Genome Analysis

Current Procedures and Applications

Edited by

Maria S. Poptsova

Weill Cornell Medical College
New York, USA
and
Moscow State University
Russia

Caister Academic Press

Caister Academic Press
Norfolk, UK

www.caister.com

British Library Cataloguing-in-Publication Data
A catalogue record for this book is available from the British Library

ISBN: 978-1-908230-29-4 (hardback)

ISBN: 978-1-908230-68-3 (ebook)

Cover design adapted from Figure 7.3

Printed and bound in Great Britain

Contents

Contributors

Sahar Al Seesi
Department of Computer Science & Engineering
University of Connecticut
Storrs, CT
USA

sahar.alseesi@engr.uconn.edu

James W. J. Anderson
Department of Statistics
University of Oxford
Oxford
United Kingdom

anderson@stats.ox.ac.uk

Alex Artyomenko
Department of Computer Science
Georgia State University
Atlanta, GA
USA

artyomenkoav@gmail.com

Irina Astrovskaya
Center for Bioinformatics and Computational Biology
University of Maryland
College Park, MD
USA

iastrovskaya@gmail.com

Michael A. Beer
Department of Biomedical Engineering and McKusick–Nathans Institute of Genetic Medicine
School of Medicine
Johns Hopkins University
Baltimore, MD
USA

mike.beer3@gmail.com

Eckart Bindewald
Basic Science Program
SAIC-Frederick Inc.
Center for Cancer Research Nanobiology Program
Frederick National Laboratory for Cancer Research
Frederick, MD
USA

eckart@mail.nih.gov

Adrian Caciula
Department of Computer Science
Georgia State University
Atlanta, GA
USA

adrian.caciula@gmail.com

Don A. Cowan
Centre for Microbial Ecology and Genomics
Department of Genetics
University of Pretoria
Pretoria
South Africa

Don.Cowan@up.ac.za

Pieter De Maayer
Centre for Microbial Ecology and Genomics
Department of Genetics
University of Pretoria
Pretoria
South Africa

Pieter.DeMaayer@up.ac.za

Lilia Ganova-Raeva
Division of Viral Hepatitis
Centers for Disease Control and Prevention
Atlanta, GA
USA

lkg7@cdc.gov

Thomas Hardcastle
Department of Plant Sciences
University of Cambridge
Cambridge
United Kingdom

tjh48@cam.ac.uk

Ilya Ioshikhes
Ottawa Institute of Systems Biology (OISB) and Department of Biochemistry, Microbiology and Immunology (BMI)
Faculty of Medicine
University of Ottawa
Ottawa, ON
Canada

iioschik@uottawa.ca

Ivan V. Kulakovskiy
Department of Computational Systems Biology
Vavilov Institute of General Genetics
Russian Academy of Sciences;
Laboratory of Bioinformatics and Systems Biology
Engelhardt Institute of Molecular Biology
Moscow
Russia

ivan.kulakovskiy@gmail.com

Dongwon Lee
Department of Biomedical Engineering
School of Medicine
Johns Hopkins University
Baltimore, MD
USA

dwlee@jhu.edu

Emmanuelle Lerat
Laboratoire Biométrie et Biologie Evolutive
UMR CNRS 5558
Université Claude Bernard Lyon 1
Villeurbanne
France

emmanuelle.lerat@univ-lyon1.fr

Vsevolod J. Makeev
Department of Computational Systems Biology
Vavilov Institute of General Genetics
Russian Academy of Sciences;
Faculty of Molecular Biology
Moscow Institute of Physics and Technology
Moscow
Russia

vsevolod.makeev@vigg.ru

Nicholas Mancuso
Department of Computer Science
Georgia State University
Atlanta, GA
USA

nick.mancuso@gmail.com

Ion Mandoiu
Department of Computer Science & Engineering
University of Connecticut
Storrs, CT
USA

ion@engr.uconn.edu

Serghei Mangul
Department of Computer Science
University of California
Los Angeles, CA
USA

serghei@cs.ucla.edu

Christopher E. Mason
Department of Physiology and Biophysics and the Institute for Computational Biomedicine
Weill Cornell Medical College
New York, NY
USA

christopher.e.mason@gmail.com

Laurent Modolo
Laboratoire Biométrie et Biologie Evolutive
UMR CNRS 5558
Université Claude Bernard Lyon 1
Villeurbanne
France

laurent.modolo@univ-lyon1.fr

Gabriel Monteiro
Centre for Biological and Chemical Engineering
Department of Bioengineering
Instituto Superior Técnico
Institute for Biotechnology and Bioengineering (IBB)
Lisbon
Portugal

gabmonteiro@ist.utl.pt

Markus Nebel
Department of Computer Science
Kaiserslautern University
Kaiserslautern
Germany

nebel@informatik.uni-kl.de

Atsushi Ogura
Department of Computer Bioscience
Nagahama Institute of Bio-Science and Technology
Shiga
Japan

aogu@whelix.info

Pedro H. Oliveira
Centre for Biological and Chemical Engineering
Department of Bioengineering
Instituto Superior Técnico
Institute for Biotechnology and Bioengineering (IBB)
Lisbon
Portugal

pcphco@gmail.com

Miguel Prazeres
Centre for Biological and Chemical Engineering
Department of Bioengineering
Instituto Superior Técnico
Institute for Biotechnology and Bioengineering (IBB)
Lisbon
Portugal

miguelprazeres@ist.utl.pt

Benjamin J. Raphael
Department of Computer Science
Brown University
Providence, RI
USA

braphael@cs.brown.edu

Marta Sánchez-Carbayo
Proteomics Unit
CIC BioGUNE
Bizkaia Technology Park
Derio
Bizkaia
Spain

mscarbayo@cicbiogune.es

Anika Schulz
Department of Computer Science
Kaiserslautern University
Kaiserslautern
Germany

anika.schulz@cs.uni-kl.de

Bruce A. Shapiro
Center for Cancer Research Nanobiology Program
National Cancer Institute
Frederick, MD
USA

shapirbr@mail.nih.gov

Suzanne S. Sindi
Applied Mathematics
School of Natural Sciences
University of California
Merced, CA
USA

ssindi@ucmerced.edu

Pavel Skums
Division of Viral Hepatitis
Centers for Disease Control and Prevention
Atlanta, GA
USA

skumsp@gmail.com

Bassam Tork
Department of Computer Science
Georgia State University
Atlanta, GA
USA

basamt@gmail.com

Angel Valverde
Centre for Microbial Ecology and Genomics
Department of Genetics
University of Pretoria
Pretoria
South Africa

Angel.Valverde@up.ac.za

Alex Zelikovsky
Department of Computer Science
Georgia State University
Atlanta, GA
USA

alexz@cs.gsu.edu

Paul Zumbo
Department of Physiology and Biophysics and the Institute for Computational Biomedicine
Weill Cornell Medical College
New York, NY
USA

pzumbo@zoho.com

Preface

When I was approached with the idea of editing a volume on the current procedures of genome analysis I was excited at the prospect of learning more about the latest achievements in the various research areas which have burgeoned in recent years due to next-generation sequencing (NGS) technology, as well as about the algorithmic challenges and pitfalls, the open problems in the various *-omics* disciplines and the future trends as seen by the experts in the field. As editor of this volume I have been fortunate to have authors as forthcoming as they are expert in their subject. Indeed, the contributed chapters have exceeded my expectations and this volume is intended to fulfil the tangible need in modern bioinformatics to elucidate the advances in current methods of genome analysis.

There is no doubt that NGS technologies made possible many types of whole-genome analysis that were deemed impossible just ten years ago. Another name for NGS is high-throughput sequencing, in which high performance is achieved by parallelizing millions of sequencing processes on a large DNA (RNA) sequence split into millions of short pieces, the so-called reads. Starting with DNA sequencing, or DNA-Seq, this technology gave birth to other, no less important high-throughput techniques, such as RNA-sequencing (RNA-seq), chromatin immunoprecipitation (ChIP) and methylated DNA immunoprecipitation (MeDIP) followed by sequencing (ChIP-Seq and MeDIP-seq). The analyses available with the help of these technologies, beyond genome sequencing *per se*, are numerous. RNA-seq permitted the whole transcriptome retrieval and identification of various types of RNA and RNA-editing. ChIP-Seq afforded the genome-wide mapping and sequencing of binding regions of transcriptional factors, nucleosomes or other protein of interest. MeDIP-seq made possible the genome-wide retrieval and sequencing of methylated regions.

However speed comes at the expense of quality, as the burden is shifted to computational processing of terabytes of data. For example, in order to reduce errors the genome is sequenced at least 30 times, i.e. it is 30-fold covered by reads providing the so-called 30× coverage. For a human genome of roughly 3 gigabases (Gb) this makes up 90 Gb of raw data that need to be stored, pre-processed and processed. Indeed, it is not merely the data volumes, which are scary, but also the abundance of noisy reads which have to be filtered out, as the chaff is from the wheat, before inferring any meaningful conclusions. However, the excitement of a scientist about the potential and actual achievements outmatches the fear of tedium and the first results from NGS experiments have opened new horizons in our understanding of the complex and multilayer processes of genome functioning.

The purpose of this book is to give an overview of the methods currently employed for NGS data analysis, highlight their problems and limitations, demonstrate the applications and point to the developing trends in various fields of genome research. The book is logically divided into two parts. The first part is devoted to the methods and applications that arose from, or were significantly advanced by, NGS technologies: identification of structural variation from DNA-seq data; whole-transcriptome analysis and discovery of small interfering RNAs (siRNAs) from RNA-seq data; motif finding in promoter regions, enhancer prediction, and nucleosome sequence code discovery from ChIP-Seq data; identification of methylation patterns in cancer from MeDIP-seq data; transposon identification in NGS data; metagenomics and metatranscriptomics; NGS of viral communities; causes and consequences of genome instabilities. The second part is devoted to the field of RNA biology, predicted to develop rapidly in the coming years with NGS experiments unveiling the role of RNA. The last three chapters are devoted to computational methods of RNA structure prediction including context-free grammar applications. Let me briefly introduce each of the chapters.

DNA-seq technologies promised a fast and relatively cheap method of obtaining a whole-genome sequence on the table. But the path from millions of reads generated by a sequencer to the continuous nucleotide sequence of a chromosome is still long and frustrating. That is why the first results from 1000 genome project represent not the complete genome sequences of 1000 individuals, but the differences between them, or variations, such as SNPs, indels and structural variants (SVs). The book begins with a chapter on a major task in DNA-seq data processing, that of algorithmic approaches to SVs discovery. In Chapter 1 Suzanne Sindi of the University of California and Benjamin Raphael of Brown University give an overview of 'increasingly sophisticated sequence-based' methods. They describe how a combination of multiple signals is performed in order to improve detection quality and discuss future trends of targeted sequencing approaches and personalized medicine.

Following DNA-seq, RNA-seq technologies are no less powerful and no less simple in terms of data processing. The applications are numerous from whole-genome transcriptome analysis, to short regulatory RNA and RNA editing discovery. In Chapter 2, Paul Zumbo and Christopher Mason of Cornell University provide a comprehensive review of the methods for RNA isolation, characterization, and sequencing. The reader will find a brief history of the key RNA discoveries, learn about the principles of RNA isolation and the methodology of RNA sequencing methods. The authors also provide an overview of recent RNA-Seq studies aimed at characterization of the so-called transcriptional landscape and discovery of RNA modifications, of which more than 100 types have been reported to date, emphasizing their emerging role in epigenetics and in particular in epitranscriptomics.

In Chapter 3 Ion Măndoiu's group of the University of Connecticut continues with the topic of RNA-seq applications and present methods for whole-genome transcriptome analyses. The authors first demonstrate how the Integer Programming method improves on transcriptome reconstruction, especially on alternative splicing and novel transcripts detection. Then the authors present expectation maximization (EM) algorithms that can be applied to both RNA-Seq and digital gene expression (DGE) sequencing protocols in order to quantify levels of RNA expression, the methods, currently replacing expression microarrays. Among the future trends is the problem of integrating the existing transcriptome analysis pipelines to a new type of transcriptomic data, such as promoter and PolyA profiling. The reader will learn about recent studies made possible with RNA-Seq whole-genome transcriptome analysis.

Another analysis made possible thanks to RNA-seq technology on the genome-wide scale is the discovery of small interfering RNAs (siRNAs), which play 'a crucial role in the regulation of transcriptomic and epigenetic factors'. In Chapter 4 Thomas Hardcastle of Cambridge University reviews the currently available tools for the analysis of the so-called siRNA-seq data, discusses experimental design and processing techniques. The reader will learn about the ways of identification of siRNA differential expression, phased siRNAs, unexplored areas in target finding, small RNA networks, as well as about the major studies in siRNA field induced by high-throughput sequencing.

ChIP-Seq is the next revolutionizing NGS technology that allows for precise identification of protein-DNA binding sites and corresponding DNA sequences that participate in binding at a genome-wide scale. Mapping of transcription factor binding sites (TFBS) with motif finding and motif discovery in ChIP-Seq data is the topic of Chapter 5, written by Ivan Kulakovskiy of the Institute of Molecular Biology and Vsevolod Makeev of the Institute of General Genetics, Moscow. The authors review the methods for TFBS motif discovery in the pregenomic era followed by the possibilities that arose from ChIP-Seq technology. They discuss the methods of data pre-processing, provide *cooking recipes* for motif analysis of ChIP-Seq data and discuss recent applications in this area. An improved modification of ChIP-Seq was developed under the name of ChIP-exo.

Mammalian regulation cannot be understood without understanding the role of enhancers. Their experimental identification was boosted by ChIP-seq technology that allowed mapping the genomic occupancies of enhancer specific protein markers. Recent genome-wide association studies (GWAS) provided evidence for enhancer association with human disease. In Chapter 6 Dongwon Le and Michael Beer of Johns Hopkins University describe the state-of-the-art computational methods to predict enhancers and specifically how support vector machine (SVM) framework, a machine learning technique, was successfully applied to mammalian enhancer prediction problem.

The other area that profited considerably from ChIP-Seq technology is the genome-wide mapping of nucleosome positions, which proved to be crucial for gene regulation to the extent that a change in 1 bp can disrupt gene function. With nucleosomes covering 70% of the eukaryotic genome it is natural to search for a sequence-specific pattern. In Chapter 7 Ilya Ioshikhes of the University of Ottawa tells the story of this scientific endeavour, which started in the pre-genomic era and was considerably advanced in the NGS era. The reader will learn about the discovery of different patterns, their biological implications, and the progress made in the formulation of a universal nucleosome pattern given the fact that nucleosome positioning differs between different tissues of the same species, while the DNA sequence remains the same.

MeDIP-seq technology promoted studies of important epigenetic modifications such as DNA methylation. Changes in the DNA methylation landscape – hypermethylation of some genes and hypomethylation of others – have been recognized as important events in many diseases, including cancer. In many tumours, silencing of tumour suppressor genes was reported to be associated with hypermethylation of promoter regions. Hypermethylation in cancer is the topic of Chapter 8, written by Marta Sánchez-Carbayo of the Spanish National Cancer Center, Madrid. The reader will learn about types of DNA methylation, hypermethylation machineries, biological pathways of frequently methylated genes in cancer, and how high-throughput technologies accelerated discoveries of epigenetic events.

One of the major problems with the analysis of NGS-type data is the presence of repetitive elements. Given the fact that almost half (45%) of the human genome is composed of transposable elements (TEs), their correct identification is understandably an important task. In Chapter 9, Laurent Modolo and Emmanuelle Lerat of the Université de Lyon give a review of classic, pre-NGS, TEs detection methods in genome sequences and then discuss how to handle TEs in the NGS data. All types of NGS technologies (DNA-seq, RNA-seq, ChIP-Seq, MeDIP-Seq) contributed to the understanding of TE functioning in the genome, but there is a great need to improve the existing TE identification pipelines.

While getting used to the host of applications which became possible with the emergence of NGS technology, the advances of metagenomics and the speed of data accumulation from various environmental niches including human body and upper troposphere are still mind-boggling. In Chapter 10, the group of Don Cowan of the University of Pretoria, the director of the Centre for Microbial Ecology and Genomics, gives a comprehensive review of the current state of metagenomic analysis. The reader will learn about three generations of sequencers used for metagenomic study, including the single-molecule, or third generation, sequencers in the market, methods and tools for metagenome assembly, contig binning, metagenome annotation, reconstruction of complete genomes of 'unculturable' organisms, all possible *meta-omics* approaches, bioprospecting methodology, and much more.

In Chapter 11, Otushi Ogura from the University of Tokushima writes specifically about metatranscriptomics studies, their initial delay due to technical and financial difficulties and quick progress made over the last few years. The author describes methods and tools that may be accommodated specifically to the transcriptomics and presents a case-study of marine microbiome from the Inland Sea of Japan where phytoplankton monitoring has been done over a 35-year period and changes in the dominance of the species were observed back in the mid-1980s. Nowadays, with metagenomics approaches one can monitor activities of functional genes in special conditions such as, for example, before and after red tides. There is an urgent need for metatranscriptome-oriented tools.

NGS provided an unprecedented opportunity to study fast evolving viral communities. Compared to all other species, it is very inexpensive to sequence a viral genome with a good coverage due to its small size. Highly infectious viruses can produce 10^{10}–10^{12}, an astonishing number, of variants per day. The strategies for a single genome assembly are different from the reconstruction of multiple, but closely related variants. In Chapter 12 the group of Alex Zelikovsky of Georgia State University reviews the methods and strategies of reconstruction of all co-existing viral variants, the so-called quasi-species, from the NGS data. The authors discuss computational challenges associated with each strategy, state-of-the-art software tools and demonstrate examples of the analysis for hepatitis C (HCV) and human immunodeficiency (HIV) viruses.

One of the first discoveries made from NGS experiments in comparing normal and tumour genomes was that genome rearrangements are a common phenomenon in cancer. However genome instability is a property inherent to genomes of all living organisms, which plays an essential yet not fully understandable role in evolution. The first part of the book ends with Chapter 13 by Gabriel Monteiro's group from the Instituto Superior Técnico, Lisbon, which expounds the causes and consequences of instabilities of bacterial genomes. The reader will learn about spontaneous and stress-induced mutagenesis, the role of non-B structures in genome instability, and modern trends in biotechnology to produce stable genome to fill-in 'microbial power-horses with streamlined pathways for production of high-value bioproducts'.

The last three chapters are devoted to RNA, considered a molecule secondary to DNA, which is nevertheless gaining ever-increasing attention. As reviewed in Chapter 3, over the last few years many types of RNA were found to play a role in nearly all of the cellular processes: transcription, translation, regulation and epigenetics. 'In hindsight it appears that the initially under-appreciated role seen in RNAs has been one of the great oversights in the field of molecular biology' – write Eckart Bindewald and Bruce Shapiro of the Frederick National Laboratory for Cancer Research in Chapter 14, which covers comparative methods for RNA structure prediction. The authors review RNA folding problems and conceptual building blocks for computational RNA structure prediction, followed by a review of the software and databases currently available. Specifically they discuss genome-wide scans for conserved RNA structural elements and present prospects of how models will develop with experimental data from NGS.

Chapter 15, authored by Markus Nebel and Anika Schulz of the University of Kaiserslautern, can be considered both as a review and as a guidebook on context-free grammars (CFGs) employed for RNA secondary structure prediction. The formalism of CFGs was developed by Noam Chomsky in the 1950s and in the 1990s Eddy and Durbin applied it to the RNA structure problem. Competing with the energy-based model, CFG models naturally developed, especially after the success of hidden Markov models (HMMs), into probabilistic models, such as stochastic CFGs (SCFGs), length-dependent SCFGs (LSCFGs), conditional log-linear models (CLLMs) and multiple context-free grammars, MCFGs, that unlike SCFGs can deal with pseudo-knots.

Finally, in Chapter 16, James Anderson of Oxford University reviews the latest achievements specifically in the area of SCFGs for RNA structure prediction and, in particular, how SCFGs can be combined with molecular evolution models and how information entropy, or Shannon entropy, is employed in calculating measures of SCFG variability. Concluding with a phrase 'RNA is as interesting as ever', the author also believes that for the single-sequence RNA structure prediction problem the future is for SCFGs rather than for the energy-based models or the machine-learning approaches. With the RNA-seq data released at an unprecedented speed 'RNA folding methods will want to be used in genome-wide studies, considering RNA structure-based effects on gene expression and regulation'.

In conclusion, with the advances of NGS technology we are definitely at a stage when experimental data accumulate much faster than the algorithmic and computational methods intended to process them. Our understanding of genome functioning before the era of high-throughput sequencing was that of a complex multilayer machinery. With the first results of the NGS-era, it has progressed to an ultra-complex multilayer machinery. Previously invisible joints, which interconnect different genome organization layers, are discovered every day and one cannot fully characterize genome regulatory processes by merely mapping transcription factor binding sites and disregarding nucleosome positioning and RNA interference. System biology aims to create this integrative view of the problem and move further from a cybernetic approach to study mechanisms of multilayer interactions. This may take us to the next level, where the laws of self-organization govern the system behaviour.

Meanwhile, the time is right to design and conduct NGS experiments, process results, collect them in the databases, develop predictive computational algorithms and to discover new type of genomic elements, connections or functioning. The major disappointment from early NGS experiments was in the field of cancer research. Sequencing of tumour and non-tumour genomes showed that cancer is extremely heterogeneous, changes at the genome level are numerous, but none can serve even as a descriptive biomarker. However, the sequencing stage cannot be omitted, and the *1000 Genome Project*, the *International HapMap* project, and *The Cancer Genome Atlas* are some major NGS projects which make their data publicly available. However, as of today, cancer, bacterial antibiotic-resistance, and evolvability of the lethal viruses, all caused by changes at the genome level, have not been fully understood and should, we believe, become a part of what would become 'Hilbert's 23 problems of bioinformatics' in the 21st century, much to the scrutiny of the contemporary and future explorers of science.

Maria S. Poptsova

Identification of Structural Variation

1

Suzanne S. Sindi and Benjamin J. Raphael

Abstract

Structural variants, rearrangements of DNA sequences, had historically been observed by low-resolution chromosomal assays. However, recent advances in high-throughput DNA sequencing have greatly improved the ability to identify and localize structural variation (SV). Today, sequence-based methods for identifying SVs have greatly improved our knowledge of genomic variation in humans and other species. DNA sequencing data from an individual genome contains multiple possible signals that indicate SVs in this genome, and these signals must be analysed and integrated using various computational techniques. Here we give an overview of SV discovery methods from sequencing data and remark on the challenges remaining.

Introduction

The genome of an organism is the source of an organism's hereditary information and ranges from a few thousand bases for viruses to billions of bases in mammalian genomes. As DNA sequencing technologies improved, reference genomes have been assembled for a variety of species such as worm, fruit fly, cow, mouse and human. These high-quality reference sequences facilitated comparisons between genomes from different species for large structural rearrangements. Some of these rearrangements are hypothesized to have contributed to speciation events (Noor *et al.*, 2001).

However until recently, it was believed that genetic diversity between members of the same species consisted primarily of small-scale changes such as single nucleotide polymorphisms (SNPs). Structural variants, rearrangements larger than a single nucleotide, were also known to occur. For example, in the 1920s chromosomal inversions were inferred in *Drosophila* from their inhibiting effect on genetic recombination (Sturtevant, 1920). In humans, rare structural variants were associated with various genetic disorders (Karayiorgou *et al.*, 2010; Stankiewicz and Lupski, 2010; Beck *et al.*, 2011; Kim *et al.*, 2012) and cancer genomes were known to exhibit many rearrangements (Bamford *et al.*, 2004; Greenman *et al.*, 2007; Bashir *et al.*, 2008; Zang *et al.*, 2011). Yet, structural variants were thought to be relatively rare events within the human population.

Most studies of genetic diversity in humans, such as the landmark International HapMap Project (Gibbs *et al.*, 2003), focused on SNPs. However, evidence for the prevalence of structural variants (SVs), such as deletions, duplications and inversions, began to accumulate. First, comparative genomic hybridization revealed the presence of hundreds of copy-number variants (CVNs). Then other methods, taking advantage of the abundance of SNP data, were developed to discover CNVs (Conrad *et al.*, 2005; McCarroll and Altshuler, 2007) and later, to a lesser extent, inversion variants (Bansal *et al.*, 2007; Sindi and Raphael, 2010).

More recently, improvements in next-generation sequencing (NGS) technologies greatly advanced structural variation detection. Over the past decade, the number of known structural variants has grown exponentially as the genomes of more individuals have been sequenced. In Tuzun *et al.* (2005), sequenced genomic clones from a single individual and identified nearly

300 structural variants. Less than a decade later, the first phase of the 1000 Genomes Project has identified over 28,000 variants from sequencing nearly 200 individuals (Mills *et al.*, 2011). In addition, data from recent studies suggests that CNVs account for more differences among individuals, covering as much as 30% of the genome, than SNPs (Zhang *et al.*, 2009). As of November 2012, The Database of Genomic Variants (DGV) contains nearly 900,000 entries accumulated from over 40 studies of SVs ranging from 50bp to over 1Mb.

While most known SV studies have focused on human genomes, SVs have also been identified in other organisms such as mouse, dog and cow (Higuchi *et al.*, 2010; Quinlan *et al.*, 2010; Yalcin *et al.*, 2011, 2012) suggesting SVs are likely to be common in mammals. Here we discuss computational and algorithmic approaches to the identification of structural variation from sequencing data and continued challenges for the field.

Defining structural variants

The term structural variation has come to refer to any genetic rearrangement that is larger than a single nucleotide but often is restricted further to rearrangements larger than 1 kb. This definition encompasses many types of events such as insertions of novel sequence, duplications and deletions, and copy neutral variants such as inversions and translocations. More general events are possible where more than one rearrangement occurs as the locus; these complex variants can be problematic to classify and, as such it is important to define the notion of a structural variant in a precise manner.

More formally, an arbitrary rearrangement event can be described as a set of one or more novel adjacencies. Given a specified reference genome and a test genome, a novel adjacency is two positions that are adjacent in the test genome, but not in the reference (see Fig. 1.1). In this context a deletion results in a single novel adjacency while an inversion results in two novel adjacencies. Each novel adjacency joins together two breakends, positions in the reference genome, which are mated. Thus, identifying structural variants in a test genome corresponds to identifying all novel adjacencies. Note that the concept of a novel adjacency is applicable to more general structural variants such as complex variants; for example, an inversion with a partial deletion of the region surrounding one end.

Causes of structural variation

Given that many human structural variants were only recently identified at the single-nucleotide resolution, the processes responsible for creating SVs are still not well understood. Recently, analysis of the surrounding DNA sequence has revealed several mechanisms likely to be responsible for much observed variation (Kidd *et al.*, 2008; Hajirasouliha *et al.*, 2010; Mani and Chinnaiyan, 2010; Mills *et al.*, 2011). First, recombination occurring between repetitive sequences, non-allelic homologous recombination (NAHR) can create duplications, deletions, inversions and even translocations depending on the placement and relative orientation of the repeats. Segmental

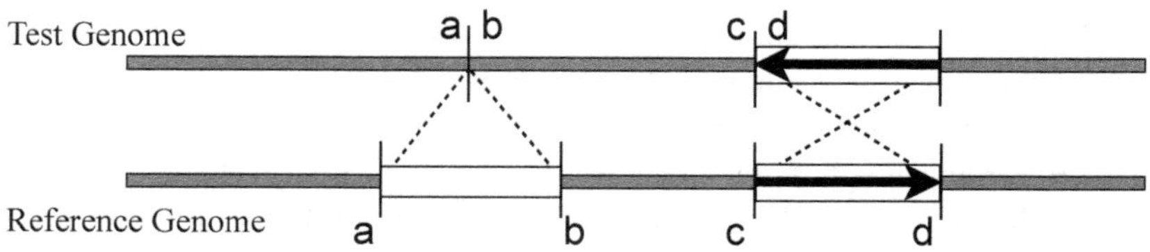

Figure 1.1 Structural variants can be described as a set of novel adjacencies, locations adjacent in a test (unknown) genome but not in the reference genome. In the example above, a deletion in the test genome (left) corresponds to novel adjacency between breakends *a* and *b* and an inversion (right) corresponds to a novel adjacency between breakends *c* and *d*.

duplications are enriched at structural variation breakpoints, supporting the importance of NAHR as a prevalent SV formation mechanism. Similarly, in the repair of a double-strand break in DNA, non-homologous end joining (NHEJ) will create a rearrangement. In addition, the mechanism of fork stalling and template switching (FoSTeS) has been proposed as an explanation for complex duplication and deletion rearrangements associated with genomic disorders (Lee *et al.*, 2007). Finally, specific classes of variants such as transposable elements and tandem repeats (VNTR) have their own replication mechanisms (Mani and Chinnaiyan, 2010).

Early methods for SV identification

Today DNA sequencing has become the dominant approach for SV identification. However, the study of chromosomal variation dates back to cytogenetic approaches in the 1920s and 1930s (Painter, 1934; Sturtevant, 1920). Such approaches were used extensively to study inversion polymorphisms in *Drosophila*. Microarray technology emerged in the late 20th century and has been used to predict various types of structural variation. Before describing the use of DNA sequencing as a tool for SV discovery we briefly discuss these earlier approaches.

Microscopy/cytological methods

Large structural variants can be detected by microscopy. Karyotyping has been used to detect gains or losses of whole chromosomes or chromosome arms. Other cytogenetic approaches identified structural variants by staining chromosomes and examining the pattern of banding (Painter, 1934).

The method of fluorescence *in situ* hybridization (FISH) is an experimental protocol used to detect the presence (or absence) of specific DNA sequences on chromosomes. Fluorescently tagged DNA sequences, which bind to complementary DNA sequence, are used to as probes and microscopy is used to study the presence, absence or position of fluorescence. When potential SVs are predicted, FISH has been used to validate the variant (Kidd *et al.*, 2008). In addition, FISH is used in chromosome painting methods, such as spectral karyotyping (SKY) (Ried *et al.*, 1998), where probes to each chromosome are differently coloured to reveal large-scale rearrangements and translocations.

Although FISH provides direct evidence of the presence of variants, only large variants can be detected and the resolution of breakpoints is quite low. However, FISH has been used as a complement to sequencing approaches to determine the presence of SVs whose endpoints cannot be well defined by sequencing approaches (Kidd *et al.*, 2008).

Array-based methods

As in FISH, DNA microarrays use hybridization between complementary DNA sequences as evidence for the presence or absence of probe sequences, but in a more high-throughput fashion. Microarrays contain anywhere from tens of thousands to millions of oligonucleotide probes which themselves can be selected to tile an entire reference genome.

In array comparative genomic hybridization (aCGH), fluorescently labelled samples hybridize to a microarray. The measured level of fluorescence reflects the abundance of DNA in the sample providing a signal for detecting copy number variants (CNVs). A reference sample is used to decide between a gain in one sample versus a loss in another one. Computational methods soon emerged to identify CNVs by segmenting the hybridization profile of a genome into regions of gains and losses (see, for example, Olshen *et al.*, 2004; Lai *et al.*, 2005; Wang *et al.*, 2009; Ritz *et al.*, 2011).

Arrays designed to detect single nucleotide polymorphisms, SNP arrays, provided another method to identify variants. The abundance of SNP data from related individuals, from efforts like The International HapMap Consortium (Gibbs *et al.*, 2003), motivated studies in SV detection. Patterns of SNPs provided evidence of different types of SVs (McCarroll and Altshuler, 2007). Deletions would appear as a run of null genotypes and would not fit the expected Mendelian inheritance from parent–child trios (Conrad *et al.*, 2005; McCarroll and Altshuler, 2007) Finally, following a similar philosophy to

Sturtevant (1920), patterns of linkage were used to provide evidence of inversion SVs (Bansal *et al.*, 2007; Sindi and Raphael, 2010).

Although arrays typically offer lower breakpoint resolution than sequencing data, their low cost continues to make arrays a powerful tool in CNV discovery (Li and Olivier, 2012).

Identification of structural variation from sequencing data

Today, DNA sequencing is the dominant approach for identifying structural variants. Current DNA sequencing technologies (e.g. Illumina HiSeq, Ion Torrent, and others) produce large volumes of DNA sequence, but with limited read lengths. *De novo* assembly of these short reads into a human genome remains a challenge, and thus most SV studies employ a resequencing strategy that exploits the availability of a high-quality reference genome. In this strategy, fragments from the test genome are sequenced. Then, these sequences are mapped to the reference genome and the resulting configuration of fragments is analysed for evidence of SV as shown in Fig. 1.2. Before discussing approaches for SV identification, we describe the sequencing and mapping procedures. We then detail the three common signals of structural variation derived from the mappings. Finally, we discuss algorithmic approaches to predicting SVs from these signals.

DNA sequencing

The first sequencing based study for SV identification in a human genome was performed by Tuzun *et al* (2005). This study used traditional Sanger sequencing of fosmid-end sequences from a single individual amounting to 0.19 times sequence coverage of the human genome. The long length of Sanger reads (~500–1000 bp) resulted in low-ambiguity in aligning reads to the reference genome.

Since then, next-generation sequencing (NGS) technologies such as 454 Life Sciences, and Illumina, and Life Technologies have become the standard approach for sequencing studies. The reads produced by NGS technologies are shorter than Sanger reads (~30–200bp) with reads longer than 100bp only recently becoming common. These short reads complicates determining their placement (or mapping) to the repeat rich human genome; however, the decreased cost of NGS facilitates higher sequence coverage. (For a detailed review of NGS sequencing technologies refer to Metzker (2009)). The first high (>40-fold) coverage sequencing of a human genome with NGS technologies was in 2008 (Bentley *et al.*, 2008). Since then, the number of whole genomes

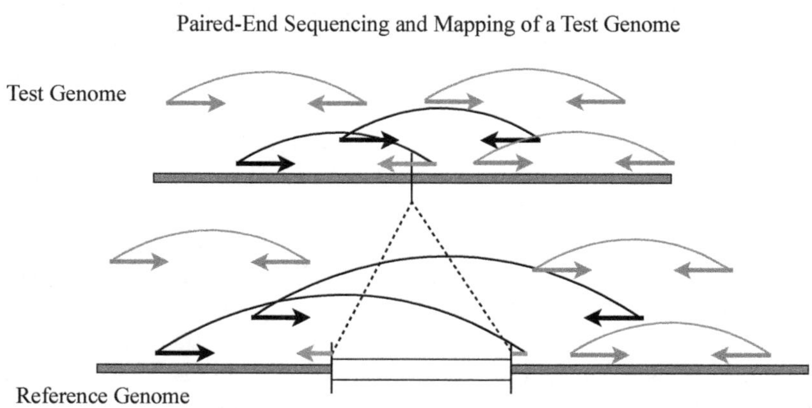

Figure 1.2 To identify structural variants (SVs) from a test genome, fragments from a test genome are sequenced from both ends and the paired-reads mapped to a reference. Evidence of a structural variant (in this case a deletion) comes from three signals in the mappings: (1) changes in read depth (RD), the depth of mapped reads (black and grey); (2) discordant paired-read mappings (PR) (black reads); and (3) split-read mappings (SR) (red read). A colour version of this figure is located in the plate section at the back of the book.

sequenced has increased culminating in the recent 1000 Genomes Project, which sequenced numerous individuals of varying coverage, some upwards of 30-fold coverage.

More recently, technologies that can sequence single molecules, such as those from Pacific Biosciences, offer the promise of substantially longer reads than NGS technologies. However, these longer reads come at the cost of higher sequencing error. For a more detailed discussion of these technologies see Gupta (2008).

Most SV identification methods use similar approaches regardless of the underlying sequencing technology. However, one important distinction is whether the DNA sequencing technology yields only reads of a certain length, or whether the technology also produces paired-reads from both ends of a longer fragment (see Fig. 1.2). The length of the larger fragment depends on the protocol followed, from 300 to 500 bp for *paired-end libraries* and from 1.5 to 20 kb for mate-pair libraries (Mardis, 2011). Such paired reads increase the effective read length for SV discovery, as discussed below.

Read mapping

Determining the read mappings, placement of each read in the reference genome, is the important first part of the SV identification process. Algorithmically, mapping is closely related to the sequence alignment problem where we want to determine the best alignment, or set of alignments between two sequences. Because of the size of the reference and number of sequenced reads involved, an exhaustive search is not usually favoured.

Early approaches for SV identification in humans, which used Sanger sequencing (Tuzun *et al.* 2005, Kidd *et al.*, 2008) sequenced about 1 million fragments for a single genome. They used Megablast (Zhang *et al.*, 2000) to determine the alignment of reads to the reference genome and worked only with reads that had a high-quality unique placement. While this mapping approach works well for the relatively long (>400 bp) reads, next-generation sequencing produces shorter reads and today SV studies consider substantially higher coverage. Current read mapping algorithms have been developed to take into account the specific challenges presented by NGS data.

Mapping algorithms begin looking for potential mapping locations in the genome based on high agreement between the reads and the reference sequence and then extending the alignment. Most read mapping algorithms employ data either hash tables or suffix trees to identify the possible set of locations. See (Li and Homer, 2010) for a more detailed review.

Hash table

A hash table is a data structure with values indexed by subsequences of length *k*, *k*-mers. The table stores the position of each *k*-mer in either the reference genome (Bfast (Homer *et al.*, 2009), Novoalign (Novocraft), SOAP (Li *et al.*, 2008b), mrFAST (Alkan *et al.*, 2009)) the reads themselves, such as MAQ (Li *et al.*, 2008a) or in some cases both, such as mrsFAST (Hach *et al.*, 2010). Read alignment begins by identifying positions in the genome with matching *k*-mer seeds, or more generally spaced seeds, which allow substitution mismatches, and extending the partial alignment to an inexact match. For example, Novoalign uses a hash-table to store exact *k*-mers in the genome and extends matches with the Smith–Waterman alignment algorithm (Smith and Waterman, 1981).

Because the number of *k*-mers is exponential in *k*, many algorithms use further restrictions to improve speed. The MAQ algorithm focuses on seeds at the beginning of the read, because the sequencing error rate of Illumina reads increases as bases are sequenced. Allowing gaps in a hash table context is more problematic. A q-gram filter, used for example by SHRiMP (Rumble *et al.*, 2009), allows an index with gaps to be built retaining the ability to quickly look up positions in a table.

Suffix trees

Suffix trees are data structures that store all suffixes of a string. Algorithms employ suffix trees to identify the locations of exact matches between substrings in the reads and the reference genome, and extend these matches. Compared to hash tables, which require storing each *k*-mer separately, identical substrings are reduced to a single entry in a suffix tree, reducing memory load. Many methods, such as BWA (Li and Durbin, 2009), Bowtie (Langmead *et al.*, 2009), SOAP2 (Li *et al.*, 2009) use the Burrows–Wheeler transform (BWT), originally developed for data compression (Burrows and Wheeler, 1994), to obtain compressed suffix arrays with lower memory requirements than suffix trees. For example Bowtie uses only 2.2 Gb of memory to store an index for a human reference genome (Langmead *et al.*, 2009). Although earlier methods (Bowtie, SOAP) did not allow gapped alignment, most recent methods do (BWA, Bowtie2 (Langmead and Salzberg, 2012), SOAP2).

Advances in read mapping

Since the error rate of NGS reads is relatively low, many methods favour heuristics guaranteed to find high quality alignments. Earlier approaches focused on quickly identifying the single best mapping for a read (BWA, MAQ, Bowtie), other methods are willing to trade speed for increased sensitivity by considering all mappings up to a particular edit distance or number (Novoalign, SOAP). In addition, some methods consider only 'full' or 'end-to-end' alignment of reads (MAQ, Botwie) while others allow local or partial alignments (Novoalign, BWA, Botwie2). As we will discuss below, such partially aligned reads are a potential source of reads containing novel adjacencies.

With evolving requirements from sequencing data such as higher coverage NGS data, longer reads with more sequencing errors, and the emergence of single molecule sequencing, read mapping programs become increasingly sophisticated. Recent programs utilize parallel GPU cores to accelerate the mapping process such as SOAP3 (Liu *et al.*, 2012) and BarraCUDA (Klus *et al.*, 2012). In addition, mapping tools are emerging specifically designed for longer reads such as BWA-SW (Li and Durbin, 2010), BLASR (Chaisson and Tesler, 2012) and YAHA (Faust and Hall, 2012).

Signals of structural variation from next generation sequencing data

Three common signals have been used to derive the presence of SVs from the resulting configuration of mapped reads. First, the number of sequenced reads mapping to locations in the genome, the read depth (RD) signal, indicates changes in copy number (Fig. 1.4). Second, although the length of each fragment is unknown, the underlying fragment length distribution depends on the sequencing protocol followed and, since most of the genome is assumed to be consistent with the reference, is often approximated from the empirical distribution of lengths resulting from the mapping procedure itself. Most fragments will map with length and orientation suggesting concordance between the test and reference

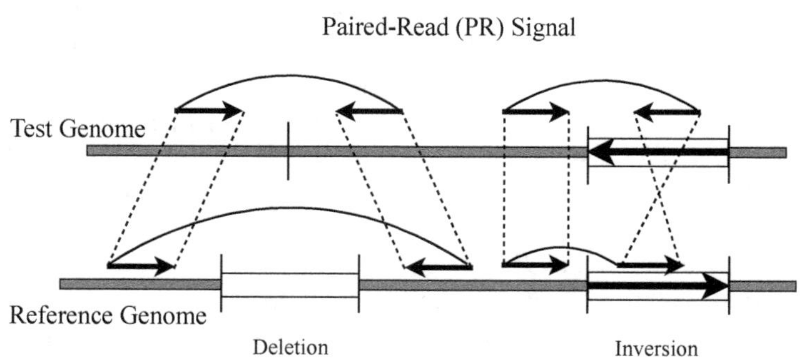

Figure 1.3 The paired-read (PR) signal of a structural variant, namely the presence of discordantly mapped fragments, indicates a potential novel adjacency between the test and reference genomes. A deletion (left) is implied by reads mapping to a distance larger than the expected length of a sequenced fragment. An inversion (right) is indicated by reads in a fragment mapping with incorrect orientation. (Note: The definition of 'incorrect orientation' varies by sequencing technology and protocols).

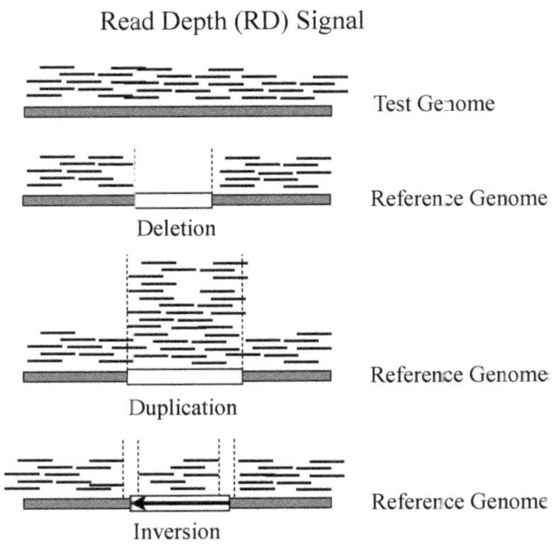

Figure 1.4 The read depth (RD) signal of a structural variant is used for predicting copy number variants (CNVs) with fewer or more reads than expected indicating a deletion or duplication respectively (top/middle). A local change in read depth at the breakend, breakend read depth (beRD), may indicate copy neutral variants such as inversions (bottom), or other types of novel adjacencies.

genome (see Fig. 1.2). As such, fragments with discordant paired read mappings (PR), indicate potential deletions, inversions, translocations and tandem duplications (see Fig. 1.3). Finally, reads without a full alignment to the reference genome may contain a SV breakpoint (novel adjacency) and are potential split reads (SR) (see Fig. 1.5). Although each of these signals can indicate the presence of rearrangements, no signal alone is sufficient to identify all types adjacencies. The RD signal is less evident for copy neutral events and may be affected by sequencing biases. The PR signal can be confounded by mis-mappings and cannot detect insertions larger than the fragment length. The SR signal usually requires one-end of the fragment be anchored, which is not possible in repetitive regions, and that each of the two sides of the read can be aligned to the reference, which is complicated by the short read length of NGS reads.

Early methods for SV identification focused on only one of these signals. More recently, algorithms have demonstrated increased sensitivity and specificity by combining more than one of these signals and exploiting the specific signature of each type of structural variant in the mapped

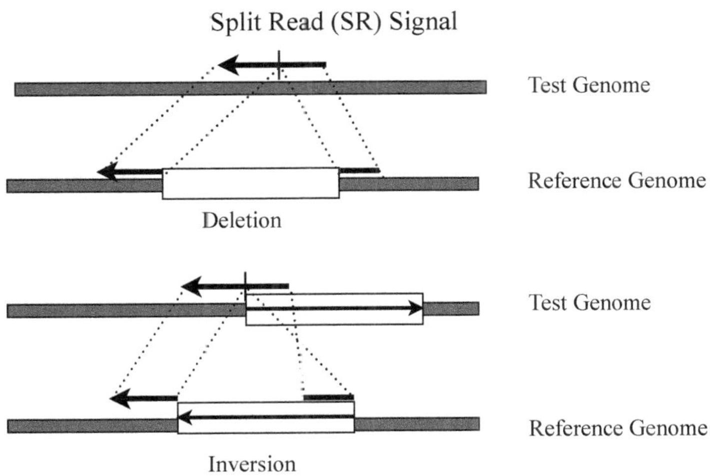

Figure 1.5 The split-read (SR) signal of a structural variant is used to detect any type of novel adjacency in the test genome. A read containing a novel adjacency will map to the reference genome as a split-alignment with two partial alignments on either side of the breakends. The orientation of the split alignments may vary depending on the variant. For example, reads containing a novel adjacency created by a deletion will have alignments are in the same direction (top), while the alignment orientations are reversed for an inversion (bottom).

reads (see Figs. 1.3, 1.4 and 1.5 and Table 1.1). In addition, with decreasing sequencing costs and increasing sequencing coverage, it has become possible to assemble the reads from the test genome into larger contiguous sequences. Ideally, these assembled sequences contain the novel adjacency and surrounding region in the test genome. Thus, the variant itself will be evident by alignment to the reference genome.

Algorithmic approaches to structural variant detection

In recent years numerous algorithms have been developed for SV detection. Rather, than exhaustively list all available methods, we focus on detailing the common approaches followed. We first describe early methods, which used only a single signal of SV to make predictions, then more recent methods that integrate multiple signals. Finally, we discuss the emergence of assembly based methods, which leverage increasingly high sequencing coverage to assemble SVs in the target genome.

Paired-read mappings

One approach taken by early methods of SV prediction was to consider only the discordant paired-read (PR) signal. After read mapping, fragments with discordant mappings were separated by the type of variant suggested: deletions were indicated by fragments whose mapped distance on the reference was larger than expected; insertions (smaller than the fragment length) were indicated by fragments whose mapped distance was smaller than expected; inversions corresponded to reads with discordant orientations and translocations corresponded to paired-reads mapping to different chromosomes (Raphael *et al.*, 2003; Volik *et al.*, 2003; Tuzun *et al.*, 2005). Other methods, such as PEMer (Korbel *et al.*, 2009), use additional linking signals between clusters to predict insertions of subsequences from elsewhere in the genome larger than the fragment length. Following a parsimony approach, fragments whose mappings were consistent with a single rearrangement were clustered to minimize the number of potential SV events.

Broadly speaking, approaches for detecting SVs from the PR signal can be grouped into two categories; those that allowed only a single mapping for a fragment, such as GASV (Sindi *et al.*, 2009), Breakdancer (Chen *et al.*, 2009), PEMer (Korbel *et al.*, 2009) and those that considered multiple alignments for each fragment, such as the method of Lee *et al.*, (2008), VariationHunter (Hormozdiari *et al.*, 2009) and Hydra (Quinlan *et al.*, 2010). However, some methods, such as Hydra, an operate in both modes. Considering multiple read alignments substantially increases the number of possible SV events, but ideally would increase sensitivity in detection of SVs in repetitive regions. Typically, methods considering

Table 1.1 Signals in read alignments that suggest common types of structural variation. Each possible variant corresponds to a distinct combination of signals

Type of structural variant	Paired read (PR) signal	Read depth (RD) signal	Split read (SR) signal
Deletion	Mapped distance too large	Decreased	Read prefix and suffix align to opposite ends of variant
Tandem duplication	Discordant read orientations	Increased	Read prefix and suffix aligns to same region
Novel insertion (small)	Mapped distance too small (only insertions smaller than fragment length)	N/A	Read contains insertion (only insertions smaller than read length)
Inversion	Discordant read orientations	Decreased only at the breakend junction	Read prefix and suffix align in opposite orientation
Reciprocal translocation	Reads map to different chromosomes	Decreased only at the breakend junction	Read prefix and suffix align to distinct chromosomes

multiple alignments select a single mapping for each fragment before making a final set of predictions. Algorithmically, this is usually accomplished by minimizing the total number of predicted variants, as in the set cover mode from Variation Hunter-SC, Hydra or Ritz *et al.*, (2010), or reporting only clusters above a specified probability distribution (VariationHunter-Pr) (Hormozdiari *et al.*, 2009).

Importantly, algorithms differ in their criteria for selecting discordant fragments and determining when two fragments were consistent with the same SV. Most methods, such as Variation Hunter, Hydra and GASV, select a fixed set of discordant mappings before making predictions. Other methods, such as Modil (Lee *et al.*, 2009) and Breakdancer-Mini (Chen *et al.*, 2009), detect insertions and deletions by comparing the observed length distribution of fragments at a particular location with the expected length distribution genome-wide. Insertions (smaller than the average fragment length) and deletions (of any size) represent statistically significant deviations from the expected length distribution. In addition, by considering a mixture of distributions Modil is able to classify an event as homozygous or heterozygous.

As for all methods which depend on mapped reads, methods using the PR signal are highly dependent on the quality of the mappings provided as well as the fragments themselves. The fragment length distribution impacts the length of variants that can be reported by PR-based methods. That is, only deletions that correspond to a statistically significant change in the mapped fragment length are detectable with the PR signal. The distribution also affects the uncertainty in SV location; namely the larger the variance of the fragment length distribution, the larger the region of uncertainty. However, as GASV (Sindi *et al.*, 2009) and Bashir *et al.* (2008) have shown, combining multiple independent discordant mappings can provide surprisingly high breakpoint resolution even with the high variance in lengths of fragments from Sanger sequencing.

Read depth

Following a similar approach to CNV detection with aCGH, methods using read depth (RD) signals identify copy number variants based on the depth of coverage of reads throughout the genome (Fig. 1.4). Deletions correspond to regions of reduced read depth while duplications will have increased read depth. The expected depth is frequently modelled as a Poisson distribution, following the approach of Lander and Waterman (1988), but more recently the negative binomial distribution has been suggested to provide a better approximation for read counts (Miller *et al.*, 2011). If the coverage is high enough, then a normal approximation has also been used (Yoon *et al.*, 2009). Regions of the genome with statistically significant departures from the expected RD signal are reported as CNVs.

Some methods, such as those described by Yoon *et al.* (2009) and Abyzov *et al.* (2011), begin by splitting the genome into short non-overlapping windows and determining counts of reads in these windows while others (Xie and Tammi, 2009) use a sliding window. Once the genome-wide read depth has been determined, the genome can be partitioned into intervals of gains or losses. A variety of techniques have been applied to analysis of the RD signal; for example, the circular binary segmentation algorithm (Olshen *et al.*, 2004; Campbell *et al.*, 2008; Chiang *et al.*, 2008), the mean-shift technique from image processing (Fukunaga and Hostetler, 1975; Abyzov *et al.*, 2011), hidden Markov models (Simpson *et al.*, 2010) and the expectation maximization algorithm (Wang *et al.*, 2012) have all been used to predict CNVs. An alternative approach to genome-wide segmentation was taken by the event-wise testing (EWT) method (Yoon *et al.*, 2009). First small regions exceeding statistically significant thresholds for gains and losses are identified. Later steps iteratively merge these small events into larger ones.

The RD signal is detectable from both single and paired-reads making it more broadly applicable than the PR signal; however, this signal has several limitations. First, the read depth signal, in isolation, is only effective at detecting CNVs and not copy-neutral SVs such as inversions, translocations and novel insertions. Second, the statistical power of RD based prediction methods will be directly related to the sequencing-coverage. In addition, since the length of the variant will impact the statistical significance of changes in coverage, longer CNVs are more reliably detected than short CNVs. Third, there are known sequencing biases in NGS sequencing platforms, such as the CG bias in Illumina. These biases need to be properly corrected before CNVs are predicted (Yoon *et al.*, 2009). Finally, because of difficulties mapping reads to repetitive regions of the genome, the ability to call CNVs in these regions is compromised.

Split reads

While methods based on paired-read PR or read depth RD provide an approximate location of a novel adjacency, sequenced reads containing the novel adjacency provide unambiguous resolution of SV location. However, since these reads will not have a full-alignment to the reference genome, additional steps must be taken to determine the appropriate paired-partial mappings to the reference genome (see Fig. 1.5).

Pindel (Ye *et al.*, 2009) was the first method to use split-read alignments to detect SVs from high-throughput sequencing data. Pindel relies on paired-reads, and looks for a split alignment of one read in the pair. Pindel first identifies candidate split reads by finding fragments where only one end has a high-quality mapping to the reference genome. These mapped reads provide an anchor for the mapping and, as such, are termed one end anchors, OEAs. Next, a pattern growth approach is used to determine the optimal split alignment of the second read within a bounded region of the anchor. In this approach, positions of all prefixes and suffixes are determined for each potential split read. A split read alignment corresponds to a choice for the prefix and suffix of a read. As originally written, Pindel did not allow gaps in the split read alignment and could only detect insertions and deletions. However, a more recent version of Pindel addresses both these limitations. SplazerS (Emde *et al.*, 2012), follows a similar approach to Pindel, but allows gaps in the alignment and does not require reads to be anchored; thus SplazerS can be applied to unanchored paired or even single read data.

Rather than consider the alignment of all possible prefixes and suffixes, other split-read approaches align only a subset of substrings from each read. GSNAP uses a hash approach to index every third 12-mer in the genome; these 12-mers are then used these as seeds for read alignment (Wu and Nacu, 2010). While Splitread (Karakoc *et al.*, 2011) considers each half of a candidate split read by splitting the read in the middle. Regardless of where the novel adjacency occurs within the read, at least one of these substrings has a full alignment to the reference genome. The authors distinguish between two types of split reads: a *balanced split*, in which case both halves of the read should align to the genome and an *unbalanced split*, in which only one of the subsequences will have a full alignment. These split-reads are then clustered into reads supporting the same variant using the balanced splits as seeds. A greedy set-cover approach is used to select a single alignment for each read. Because GSNAP and Splitread approaches are not based on local realignment, their algorithms are not limited to a maximum variant length.

Finally, with the development of more sophisticated mapping tools, such as BWA-SW, partial mappings of reads are reported. Thus, methods are emerging which cluster these partial alignments as a special case of a discordant alignment (Zhang and Wu, 2011).

Using multiple signals for SV detection

Recently, algorithms have been developed to predict SVs by combining two or more of the three common signals (RP, PR and SR). By combining signals when predicting SVs, methods are more robust to potential false-positive signals impacting a single signal type; for example sequencing bias is likely to have a larger bias on the RD signal. We describe the recent approaches taken when combining SV signals and discuss possible improvements.

Combining PR with SR

Most algorithms, thus far, have combined discordant paired-read mappings (PR) with split-read mappings (SR) with a two-step approach for SV prediction (Jiang *et al.*, 2012; Rausch *et al.*, 2012; Zhang *et al.*, 2011, 2012; Pindel 0.20 unpublished). First, discordant paired-read mappings are used to determine regions of the test genome likely to contain a SV and then these regions are examined for evidence of SR by identifying OEAs or soft-clipped reads, indicating potential split-reads. For example, a simple two-tiered approach is used by a new version of Pindel (v 0.2.0) (unpublished); Breakdancer predictions (PR signal) are used as a pre-processing step to determine regions of the genome to analyse for split reads. Since the PR signal isolates potential breakends to relatively small intervals, a more careful alignment of split-reads is possible than in a genome-wide search and there is no need for an upper limit on variant size.

The pair-read informed split mapping (PRISM) method (Jiang *et al.*, 2012) adapts the Needleman–Wunsch algorithm for determining split-read alignments where two dynamic programming matrices, one for each side of the split, are populated. DELLY (Rausch *et al.*, 2012) follows a similar approach to PRISM, but uses a two-tiered approach to the SR analysis to avoid computing a split-alignment for each read. First, a *k*-mer index is determined for the regions and *k*-mer counts are used to identify candidate split-reads. Next a consensus sequence is built based on the split-reads and re-aligned to the reference genome with a similar double dynamic programming approach to DELLY.

Combining PR with RD

Alternative approaches to SV prediction have been taken when combining PR with RD. For example, the CNVer (Medvedev *et al.*, 2010) method uses the discordant paired-read signal to supplement CNV predictions made from read depth while GASVPro (Sindi *et al.*, 2012) uses read depth to refine predictions made from discordantly mapped paired-reads. As for methods using the RD signal alone, biases in the sequencing process must be considered by approaches combining RD with other signals.

The CNVer method (Medvedev *et al.*, 2010) begins by partitioning the genome into maximal distinct intervals by self-alignment of the reference and then determines which of these intervals correspond to CNVs. CNVer constructs a donor graph, a graph representing the test genome, where vertices are these distinct intervals and edges correspond to either known adjacencies in the reference or novel adjacencies predicted by clustering discordant fragments. Each interval is then associated with its read depth (RD) and CNVs are predicted by solving a minimum cost flow problem. Thus, in addition to predicting CNVs, CNVer estimates the copy count in the test genome. A similar approach was used in the Zinfandel method (Shen *et al.*, 2011), which used a hidden Markov model (HMM) and modelled transitions between normal, elevated and decreased copy number genome-wide.

Alternatively, GASVPro uses a generative probabilistic model, based on Poisson coverage (Lander and Waterman, 1988), to determine the likelihood for each potential SV identified with the PR signal. GASVPro first clusters all discordant paired-reads and then uses the read depth signal appropriate for the type of variant suggested by the PR signal. For deletions, GASVPro determines the read depth in the minimally deleted interval, a true deletion should have reduced read depth throughout this interval. For inversions and translocations, GASVPro considers a localized read depth only at a novel adjacency, the breakend read depth (beRD). If there is a novel adjacency in the test genome between positions *a* and *b*, no reads containing these adjacencies should map concordantly to the reference. For each prediction, GASVPro determines the probability that the configuration of reads represents a true variant, as opposed to an error, and reports the log-likelihood of a variant, the log of the ratio of the two probabilities.

Combining RD with SR

While most methods using multiple signals have relied on the presence of discordant paired-reads, a recent approach demonstrates the use of combined read depth and split read signals to determine CNVs (Nord *et al.*, 2011). First a straightforward method is applied to call CNVs by analysing the read depth in a sliding window. As in the EWT approach (Yoon *et al.*, 2009), a normal approximation is used for the read depth in a window; small regions with aberrant *Z*-scores are identified and merged into larger variants. The CNVs predicted by RD are then examined for a signature of split-reads whose alignments did not extend beyond the breakpoint. Taking advantage of recent alignment tools that report partial alignments of reads, through soft-clipping the read sequence, this method identifies all soft-clipped reads as potential split-reads.

Local assembly for SV prediction

In genome assembly, an unknown test genome is reconstructed from short sequenced fragments. Recently, assembly has emerged as a powerful method in SV identification. In contrast to *de novo* assembly of the entire genome sequence, these approaches perform local assembly from a subset of reads that are determined as likely to contain a novel adjacency. The resulting contiguous sequence from the local assembly provides a reconstruction of the variant itself and can be compared to the reference genome. Importantly, this is the only technique thus far discussed that is capable of determining all SV types and that also has the promise to reconstruct complex variants.

Local assembly was first used to predict novel insertions. Since novel insertions represent sequence not present in the reference genome, assembly was the only way to determine these sequences. The use of local assembly to determine novel insertions was applied to studies using Sanger sequencing (Kidd *et al.*, 2008). Researchers identified fragments where only one end read mapped to the reference genome. These one-ended anchors and their corresponding unmapped mated reads were assembled into a contiguous sequence, a contig, with the TIGR assembler (Sutton *et al.*, 1995). This approach has since been adapted for NGS data in a high-throughput fashion with NovelSeq (Hajirasouliha *et al.*, 2010).

Later, local assembly was employed as a step in the validation of SV predictions (Tuzun *et al.*, 2005; TIGRA Assembler, unpublished) but due to the increasing coverage in sequencing studies, assembly has developed into an independent tool for the discovery of all types of SVs. More recent methods predict SVs by aligning the contigs to the reference genome (Abyzov and Gerstein, 2011; Wang *et al.*, 2011).

CREST (Wang *et al.*, 2011) detects SVs by assembling contigs starting from reads having only partial alignments to the reference genome. After pulling in all possible reads believed to originate from the same genomic region, the contig is assembled and aligned to the reference. However, aligning these contigs to the reference genome is complicated by the presence of the SV itself. AGE (Assembly with Gap Excision) (Abyzov and Gerstein, 2011) addresses the problem of aligning in the presence of a structural variant by developing a double dynamic programming method for sequence alignment of a contig to the reference by allowing a single large jump in the alignment. This allows a contig containing a copy number variant to be aligned to the reference; a slight modification to the algorithm is used to align a contig containing an inversion relative to the reference.

While still in their infancy, the value of assembly-based methods for SV detection is likely to increase with increasing sequencing coverage.

Discussion

There have been tremendous achievements made in DNA sequencing technologies and corresponding computational techniques for SV prediction, with many recent techniques integrating multiple signals of potential SVs in a genome of interest. However, there remains room to improve upon current approaches. Moreover, many open questions remain surrounding the generation of SVs and their role in human genetics.

First, as described in the previous sections, nearly all of the current methods for SV identification that use multiple signals are not truly integrative. Most use only two of the three common signals for detection and primarily use multiple signals to refine predictions made by a single dominant signal. This two-tiered approach means that, while the specificity of prediction may be increased, the full sensitivity of the signals has not been realized. Complicating the process of full signal integration is that not all variants are detectable with all signals (see Table 1.1). Analysis from the 1000 Genomes Project (Mills *et al.*, 2011) has shown that, while many variants were detectable with multiple signals, there were differences in the signals based on genomic region and accessible size-range. Thus, a more feasible approach would be to use all sequenced data from a particular locus as evidence to assess the likelihood of a variant.

Second, most methods restrict the possible mappings for each fragment. Many methods consider only a single mapping for each read/fragment, thereby limiting the ability to predict SVs in repetitive regions. Even methods that consider multiple mappings for reads usually restrict these mappings to one particular category of signal. For example, methods such as VariationHunter or Hydra consider multiple possible discordant alignments for fragments, but do not simultaneously examine the one-end anchors that Pindel analyses. By allowing only a subset of possible alignments at the outset, incorrect mappings of a particular read or fragment will percolate through an analysis. Rather than treating a mapping as an immutable part of the data, a truly integrative approach would allow for the same fragment to contribute to multiple signals throughout the analysis. By integrating the read alignment as part of the SV prediction pipeline, a single read could be simultaneously considered for both full and split-alignments or discordant and concordant alignments without classifying the mapping before SV prediction.

Third, while the number of SV detection methods has increased considerably, the ability to compare the relative performance of methods is complicated. Typically each method publishes their own evaluation, perhaps only on simulated data where variants were randomly added to a genome. Often the results presented reflect pre-processing of read alignments or post-processing of predictions that may be poorly documented making it difficult to reproduce results. One trivial example is the use of different thresholds for mapping quality or maximum size of a predicted variant. Greater care in documenting SV analysis pipelines would aid in method development as well as better guide researchers using available SV software. One step that would ease both method comparison as well as guide further studies would be the establishment of standard datasets. Recent efforts like the 1000 Genomes Project (Mills *et al.*, 2011) and assembly of the Venter genome (Levy *et al.*, 2007) have collected and characterized over 20,000 structural variants. These repositories could serve as a standard for benchmarking future SV methods or as a resource for generating simulations of SVs.

Lastly, an important aspect of the analysis and identification of structural variants is how to represent the location of a variant itself. Often, the breakends of a structural variant are not known to the resolution of a single nucleotide. The level of uncertainty in the location of the breakend typically corresponds to the method of identification. For example, FISH identifies variants at relatively low resolution; arrayCGH, at best, identifies variants to a position between two probes on the array; DNA sequencing identifies variants at a higher resolution, but typically to a region larger than a nucleotide that depends on the size of the sequenced fragments. Despite the inherent ambiguity in localizing the breakends in a predicted variant, structural variants are often reported to the specific nucleotide. For example, although the Database of Genomic Variants annotates each entry with its originating study, the coordinates are often specified to the nucleotide resolution. In addition, methods for SV prediction from sequencing data often do not specify the uncertainty of their predictions, as discussed in Sindi *et al.* (2012). The loss of uncertainty makes it difficult to compare sets of predicted variants to one another, and determine if a predicted variant is merely located near previously identified variants or represents an entirely novel structure.

From the 1000 Genomes Project an effort has emerged to standardize the representation of SVs. The variant calling format (VCF) was established as a standard to represent structural variants, from single nucleotide SNPs to large rearrangements in an unambiguous fashion. The format supports exact precision of breakpoints, when they are known, but also allows uncertainty in location. In addition, the format supports annotation with additional information such as small gains/losses, which may occur at breakend junctions. As the VCF continues to be used, comparisons between predictions and previously identified variants will be greatly simplified.

Future trends

Structural variation with mixtures of data

Much of the original effort in analysing structural variation has been focused on identifying variants in a single genome from a single sequencing effort. However, as DNA sequencing continues to change, it becomes possible to simultaneously analyse data from multiple individuals and sample preparations. A special case of data mixtures arises in cancer genomics. To isolate cancer-specific variants, a common procedure is to sequence both the DNA from cancer cells and normal cells in the same individual (Campbell *et al.*, 2008; Quinlan and Hall, 2012; Oesper *et al.*, 2012). Algorithms for SV detection based on a single signal could be run on fragments sequenced from many individuals (Sindi *et al.*, 2009; Hormozdiari *et al.*, 2011). However, more sophisticated methods considering multiple signals from mixtures of data have only begun to emerge (Magi *et al.*, 2011).

Finally, distinct types of data from the same individual may together implicate a structural variant. Thus far, few methods have been developed that can combine, for example, aCGH and paired-read data from the same individual (Sindi *et al.*, 2009). However, as knowledge of structural variation increases, integrating data from prior analyses or prior knowledge will continue to be valuable. In a call back to early work in *Drosophila*, a recent study combined discordant paired-reads with high F_{st}, a signature of nucleotide variation expected to be associated with inversion breakpoints, to detect and assemble the breakpoints of inversion polymorphisms (Corbett-Detig *et al.*, 2012).

Detection of complex variants

Current SV identification methods have focused on specific classes of SVs when making predictions. However, there are many rearrangement events that do not fall into one of the common categories. Complex structural variants (Quinlan and Hall, 2012) cannot be distinguished by a single rearrangement event and are problematic to detect with current SV identification approaches. While these events are primarily associated with cancer genomes (Stephens *et al.*, 2011), SVs occur in hot-spots in the genome (Mills *et al.*, 2011) causing variants to co-localize and interact. Thus, even for noncancerous genomes there are difficulties in decomposing the evidence at a single locus into simple rearrangement events.

To address the challenges of complex variants, methods will need to be developed that consider detecting more than a novel adjacency in isolation, but characterizing the set of adjacencies interacting with one another. Some efforts along this line have recently been proposed for complex rearrangements in cancer genomes (Greenman *et al.*, 2012; McPherson *et al.*, 2012; Oesper *et al.*, 2012). A combination of such linking approaches, with a local assembly of the test genome, should provide insight into the rearrangements creating a complex event. In addition, the longer reads emerging from single-molecule sequencing approaches will further aid this analysis.

Targeted sequencing approaches

While there are still questions to be answered about SVs identified from genome-wide sequencing data, many new kinds of questions are emerging from recent trends in sequencing. Targeted sequencing approaches, such as Exome-Seq and RNA-Seq data motivate the analysis of sequences for splice junctions and alternative splicing events (Karakoc *et al.*, 2011; Krumm *et al.*, 2012). Although methods have begun to be developed to address these questions, most are again based on a single signal of detection, such as read depth with CoNIFER (Krumm *et al.*, 2012) and SR with SpliceMap (Au *et al.*, 2010). It is likely the field will undergo a similar evolution to SV detection methods, and combine signals for detection.

Personalized medicine

A major motivation behind SV identification has been the promise of personalized medicine (Lee and Morton, 2008; Valencia and Hidalgo, 2012). Although knowledge about SVs has increased tremendously in recent years, there are still major technical and scientific obstacles towards integrating SV discovery into a medical setting. First, the costs of genome sequencing, while decreasing, remain high enough to limit its use in common medical setting. Second, although numerous structural variants have been associated with diseases (Stankiewicz and Lupski, 2010), most variants need to be better characterized.

Evolutionary significance

From an evolutionary perspective, the implications of SVs are fascinating. Previous work has postulated that the presence of variation can contribute to speciation. Large variants can act to restrict recombination between chromosomes, which can contribute to reproductive isolation between haplotypes (Noor *et al.*, 2001). Intriguingly, Feuk *et al* 2006 showed that several of the inversions distinguishing human from chimp are still polymorphic in humans. Other studies have identified micro-inversions between human and chimp (Chaisson *et al.*, 2006; Hou *et al.*, 2011).

Little work has been done in the population genetics of human structural variants and the possible evolutionary roles they may have had. The importance of SVs, particularly inversions, has been demonstrated through detailed studies in *Drosophila*. Thus, the overall role of SVs as drivers of evolution, or more recently emerged genomic events is still largely unknown. But, efforts such as the 1000 Genomes Project and recent SV and CNV studies in related species (Quinlan *et al.*, 2010; Alvarez and Akey, 2012; Liu and Bickhart, 2012; Yalcin *et al.*, 2012) may represent a transition point in the evolutionary analysis of SVs.

Web resources

Database of Genomic Variants (http://projects.tcag.ca/variation/)

NCBI DBVar (http://www.ncbi.nlm.nih.gov/dbvar/)

NHGRI Structural Variation Project (http://www.ncbi.nlm.nih.gov/projects/genome/StructuralVariation/NHGRIStructuralVariation.shtml)

Variant Calling format (http://vcftools.sourceforge.net/specs.html)

1000 Genomes Site (http://www.1000genomes.org/)

Human Genome Browser (http://genome.ucsc.edu/)

References

Abyzov, A., and Gerstein, M. (2011). AGE: defining breakpoints of genomic structural variants at single-nucleotide resolution, through optimal alignments with gap excision. Bioinformatics 27, 595–603.

Abyzov, A., Urban, A.E., Snyder, M., and Gerstein, M. (2011). CNVnator: An approach to discover, genotype, and characterize typical and atypical CNVs from family and population genome sequencing. Genome Res. *21*, 974–984.

Alkan, C., Kidd, J.M., Marques-Bonet, T., Aksay, G., Antonacci, F., Hormozdiari, F., Kitzman, J.O., Baker, C., Malig, M., and Mutlu, O. (2009). Personalized copy number and segmental duplication maps using next-generation sequencing. Nat. Genet. *41*, 1061–1067.

Alvarez, C.E., and Akey, J.M. (2012). Copy number variation in the domestic dog. Mamm. Genome 1–20.

Au, K.F., Jiang, H., Lin, L., Xing, Y., and Wong, W.H. (2010). Detection of splice junctions from paired-end RNA-seq data by SpliceMap. Nucleic Acids Res. *38*, 4570–4578.

Bamford, S., Dawson, E., Forbes, S., Clements, J., Pettett, R., Dogan, A., Flanagan, A., Teague, J., Futreal, P., and Stratton, M. (2004). The COSMIC (Catalogue of Somatic Mutations in Cancer) database and website. Br. J. Cancer *91*, 355–358.

Bansal, V., Bashir, A., and Bafna, V. (2007). Evidence for large inversion polymorphisms in the human genome from HapMap data. Genome Res. *17*, 219–230.

Bashir, A., Volik, S., Collins, C., Bafna, V., and Raphael, B.J. (2008). Evaluation of paired-end sequencing strategies for detection of genome rearrangements in cancer. PLoS Comput. Biol. *4*, e1000051.

Beck, C.R., Garcia-Perez, J.L., Badge, R.M., and Moran, J.V. (2011). LINE-1 elements in structural variation and disease. Annu. Rev. Genomics Hum. Genet. *12*, 187–215.

Bentley, D.R., Balasubramanian, S., Swerdlow, H.P., Smith, G.P., Milton, J., Brown, C.G., Hall, K.P., Evers, D.J., Barnes, C.L., and Bignell, H.R. (2008). Accurate whole human genome sequencing using reversible terminator chemistry. Nature *456*, 53–59.

Burrows, M., and Wheeler, D.J. (1994). A block-sorting lossless data compression algorithm. Technical report 124. Palo Alto, CA: Digital Equipment Corporation.

Campbell, P.J., Stephens, P.J., Pleasance, E.D., O'Meara, S., Li, H., Santarius, T., Stebbings, L.A., Leroy, C., Edkins, S., and Hardy, C. (2008). Identification of somatically acquired rearrangements in cancer using genome-wide massively parallel paired-end sequencing. Nat. Genet. *40*, 722–729.

Chaisson, M.J., and Tesler, G. (2012). Mapping single molecule sequencing reads using basic local alignment with successive refinement (BLASR): application and theory. BMC Bioinformatics *13*, 238.

Chaisson, M.J., Raphael, B.J., and Pevzner, P.A. (2006). Microinversions in mammalian evolution. Proc. Natl. Acad. Sci. U.S.A. *103*, 19824–19829.

Chen, K., Wallis, J.W., McLellan, M.D., Larson, D.E., Kalicki, J.M., Pohl, C.S., McGrath, S.D., Wendl, M.C., Zhang, Q., and Locke, D.P. (2009). BreakDancer: an algorithm for high-resolution mapping of genomic structural variation. Nat. Methods *6*, 677–681.

Chiang, D.Y., Getz, G., Jaffe, D.B., O'Kelly, M.J.T., Zhao, X., Carter, S.L., Russ, C., Nusbaum, C., Meyerson, M., and Lander, E.S. (2008). High-resolution mapping of copy-number alterations with massively parallel sequencing. Nat. Methods *6*, 99–103.

Church, D.M., Lappalainen, I., Sneddon, T.P., Hinton, J., Maguire, M., Lopez, J., Garner, J., Paschall, J., DiCuccio, M., and Yaschenko, E. (2010). Public data archives for genomic structural variation. Nat. Genet. *42*, 813–814.

Conrad, D.F., Andrews, T.D., Carter, N.P., Hurles, M.E., and Pritchard, J.K. (2005). A high-resolution survey of deletion polymorphism in the human genome. Nat. Genet. *38*, 75–81.

Corbett-Detig, R.B., Cardeno, C., and Langley, C.H. (2012). Sequence-based detection and breakpoint assembly of polymorphic inversions. Genetics *192*, 131–137.

Database of Genomic Variants (2012). Available at: http://projects.tcag.ca/variation/

Emde, A.K., Schulz, M.H., Weese, D., Sun, R., Vingron, M., Kalscheuer, V.M., Haas, S.A., and Reinert, K. (2012). Detecting genomic indel variants with exact breakpoints in single- and paired-end sequencing data using SplazerS. Bioinformatics *28*, 619–627.

Faust, G.G., and Hall, I.M. (2012). YAHA: fast and flexible long-read alignment with optimal breakpoint detection. Bioinformatics *28*, 2417–2424.

Fukunaga, K., and Hostetler, L. (1975). The estimation of the gradient of a density function, with applications in pattern recognition. IEEE Trans. Inf. Theory *21*, 32–40.

Gibbs, R.A., Belmont, J.W., Hardenbol, P., Willis, T.D., Yu, F., Yang, H., Ch'ang, L.Y., Huang, W., Liu, B., and Shen, Y. (2003). The international HapMap project. Nature *426*, 789–796.

Greenman, C., Stephens, P., Smith, R., Dalgliesh, G.L., Hunter, C., Bignell, G., Davies, H., Teague, J., Butler, A., and Stevens, C. (2007). Patterns of somatic mutation in human cancer genomes. Nature *446*, 153–158.

Greenman, C.D., Pleasance, E.D., Newman, S., Yang, F., Fu, B., Nik-Zainal, S., Jones, D., Lau, K.W., Carter, N., and Edwards, P.A.W. (2012). Estimation of rearrangement phylogeny for cancer genomes. Genome Res. *22*, 346–361.

Gupta, P.K. (2008). Single-molecule DNA sequencing technologies for future genomics research. Trends Biotechnol. *26*, 602–611.

Hach, F., Hormozdiari, F., Alkan, C., Hormozdiari, F., Birol, I., Eichler, E.E., and Sahinalp, S.C. (2010). mrsFAST: a cache-oblivious algorithm for short-read mapping. Nat. Methods *7*, 576–577.

Hajirasouliha, I., Hormozdiari, F., Alkan, C., Kidd, J.M., Birol, I., Eichler, E.E., and Sahinalp, S.C. (2010). Detection and characterization of novel sequence insertions using paired-end next-generation sequencing. Bioinformatics *26*, 1277–1283.

Higuchi, D.A., Cahan, P., Gao, J., Ferris, S.T., Poursine-Laurent, J., Graubert, T.A., and Yokoyama, W.M. (2010). Structural variation of the mouse natural killer gene complex. Genes Immun. *11*, 637–648.

Homer, N., Merriman, B., and Nelson, S.F. (2009). BFAST: an alignment tool for large scale genome resequencing. PLoS One *4*, e7767.

Hormozdiari, F., Alkan, C., Eichler, E.E., and Sahinalp, S.C. (2009). Combinatorial algorithms for structural variation detection in high-throughput sequenced genomes. Genome Res. *19*, 1270–1278.

Hormozdiari, F., Hajirasouliha, I., McPherson, A., Eichler, E.E., and Sahinalp, S.C. (2011). Simultaneous structural variation discovery among multiple paired-end sequenced genomes. Genome Res. *21*, 2203–2212.

Hou, M., Yao, P., Antonou, A., and Johns, M.A. (2011). Pico-inplace-inversions between human and chimpanzee. Bioinformatics *27*, 3266–3275.

Jiang, Y., Wang, Y., and Brudno, M. (2012). PRISM: Pair read informed split read mapping for base pair level detection of insertion, deletion and structural variants. Bioinformatics *28*, 2576–2583.

Karakoc, E., Alkan, C., O'Roak, B.J., Dennis, M.Y., Vives, L., Mark, K., Rieder, M.J., Nickerson, D.A., and Eichler, E.E. (2011). Detection of structural variants and indels within exome data. Nat. Methods *9*, 176–178.

Karayiorgou, M., Simon, T.J., and Gogos, J.A. (2010). 22q11. 2 microdeletions: linking DNA structural variation to brain dysfunction and schizophrenia. Nat. Rev. Neurosci. *11*, 402–416.

Kidd, J.M., Cooper, G.M., Donahue, W.F., Hayden, H.S., Sampas, N., Graves, T., Hansen, N., Teague, B., Alkan, C., and Antonacci, F. (2008). Mapping and sequencing of structural variation from eight human genomes. Nature *453*, 56–64.

Kim, R.N.A.M., Kim, A., Kim, D.W., Choi, S.H., Kim, D.A.E.S.O.O., Nam, S.H., Kang, A., Kim, M.I.N.Y., Park, K.U.N.H., and Yoon, B.H.A. (2012). Analysis of indel variations in the human disease-associated genes CDKN2AIP, WDR66, USP20 and OR7C2 in a Korean population. J. Genet. 1–11.

Korbel, J.O., Abyzov, A., Mu, X.J., Carriero, N., Cayting, P., Zhang, Z., Snyder, M., and Gerstein, M.B. (2009). PEMer: a computational framework with simulation-based error models for inferring genomic structural variants from massive paired-end sequencing data. Genome Biol. *10*, R23.

Krumm, N., Sudmant, P.H., Ko, A., O'Roak, B.J., Malig, M., Coe, B.P., Quinlan, A.R., Nickerson, D.A., and Eichler, E.E. (2012). Copy number variation detection and genotyping from exome sequence data. Genome Res. *22*, 1525–1532.

Klus, P., Lam, S., Lyberg, D., Cheung, M.S., Pullan, G., McFarlane, I., Yeo, G.S.H., and Lam, B.Y.H. (2012). BarraCUDA-a fast short read sequence aligner using graphics processing units. BMC Res. Notes *5*, 27.

Lai, W.R., Johnson, M.D., Kucherlapati, R., and Park, P.J. (2005). Comparative analysis of algorithms for identifying amplifications and deletions in array CGH data. Bioinformatics *21*, 3763–3770.

Lander, E.S., and Waterman, M.S. (1988). Genomic mapping by fingerprinting random clones: a mathematical analysis. Genomics *2*, 231–239.

Langmead, B., and Salzberg, S.L. (2012). Fast gapped-read alignment with Bowtie 2. Nat. Methods *9*, 357–359.

Langmead, B., Trapnell, C., Pop, M., and Salzberg, S.L. (2009). Ultrafast and memory-efficient alignment of short DNA sequences to the human genome. Genome Biol. *10*, R25.

Lee, C., and Morton, C.C. (2008). Structural genomic variation and personalized medicine. N. Engl. J. Med. *358*, 740–741.

Lee, J.A., Carvalho, C., and Lupski, J.R. (2007). A DNA replication mechanism for generating nonrecurrent rearrangements associated with genomic disorders. Cell *131*, 1235–1247.

Lee, S., Cheran, E., and Brudno, M. (2008). A robust framework for detecting structural variations in a genome. Bioinformatics *24*, i59-i67.

Lee, S., Hormozdiari, F., Alkan, C., and Brudno, M. (2009). MoDIL: detecting small indels from clone-end sequencing with mixtures of distributions. Nat. Methods *6*, 473–474.

Levy, S., Sutton, G., Ng, P.C., Feuk, L., Halpern, A.L., Walenz, B.P., Axelrod, N., Huang, J., Kirkness, E.F., and Denisov, G. (2007). The diploid genome sequence of an individual human. PLoS Biol. *5*, e254

Li, H., and Durbin, R. (2009). Fast and accurate short read alignment with Burrows–Wheeler transform. Bioinformatics *25*, 1754–1760.

Li, H., and Durbin, R. (2010). Fast and accurate long-read alignment with Burrows–Wheeler transform. Bioinformatics *26*, 589–595.

Li, H., and Homer, N. (2010). A survey of sequence alignment algorithms for next-generation sequencing. Brief. Bioinform. *11*, 473–483.

Li, H., Ruan, J., and Durbin, R. (2008a). Mapping short DNA sequencing reads and calling variants using mapping quality scores. Genome Res. *18*, 1851–1858.

Li, R., Li, Y., Kristiansen, K., and Wang, J. (2008b). SOAP: short oligonucleotide alignment program. Bioinformatics *24*, 713–714.

Li, R., Yu, C., Li, Y., Lam, T.W., Yiu, S.M., Kristiansen, K., and Wang, J. (2009). SOAP2: an improved ultrafast tool for short read alignment. Bioinformatics *25*, 1966–1967.

Li, W., and Olivier, M. (2012). Current analysis platforms and methods for detecting copy number variation. Physiol. Genomics *45*, 1–16.

Liu, G.E., and Bickhart, D.M. (2012). Copy number variation in the cattle genome. Funct. Integr. Genomics 1–16.

Liu, C.M., Wong, T., Wu, E., Luo, R., Yiu, S.M., Li, Y., Wang, B., Yu, C., Chu, X., and Zhao, K. (2012). SOAP3: ultra-fast GPU-based parallel alignment tool for short reads. Bioinformatics *28*, 878–879.

Magi, A., Benelli, M., Yoon, S., Roviello, F., and Torricelli, F. (2011). Detecting common copy number variants in high-throughput sequencing data by using JointSLM algorithm. Nucleic Acids Res. *39*, e65–e65.

Mani, R.S., and Chinnaiyan, A.M. (2010). Triggers for genomic rearrangements: insights into genomic, cellular and environmental influences. Nat. Rev. Genet. *11*, 819–829.

Mardis, E.R. (2011). A decade's perspective on DNA sequencing technology. Nature *470*, 198–203.

McCarroll, S.A., and Altshuler, D.M. (2007). Copy-number variation and association studies of human disease. Nat. Genet. *39*, S37-S42.

McPherson, A.W., Wu, C., Wyatt, A., Shah, S.P., Collins, C., and Sahinalp, S.C. (2012). nFuse: Discovery of complex genomic rearrangements in cancer using high-throughput sequencing. Genome Res.

Medvedev, P., Fiume, M., Dzamba, M., Smith, T., and Brudno, M. (2010). Detecting copy number variation with mated short reads. Genome Res. *20*, 1613–1622.

Metzker, M.L. (2009). Sequencing technologies – the next generation. Nat. Rev. Genet. *11*, 31–46.

Miller, C.A., Hampton, O., Coarfa, C., and Milosavljevic, A. (2011). ReadDepth: a parallel R package for detecting copy number alterations from short sequencing reads. PLoS One *6*, e16327.

Mills, R.E., Walter, K., Stewart, C., Handsaker, R.E., Chen, K., Alkan, C., Abyzov, A., Yoon, S.C., Ye, K., and Cheetham, R.K. (2011). Mapping copy number variation by population-scale genome sequencing. Nature *470*, 59–65.

Noor, M.A.F., Grams, K.L., Bertucci, L.A., and Reiland, J. (2001). Chromosomal inversions and the reproductive isolation of species. Proc. Natl. Acad. Sci. *98*, 12084–12088.

Nord, A.S., Lee, M., King, M.C., and Walsh, T. (2011). Accurate and exact CNV identification from targeted high-throughput sequence data. BMC Genomics *12*, 184.

Novocraft: Novoalign. Available at: http://www.novocraft.com/main/index.php.

Oesper, L., Ritz, A., Aerni, S.J., Drebin, R., and Raphael, B.J. (2012). Reconstructing cancer genomes from paired-end sequencing data. BMC Bioinformatics 13(Suppl 6), S10.

Olshen, A.B., Venkatraman, E., Lucito, R., and Wigler, M. (2004). Circular binary segmentation for the analysis of array-based DNA copy number data. Biostatistics *5*, 557–572.

Painter, T.S. (1934). A new method for the study of chromosome aberrations and the plotting of chromosome maps in Drosophila melanogaster. Genetics *19*, 175.

Pindel. (2012). Available at: https://trac.nbic.nl/pindel/

Quinlan, A.R., and Hall, I.M. (2012). Characterizing complex structural variation in germline and somatic genomes. Trends Genet. *28*, 43–53.

Quinlan, A.R., Clark, R.A., Sokolova, S., Leibowitz, M.L., Zhang, Y., Hurles, M.E., Mell, J.C., and Hall, I.M. (2010). Genome-wide mapping and assembly of structural variant breakpoints in the mouse genome. Genome Res. *20*, 623–635.

Raphael, B.J., Volik, S., Collins, C., and Pevzner, P.A. (2003). Reconstructing tumor genome architectures. Bioinformatics *19*, ii162-ii171.

Ried, T., Schröck, E., Ning, Y., and Wienberg, J. (1998). Chromosome painting: a useful art. Hum. Mol. Genet. *7*, 1619–1626.

Ritz, A., Bashir, A., and Raphael, B.J. (2010). Structural variation analysis with strobe reads. Bioinformatics *26*, 1291–1298.

Ritz, A., Paris, P.L., Ittmann, M.M., Collins, C., and Raphael, B.J. (2011). Detection of recurrent rearrangement breakpoints from copy number data. BMC Bioinformatics *12*, 114.

Rausch, T., Zichner, T., Schlattl, A., Stütz, A.M., Benes, V., and Korbel, J.O. (2012). DELLY: structural variant discovery by integrated paired-end and split-read analysis. Bioinformatics *28*, i333-i339.

Rumble, S.M., Lacroute, P., Dalca, A.V., Fiume, M., Sidow, A., and Brudno, M. (2009). SHRiMP: accurate mapping of short color-space reads. PLoS Comput. Biol. *5*, e1000386.

Shen, Y., Gu, Y., and Pe'er, I. (2011). A Hidden Markov Model for Copy Number Variant prediction from whole genome resequencing data. BMC Bioinformatics *12*, S4.

Simpson, J.T., McIntyre, R.E., Adams, D.J., and Durbin, R. (2010). Copy number variant detection in inbred strains from short read sequence data. Bioinformatics *26*, 565–567.

Sindi, S.S., and Raphael, B.J. (2010). Identification and frequency estimation of inversion polymorphisms from haplotype data. J. Comput. Biol. *17*, 517–531.

Sindi, S., Helman, E., Bashir, A., and Raphael, B.J. (2009). A geometric approach for classification and comparison of structural variants. Bioinformatics *25*, i222–30.

Sindi, S.S., Onal, S., Peng, L.C., Wu, H.T., and Raphael, B.J. (2012). An integrative probabilistic model for identification of structural variation in sequencing data. Genome Biol. *13*, R22.

Smith, T., and Waterman, M. (1981). Identification of common molecular subsequences. J. Mol. Biol. *147*, 195–197.

Stankiewicz, P., and Lupski, J.R. (2010). Structural variation in the human genome and its role in disease. Annu. Rev. Med. *61*, 437–455.

Stephens, P.J., Greenman, C.D., Fu, B., Yang, F., Bignell, G.R., Mudie, L.J., Pleasance, E.D., Lau, K.W., Beare, D., and Stebbings, L.A. (2011). Massive genomic rearrangement acquired in a single catastrophic event during cancer development. Cell *144*, 27–40.

Sturtevant, A. (1920). Genetic studies on Drosophila simulans. I. Introduction. Hybrids with *Drosophila melanogaster*. Genetics *5*, 488.

Sutton, G.G., White, O., Adams, M.D., and Kerlavage, A.R. (1995). TIGR Assembler: A new tool for assembling large shotgun sequencing projects. Genome Sci. Technol. *1*, 9–19.

T.I.G.R.A_S.V., The Genome Institute at Washington University. (2012). Available at: http://genome.wustl.edu/software/tigra_sv.

Tuzun, E., Sharp, A.J., Bailey, J.A., Kaul, R., Morrison, V.A., Pertz, L.M., Haugen, E., Hayden, H., Albertson, D., and Pinkel, D. (2005). Fine-scale structural variation of the human genome. Nat. Genet. *37*, 727–732.

Valencia, A., and Hidalgo, M. (2012). Getting personalized cancer genome analysis into the clinic: the challenges in bioinformatics. Genome Med. *4*, 61.

Variant Calling Format. Available at: http://vcftools.sourceforge.net/specs.html.

Volik, S., Zhao, S., Chin, K., Brebner, J.H., Herndon, D.R., Tao, Q., Kowbel, D., Huang, G., Lapuk, A., and Kuo, W.L. (2003). End-sequence profiling: sequence-based analysis of aberrant genomes. Proc. Natl. Acad. Sci. U.S.A. *100*, 7696–7701.

Wang, L., Abyzov, A., Korbel, J.O., Snyder, M., and Gerstein, M. (2009). MSB: A mean-shift-based approach for the analysis of structural variation in the genome. Genome Res. *19*, 106–117.

Wang, J., Mullighan, C.G., Easton, J., Roberts, S., Heatley, S.L., Ma, J., Rusch, M.C., Chen, K., Harris, C.C., and Ding, L. (2011). CREST maps somatic structural variation in cancer genomes with base pair resolution. Nat. Methods *8*, 652–654.

Wang, Z., Hormozdiari, F., Yang, W.Y., Halperin, E., and Eskin, E. (2012) E. CNVeM: Copy number variation detection using uncertainty of read mapping. In Research in Computational Molecular Biology. (Springer, Berlin) pp. 326–340.

Wu, T.D., and Nacu, S. (2010). Fast and SNP-tolerant detection of complex variants and splicing in short reads. Bioinformatics *26*, 873–881.

Xie, C., and Tammi, M.T. (2009). CNV-seq, a new method to detect copy number variation using high-throughput sequencing. BMC Bioinformatics *10*, 80.

Yalcin, B., Wong, K., Agam, A., Goodson, M., Keane, T.M., Gan, X., Nellåker, C., Goodstadt, L., Nicod, J., and Bhomra, A. (2011). Sequence-based characterization of structural variation in the mouse genome. Nature *477*, 326–329.

Yalcin, B., Wong, K., Bhomra, A., Goodson, M., Keane, T.M., Adams, D.J., and Flint, J. (2012). The fine-scale architecture of structural variants in 17 mouse genomes. Genome Biol. *13*, R18.

Ye, K., Schulz, M.H., Long, Q., Apweiler, R., and Ning, Z. (2009). Pindel: a pattern growth approach to detect break points of large deletions and medium sized insertions from paired-end short reads. Bioinformatics *25*, 2865–2871.

Yoon, S., Xuan, Z., Makarov, V., Ye, K., and Sebat, J. (2009). Sensitive and accurate detection of copy number variants using read depth of coverage. Genome Res. *19*, 1586–1592.

Zang, Z.J., Ong, C.K., Cutcutache, I., Yu, W., Zhang, S.L., Huang, D., Ler, L.D., Dykema, K., Gan, A., and Tao, J. (2011). Genetic and structural variation in the gastric cancer genome revealed through targeted deep sequencing. Cancer Res. *71*, 29–39.

Zhang, F., Gu, W., Hurles, M.E., and Lupski, J.R. (2009). Copy number variation in human health, disease, and evolution. Annu. Rev. Genomics Hum. Genet. *10*, 451–481.

Zhang, J., and Wu, Y. (2011). SVseq: an approach for detecting exact breakpoints of deletions with low-coverage sequence data. Bioinformatics *27*, 3228–3234.

Zhang, J., Wang, J., and Wu, Y. (2012). An improved approach for accurate and efficient calling of structural variations with low-coverage sequence data. BMC Bioinformatics *13*, S6.

Zhang, Z., Schwartz, S., Wagner, L., and Miller, W. (2000). A greedy algorithm for aligning DNA sequences. J. Comput. Biol. *7*, 203–214.

Zhang, Z.D., Du, J., Lam, H., Abyzov, A., Urban, A.E., Snyder, M., and Gerstein, M. (2011). Identification of genomic indels and structural variations using split reads. BMC Genomics *12*, 375.

Methods for RNA Isolation, Characterization and Sequencing (RNA-Seq)

2

Paul Zumbo and Christopher E. Mason

Abstract

Ribonucleic acid (RNA) is a key substrate for storing and transmitting biological information in cells, along with deoxyribonucleic acid (DNA), proteins, and other small molecules and metabolites. Since the discovery of nucleic acids by Friedrich Miescher in 1869 (Dahm *et al.*, 2008), RNA has been observed in an expanded range of functions within and between cells, tissues, and even between generations. In 1958, Francis Crick proposed the Central Dogma of Molecular Biology (Crick, 1958), and he placed RNA as a simple intermediary of unidirectional information transfer between DNA and proteins (Fig. 2.1). Yet, today, we know that a wide range of activities surround and violate this dogma, and recent work has shown that RNAs come in many varieties, serve essential regulatory and catalytic roles, and that RNA bases can harbour many small, chemical modifications that can also change its function. This chapter will review the history of RNA's expansion as a mediator and as a catalytic molecule in cells, the new methods developed to characterize and sequence RNA, and the means for contextualizing the roles of RNA.

A brief history of RNA

Methods for testing and analysing RNA date to the beginning of the 20th century, but focused work on teasing out the code of messenger RNA (mRNA) did not begin until the 1950s, with the isolation of mRNA in the laboratories of Francois Jacob, Jacques Monod, Elliot Volkin and Lazarus Astrachan, Sydney Brenner, and Francis Crick. Then, a race to elucidate the three-base genetic code started in 1961, when it was shown that UUU encoded phenylalanine (Nirenberg and Matthaei, 1961), and the race completed in 1966 with the role of each codon's targets in proteins, as well as the discernment of the three stop codons (UGA, UAG, UAA). Owing to the flexibility of RNA and its central role in information transfer, some scientists (Crick, Woese and Orgel) also proposed an 'RNA world' hypothesis, wherein the primordial molecules on Earth that led to the formation of complex life were RNA, rather than DNA or proteins (Woese, 1967).

However, methods for extracting RNA and characterizing these essential molecules were still limited. In 1970, the discovery of reverse transcriptase by Howard Temin (Temin and Mizutani, 1970) and David Baltimore (Baltimore, 1970) allowed for RNA to be converted into complementary DNA (cDNA), which transformed RNA into a more stable molecule and which enabled the use of methods that had been previously created for DNA sequencing. Once cDNA could be profiled and sequenced, the exact arrangement of genes' sequence outputs were constructed for the first time. This led to the discovery about adenoviruses in 1977 that showed RNAs are not simple, linear copies of their genomic template DNA, but rather that they can be rearranged, or 'spliced,' to create alternative isoforms of the original gene (Chow *et al.*, 1977). Interestingly, alternative splicing (AS) was initially thought to be a very rare event, and perhaps limited to viruses or only a few organisms, but it is now known that AS occurs in all eukaryotic organisms and to at least 92% of all [Refseq] genes (Wang *et al.*, 2008).

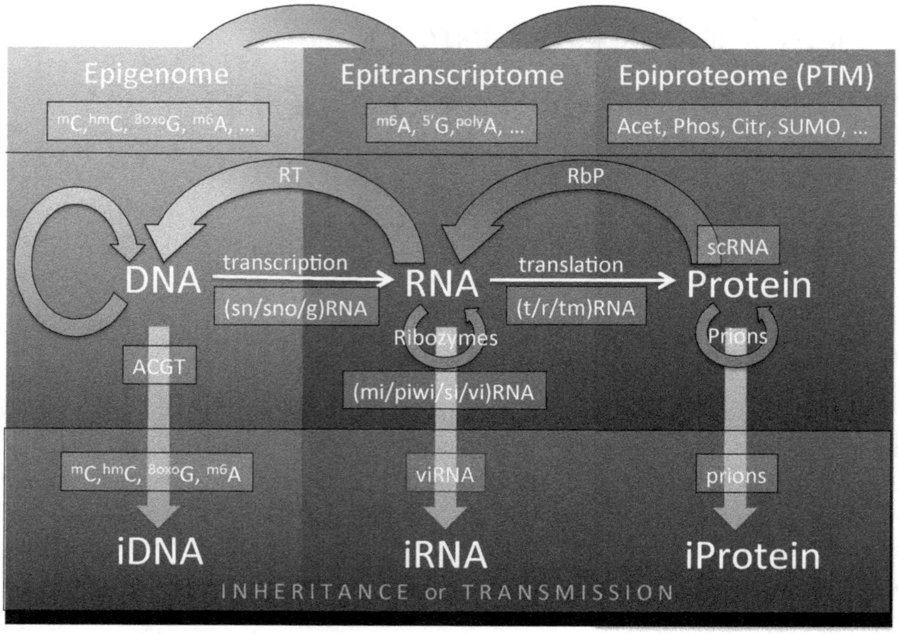

Adapted from Saletore *et al.*, Genome Biology, 2012

Figure 2.1 The increasing complexity of the central dogma of molecular biology. (Top) The originally proposed central dogma, with unidirectional information flow. (Bottom) The current view of the central dogma, wherein information content can flow 'backwards' or 'sideways' with reverse transcriptases (RT), RNA-binding proteins (RbPs) and RNA editing. Also, information can be copied within each of the three realms: genetic (blue) copying such as with transposable elements, transcriptional (red) copying with ribozymes and rich levels of RNA regulation using small RNAs (micro, piwi, si, vi RNAs), and proteomic (green) copying using prions. A colour version of this figure is located in the plate section at the back of the book.

Additional complexity of RNA was revealed in 1982, with the discovery of 'ribozymes,' a class of self-splicing, catalytic RNA molecules (Kruger *et al.*, 1982). The work on ribozymes demonstrated that RNAs can function independently and utilize their secondary structure to serve as enzymes for catalysing chemical reactions. Indeed, catalytic RNAs in large complexes of RNA (called ribosomes) were later shown to serve as the critical mediator for the formation of peptide bonds in proteins. When these complexes or ribosomes merge together to form polysomes, they enable efficient and dynamic regulation of protein translation.

However, until the 1990s, only a few classes of RNAs were known (tRNA, rRNA, mRNA) and most of the known RNAs were relatively large (>150 bp). In 1986, the observation that gene expression in transgenic plants could be modified by small, anti-sense RNA created speculation that a 'virus-induced gene silencing (VIGS)' or 'post-transcriptional gene silencing (PTGS)' mechanism was possible in transgenic plants. These results continued to be replicated in other plant species, which led researchers to hypothesize that plants used VIGS/PTGS as part of a viral defence system that targeted the alien RNA, with small RNAs as the mediator. In 1998, Andrew File and Craig Mello adapted these ideas for work in *Caenorhabditis elegans*, the nematode, and dramatically changed our understanding of RNA interference (RNAi) and the functions of small RNAs. They showed that the addition of double-stranded RNA (dsRNA) that was complementary to a specific mRNA could 'silence' the gene because the small, dsRNAs would bind to the mRNA, and the transcript would then be targeted for degradation – thus never becoming a protein.

RNAi was subsequently demonstrated in other invertebrates and then mammals, and it is now applied as a standard method in functional genomics to understand gene function and in pharmacological products for gene silencing. Also, the presence of these micro RNAs (miRNAs) has been shown in many organisms and demonstrated as a key mechanism of gene regulation that primarily targets the 3′UTR of genes. As a result of this, changes in the length of 3′UTRs and the polyadenylation sites (PASs) of genes can alter the number and type of miRNAs that can target a gene, which has led the competing endogenous RNA (ceRNA) hypothesis of gene regulation, whereby all potential miRNA binding sites can serve as 'sponges' to absorb the targets of miRNAs, which often are shared between many genes. Thus, when modelling gene expression changes that may be controlled by the RNA-induced silencing complex (RISC), one must understand all the potential targets of a miRNA and use this information in a model of gene expression.

However, other types of RNAs and RNA function have also recently emerged. Piwi-interacting RNAs (piwi-RNAs) are another class of small, noncoding RNAs that are slightly larger (26–31 nt) than most miRNAs (18–26nt), and they are believed to perform essential functions for germline tissues. In particular, piwiRNAs are required for spermatogenesis, germ-cell development, and stem cell development (Ruby *et al.*, 2006). Also, miRNAs have recently been demonstrated to be a non-Mendelian inheritance mechanism, carried forward as a viral-induced miRNA (viRNA) response that moves between generations (Rechavi *et al.*, 2011). Finally, long, intergenic noncoding RNAs (lincRNAs) and long, noncoding RNAs (lncRNAs) have been shown to be hallmarks of active and inactive regions of the genome, required for metastasis in some cancers, and important in overall epigenetic regulation (Gupta *et al.*, 2010). Indeed, in the most recent total of all types of genes (55,123) present in the human genome from the ENCODE (Encyclopaedia of DNA Elements) Project, there are now more non-coding RNAs present (21,566) than coding RNAs (20,070), which is a complete reversal from the expected roles of RNAs from the 1980s.

Currently, there are 27 known types of RNA across the Archaea, Bacteria and Eukaryotes (Table 2.1). As described above, these various classes of RNA can serve as substrates for biochemical reactions, catalytic entities, self-splicing RNAs, and also mediators of information between cells and generations. However, an experiment that aims to understand or characterize any RNA must first be designed with the appropriate biochemical approach, since there are multiple ways to isolate, purify, and characterize RNA. Although there is a long history of Northern blots, RNase protection assays, and microarray work for RNA characterization, we will focus this chapter on the methods and considerations for sequencing RNA.

Principles of RNA isolation

In order to experiment with RNA, one must first extract and purify it from the cellular milieu. Although there are a variety of methods for nucleic acid isolation, the most prevalent method for RNA extraction is the phenol–chloroform extraction using guanidinium salts. Volkin and Carter reported the first use of guanidinium chloride in the isolation of RNA in 1951 (Volkin *et al.*, 1951). In 1953, Grassmann and Defner described the efficacy of phenol at extracting proteins from aqueous solution (Grassman *et al.*, 1953). Utilizing this find, Kirby demonstrated the use of phenol to separate nucleic acids from proteins in 1956 (Kirby, 1956). Cox and others renewed interest in the use of guanidinium chloride in the isolation of RNA from ribonucleoproteins in the 1960s (Cox *et al.*, 1963; Cox, 1968, 1996). From then on, guanidinium extractions were the method of choice for RNA purification, replacing phenol extraction. The use of guanidinium thiocyanate instead of guanidinium chloride was first briefly mentioned by Ullrich *et al.* in 1977, and later successfully employed by Chirgwin *et al.* in 1979. Chirgwin *et al.* used guanidinium thiocyanate to isolate undegraded RNA from ribonuclease-rich tissues like pancreas. Feramisco *et al.* (1982) reported a combination of guanidinium thiocyanate and hot phenol for RNA isolation in 1981. In 1987, Chomczynski and Sacchi combined guanidinium thiocyanate with phenol–chloroform extraction under acidic conditions (Chomczynski *et al.*,

Table 2.1 Types of RNA

Type	Abbreviation	Function	Organisms
7SK RNA	7SK	Negatively regulating CDK9/cyclin T complex	Metazoans
Signal recognition particle RNA	7SRNA	Membrane integration	All organisms
Antisense RNA	aRNA	Regulatory	All organisms
CRISPR RNA	crRNA	Resistance to parasites	Bacteria and Archaea
Guide RNA	gRNA	mRNA nucleotide modification	Kinetoplastid mitochondria
Long noncoding RNA	ln cRNA	XIST (dosage compensation), HOTAIR (cancer)	Eukaryotes
MicroRNA	miRNA	Gene regulation	Most eukaryotes
Messenger RNA	mRNA	Codes for protein	All organisms
Piwi-interacting RNA	piRNA	Transposon defence, maybe other functions	Most animals
Repeat associated siRNA	rasiRNA	Type of piRNA; transposon defence	Drosophila
Retrotransposon	retroRNA	self-propagation	Eukaryotes and some bacteria
Ribonuclease MRP	RNase MRP	rRNA maturation, DNA replication	Eukaryotes
Ribonuclease P	RNase N	tRNA maturation	All organisms
Ribosomal RNA	rRNA	Translation	All organisms
Small Cajal body-specific RNA	scaRNA	Guide RNA to telomere in active cells	Metazoans
Small interfering RNA	siRNA	Gene regulation	Most eukaryotes
SmYRNA	SmY	mRNA trans-splicing	Nematodes
Small nucleolar RNA	snoRNA	Nucleotide modification of RNAs	Eukaryotes and Archaea
Small nuclear RNA	snRNA	Splicing and other functions	Eukaryotes and Archaea
Trans-acting siRNA	tasiRNA	Gene regulation	Land plants
Telomerase RNA	telRNA	Telomere synthesis	Most eukaryotes
Transfer-messenger RNA	tmRNA	Rescuing stalled ribosomes	Bacteria
Transfer RNA	tRNA	Translation	All organisms
Viral Response RNA	viRNA	Antiviral immunity	C elegans
Vault RNA	vRNA	self-propagation	Expulsion of xenobiotics
Y RNA	yRNA	RNA processing, DNA replication	Animals

1987). Since its inception, the Chomczynski and Sacchi method has been the method of choice to isolate RNA from cultured cells and most animal tissues (Chomczynski *et al.*, 2006). Here, we will review the principles behind this widely used extraction.

A phenol–chloroform extraction is a liquid–liquid extraction. A liquid–liquid extraction is a method that separates mixtures of molecules based on the differential solubilities of the individual molecules in two different immiscible liquids. Liquid–liquid extractions are widely used to isolate RNA, DNA, or proteins. In the extraction of RNA, an equal volume of acidic phenol–chloroform is added to an aqueous solution of lysed cells or homogenized tissue, the two phases are mixed and then separated by centrifugation.

Nucleic acids (RNA and DNA) are polymers of nucleotides that are composed of a nucleobase (a nitrogenous base), a five-carbon sugar (either ribose or 2-deoxyribose, for RNA or DNA, respectively), and a phosphate group. Nucleotides within a nucleic acid polymer are joined together by the formation of phosphodiester bonds; the oxygen on the 5′ carbon of one nucleotide is connected to the oxygen on the 3′ carbon of its neighbouring nucleotide. This chain of phosphodiester bonds is often referred to as the backbone chain.

Because of the oxygen atoms and the nitrogen atoms in the backbone, nucleic acids are polar molecules (the oxygen and nitrogen atoms act as hydrogen bond acceptors, and the various protons act as hydrogen bond donors). Nucleic acids are therefore soluble in water. Unlike salt compounds, nucleic acids do not dissociate in water (because the intramolecular forces linking nucleotides together are stronger than the intermolecular forces between nucleic acids and water). Instead, water molecules form solvation shells with nucleic acids through dipole–dipole interactions, which separates different polymers of nucleic acids from each other, effectively dissolving the molecules in solution.

In general, a solute dissolves best in a solvent that is most similar in chemical structure to itself (this is because bonds between a solute particle and solvent molecule substitute for the bonds between the solute particle itself). With respect to solvation, whether or not a solute and a solvent are similar in chemical structure to each other depends primarily on each substance's polarity. Typically, polar solutes will dissolve *only* in polar solvents, and non-polar solutes will dissolve *only* in non-polar solvents. Accordingly, a very polar solute such as urea is very soluble in highly polar water, less soluble in somewhat polar methanol, and almost insoluble in non-polar solvents such as chloroform and ether.

Because nucleic acids are polar, nucleic acids are soluble in the upper aqueous phase instead of the lower organic phase (because water is more polar than phenol). Conversely, proteins contain varying proportions of charged and uncharged domains, producing hydrophobic and hydrophilic regions. In the presence of phenol, the hydrophobic cores interact with phenol, causing precipitation of proteins and polymers (including carbohydrates) that collect at the interface between the two phases (often as a white flocculent) or causing lipids to dissolve in the lower organic phase. Centrifugation of a mixture of phenol–chloroform and lysed cells accordingly yields two discrete phases: the lower organic phase and the upper aqueous phase containing purified RNA.

Chloroform mixed with phenol is more efficient at denaturing proteins than either reagent is alone. The phenol–chloroform combination reduces the partitioning of poly(A)+ mRNA into the organic phase and reduces the formation of insoluble RNA–protein complexes at the interphase (Perry *et al.*, 1972). However, phenol retains about 10–15% of the aqueous phase, which results in a corresponding loss of RNA; chloroform prevents this retention of water and thus improves RNA yield (Palmiter, 1974). Typical mixtures of phenol to chloroform are 1:1 and 5:1 (v/v). At acidic pH, a 5:1 ratio results in the absence of DNA from the upper aqueous phase; whereas a 1:1 ratio, while providing maximal recovery of all RNAs, maintains some DNA present in the upper aqueous phase (Kedzierzki *et al.*, 1991). Isoamyl alcohol is sometimes added to prevent foaming (typically in a ratio of 24 parts chloroform to 1 part isoamyl alcohol), and the guanidinium salts are used to reduce the effect of nucleases. Thus, varying the ratio of phenol is an essential factor for extracting the types, and purity, of RNA that one desires to use for experiments.

Purified phenol has a density of 1.07 g/cm^3 and therefore forms the lower phase when mixed with water (1.00 g/cm^3) (21). Chloroform ensures phase separation of the two liquids because chloroform is miscible with phenol and it has a higher density (1.47 g/cm^3) than phenol; it thereby forces a sharper separation of the organic and aqueous phases, assisting in the removal of the aqueous phase with minimal cross contamination from the organic phase.

The pH of phenol determines the partitioning of DNA and RNA between the organic phase and the aqueous phase (Brawerman *et al.*, 1972; Perry *et al.*, 1972). At neutral or slightly alkaline pH (pH 7–8), the phosphate diesters in nucleic acids are negatively charged, and thus DNA and RNA both partition into the aqueous phase. DNA is removed from the aqueous layer with increasing efficiency as the pH is lowered with a maximum efficiency at pH 4.8. At this acidic pH, most proteins and small DNA fragments (<10 kb) fractionate into the organic phase and large DNA fragments and some proteins remain at the interphase between the organic and aqueous phases (Brawerman *et al.*, 1972; Chomczynski *et al.*, 1987; Puissant *et al.*, 1990). Acidic phenol retains RNA in the aqueous phase, but moves DNA into the phenol phase, because the phosphate groups on the DNA are more easily neutralized than those in RNA (i.e. DNA is less acidic/has a greater pK_a than RNA) (Bradley *et al.*, 2001). An acid pH also minimizes RNase activity.

Methods of RNA sequencing

The transcriptome is the set of all ribonucleic acid (RNA) molecules produced in a particular cell or in a population of cells. Understanding the transcriptome is crucial to understanding the relationship between the genome and the functioning of cells, i.e. the spatial, temporal, developmental, and physiological conditions under which a gene is expressed can provide clues about its biological role.

A bacterial cell contains 0.05–0.10 pg of RNA, whereas a mammalian cell contains 20–30 pg of RNA. There are two major types of cellular RNA: coding RNA and non-coding RNA. Coding RNA consists entirely of messenger RNA (mRNA) and accounts for 2–4% of total cellular RNA. Non-coding RNAs are more diverse than coding RNAs. There are two predominant classes of non-coding RNA: ribosomal RNA (rRNA) and transfer RNA (tRNA). Ribosomal RNAs are the most abundant RNA in the cell, accounting for at least 80% of all RNAs. There are other non-coding RNAs, including small nuclear RNA (snRNA), small nucleolar RNA (snoRNA), small cytoplasmic RNA (scRNA), and transfer messenger RNA (tmRNA), but they either do not make up a significant portion of cellular RNA or are limited in their distribution (Table 2.1).

Messenger RNAs are protein-coding genes and are eventually translated into protein. Although accounting for no more than 4% of total cellular RNA, mRNAs are traditionally considered to be the most significant RNA because they specify the composition of the proteome, and proteins and their interactions are thought to be the ultimate causative forces in the cell. Because it is widely assumed that the amount of mRNA in a cell is proportional to the amount of protein in a cell, researchers preferentially study mRNA instead of proteins themselves, hoping to infer protein abundance levels or biochemical capacity of cells from mRNA experiments. Additionally, mRNAs are amenable to large-scale experimentation, whereas proteins are not; and mRNAs can, by themselves, sometimes serve as biomarkers for cancer, disease, and other phenotypic manifestations. The belief that changes in mRNA correspond to changes in protein levels has, however, been significantly challenged: most studies that have investigated the relationship between mRNA and protein levels have found limited, at best, correlations (Greenbaum *et al.*, 2003). Recent reports indicate that this is likely because translational and post-translation regulation play more important roles than transcriptional regulation (Lu *et al.*, 2009); that is, protein abundance may be predominantly regulated at the ribosome (Schwanhaeusser *et al.*, 2011).

Previous studies have interrogated the transcriptome through Northern blots, RNase protection assays, reverse-transcription quantitative PCR (RT-qPCR), serial analysis of gene expression (SAGE), and DNA microarrays. High-throughput transcriptome sequencing (RNA-Seq) is gradually supplanting microarrays and is becoming the method of choice for studies aimed at mapping and quantifying the transcriptome. Although there are different high-throughput sequencing platforms, the Illumina Genome Analysis systems have become the most ubiquitous one.

RNA-Seq offers a few key advantages over previous technologies. Unlike hybridization-based approached (DNA microarrays), RNA-Seq is capable of detecting transcripts which do not correspond to previously annotated genomic sequence. Moreover, RNA-Seq (especially at longer read lengths) is capable of quantifying differential isoforms, distinguishing allelic expression, and of revealing novel connectivity between multiple exons. RNA-Seq can also reveal, at single-base resolution, sequence variations (SNPs) in the transcribed regions relative to the reference genome, which can be utilized to identify both genomic SNPs and RNA editing events (Park 2012). Other advantages of RNA-Seq include an ability to find fusion gene products, a low background noise and boundless upper limit for quantification (and therefore a large dynamic range of expression levels over which transcripts can be detected). RNA-Seq is consequently the first method that allows the entire transcriptome to be surveyed in a very high-throughput and quantitative manner for a relatively low cost.

RNA-Seq, in general, consists of the following steps: isolation of RNA (examined above), removal of ribosomal RNA, creation of cDNA, ligation of platform-specific adapters, PCR amplification, and sequencing. As methods for RNA-Seq preparation continue to develop, it is becoming clear that sample preparation is the largest influence on the quality and scope of the results. Here, we will review a few important steps in RNA-Seq library preparation that delineate the range of possible results depending upon the method that is selected.

Removal of ribosomal RNA

RNA-Seq typically begins with the removal of rRNAs, because these transcripts would otherwise saturate total RNA preparations (because of their high abundance). In eukaryotes, primary or precursor mRNA transcripts are *processed* into mature ones, which is comprised of three steps: the covalent linkage of a 3′ polyadenylate-tail, the addition of a 5′-cap, and splicing (the removal of introns and the consequent ligation of multiple exons together). In eukaryotes, poly(A) tails are important for mRNA stability and translation initiation, and are added to most precursor RNAs, but typically not added to rRNAs (a few non-abundant poly(A)+ transcripts of the 18S and 28S rRNAs have been detected in eukaryotic cells (Slomovic *et al.*, 2006)). In prokaryotes, on the other hand, polyadenylation promotes RNA degradation. Ribosomal RNAs are removed either by enrichment for poly(A)+ fractions or by direct depletion of rRNAs.

Enrichment of poly(A)+ fractions is the most popular method for the removal of rRNAs to date. This is partially because of the ubiquitous belief that protein-coding genes (mRNAs) are the most significant RNA for predicting biological function, but also because poly(A)+ selection is technically easier and more efficient at removing rRNA than direct depletion of rRNAs. Poly(A)+ isolation of mRNA is typically achieved by using biotinylated oligo(dT) and streptavidin-conjugated beads. This procedure relies on base pairing between the poly(A)+ tail of mRNA and the oligo(dT) sequences bound to the surface of the beads. After annealing of poly(A)+ transcripts to the beads, the bead–RNA mixture is placed on a magnet, pulling the beads and whatever is bound to them towards the side of the tube. Contaminants are discarded by the removal of the supernatant, which contains any transcript or biological material which has not bound to the oligo(dT) beads. The purified poly(A)+ RNA is eluted from the beads and library preparation commences. Depletion of rRNA is achieved in a similar procedural manner.

Total RNA can be depleted by the removal of ribosomal RNA directly instead of by the exclusion of ribosomal RNA by the selection of transcripts containing poly(A)+ tails. This is accomplished by using oligonucleotide probes bound to magnetic beads. These oligonucleotide probes are complimentary to ribosomal RNA sequences, and, in the case of eukaryotes, typically include sequences for 28S, 18S, 5.8S and 5S ribosomal RNA. Ribosomal RNA, subject to evolutionary differences, mandates that these probes be tailored to particular species. The probes are mixed with total RNA, and the ribosomal RNA is allowed to anneal. Afterward, the beads are collected with magnets, and the supernatant, containing the purified, rRNA-free fraction, is removed and used for subsequent library preparation.

A less commonly used method to reduce the abundance of rRNA is cDNA normalization. The method receiving the most attention is normalization by duplex-specific nucleases (DSN). This method works by exploiting nucleic acid hybridization kinetics; in particular, that abundant transcripts convert to the double-stranded cDNA (ds-cDNA) form more effectively than those that are less abundant. DSN normalization is typically performed prior to PCR amplification. Double-stranded cDNA is denatured, and then allowed to reanneal; since the most abundant RNAs re-anneal more effectively than those that are less frequent, two fractions are formed:

single-stranded cDNA (ss-cDNA) and ds-cDNA arising from highly abundant transcripts. The ds-cDNA is then digested by duplex-specific nuclease. This results in 'normalized cDNA, or cDNA significantly depleted of highly expressed RNAs (see Christodoulou *et al.*, 2012). The main benefit to this method is that it facilitates the assessment of the full spectrum of the transcriptome with less sequencing depth by reducing the dynamic range amongst transcripts. A con, however, is that the quantification of highly expressed transcripts would be affected (Yi *et al.*, 2011).

The method used to remove ribosomal RNA results in the querying of different transcriptional landscapes. Recent studies have suggested that there exists important, functional non-coding and protein-coding RNAs which do not possess poly(A)+ tails. These include small RNAs generated by RNA polymerase III, replication-dependent histone mRNAs, and long-noncoding RNAs (lncRNAs) transcribed by RNA polymerase II. Histone proteins, for example, are essential for cell proliferation and required for the packaging of DNA into chromatin (Keall, 2007); while the brain-associated *BC200* RNA has been found to be related to tumour induction and progression (Chen *et al.*, 1997). In some cases, poly(A)– transcripts have been found to be two times more abundant (36%) than poly(A)+ transcripts in cytosolic RNAs in human cells (Cheng *et al.*, 2005), while other studies more conservatively indicated as low as 2.5–2.7% of the annotated transcripts are poly(A)– in HeLa and H9 cells, respectively (Yang *et al.*, 2011). Whatever the case may be, it is clear that, the transcribed portion of the genome is composed of both poly(A)+ and poly(A)– transcripts; and therefore that, for a more complete view of the transcriptome, the direct removal of ribosomal RNA is necessary.

Another important consideration when deciding which method to use for removing ribosomal RNA is the integrity of the starting RNA material. Integrity of RNA is most often determined with respect to the ribosomal RNA: for mammalian samples, for example, RNA is often visualized on an agarose gel or through capillary electrophoresis to determine if the ratio between the 28S:18S bands is 2:1, a traditional benchmark for the of intactness of RNA. It should be noted, however, that is isn't entirely clear that intact RNA by this measure is an accurate reflection of the integrity of the underlying mRNA (Vermeulen *et al.*, 2011). If the starting sample is severely degraded, then removal of rRNA by poly(A)+ selection would result in severely 3′ biased sequencing libraries. In the case of direct removal of rRNA, however, fragments of mRNA from degraded templates would be preserved.

Fragmentation

Fragmentation of RNA is standard practice in most RNA sequencing protocols. This is partially because most next-generation sequencers are short read sequencers; Illumina currently supports 250×250 paired-end reads while 454 can deliver up to 1000 bp in length. However, most human transcripts are greater than 1000 nucleotides (nt) in length, meaning that the reconstruction of the transcripts after sequencing becomes a computational challenge. Ideally, all sequences would be full-length cDNAs, but pending sequencing advances aside, these have traditionally been a challenge to create *in vitro*. A full-length cDNA depends on both capturing full-length mRNA (that is, starting with material that is perfectly intact) and on effective reverse transcriptase that goes to the extreme end of the mRNA.

Researchers fragment RNA either immediately after removal of ribosomal RNA or immediately after synthesis of cDNA. Because of the challenges involved in synthesizing full-length cDNA, fragmentation of cDNA accordingly results in coverage across the gene body that is 3′ biased. Fragmentation of the RNA itself, on the other hand, results in more even coverage across the gene body, save for the 3′ and 5′ ends of the transcript, which are relatively depleted (Wang *et al.*, 2010). It is important to consider which areas of the transcript are of most interest when deciding on which fragmentation method to use. Some researchers, for example, are interested in unearthing 3′ UTR extensions, since the 3′ UTR often contains microRNA binding sites and other stability determining regions thought to be important in translation control or development. In this case, cDNA fragmentation, not RNA fragmentation, may be best suited, particularly coupled with a poly(A)+ priming method.

Directionality

One of the drawbacks of initial RNA-Seq preparation methods was the lack of strand specificity in the data. Standard RNA-Seq libraries mask the directionality of transcription, as these protocols sequence both strands of double-stranded cDNA (ds-cDNA), which consequently reveals during downstream analysis an equal number of reads mapping to both the forward and the reverse strands of genomic features regardless of transcriptional direction. Various protocols have since been devised to overcome this limitation, effectively allowing for the deduction of strand information with high accuracy.

One of the most popular and well-balanced methods for creating strand-specific libraries is the 'dUTP method' (Parkhomchuk *et al.*, 2009; Levin *et al.*, 2010). This protocol works by incorporating deoxy-UTP (dUTP) during second-strand cDNA synthesis instead of incorporating the standard deoxy-TTP (dTTP). When dUTP replaces dTTP as a dNTP substrate in second-strand cDNA synthesis, it can subsequently be targeted for degradation. Uracil-N-glycosylase (UNG) is used to target the uridine-containing strand in the sequencing library, successfully destroying it and allowing for the identification of the orientation of transcripts during secondary analysis of the data. UNG works by removing uracil incorporated into any containing molecules, leaving apyrimidinic sites; the offending molecules are subsequently decomposed into a nick or destroyed by hydrolysis of the phosphate backbone at the abasic sites (see Grossman *et al.*, 1982).

Strand information can enhance an RNA-Seq experiment in a few ways. Practically, strand information helps resolve transcription from signal that arises from regions of the genome which contain overlapping genes on opposite strands. It also aids in the completeness of annotation when discovering novel genes. More interestingly, strand information can help identify cases of anti-sense transcription, which may play an important regulatory role. Concerns that antisense transcription may represent transcriptional noise rather than a regulatory mechanism have been weakened by numerous studies. One study, for example, determined a relationship between alternative splicing and antisense transcription in the human genome; in particular, instances of antisense transcription were strongly correlated to sense gene splicing, suggesting an antisense transcription-mediated mechanism of splicing regulation (Morrissy *et al.*, 2011). Other studies have implicated anti-sense transcripts in RNA degradation of the corresponding sense transcripts (RNA interference), as well as gene silencing at the chromatin level.

Using RNA sequencing to define transcriptional landscapes

A major goal in genomics is to identify all genomic features, including coding regions, introns, and untranslated regions. Because RNA-seq is high-throughput, cost-effective, and single-base resolution, it has the potential to help achieve this goal. Nagalakshmi *et al.* (2008) demonstrated the efficacy of RNA-Seq at mapping a transcriptional landscape by applying it to generate a high-resolution transcriptome map of the yeast genome, a genome whose 5′ and 3′ boundaries of most genes had previously been undefined.

Nagalakshmi *et al.* (2008) prepared four fragmented, poly(A)+ RNA-seq libraries from *Saccharomyces cerevisiae*; and subjected the libraries to single-end, high-throughput Illumina sequencing, in which 35 base pairs of sequence was determined from one end of each fragment. The reads were mapped to the yeast reference genome, pruned of adapters, and filtered of low-quality reads. The processed reads were then used to ascertain the transcriptional landscape of the yeast genome.

The 5′ and 3′ boundaries of annotated genes were mapped to within 10–50 nt by searching the data for a sharp reduction in signal in the transcribed regions. The 3′ boundaries were more precisely mapped by searching for poly(A) tails, and the introns were mapped by searching for tags that span GT–AG splicing consensus sites. Using these methods, it was determined that 74.5% of the yeast genome and 91.5% of annotated ORFs were expressed. Moreover, the 5′ and 3′ boundaries of 80% and 85% of all annotated genes, respectively, were able to be determined; the median length of the 5′ UTRs was 50 nt (range 0–990 nt), while the median length of the 3′ UTRs was 104 nt (range 0–1461 nt).

Djebali *et al.* (2012), in participation of the ENCODE project, also utilized RNA-Seq to define transcriptional landscape, but in human cells. RNA-Seq libraries were created from 15 different cells lines, fractionated by their nuclear and cytosolic subcompartments. RNAs from each of these subcompartments were further subdivided by length into short (<200 nt long) and long (>200 nt) subgroups, with the latter group ultimately fractionated into polyadenylated and non- polyadenylated transcripts. Using the long RNA transcripts, Djebali *et al.* (2012) cumulatively detected 70% of annotated splice junctions, transcripts, and genes; and 85% of all previously annotated GENCODE exons. Novel elements were discovered which further extended the GENCODE collection of exon, splice sites, transcripts, and genes by 19%, 22%, 45%, and 80%, respectively. The prominence of alternative splicing was also analysed, and it was determined that most isoform expression does not follow a minimalistic strategy, i.e. genes tend to express many isoforms simultaneously, with a plateau at about 10–12 expressed isoforms per gene, and with one isoform typically dominating.

RNA sequencing to discover RNA modifications

The processing of RNA was originally thought to be a perfectly faithful copy of the genomic DNA template. With the discovery of alternative splicing in 1977, this notion began to fade. Indeed, subsequent research has shown that there are at least two other common means by which RNA is modified: RNA editing and base modification. From profiling work in a wide variety of organisms, three RNA modifications (splicing, editing, and base modification) are now anticipated to be as common as post-translational modifications (PTMs) of proteins. Moreover, just as protein modifications (such as phosphorylation and acetylation) are dynamically regulated to alter a protein's function, folding, stability, or binding partners; RNA modifications also show dynamic changes in development, tissue-specificity, and relevance for disease.

'RNA editing' was first discovered in 1986 through the comparison of cDNA and DNA from a single trypanosome, wherein certain bases that did not exist in the DNA were appearing in the cDNA. These data indicated that not only could the *order* of the information of a gene be changed (i.e. splicing), but also the *content* of a transcript could be modified in a post-transcriptional process. RNA editing can occur in two forms: single-base substitution (such as a A→G transition) or an insertion/deletion (indel) of several bases into the transcript.

The implications for such a modification quickly appeared as researchers used cDNA sequencing more frequently. For example, the human apolipoprotein B gene (APOB) was reported to have an RNA edit at nucleotide 6666 (CAA to UAA) that resulted in a stop codon for the gene (Tennyson, 1989). The APO genes are important for lipid metabolism, and they are also the family of proteins implicated in Alzheimer's disease. However, this modification was surprising not only for its confirmation of the existence of RNA editing in humans in a disease-related gene, but also because the RNA editing event was observed to be tissue-specific. In the liver, the transcript remains un-edited, creating a longer protein that can function in cholesterol transport; whereas in the intestine, the transcript is edited (and truncated) to change its role to a protein that can help absorb lipids from the intestine.

Since those early discoveries, both single-base and indel RNA editing have been observed in dozens of species and tissues, and it is now estimated to occur in thousands of genes (Li *et al.*, 2011) and even confer specific roles for potassium channel function in different environments and indicate that RNA editing can be a phenotype under evolutionary selection (Garrett and Rosenthal, 2012). However, even though high-throughput RNA-Sequencing has created an unprecedented depth of data with which to examine RNA editing, there are many ways in which to mis-call an RNA edit (Kleinman 2012, Pickrell, 2012), and care must be taken when aligning short reads from RNA-Seq to the genome for discerning RNA editing events.

Nonetheless, RNA editing has emerged as an essential component of eukaryotic biology that is hypothesized to serve several roles. The editing can (1) expand the transcriptional complexity of an organism without adding to the DNA content, (2) engineer tissue- or cell-specific functions for a single gene, (3) protect the genome against viruses or retrotransposons, (4) prevent dsRNA formation in transcripts to avoid accidental mRNA degradation, and (5) correct mutations that arise in the native DNA. In any one cell, some or all of these functions may be utilized and give additional fitness and function for a single set of genes from a single organism.

RNA base modifications and epitranscriptomics

RNA can also be modified at the base moiety, just as DNA can be modified in epigenetics. Modifications of the bases of RNA were first described in 1974 for a single mRNA (Desrosiers, 1974), but many subsequent studies confirmed these results in rRNA, snRNAs, and tRNAs (Desrosiers, 1975; Wei, 1976; Levis, 1978; Chen-Kiang, 1979; Horowitz, 1984; Bringmann, 1987). However, most of these modifications were not observed in mRNA, and, until recently, there were no high-throughput means to examine the distribution or localization of RNA modifications.

With the advent of RNA-Seq, researchers could begin to use the immunoprecipitation (IP) techniques used in ChIP-Seq (chromatin immunoprecipitation and sequencing) for enriching transcripts that contain modified bases. For example, methyl-6-adenosine (m^6A) is found in E. coli as a DNA modification, but an antibody to m^6A recognizes the adenine and not the ribose moiety of the nucleotide, and therefore should be able to bind to m^6A in both RNA and DNA. This technique (Methyl-RNA Immunoprecipitation Sequencing, or MeRIP-Seq) was recently used to map the distribution and localization of m^6A across the mouse and human transcriptomes (Meyer 2012, Domminissini 2012), effectively opening up the study of 'epitranscriptomics.' Interestingly, m^6A showed a strong enrichment for the beginning of the 3′UTR, implicating RNA modifications in translation termination.

The exact role or roles of epitranscriptomic marks is not yet known. There is some evidence of increased m^6A in alternatively spliced exons, indicating that epitranscriptomic marks could flag certain exons for inclusion or exclusion during splicing. Also, there is some enrichment of m^6A in highly conserved regions and an anti-correlation to miRNA binding site, and m^6A may contribute to the miRNA signal, or the polyadenylation or AU-rich signals near the end of a transcript. Indeed, m^6A may affect the degradation of the transcripts with AU-rich sequences, and the methyl marks of m^6A may serve to attenuate the degradation dynamics of genes.

Taken together, these modifications could create a 'code' of the transcriptome that is read by a variety of RNA binding proteins to respond to the specific modified nucleic acid, and then move to change the function of the RNA, just as various proteins respond to DNA and histone changes in epigenetics. Indeed, it is possible that RNA base modifications can also impact RNA editing (Fig. 2.2).

Currently, there are at least 107 types of RNA modifications that have been observed in organisms across the phylogenetic tree (Table 2.2, from the RNA Modification Database). Many of them are currently believed to be specific to one clade or to one type of RNA, but their study has only recently begun. It is noteworthy that the gene for de-methylating RNA (FTO) is implicated in obesity and its targets are also implicated in Alzheimer's Disease, and that recent work has shown the m^6A modifications to be very dynamic in development, cancer, and in response to cellular stress. These data indicate that RNA has emerged as a critical regulatory layer in disease, development, and cellular function.

Conclusions

The changes in RNA concentrations and RNA types within different cellular compartments and tissues is a key means through which epigenetic and organismal programs emerge and how cellular lineages are decided during development. Indeed, the importance of gene regulation was hypothesized as early as 1975, when the first comparisons of human and chimps showed extremely high similarity at the protein level (98–100%) for 44 genes. The early work of King and Wilson

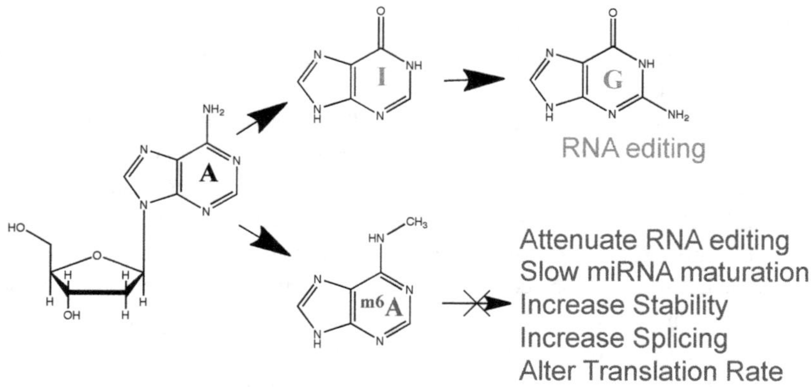

Figure 2.2 The Interplay of RNA base modifications and RNA editing. Adenosine (left) can be de-animated to become Inosine (upper, middle) and then Guanosine (upper right) in RNA editing, but the presence of m6A can prevent such a transition (below).

Table 2.2 RNA base modifications

Abbreviation	Chemical name
m1acp3Y	1-Methyl-3-(3-amino-3-carboxypropyl) pseudouridine
m1A	1-Methyladenosine
m1G	1-Methylguanosine
m1I	1-Methylinosine
m1Y	1-Methylpseudouridine
m1Am	1,2′-O-Dimethyladenosine
m1Gm	1,2′-O-Dimethylguanosine
m1Im	1,2′-O-Dimethylinosine
m2A	2-Methyladenosine
ms2io6A	2-Methylthio-N6-(cis-hydroxyisopentenyl) adenosine
ms2hn6A	2-Methylthio-N6-hydroxynorvalyl carbamoyladenosine
ms2i6A	2-Methylthio-N6-isopentenyladenosine
ms2m6A	2-Methylthio-N6-methyladenosine
ms2t6A	2-Methylthio-N6-threonyl carbamoyladenosine
s2Um	2-Thio-2′-O-methyluridine
s2C	2-Thiocytidine
s2U	2-Thiouridine
Am	2′-O-Methyladenosine
Cm	2′-O-Methylcytidine
Gm	2′-O-Methylguanosine
Im	2′-O-methylinosine
Ym	2'-O-methylpseudouridine
Um	2′-O-methyluridine
Ar(p)	2′-O-ribosyladenosine (phosphate)
Gr(p)	2′-O-ribosylguanosine (phosphate)
acp3U	3-(3-amino-3-carboxypropyl)uridine
m3C	3-Methylcytidine

Abbreviation	Chemical name
m3Y	3-Methylpseudouridine
m3U	3-Methyluridine
m3Um	3,2′-O-dimethyluridine
imG-14	4-Demethylwyosine
s4U	4-Thiouridine
chm5U	5-(Carboxyhydroxymethyl)uridine
mchm5U	5-(Carboxyhydroxymethyl)uridine methyl ester
inm5s2U	5-(Isopentenylaminomethyl)- 2-thiouridine
inm5Um	5-(Isopentenylaminomethyl)- 2′-O-methyluridine
inm5U	5-(Isopentenylaminomethyl)uridine
nm5s2U	5-Aminomethyl-2-thiouridine
ncm5Um	5-Carbamoylmethyl-2′-O-methyluridine
ncm5U	5-Carbamoylmethyluridine
cmnm5Um	5-Carboxymethylaminomethyl- 2′-O-methyluridine
cmnm5s2U	5-Carboxymethylaminomethyl-2-thiouridine
cmnm5U	5-Carboxymethylaminomethyluridine
cm5U	5-Carboxymethyluridine
f5Cm	5-Formyl-2′-O-methylcytidine
f5C	5-Formylcytidine
hm5C	5-Hydroxymethylcytidine
ho5U	5-Hydroxyuridine
mcm5s2U	5-Methoxycarbonylmethyl-2-thiouridine
mcm5Um	5-Methoxycarbonylmethyl-2′-O-methyluridine
mcm5U	5-Methoxycarbonylmethyluridine
mo5U	5-Methoxyuridine
m5s2U	5-Methyl-2-thiouridine
mnm5se2U	5-Methylaminomethyl-2-selenouridine
mnm5s2U	5-Methylaminomethyl-2-thiouridine
mnm5U	5-Methylaminomethyluridine
m5C	5-Methylcytidine
m5D	5-Methyldihydrouridine
m5U	5-Methyluridine
tm5s2U	5-Taurinomethyl-2-thiouridine
tm5U	5-Taurinomethyluridine
m5Cm	5,2′-O-Dimethylcytidine
m5Um	5,2′-O-dimethyluridine
preQ1	7-Aminomethyl-7-deazaguanosine
preQ0	7-Cyano-7-deazaguanosine
m7G	7-Methylguanosine
G+	Archaeosine
D	Dihydrouridine
oQ	Epoxyqueuosine
galQ	Galactosyl-queuosine

Table 2.2 RNA base modifications (*continued*)

Abbreviation	Chemical name
OHyW	Hydroxywybutosine
I	Inosine
imG2	Isowyosine
k2C	Lysidine
manQ	Mannosyl-queuosine
mimG	Methylcytosine
m2G	N2-methylguanosine
m2Gm	N2,2′-O-dimethylguanosine
m2,7G	N2,7-dimethylguanosine
m2,7Gm	N2,7,2'-O-trimethylguanosine
m22G	N2,N2-dimethylguanosine
m22Gm	N2,N2,2′-O-trimethylguanosine
m2,2,7G	N2,N2,7-trimethylguanosine
ac4Cm	N4-acetyl-2′-O-methylcytidine
ac4C	N4-acetylcytidine
m4C	N4-methylcytidine
m4Cm	N4,2′-O-dimethylcytidine
m42Cm	N4,N4,2′-O-trimethylcytidine
io6A	N6-(cis-hydroxyisopentenyl)adenosine
ac6A	N6-acetyladenosine
g6A	N6-glycinylcarbamoyladenosine
hn6A	N6-hydroxynorvalylcarbamoyladenosine
i6A	N6-isopentenyladenosine
m6t6A	N6-methyl-N6-threonylcarbamoyladenosine
m6A	N6-methyladenosine
t6A	N6-threonylcarbamoyladenosine
m6Am	N6,2′-O-dimethyladenosine
m62A	N6,N6-dimethyladenosine
m62Am	N6,N6,2′-O-trimethyladenosine
o2yW	Peroxywybutosine
Y	Pseudouridine
Q	Queuosine
OHyW*	Undermodified hydroxywybutosine
cmo5U	Uridine 5-oxyacetic acid
mcmo5U	Uridine 5-oxyacetic acid methyl ester
yW	Wybutosine
imG	Wyosine

	Aceview	UCSC	Vega	ENSEMBL	Refseq
Aceview	178				
UCSC	76	81			
Vega	51	42	58		
ENSEMBL	64	60	43	70	
RefSeq	60	61	37	57	62

Figure 2.3 Proportion of the human genome considered 'exonic' (in Mb). An overlap chart of five commonly used gene annotations show that the five ranges of total exonic space ranges from 62178Mb, with a varying degree of overlap between them.

indicated that not only are protein changes important, but also that regulation of genes and its adjacent regulatory sequences could themselves be critical and targets of evolutionary selection. Namely, it is not simply the presence or absence of a gene in the genome that is important, but also *if*, *when*, and *how* a gene is used.

Yet, our understanding of gene and RNA regulation has only recently begun to be illuminated, through targeted experiments or large-scale profiling work from groups such as the ENCODE group. Also, even the relatively simple question of 'what per cent of the human genome is exonic?' is rife with disagreement between gene annotation sets (Fig. 2.3). As the technology advances, our understanding of the changes in the transcriptome and epitranscriptome will improve, including single-molecule sequencing and detection of base modifications (Saletore *et al.*, 2012). Eventually, the small 'RNA world' and all of its informational content that is present in each cell will be leveraged for our understanding of disease, life, and evolution.

References

Baltimore, D. (1970). Viral RNA-dependent DNA polymerase: RNA-dependent DNA polymerase in virions of rna tumour viruses. Nature *226*, 1209–1211.

Bradley, J., Bradley, J., Johnson, D., and Rubenstein, D. (2001). Lecture Notes on Molecular Medicine. 2nd Edition (Oxford, Blackwell).

Brawerman, G., Mendecki, J., and Lee, S.Y. (1972) A procedure for the isolation of mammalian messenger ribonucleic acid. Biochemistry *11*, 637–641.

Bringmann, P., and Lührmann, R. (1987). Antibodies specific for N6-methyladenosine react with intact snRNPs U2 and U4/U6. FEBS Lett. *213*, 309–315

Chen, W., Böcker, W., Brosius, J., and Tiedge, H. (1997). Expression of neural BC200 RNA in human tumours. J. Pathol. *183*, 345–51.

Chen-Kiang, S., Nevins, J.R., and Darnell Jr, J.E. (1979). N-6-methyl-adenosine in adenovirus type 2 nuclear RNA is conserved in the formation of messenger RNA. J. Mol. Biol. *135*, 733–752.

Cheng, J., Kapranov, P., Drenkow, J., Dike, S., Brubaker, S., Patel, S., Long, J., Stern, D., Tammana, H., Helt, G., *et al.* (2005). Transcriptional maps of 10 human chromosomes at 5-nucleotide resolution. Science *308*, 1149–54.

Chirgwin, J.M., Przybyla, A.E., MacDonald, R.J., and Rutter, W.J. (1979). Isolation of biologically active ribonucleic acid from sources enriched in ribonuclease. Biochemistry *18*, 5294–5299.

Chomczynski, P., and Sacchi, N. (1987). Single-step method of RNA isolation by acid guanidinium thiocyanate-phenol–chloroform extraction. Anal. Biochem. *162*, 156–159.

Chomczynski, P., and Sacchi, N. (2006). The single-step method of RNA isolation by acid guanidinium thiocyanate–phenol–chloroform extraction: twenty-something years on. Nat. Protoc. *1*, 581–585.

Chow, L.T., Gelinas, R.E., Broker, T.R., and Roberts, R.J (1977). An amazing sequence arrangement at the 5' ends of adenovirus 2 messenger RNA. Cell *12*, 1–8.

Christodoulou, D.C., Gorham, J.M., Herman, D.S., and Seidman, J.G. (2011). Construction of normalized RNA-seq libraries for next-generation sequencing using the crab duplex-specific nuclease. Curr. Protoc. Mol. Biol. Chapter *4*, Unit4.12.

Cox, R.A. (1968). The use of guanidinium chloride in the isolation of nucleic acids. In Methods in Enzymology, Grossman, L., and Moldave, K., eds. (Orlando, FL, Academic Press), 12-B, pp. 120–129.

Cox, R.A. (1996). Ribonucleic acid from ribonucleoprotein particles. Biochem. Prep. *11*, 104–109.

Cox, R.A., and Arnstein, H.R.V. (1963). The isolation, characterization and acid–base properties of ribonucleic acid from rabbit-reticulocyte ribosomes. J. Biochem. *89*, 574.

Crick, F.H.C. (1958). On protein synthesis. Symp. Soc. Exp. Biol. XII, 139–163.

Dahm, R. (2008). Discovering DNA: Friedrich Miescher and the early years of nucleic acid research. Hum Genet. *122*, 565–81.

Djebali, S., Davis, C.A., Merkel, A., Dobin, A., Lassmann, T., Mortazavi, A., Tanzer, A., Lagarde, J., Lin, W., Schlesinger, F., *et al.* (2012). Landscape of transcription in human cells. Nature *489*, 101–108.

Desrosiers, R., Friderici, K., and Rottman, F. (1974). Identification of methylated nucleosides in messenger RNA from Novikoff hepatoma cells. PNAS *71*, 3971–3975.

Desrosiers, R.C., Friderici, K.H., and Rottman, F.M. (1975). Characterization of Novikoff hepatoma mRNA methylation and heterogeneity in the methylated 5′ terminus. Biochemistry *14*, 4367–4374.

Dominissini, D., Moshitch-Moshkovitz, S., Schwartz, S., Salmon-Divon, M., Ungar, L., Osenberg, S., Cesarkas, K., Jacob-Hirsch, J., Amariglio, N., Kupiec, M., *et al.* (2012). Topology of the human and mouse m6A RNA methylomes revealed by m6A -seq. Nature *485*, 201–206.

Feramisco, J.R., Smart, J.E., Burridgel, K., Helfmanll, D.M., and Thomas, G.P. (1982). Co-existence of Vinculin and a Vinculin-like Protein of Higher Molecular Weight in Smooth Muscle. J. Biol. Chem. *257*, 11024–11031.

Garrett, S., and Rosenthal, J.J. (2012). RNA editing underlies temperature adaptation in K+ channels from polar octopuses. Science *335*, 848–51.

Grassman, W., and Deffner, G. (1953). Verteilungschromatographisches Verhalten von Proteinen und Peptiden in phenolhaltigen Lösungsmitteln. Hoppe-Seylor's Z. Physiol. Chem. *293*, 89–98.

Greenbaum, D., Colangelo, C., Williams, K., and Gerstein M. (2003). Comparing protein abundance and mRNA expresssion levels on a genomic scale. Genome Biol. *4*, 117.

Grossman, L., and Grafstrom, R. (1982). AP sites and AP endonucleases. Biochimie. *64*, 577–80.

Gupta, R.A., Shah, N., Wang, K.C., Kim, J., Horlings, H.M., Wong, D.J., Tsai, M.C., Hung, T., Argani, P., Rinn, J.L., *et al.* (2010). Long non-coding RNA HOTAIR reprograms chromatin state to promote cancer metastasis. Nature *464*, 1071–6.

Horowitz, S., Horowitz, A., and Nilsen, T.W. (1984). Mapping of N6-methyladenosine residues in bovine prolactin mRNA. PNAS *81*, 5667–5671.

Keall, R., Whitelaw, S., Pettitt, J., and Müller, B. (2007). Histone gene expression and histone mRNA 3′ end structure in *Caenorhabditis elegans*. BMC Mol. Biol. *8*, 51.

Kedzierzki, W., and Porter, J.C. (1991). A novel non-enzymatic procedure for removing DNA template from RNA transcription mixtures. Biotechniques *10*, 210–214.

Kirby, K.S. (1956). A new method for the isolation of ribonucleic acids from mammalian tissues. Biochem. J. *64*, 405.

Kleinman, C.L., and Majewski, J. (2012). Comment on 'Widespread RNA and DNA sequence differences in the human transcriptome'. Science *335*, 1302; author reply 1302.

Kruger, K., Grabowski, P.J., Zaug, A.J., Sands, J., Gottschling, D.E., and Cech, T.R. (1982). Self-splicing RNA: autoexcision and autocyclization of the ribosomal RNA intervening sequence of Tetrahymena. Cell *31*, 147–57.

Levin, J.Z., Yassour, M., Adiconis, X., Nusbaum, C., Thompson, D.A., Friedman, N., Gnirke, A., and Regev, A. (2010). Comprehensive comparative analysis of strand-specific RNA sequencing methods. Nat. Methods *7*, 709–15.

Levis, R., and Penman, S. (1978). 5′-Terminal structures of poly(A)+ cytoplasmic messenger RNA and of poly(A)+ and poly(A)− heterogeneous nuclear RNA of cells of the dipteran Drosophila melanogaster. J. Mol. Biol. *120*, 487–515.

Li, M., Wang, I.X., Li, Y., Bruzel, A., Richards, A.L., Toung, J.M., and Cheung, V.G. (2011). Widespread RNA and DNA sequence differences in the human transcriptome. Science *333*, 53–58.

Lu, R., Markowetz, F., Unwin, R.D., Leek, J.T., Airddi, E.M., MacArthur, B.D., Lachmann, A., Rozov, R., Ma'ayan, A., Bayer, L.A., *et al.* (2009). Systems-level dynamic analyses of fate change in murine embryonic stem cells. Nature *62*, 358–362.

Meyer, K.D., Saletore, Y., Zumbo, P., Elemento, O., Mason, C.E., and Jaffrey, S.R. (2012). Comprehensive analysis of mRNA methylation reveals enrichment in 3′ UTRs and near stop codons. Cell *149*, 1635–1646.

Morrissy, A.S., Griffith, M., and Marra, M.A. (2011). Extensive relationship between antisense transcription and alternative splicing in the human genome. Genome Res. *21*, 1203–12.

Nagalakshmi, U., Wang, Z., Waern, K., Shou, C., Raha, D., Gerstein, M., and Snyder, M. (2008). The transcriptional landscape of the yeast genome defined by RNA sequencing. Science *320*, 1344–1349.

Nirenberg, M.W., and Matthaei, J.H. (1961). The dependence of cell-free protein synthesis in E. coli upon naturally occurring or synthetic polyribonucleotides. PNAS 47, 1588–1602.

Palmiter, R.D. (1974). Magnesium precipitation of ribonucleoprotein complexes. Expedient techniques for the isolation of undegraded polysomes and messenger ribonucleic acid. Biochemistry *13*, 3606–3615.

Park, E., Williams, B., Wold, B.J., and Mrtazavi, A. (2012). RNA editing in the human ENCODE RNA-seq data. Genome Res. *22*, 1626–1633.

Parkhomchuk, D., Borodina, T., Amstislavskiy, V., Banaru, M., Hallen, L., Krobitsch, S., Lehrach, H., and Soldatov, A. (2009). Transcriptome analysis by strand-specific sequencing of complementary DNA. Nucleic Acids Res. *37*, e123.

Perry R.P and Kelley D.E. (1972). The production of ribosomal RNA from high molecular weight precursors. J. Mol. Biol. *70*, 265–279.

Perry, R.P., La Torre, J., Kelley, D.E., and Greenberg, J.R. (1972). On the lability of poly(A) sequences during extraction of messenger RNA from polyribosomes. Biochim. Biophys. Acta. *262*, 220–6.

Pickrell, J.K., Gilad, Y., and Pritchard, J.K. (2012). Comment on 'Widespread RNA and DNA sequence differences in the human transcriptome'. Science *335*, 1302; author reply 1302.

Puissant, C., and Houdebine, L.M. (1990). An improvement of the single-step method of RNA isolation by acid guanidinium thiocyanate-phenol–chloroform extraction. Biotechniques *8*, 148–149.

Rechavi, O., Minevich, G., and Hobert, O. (2011). Transgenerational inheritance of an acquired small RNA-based antiviral response in C. elegans. Cell *147*, 1248–56.

Ruby, J.G., Jan, C., Player, C., Axtell, M.J., Lee, W., Nusbaum, C., Ge, H., and Bartel, D.P., (2006). Large-scale sequencing reveals 21U-RNAs and additional microRNAs and endogenous siRNAs in *C. elegans*. Cell *127*, 1193–1207.

Saletore, Y., Meyer, K., Korlach, J., Vilfan, I.D., Jaffrey, S., and Mason, C.E. (2012). The birth of Epitranscriptome: deciphering the function of RNA modifications. Genome Biol. *13*, 175.

Schwanhaeusser, B., Busse, D., Li, N., Dittmar, G., Schuchhardt, J., Wolf, J., Chen, W., and Selbach, M. (2011). Global quantification of mammalian gene expression control. Nature *495*, 126–127.

Slomovic, S., Laufer, D., Geiger, D., and Schuster, G. (2006). Polyadenylation of ribosomal RNA in human cells. Nucleic Acids Res. *34*, 2966–2975.

Temin, H.M., and Mizutani, S. (1970). Viral RNA-dependent DNA polymerase: RNA-dependent DNA polymerase in virions of Rous sarcoma virus. Nature *226*, 1211–1213.

Tennyson, G.E., Sabatos, C.A., Eggerman, T.L., and Brewer, H.B. (1989). Characterization of single base substitutions in edited apolipoprotein B transcripts. Nucleic Acids Res. *17*, 691–8.

Ullrich, A., Shine, J., Chirgwin, J., Pictet, R., Tischer, E., Rutter, W.J., and Goodman, H.M. (1977). Rat insulin genes: construction of plasmids containing the coding sequences. Science *196*, 1313–1319.

Vermeulen, J., De Preter, K., Lefever, S., Nuytens, J., De Vloed, F., Derveaux, S., Hellemans, J., Speleman, F., and Vandesompele, J. (2011). Measurable impact of RNA quality on gene expression results from quantitative PCR. Nucleic Acids Res. *39*, e63.

Volkin, E., and Carter, C.E. (1951). The preparation and properties of mammalian ribonucleic acids. J. Am. Chem. Soc., *73*, 1516–1519.

Wang, E.T., Sandberg, R., Luo, S., Khrebtukova, I., Zhang, L., Mayr, C., Kingsmore, S.F., Schroth, G.P., and Burge, C.B. (2008). Alternative isoform regulation in human tissue transcriptomes. Nature *456*, 470–6.

Wang, Z., Gerstein, M., and Snyder, M. (2009). RNA-Seq: a revolutionary tool for transcriptomics. Nat. Rev. Genetics *10*, 57–63.

Wei, C.M., Gershowitz, A., and Moss, B. (1976). 5′-Terminal and internal methylated nucleotide sequences in HeLa cell mRNA. Biochemistry *15*, 397–401.

Woese, C.R. (1967). The Genetic Code: The Molecular Basis for Genetic Expression. (New York, Harper & Row), p. 186.

Yang, L., Duff, M.O., Graveley, B.R., Carmichael, G.G., and Chen, L.L. (2011). Genomewide characterization of non-polyadenylated RNAs. Genome Biol. *12*, R16.

Yi, H., Cho, Y.J., Won, S., Lee, J.E., Jin Yu, H., Kim, S., Schroth, G.P., Luo, S., and Chun, J. (2011). Duplex-specific nuclease efficiently removes rRNA for prokaryotic RNA-Seq. Nucleic Acids Res. *39*, e140.

Transcriptome Reconstruction and Quantification from RNA Sequencing Data

3

Sahar Al Seesi, Serghei Mangul, Adrian Caciula, Alex Zelikovsky and Ion Măndoiu

Abstract

Massively parallel whole-transcriptome sequencing has become the technology of choice for transcriptome analysis since it supports a wider range of problems than the previously popular microarray technology. In this chapter we focus on two of these applications, namely transcriptome reconstruction and quantification. We discuss the key computational problems related with these applications and describe some of the best-performing algorithms available for each. For transcriptome reconstruction, we present in detail a statistical genome-guided method called transcriptome reconstruction using integer programming (TRIP) that incorporates fragment length distribution into novel transcript reconstruction from paired-end RNA-Seq reads. Experimental results on both real and synthetic datasets show that TRIP is more accurate than methods ignoring fragment length distribution information. For transcriptome quantification, we focus on two expectation maximization (EM) algorithms for both RNA-Seq and digital gene expression (DGE) sequencing protocols. Both algorithms take into account alternative splicing and mapping ambiguities. We present experimental results on real datasets comparing the two protocols as well as methods for each protocol. Results show that the EM algorithms outperform other available methods for both RNA-Seq and DGE, and that they yield comparable quantification accuracy on real data generated using the RNA-Seq and DGE protocols.

Introduction

Massively parallel whole-transcriptome sequencing and its ability to probe the complete transcriptome at single base resolution provides a powerful tool with multiple interrelated applications, including transcriptome reconstruction (Guttman *et al.*, 2010; Trapnell *et al.*, 2010; Li *et al.*, 2011b; Mangul *et al.*, 2012), gene/isoform expression estimation (Mortazavi *et al.*, 2008; Wang *et al.*, 2008; Li *et al.*, 2010; Nicolae *et al.*, 2011), also known as transcriptome quantification, studying trans- and cis-regulatory effects (McManus *et al.*, 2010), studying parent-of-origin effects (Degner *et al.*, 2009; Gregg *et al.*, 2010; McManus *et al.*, 2010), and calling expressed variants (Duitama *et al.*, 2012). As a result, whole transcriptome sequencing has become the technology of choice for performing transcriptome analysis, rapidly replacing array-based technologies (Wang *et al.*, 2009). To date, two main transcriptome sequencing protocols have been proposed in the literature. The most commonly used one, referred to as RNA-Seq, generates short (single or paired) sequencing tags from the ends of randomly generated cDNA fragments. An alternative protocol is the 3′-tag digital gene expression (DGE), also known as high-throughput sequencing-based serial analysis of gene expression (SAGE-Seq).

Most current research, using transcriptome sequencing data, employs methods that depend on existing transcriptome annotations. Unfortunately, as shown by recent studies (Mercer *et al.*, 2012), existing transcript libraries still miss large numbers of transcripts. The incompleteness of annotation libraries poses a serious limitation to using this powerful technology since accurate normalization of data critically requires knowledge of expressed transcript sequences (Mortazavi *et al.*, 2008; Wang *et al.*, 2008; Li *et al.*, 2010; Nicolae *et al.*, 2011). Another challenge in transcriptomic analysis comes from the ambiguities in read/tag mapping to transcripts. Ubiquitous regulatory mechanisms such as the use of alternative transcription start and polyadenylation sites, alternative splicing, and RNA editing result in multiple messenger RNA (mRNA) isoforms being generated from a single genomic locus. Most prevalently, alternative splicing is estimated to take place for over 90% of the multi-exon human genes across diverse cell types (Wang *et al.*, 2008), with as much as 68% of multi-exon genes expressing multiple isoforms in a clonal cell line of colorectal cancer origin (Griffith *et al.*, 2010). The ability to reconstruct full-length isoform sequences and accurately estimate their expression levels is widely believed to be critical for unravelling gene functions and transcription regulation mechanisms (Ponting and Belgard, 2010). In this chapter, we focus on two main problems in transcriptome data analysis, namely, transcriptome reconstruction and expression level quantification, and describe recent techniques developed to address these challenges. Transcriptome reconstruction, also referred to as novel transcript discovery, is the problem of reconstructing the transcript sequences from the sequencing data. Reconstruction can be done *de novo* or it can be assisted by existing genome and transcriptome annotations. Transcriptome quantification refers to the problem of estimating the expression level of each transcript. In the remainder of this section we give a brief description of the two main protocols for mRNA sequencing. The rest of this chapter presents computational methods for transcriptome reconstruction from RNA-Seq data and transcriptome quantification from both RNA-Seq and DGE data.

RNA-seq protocol

RNA-Seq uses next-generation sequencing technologies, such as SOLiD (Pandey *et al.*, 2008), 454 (Thomas *et al.*, 2006), Illumina (Bentley *et al.*, 2008), or ION Torrent (Rothberg *et al.*, 2011). Fig. 3.1a depicts the main steps in an RNA-Seq experiment, ending with the first step of analysis which is typically mapping the data to a reference. The mRNA extracted from a sample is converted to cDNA using reverse transcription and sheared into fragments. Fragments with lengths within a certain range are selected, and ligated with sequencing adapters. This is usually followed by an amplification step after which one or both ends of the cDNA fragments are sequenced to produce either single or paired-end reads. cDNA synthesis and adapter ligation can be done in a strand-specific manner, in which case the strand of each read is known; this is commonly referred to as directional sequencing. In the more common non-directional RNA-Seq protocols strand specificity is not maintained. The specifics of the sequencing protocols vary from one technology to the other. In particular, the length of produced reads varies depending on the technology, with newer high-throughput technologies typically producing longer reads.

DGE protocol

The DGE protocol generates short cDNA tags from an mRNA population in several steps (as shown in Fig. 3.1b). First, PolyA+ mRNA is captured from total RNA using oligo-dT magnetic beads and used as template for cDNA synthesis. The double stranded cDNA is then digested with a first restriction enzyme, called Anchoring Enzyme (AE), with known sequence specificity (e.g. the NlaIII enzyme cleaves cDNA at sites at which the four nucleotide motif CATG appears). We refer to the cDNA sites cleaved by the anchoring enzyme as AE sites. The recognition site of a second restriction enzyme, called tagging enzyme (TE) is ligated to the fragments of cDNA that remain attached to the beads after cleavage with the AE, immediately upstream of the AE site. The cDNA fragments are then digested with TE, which cleaves several bases away from its recognition site. This results in very short cDNA tags (10 to 26 bases long, depending on the TE used), which are then sequenced using any of the available high-throughput technologies. Since the recognition site of AE is only 4 bases long, most transcripts contain multiple AE sites. Under perfect experimental conditions, full digest by AE would ensure that DGE tags are

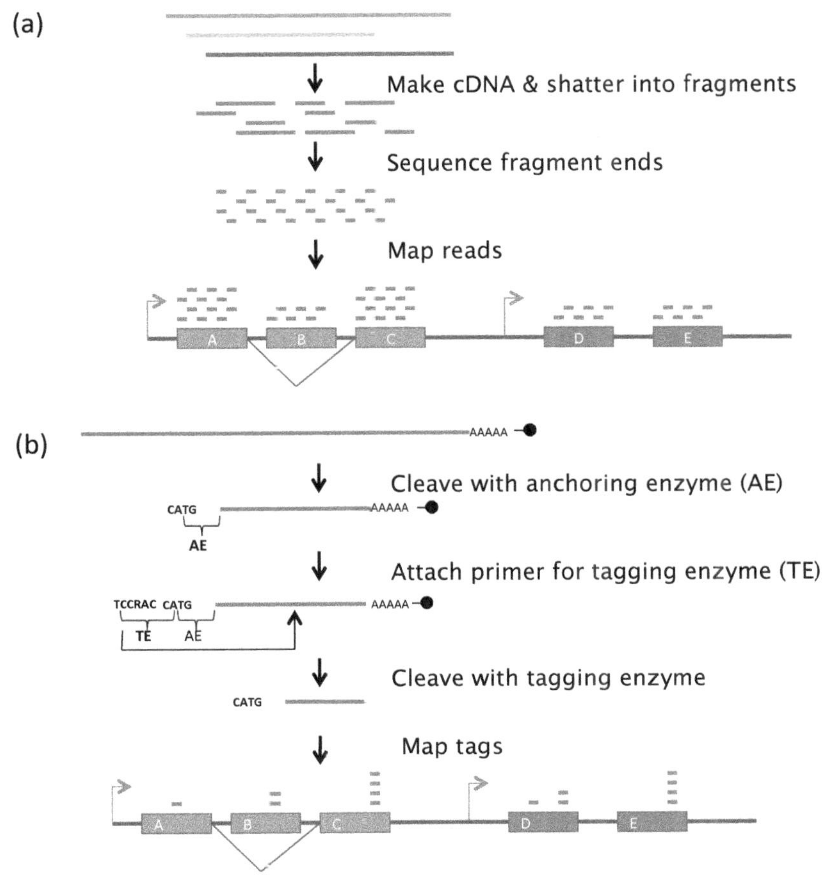

Figure 3.1 A schematic representation of the RNA sequencing protocols. (a) RNA-Seq (b) DGE. A colour version of this figure is located in the plate section at the back of the book.

generated only from the most 3′ AE site of each transcript. In practice, some mRNA molecules release tags from other AE sites, or no tag at all. The cleavage probability of the *AE*, denoted by *p*, is assumed to be the same for all *AE* sites of all transcripts. Since only the most 3′ cleaved *AE* site of a transcript releases a *DGE* tag, the probability of generating a tag from site $i = 1, \ldots, k$ follows a geometric distribution with ratio $1 - p$ as shown in Fig. 3.2, where sites are numbered starting from the 3′ end. Note that splicing isoforms of a gene are likely to share many *AE* sites. However, the probability of generating a tag from a site is isoform specific since it depends on the number downstream *AE* sites on each isoform. Thus, in order for any algorithm to properly allocate ambiguous tags among *AE* sites, it must take into account alternative splicing isoforms.

Transcriptome reconstruction

Transcriptome reconstruction from RNA-Seq has been the focus of much research in recent years. The sequences of novel transcripts can be reconstructed from deep RNA-Seq data, but this is computationally challenging due to the short length of the reads, high percentage of sequencing errors, uneven coverage of expressed transcripts, and the need to distinguish between highly similar transcripts produced by alternative splicing. A number of methods address the problem of transcriptome reconstruction from RNA-Seq reads. These methods fall into three categories: 'genome-guided', 'genome-independent' and 'annotation-guided' methods (Garber *et al.*, 2011). Genome-independent methods such as Trinity (Grabherr *et al.*, 2011) or transAbyss (Robertson *et al.*, 2010) directly assemble reads

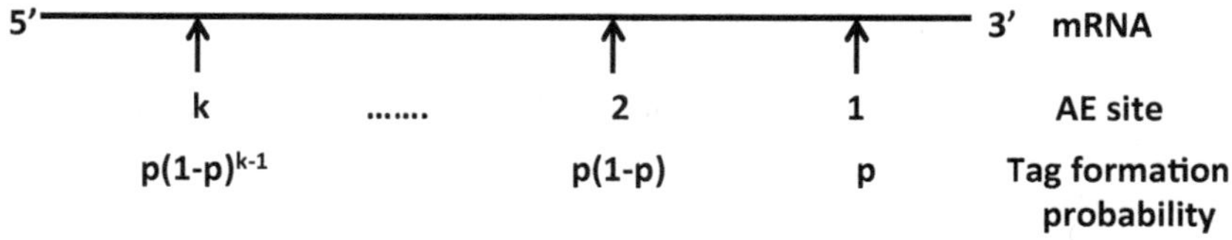

Figure 3.2 Tag formation probability; p denotes the AE cutting probability.

into transcripts. A commonly used approach for such methods is the de Bruijn graph (Pevzner, 1989) utilizing k-mers. The use of genome-independent methods becomes essential when there is no trusted genome reference that can be used to guide reconstruction. At the other end of the spectrum, annotation guided methods (Feng *et al.*, 2011; Mangul *et al.*, 2011; Roberts *et al.*, 2011a) make use of available information in existing transcript annotations to aid in the discovery of novel transcripts. RNA-Seq reads can be mapped onto the reference genome, reference transcript libraries, exon–exon junction libraries, or combinations thereof, and the resulting alignments are used to reconstruct novel transcripts.

Most transcriptome reconstruction methods fall in the genome-guided category. They typically start by mapping sequencing reads onto the reference genome, using spliced alignment tools, such as TopHat (Trapnell *et al.*, 2009) or SpliceMap (Au *et al.*, 2010). The spliced alignments are used to identify putative exons and transcripts that explain the alignments. While some methods aim to achieve the highest sensitivity, others work to predict the smallest set of transcripts explaining the given input reads. Furthermore, some methods aim to reconstruct the set of transcripts that would insure the highest quantification accuracy. Scripture (Guttman *et al.*, 2010) constructs a splicing graph from the mapped reads and reconstructs isoforms corresponding to all possible paths in this graph. It then uses paired-end information to filter out some transcripts. Although Scripture achieves very high sensitivity, it may predict a large number of incorrect isoforms. The method of Trapnell *et al.* (2010) and Roberts *et al.* (2011b), referred to as Cufflinks, constructs a read overlap graph and generates candidate transcripts by finding a minimum size path cover via a reduction to maximum matching in a weighted bipartite graph. TRIP (Mangul *et al.*, 2012) uses an integer programming model where the objective is to select the smallest set of putative transcripts that yields a good statistical fit between the fragment length distribution empirically determined during library preparation and fragment lengths implied by mapping read pairs to selected transcripts. Cufflinks, Scripture, and TRIP do not target the quantification accuracy. IsoLasso (Li *et al.*, 2011b) uses the LASSO (Tibshirani, 2011) algorithm and aims to achieve a balance between quantification accuracy and predicting the minimum number of isoforms. It formulates the problem as a quadratic program, with additional constraints to ensure that all exons and junctions supported by the reads are included in predicted isoforms. SLIDE (Li *et al.*, 2011a) uses statistical modelling under sparsity assumption while CLIIQ (Lin *et al.*, 2012) uses an integer linear programming solution that minimizes the number of predicted isoforms explaining the RNA-Seq reads while minimizing the difference between estimated and observed expression levels of exons and junctions within the predicted isoforms. Table 3.1 compares the features of these methods for transcriptome reconstruction. Experimental results on both real and synthetic datasets generated with various sequencing parameters and distribution assumptions show that the TRIP method of (Mangul *et al.*, 2012) has increased transcriptome reconstruction accuracy compared to other methods.

In the remainder of this section we describe TRIP in more detail. The first step of TRIP is to infer exon boundaries and exon–exon junctions from spliced genome alignments of the reads. Then, a splice graph is created based on inferred exon boundaries and junction information. The third step enumerates all maximal paths in the

Table 3.1 Transcriptome reconstruction methods

Methods	Support paired-end reads	Consider fragment length distribution	Require annotations
TRIP	Yes	Yes	No
IsoLasso	Yes	No	No
IsoInfer	No	No	TES/TSS
Cufflinks	Yes	Yes	No
CLIQ	No	No	No
Scripture	Yes	No	No
SLIDE	Yes	No	Gene/exon boundaries

splice graph corresponding to putative transcripts. The problem of selecting true transcripts is formulated as an integer program (IP) which minimizes the number of selected transcripts subject to achieving a good statistical fit between the fragment length distribution (empirically determined during library preparation) and fragment lengths implied by mapped read pairs.

Read mapping

As with many RNA-Seq analysers, the first step of TRIP is to map the reads. Reads are mapped onto the genome reference using any of the available spliced alignment tools ((Mangul *et al.*, 2012) used TopHat (Trapnell *et al.*, 2009) with default parameters). Note that a paired read consists of two reads flanking a fragment whose length usually follows normal distribution. The mean and variance of fragment length distribution are usually known from library preparation or can be inferred from alignments of reads coming from single transcript genes.

Construction of the splice graph and enumeration of candidate transcripts

Typically, a gene can express multiple mRNA transcripts due to alternative transcriptional or splicing events including alternative first exon, alternative last exon, exon skipping, intron retention, alternative 5′ splice site (A5SS), and alternative 3′ splice site (A3SS) (Pal *et al.*, 2011). To represent such alternative transcripts, a gene is processed as a set of so called 'pseudo-exons' based on alternative variants obtained from aligned RNA-Seq reads. A pseudo-exon is a region of a gene between consecutive transcriptional or splicing events, i.e. starting or ending of an exon, as shown in Fig. 3.3. Hence every gene has a set of non-overlapping pseudo-exons, from which it is possible to reconstruct the set of candidate transcripts.

The splice graph is a directed acyclic graph (see Fig. 3.4), whose vertices represent pseudo-exons and edges represent pairs of pseudo-exons immediately following one another in at least one transcript (which is witnessed by at least

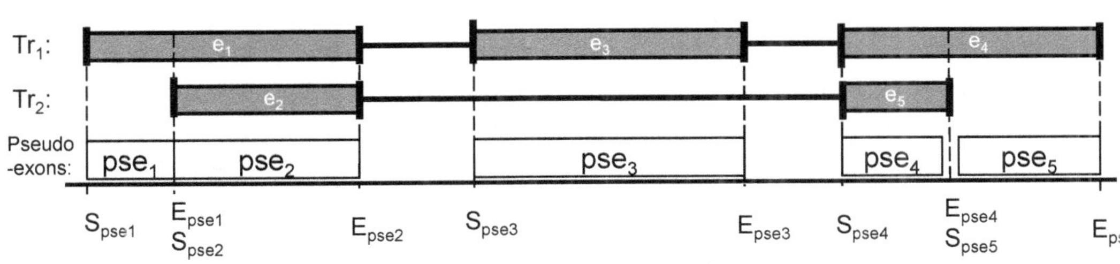

Figure 3.3 Pseudo-exons (white boxes) are regions of a gene between consecutive transcriptional or splicing events. An example of two transcripts, Tr_1 and Tr_2, each sharing exons (blue boxes, denoted by e_i). S_{psej} and E_{psej} represent the starting and ending position of pseudo-exon j, respectively.

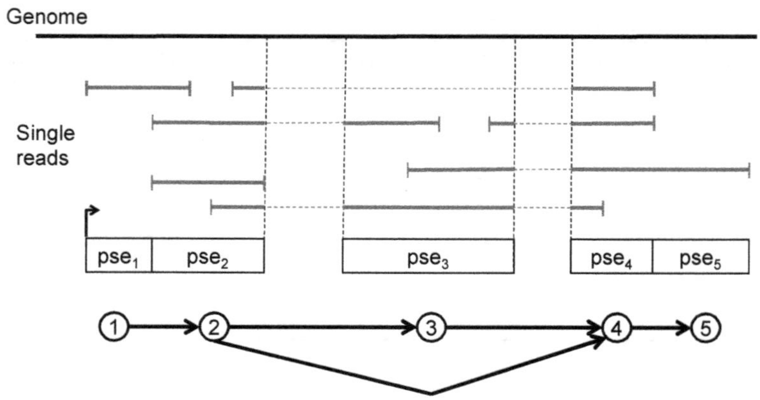

Figure 3.4 Splice graph. The horizontal lines represent single reads. Reads interrupted by dashed lines are spliced reads. Each vertex of the splice graph corresponds to a pseudo-exon and each directed edge corresponds to a (splice) junction between two pseudo-exons.

one spliced read). TRIP enumerates all maximal paths in the splice graph using a depth-first-search algorithm. These paths correspond to candidate transcripts. Note that a gene with n pseudo-exons may have as many as 2^n-1 possible candidate transcripts, each composed of a subset of the n pseudo-exons. The next subsection presents an integer program that selects a minimal subset of candidate transcripts that statistically fits well-observed paired reads.

Integer program formulation

The following notations are used in the Integer Program (IP) formulation:

- N total number of read pairs;
- J_l lth splice junction;
- p_j paired-end read pair j, $1 \le j \le N$;
- t_k kth candidate transcript, $1 \le k \le K$;
- s_i expected portion of reads mapped within i standard deviations ($s_1 \approx 68\%$, $s_2 \approx 95\%$, $s_3 \approx 99.7\%$);
- ε allowed deviation from the rule ($\varepsilon = 0.05$);
- $T_i(p_j)$ set of candidate transcripts where p_j can be mapped with a fragment length between $i-1$ and i standard deviations; $1 \le i \le 3$;
- $T_4(p_j)$ set of candidates transcripts where p_j can be mapped with a fragment length within more than 3 standard deviations;

The IP uses two sets of Boolean variables:

- $y(t_k) = 1$ if candidate transcript t_k is selected, and 0 otherwise;
- $x_i(p_j) = 1$ if the read pair p_j is mapped between $i-1$ and i standard deviations, and 0 otherwise;

Using these variables, the transcriptome reconstruction problem is formulated as follows:

$$\sum_{t_k T} y(t) \rightarrow \min$$

subject to:

$$(1) \sum_{t_k T_i(p)} y(t) \ge x_i(p), \forall p, i = \overline{1,4}$$

$$(2) N(s_i - \varepsilon) \le \sum_j x_i(p_j) \le N(s_i + \varepsilon), i = \overline{1,4}$$

$$(3) \sum_i x_i(p) \le 1, \forall p$$

$$(4) \sum_{t_k J_l} y(t) \ge 1, \forall J_l$$

The objective of the IP is to minimize the number of candidate transcripts subject to four constraints. Constraint (1) ensures that a paired-end read p is counted as being mapped to within i standard deviations only if at least one transcript $t \in T_i(p)$ is selected. Constraint (2) restricts the number of paired-end reads mapped within every category of standard deviation. Constraint (3) ensures that each paired-end read p gets counted towards at most one category of standard deviations. Finally, constraint (4) ensures that every splice junction supported by reads is present in the set of selected transcripts at least once.

Simulation setup and matching criteria

TRIP was tested by Mangul *et al.* (2012) on simulated human RNA-Seq data. The human genome sequence (hg*18*, NCBI build *36*) was downloaded from UCSC together with the *KnownGenes* transcripts annotation table. Genes were defined as clusters of known transcripts defined by the GNFAtlas2 table. The dataset contains a total of 66,803 transcripts pertaining to 19,372 genes.

Error-free paired-end reads of length 100 base pairs were randomly generated for each gene by sampling fragments from known transcripts at a coverage of 100× per transcript. Fragment lengths were simulated from a normal probability distribution with a mean of 500 and standard deviations of 50 and 500, respectively.

All reconstructed transcripts were matched against annotated transcripts. As in Trapnell *et al.* (2010) and Li *et al.* (2011b), two transcripts were assumed to match if and only if internal pseudo-exon boundaries coordinates (i.e. all pseudo-exons coordinates except the beginning of the first pseudo-exon and the end of the last pseudo-exon) were identical. Sensitivity, positive predictive value (PPV) and the *F*-score were used to evaluate the performance of different methods. Sensitivity is defined as the proportion of annotated transcript sequences that match reconstructed sequences, i.e. sensitivity = TP/(TP + FN). PPV is defined as the proportion of reconstructed sequences that match annotated transcript sequences, i.e. PPV = TP/(TP + FP) and the *F*-score is defined as the harmonic mean of sensitivity and PPV, i.e. *F*-score = 2 × (PPV × Sens)(PPV + Sens).

Comparison of methods on simulated data

TRIP was compared in Mangul *et al.* (2012) to Cufflinks version 2.0.0. The integer programs for TRIP were solved by IBM ILOG CPLEX (version 12.2.0.0). The comparison in Mangul *et al.* (2012) also included a method that reports all candidate transcripts in order to illustrate the effectiveness of selection produced by the integer program (IP) in TRIP. A variant of TRIP that is given the transcription start sites (TSS) and transcription end sites (TES) was also compared; this is useful for assessing the benefits of complementing RNA-Seq data with TSS/TES data generated by specialized protocols such as the RAMPAGE (Batut *et al.*, 2012) and PolyA-Seq (Derti *et al.*, 2012).

Figs. 3.5–3.7 compare the performance of these four methods (Cufflinks, all candidate transcripts, TRIP with and without TSS/TES) on simulated data with respect to the number of transcripts per gene. Note that sensitivity (see Fig. 3.5) for single-transcript genes is 100% for all methods. With the growth in number of transcripts per gene. TRIP's sensitivity gain over Cufflinks gradually improves while sensitivity of candidate transcripts stays almost 100%. The advantage of TRIP over

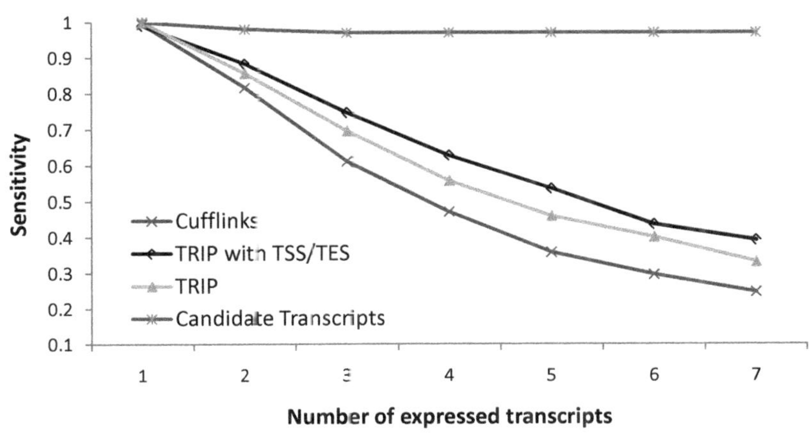

Figure 3.5 Sensitivity comparison between methods for groups of genes with *n* transcripts (*n*=1,..., ≥7) on simulated datasets with mean fragment length 500, standard deviation 50, and read length of 100×2.

Cufflinks can be explained by extra statistical constraints in the *IP* that are not taken into account by Cufflinks.

Fig. 3.6 shows that Cufflinks has an advantage over TRIP in the portion of correctly predicted transcripts but overall comparison using *F*-score (see Fig. 3.7) shows that TRIP improves over Cufflinks.

Influence of fragment length distribution

Although high-throughput technologies allow users to make trade-offs between read length and the number of generated reads, very little has been done to determine optimal choice of fragment length. Additionally, novel Next Generation Sequencing (*NGS*) technologies such as Ion Torrent may allow learning exact fragment length. For the case when fragment length is known, TRIP-L is a variation of TRIP that takes advantage of known exact fragment length. Figs. 3.8 and 3.9 compare sensitivity and PPV of TRIP, TRIP-L, and Cufflinks when run on simulated data generated with mean fragment length of 500 bp and standard deviation of either 50 bp or 500 bp. The results show that TRIP has a better sensitivity than Cufflinks, while TRIP-L further improves it. Interestingly, higher variation in fragment lengths actually improves performance of all methods.

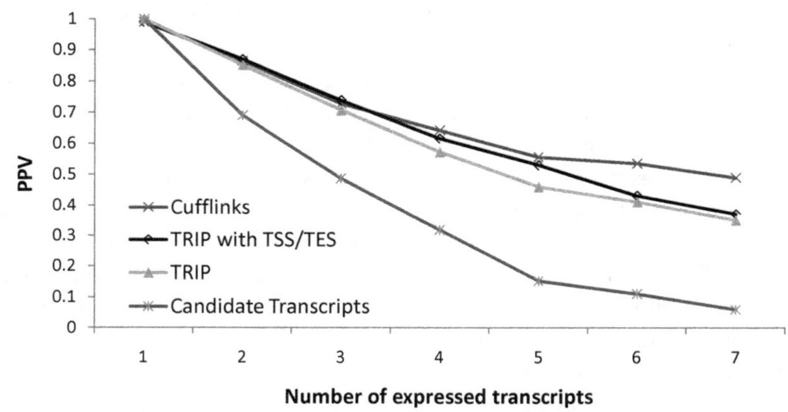

Figure 3.6 PPV comparison between methods for groups of genes with *n* transcripts (n=1 ,...,≥7) on simulated datasets with mean fragment length 500, standard deviation 50, and read length of 100×2.

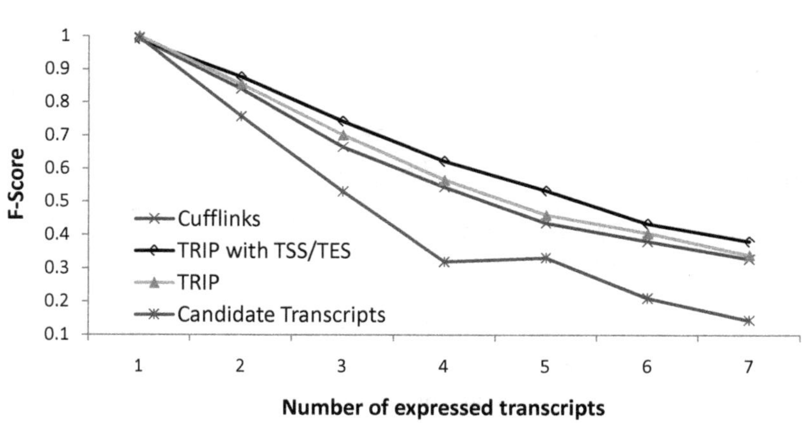

Figure 3.7 F-Score comparison between methods for groups of genes with *n* transcripts (n=1,..., ≥7) on simulated datasets with mean fragment length 500, standard deviation 50, and read length of 100×2.

Results on real RNA-seq data

TRIP was also tested on real RNA-Seq data that was sequenced from a CD1 mouse retina RNA samples. Mangul *et al.* (2012) selected a specific gene that has 33 annotated transcripts in Ensembl. The gene was picked and validated experimentally due to interest in its biological function. The read alignments falling within the genomic locus of the selected gene were used to construct a splicing graph; then candidate transcripts were selected using TRIP. The dataset used consists of 46,906 alignments for 22,346 read pairs with read length of 68. TRIP was able to infer 5 out of 10 transcripts that we confirmed using qPCR. For comparison, the same experiment was executed using Cufflinks which was able to infer only 3 out of the 10 experimentally confirmed transcripts.

Transcriptome quantification

Transcriptome quantification is usually done on one of two levels: gene expression-level estimation (GE) and isoform expression-level estimation (IE). Targeted GE using methods such as quantitative PCR has long been a staple of genetic studies. The completion of the human genome has been a key enabler for genome-wide GE performed using expression microarrays. Since expression microarrays have limited capability of detecting alternative splicing events, specialized splicing arrays have been developed for genome-wide interrogation of both annotated exons and exon–exon junctions. However, despite sophisticated deconvolution algorithms (Anton *et al.*, 2008; She *et al.*, 2009), the fragmentary information provided by splicing arrays is typically insufficient for unambiguous identification

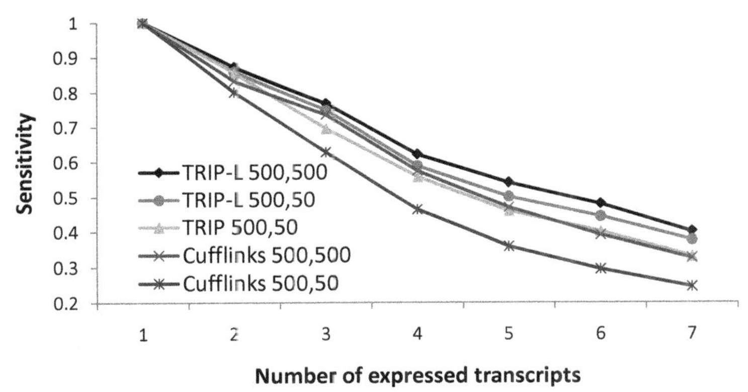

Figure 3.8 Sensitivity comparison between methods on simulated datasets with different fragment length distribution assumptions.

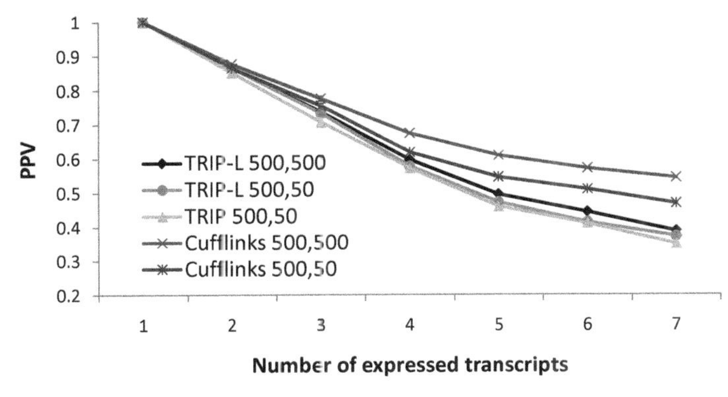

Figure 3.9 Positive predictive value (PPV) comparison between methods on simulated datasets with different fragment length distribution assumptions.

of full-length transcripts (Lacroix *et al.*, 2008; Hiller *et al.*, 2009). Massively parallel transcriptome sequencing is quickly replacing microarrays as the technology of choice for performing gene expression profiling due to its wider dynamic range and digital quantitation capabilities. However, accurate estimation of expression levels from sequencing data remains challenging due to the short read length delivered by current sequencing technologies and still poorly understood protocol- and technology-specific biases. In this section we discuss methods for transcriptome quantification using mRNA sequencing generated using the RNA-Seq and DGE protocols. We focus on two expectation maximization (EM) models, one for RNA-Seq called IsoEM (Nicolae *et al.*, 2011) and another for DGE, called DGE-EM (Nicolae and Măndoiu, 2011) as they have been shown to outperform other commonly used methods.

RNA-seq methods

Quantitative analyses of mRNA sequencing data typically start by mapping sequencing reads onto the reference genome, reference transcriptome, exon–exon junction libraries, or combinations thereof. Early RNA-Seq studies have recognized that limited read lengths result in a significant percentage of so called multireads, i.e. reads that map equally well at multiple locations in the genome. A simple approach is to discard multireads, and estimate expression levels using only the so called unique reads. Mortazavi *et al.* (2008) proposed a multiread '*rescue*' method whereby initial gene expression levels are estimated from unique reads and used to fractionally allocate multireads, with final expression levels obtained by re-estimation based on total counts obtained after multiread allocation. An EM algorithm that extends this scheme by repeatedly alternating between fractional read allocation and re-estimation of gene expression levels was proposed in Pasaniuc *et al.* (2011). A number of recent works have addressed the IE problem, namely isoform expression level estimation from RNA-Seq reads. Under a simplified 'exact information' model, Lacroix *et al.* (2008) showed that neither single nor paired read RNA-Seq data can theoretically guarantee unambiguous inference of isoform expression levels, although paired reads may be sufficient to deconvolute expression levels for the majority of annotated isoforms. The key challenge in IE is accurate assignment of ambiguous reads to isoforms. Compared with the GE context, read ambiguity is much more significant, since it affects not only multireads, but also reads that map at a unique genome location expressed in multiple isoforms. Estimating isoform expression levels based solely on unambiguous reads, as suggested, e.g. in Griffith *et al.* (2010), results in splicing-dependent biases similar to the transcript-length bias noted in Oshlack and Wakefield (2009), further complicating the design of differential expression tests based on RNA-Seq data. To overcome this difficulty, Jiang and Wong (2009) proposed a Poisson model of single-read RNA-Seq data explicitly modelling isoform frequencies. Under their model, maximum likelihood estimates are obtained by solving a convex optimization problem, and uncertainty of estimates is obtained by importance sampling from the posterior distribution. The IE problem for single reads is also tackled in Richard *et al.* (2010), who propose an EM algorithm for inferring isoform expression levels from the read coverage of exons (reads spanning exon junctions are ignored).

Three of the most commonly used algorithms for IE from RNA-Seq data are RSEM, Cufflinks, and IsoEM. RSEM is an expectation maximization (EM) algorithm that works on the isoform level. The initial version of RSEM (Li *et al.*, 2010) only handled single-end reads, however, the latest version (Li and Dewey, 2011) has been extended to support paired-end reads, variable-length reads, and incorporates fragment length distribution and quality scores in its modelling. In addition to the maximum likelihood estimates of isoform expressions, RSEM also calculates a 95% confidence intervals and posterior mean estimates. As part of transcriptome reconstruction and quantification, Cufflinks (Trapnell *et al.*, 2010) employs a likelihood function to estimate isoform expression from read alignments, incorporating the fragment length distribution. They use a Bayesian inference procedure based on importance sampling to calculate the maximum a posteriori (MAP) estimate. They also report a 95% CI for the MAP estimates. In its latest version (Roberts *et al.*, 2011b), Cufflinks handles sequence bias and positional bias (non-uniformity of fragment positions). IsoEM is an EM algorithm for single and paired-end reads that incorporates fragment length distribution and other available information to handle read mapping ambiguities. Both RSEM and IsoEM have the ability to infer expression levels for a given set of transcripts without the need for a gene annotation file (GTF). In the remainder of this subsection we detail the IsoEM algorithm. Experimental results shown later in the section show that IsoEM significantly outperforms Cufflinks. These results agree with those in Li and Dewey (2011), where the authors find that RSEM and IsoEM have comparable accuracy and both outperform Cufflinks in terms of estimation accuracy. (Li and Dewey, 2011) also shows that IsoEM is the fastest among the three methods; although it also has the highest memory requirements.

Additional statistical models for transcriptome quantification are reviewed by Pachter (2011), who classifies the available models based on the different features they support. Table 3.2 includes a list of these features, and details which of the algorithms discussed above support each feature.

IsoEM (Nicolae *et al.*, 2011) estimates isoform expression levels (interchangeably referred to as frequencies) from RNA-Seq reads, under the assumption that a complete list of candidate isoforms is available. It is an expectation maximization algorithm for isoform frequency estimation from any mixture of single and paired RNA-Seq reads. A key feature of the algorithm is that it exploits information provided by the distribution of insert sizes, which is tightly controlled during sequencing library preparation under current RNA-Seq protocols. Such information is not modelled in the 'exact' information models of Lacroix *et al.* (2008) and Hiller *et al.* (2009), challenging the validity of their negative results. Nicolae *et al.* (2011) show that modelling insert sizes is highly beneficial for IE even for RNA-Seq data consisting of single reads. Insert sizes contribute to increased estimation accuracy in two different ways. On one hand, they can help disambiguating the isoform of origin for the reads. Insert lengths are combined with base quality scores, and, if available, read pairing and strand information to probabilistically allocate reads to isoforms during the expectation step of the algorithm. The genomic locations of multireads are also resolved probabilistically in this step, further contributing to overall accuracy compared to methods that ignore or fractionally pre-allocate multireads. On the other hand, insert size distribution is used to accurately adjust isoform lengths during frequency re-estimation in the maximization step of the IsoEM algorithm. Additionally, IsoEM corrects for sequence bias introduced by random hexamers used to prime reverse transcription by implementing a simple reweighting scheme introduced in Hansen *et al.* (2010). To avoid biases from incorrectly mapped reads originating from repetitive regions, IsoEM has the option to discard reads that overlap annotated repeats. When applying this correction, isoform lengths are automatically adjusted by subtracting the number of positions resulting in reads that would be discarded.

The first step of IsoEM is to map the reads. The IsoEM pipeline suggests to map them onto the library of known isoforms using any one of the

Table 3.2 Main features of transcriptome quantification methods

Feature	Methods
Paired-end reads	IsoEM, RSEM, Cufflinks
Handling read ambiguity	IsoEM, RSEM, Cufflinks, (Jiang and Wong, 2009; Richard *et al.*, 2010; Pasaniuc *et al.*, 2011)
Sequence bias	IsoEM, Cufflinks
Positional bias	Cufflinks, RSEM
Independence of transcript annotations	IsoEM*, RSEM
Confidence intervals	Cufflinks, RSEM

*Feature available in IsoEM through SEQ_FRQ script available for download at the IsoEM website http://dna.engr.uconn.edu/?page_id=105

many available ungapped aligners, such as Bowtie (Langmead *et al.*, 2009). An alternative strategy is to map the reads onto the genome using a spliced alignment tool such as TopHat (Trapnell *et al.*, 2009). The IsoEM implementation is compatible with both mapping approaches. It always converts read alignments to genome coordinates and performs all read-isoform compatibility calculations in genome space.

The candidate set of isoforms for each read is obtained by combining all genome coordinates of reads and isoforms, sorting them and using a line sweep technique to detect read-isoform compatibilities. During the line sweep reads are grouped into equivalence classes defined by their isoform compatibility sets; this speeds up the E-steps of the IsoEM algorithm by allowing the processing of an entire read class at once.

Some of the reads match multiple positions in the genome, which we refer to as *alignments* (for paired-end reads, an alignment consists of the positions where the two reads in the pair align with the genome). Each alignment a can in turn be compatible with multiple isoforms that overlap at that position of the genome. During the line sweep, the relative 'weight' of assigning a given read/pair r to isoform j is computed as $w(r j) = \Sigma Q_a F_a O_a$ where the sum is over all alignments of r compatible with j, and the factors of the summed products are defined as follows:

- Q_a represents the probability of observing the read from the genome locations described by the alignment. This is computed from the base quality scores as

$$Q_a = \prod_{k=1}^{|r|} [(1-\varepsilon_k) M_{a_k} + \frac{\varepsilon_k}{3}(1-M_{a_k})]$$

  where $M_{a_k} = 1$ if position k of alignment a matches the reference genome sequence and 0 otherwise, while ε_k denotes the error probability of kth base of r.
- For paired-end reads, F_a represents the probability of the fragment length needed to produce alignment a from isoform j; note that the length of this fragment can be inferred from the genome coordinates of the two aligned reads and the available isoform annotation. For single reads, only an upper bound u on the fragment length can only be estimated: if the alignment is on the same strand as the isoform then u is the number of isoform annotated bases between the 5′ end of the aligned read and the 3′ end of the isoform, otherwise u is the number of isoform annotated bases between the 5′ end of the aligned read and the 5′ end of the isoform. In this case F_a is defined as the probability of observing a fragment with length of u bases or fewer.
- O_a is 1 if alignment a of r is consistent with the orientation of isoform j, and 0 otherwise. Consistency between the orientations of r and j depends on whether or not the library preparation protocol preserves the strand information. For single reads $O_a = 1$ when reads are generated from fragment ends randomly or, for directional RNA-Seq, when they match the known isoform orientation. For paired-end reads, $O_a = 1$ if the two reads come from different strands, point to each other, and, in the case of directional RNA-Seq, the orientation of the first read matches the known isoform orientation.

The algorithm starts with the set of N known isoforms. For each isoform $l(j)$ denotes its length and $f(j)$ denotes its (unknown) frequency. If we denote by $n(j)$ the number of reads coming from isoform j and let $p(k)$ denote the probability of a fragment of length k, then

$$E[n(j)] \propto \sum_{k \le l(j)} p(k)(l(j)-k-1)$$

since the number of fragments of length k is expected to be proportional to the number of valid starting positions for a fragment of that length in the isoform. Thus, if the isoform of origin is known for each read, the maximum likelihood estimator for $f(j)$ is given by $c(j)/(c(1)+\ldots+c(N))$, where

$$c(j) = n(j) \Big/ \sum_{k \le l(j)} p(k)(l(j)-k-1)$$

denotes the length-normalized fragment coverage. Note that the length of most isoforms is significantly larger than the mean fragment length typical of current sequencing libraries; for such isoforms $c(j)$ can be approximated by $n(j)/(l(j) - \mu + 1)$.

Since some reads match multiple isoforms, their isoform of origin cannot be established unambiguously. The IsoEM algorithm overcomes this difficulty by simultaneously estimating the frequencies and imputing the missing read origin within an iterative framework. After initializing frequencies $f(j)$ at random, the algorithm repeatedly performs the next two steps until convergence:

- E-step: Compute the expected number $n(j)$ of reads that come from isoform j under the assumption that isoform frequencies $f(j)$ are correct, based on weights $w(r,j)$ computed as described earlier.
- M-step: For each j, set the new value of $f(j)$ to $c(j)/(c(1) + \ldots + c(N))$, where normalized coverages $c(j)$ are based on expected counts computed in the prior E-step.

Below we describe two implementation optimizations that significantly improve the performance of IsoEM by reducing both runtime and memory usage. The first optimization consists of partitioning the input into compatibility components. The compatibility between reads and isoforms naturally induces a bipartite read-isoform compatibility graph, with edges connecting each isoform with all reads that can possibly originate from it. Connected components of the compatibility graph can be processed independently in IsoEM since the frequencies of isoforms in one connected component do not affect the frequencies of isoforms in any other connected component.

Although this optimization can be applied to any EM algorithm, its impact is particularly significant in IsoEM. Indeed, in this context the compatibility graph is decomposed in numerous small components (see Fig. 3.10a for a typical distribution of component sizes; a similar distribution of component sizes is reported for *Arabidopsis* gene models in Howard and Heber (2010). The resulting speed-up comes from the fact that, in each iteration of IsoEM, frequencies of isoforms in a single compatibility component are updated, avoiding needless updates for other isoforms.

The second IsoEM optimization consists of partitioning the set of reads within each compatibility component into equivalence classes. Two reads are equivalent for IsoEM if they are compatible with the same set of isoforms and their compatibility weights to the isoforms are proportional. Keeping only a single representative from each read class (with appropriately adjusted frequency) drastically reduces the number of reads kept in memory (see Fig. 3.10b). As the number of reads increases, the number of read classes increases much slower. Eventually this reaches saturation and no new read classes appear – at which point the runtime of IsoEM becomes virtually independent of the number of reads. Indeed, in practice the runtime bottlenecks are parsing the reads, computing the compatibility graph and detecting equivalent reads.

Once read classes are constructed, the E-step of IsoEM is slightly modified to use read classes instead of reads. For details on the union-find algorithm used for efficiently finding compatibility components and read classes the reader is directed to Nicolae *et al.* (2011).

DGE methods

While computational methods for accurate inference of gene (and isoform) specific expression levels from RNA-Seq data have attracted much attention recently, analysis of DGE data still relies on direct estimates obtained from counts of uniquely mapped DGE tags (Asmann *et al.*, 2009; Zaretzki *et al.*, 2010). In part this is due to salient features of the DGE protocol, which, unlike RNA-Seq, guarantees that each mRNA molecule in the sample generates at most one tag and obviates the need for length normalization. Nevertheless, ignoring ambiguous DGE tags (which, due to the severely restricted tag length, can represent a sizeable fraction of the total) may result in systematic inference biases. A simple heuristic redistribution scheme for rescuing some of the ambiguous tags was proposed by (Wu *et al.*, 2010). In their approach, the rightmost site in each transcript is identified as a 'best' site. If a tag matches several locations, but only one of them is a best site, then the tag is assigned to that site. If a tag matches

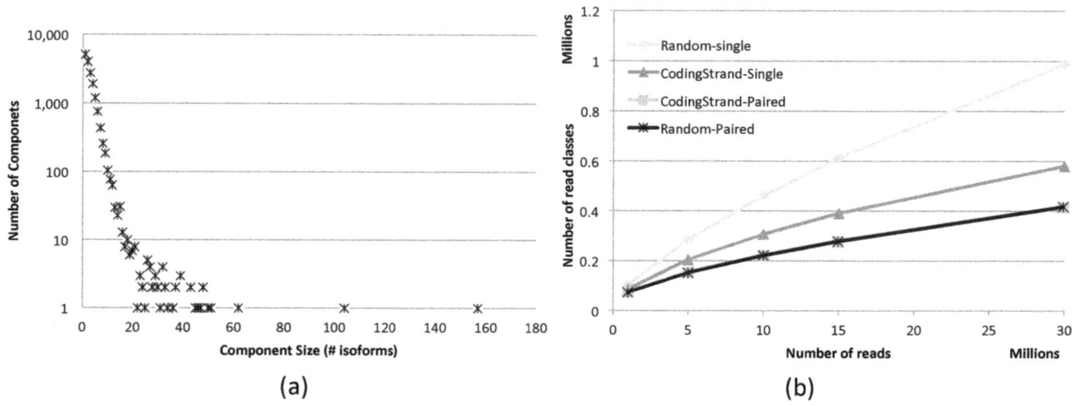

Figure 3.10 Distribution of compatibility component sizes (defined as the number of isoforms) for 10 million single reads of length 75 (a) and number of read classes for 1 to 30 million single reads or pairs of reads of length 75 (b).

multiple locations, none of which is a best site, the tag is equally split between these locations.

In this section we detail an EM algorithm, referred to as DGE-EM (Nicolae and Măndoiu, 2011), which probabilistically assigns DGE tags to candidate AE sites in different genes, different isoforms of the same gene, as well as different sites within the same isoform. In a pre-processing step, a weight is assigned to each (DGE tag, AE site) pair, reflecting the conditional probability of the tag given the site that releases it. This probability is computed from base quality scores assuming that sequencing errors at different tag positions arise independently of one another. Formally, the weight for the alignment of tag t with the jth rightmost AE site in isoform i is

$$w_{t,i,j} \propto \prod_{k=1}^{|t|}\left[(1-\varepsilon_k)M_{t_k} + \frac{\varepsilon_k}{3}(1-M_{t_k})\right],$$

where M_{t_k} is 1 if position k of tag t matches the corresponding position at site j in the transcript, 0 otherwise, while ε_k denotes the error probability of the kth base of t, derived from the corresponding quality score reported by the sequencing machine. In practice we only compute these weights for sites at which a tag can be mapped with a small (user selected) number of mismatches, and assume that remaining weights are 0. With each tag t, we associate a 'tag class' y_t which consists of the set of triples (i, j, w), where i is an isoform, j is an AE site in isoform i, and $w > 0$ is the weight associated as above with tag t and site j in isoform i. The collection of tag classes, $y = (y_t)_t$, represents the observed DGE data.

Let m be the number of isoforms. The parameters of the model are the relative frequencies of each isoform, $\theta = (f_i)_{i=1,\dots,m}$. Let $n_{i,j}$ denote the (unknown) number of tags generated from AE site j of isoform i. Thus, $x = (n_{i,j})_{i,j}$ represents the complete data. Denoting by k_i the number of AE sites in isoform i, by

$$N_i = \sum_{j=1}^{k_i} n_{i,j}$$

the total number of tags from isoform i, and by

$$N = \sum_{i=1}^{m} N_i$$

the total number of tags overall, we can write the complete data likelihood as

$$g(x|\theta) = \prod_{i=1}^{m} \prod_{j=1}^{k_i}\left[\frac{f_i(1-p)^{j-1}p}{S}\right]^{n_{i,j}}$$

where

$$S = \sum_{i=1}^{m} \sum_{j=1}^{k_i} f_i(1-p)^{j-1}p = \sum_{i=1}^{m} f_i(1-(1-p)^{k_i}).$$

Put into words, the probability of observing a tag from site j in isoform i is the frequency of that isoform (f_i) times the probability of not cutting at any of the first $j-1$ sites and cutting at the jth, i.e. $[(1-p)^{j-1}p]$. Notice that the algorithm effectively down-weights the matching AE sites far from the 3′ end based on the site probabilities shown in Fig. 3.2. Since for each transcript there is a probability that no tag is actually generated, in order for the above formula to yield a proper probability distribution, we have to normalize by the sum S over all observable AE sites.

Taking logarithms gives the complete data log-likelihood:

$$\begin{aligned}\log g(x|\theta) &= \sum_{i=1}^{m}\sum_{j=1}^{k_i} n_{i,j}[\log f_i + (j-1)\log(1-p) \\ &\quad + \log p - \log S] \\ &= \sum_{i=1}^{m}\sum_{j=1}^{k_i} n_{i,j}[\log f_i + (j-1)\log(1-p) \\ &\quad + \log p - \log S] + N \log p \\ &\quad - N \log\left(\sum_{i-1}^{m} f_i\left(1-(1-p)^{k_i}\right)\right) + constant\end{aligned}$$

E-step

Let

$$c_{i,j} = \{\, y_t \mid \exists w \ s.t.\ (i,j,e) \in y_t \}$$

be the collection of all tag classes that are compatible with AE site j in isoform i. The expected number of tags from each cleavage site of each isoform, given the observed data and the current parameter estimates, can be computed as

$$\begin{aligned}n_{i,j}^{(r)} &= E(n_{i,j}|y, \theta^{(r)}) \\ &= \sum_{y_t \in c_{i,j}, (i,j,w) \in y_t} \frac{f_i(1-p)^{i-1} pw}{\sum_{(w,q,z)\in y_t} f_l(1-p)^{q-1} p_z}\end{aligned}$$

This means that each tag class is fractionally assigned to the compatible isoform AE sites based on the frequency of the isoform, the probability of cutting at the cleavage sites where the tag matches, and the confidence that the tag comes from each location.

M-step

In this step we want to select parameters that maximize the Q function,

$$\begin{aligned}Q(\theta|\theta^{(r)}) &= E\left[\log g(x|\theta)|y,\theta^{(r)}\right] \\ &= \sum_{i=1}^{m}\sum_{j=1}^{k_i} n_{i,j}^{(r)}[\log f_i + (j-1)\log(1-p)] \\ &\quad + N \log p \\ &\quad - N \log\left(\sum_{i-1}^{m} f_i\left(1-(1-p)^{k_i}\right)\right) \\ &\quad + constant\end{aligned}$$

Partial derivatives of the Q function are

$$\begin{aligned}\frac{\delta Q(\theta|\theta^{(r)})}{\delta f_i} &= \frac{1}{f_i}\sum_{j=1}^{k_i} n_{i,j}^{(r)} \\ &\quad + N\frac{1-(1-p)^{k_i}}{\sum_{l=1}^{m} f_l(1-(1-p)^{k_l})}\end{aligned}$$

Letting $C = N/\left(\sum_{l=1}^{m} f_l\left(1-(1-p)^{k_l}\right)\right)$ and equating partial derivatives to 0 gives

$$\begin{aligned}&\frac{N_i^{(r)}}{f_i} + C\left(1-(1-p)^{k_i}\right) \\ &= 0 \Rightarrow f_i = -\frac{N_i^{(r)}}{C\left(1-(1-p)^{k_i}\right)}\end{aligned}$$

since $\sum_{i=1}^{m} f_i = 1$ it follows that

$$f_i = \frac{N_i^{(r)}}{1-(1-p)^{k_i}}\left(\sum_{l=1}^{m}\frac{N_l^{(r)}}{1-(1-p)^{k_l}}\right)^{-1}$$

Inferring p

In the above calculations we assumed that p is known, which may not be the case in practice. Assuming the geometric distribution of tags to sites, the observed tags of each isoform provide an independent estimate of p (Zaretski *et al.*, 2011). However, the presence of ambiguous tags complicates the estimation of p on an isoform by isoform basis. In order to globally capture the value of p, the DGE-EM algorithm incorporates it as a hidden variable and iteratively re-estimate it as the distribution of tags to isoforms changes from iteration to iteration.

The value of p is estimated as N^1/D, where denotes the total number of RNA molecules with at least one AE site, and

$$N^1 = \sum_{i=1}^{m} n_{i1}$$

denotes the total number of tags coming from first AE sites. The total number of RNA molecules representing an isoform is computed as the number of tags coming from that isoform divided by the probability that the isoform is cut. This gives

$$D = \sum_{i=1}^{m} N_i / (1-(1-p)^{k_i}),$$

which happens to be the normalization term used in the M step of the algorithm.

Comparison of transcriptome quantification methods and protocols

Due to the availability of both DGE and RNA-Seq data generated from two commercially available reference RNA samples that have been well characterized by quantitative real time PCR as part of the MicroArray Quality Control Consortium (MAQC), it is possible to directly compare estimation performance of inference methods both within and between protocols. While RNA-Seq is clearly more powerful than DGE at detecting alternative splicing and novel transcripts, previous studies have suggested that for gene expression profiling DGE may yield accuracy comparable to that of RNA-Seq at a fraction of the cost ('t Hoen *et al.*, 2008). The results presented below show that indeed the two protocols achieve similar cost-normalized accuracy on the MAQC samples when using state-of-the-art estimation methods. For RNA-Seq, we compare IsoEM algorithm with the widely used Cufflinks algorithm. Estimation accuracy is assessed using the correlation coefficient (R^2) and the median per cent error (MPE), which gives the median value of the relative errors (in percentage) over all genes.

The comparison includes results from real DGE datasets consisting of nine libraries that were used in Asmann *et al.* (2009). These libraries were independently prepared and sequenced at multiple sites using six flow cells on Illumina Genome Analyser (GA) I and II platforms, for a total of 35 lanes. The first eight libraries were prepared from the Ambion Human Brain Reference RNA, (Catalogue #6050), henceforth referred to as HBRR and the ninth was prepared from the Stratagene Universal Human Reference RNA (Catalogue #740000) henceforth referred to as UHRR. DpnII, with recognition site GATC, was used as anchoring enzyme and MmeI as tagging enzyme, resulting in approximately 238 million tags of length 20 across the 9 libraries. Unless otherwise indicated, all tags mapped with at most 1 mismatch (83% of all tags) were used.

For RNA-Seq, we compare IsoEM with Cufflinks, on real datasets generated by two technologies, Illumina and ION Torrent. Two Illumina datasets for the HBRR sample and six datasets for the UHRR sample (SRA study SRP001847 (Bullard *et al.*, 2010)) were used in the comparison. Illumina RNA-Seq datasets, each containing between 47 and 92 million reads of length 35, were mapped onto Ensembl known isoforms using Bowtie (Langmead *et al.*, 2009) after adding a polyA tail of 200 bases to each transcript. Allowing for up to two mismatches, between 65% and 72% of the reads were mapped. IsoEM and Cufflinks were then run, assuming a mean fragment length of 200 bases with standard deviation 50. The ION Torrent datasets used in the comparison consisted of five HBRR datasets and five UHRR datasets. The ION Torrent reads were mapped onto Ensembl known isoforms, also with polyA tail of 200 bases added, using the Ion Torrent mapper, tmap, with default parameters. Mapping statistics for the HBRR and UHRR datasets are given in Tables 3.3 and 3.4, respectively.

To assess accuracy, gene expression levels estimated from real DGE and RNA-Seq datasets were compared against TaqMan qPCR measurements (GEO accession GPL4097) collected by the MicroArray Quality Control Consortium (MAQC). As described in MAQC (2006), each TaqMan Assay was run in four replicates for each measured gene. POLR2A (ENSEMBL id ENSG00000181222) was chosen as the reference gene and each replicate CT was subtracted from the average POLR2A CT to give the log2 difference (delta CT). For delta CT calculations, a CT value of 35 was used for any replicate that had CT > 35. Normalized expression values were computed as $2^{(\text{CT of POLR2A})-(\text{CT of the tested gene})}$, and the average of the qPCR expression values in the four replicates was used as the ground truth.

Table 3.3 Read mapping statistics and correlation between gene expression levels estimated by IsoEM and qPCR measurements for ION-Torrent HBRR datasets

Run	Number of reads	Number of mapped reads	R^2
LUC-140_265	1,588,375	1,142,306	0.728274492
POZ-124_266	1,495,151	1,066,809	0.72914521
DID-143_282	1,703,169	1,215,093	0.732123232
GOG-139_281	1,621,848	1,208,950	0.736479778
LUC-141_267	1,390,667	1,039,816	0.747284176
All runs	7,799,210	5,672,974	0.755807674

Table 3.4 Read mapping statistics and correlation between gene expression levels estimated by IsoEM and qPCR measurements for ION-Torrent UHRR datasets

Run	Number of reads	Number of mapped reads	R^2
POZ-125_268	1,601,962	1,103,357	0.488817714
DID-144_283	1,990,213	1,368,073	0.486842503
POZ-126_269	1,800,034	1,291,935	0.469138694
GOG-140_284	2,052,587	1,452,006	0.498807425
POZ-127_270	2,263,519	1,615,623	0.484335267
All Runs	9,708,315	6,830,990	0.484644315

Mapping gene names to Ensembl gene IDs using the HUGO Gene Nomenclature Committee (HGNC) database resulted in TaqMan qPCR expression levels for 832 Ensembl genes. Expression levels inferred from DGE and RNA-Seq data were similarly divided by the expression level inferred for POLR2A prior to computing accuracy.

Fig. 3.11 shows the comparison between DGE and RNA-Seq Illumina datasets using the EM algorithms, IsoEM and DGE-EM. The AE cutting probability p inferred by DGE-EM is almost the same for all libraries, with a mean of 0.8837 and standard deviation 0.0049. This is slightly higher than the estimated value of 70–80% suggested in the original study (Wang *et al.*, 2008), possibly due to their discarding of non-unique or non-perfectly matched tags. Normalized for sequencing cost, DGE performance is comparable to that of RNA-Seq estimates obtained by IsoEM, with accuracy differences between libraries produced using different protocols within the range of library-to-library variability within each of the two protocols. Fig. 3.12, which compares the accuracy of IsoEM and Cufflinks on RNA-Seq Illumina data, shows that the MPE of estimates generated from RNA-Seq data by Cufflinks is significantly higher than that of IsoEM while the correlation with qPCR estimates is lower.

We also compared IsoEM and Cufflinks on ION Torrent datasets. Tables 3.2 and 3.3 show the coefficient of correlation (R^2) between estimates obtained from IsoEM and qPCR values. IsoEM estimates correlate better with qPCR measurements compared to Cufflinks (Fig. 3.13). Additionally, IsoEM has a consistent accuracy, across datasets, shown by the tight range of R^2 compared with the more spread out Cufflinks R^2 values.

Future trends

Advances in sequencing protocols continue to generate additional types of transcriptomic data, such as promoter (Batut *et al.*, 2012) and PolyA profiling (Derti *et al.*, 2012) at a rapid pace. Integrating these new types of data in existing transcriptome analysis pipelines will continue to generate challenging statistical and computational problems. Dealing with the ever increasing amounts of data also raises important scalability challenges, both in terms of running time and

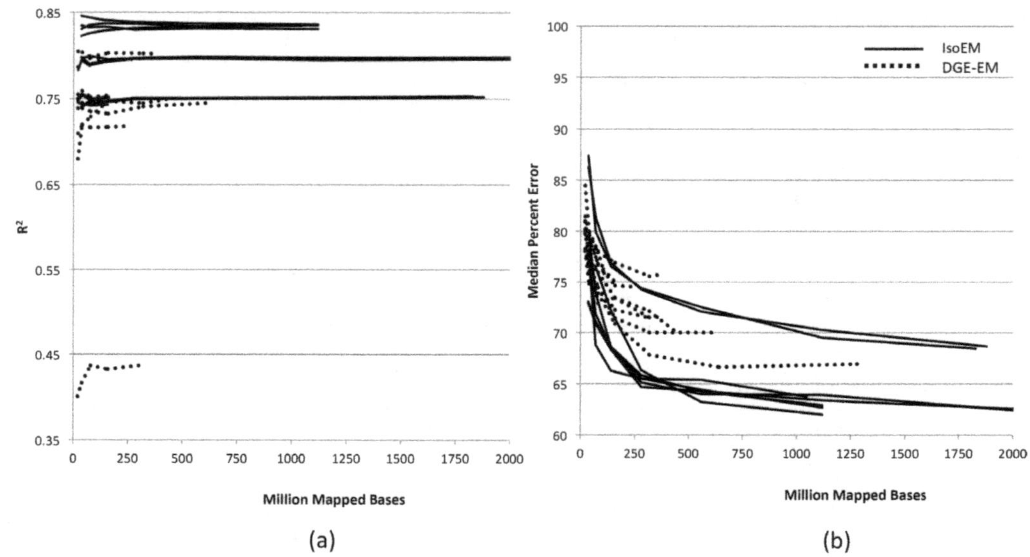

Figure 3.11 Comparison between RNA-Seq and DGE. (a) Correlation coefficient (R^2) as a function of the number of mapped bases. (b) Median Percent Error as a function of the number of mapped bases. Sold lines: IsoEM (RNA-Seq); dotted lines: DGE-EM (GDE).

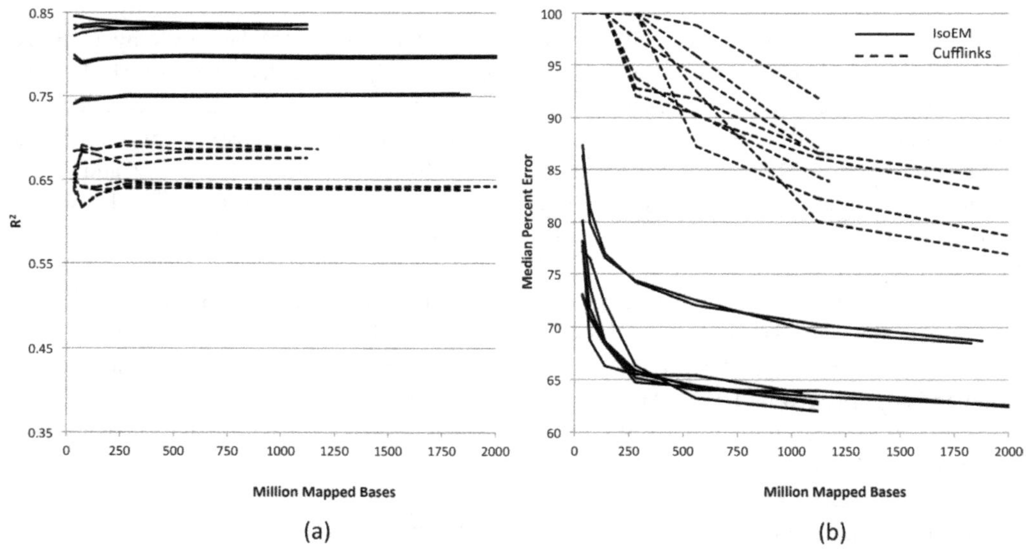

Figure 3.12 Comparison between IE methods for RNA-Seq protocol. (a) Correlation coefficient (R^2) as a function of the number of mapped bases. (b) Median percent error as a function of the number of mapped bases. Sold lines: IsoEM; dashed lines: Cufflinks.

memory requirements. Recently, online estimation algorithms have been successfully employed for the problem of base calling in Illumina sequencing-by-synthesis reads (Das and Vikalo, 2012) as well as transcriptome quantification (Roberts and Pachter, 2012). We expect that need for efficient online (also referred to as streaming, or adaptive) algorithms for transcriptome analysis will continue to increase in the future.

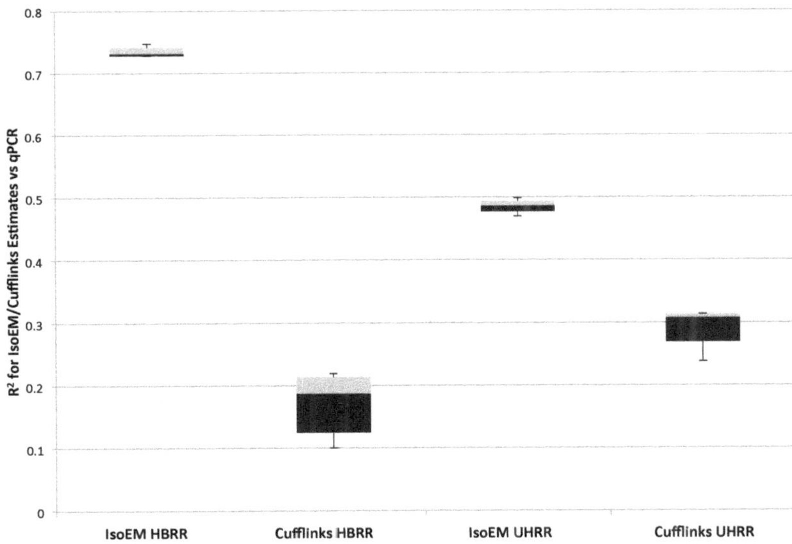

Figure 3.13 Correlation of estimates obtained by both IsoEM and Cufflinks with qPCR measurements for HBRR and UHRR datasets.

Many biological samples, including developing embryos, tumours, as well as many normal tissues are complex mixtures consisting of numerous cell types interacting with each other. Understanding such complex biological systems would greatly benefit from measuring whole-genome expression profiles of individual cell types, yet this is difficult to achieve since existing whole-transcriptome profiling technologies require large amounts of RNA. While sequencing of single cell transcriptomes has been demonstrated [see for example, Kalisky *et al.* (2011) for a recent review], PCR-based amplification of cDNA produced from single cell quantities may result in non-uniform amplification of different transcripts and chimeric amplification products. Computational deconvolution of RNA-Seq data generated from heterogeneous mixtures into individual cell-type expression profiles is an important future research direction.

Conclusions

In this chapter we discussed different methods for transcriptome reconstruction and quantification form RNA sequencing data. Graveley (2001) emphasized the role of alternative splicing in increasing the diversity of the proteome, and recent research has shown how high-throughput RNA sequencing is a powerful tool to study alternative splicing in human populations and the role it plays in human diseases and drug responses (Lu *et al.*, 2012). Using RNA-Seq for isoform expression estimation was used in studying trans- and cis regulatory effect (McManus *et al.*, 2010), and studying parent of origin effect (Degner *et al.*, 2009, Gregg *et al.*, 2010, McManus *et al.*, 2010). It also proved to be more accurate in studying expression quantitative trait loci (eQTLs) variations (Pickrell *et al.*, 2010; Majewski and Pastinen, 2011).

We focused on three methods for transcriptome reconstruction and quantification. TRIP is an Integer Programming method for transcriptome reconstruction from paired-end RNA-Seq reads. TRIP critically exploits the distribution of fragment lengths, and can take advantage of additional experimental data such as TSS/TES and individual fragment lengths estimated, e.g. from ION Torrent (Rothberg *et al.*, 2011) flowgram data. Experimental results on both real and synthetic datasets generated with various sequencing parameters and distribution assumptions show that this IP approach is scalable and has increased transcriptome reconstruction accuracy compared to previous methods that ignore information about fragment length distribution. IsoEM is an expectation maximization algorithm for isoform expression level estimation for single and paired RNA-Seq data. IsoEM explicitly models insert size distribution, base quality scores, strand and read pairing information. Experiments on both real Illumina and ION Torrent data from the MAQC project show that IsoEM consistently outperforms the widely used Cufflinks algorithm. DGE-EM is an EM algorithm for inference of gene-specific expression levels from DGE tags. It takes into account alternative splicing isoforms and tags that map at multiple locations in the genome within a unified statistical model, and can further correct for incomplete digestion and sequencing errors. Experimental results show that DGE-EM has cost-normalized accuracy comparable to that achieved by state-of-the-art RNA-Seq estimation algorithms on the tested real datasets.

Web resources

Links to download different tools:

- TRIP (Mangul *et al.*, 2012): http://alan.cs.gsu.edu/NGS/?q=trip
- Cufflinks (Roberts *et al.*, 2011b): http://cufflinks.cbcb.umd.edu/
- IsoLasso (Li *et al.*, 2011b): http://www.cs.ucr.edu/~liw/isolasso.html
- IsoInfer (Feng *et al.*, 2011): http://www.cs.ucr.edu/~jianxing/IsoInfer0.9.1.html
- Scripture (Guttman *et al.*, 2010): http://www.broadinstitute.org/software/scripture/
- SLIDE (Li *et al.*, 2011a): https://sites.google.com/site/jingyijli/SLIDE.zip
- IsoEM (Nicolae *et al.*, 2011): http://dna.engr.uconn.edu/?page_id=105
- DGE-EM (Nicolae and Măndoiu, 2011): http://dna.engr.uconn.edu/?page_id=163
- RSEM (Li and Dewey, 2011): http://deweylab.biostat.wisc.edu/rsem/
- Bowtie (Langmead *et al.*, 2009): http://bowtie-bio.sourceforge.net/index.shtml
- Tophat (Trapnell *et al.*, 2009): http://tophat.cbcb.umd.edu/

References

Anton, M.A., Gorostiaga, D., Guruceaga, E., Segura, V., Carmona-Saez, P., Pascual-Montano, A., Pio, R., Montuenga, L.M., and Rubio, A. (2008). SPACE: an algorithm to predict and quantify alternatively spliced isoforms using microarrays. Genome Biol. *9*, R46.

Asmann, Y., Klee, E., Thompson, E.A., Perez, E., Middha, S., Oberg, A., Therneau, T., Smith, D., Poland, G., Wieben, E., *et al.* (2009). 3′ tag digital gene expression profiling of human brain and universal reference RNA using Illumina Genome Analyser. BMC Genomics *10*, 531.

Au, K.F., Jiang, H., Lin, L., Xing, Y., and Wong, W.H. (2010). Detection of splice junctions from paired-end RNA-seq data by SpliceMap. Nucleic Acids Res. *38*, 4570–4578.

Batut, P., Dobin, A., Plessy, C., Carninci, P., and Gingeras, T.R. (2012). High-fidelity promoter profiling reveals widespread alternative promoter usage and transposon-driven developmental gene expression. Genome Res. *23*, 169–180.

Bentley, D.R., Balasubramanian, S., Swerdlow, H.P., Smith, G.P., Milton, J., Brown, C.G., Hall, K.P., Evers, D.J., Barnes, C.L., Bignell, H.R., *et al.* (2008). Accurate whole human genome sequencing using reversible terminator chemistry. Nature *456*, 53–59.

Bullard, J., Purdom, E., Hansen, K., and Dudoit, S. (2010). Evaluation of statistical methods for normalization and differential expression in mRNA-Seq experiments. BMC Bioinformatics *11*, 94.

Das, S., and Vikalo, H. (2012). OnlineCall: fast online parameter estimation and base calling for illumina's next-generation sequencing. Bioinformatics *28*, 1677–1683.

Degner, J.F., Marioni, J.C., Pai, A.A., Pickrell, J.K., Nkadori, E., Gilad, Y., and Pritchard, J.K. (2009). Effect of read-mapping biases on detecting allele-specific expression from RNA-sequencing data. Bioinformatics *25*, 3207–3212.

Derti, A., Garrett-Engele, P., MacIsaac, K.D., Stevens, R.C., Sriram, S., Chen, R., Rohl, C.A., Johnson, J.M., and Babak, T. (2012). A quantitative atlas of polyadenylation in five mammals. Genome Res. *22*, 1173–1183.

Duitama, J., Srivastava, P., and Mandoiu, I. (2012). Towards accurate detection and genotyping of expressed variants from whole transcriptome sequencing data. BMC Genomics *13*, S6.

Feng, J., Li, W., and Jiang, T. (2011). Inference of isoforms from short sequence reads. J. Comput. Biol. *18*, 305–321.

Garber, M., Grabherr, M.G., Guttman, M., and Trapnell, C. (2011). Computational methods for transcriptome annotation and quantification using RNA-seq. Nat. Methods *8*, 469–477.

Grabherr, M.G., Haas, B.J., Yassour, M., Levin, J.Z., Thompson, D.A., Amit, I., Adiconis, X., Fan, L., Raychowdhury, R., Zeng, Q., *et al.* (2011). Full-length transcriptome assembly from RNA-Seq data without a reference genome. Nat. Biotechnol. *29*, 644–652.

Graveley, B.R. (2001). Alternative splicing: increasing diversity in the proteomic world. Trends Genet. *17*, 100–107.

Gregg, C., Zhang, J., Butler, J.E., Haig, D., and Dulac, C. (2010). Sex-specific parent-of-origin allelic expression in the mouse brain. Science *329*, 682–685.

Griffith, M., Griffith, O.L., Mwenifumbo, J., Goya, R., Morrissy, A.S., Morin, R.D., Corbett, R., Tang, M.J., Hou, Y.C., Pugh, T.J., *et al.* (2010). Alternative expression analysis by RNA sequencing. Nat. Methods *7*, 843–847.

Guttman, M., Garber, M., Levin, J.Z., Donaghey, J., Robinson, J., Adiconis, X., Fan, L., Koziol, M.J., Gnirke, A., Nusbaum, C., *et al.* (2010). Ab initio reconstruction of cell type-specific transcriptomes in mouse reveals the conserved multi-exonic structure of lincRNAs. Nature Biotechnol. *28*, 503–510.

Hansen, K.D., Brenner, S.E., and Dudoit, S. (2010). Biases in Illumina transcriptome sequencing caused by random hexamer priming. Nucleic Acids Res. *38*, e131.

Hiller, D., Jiang, H., Xu, W., and Wong, W.H. (2009). Identifiability of isoform deconvolution from junction arrays and RNA-Seq. Bioinformatics 25, 3056–3059.

Howard, B., and Heber, S. (2010). Towards reliable isoform quantification using RNA-SEQ data. BMC Bioinformatics *11*, S6.

Jiang, H., and Wong, W.H. (2009). Statistical inferences for isoform expression in RNA-Seq. Bioinformatics *25*, 1026–1032.

Kalisky, T., Blainey, P., and Quake, S.R. (2011). Genomic Analysis at the Single-Cell Level. Annu. Rev. Genet. *45*, 431–445.

Lacroix, V., Sammeth, M., Guigo, R., and Bergeron, A. (2008). Exact transcriptome reconstruction from short sequence reads. In Proceedings of the 8th International Workshop on Algorithms in Bioinformatics (Karlsruhe, Germany, Springer-Verlag), pp. 50–63.

Langmead, B., Trapnell, C., Pop, M., and Salzberg, S. (2009). Ultrafast and memory-efficient alignment of short DNA sequences to the human genome. Genome Biol. *10*, R25.

Li, B., and Dewey, C.N. (2011). RSEM: accurate transcript quantification from RNA-Seq data with or without a reference genome. BMC Bioinformatics *12*, 323.

Li, B., Ruotti, V., Stewart, R.M., Thomson, J.A., and Dewey, C.N. (2010). RNA-Seq gene expression estimation with read mapping uncertainty. Bioinformatics *26*, 493–500.

Li, J.J., Jiang, C.R., Brown, J.B., Huang, H., and Bickel, P.J. (2011a). Sparse linear modelling of next-generation mRNA sequencing (RNA-Seq) data for isoform discovery and abundance estimation. Proc. Natl. Acad. Sci. U.S.A. *108*, 19867–19872.

Li, W., Feng, J., and Jiang, T. (2011b). IsoLasso: a LASSO regression approach to RNA-Seq based transcriptome assembly. J. Comput. Biol. *18*, 1693–1707.

Lin, Y.-Y., Dao, P., Hach, F., Bakhshi, M., Mo, F., Lapuk, A., Collins, C., and Sahinalp, S.C. (2012). CLIIQ: Accurate comparative detection and quantification of expressed isoforms in a population. In Algorithms in Bioinformatics, Raphael, B., and Tang, J., eds. (Springer, Berlin Heidelberg), pp. 178–189.

Lu, Z.X., Jiang, P., and Xing, Y. (2012). Genetic variation of pre-mRNA alternative splicing in human populations. Wiley Interdiscip. Rev. RNA *3*, 581–592.

Majewski, J., and Pastinen, T. (2011). The study of eQTL variations by RNA-seq: from SNPs to phenotypes. Trends Genet. *27*, 72–79.

Mangul, S., Caciula, A., Mandoiu, I., and Zelikovsky, A. (2011). RNA-Seq based discovery and reconstruction of unannotated transcripts in partially annotated genomes. Paper presented at: IEEE International Conference on Bioinformatics and Biomedicine Workshops (BIBMW), Atlanta, GA.

Mangul, S., Caciula, A., Al Seesi, S., Brinza, D., Banday, A.R., Kanadia, R., Mandoiu, I., and Zelikovsky, A. (2012). An integer programming approach to novel transcript reconstruction from paired-end RNA-seq reads. In ACM Conference on Bioinformatics, Computational Biology and Biomedicine. ACM Press, New York, NY, 369–376.

M.A.Q.C. (2006). The MicroArray Quality Control (MAQC) project shows inter- and intraplatform reproducibility of gene expression measurements. Nat. Biotechnol. *24*, 1151–1161.

McManus, C.J., Coolon, J.D., Duff, M.O., Eipper-Mains, J., Graveley, B.R., and Wittkopp, P.J. (2010). Regulatory divergence in Drosophila revealed by mRNA-seq. Genome Res. *20*, 816–825.

Mercer, T.R., Gerhardt, D.J., Dinger, M.E., Crawford, J., Trapnell, C., Jeddeloh, J.A., Mattick, J.S., and Rinn, J.L. (2012). Targeted RNA sequencing reveals the deep complexity of the human transcriptome. Nat. Biotechnol. *30*, 99–104.

Mortazavi, A., Williams, B.A., McCue, K., Schaeffer, L., and Wold, B. (2008). Mapping and quantifying mammalian transcriptomes by RNA-Seq. Nat. Methods *5*, 621–628.

Nicolae, M., and Măndoiu, I. (2011). Accurate estimation of gene expression levels from dge sequencing data. In Bioinformatics Research and Applications, Chen, J., Wang, J., and Zelikovsky, A., eds. (Springer, Berlin Heidelberg), pp. 392–403.

Nicolae, M., Mangul, S., Mandoiu, I., and Zelikovsky, A. (2011). Estimation of alternative splicing isoform frequencies from RNA-Seq data. Algorithms Mol. Biol. *6*, 9.

Oshlack, A., and Wakefield, M.J. (2009). Transcript length bias in RNA-seq data confounds systems biology. Biology direct *4*, 14.

Pachter, L. (2011). Models for transcript quantification from RNA-Seq. arXiv:11043889 [q-bioGN].

Pal, S., Gupta, R., Kim, H., Wickramasinghe, P., Baubet, V., Showe, L.C., Dahmane, N., and Davuluri, R.V. (2011). Alternative transcription exceeds alternative splicing in generating the transcriptome diversity of cerebellar development. Genome Res. *21*, 1260–1272.

Pandey, V., Nutter, R.C., and Prediger, E. (2008). Applied biosystems SOLiD™ system: ligation-based sequencing. In Next Generation Genome Sequencing (Wiley-VCH Verlag GmbH & Co. KGaA), pp. 29–42.

Pasaniuc, B., Zaitlen, N., and Halperin, E. (2011). Accurate estimation of expression levels of homologous genes in RNA-seq experiments. J. Comput. Biol. *18*, 459–468.

Pevzner, P.A. (1989). 1-Tuple DNA sequencing: computer analysis. J. Biomol. Struct. Dyn. *7*, 63–73.

Pickrell, J.K., Marioni, J.C., Pai, A.A., Degner, J.F., Engelhardt, B.E., Nkadori, E., Veyrieras, J.B., Stephens, M., Gilad, Y., and Pritchard, J.K. (2010). Understanding mechanisms underlying human gene expression variation with RNA sequencing. Nature *464*, 768–772.

Ponting, C.P., and Belgard, T.G. (2010). Transcribed dark matter: meaning or myth? Hum. Mol. Genet. *19*, R162–168.

Richard, H., Schulz, M.H., Sultan, M., Nurnberger, A., Schrinner, S., Balzereit, D., Dagand, E., Rasche, A., Lehrach, H., Vingron, M., *et al.* (2010). Prediction of alternative isoforms from exon expression levels in RNA-Seq experiments. Nucleic Acids Res. *38*, e112.

Roberts, A., and Pachter, L. (2012). Streaming fragment assignment for real-time analysis of sequencing experiments. Nat. Methods [Epub ahead of print].

Roberts, A., Pimentel, H., Trapnell, C., and Pachter, L. (2011a). Identification of novel transcripts in annotated genomes using RNA-Seq. Bioinformatics *27*, 2325–2329.

Roberts, A., Trapnell, C., Donaghey, J., Rinn, J.L., and Pachter, L. (2011b). Improving RNA-Seq expression estimates by correcting for fragment bias. Genome Biol. *12*, R22.

Robertson, G., Schein, J., Chiu, R., Corbett, R., Field, M., Jackman, S.D., Mungall, K., Lee, S., Okada, H.M., Qian, J.Q., *et al.* (2010). De novo assembly and analysis of RNA-seq data. Nat. Methods *7*, 909–912.

Rothberg, J.M., Hinz, W., Rearick, T.M., Schultz, J., Mileski, W., Davey, M., Leamon, J.H., Johnson, K., Milgrew, M.J., Edwards, M., *et al.* (2011). An integrated semiconductor device enabling non-optical genome sequencing. Nature *475*, 348–352.

She, Y., Hubbell, E., and Wang, H. (2009). Resolving deconvolution ambiguity in gene alternative splicing. BMC Bioinformatics *10*, 237.

't Hoen, P.A., Ariyurek, Y., Thygesen, H.H., Ureugdenhil, E., Uossen, R.H., de Menezes, R.X., Boer, J.M., van Ommen, G-J.J., and den Dunnen, J.T. (2008). Deep sequencing-based expression analysis shows major advances in robustness, resolution and inter-lab portability over five microarray platforms. Nucleic Acids Res. *36*, e141.

Thomas, R.K., Nickerson, E., Simons, J.F., Janne, P.A., Tengs, T., Yuza, Y., Garraway, L.A., LaFramboise, T., Lee, J.C., Shah, K., *et al.* (2006). Sensitive mutation detection in heterogeneous cancer specimens by massively parallel picoliter reactor sequencing. Nat. Med. *12*, 852–855.

Tibshirani, R. (2011). Regression shrinkage and selection via the lasso: a retrospective. J. R. Stat. Soc. Series B Stat. Methodol. *73*, 273–282.

Trapnell, C., Pachter, L., and Salzberg, S.L. (2009). TopHat: discovering splice junctions with RNA-Seq. Bioinformatics *25*, 1105–1111.

Trapnell, C., Williams, B.A., Pertea, G., Mortazavi, A., Kwan, G., van Baren, M.J., Salzberg, S.L., Wold, B.J., and Pachter, L. (2010). Transcript assembly and quantification by RNA-Seq reveals unannotated transcripts and isoform switching during cell differentiation. Nat. Biotechnol. *28*, 511–515.

Wang, E.T., Sandberg, R., Luo, S., Khrebtukova, I., Zhang, L., Mayr, C., Kingsmore, S.F., Schroth, G.P., and Burge, C.B. (2008). Alternative isoform regulation in human tissue transcriptomes. Nature *456*, 470–476.

Wang, Z., Gerstein, M., and Snyder, M. (2009). RNA-Seq: a revolutionary tool for transcriptomics. Nat. Rev. Genet. *10*, 57–63.

Wu, Z.J., Meyer, C.A., Choudhury, S., Shipitsin, M., Maruyama, R., Bessarabova, M., Nikolskaya, T., Sukumar, S., Schwartzman, A., Liu, J.S., *et al.* (2010). Gene expression profiling of human breast tissue samples using SAGE-Seq. Genome Res. *20*, 1730–1739.

Zaretzki, R., Gilchrist, M., Briggs, W., and Armagan, A. (2010). Bias correction and Bayesian analysis of aggregate counts in SAGE libraries. BMC Bioinformatics *11*, 72.

Identification of Small Interfering RNA from Next-generation Sequencing Data

4

Thomas J. Hardcastle

Abstract

Small interfering RNAs (siRNAs) play a crucial role in the regulation of transcriptomic and epigenetic factors. Next generation sequencing technologies allow the identification and quantification of siRNAs on a genome-wide scale. Given the proper tools for analysis, biologically meaningful inferences can be made about the biogenesis and patterns of expression of these key components of biological systems. This review discusses the application of sequencing technologies to siRNAs and the currently available tools for the analysis of the data thus generated.

Introduction

Eukaryotic small interfering RNAs (siRNAs) play a key role in the regulatory mechanisms of many fundamental cellular processes. Genome-wide analyses of siRNAs have only recently become possible with the development of next-generation sequencing methods. These are well suited to the analysis of siRNAs and have been instrumental in revealing a great amount of detail about the mechanisms of both production and action of these important regulatory elements.

Small interfering RNAs first gained attention for their role in post-transcriptional modification (Hamilton and Baulcombe, 1999), in which they act through sequence complementarity to guide RNA-induced transcriptional silencing complexes (RISCs), of which an Argonaute protein forms the central part. These RISCS act to cleave the messenger RNA (mRNA) at the approximate centre of the region of complementarity (Martinez *et al.*, 2002; Martinez and Tuschl, 2004). This may result in the production of secondary small RNAs (see below) but is primarily a mechanism for preventing translation of the mRNA, which is usually rapidly degraded following cleavage.

More recently, the role of siRNAs in chromatin modification (Moazed, 2009) has been recognized as of increasing importance, especially since the association of siRNAs with heritable modifications (Jones *et al.*, 2001; Vastenhouw *et al.*, 2006; Mosher and Melnyk, 2010). SiRNAs have been reported as playing a role in both the establishment and maintenance of cytosine methylation (Zilberman *et al.*, 2003; Henderson *et al.*, 2006) and histone methylation (Reinhart and Bartel, 2002; Volpe *et al.*, 2002), as well as heterochromatin formation (Pal-Bhadra *et al.*, 2004; Deshpande *et al.*, 2005; Kim *et al.*, 2005). These modifications usually act to repress transcription; however, it has been suggested that siRNAs targeting the promoter region of a gene may have an activating effect (Li *et al.*, 2006). SiRNAs are also able to act as a signalling mechanism in which silencing effects are able to spread from cell to cell through the movement of siRNAs (Voinnet and Baulcombe, 1997). In plant systems, siRNAs have been shown to move over greater distances, presumably through the vascular systems (Molnar *et al.*, 2010).

SiRNAs are canonically processed from long double-stranded RNA (dsRNA) precursor molecules by a Dicer or Dicer-like (DCL) nuclease protein (Meister and Tuschl, 2004) into 19–28 nucleotide dsRNA duplexes with a two nucleotide 3′ overhang (Elbashir *et al.*, 2001). From these RNA duplexes the small RNA strands with lower thermodynamic stability at the 5′ ends are retained by an Argonaute (AGO) (Khvorova *et al.*, 2003; Schwarz *et al.*, 2003), the effector protein of RNA silencing. This process is influenced by the 5′ nucleotide of siRNAs (Mi *et al.*, 2008). SiRNAs guide the AGO silencing complex to complementary nucleic acids to induce RNA silencing at the transcriptional level through DNA or histone modification, and at the post-transcriptional level by mRNA cleavage and degradation or translational repression. The genes encoding DCL and AGO proteins are often multiplied and diversified in eukaryotes which results in diverse silencing pathways in many organisms to control the expression of endogenous genes, transposons, transgenes and viruses (Lee *et al.*, 2004; Margis *et al.*, 2006; Hutvagner and Simard, 2008).

In addition to the canonical formation of siRNAs described above, a variety of endogenous and exogenous dsRNA molecules have been implicated in their production. Endogenous siRNAs can be produced from transcribed gene pairs possessing complementarity *in cis* (*cis* natural antisense transcripts; (Zhang *et al.*, 2012b)). SiRNAs are often generated from double-stranded fold-back transcripts of inverted repeat DNA (Okamura *et al.*, 2008). In addition, small RNAs can also be processed from a partially double-stranded region of imperfectly matched foldback RNA. In this case, the small RNA that is incorporated into AGO is called the microRNA (miRNA), and generally considered as a separate class to true siRNAs.

In plants, miRNA and siRNA targeted transcripts can be converted into dsRNA by RNA-dependent RNA polymerases (RDRs). This process generates further substrates for DCL processing, thereby allowing the amplification of silencing RNAs and the production of secondary siRNAs (Voinnet, 2008). RDRs are also involved in primary siRNA production by converting the transcripts of a plant-specific DNA-dependent RNA polymerase IV (POL IV) (Pontes *et al.*, 2006) into dsRNA. Owing to the nature of their production, these secondary siRNAs may demonstrate particular features which can be identified through high-throughput sequencing (see Phasing discovery, below).

Naturally occurring exogenous siRNAs are primarily derived from viral RNA. The mechanisms of production of these siRNAs are not fully characterized but it is likely that the primary pathways involve either the action of a Dicer protein on the double-stranded intermediates of viral replication, or from the formation of hairpins in the viral RNA (Ding and Voinnet, 2007). These virus-derived siRNAs (viRNAs) play an important role in antiviral response (Molnár *et al.*, 2005).

Applying sequencing technologies to siRNAs

The application of sequencing technologies to siRNAs usually begins with the extraction of total RNA through deproteinization of RNA followed by ethanol-induced precipitation (Zhuang *et al.*, 2012). Until relatively recently, sequencing required the siRNAs to be isolated from a total RNA extraction through size profiling alone. In this process, polyacrylamide gel electrophoresis is used to separate the RNA by size class and the appropriate bands excised from the gel (Chappell *et al.*, 2006; Ho *et al.*, 2008). These methods have since been enhanced by the development of adaptors designed to attach specifically to small RNAs (Applied Biosystems, 2008; Illumina, 2009) through targeting of the 3′-hydroxyl group that is produced by the action of Dicer (Martinez and Tuschl, 2004). Purification by size fractionation can be carried out either before or after adaptor ligation and PCR amplification, depending on the specific sequencing protocol being used.

The small length of siRNAs means that the choice of sequencing technology tends to be influenced by the number of reads that can be sequenced, rather than the maximum read length. For this reason, Illumina technology has been by far the most commonly used in siRNA-seq applications. In this technology, adapters are ligated to both the 5′ and 3′ ends of extracted small RNAs, allowing the production of cDNA and subsequent amplification and sequencing through the standard Illumina protocols (Bentley *et al.*, 2008). The number of reads generated by HiSeq technology (Minoche *et al.*, 2011) will generally be in excess of that required in siRNA-seq experiments unless barcoding is being used to increase the number of samples concurrently sequenced, or unless extreme sensitivity is required (see below for a discussion of sequencing depths).

The ABI SOLiD technology (Valouev *et al.*, 2008) has also been used for the sequencing of small RNAs (Calviño *et al.*, 2011; Schopman *et al.*, 2012). As with Illumina's HiSeq, this platform is also capable of generating very high numbers of short reads. Widespread use of this platform has perhaps been limited by difficulties in the processing of the raw 'colourspace' data, although the development of the Small RNA Analysis Pipeline Tool has simplified this process.

Of the newer breed of 'desktop sequencers' (Quail *et al.*, 2012), the Ion Torrent (Rothberg *et al.*, 2011) and Illumina MiSeq have protocols suitable for sequencing of small RNAs. However, little information is currently available with which to compare the performance of these technologies for siRNA-seq experiments.

Argonaute immunoprecipitation

An interesting variant on conventional siRNA-seq experiments is that of Argonaute immunoprecipitation, in which individual Argonautes are immunoprecipitated before RNA extraction from the eluted protein (Goff *et al.*, 2009; Hong *et al.*, 2009; Havecker *et al.*, 2010) as before. Sequencing the extracted siRNAs allows comparisons between Argonaute binding propensities and may indicate the functional pathways of individual siRNAs by association with particular Argonautes (Carbonell *et al.*, 2012)

Experimental designs

In any application of high-throughput sequencing that aims to quantify expression levels, care in experimental design is needed in order to produce meaningful and robust results. This is required both to identify and reduce the inevitable failures of sample preparation and sequencing and, perhaps more importantly, to assess the level of biological variation within a population. The expression of siRNAs shows not inconsequential variation even between genetically identical individuals in controlled and identical environmental conditions. Some of this variation may be attributed to minor, uncontrolled (and largely uncontrollable) variation in environmental factors. However, the effects of epiallelic variation in which methylation (Becker *et al.*, 2011; Schmitz *et al.*, 2011), and potentially other chromatin modifications can spontaneously and heritably alter may also require consideration. The effect of this variation on siRNA expression may be significant, particularly in experimental designs that span several generations.

Replication and sequencing depths

Accounting for biological variation requires adequate replication within the experiment. In most applications of next-generation sequencing, technical variation is of low magnitude and relatively easy to model in the absence of serious technical errors. In general, such events can be identified relatively easily through various quality control checks (see below). Technical replicates, in which the same samples are sequenced multiple times, are thus generally of little value. An exception to this rule arises when high variation is expected between individuals, and consequently true biological replication is difficult to identify. In this case, it may be difficult to distinguish whether an unusual sequencing result is the result of technical error or a result of true biological variation. In this case, technical replicates may be of value to discriminate between these alternatives.

In most cases, however, it is biological variation that is of significance in making meaningful statements about the expression of siRNAs. Sufficient numbers of biological replicates are therefore required to accurately assess this variability. In practice, this suggests that a minimum of two biological replicates should be produced for each condition within an experiment, as this allows some estimate of the biological variability to be calculated. Such a minimum is suitable for well-characterized systems in which the biological variation is known to be low for all, or almost all siRNAs. In many cases, and in particular when novel organisms or experimental conditions are being considered for siRNA sequencing, the level of replication should be much higher in order to establish the level of variability.

The depth to which individual samples are sequenced is usually much less significant than the availability of biological replicates. Depth of sequencing will affect both the number of lowly expressed siRNAs discovered, and the sensitivity with which differential expression between experimental conditions can be detected. Except where whole classes of siRNAs are both biologically significant and lowly expressed, depth of sequencing is unlikely to affect any genome-wide inferences about the data (for example, that some subclass of siRNAs shows association with a particular genomic feature). Beyond a certain depth of sequencing for any individual siRNA, increased depth grants no significant advantage in estimating the true expression profile of that siRNA in an individual sample. However, the biological variability of that profile will be uncertain without sufficient biological replication. By prioritizing biological replication over sequencing depth, it is thus usually possible to acquire a more robust set of conclusions from the data.

Exceptions to these guidelines may arise when lowly expressed siRNAs are of importance. Such a case is probably most likely to arise when examining viral siRNAs (Schopman *et al.*, 2012), whose population may be much smaller than the total small RNA population within each sample. High sequencing depths may also be advantageous when considering genetically diverse individuals; in this case depth will be required to distinguish between single nucleotide polymorphisms (SNPs) in the sequenced siRNAs and sequencing error at the alignment stage of analysis (see below).

Pooling samples

Pooling the RNA extracted from biological replicates can be a useful technique in siRNA-seq experiments in reducing biological variability without increasing the sequencing costs. When the quantity of material that can be collected from any individual sampling is small, pooling allows the amount of RNA available for sequencing to be increased to levels sufficient for the current generation of sequencing technology. Indeed, almost all sequencing currently involves at least a pooling of data from multiple cells, usually a pooling of multiple cells from different tissue types, and often a pooling of data from individuals. It is, however, important to recognize that pooling will not remove all biological variability from the data. Only by sequencing biological replicates can estimations of the biological variability be made and hence statistically meaningful analyses be carried out. Pooling does not therefore remove the need to sequence biological replicates, although these may now be replicate pools, containing different sets of biological replicates.

A further drawback of pooling is that biological variability can be significant in the interpretation of the results. Consider the case where a siRNA does show some change in average expression between experimental conditions, but has a variability between replicate samples orders of magnitude greater than this change in expression. Pooling may lead to the identification of this siRNA as significantly differentially expressed; however, the variability of this siRNA means that such a difference is unlikely to have an important functional effect. The pools used should thus not contain an excessive number of samples, and interpretations of the data thus acquired should allow for the greater variability in expression levels between individuals than that seen in the sequencing data.

Barcoding

In next-generation sequencing, barcoding (Smith *et al.*, 2010) refers to the tagging of individual samples or pools with some short identifying sequence. These barcodes can be introduced at the ligation stage, in the primer for cDNA synthesis, or during PCR amplification. The samples identified by unique barcodes can subsequently be pooled and sequenced in a single lane. Given properly constructed barcodes, it is then possible to unambiguously separate the data acquired for each sample based on these sequences, and these techniques have been applied to siRNA sequencing experiments (Pang *et al.*, 2009; Qi *et al.*, 2009).

In theory, barcoding provides the ideal solution to minimizing the cost of high-throughput sequencing whilst keeping high numbers of biological replicates within the experiment. The latest generations of sequencing machines are capable of producing on the order of two hundred million reads in a single lane, which will generally be well in excess of that needed to characterize the majority of siRNA expression in a single sample. The use of barcodes allows multiple samples to be characterized within a single lane.

However, barcodes have been shown to introduce bias into high-throughput sequencing of small RNAs (Alon *et al.*, 2011). This is likely to arise from interactions between the barcode sequence and the sequence of individual small RNAs, presumably by altering the secondary structure of the combined sequence. These changes are likely to affect the ligation efficiency of the individual reads and consequently it may be advantageous to add the barcodes at a later stage of sample preparation. However, PCR amplification is also susceptible to sequence based biases (Alon *et al.*, 2011), and the sequencing technologies themselves have also been reported to show sequence based biases (Dohm *et al.*, 2008). Barcode induced bias is thus unlikely to be removed completely by any method.

The biases that are currently inherent in barcoding make the use of this technology problematic for most siRNA-seq experiments. Careful construction and placement of the barcodes may be able to reduce these effects, though it is unlikely to remove them completely. In large siRNA-seq experiments where sufficiently many biological replicates are sought, it may be possible to assign the barcodes to individual samples in a randomized block design such that the effects of individual barcodes can be accounted for within the statistical models used to analyse these data. However, computational methods for such an analysis are currently not well developed. At present, therefore, barcoding remains a technology to be used with caution in the sequencing of siRNAs.

Available tools for analysis of siRNA-seq data

A diverse range of freely available tools have been developed and released for the analysis of high-throughput sequencing data, and many of these are applicable to siRNA-seq data. A number of packages have also been developed specifically for the analysis of siRNA-seq data. Methods suitable for individual steps in analysis are discussed at length below, however, there also exist packages intended as a single solution to siRNA-seq analysis.

Available methods differ substantially in their accessibility to users. The most accessible and user-friendly methods are generally those available as web services, which provide easy-to-run front-ends to both existing and novel tools. One of the most complete analysis suites for siRNA analysis (and analysis of small RNAs in general) is the UEA sRNA Toolkit (Moxon *et al.*, 2008) (http://srna-tools.cmp.uea.ac.uk/). Tools from this web-service are particularly useful for early processing of the siRNA-seq data, allowing for adaptor removal, alignment and visualization of siRNAs, filtration of sequences matching degradation products of tRNAs and rRNAs. These tasks are relatively standard in analysis of sRNA data and do not require especially high computational resources, and so are ideal candidates for a web-based approach. The DSAP web-service (Huang *et al.*, 2010) offers a generally more limited functionality than the UEA toolkit, but allows some alternative filtering mechanisms such as poly-A removal (see below).

The convenience of web-based tools comes with some costs, of which the user should be aware. Firstly, the tools used are of necessity of general application, and may not address specific issues within a particular dataset. In selecting the underlying algorithms of the analysis, greater emphasis is likely to be placed on computational efficiency over performance in order to reduce the load placed on the host's servers. Greater care must be taken to ensure consistency of analysis of the data over time, as the parameters (and even the algorithms used) may change without obvious warning. Finally, for particularly sensitive data (e.g. medical data) it may not be appropriate to export sequencing data onto servers outside the immediate control of the user. These last two concerns can often be addressed by the local installation of web-based methods onto servers within the user's control, however, this requires a level of expertise and maintenance which may be prohibitive.

Less convenient than web-based packages but allowing more fine control, several downloadable packages for pipeline analysis of siRNA-seq data exist. The *shortran* package (Gupta *et al.*, 2012) consists of a set of Python scripts (released under the GPL-3 software licence) that carry out the initial processing of raw siRNA-seq data, producing graphical summaries for quality control and initial analysis, and load the data into a MySQL database for subsequent downstream analysis. The ADAAPTS package (http://www.plantsci.cam.ac.uk/Bioinformatics/addapts.html) offers similar functionality through a set of Perl scripts, also released under the GPL-3 licence. These tools allow more control over the processing of the data than web-based methods, and may provide a useful foundation for further development.

A large repository of tools is also available for analysis of high-throughput sequencing in the **R** (R Development Core Team, 2012) programming language through the Bioconductor project (Gentleman *et al.*, 2004). **R** is a complete programming language and hence offers complete control over the analysis of high-throughput sequencing data, although considerable expertise is required to use this to the full. The community produced packages available through Bioconductor and elsewhere simplify the analysis process considerably and provide extensive documentation.

The most in-depth analyses of siRNA-seq data are likely to require selection of individual tools from a wide-range of sources, combined with novel tool development. Such analyses are clearly likely to be the most time-consuming and require a high level of bioinformatics expertise, while potentially providing the most detailed and novel results from the data. However, it is highly unlikely that this approach will be needed for all aspects of analysis, but rather, will be applied to a few key areas of particular significance to the experiments being conducted.

Processing of siRNA-seq data

Trimming and de-multiplexing

The initial step in processing siRNA-seq data is the removal of adaptor sequences from the sequenced reads, allowing the siRNA sequence alone to be considered. This is a relatively straightforward task that may nevertheless be approached in two different ways. The FASTX Toolkit (Pearson *et al.*, 1997), for example, simply trims a defined number of bases from each end of each sequenced read. A better approach is likely to be the identification of the primer sequences in the correct locations in the sequenced reads. This approach is taken by the UEA toolkit, although perfect matching is required between the given primers and the ends of the sequenced reads, with imperfectly matching reads being discarded. The degree of matching of both primer and, if used, barcode sequences for an individual read can be used as a filter on poor-quality sequences. As a general rule, sequenced reads with a high error rate at the 5′-end will in general be unreliable throughout. However, some errors may be permissible, especially if long primers are used. The TagCleaner resource (Schmieder *et al.*, 2010) offers far more options for primer removal, including ambiguity codes, mismatches in the primer sequence and removal of repetitive primer sequences.

Tools for demultiplexing of barcoded sequence libraries, by which sequenced reads are split into separate files according to their barcode, are less widespread. The FASTX Toolkit offers a fairly complete method which allows detection of barcodes at either end of the sequenced read and permits mismatch detection. Complications in adaptor removal and de-multiplexing can occur depending on the location of the barcode within the sequenced read. Depending on the process used to introduce the barcode, the barcode in siRNA sequencing data can appear in front of, behind, or within either the 5′- or 3′-adaptor sequence. It may thus be necessary first to remove the adaptors, then demultiplex based on the barcode sequence, or first to demultiplex before removing the adaptors. If the barcode is contained within the adaptor sequence, the ambiguity codes allowed by the TagCleaner resource may be useful.

Alignment

The short length of siRNAs affects the choice of alignment method in two ways. Firstly, with short read lengths there is a greatly reduced likelihood of sequencing error in any individual sequenced read, in part because sequencing errors tend to accumulate towards the end of a sequencing run (Cox *et al.*, 2010; Nakamura *et al.*, 2011; McElroy *et al.*, 2012) and in part because there are simply fewer opportunities for an error to occur within a shorter read. However, the short read length also increases the potential ambiguity of a mapping to a reference genome. In combination, these two factors suggest that, given a suitable reference genome, it is preferable to align based on a perfect matching of the sequenced read to the reference sequence. This removes the need to incorporate quality scores within the alignment, and so a wide range of alignment methods are suitable for aligning siRNA-seq data. Bowtie (Langmead *et al.*, 2009), perhaps the most widely used alignment method, is suitable for alignment of siRNA-seq data, although the parameters of alignment will require modification from the defaults if perfect matching is required. The PatMaN aligner (Prüfer *et al.*, 2008) has also been used extensively for siRNA analyses, and is well-suited for perfect matching.

Several factors may make the approach of requiring perfect alignment to the reference genome problematic. In the case of genomic differences between the sequenced organism and the reference, a number of reads will be unnecessarily discarded. Post-transcriptional modification of the siRNAs (Kim *et al.*, 2010) may also lead to imperfect matching of true siRNAs. In many cases the small number of reads such as these may be insignificant to the goals of the experiment, and allowing only perfect matching is likely to be sufficient. Where such reads are likely to be of importance, a viable strategy may be to separate the sequenced reads into those which match perfectly to the reference and those which do not. The alignment of the imperfectly matching reads might then be considered, requiring strict controls on sequence quality, abundance of particular sequences, and perhaps the location of the mismatching bases.

In the absence of a high-quality reference genome, analysis of siRNA-seq data is much more problematic. The short and (due to accumulation biases) largely non-overlapping sequenced reads are not suitable for *de novo* assembly. If some sufficiently close reference genome exists, this may be used, although in this case, it may be necessary to consider non-perfect matches to the reference. The ' -- best' option of Bowtie, which returns the potential alignments of each sequenced read ranked by quality of alignment to the reference, may be of use in such a case. Failing the existence of such a reference, the best that can be done is an analysis of global properties of the siRNAs (see below) and analyses of the expression levels of individual reads (see below).

Visualization and quality control

In analysing siRNA-seq data there are several indicators of data quality to be considered. Firstly, the percentage of reads from a given sequencing library which align to the genome, while dependent on the quality of the reference, should not be too low when compared to other libraries from the same genome. Initial visualisations of the aligned reads should then be carried out as an essential stage of quality control and preliminary analyses of the data.

Visualization of the abundance of the sequenced reads by length (Fig. 4.1) will be of importance in almost all siRNA-seq experiments. This shows firstly that the sequenced reads are in the correct length range for siRNAs; that is, predominantly nineteen to twenty-eight bases in length. However, more detailed patterns are often present in the data. In the case shown in Fig. 4.1a, the small RNAs are sequenced from the rootstock of *Arabidopsis thaliana*, for which it is known that there is a peak of production of small RNAs of 21 bases in length, as well as a much larger peak at twenty-four bases in length. This can be observed in the data, and an absence of this bimodality might suggest problems with the extraction or sequencing of the small RNAs. This can be compared to the sample shown in Fig. 4.1b, where the only peak in abundance is at twenty-one bases. In this case, the sample is, again of small RNAs extracted from the rootstock of *Arabidopsis thaliana*, but of a triple mutant in which Dicer-like proteins, 2, 3 and 4 have been knocked out (Molnar *et al.*, 2010). In such a mutant, the production of siRNAs of twenty-four bases in length is suppressed, and so such a result is expected in this case.

Visualisations of this type can also be useful in some circumstances in inferring general results from the data. Fig. 4.1c shows the abundance of sequenced reads of small RNAs, again from a rootstock of a triple mutant in which Dicer-like proteins, 2, 3 and 4 have been knocked out, but now grafted to a non-mutant shoot. In this case, the peak in abundance of small RNAs of 24 bases is restored, indicating that a proportion of siRNAs of twenty-four bases in length are able to migrate from the non-mutant shoot into the mutant root-stock.

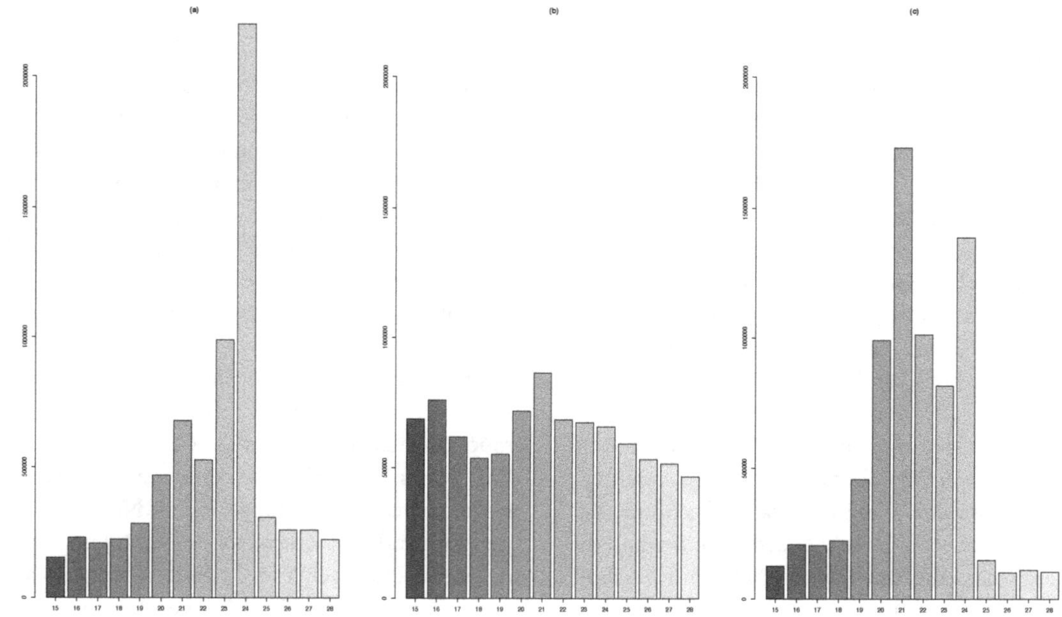

Figure 4.1 Number of sequenced reads split by adaptor trimmed read length from three Illumina libraries of small RNAs in *Arabidopsis thaliana* rootstock. The distribution of in a wild-type sample (a) shows a major peak in the number of reads sequenced of 24 bases in length, and a minor peak at 21 bases. In a triple mutant of Dicer-like 2, 3 and 4 proteins (b) the peak at 24 bases has been removed. In another triple mutant of Dicer-like 2, 3 and 4 proteins, now grafted to a wild-type shoot of *Arabidopsis thaliana*, the peak at 24 bases has been largely recovered. Note that in (b) where the majority of reads of 24 bases in length have been removed, there is a greater depth of sequencing at all other lengths, reflecting the competitive nature of the sequencing process.

The data should also be examined for the numbers of unique reads. In some cases, the majority of the sequenced reads will be of a very few sequences. There are multiple technical failures that can lead to such an outcome. Amongst the most common are a failure to properly extract the small RNAs that leads to contamination of the sequenced reads from tRNA and rRNA degradation products, and errors in primer ligation that lead to very high numbers of technical artefacts such as dimer–dimer reads. The situation can also occur as a result of highly expressed miRNAs, which are often expressed at orders of magnitude higher than siRNAs and may saturate the sequencing library. A further common source of error is the formation of poly-A reads (see below). These will be present in almost all sequencing data, but if too abundant may drown out the true signal.

Removing tRNAs, rRNAs and miRNAs

Despite the methods now available for isolation of short RNAs from total RNA, some degradation products of transfer RNAs (tRNAs) and ribosomal RNAs (rRNAs) are likely to be included in the sequencing libraries. These can be highly abundant and are likely to show high variability between samples. It is therefore generally considered advisable to remove reads whose sequences show perfect alignment to these species of RNA, although it has been suggested that at least some of these may represent true small RNA populations (Cole *et al.*, 2009).

MicroRNAs are extracted from total RNA samples by the same methods as siRNAs, and so will be sequenced alongside the siRNA population. The mechanisms by which miRNAs are produced from a stem–loop structure are far better understood than those that produce siRNAs and, in consequence, the models used to identify miRNAs can be considerably more specific than those for siRNAs. For many organisms, more or less complete lists of miRNAs exist (Griffiths-Jones *et al.*, 2006), and methods exist to exploit the stem–loop formation of miRNAs to discover novel miRNAs from sequenced small RNA data from both plant (Yang and Li, 2011) and animal (Friedländer *et al.*, 2008) sources. It is therefore usually possible to remove reads whose sequence matches that of miRNAs. This is recommended in order to avoid confusion between the two species of small RNA.

Poly-A reads

A common form of sequencing error in Illumina sequencing is the presence of poly-A reads, in which a read is falsely reported as consisting almost wholly of adenines. Errors in the reporting may lead to a few of the bases in this read to be reported not as adenines but as some other base. Since poly-A regions are also common features in genomic sequences, these reads are likely to align across a relatively large regions of the genome, giving rise to regions that superficially appear to be siRNA loci (see below). The removal from the data of any sequenced read containing long poly-A sequences is thus recommended.

Library scaling factors

'Normalization' by library scaling factor is a common requirement in analyses of expression of high-throughput sequencing. The requirement arose initially from the variation in the number of sequenced reads produced in each sequencing library. If some sample is sequenced by constructing two libraries, A and B, but is sequenced to twice the depth in library A compared to library B, then the number of copies of each sequence will be approximately doubled in the data from library A compared with that in library B. Since the sample used is identical in both libraries, a normalization factor based on the total number of reads sequenced in each library should be used in any analysis of the two sets of data.

Early suggestions for a normalization factor were based on the total number of reads sequenced in each library (Mortazavi *et al.*, 2008). However, these neglected to take into account that next-generation sequencing is a competitive process. If a given biological sample sequenced contains a few very highly expressed sequences, these will account for a large proportion of the sequenced reads from that sample. Consequently, the number of sequenced reads from more lowly expressed sequences will be diminished, even though the true expression of these sequences may not be altered. This is likely to be a particular problem in sequencing siRNAs, both because degradation products from tRNAs and rRNAs may be present in high abundance, but also because individual small RNAs may be present at extremely high levels. More sophisticated methods of calculating the library scaling factor used for normalization are thus essential for siRNA-seq data.

The method of normalization by 75th percentile (Bullard *et al.*, 2010) is usually sufficient for most siRNA data. In this method, the sequences from a given library are ranked by the number of copies of this sequence that appear in the data. The sum of the number of copies of the lower seventy-five per cent of these sequences is taken as the library scaling factor. This removes those sequences for which the greatest number of copies are sequenced from an estimation of the library scaling factor; thus, if any library is dominated by a few very highly expressed sequences, the scaling factor of this library will be small. More complex methods such as the trimmed mean of M-values (Robinson and Oshlack, 2010), or the median ratio method (Anders and Huber, 2010) may also be used to estimate library scaling factors. These methods trim both the most and least expressed sequences from the data. This approach may offer advantages in the case of mRNA-seq data, where very lowly expressed genes are relatively rare, however, in siRNA-seq data the majority of sequences appear infrequently and so these methods may exclude too much data to be appropriate for analyses of siRNA-seq data.

To reduce the effect of contamination or sequencing error in calculating the library scaling factors, it is usually advisable to consider only those reads which align to the reference genome when calculating the library size. This removes the effect of contamination by, for example, viral siRNAs, which if present in large numbers will reduce the perceived expression of siRNAs from the reference organism, as above. It is also likely to be beneficial to calculate library sizes after first removing poly-A reads, potential tRNA and rRNA degradation products and miRNAs, for the same reasons.

Locus finding

A primary challenge in the bioinformatic analysis of siRNA-seq data is in the discovery of siRNA 'loci' on the genome. These loci are genomic regions associated with a high density of sequenced siRNA reads, which approximate the locations of the double-stranded precursors from which the siRNAs derive. These precursors are highly transient and consequently not susceptible to sequencing; however, they can be inferred from an analysis of the relatively easily sequenced siRNAs. Since the siRNAs deriving from a common precursor are at least to some extent regulated by the same factors, the abundances and functions of these siRNAs are also likely to be correlated. The identification of siRNA loci from high-throughput sequencing thus gives valuable information about the interactions and functions of individual siRNAs, and represents an important biological question.

Where a set of siRNAs derive from the same precursor, we expect to see the sequenced reads align to the genome in close proximity and with non-independent abundances, and this allows us to define a set of models for the identification of the siRNA loci. However, several challenges make this a non-trivial task.

Accumulation biases

The greatest difficulty in identifying siRNA loci is the problem of accumulation bias. From a single siRNA precursor, multiple siRNAs are produced. Of these, some will be stabilized by association with a protein complex and thus will be available for sequencing. However, other potential siRNAs diced from the precursor will be rapidly degraded, and thus will not be sequenced. The mechanisms for selection of individual siRNAs for association with a stabilizing protein complex are at present not well understood, although there is likely to be a dependence on the presence or absence of particular Argonaute proteins (Havecker *et al.*, 2010), tissue type, and environmental factors. Of those siRNAs from a given locus that do associate with some stabilizing complex, they may do so with either greatly varying degrees, or be degraded at greatly varying rates.

The consequence of these accumulation biases is that siRNA reads sequenced from an individual locus show much higher variability than might be expected, with most siRNA loci containing multiple regions within the locus where no sequenced read appears at all. Fig. 4.2 shows an example of this; the coverage of any individual base within the locus ranges from zero to several hundred covering sequenced reads. This variation precludes the use of techniques such as employed in *de novo* transcriptome assembly of mRNA-seq data (Zerbino and Birney, 2008; Birol *et al.*, 2009; Grabherr *et al.*, 2011) or peak-calling in ChIP-seq data (Zhang *et al.*, 2008). An alternative approach is thus required for locus discovery from siRNA-seq data.

Multireads

Multireads, in which a sequenced read aligns to multiple locations on the genome, are a common problem in high-throughput sequencing. However, the problem tends to be more severe in siRNA-seq data. This occurs in part due to the short nature of siRNA reads, which are consequently more likely to align to multiple locations, but primarily due to the known associations (Sunkar *et al.*, 2005; Ghildiyal *et al.*, 2008) of siRNAs with repetitive genome elements. It is thus not uncommon for entire siRNA loci to be duplicated in multiple locations on the genome.

Methods that have been successfully applied to the problem of multireads in mRNA-seq data (Kim *et al.*, 2011; Li and Dewey, 2011) are thus unfortunately unsuitable for analysis of siRNA-seq data. For multireads occurring in mRNA-seq data, there will usually exist some uniquely mapping reads within each gene or isoform which are able to guide the placement of the multireads based on a model of consistent sequencing of reads from each location within that gene. However, when whole siRNA loci are duplicated this is not a viable approach. Moreover, the accumulation biases described above make assumptions

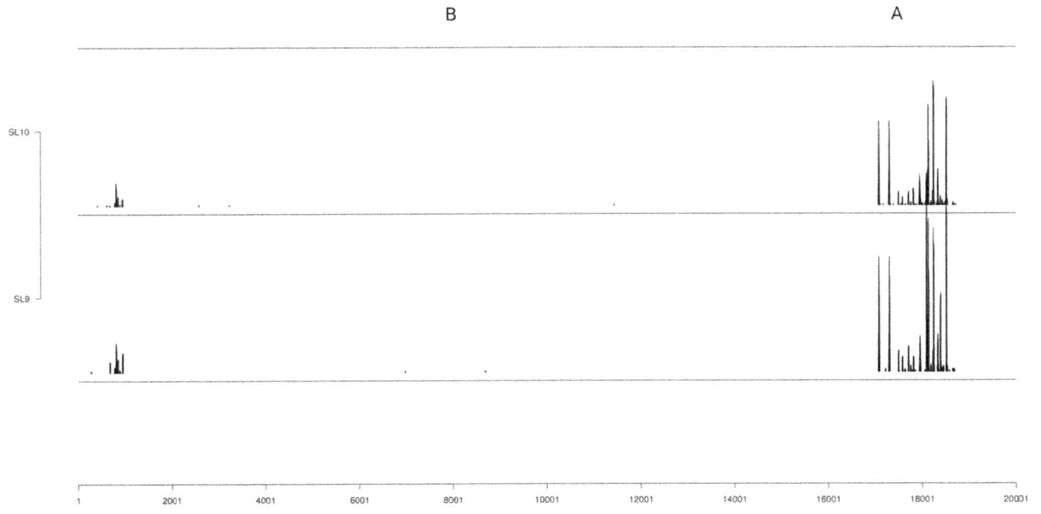

Figure 4.2 Plot of observed base coverage for the first 20,000 bases of *Arabidopsis thaliana* in two replicated sequencing libraries. Note the highly variable coverage of the siRNA locus at region A and the non-replicated background noise at region B.

of consistent expression within a siRNA locus questionable.

One approach that has been suggested for multireads in siRNA data is to discard them from the data. This has the advantage that any locus discovered can be assigned unambiguously to a particular location. However, given the tendency of siRNAs to associate with repetitive elements, multireads can form a large proportion of the sequenced data and to discard them is consequently to lose significant volumes of data. More seriously, in discarding multireads a bias is introduced into the classes of siRNA loci that will be discovered; specifically, only those siRNA loci which are not associated with repetitive elements will be found. Any general statements about the mechanisms of production and association of siRNAs on the basis of the sequencing data will be thus limited in their application. A better approach is thus to include the multireads within the data and account for duplication of loci in downstream analyses.

Inclusion of multireads within the data must be done with some care. The expression of a locus should be estimated based on the total number of reads that map to that locus. Where a read maps multiple times within a single locus, it is clearly erroneous to count that read multiple times when estimating the expression of that locus. It is therefore necessary, if multireads are included within the data, to track their locations carefully (Hardcastle *et al.*, 2012)

Background noise

Background noise arises in high-throughput sequencing of siRNAs from a number of causes. As a consequence, not every sequenced read need be associated with a siRNA locus. The most prevalent, and easily identified sources of noise are sequencing errors and the sequencing of degraded longer molecules such as rRNAs, tRNAs, and mRNAs. Reads that derive from such cases are unlikely to be seen in multiple biological replicates and so can be identified as background noise. Fig. 4.2 shows examples of this, where non-replicated sequenced reads can clearly be identified as non-locus associated sequenced reads.

Naïve locus discovery

Early methods for locus discovery from siRNA reads (Moxon *et al.*, 2008; MacLean *et al.*, 2010b) in general simply looked for genomic regions in which the number of sequenced reads exceeded some minimum value and no sufficiently large gap existed within the region. Such a model accounts for some of the characteristics of siRNA loci, predominantly, the association of siRNA loci with high numbers of sequenced reads. Accumulation bias is permitted to create gaps, or regions of the genome to which no sequenced read maps, within a siRNA locus. However, should these gaps be of too great a length, it is assumed that this is the result of the termination of the locus rather than accumulation bias.

These early models, while accounting for some of the features of siRNA loci, were flawed in a number of important ways. The thresholds chosen for siRNA loci are largely arbitrary. Moreover, the choice of these thresholds must vary between organisms and with sequencing depth as the characteristics of the siRNA loci change, however, no clear mechanism has been proposed for making these choices in such a way as to make results comparable between different sequencing experiments. Furthermore, these models neglect to account for the evidence provided by biological replicates of the data; that is, if a siRNA locus discovered through one set of sequencing data is to be regarded as a true finding then it should appear in all biological replications of these data. This failure to account for replication limits these naïve models to the concurrent analysis of siRNA-seq data from at most two samples.

Empirical Bayesian locus discovery

The problems identified for siRNA locus discovery through a naïve analysis of siRNA-seq data have been addressed by an empirical Bayesian approach (Hardcastle *et al.*, 2012) implemented as the R package segmentSeq. In applying this method, a set of *candidate loci* are defined on the genome consisting of all those genomic regions which, on the basis of aligned sequenced reads in those regions, might represent a siRNA locus. A region cannot be considered a true siRNA locus if it contains a 'null' region, that is, a region to which sequenced reads align only at background levels, and so a set of *candidate nulls* must also be defined. Based on the density of reads aligning within these candidate regions, classifications are made as to which represent true loci and nulls. Given these classifications, a locus map is constructed by combining the sets of true loci and nulls from multiple replicate groups.

To acquire the classifications, models are constructed for locus and null regions based on a negative binomial distribution for the number of sequenced reads that derive from a particular biological sample and align to some region of the genome. In order to evaluate the likelihoods of these models, a set of parameters on the distributions must be defined. However, the primary advantage of an empirical Bayesian approach is that the parameters of the underlying negative binomial distributions are estimated from the complete dataset. By borrowing power in this manner, it is no longer necessary to specify an arbitrary set of parameters in order to define the loci. The models are designed to account for biological replicates and there is no theoretical limit on the number of samples that can be considered.

The primary challenge in applying empirical Bayesian methods to high-throughput sequencing data is that in order to estimate the model parameters from the data, some elements of the data which approximate the model must be known. In the case of siRNA locus discovery, this means that the locations of at least some siRNA loci must be known, as well as some regions of the genome which are known not to contain a siRNA locus (the null regions). Fortunately, in the application of empirical Bayesian methods to high-throughput sequencing data, the initial approximation need not be of especially high quality. Consequently, heuristic methods can be used to give an initial approximation to the true loci and nulls, and this approximation can be used to inform an empirical Bayesian analysis. The use of heuristic methods does require a set of initial parameters to be chosen in order to construct the approximation; however, these have far less impact on the final locus definitions than upon the initial approximation.

Fig. 4.3 shows the application of these methods on sequencing libraries of siRNAs from *Arabidopsis thaliana*. The initial approximation can be seen to identify many small segments of the genome as individual loci in close proximity to one another rather than as one contiguous locus, and does not account for low reproducibility of the data between biological replicates. Adjusting the parameters of the heuristic method

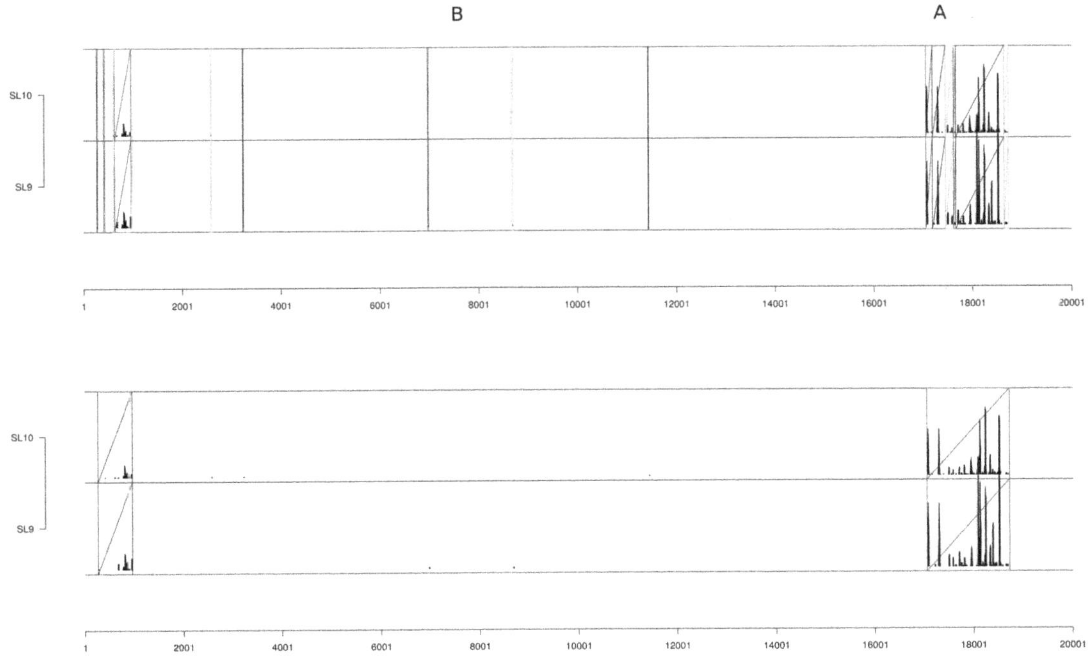

Figure 4.3 Plot of observed base coverage for the first 20,000 bases of *Arabidopsis thaliana* in two replicated sequencing libraries showing estimated siRNA loci (grey scale rectangles) based on a naïve heuristic (top) and a set of empirical Bayesian methods (bottom) initiated by the naïve heuristic loci. The heuristic approach identifies regions of non-replicated low abundance base coverage as representing loci (region B), and splits the large locus (region A) into multiple separate loci. The empirical Bayesian method identifies region A as a single locus and discards the non-replicated cases in region B.

might resolve one or both of these flaws, but a single set of parameters is unlikely to give good results in all such cases for all experiments. However, the empirical Bayesian analysis based on this initial approximation resolves both of these issues without requiring parameter adjustments.

The principal drawback in applying an empirical Bayesian approach to locus discovery is that the methods are extremely computationally intensive. Fortunately, these methods are also readily parallelisable, allowing the computational load to be distributed over a large number of machines. Since there exists no theoretical limit to the number of sequencing libraries and replicate groups that can be considered by these methods, it may also be possible to construct a robust set of siRNA loci from sufficiently many sequencing libraries. For many siRNA-seq experiments, the loci identified from such an analysis may be sufficient for an initial analysis.

Association of siRNA loci with genomic features

It is often of interest to explore how a defined set of siRNA loci might interact with other annotation features on the genome. A straightforward approach to this question is to consider the number of cases where the genomic coordinates of a siRNA locus overlaps those of an annotation feature of interest. By comparing this number with the total number of siRNA loci, and the total number of times that our annotation feature appears, the likelihood of the observed number of overlaps can be calculated either through Fisher's test or based on a random permutation of siRNAs on the genome.

However, this naïve approach is unlikely to give reliable estimates as the distribution of siRNA loci upon the genome tends to be highly non-uniform. SiRNA loci show a strong tendency to form clusters on the genome, as do important classes of annotation features with which siRNAs associate. If one of these clusters overlaps with a particular annotation feature, then many cases of overlap for individual siRNA loci will be identified. However, this derives from the clustering of the siRNA loci, rather than a true association of the siRNA loci with a particular feature. A solution to this problem has been suggested based on a block-bootstrap approach (Bickel *et al.*, 2010). This approach divides the genome into 'blocks' in order to preserve the clustering structures of siRNAs. By randomly permuting the position of these blocks upon the genome, a null distribution that accounts for clustering can be used to discover whether the number of observed overlaps on the genome is significant. This approach is of general application to analyses of overlap between genomic features but is of particular relevance to the siRNA world.

Differential expression in siRNAs

Differential expression for siRNAs can be identified by much the same methods as for other sequencing data. Several methods have been published in recent years for the analysis of differential expression, and these can be divided into three main groups. By far the largest set of methods is that consisting of primarily 'classical' statistical approaches. These methods impose a null model on the data (no differential expression) and evaluate the likelihood of the observed data given this null model. The implementation of these methods to high-throughput sequencing has focused primarily on stabilizing estimates of the parameters of the null model for an individual gene or siRNA by borrowing information from the large dataset. Widely used examples of methods adopting this approach are the edgeR (Robinson, 2010) and DESeq (Anders and Huber, 2010) R packages. These methods were originally designed for analysis of pairwise differential expression only; that is, to discover differential expression between two experimental groups. Recent developments have allowed a generalized linear models approach (McCarthy et al., 2012) to be applied to high-throughput sequencing data based on the same principles, although at some cost in performance.

An alternative to the classical approach is an empirical Bayesian method released as the R package baySeq (Hardcastle and Kelly, 2010). In this approach, multiple models can be examined on the data (including the model of no differential expression). The posterior likelihoods of these models are evaluated simultaneously for an individual gene or siRNA based on the observed data and the underlying parameters of the models, which in this case are estimated by sampling from the whole dataset. For pairwise differential expression, the interpretation of an analysis with this method is similar to that of the classical methods described above. Independent comparisons of differential expression analysis methods (Cordero et al., 2012; Kvam et al., 2012) for pairwise analysis methods, suggest that the empirical Bayesian methods perform as well or better than the classical alternatives currently available. Where multiple experimental conditions are simultaneously being considered, the models defined by the baySeq package differ from those available in a generalized linear models approach (McCarthy et al., 2012). Depending on the particular experimental design and the questions being asked of the data, either method may be more or less appropriate.

The third class of differential expression analysis methods are non-parametric in nature. Both the classical and empirical Bayesian approaches assume that the data are distributed as an over-dispersed Poisson, usually the negative binomial. This is appropriate in many situations, as the technical variation that arises in high-throughput sequencing is Poisson distributed (Lee *et al.*, 2008; Marioni *et al.*, 2008) and the biological variation is generally small relative to the technical variation (that the biological variation is non-zero leads to an assumption of over-dispersion). In some cases, however, the biological variation is likely to dominate and the assumption of an over-dispersed Poisson is no longer appropriate. In this case, it may not be apparent what distribution will model the data and a non-parametric method becomes necessary. This is perhaps more likely to occur in analyses of siRNA data than other forms of sequencing data, as siRNAs can be involved in positive feedback loops that cause very high expression levels in individual samples. The SAMSeq method (Li and Tibshirani, 2011), in which differential expression is calculated based on a ranking of expression levels, is a good example of a robust analysis method for high-throughput sequencing data. However, non-parametric methods will inevitably require substantially higher numbers of replicates in order to provide reliable results, as they make their inferences without the benefit of a guiding model. In cases where large numbers of replicates are not available, it is likely that better results will be obtained through careful removal of outliers before using one of the classical or empirical Bayesian methods described above.

Methods for detection of differential expression can be applied to the observed abundances of individual siRNAs or to the combined abundances of siRNAs mapping to an individual siRNA locus. In general, it is probably true to say that an analysis of differential expression of siRNA loci will give most information about the mechanisms acting to produce siRNAs, and an analysis of differential expression of individual siRNAs will give most information about potential effects of siRNAs; however, this is a largely unexplored model of interpretation. In analyses of differential expression in individual siRNAs, of which there may be several million within the data from even a small sequencing experiment, it is often worth imposing filters on the data to reduce the computational costs of such an analysis. These can often be fairly crude in form; a good rule of thumb is to discard any siRNA which is seen on average less than once in any sequenced library as it will usually not be practical to identify differential expression in such cases.

Phased siRNAs

Phased siRNAs form a particular class due both to their biogenesis and the association with *trans*-acting siRNAs (ta-siRNAs). These ta-siRNAs are formed as a result of the targeted cleavage of some long RNA transcript by the action of either a miRNA or siRNA. Following this, the remaining transcript is converted to dsRNA by some RNA-dependent RNA polymerase, rather than being degraded. The mechanisms by which some transcripts are identified for conversion to double-stranded RNA rather than degradation are not clear, although multiple hypotheses have been suggested (Axtell *et al.*, 2006; Chen *et al.*, 2010; Cuperus *et al.*, 2010; Manavella *et al.*, 2012) These hypotheses tend to centre on the lengths of the small RNAs guiding cleavage; however, no single hypothesis has fully explained all cases of phased siRNA production.

Given the formation of the dsRNA, a Dicer or Dicer-like protein then acts to cleave this dsRNA. The action of the Dicer protein is such that the cleavage occurs at regularly spaced intervals (usually every twenty-one bases) on this dsRNA. Since the initial targeted cleavage occurs at the same location for each copy of the initial long RNA transcript, this process leads to 'phased' siRNAs.

These phased siRNAs are of particular interest because in many cases they have been shown to be trans-acting and hence have the capacity to form hubs of regulatory activity. Phased siRNAs have been shown to initiate cascades of secondary siRNA production (Daxinger *et al.*, 2009; Chen *et al.*, 2010; Zhang *et al.*, 2012a), in which further long RNA transcripts are targeted by phased siRNAs to initiate further production of siRNAs. Phased siRNAs have been identified in a number of plant species, including *Arabidopsis thaliana* (Howell *et al.*, 2007), *Chlamydomonas reinhardtii* (Zhao *et al.*, 2007), rice (Johnson *et al.*, 2009) and grapevine (Zhang *et al.*, 2012a). Although the majority of known phasing occurs with an interval of 21 bases, other phasing intervals have been reported (Johnson *et al.*, 2009). The number of known locations of phased loci varies substantially between species, from over eight hundred in rice to less than a dozen in *Arabidopsis thaliana*.

Identification of phased siRNA loci

Analyses and identification of phased siRNAs can take advantage of the additional structure of the data to build more sophisticated models than those available in the siRNA locus detection described above. SiRNAs from a phased locus should map to the genome in a regularly spaced manner, usually every 21 bases, with non-independent abundances. In practice, several factors complicate this model. It is not uncommon for production of siRNAs by other mechanisms to occur at the same location as that described for phased loci, and so there may be a number of reads that occur away from the phasing sites. Secondly, the action of the cleaving Dicer or Dicer-like protein is not always precise, and, while most dicing occurs every 21 bases, it is possible for individual cuts to occur at 20 or 22 bases. These errors in position accumulate with distance from the initial cleavage site, and so at sufficient distance from the cleavage site there is no clear association of siRNAs accumulating at a particular location. Finally, the accumulation biases described for siRNAs in general also apply to phased siRNAs, with some sequences of phased siRNAs failing to associate with an Ago protein and hence not appearing in the sequencing data.

Despite these difficulties, a number of methods have been applied to identify phased siRNAs from next-generation sequencing data with reasonable success. Heuristic methods have been applied with some success to identify phased loci in Arabidopsis thaliana (Howell *et al.*, 2007) and *Brachypodium distachyon* (Vogel *et al.*, 2010). These methods generate a phasing score P for a window on the genome such that, for the analysis of *Brachypodium distachyon:*

$$P = \log\left[\left(1 + 10\sum_{i=1}^{10} \frac{P_i}{1+U}\right)^{n-2}\right], n > 3$$

where n gives the total number of occupied phased positions, P_i is the total number of sequenced reads mapping to the ith phased location and U is the total number of sequenced reads mapping to an out of phase location in the genome. This model attempts to account for the primary features of phased siRNA loci, in that a region for which many locations of phasing are occupied by the majority of reads in that window will result in a high score. In order to account for imprecision in the action of the Dicer proteins, phasing locations are defined not as single bases separated by the phasing distance but as a region of three bases in length centred on these locations.

This approach, and the simplified version of it applied to *Arabidopsis thaliana* have been successful in discovering likely candidates for phased loci. However, the model is largely arbitrary and it is difficult to interpret the scores acquired. A more statistically rigorous method has been proposed based on the hypergeometric distribution (Chen *et al.*, 2007), and implemented (with minor variations) in the UEA Small RNA Toolkit. Within a window on the genome, each location is examined for occupancy by some sequenced read. The ratio of occupied to non-occupied positions in those locations potentially associated with phasing can be compared to that ratio for those locations not associated with phasing through Fisher's test. This gives a statistically meaningful indication of whether there is greater association of sequenced reads with phasing locations than would be expected by chance. However, the method does neglect to take into account the abundance of the reads matching to each location; a site is considered occupied based on a single read. This makes it more difficult to identify phasing in the presence of a diverse population of non-phased small RNAs using this approach.

Post-analysis visualization

Visualization of the siRNA data following analysis is an important tool both to validate and explore the reported results. A valuable tool for such visualization is the Generic Genome Browser (Gbrowse) (Stein *et al.*, 2002), which allows detailed visualization of aligned siRNA reads together with annotation tracks. Fig. 4.4 shows an example of such a visualization in which a differentially expressed siRNA locus can be seen to exist close to a coding region of the genome. Visualization in this way also allows strand biases, in which the sequenced siRNAs are predominantly from one strand of the genome, and length biases, in which the sequenced siRNAs are predominantly of a particular length, to be easily visualized. Publicly available implementations of the browser which allow user upload of data are available for many model organisms.

Target finding and small RNA networks

High-throughput sequencing is likely to play an important role in establishing targeting rules operating within the various regulatory pathways of siRNAs (Asikainen *et al.*, 2008). This is at present a relatively unexplored area of analysis; while it is known that some degree of sequence complementarity is required between a siRNA and its target,

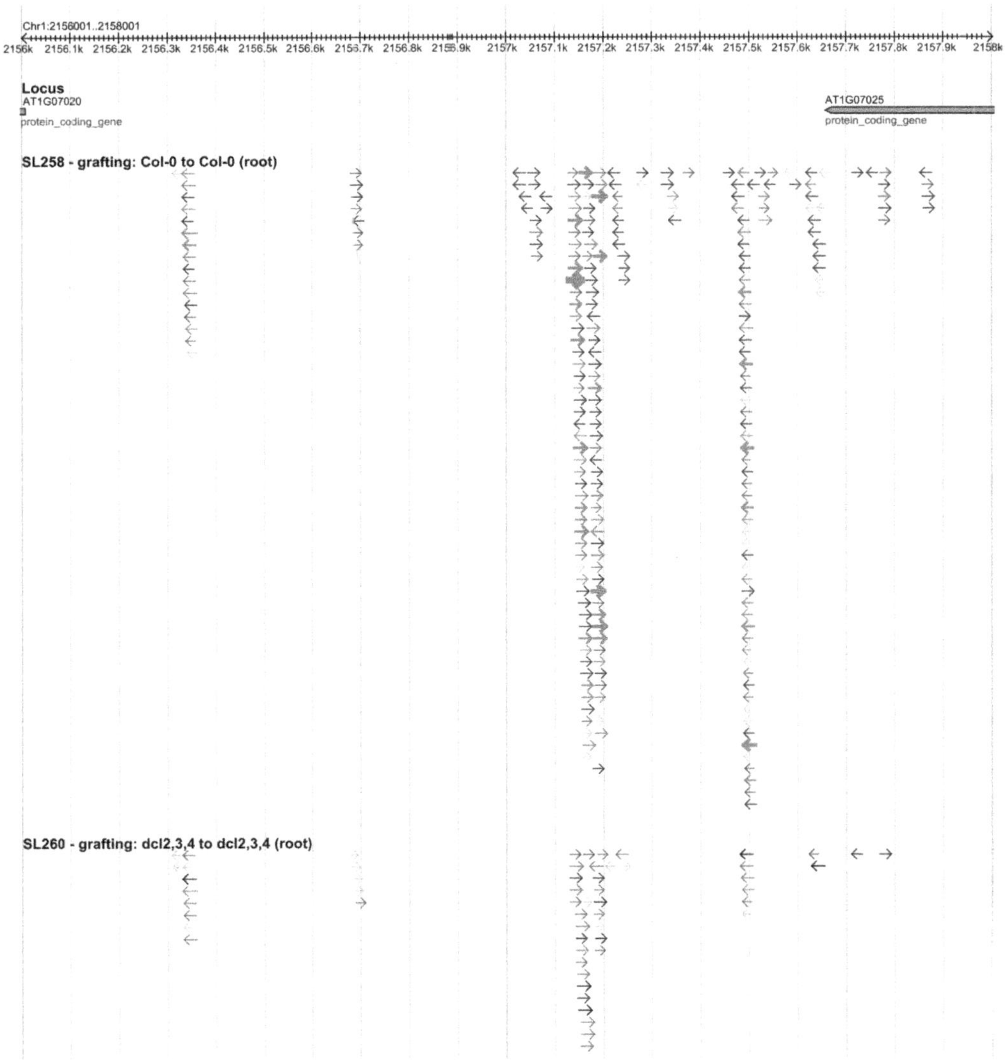

Key: Pink 15-19 nt;Red 20-21 nt; Green 22 nt; Blue 23-24 nt;Grey 25-27 nt;Black 28-35 nt; Orange other;

Figure 4.4 Post-analysis visualization of a locus in Gbrowse showing differential expression between two sequencing libraries. Visualization in this way allows comparison with nearby annotation features; in this case, a known coding sequence, identification of strandedness of individual siRNAs, and, if properly configured, the length of sequenced siRNAs in the region. A colour version of this figure is located in the plate section at the back of the book.

it is not clear what level of complementarity is required, nor if there are particular rules guiding acceptable mismatches, as for miRNAs (Jones-Rhoades and Bartel, 2004; Rajewsky, 2006). In combination with other high-throughput sequencing data, classes of siRNAs which directly influence chromatin modifications and induce mRNA cleavage can be identified. Of particular interest in the latter case is degradome sequencing (Addo-Quaye *et al.*, 2008), in which the mRNA cleavage products can be isolated from the data and sequenced. In combination with studies of Argonaute associations and siRNA sequence properties, these analyses are of relevance not only to the endogenous action of siRNAs but in identifying off-target effects in RNA interference

(Jackson and Linsley, 2010), a significant barrier to the use of this technology in therapeutic applications.

A related question is the interaction of siRNAs as self-regulatory elements (MacLean *et al.*, 2010a). Evidence for this hypothesis is currently limited; however, it does appear that networks of siRNA loci linked by sequence similarity do possess characteristics usually associated with networks of genomic features. Extensive sequencing of siRNAs may help to validate this suggestion by identifying under which circumstances two siRNA loci sharing sequence similarity also show correlated expression profiles. In combination with the target finding analyses described above, these novel data and analysis techniques will place siRNAs in a systems biology framework in which their interactions with diverse regulatory features may be considered (Baulcombe, 2006).

Discussion

Sequencing of siRNAs allows the analysis of these key components of regulatory networks on a genome-wide scale. A wide range of tools has been developed to assist in the initial processing of the data generated in such experiments. A particularly interesting and challenging problem is the identification of the precursor elements of the siRNAs, and methods have been successfully applied to such analyses.

High-throughput sequencing has already been instrumental in describing many of the characteristics of small interfering RNAs. The manner in which Argonaute association with siRNAs is influenced by the 5′ nucleotide (Mi *et al.*, 2008; Takeda *et al.*, 2008) and the influence of tissue type (Havecker *et al.*, 2010) both relied on high-throughput sequencing data to make the genome-wide assessments of association required. The capacity of siRNAs to move over significant distances within an organism (Dunoyer *et al.*, 2010; Molnar *et al.*, 2010) also required high-throughput sequencing in order to establish the high numbers of endogenous siRNAs involved.

Small interfering RNAs play a key role in viral defence mechanisms in plants and this has been a fruitful area in which to apply high-throughput sequencing technologies. The characterisation of virus (Molnár *et al.*, 2005) and viroid (Di Serio *et al.*, 2009; Bolduc *et al.*, 2010) derived siRNAs indicates hotspots of generation together with sequence biases. These analyses allow inferences to be made regarding the mechanisms of production of such siRNAs.

Next-generation sequencing data allows the quantification of expression of both individual siRNAs and the siRNA precursor elements, allowing patterns of expression to be identified within experimental groups and tissue types. This is of particular importance in an interpretation of siRNA function within a systems framework in which the associations between siRNAs and epigenetic factors (Henderson and Jacobsen, 2007) can be identified. Recent work in identifying cascades of secondary small RNA production (Daxinger *et al.*, 2009; Chen *et al.*, 2010; Zhang *et al.*, 2012a) indicates the significance of such a systems approach and the necessity for genome-wide analyses in order to make such discoveries. Consequently, as the volumes of data generated by next-generation sequencing technologies increases, this is likely to lead to new insights as to the mechanisms by which siRNAs act as regulators and are themselves regulated.

References

Addo-Quaye, C., Eshoo, T.W., Bartel, D.P., and Axtell, M.J. (2008). Endogenous siRNA and miRNA targets identified by sequencing of the *Arabidopsis* degradome. Curr. Biol. *18*, 758–762.

Alon, S., Vigneault, F., Eminaga, S., Christodoulou, D.C., Seidman, J.G., Church, G.M., and Eisenberg, E. (2011). Barcoding bias in high-throughput multiplex sequencing of miRNA. Genome Res. *21*, 1506–1511.

Anders, S., and Huber, W. (2010). Differential expression analysis for sequence count data. Genome Biol. *11*, R106.

Applied Biosystems (2009). Whole Genome discovery and profiling of small RNAs using the SOLiD System. BioTechniques *46*, 232–234.

Asikainen, S., Heikkinen, L., Wong, G., and Storvik, M. (2008). Functional characterization of endogenous siRNA target genes in *Caenorhabditis elegans*. BMC Genomics *9*, 270.

Axtell, M.J., Jan, C., Rajagopalan, R., and Bartel, D.P. (2006). A two-hit trigger for siRNA biogenesis in plants. Cell *127*, 565–577.

Baulcombe, D.C. (2006). Short silencing RNA: the dark matter of genetics? Cold Spring Harb. Symp. Quant. Biol. *71*, 13–20.

Becker, C., Hagmann, J., Müller, J., Koenig, D., Stegle, O., Borgwardt, K., and Weigel, D. (2011). Spontaneous epigenetic variation in the Arabidopsis thaliana methylome. Nature *480*, 245–249.

Bentley, D.R., Balasubramanian, S., Swerdlow, H.P., Smith, G.P., Milton, J., Brown, C.G., Hall, K.P., Evers, D.J., Barnes, C.L., Bignell, H.R., *et al.* (2008). Accurate whole human genome sequencing using reversible terminator chemistry. Nature *456*, 53–59.

Bickel, P.J., Boley, N., Brown, J.B., Huang, H., and Zhang, N.R. (2010). Subsampling methods for genomic inference. Ann. Appl. Stat. *4*, 1660–1697.

Birol, I., Jackman, S.D., Nielsen, C.B., Qian, J.Q., Varhol, R., Stazyk, G., Morin, R.D., Zhao, Y., Hirst, M., Schein, J.E., *et al.* (2009). De novo transcriptome assembly with ABySS. Bioinformatics *25*, 2872–2877.

Bolduc, F., Hoareau, C., St-Pierre, P., and Perreault, J.-P. (2010). In-depth sequencing of the siRNAs associated with peach latent mosaic viroid infection. BMC Mol. Biol. *11*, 16.

Bullard, J.H., Purdom, E., Hansen, K.D., and Dudoit, S. (2010). Evaluation of statistical methods for normalization and differential expression in mRNA-Seq experiments. BMC Bioinformatics *11*, 94.

Calviño, M., Bruggmann, R., and Messing, J. (2011). Characterization of the small RNA component of the transcriptome from grain and sweet sorghum stems. BMC Genomics *12*, 356.

Carbonell, A., Fahlgren, N., Garcia-Ruiz, H., Gilbert, K.B., Montgomery, T.A., Nguyen, T., Cuperus, J.T., and Carrington, J.C. (2012). Functional Analysis of Three Arabidopsis ARGONAUTES Using Slicer-Defective Mutants. Plant Cell *24*, 3613–3629.

Chappell, L., Baulcombe, D., and Molnár, A. (2006). Isolation and cloning of small RNAs from virus-infected plants. Curr. Protoc. Microbiol *Chapter 16*, Unit 16H.2.

Chen, H.-M., Li, Y.-H., and Wu, S.-H. (2007). Bioinformatic prediction and experimental validation of a microRNA-directed tandem trans-acting siRNA cascade in Arabidopsis. Proc. Natl. Acad. Sci. U.S.A.*104*, 3318–3323.

Chen, H.-M., Chen, L.-T., Patel, K., Li, Y.-H., Baulcombe, D.C., and Wu, S.-H. (2010). 22-Nucleotide RNAs trigger secondary siRNA biogenesis in plants. Proc. Natl. Acad. Sci. U.S.A.*107*, 15269–15274.

Cole, C., Sobala, A., Lu, C., Thatcher, S.R., Bowman, A., Brown, J.W.S., Green, P.J., Barton, G.J., and Hutvagner, G. (2009). Filtering of deep sequencing data reveals the existence of abundant Dicer-dependent small RNAs derived from tRNAs. RNA *15*, 2147–2160.

Cordero, F., Beccuti, M., Arigoni, M., Donatelli, S., and Calogero, R.A. (2012). Optimizing a massive parallel sequencing workflow for quantitative miRNA expression analysis. PLoS One *7*, e31630.

Cox, M.P., Peterson, D.A., and Biggs, P.J. (2010). SolexaQA: At-a-glance quality assessment of Illumina second-generation sequencing data. BMC Bioinformatics *11*, 485.

Cuperus, J.T., Carbonell, A., Fahlgren, N., Garcia-Ruiz, H., Burke, R.T., Takeda, A., Sullivan, C.M., Gilbert, S.D., Montgomery, T.A., and Carrington, J.C. (2010). Unique functionality of 22-nt miRNAs in triggering RDR6-dependent siRNA biogenesis from target transcripts in Arabidopsis. Nat. Struct. Mol. Biol. *17*, 997–1003.

Daxinger, L., Kanno, T., Bucher, E., Van der Winden, J., Naumann, U., Matzke, A.J.M., and Matzke, M. (2009). A stepwise pathway for biogenesis of 24-nt secondary siRNAs and spreading of DNA methylation. EMBO J. *28*, 48–57.

Deshpande, G., Calhoun, G., and Schedl, P. (2005). Drosophila argonaute-2 is required early in embryogenesis for the assembly of centric/centromeric heterochromatin, nuclear division, nuclear migration, and germ-cell formation. Genes Dev. *19*, 1680–1685.

Ding, S.-W., and Voinnet, O. (2007). Antiviral immunity directed by small RNAs. Cell *130*, 413–426.

Dohm, J.C., Lottaz, C., Borodina, T., and Himmelbauer, H. (2008). Substantial biases in ultra-short read datasets from high-throughput DNA sequencing. Nucleic Acids Res. *36*, e105.

Dunoyer, P., Brosnan, C.A., Schott, G., Wang, Y., Jay, F., Alioua, A., Himber, C., and Voinnet, O. (2010). An endogenous, systemic RNAi pathway in plants. EMBO J. *29*, 1699–1712.

Elbashir, S.M., Lendeckel, W., and Tuschl, T. (2001). RNA interference is mediated by 21- and 22-nucleotide RNAs. Genes Dev. *15*, 188–200.

Friedländer, M.R., Chen, W., Adamidi, C., Maaskola, J., Einspanier, R., Knespel, S., and Rajewsky, N. (2008). Discovering microRNAs from deep sequencing data using miRDeep. Nat. Biotechnol. *26*, 407–415

Gentleman, R.C., Carey, V.J., Bates, D.M., Bolstad, B., Dettling, M., Dudoit, S., Ellis, B., Gautier, L., Ge, Y., Gentry, J., *et al.* (2004). Bioconductor: open software development for computational biology and bioinformatics. Genome Biol. *5*, R80.

Ghildiyal, M., Seitz, H., Horwich, M.D., Li, C., Du, T., Lee, S., Xu, J., Kittler, E.L.W., Zapp, M.L., Weng, Z., *et al.* (2008). Endogenous siRNAs derived from transposons and mRNAs in Drosophila somatic cells. Science *320*, 1077–1081.

Goff, L.A., Davila, J., Swerdel, M.R., Moore, J.C., Cohen, R.I., Wu, H., Sun, Y.E., and Hart, R.P. (2009). Ago2 immunoprecipitation identifies predicted microRNAs in human embryonic stem cells and neural precursors. PLoS One *4*, e7192.

Grabherr, M.G., Haas, B.J., Yassour, M., Levin, J.Z., Thompson, D.A., Amit, I., Adiconis, X., Fan, L., Raychowdhury, R., Zeng, Q., *et al.* (2011). Full-length transcriptome assembly from RNA-Seq data without a reference genome. Nat. Biotechnol. *29*, 644–652.

Griffiths-Jones, S., Grocock, R.J., Van Dongen, S., Bateman, A., and Enright, A.J. (2006). miRBase microRNA sequences, targets and gene nomenclature. Nucleic Acids Res. *34*, D140–4.

Gupta, V., Markmann, K., Pedersen, C.N.S., Stougaard, J., and Andersen, S.U. (2012). shortran: A pipeline for small RNA-seq data analysis. Bioinformatics *28*, 2698–2700.

Hamilton, A.J., and Baulcombe, D.C. (1999). A species of small antisense RNA in posttranscriptional gene silencing in plants. Science *286*, 950–952.

Hardcastle, T.J., and Kelly, K.A. (2010). baySeq: empirical Bayesian methods for identifying differential expression in sequence count data. BMC Bioinformatics *11*, 422.

Hardcastle, T.J., Kelly, K.A., and Baulcombe, D.C. (2012). Identifying small interfering RNA loci from high-throughput sequencing data. Bioinformatics *28*, 457–463.

Havecker, E.R., Wallbridge, L.M., Hardcastle, T.J., Bush, M.S., Kelly, K.A., Dunn, R.M., Schwach, F., Doonan, J.H., and Baulcombe, D.C. (2010). The Arabidopsis RNA-directed DNA methylation argonautes functionally diverge based on their expression and interaction with target loci. Plant Cell *22*, 321–334.

Henderson, I.R., and Jacobsen, S.E. (2007). Epigenetic inheritance in plants. Nature *447*, 418–424.

Henderson, I.R., Zhang, X., Lu, C., Johnson, L., Meyers, B.C., Green, P.J., and Jacobsen, S.E. (2006). Dissecting Arabidopsis thaliana DICER function in small RNA processing, gene silencing and DNA methylation patterning. Nat. Genet. *38*, 721–725.

Ho, T.X., Rusholme, R., Dalmay, T., and Wang, H. (2008). Cloning of short interfering RNAs from virus-infected plants. Methods Mol. Biol. *451*, 229–241.

Hong, X., Hammell, M., Ambros, V., and Cohen, S.M. (2009). Immunopurification of Ago1 miRNPs selects for a distinct class of microRNA targets. Proc. Natl. Acad. Sci. U.S.A.*106*, 15085–15090.

Howell, M.D., Fahlgren, N., Chapman, E.J., Cumbie, J.S., Sullivan, C.M., Givan, S.A., Kasschau, K.D., and Carrington, J.C. (2007). Genome-wide analysis of the RNA-dependent RNA polymerase6/Dicer-like4 pathway in *Arabidopsis* reveals dependency on miRNA- and tasiRNA-directed targeting. Plant Cell *19*, 926–942.

Huang, P.-J., Liu, Y.-C., Lee, C.-C., Lin, W.-C., Gan, R.R.-C., Lyu, P.-C., and Tang, P. (2010). DSAP: deep-sequencing small RNA analysis pipeline. Nucleic Acids Res. *38*, W385–91.

Hutvagner, G., and Simard, M.J. (2008). Argonaute proteins: key players in RNA silencing. Nat. Rev. Mol. Cell Biol. *9*, 22–32.

Illumina (2009). Preparing Samples for Small RNA Sequencing Using the Alternative v1.5 Protocol.

Jackson, A.L., and Linsley, P.S. (2010). Recognizing and avoiding siRNA off-target effects for target identification and therapeutic application. Nat. Rev. Drug Discov. *9*, 57–67.

Johnson, C., Kasprzewska, A., Tennessen, K., Fernandes, J., Nan, G.-L., Walbot, V., Sundaresan, V., Vance, V., and Bowman, L.H. (2009). Clusters and superclusters of phased small RNAs in the developing inflorescence of rice. Genome Res. *19*, 1429–1440.

Jones, L., Ratcliff, F., and Baulcombe, D.C. (2001). RNA-directed transcriptional gene silencing in plants can be inherited independently of the RNA trigger and requires Met1 for maintenance. Curr. Biol. *11*, 747–757.

Jones-Rhoades, M.W., and Bartel, D.P. (2004). Computational identification of plant microRNAs and their targets, including a stress-induced miRNA. Mol. Cell *14*, 787–799.

Khvorova, A., Reynolds, A., and Jayasena, S.D. (2003). Functional siRNAs and miRNAs exhibit strand bias. Cell *115*, 209–216.

Kim, H., Bi, Y., Pal, S., Gupta, R., and Davuluri, R.V. (2011). IsoformEx: isoform level gene expression estimation using weighted non-negative least squares from mRNA-Seq data. BMC Bioinformatics *12*, 305.

Kim, J.K., Gabel, H.W., Kamath, R.S., Tewari, M., Pasquinelli, A., Rual, J.-F., Kennedy, S., Dybbs, M., Bertin, N., Kaplan, J.M., *et al.* (2005). Functional genomic analysis of RNA interference in C. elegans. Science *308*, 1164–1167.

Kim, Y.-K., Heo, I., and Kim, V.N. (2010). Modifications of small RNAs and their associated proteins. Cell *143*, 703–709.

Kvam, V.M., Liu, P., and Si, Y. (2012). A comparison of statistical methods for detecting differentially expressed genes from RNA-seq data. Am. J. Bot. *99*, 248–256.

Langmead, B., Trapnell, C., Pop, M., and Salzberg, S.L. (2009). Ultrafast and memory-efficient alignment of short DNA sequences to the human genome. Genome Biol. *10*, R25.

Lee, A., Hansen, K.D., Bullard, J., Dudoit, S., and Sherlock, G. (2008). Novel low abundance and transient RNAs in yeast revealed by tiling microarrays and ultra high-throughput sequencing are not conserved across closely related yeast species. PLoS Genet. *4*, e1000299.

Lee, Y.S., Nakahara, K., Pham, J.W., Kim, K., He, Z., Sontheimer, E.J., and Carthew, R.W. (2004). Distinct roles for Drosophila Dicer-1 and Dicer-2 in the siRNA/miRNA silencing pathways. Cell *117*, 69–81.

Li, B., and Dewey, C.N. (2011). RSEM: accurate transcript quantification from RNA-Seq data with or without a reference genome. BMC Bioinformatics *12*, 323.

Li, J., and Tibshirani, R. (2011). Finding consistent patterns: a nonparametric approach for identifying differential expression in RNA-Seq data. Stat. Methods Med. Res. (Epub ahead of print).

Li, L.-C., Okino, S.T., Zhao, H., Pookot, D., Place, R.F., Urakami, S., Enokida, H., and Dahiya, R. (2006). Small dsRNAs induce transcriptional activation in human cells. Proc. Natl. Acad. Sci. U.S.A.*103*, 17337–17342.

MacLean, D., Elina, N., Havecker, E.R., Heimstaedt, S.B., Studholme, D.J., and Baulcombe, D.C. (2010a). Evidence for large complex networks of plant short silencing RNAs. PLoS One *5*, e9901.

MacLean, D., Moulton, V., and Studholme, D.J. (2010b). Finding sRNA generative locales from high-throughput sequencing data with NiBLS. BMC Bioinformatics *11*, 93.

Manavella, P.A., Koenig, D., and Weigel, D. (2012). Plant secondary siRNA production determined by microRNA-duplex structure. Proc. Natl. Acad. Sci. U.S.A.*109*, 2461–2466.

Margis, R., Fusaro, A.F., Smith, N.A., Curtin, S.J., Watson, J.M., Finnegan, E.J., and Waterhouse, P.M. (2006). The evolution and diversification of Dicers in plants. FEBS Lett. *580*, 2442–2450.

Marioni, J.C., Mason, C.E., Mane, S.M., Stephens, M., and Gilad, Y. (2008). RNA-seq: an assessment of technical reproducibility and comparison with gene expression arrays. Genome Res. *18*, 1509–1517.

Martinez, J., and Tuschl, T. (2004). RISC is a 5′ phosphomonoester-producing RNA endonuclease. Genes Dev. *18*, 975–980.

Martinez, J., Patkaniowska, A., Urlaub, H., Lührmann, R., and Tuschl, T. (2002). Single-stranded antisense siRNAs guide target RNA cleavage in RNAi. Cell *110*, 563–574.

McCarthy, D.J., Chen, Y., and Smyth, G.K. (2012). Differential expression analysis of multifactor RNA-Seq experiments with respect to biological variation. Nucleic Acids Res. *40*, 4288–4297.

McElroy, K.E., Luciani, F., and Thomas, T. (2012). GemSIM: general, error-model based simulator of next-generation sequencing data. BMC Genomics *13*, 74.

Meister, G., and Tuschl, T. (2004). Mechanisms of gene silencing by double-stranded RNA. Nature *431*, 343–349.

Mi, S., Cai, T., Hu, Y., Chen, Y., Hodges, E., Ni, F., Wu, L., Li, S., Zhou, H., Long, C., *et al.* (2008). Sorting of small RNAs into Arabidopsis argonaute complexes is directed by the 5′ terminal nucleotide. Cell *133*, 116–127.

Minoche, A.E., Dohm, J.C., and Himmelbauer, H. (2011). Evaluation of genomic high-throughput sequencing data generated on Illumina HiSeq and genome analyzer systems. Genome Biol. *12*, R112.

Moazed, D. (2009). Small RNAs in transcriptional gene silencing and genome defence. Nature *457*, 413–420.

Molnár, A., Csorba, T., Lakatos, L., Várallyay, E., Lacomme, C., and Burgyán, J. (2005). Plant virus-derived small interfering RNAs originate predominantly from highly structured single-stranded viral RNAs. J. Virol. *79*, 7812–7818.

Molnar, A., Melnyk, C.W., Bassett, A., Hardcastle, T.J., Dunn, R., and Baulcombe, D.C. (2010). Small silencing RNAs in plants are mobile and direct epigenetic modification in recipient cells Science *328*, 872–875.

Mortazavi, A., Williams, B.A., McCue, K., Schaeffer, L., and Wold, B. (2008). Mapping and quantifying mammalian transcriptomes by RNA-Seq. Nat. Methods *5*, 621–628.

Mosher, R.A., and Melnyk, C.W. (2010). siRNAs and DNA methylation: seedy epigenetics. Trends Plant Sci. *15*, 204–210.

Moxon, S., Schwach, F., Dalmay, T., Maclean, D., Studholme, D.J., and Moulton, V. (2008). A toolkit for analysing large-scale plant small RNA datasets. Bioinformatics *24*, 2252–2253.

Nakamura, K., Oshima, T., Morimoto, T., Ikeda, S., Yoshikawa, H., Shiwa, Y., Ishikawa, S., Linak, M.C., Hirai, A., Takahashi, H., *et al.* (2011). Sequence-specific error profile of Illumina sequencers. Nucleic Acids Res. *39*, e90.

Okamura, K., Chung, W.-J., and Lai, E.C. (2008). The long and short of inverted repeat genes in animals: microRNAs, mirtrons and hairpin RNAs. Cell Cycle *7*, 2840–2845.

Pal-Bhadra, M., Leibovitch, B.A., Gandhi, S.G., Rao, M., Bhadra, U., Birchler, J.A., and Elgin, S.C.R. (2004). Heterochromatic silencing and HP1 localization in Drosophila are dependent on the RNAi machinery. Science *303*, 669–672.

Pang, M., Woodward, A.W., Agarwal, V., Guan, X., Ha, M., Ramachandran, V., Chen, X., Triplett, B.A., Stelly, D.M., and Chen, Z.J. (2009). Genome-wide analysis reveals rapid and dynamic changes in miRNA and siRNA sequence and expression during ovule and fiber development in allotetraploid cotton (Gossypium hirsutum L.). Genome Biol. *10*, R122.

Pearson, W.R., Wood, T., Zhang, Z., and Miller, W. (1997). Comparison of DNA sequences with protein sequences. Genomics *46*, 24–36.

Pontes, O., Li, C.F., Costa Nunes, P., Haag, J., Ream, T., Vitins, A., Jacobsen, S.E., and Pikaard, C.S. (2006). The Arabidopsis chromatin-modifying nuclear siRNA pathway involves a nucleolar RNA processing center. Cell *126*, 79–92.

Prüfer, K., Stenzel, U., Dannemann, M., Green, R.E., Lachmann, M., and Kelso, J. (2008). PatMaN: rapid alignment of short sequences to large databases. Bioinformatics *24*, 1530–1531.

Qi, X., Bao, F.S., and Xie, Z. (2009). Small RNA deep sequencing reveals role for Arabidopsis thaliana RNA-dependent RNA polymerases in viral siRNA biogenesis. PLoS One *4*, e4971.

Quail, M.A., Smith, M., Coupland, P., Otto, T.D., Harris, S.R., Connor, T.R., Bertoni, A., Swerdlow, H.P., and Gu, Y. (2012). A tale of three next generation sequencing platforms: comparison of Ion Torrent, Pacific Biosciences and Illumina MiSeq sequencers. BMC Genomics *13*, 341.

R Development Core Team (2012). R: A Language and Environment for Statistical Computing.

Rajewsky, N. (2006). microRNA target predictions in animals. Nat. Genet. *38*, S8–13.

Reinhart, B.J., and Bartel, D.P. (2002). Small RNAs correspond to centromere heterochromatic repeats. Science *297*, 1831.

Robinson, M.D., and McCarthy D.J. (2010). edgeR: a Bioconductor package for differential expression analysis of digital gene expression data. Bioinformatics *26*, 139–140.

Robinson, M.D., and Oshlack, A. (2010). A scaling normalization method for differential expression analysis of RNA-seq data. Genome Biol. *11*, R25.

Rothberg, J.M., Hinz, W., Rearick, T.M., Schultz, J., Mileski, W., Davey, M., Leamon, J.H., Johnson, K., Milgrew, M.J., Edwards, M., *et al.* (2011). An integrated semiconductor device enabling non-optical genome sequencing. Nature *475*, 348–352.

Schmieder, R., Lim, Y.W., Rohwer, F., and Edwards, R. (2010). TagCleaner: Identification and removal of tag sequences from genomic and metagenomic datasets. BMC Bioinformatics *11*, 341.

Schmitz, R.J., Schultz, M.D., Lewsey, M.G., O'Malley, R.C., Urich, M.A., Libiger, O., Schork, N.J., and Ecker, J.R. (2011). Transgenerational epigenetic instability is a source of novel methylation variants. Science *334*, 369–373.

Schopman, N.C.T., Willemsen, M., Liu, Y.P., Bradley, T., Van Kampen, A., Baas, F., Berkhout, B., and Haasnoot, J. (2012). Deep sequencing of virus-infected cells reveals HIV-encoded small RNAs. Nucleic Acids Res. *40*, 414–427.

Schwarz, D.S., Hutvágner, G., Du, T., Xu, Z., Aronin, N., and Zamore, P.D. (2003). Asymmetry in the assembly of the RNAi enzyme complex. Cell *115*, 199–208.

Di Serio, F., Gisel, A., Navarro, B., Delgado, S., Martínez de Alba, A.-E., Donvito, G., and Flores, R. (2009). Deep sequencing of the small RNAs derived from two symptomatic variants of a chloroplastic viroid: implications for their genesis and for pathogenesis. PLoS One *4*, e7539.

Smith, A.M., Heisler, L.E., St Onge, R.P., Farias-Hesson, E., Wallace, I.M., Bodeau, J., Harris, A.N., Perry, K.M., Giaever, G., Pourmand, N., *et al.* (2010). Highly-multiplexed barcode sequencing: an efficient method for parallel analysis of pooled samples. Nucleic Acids Res. *38*, e142.

Stein, L.D., Mungall, C., Shu, S., Caudy, M., Mangone, M., Day, A., Nickerson, E., Stajich, J.E., Harris, T.W., Arva, A., *et al.* (2002). The generic genome browser: a building block for a model organism system database. Genome Res. *12*, 1599–1610.

Sunkar, R., Girke, T., and Zhu, J.-K. (2005). Identification and characterization of endogenous small interfering RNAs from rice. Nucleic Acids Res. *33*, 4443–4454.

Takeda, A., Iwasaki, S., Watanabe, T., Utsumi, M., and Watanabe, Y. (2008). The mechanism selecting the guide strand from small RNA duplexes is different among argonaute proteins. Plant Cell Physiol *49*, 493–500.

Valouev, A., Ichikawa, J., Tonthat, T., Stuart, J., Ranade, S., Peckham, H., Zeng, K., Malek, J.A., Costa, G., McKernan, K., *et al.* (2008). A high-resolution, nucleosome position map of C. elegans reveals a lack of universal sequence-dictated positioning. Genome Res. *18*, 1051–1063.

Vastenhouw, N.L., Brunschwig, K., Okihara, K.L., Müller, F., Tijsterman, M., and Plasterk, R.H.A. (2006). Gene expression: long-term gene silencing by RNAi. Nature *442*, 882.

Vogel, J.P., Garvin, D.F., Mockler, T.C., Schmutz, J., Rokhsar, D., Bevan, M.W., Barry, K., Lucas, S., Harmon-Smith, M., Lail, K., *et al.* (2010). Genome sequencing and analysis of the model grass Brachypodium distachyon. Nature *463*, 763–768.

Voinnet, O. (2008). Use, tolerance and avoidance of amplified RNA silencing by plants. Trends Plant Sci. *13*, 317–328.

Voinnet, O., and Baulcombe, D.C. (1997). Systemic signalling in gene silencing. Nature *389*, 553.

Volpe, T.A., Kidner, C., Hall, I.M., Teng, G., Grewal, S.I.S., and Martienssen, R.A. (2002). Regulation of heterochromatic silencing and histone H3 lysine-9 methylation by RNAi. Science *297*, 1833–1837.

Yang, X., and Li, L. (2011). miRDeep-P: a computational tool for analyzing the microRNA transcriptome in plants. Bioinformatics *27*, 2614–2615.

Zerbino, D.R., and Birney, E. (2008). Velvet: algorithms for *de novo* short read assembly using de Bruijn graphs. Genome Res. *18*, 821–829.

Zhang, C., Li, G., Wang, J., and Fang, J. (2012a). Identification of trans-acting siRNAs and their regulatory cascades in grapevine. Bioinformatics *28*, 2561–2568.

Zhang, W., Zhou, X., Xia, J., and Zhou, X. (2012b). Identification of microRNAs and natural antisense transcript-originated endogenous siRNAs from small-RNA deep sequencing data. Methods Mol. Biol. *883*, 221–227.

Zhang, Z.D., Rozowsky, J., Snyder, M., Chang, J., and Gerstein, M. (2008). Modeling ChIP sequencing *in silico* with applications. PLoS Comput. Biol. *4*, e1000158.

Zhao, T., Li, G., Mi, S., Li, S., Hannon, G.J., Wang, X.-J., and Qi, Y. (2007). A complex system of small RNAs in the unicellular green alga Chlamydomonas reinhardtii. Genes Dev. *21*, 1190–1203.

Zhuang, F., Fuchs, R.T., and Robb, G.B. (2012). Small RNA expression profiling by high-throughput sequencing: implications of enzymatic manipulation. J. Nucleic Acids *2012*, 360358.

Zilberman, D., Cao, X., and Jacobsen, S.E. (2003). ARGONAUTE4 control of locus-specific siRNA accumulation and DNA and histone methylation. Science *299*, 716–719.

Motif Discovery and Motif Finding in ChIP-Seq Data

5

Ivan V. Kulakovskiy and Vsevolod J. Makeev

Abstract

Modern bioinformatics and molecular biology research are impossible to imagine without application of high-throughput DNA sequencing technologies, also called next-generation sequencing. In particular, transcriptional regulation studies determining how different genes become 'on' and 'off' in different tissues in different conditions rely heavily on next-generation sequencing. The chromatin immunoprecipitation followed by deep sequencing (ChIP-Seq) technology allows genome-wide *in vivo* studies of binding sites for different transcription factors, the proteins that can specifically facilitate or prevent proper construction of the transcription initiatory complex necessary to activate transcription of a specific gene. Transcription initiation control in higher eukaryotes is extremely complex, and its analysis is especially difficult because of the genome size and comparably short transcription factor binding sites. Availability of ChIP-Seq data provided new insights into genome-wide distribution of binding events. It was a new challenge for computational biology to handle enormous amounts of data and detect actual binding sites within DNA segments identified by ChIP-Seq. Here we focus on application and advances of DNA motif discovery and motif finding, a very well-established field in bioinformatics sequence analysis, which has been given a second birth by the ChIP-Seq technology.

Introduction

The key questions in transcriptional regulation

Our understanding of transcriptional regulatory machinery is very far from complete. One of the basic yet actively discussed topics is the tissue- and time-specific regulation of gene expression. Several regulatory mechanisms have been discovered (Lelli *et al.*, 2012), and one of them is represented by complexes of interacting proteins, the so-called transcription factors (TFs), which facilitate and control construction of transcription initiation machinery. In higher eukaryotes different complexes of transcription factors bound at different genomic locations function as activators or inhibitors of different groups of target genes in different tissues and different conditions. For a particular transcription factor (TF), several typical questions arise. What DNA binding sequence specificity is exhibited by a particular TF, i.e. is there any common sequence pattern shared by TF binding sites (TFBSs)? Which genomic segments can be considered to be binding sites? Which genes are targeted? And, finally, how are binding sites of a particular TF involved in formation of the putative regulatory code? That is, whether TFBSs of a particular TF tend to specifically co-localize with active transcription start sites, histone modifications, and binding sites of other TFs. While the chromatin immunoprecipitation followed by sequencing (ChIP-Seq) technology is applicable to study several aspects of transcriptional regulation, including mapping of transcription start sites of RNA polymerase II or various histone modifications (Wei *et al.*, 2012), we shall focus on transcription factor binding sites, the subject of numerous ChIP-Seq studies in recent years, that provide an extremely interesting challenge for downstream bioinformatics analysis, such as sequence motif discovery. We are going to discuss data on higher eukaryotes, which have long genome sequences, broad sets of transcriptional regulators and, surprisingly, comparably short transcription factor binding sites (10–15 or sometimes nearly 20 base pairs, bp). While motif discovery and motif finding by themselves constitute a wide bioinformatics field which rapidly develops especially for TFBS studies, here we try to give an overview of TFBS motif analysis advances in the context of ChIP-Seq data.

Sequence motifs: a common terminological trap

Motif discovery is a common expression in bioinformatics, especially in sequence analysis. Yet, the definition of 'TFBS motif' is not fully established. Some authors (Wang *et al.*, 2005) use the term 'motif' for the DNA sequence of a particular TFBS and 'motif family' for a sequence pattern model like the positional weight matrix (PWM) (Monteiro *et al.*, 2008). At the same time, previous well-known publications (Sinha and Tompa, 2002) used the term 'motif' as the TFBS recognition rule or TFBS model including the PWM and the degenerate consensus string. Sometimes, even within the same publication (D'haeseleer, 2006; Xie *et al.*, 2009) 'the motif' simultaneously denotes the nucleotide sequence of a particular TFBS and the TFBS pattern or model without any specific comments. No wonder that in the work by (Sandelin and Wasserman, 2004) 'binding profiles' define DNA patterns recognized by transcription factors and the term 'motif' is avoided. All in all, careless usage of the term 'motif' can be misleading.

On the other hand, 'motif discovery' and 'motif finding' (combined and referred to as 'motif analysis' below) retain their meaning for both definitions of 'the motif', either being a pattern or pattern occurrences. Thus, here 'motif discovery' will denote the discovery of common DNA patterns in DNA texts. The 'motif finding', in turn, will be used to denote search for occurrences of selected DNA patterns. The detected pattern occurrences themselves will be named 'motifs hits' or 'motif occurrences'.

Transcription factor binding sites as DNA sequence patterns

Pregenomic studies of TFBSs yielded only a limited amount of data on sequence preferences of specific protein-DNA binding. Simple models like consensus sequences (Day and McMorris, 1992) or regular expressions (Myers and Miller, 1989) were applied for motif finding and motif discovery. Surprisingly, positional weight matrix, a more general model, was presented in the earlier fundamental work (Berg and Von Hippel, 1987, 1988). Basically, a gapless multiple local alignment (GMLA) of TFBS sequences can be converted into a positional weight matrix (PWM), with nucleotides at rows and binding site positions at columns (with the length of the alignment determined by the number of columns, see Fig. 5.1). The values in the matrix show the preferences for corresponding nucleotides in each particular alignment position. Depending on the particular strategy of PWM construction, the matrix may contain probabilities of nucleotides or weights, as log-odds transformations of nucleotide frequencies (Stormo, 2000).

A DNA segment of a fixed length, the DNA word, can be then used to select a single real value corresponding to each particular nucleotide at each column. The sum of the matrix values selected for the DNA word is called 'the word score' (or 'the binding site score'). This score represents the TFBS quality as recognized by the PWM and was shown to be related to TF binding affinity (Berg and Von Hippel, 1987). The length of the alignment measured as the number of columns in the positional weight matrix is called 'the motif length' or 'the motif width'. Again, this is terminologically safe since both the pattern and its occurrences share the same length.

Many models having a larger number of parameters than PWM have been proposed (Oshchepkov *et al.*, 2004; Levitsky *et al.*, 2007) with their prediction power evaluated. Yet, positional weight matrices continue to be the most widely used model for short TFBS sequence patterns even in the era of ChIP-Seq data.

TFBS motif discovery in pregenomic era

Different experimental technologies have been designed during the long pregenomic age of experimental TFBS identification. Studies *in vivo* were limited, and only fragmentary data on TF binding specificity was assessed by low-throughput *in vitro* techniques (Galas and Schmitz, 1978; Tuerk and Gold, 1990). Such techniques were characterized by comparably low numbers and lengths of detected binding regions, resulting in 10 to 100 sequences from 10 to 20 bp.

Many computational tools were developed for motif discovery, i.e. for identification of a common sequence pattern in these small sets of comparably short sequences. At the same time, with the lack of experimental data, small sets of proximal promoter regions of genes were actively used as input for motif discovery tools, following a basic idea that putatively co-regulated genes can exhibit a common structured sequence signal (Favorov *et al.*, 2005). One of the widely cited benchmark studies concerning motif discovery in pregenomic era (Tompa *et al.*, 2005) displayed discrepant results for different computational tools tested with different datasets. But the real challenge and no less success for a computational biology of transcriptional regulation came forward with rising of high-throughput methods.

Pre-ChIP-Seq high-throughput experimental techniques for TFBS identification

One of the early attempts to look at a genome-wide landscape of TF binding was published by (Impey *et al.*, 2004) who combined chromatin immunoprecipitation (ChIP) with a modification of serial analysis of gene expression. This publication was issued just two years after the ChIP-chip technique (chromatin immunoprecipitation followed by hybridization on a DNA microarray (Horak and Snyder, 2002)) was presented as an experimental technique for high-throughput analysis of TFBSs. While there were alternative experimental methods developed in parallel, such as DamID (Van Steensel *et al.*, 2001), the ChIP-chip on tiling arrays formed a general trend in TFBS analysis being able to capture a genome-wide distribution of binding sites *in vivo*, comparing to *in vitro* high-throughput methods, such as DNA immunoprecipitation-chip, DIP-chip (Liu *et al.*, 2005), cognate site identifier, CSI (Warren *et al.*, 2006) or protein-binding microarray, PBM (Bulyk, 2006).

The wet-lab part of the ChIP-chip workflow is much the same as that of the ChIP-Seq and other similar technologies. Proteins are crosslinked to DNA *in vivo* using, for example, a chemical agent such as formaldehyde. The DNA is fragmented by ultrasound sonication or with a non-specific DNase, and DNA–protein complexes are extracted using antibodies against the protein-of-interest (the immunoprecipitation procedure). The crosslinks are then removed, and the final DNA pool is purified. DNA fragments are amplified, fluorescently labelled and subjected, in the case of ChIP-chip, to microarray hybridization. The effectiveness of a ChIP-chip experiment heavily depends on the microarray design and downstream dry-lab bioinformatics analysis. Moreover, while DNA probes on expression microarrays need to cover only a small part of the genome, tiling arrays for ChIP-chip require probes covering as greater part of the whole genome as possible, better with a density allowing the probes to overlap. The hybridization technology itself has many specific disadvantages; more details are given, for example, by (Buck and Lieb, 2004). When it comes to TFBS detection, ChIP-chip allows detecting fairly extended DNA regions where it is difficult to precisely identify TFBSs without any additional evidence or prior knowledge. Several studies were focused on ChIP-chip motif analysis (Linhart *et al.*, 2008) including the data integration strategies, which proved successful provided the pregenomic data on TFBSs was taken into account (Kulakovskiy and Makeev, 2010). All in all, ChIP-chip was a reliable technology, which made a basis for ChIP-Seq development by setting well-established experimental protocols for chromatin immunoprecipitation. The TFBS motif discovery in ChIP-chip data revealed many advantages and difficulties which were a forerunner of vital problems and solutions in ChIP-Seq data analysis.

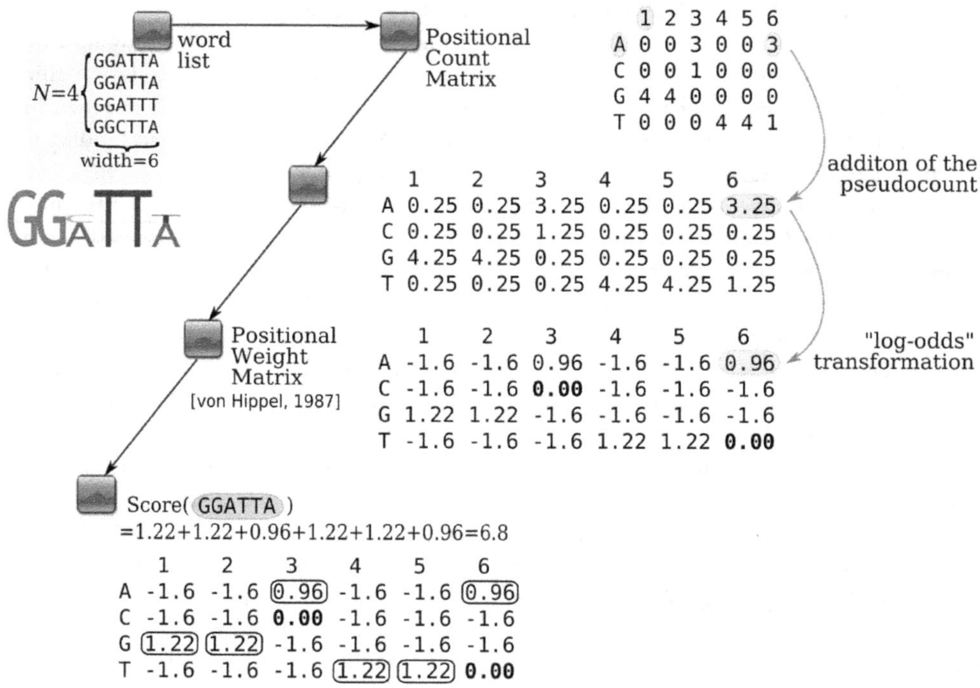

Figure 5.1 PWM as the basic TFBS model constructed from a gapless multiple local alignment. The motif logo representation (shown below the alignment) is commonly used to display importance of columns (overall column height) and particular nucleotides (letter height), see (Crooks *et al.*, 2004). The pseudo count value is added to model nucleotides of possible functional TFBS that are missing in the alignment. The log-odds transformation (Stormo, 2000; Lifanov *et al.*, 2003) is applied to make scores of individual columns additive.

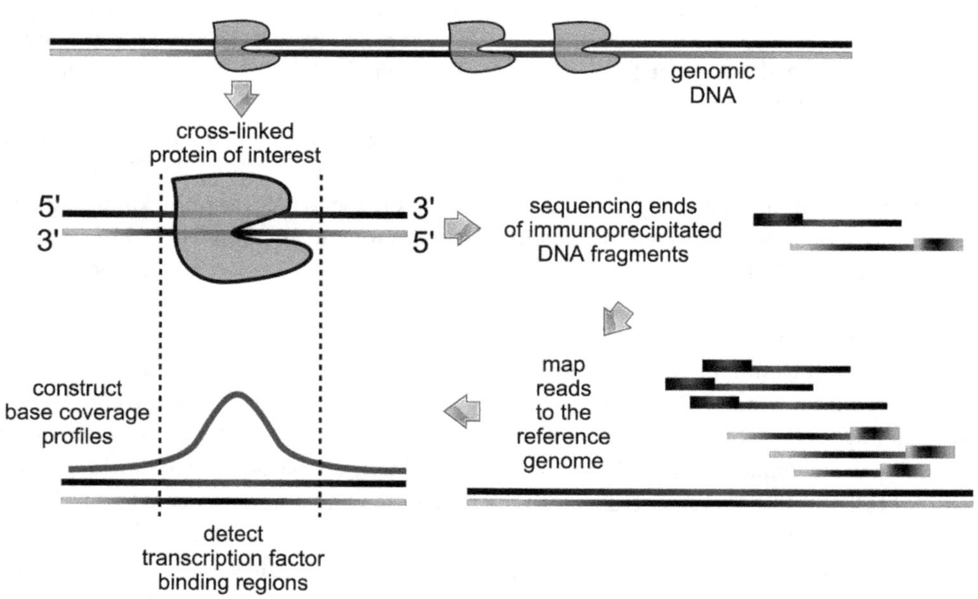

Figure 5.2 Schematic representation of the ChIP-Seq procedure.

The rise of ChIP-Seq technology

The increasing usage of next-generation sequencing brought about sequencing of the resulting DNA libraries instead of the hybridization step involved in ChIP-chip workflow. This seemingly incremental difference nearly revolutionized the field. In ChIP-Seq (Robertson *et al.*, 2007) with its sister technologies like ChIP-PET, paired-end diTag (Ng *et al.*, 2006; Wei *et al.*, 2006), next generation sequencing technologies are used to obtain sequences of terminal segments of protein-bound DNA fragments. Several highly accurate studies on genome-wide TF binding profiles were published in succession (Johnson *et al.*, 2007; Robertson *et al.*, 2007; Valouev *et al.*, 2008) revealing the genome-wide landscape of TF binding at a precise resolution and thus allowing to accurately detect target genes.

Let us have a more detailed look at the wet-lab part of a ChIP-Seq experiment. The ChIP-Seq wet-lab provides an immense amount of data delivering millions of short sequence reads, also called tags. In contrast to ChIP-chip, ChIP-Seq does not need a pre-developed tiling array, but yields a vast number of short reads that need to be unambiguously mapped to some existing genome assembly. Thus, availability and quality of the genome assembly plays a critical role in all further analyses. Moreover, different sequencing techniques have different types of characteristic sequencing errors that influence mapping quality. Lastly, the ultimate result depends on peculiarities of the bioinformatics pipeline, like read mapping and peak calling (see below), which further complicate the picture. Read mapping strategies were reviewed by (Leleu *et al.*, 2010), partially covering other topics as well, including ChIP-Seq peak finding (peak calling) discussed in the next section. Fig 5.2 shows a basic schematic representation of the ChIP-Seq workflow.

ChIP-Seq peak calling

In practice, resolution of ChIP-Seq used for identification of actual TFBS locations heavily depends on the experimental protocol, particularly on distribution of DNA fragment lengths, the sequencing depth, and the quality of antibodies. In addition, the arrangements of binding sites within DNA segments also have a striking impact. A common type of TFBS arrangement is a homotypic cluster of binding sites, a tight group of possibly overlapping TFBSs in a given DNA segment (Lifanov *et al.*, 2003; Gotea *et al.*, 2010). In this case immunoprecipitation would give a set of DNA fragments produced from different parts of the TFBS-containing DNA segment. The resultant reads are obtained from terminal parts of immunoprecipitated DNA fragments. Being mapped to the genome the reads would form a high yet wide pileup. Without additional processing it is impossible to directly segregate such pileup into groups of reads attributed to certain TFBSs. Indeed, for a homotypic cluster of binding sites the characteristic distance from one TFBS to another is around 100 bp. It is much the same (or shorter) as the DNA fragment length (~300 bp). If the single-end sequencing protocol is used, the signal becomes even more blurred due to possible uncertainty as to what direction the mapped reads should be extended to when restoring the proper DNA fragment pileup. Paired end reads help to improve the resolution to some extent, but, nevertheless, signals from the neighbouring TFBS interfere.

Having the reads mapped to the reference genome, the next step of dry-lab processing is to find the so-called peaks, the genomic regions enriched with DNA fragments resulted from immunoprecipitation. A genome region where reads of these fragments are mapped in abundance is expected to contain numerous binding sites or binding sites with higher affinity to the protein under study. Many approaches and even careful comparative studies (Wilbanks and Facciotti, 2010; Rye *et al.*, 2011) have been recently published, showing that for downstream analysis it is highly important what kind of peaks the peak finder tends to produce in practice. Peak finders are based on different statistical models and different assumptions; some of them tend to produce very short peak regions (Jothi *et al.*, 2008). In some cases, the produced peaks are directly based on 'peak shape' profiles, i.e. DNA base coverage profiles reconstructed from read pileups (Fejes *et al.*, 2008). The shape and the height of the base coverage profile are crucial for a downstream analysis. Some general suggestions for peak finders, adequate for further motif analysis, are presented in the separate 'cookbook'-like section below. In brief, for motif discovery any peak finder can be used in the case of high-quality ChIP-Seq data; in the case of experimental errors or simply the low number of mapped reads (either due to errors or differences between the studied and reference genomes) the simplest 'count-these-tags' algorithm would be the most robust. A complex example is shown in Fig. 5.3. Here three ChIP-Seq datasets for REST (Johnson *et al.*, 2007), GABPA (Valouev *et al.*, 2008) transcription factors and EWS-FLI1 Ewing's sarcoma fusion protein (Guillon *et al.*, 2009) were studied with FindPeaks (Fejes *et al.*, 2008) used to identify the peaks. While the top peaks for classic NRSF and GABP datasets were bell-shaped with a pronounced maximum, those obtained for EWS-FLI1 were fuzzier. In this particular case the reason is not obvious. Yet one should keep in mind that ChIP-Seq peaks do not always exhibit a clear single summit or even a set of well-defined summits. Thus, it is not always obvious how to truncate correctly the peak around its summit (where the actual binding sites are expected to be found).

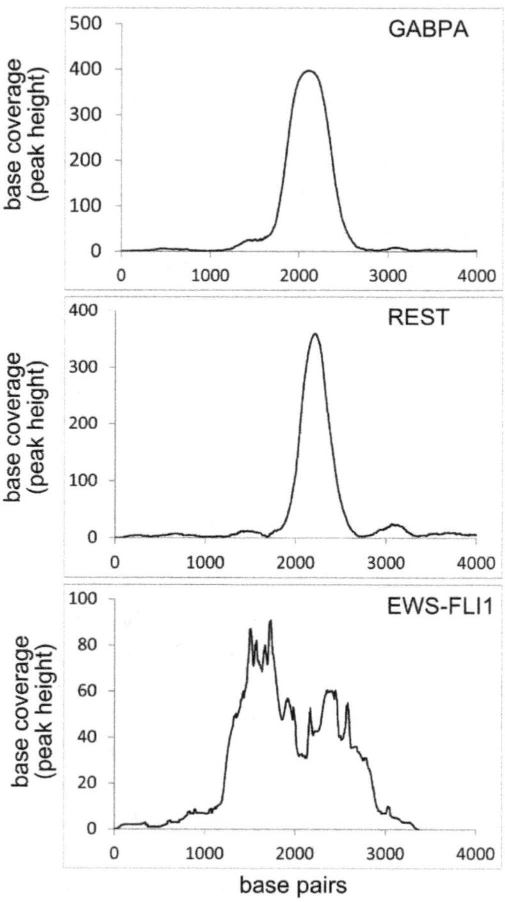

Figure 5.3 Examples of top ChIP-Seq peak shapes for three independent ChIP-Seq datasets. GABPA and REST datasets contain well-shaped peaks while EWS-FLI1 peak has oscillating shape. Note that the peak from EWS-FLI1 dataset has much lower coverage (coverage is shown at Y axis, see axis scales) and is much broader (X axis, base pairs).

Another related problem is the proper usage of ChIP-Seq control data like control DNA extracted without immunoprecipitation (Johnson *et al.*, 2007) or precipitated with non-specific antibodies (Chen *et al.*, 2008). More details on peak finding software and ChIP-Seq data processing strategies can be found in the following reviews (Wilbanks and Facciotti, 2010; Rye *et al.*, 2011).

When the peak segments are long and base coverage profiles have noisy or oscillating shapes, it becomes difficult to locate the actual binding sites of a TF under study. Likewise, it becomes difficult to properly find TF target genes or assign a peak to the appropriate category (promoter, enhancer or intergenic). In this case motif discovery and motif finding algorithms are applied for precise detection of binding sites within ChIP-Seq peaks by means of sequence analysis.

ChIP-Seq data: advantages and challenges for sequence analysis bioinformatics

In the framework of sequence analysis the ChIP-Seq analysis problem is defined as follows: given a number of peaks possibly supplied with information on the shape of the corresponding base coverage profiles, it is necessary to locate actual TFBS as precise as possible. The number of peaks is usually large and amounts to thousands or tens of thousands. The lengths of the peaks vary widely from tens to thousands of base pairs depending on the whole wet- and dry-lab processing protocol and the TF under study. Extremely short peaks made of piles of identical reads usually are PCR or read mapping artefacts. Extremely long peaks, in turn, may correspond to segments enriched with TFBSs, which makes it impossible to directly identify each single binding site within. However, PCR artefacts can be removed during either read mapping or peak finding; read mapping can be checked for tandem repeats and other ambiguous mapping segments; the longer peaks often can be split into several shorter peaks to facilitate further analysis.

Still thousands and tens of thousands of sequences are ready for motif discovery. This overcomes the problem of paucity of data for TFBS identification that existed during the pre-genome era. Moreover, ChIP-Seq produces much clearer data as compared to ChIP-chip, with comparably shorter TF binding regions of about tens to hundreds or rarely thousands of base pairs and, which is more important, with base coverage profiles pointing out the expected location of actual binding sites. One remaining difficulty is that motif discovery tools developed in the pre-genomic era usually cannot process such large numbers of data consisting of DNA segments with significantly varying lengths, whereas tools designed for ChIP-chip data do not take into consideration information on read pileup profiles.

Basic motif analysis for ChIP-Seq data

Theoretically, all available peaks can be supplied to a motif discovery pipeline. Still, the optimal overall size of the input sequence set is not obvious. While higher peaks (constructed from larger sets of reads and thus corresponding to larger sets of initial immunoprecipitation-induced DNA fragments) correspond to probably stronger binding sites or even homotypic clusters of binding sites, lower ChIP-Seq peaks still can provide important information on protein-DNA recognition or TF target genes. At the beginning of the ChIP-Seq era several simple heuristics were invented to reduce the data volume to values suitable for application of pre-genomic motif discovery tools

At first glance a subset of the highest peaks is enough to detect the TF binding pattern. Next, only a short segment centred at the peak summit can be subjected to analysis. These heuristics allowed applying many well-established motif discovery tools, such as MEME (multiple expectation maximization for motif elicitation; Bailey, 2002) or Bioprospector (Liu *et al.*, 2001), to analyse ChIP-Seq data (Jothi *et al.*, 2008; Valouev *et al.*, 2008; Guillon *et al.*, 2009; Tuteja *et al.*, 2009).

However, these heuristics have several implications: top peaks may contain only consensus binding sites with the highest affinity. In this case weak binding sites would not be included in the training set, and hence, weak positions of the TFBS alignment would not be carefully estimated. Also, as it was discussed earlier, it is not always trivial to define the peak summits correctly. Moreover, apart from the major sequence pattern of TF binding, a more detailed motif analysis of ChIP-Seq data can provide insight into more interesting issues.

Possible outcomes of motif analysis

ChIP-Seq peak data supplied with results of motif discovery approaches can yield interesting observations on TFBS positional preferences, starting from the simple question as to how strongly TFBSs tend to be located near the peak summit. Next, since ChIP-Seq provides genome-wide *in vivo* TFBS data, one can obtain essential information on specific TFBS subtypes or even completely different sequence patterns recognized by the TF under study. These TFBS subtypes can possibly be present in peaks obtained in different experiments and in different tissues or stages of cell development or even in different peak subsets of a single experiment. Next, by applying motif discovery to peak regions other than the segments covered by peak summits, one can identify additional overrepresented sequence patterns which might be characteristic to TFBSs of regulatory proteins acting as putative cofactors to the TF under study. An ultimate challenge of motif discovery applications is the construction of a TFBS model accurate enough for detection of actual location of binding sites within ChIP-Seq peaks or for reliable prediction of binding sites in other genomic sequences. With such a large amount of data, one can perform careful estimation of model parameters which, in turn, should increase both model sensitivity and specificity. Good performance of TFBS models derived from ChIP-Seq data was shown in several comparative studies, including the benchmark of different

ChIP-Seq motif discovery tools done by Ma *et al.* (2012), and comparison by Bi *et al.* (2011) that showed, in particular, how ChIP-Seq-based TFBS models perform versus known TFBS models from TRANSFAC (Matys *et al.*, 2006); and a recent comparison with PBM-derived models (Weirauch *et al.*, 2013). HOCOMOCO database (http://autosome.ru/HOCOMOCO) provides an annotated collection of TFBS models derived by integrating low-throughput data from various sources and ChIP-Seq data from ENCODE (Myers *et al.*, 2011). A comprehensive benchmark comparing HOCOMOCO TFBS models with those in TRANSFAC and JASPAR was also based on ChIP-Seq data (Kulakovskiy *et al.*, 2013b).

The importance of the base coverage profiles

Let's assume a simple task to search for the main TF binding sequence pattern based on ChIP-Seq data for this TF. Can we simply take a small set of full-length peaks and supply them to a motif discovery tool? We have checked four motif discovery tools (Fig. 5.4), two pregenomic, MEME (Bailey, 2002) and SeSiMCMC (sequence similarities by Markov chain Monte Carlo; Favorov *et al.*, 2005), and two ChIP-Seq oriented tools, ChIPMunk (Kulakovskiy *et al.*, 2010) and HMS (hybrid motif sampler; Hu *et al.*, 2010), that are capable of using base coverage profiles. ChIP-Seq datasets for REST (Johnson *et al.*, 2007), GABPA (Valouev *et al.*, 2008) and EWS-FLI1 (Guillon *et al.*, 2009) TFs were used with FindPeaks (Fejes *et al.*, 2008) as a peak finder. The striking result was that even for such a strong long pattern as REST it was impossible for all motif discovery programs to succeed on full-length peaks (as compared with the sequences cut to 10% around peak summits). The tools accounting for the base coverage profile were more successful and capable of handling the 'well-shaped' peaks; and even full-length noisy peaks of EWS-FLI1 were processed more or less successfully.

Performance of the algorithms depends on implementation details and the particular test dataset. For instance, HMS and ChIPMunk, that both use peak shapes to assign positional priors for motif discovery, still show different performance for EWS-FLI1 data. Yet, even if the peak shape is taken into account, it appears useful to reduce rationally peak lengths to consider only the regions around peak summits corresponding to the most probable binding events. Some peak calling software (Jothi *et al.*, 2008) produces

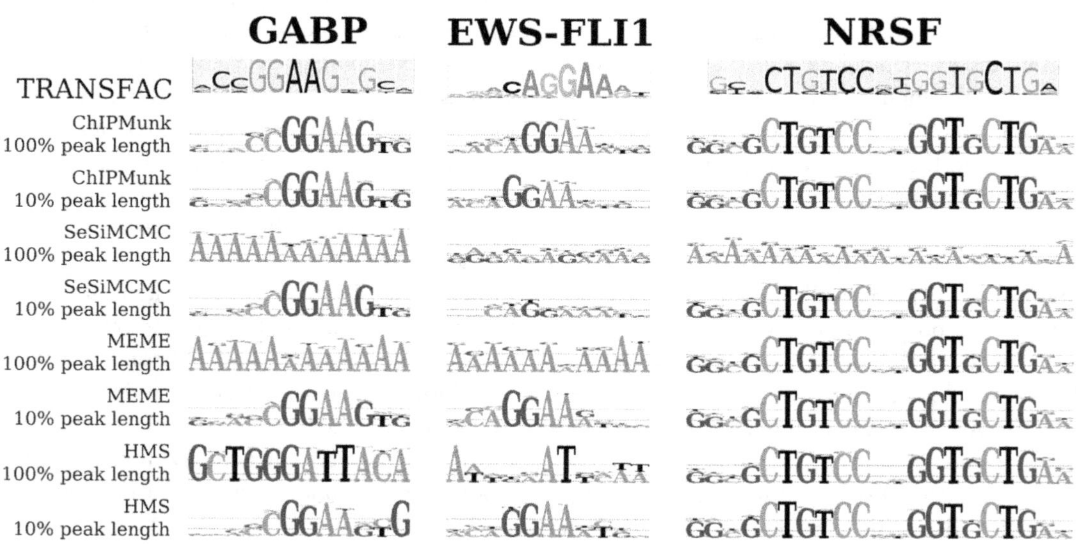

Figure 5.4 Results of motif discovery in ChIP-Seq peaks truncated around peaks summits. Known TFBS models from TRANSFAC (Matys *et al.*, 2006) are shown for REST and GABPA TFs. EWS-FLI1 TFBS model is shown according to Guillon *et al.* (2009). Note: peaks with GGAA repeats were pre-filtered for the EWS-FLI1 dataset, see the section on motif subtypes below for details.

shorter peaks, and at least theoretically, this might be more suitable for motif discovery pipelines involving conventional motif discovery tools. At the same time, for extremely short peaks there is a chance to miss or partially truncate true binding sites.

Yet, not all peak finders directly produce base coverage profiles; and not all motif discovery tools take this information into account. Thus, an additional post-processing step is often used to verify if the motif occurrences really tend to be located under the peak summits. This is done, for example, by CentriMo (Bailey and Machanick, 2012) or in the complex peak-motifs software workflow (Thomas-Chollier *et al.*, 2012).

Motif subtypes in ChIP-Seq data

The discussion whether a single TF can recognize TFBSs defined by several distinct sequence patterns remains open for quite a while (Hannenhalli and Wang, 2005). Recently it was revitalized by high-throughput data, such as ChIP-Seq (Chan *et al.*, 2012). Modern high-throughput *in vitro* methods like PBM demonstrated that for many TFs some TF-bound DNA fragments contain occurrences of secondary patterns (Badis *et al.*, 2009), which may be quite dissimilar to those contained by the majority of fragments assayed. ChIP-Seq data are appropriate for studying these 'motif subtypes' or even for constructing a complex TFBS model accounting for their concurrent recognition (Bi *et al.*, 2011). In some cases, a careful analysis of ChIP-Seq data allowed identifying completely different TFBS sequence patterns, all recognized by the same TF. An example is the EWS-FLI1 oncogenic fusion protein, for which JASPAR, a popular database of TFBS models (Portales-Casamar *et al.*, 2010) reported only a single GGAA tandem repeat pattern. Using the data published by Guillon *et al.* (2009) and applying the ChIP-Munk motif discovery tool, we can rediscover two quite different EWS-FLI1 binding patterns (Fig. 5.5). One of them is the -GGAA- tandem repeat, and the other is a common ETS-family motif. In this case the ChIPHorde extension of ChIPmunk was applied in its filtering mode (that sequentially excludes whole sequences with motif occurrences discovered during previous iterations from all subsequent motif discovery iterations).

Figure 5.5 Sequence patterns presented in different peak subsets of EWS-FLI1 ChIP-Seq data. Top panel: known -GGAA- tandem repeat pattern (strong occurrences found in approximately one-third of the peaks). Bottom panel: ETS-family like pattern (found approximately in 2/3 of the peaks).

On the way to understanding TFBS-based regulatory code

Suppose one knows what motif occurrences correspond to TFBSs for the TF under study. Naturally arises the next question as to what are the other patterns found in the ChIP-Seq peaks? Can one discover any other overrepresented signals? Most of the motif discovery tools can solve this task. But having a complete set of sequence patterns overrepresented in the peaks is not enough. To make biological conclusions one needs additional information, specifically, whether the detected patterns correspond to any known TFBS models and if there are any positional preferences concerning the observed patterns. This information would allow detecting putative cofactor proteins possibly involved in transcriptional regulation via their interaction with a ChIP-targeted protein. Searching for multiple patterns is a typical task of motif discovery software. Yet, only particular tools provide an integrated analysis of the newly found patterns by comparing them with known TFBS models stored in databases (Mercier *et al.*, 2011; Thomas-Chollier *et al.*, 2012) or by studying positional preferences of motif occurrences (Mercier *et al.*, 2011; Whitington *et al.*, 2011; Guo *et al.*, 2012). One can make interesting observations even by inspecting preferred distances between motif occurrences solely for the ChIP-Seq target TF. For example, Ridinger-Saison *et al.*, (2012) have shown dramatic differences for distances between Spi-1 binding sites when a subset of peaks characteristic for particular genomic regions was selected (Fig. 5.6). These TFBS distance preferences appear to be directly related to

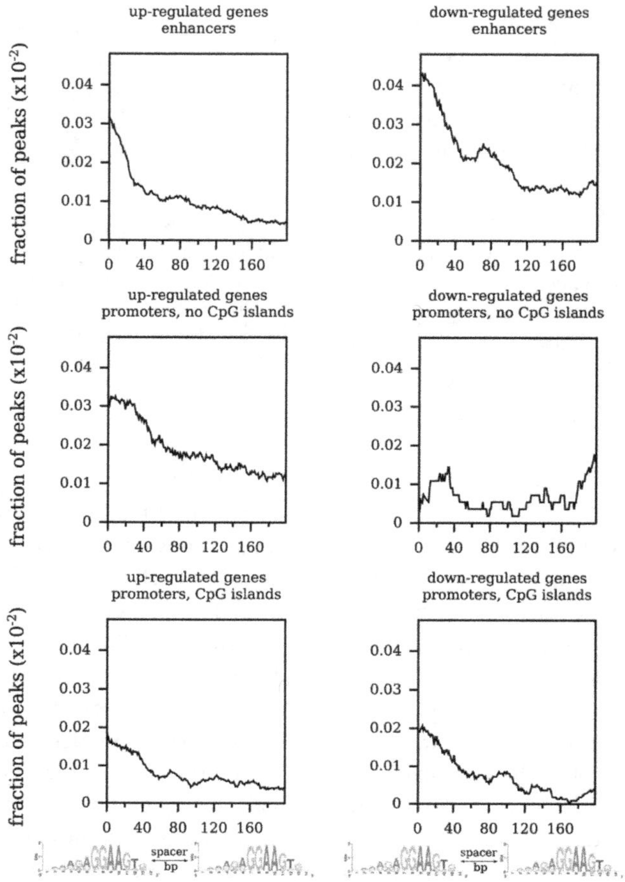

Figure 5.6 Positional preferences of Spi-1 motif occurrences in different ChIP-Seq peak categories. X axis shows the distance between Spi-1 motif occurrences in the same orientation. Y axis shows the fraction of ChIP-Seq peaks having two motif occurrences separated by a given spacer. It is notable that close homotypic pairs of Spi-1 TFBSs can have opposite effects on expression of target genes depending on (1) the category of the peak, which is defined by its location relative to the target gene; (2) the presence of the CpG islands (for the promoter peaks). The data is given according to Ridinger-Saison *et al.* (2012).

the regulatory effect of Spi-1 on the corresponding target genes.

Deriving precise TFBS models from ChIP-Seq data

The ChIP-Seq technology is not perfect, the output peak set can contain false positive peaks unrelated to the specific protein binding events, and false negatives can also happen, with a TFBS unrevealed in the experiment. For instance, 'piggy-back' binding of a cofactor protein tightly interacting with the target TF can produce a clear binding signal unrelated to the signal of target TF observed *in vitro*. Details of crosslinking, library preparation, sequencing and even bioinformatics pipeline can distort the final TF binding signal. To separate true TF binding events from such noise, a TFBS model of high specificity and selectivity is needed.

Most of the existing tools for motif discovery in ChIP-Seq data employ traditional PWMs for motif models. It should be noted that different strategies used to estimate the matrix elements, the positional weights, result in surprisingly different matrices, and the ultimate solution of the problem of constructing the 'true' optimal PWM from an experimental dataset is still unavailable. Several benchmark studies compared motif discovery tools in the context of their efficiency within integrated pipelines, for example, with the completeMOTIFs pipeline (Kuttippurathu *et al.*, 2011). Several independent benchmarks demonstrated that our ChIPMunk algorithm produced high-quality PWMs in a variety of scenarios (Bi *et al.*, 2011; Ma *et al.*, 2012; Weirauch *et al.*, 2013) if ChIP-Seq base coverage profiles were provided along with sequence data. Still, despite surprisingly high effectiveness of the PWM model in practical applications, it is based upon an oversimplified assumption of the independence of nucleotides, which occupy different binding site positions. Several studies (Siddharthan, 2010; Hooghe *et al.*, 2012) clearly demonstrated that this approximation is not generally valid. Partly, PWM owes its lasting popularity to the fact that in the pre-genomic age the data on specific DNA-protein binding was rather sparse, with a set of more than 100 TFBSs being a rare case. Hence, in the most cases the available data was insufficient for reliable estimation of all elements of PWM, not to mention models with a larger number of parameters. The abundance of DNA-protein binding data, obtained by high-throughput methods, prompted the development of more complex models. Surprisingly, in many cases the increase of accuracy of these models was only incremental (e.g. Tree-PWM in Bi *et al.*, 2011) as compared to the models based on PWM thoroughly trained on large datasets.

A very specific and important family of TFBS models with non-independent positional weights takes into account correlations of nucleotides in neighbouring positions within TFBSs and in the corresponding GMLA. The correlation of neighbouring nucleotides agrees well with their role in DNA structure formation (SantaLucia and Hicks, 2004). The most straightforward model here is a matrix similar to PWM but having matrix elements estimated for dinucleotides, which allows for interaction of neighbouring nucleotides. It has been previously shown that this dinucleotide PWM outperforms the classic mononucleotide PWM (Gershenzon *et al.*, 2005; Levitsky *et al.*, 2007) if trained on a set of sequences large enough to estimate the increased number of model parameters. Moreover, a recent reanalysis of protein-binding microarray data with the use of TFBS models, that allow for frequencies of dinucleotides, revealed a high correlation between experimentally measured and predicted biding site affinities (Zhao *et al.*, 2012). So it looked extremely interesting to try training of a similar model on ChIP-Seq data, which also can provide a large dataset of binding sequences. For example, the ChIPMunk algorithm (http://autosome.ru/ChIPMunk/) was successfully extended as diChIPMunk (http://autosome.ru/dichipmunk) to use dinucleotide PWM models (Kulakovskiy *et al.*, 2013a), but comprehensive benchmarks and downstream analysis tools for such models are only expected to arrive in the future.

Other applications of motif discovery and motif finding

Motif analysis proved to be a highly useful addition for interpretation of ChIP-Seq experimental results. For instance, low ChIP-Seq peaks containing 'strong' motif occurrences can be retained and used in subsequent functional annotation, whereas other low peaks missing motif occurrences can be considered as non-relevant noise and discarded. Several software tools were developed to profit from this idea. The first example is MICSA, motif identification for ChIP-Seq analysis (Boeva *et al.*, 2010), where motif discovery is used as an additional processing step to increase peak detection accuracy, as compared with peak finding tools that do not take into account information from the TFBS model. The recently published GEM software (Guo *et al.*, 2012) introduces precise identification of binding sites directly into the peak calling process.

Another application of motif finding is functional annotation of ChIP-Seq peaks. As it was mentioned above, two or more closely positioned TFBS can produce a single ChIP-Seq peak. The question is how the shape of a ChIP-Seq peak is related to the actual TFBS arrangements. To look into this issue, we have studied several independent experimental and bioinformatics data sources for the regulatory system of *Drosophila* early development.

We extracted (–12,000; +6000) genomic regions around transcription start sites for several genes of *Drosophila* early development, including *knirps* (shown in Fig. 5.7). We used ChIP-Seq

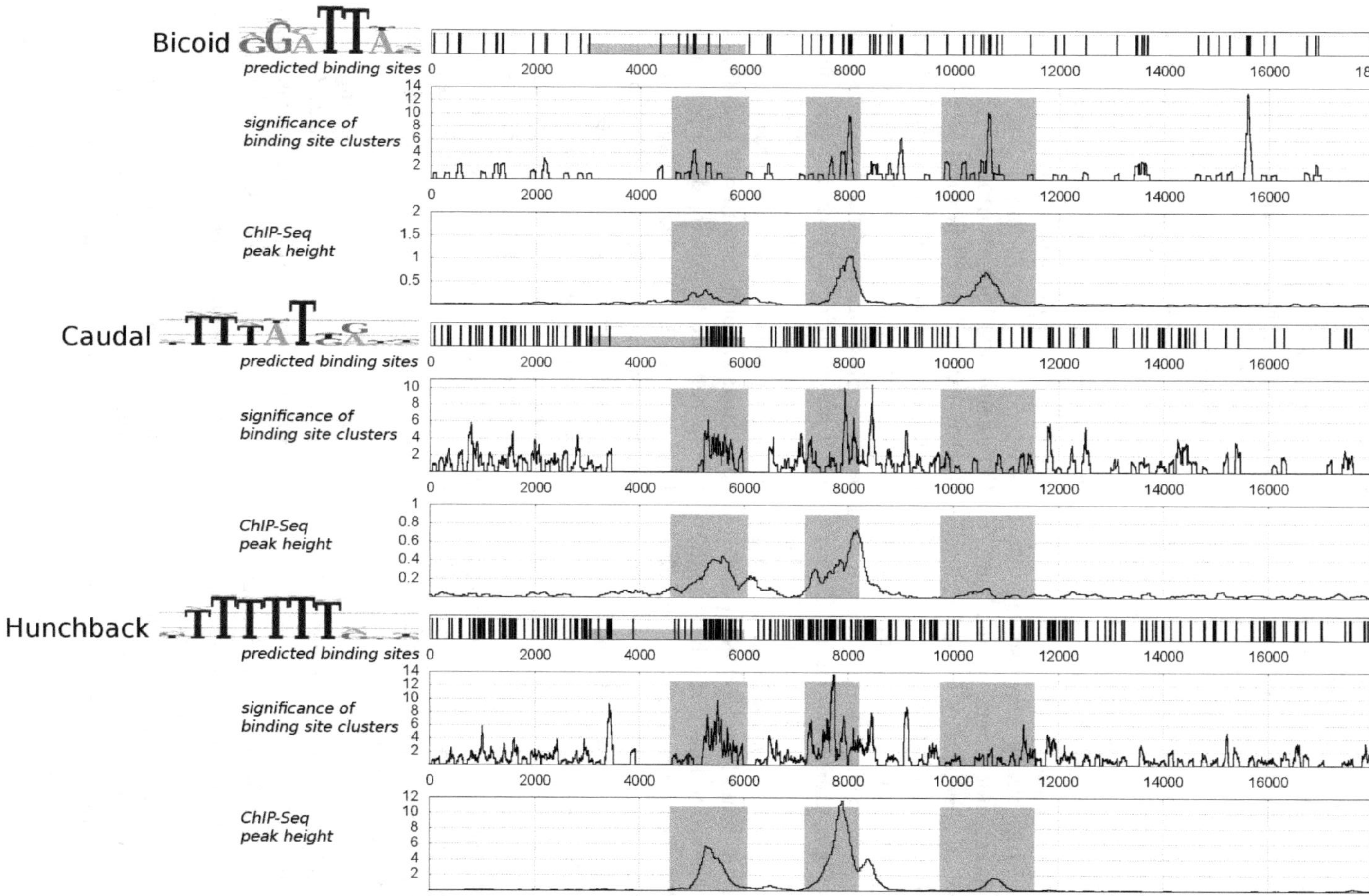

Figure 5.7 ChIP-Seq peak profiles and the predicted statistical significance of homotypic TFBS clusters for regulatory regions of the *knirps* gene (*Drosophila* early development system). Data for three TFs (Bicoid, Caudal, Hunchback) is shown. The TFBS models LOGOs are given according to Kulakovskiy and Makeev (2010). X axis corresponds to the genomic region in the vicinity of the *knirps* gene (shown as the grey box on the 'predicted binding sites' track). Note that the gene is located on the reverse strand and thus its 'upstream' region is on the right of the graph. The grey boxes on the remaining tracks correspond to DNase accessible regions. ChIP-Seq peak height is given according to Kaplan *et al.* (2011).

peaks obtained for several key TFs (Kaplan *et al.*, 2011), a collection of published TFBS models (Kulakovskiy and Makeev, 2010), and data on DNase accessibility regions (Li *et al.*, 2011), all obtained for the same fifth embryogenesis stage as the ChIP-Seq data. The putative TFBSs were predicted by applying PWMs requiring a minimal score of no less than a mean plus three standard deviations based on the overall PWM score distribution. Then we used AhoPro software (Boeva *et al.*, 2007) to estimate *P*-values in 50 bp sliding windows. The *P*-value was defined as the probability to observe no less than the given number of motif occurrences in a given window with a given nucleotide composition. Having compared the statistical significance as −log(*P*-value) with the ChIP-Seq peak profile height, we observed strikingly similar shapes of profiles for the statistical significance of motif occurrences and the ChIP-Seq base coverage in the DNase-accessible regions. This suggested that the ChIP-Seq peak shape profile actually showed not only the presence of binding sites in a particular region, but also their complex interference resulting from a specific arrangement that could be identified with the help of motif analysis.

Cooking recipes for motif analysis of ChIP-Seq data

Finally, we discuss how the ChIP-Seq analysis workflow can be configured to incorporate motif discovery and motif finding. The workflow starts from mapping of sequence reads to the target genome, and it includes careful peak finding, selection of peak subsets for motif discovery, search for the best TFBS model for the TF under study, search for additional sequence patterns present in the peaks, comparison of the sequence patterns found with those stored in databases of known TFBS models, and estimation of positional preferences of pattern occurrences relative to each other and to the ChIP-Seq peak summits. It is also practical to have an option of exporting the data in commonly used formats at any step of the analysis, allowing usage of extra visualization and analysis tools. An exhaustive benchmark of ChIP-Seq processing pipelines is not yet available. Here we offer basic 'cookbook'-style suggestions for the most advanced and actively developing tools that might be a good starting point for anybody interested in motif discovery in ChIP-Seq data.

Recipes for read mapping

Several benchmarks of read mapping tools for next-generation sequencing have been reported, see, e.g. (Flicek and Birney, 2009; Li and Homer, 2010). It is important that analysis of ChIP-Seq data, oriented to TFBS prediction sets, has less stringent requirements to read mapping because it aims at identification of peaks based on read pileups. A single missing read or an incorrectly mapped read cannot dramatically distort the profile. That is why the correct mapping of all reads is desirable but actually not necessary and sometimes can be traded for speed. Many existing pipelines use Bowtie (Langmead *et al.*, 2009) or BWA (Li and Durbin, 2009) for read mapping, which ensure reasonable trade-off between the mapping computational cost and quality and provide mapping format that can be directly used by major peak finding tools. The newest generation of read mapping tools profit from special hardware like GPGPU, general-purpose graphics processing unit. A specially optimized software, e.g. SOAP3 (Liu *et al.*, 2012), is able to produce more precise read mappings with a reasonable speed.

Recipes for peak finding

Several benchmark studies on peak finders have been published to date (Wilbanks and Facciotti, 2010; Kim *et al.*, 2011; Rye *et al.*, 2011). With the subsequent motif analysis procedure kept in mind, selection of the proper peak finding software may be a non trivial task where only very general guidelines can be given. Those requiring the graphical user interface probably should stick to CisGenome (Ji *et al.*, 2008) or Sole-Search (Blahnik *et al.*, 2010). Users of the R programming language may prefer PICS (Mercier *et al.*, 2011) that is available in the Bioconductor R package. However, the golden standard is now defined by a command-line tool MACS, Model-based Analysis of ChIP-Seq (Feng *et al.*, 2012). Other important peak callers include SPP (Kharchenko *et al.*, 2008), also suitable for R users, and PeakSeq (Rozowsky *et al.*, 2009), both used in ENCODE project (Landt *et al.*, 2012) along with MACS.

Difficult cases with noisy peaks or low sequencing depth can be resolved by more straightforward approaches, such as FindPeaks (Fejes *et al.*, 2008). Its successor, Vancouver Short Read Analysis Package (http://vancouvershortr.sourceforge.net), provides several additional tools to process and convert data, e.g. from different read-mapping tools. Selection of the peak height threshold is another non-trivial task. No general strategy is available, since different peak finding tools employ different strategies and different estimations of statistical significance with or without reference control data.

Recipes for motif discovery and motif finding in ChIP-Seq data

For developers preferring all-in-one software available on the web, generally, the peak-motifs pipeline within RSAT, Regulatory Sequence Analysis Toolbox, would be the best choice (Thomas-Chollier *et al.*, 2012). Another popular way is to apply the web-based software from MEME suite (Bailey *et al.*, 2009) that includes several tools for motif discovery in ChIP-Seq peaks, MEME-ChIP (Machanick and Bailey, 2011). Analysis of positional preferences of motif occurrences corresponding to different discovered patterns can also be done by stand-alone tools, where rGADEM, a genetic algorithm guided formation of spaced dyads, identifies overrepresented two-boxed motifs (Mercier *et al.*, 2011) and is more suitable for R users; and SpaMo, spaced motif analysis (Whitington *et al.*, 2011) would be helpful for those preferring the MEME suite. Advanced bioinformaticians building their own computational pipelines and those interested in precise TFBS models may prefer using ChIPMunk (Kulakovskiy *et al.*, 2010) or its dinucleotide progeny, diChIPMunk (Kulakovskiy *et al.*, 2013a).

Conclusion: the present and the future of motif analysis for the ChIP-Seq technology

ChIP-Seq became a technology of choice for studying genome-wide TFBS landscapes (Park, 2009). Many studies, including recent findings from ENCODE project (Myers *et al.*, 2011; Dunham *et al.*, 2012), provide large volumes of data on TFBSs and target genes of different TFs in different tissues. For those planning their own ChIP-Seq experiments or analysis, several references would be especially useful, including comprehensive studies on experimental design (Landt *et al.*, 2012), extensive motif analysis (Wang *et al.*, 2012), and studies on association between TFBSs and different chromatin features (Kundaje *et al.*, 2012). Useful reviews have been focused specifically on ChIP-Seq analysis (Pepke *et al.*, 2009) and motif discovery in the next-generation sequencing era (Zambelli *et al.*, 2012). Even more, further applications of TFBS ChIP-Seq continue to evolve. A nice recent example is given by a ChIP-Seq driven discovery of regulatory single-nucleotide polymorphisms (Ni *et al.*, 2012).

The experimental technologies are developing very rapidly. Recently an improved modification of ChIP-Seq was reported under the name 'ChIP-exo technology' (Rhee and Pugh, 2011). ChIP-exo has inherited many advantages of ChIP-Seq and, with additional DNase digestion step, possibly allows the resulting peaks to point to the TFBS location at a significantly better resolution. It is an open question whether ChIP-Seq-oriented tools, including motif analysis software, will be applicable and/or necessary for dry-lab analysis of novel data in the coming years. Still, with the large volumes of ChIP-Seq data projected and already produced, the tools and workflows described in this review have years of solid ground for application.

Web resources

Read mapping

- Bowtie: http://bowtie-bio.sourceforge.net
- BWA: http://bio-bwa.sourceforge.net/
- SOAP3: http://soap.genomics.org.cn/soap3.html

Peak calling

- FindPeaks: http://vancouvershortr.sourceforge.net
- MACS: http://liulab.dfci.harvard.edu/MACS/

- CisGenome: http://www.biostat.jhsph.edu/~hji/cisgenome/
- Sole-Search: http://havoc.genomecenter.ucdavis.edu/
- PeakSeq: http://info.gersteinlab.org/PeakSeq
- SPP : http://compbio.med.harvard.edu/Supplements/ChIP-seq/
- PICS: http://www.bioconductor.org/packages/release/bioc/html/PICS.html

Motif discovery

- ChIPMunk: http://autosome.ru/ChIPMunk/
- diChIPMunk: http://autosome.ru/diChIP-Munk/
- completeMOTIFs: http://cmotifs.tchlab.org/
- POSMO: http://cb.utdallas.edu/Posmo/index.html
- HMS: http://www.sph.umich.edu/csg/qin/HMS/

Complex analysis

- peak-motifs: http://rsat.ulb.ac.be/peak-motifs_form.cgi
- MEME-ChIP: http://meme.ebi.edu.au/meme/cgi-bin/meme-chip.cgi
- GEM: http://cgs.csail.mit.edu/gem/

Acknowledgements

The authors are pleased to personally thank Eugenia V Serebrova for her valuable comments and suggestions.

Funding

This work was supported, in part, by a Dynasty Foundation Fellowship (I.V.K.); the Russian Foundation for Basic Research (grant 12-04-32082 to I.V.K.); Program on Molecular and Cell Biology of the Presidium of RAS (V.J.M); Program 'Wildlife: Current State and Development' of the Presidium of RAS (grant 2012.3-1.5 to V.J.M.); Russian Ministry of Science and Education grant (11.G34.31.0008).

References

Badis, G., Berger, M.F., Philippakis, A.A., Talukder, S., Gehrke, A.R., Jaeger, S.A., Chan, E.T., Metzler, G., Vedenko, A., Chen, X., *et al.* (2009). Diversity and complexity in DNA recognition by transcription factors. Science *324*, 1720–1723.

Bailey, T.L. (2002). Discovering novel sequence motifs with MEME. Curr. Protoc. Bioinformatics *Chapter 2*, Unit 2.4.

Bailey, T.L., and Machanick, P. (2012). Inferring direct DNA binding from ChIP-seq. Nucleic Acids Res. *40*, e128.

Bailey, T.L., Boden, M., Buske, F.A., Frith, M., Grant, C.E., Clementi, L., Ren, J., Li, W.W., and Noble, W.S. (2009). MEME SUITE: tools for motif discovery and searching. Nucleic Acids Res. *37*, W202–8.

Berg, O.G., and Von Hippel, P.H. (1987). Selection of DNA binding sites by regulatory proteins. Statistical-mechanical theory and application to operators and promoters. J. Mol. Biol. *193*, 723–750.

Berg, O.G., and Von Hippel, P.H. (1988). Selection of DNA binding sites by regulatory proteins. Trends Biochem. Sci. *13*, 207–211.

Bi, Y., Kim, H., Gupta, R., and Davuluri, R.V. (2011). Tree-based position weight matrix approach to model transcription factor binding site profiles. PLoS One *6*, e24210.

Blahnik, K.R., Dou, L., O'Geen, H., McPhillips, T., Xu, X., Cao, A.R., Iyengar, S., Nicolet, C.M., Ludäscher, B., Korf, I., *et al.* (2010). Sole-Search: an integrated analysis program for peak detection and functional annotation using ChIP-seq data. Nucleic Acids Res. *38*, e13.

Boeva, V., Clément, J., Régnier, M., Roytberg, M.A., and Makeev, V.J. (2007). Exact *p*-value calculation for heterotypic clusters of regulatory motifs and its application in computational annotation of cis-regulatory modules. Algorithms Mol. Biol. 2, 13.

Boeva, V., Surdez, D., Guillon, N., Tirode, F., Fejes, A.P., Delattre, O., and Barillot, E. (2010). De novo motif identification improves the accuracy of predicting transcription factor binding sites in ChIP-Seq data analysis. Nucleic Acids Res. *38*, e126.

Buck, M.J., and Lieb, J.D. (2004). ChIP-chip: considerations for the design, analysis, and application of genome-wide chromatin immunoprecipitation experiments. Genomics *83*, 349–360.

Bulyk, M.L. (2006). DNA microarray technologies for measuring protein–DNA interactions. Curr. Opin. Biotechnol. *17*, 422–430.

Chan, T.-M., Leung, K.-S., Lee, K.-H., Wong, M.-H., Lau, T.C.-K., and Tsui, S.K.-W. (2012). Subtypes of associated protein-DNA (Transcription Factor-Transcription Factor Binding Site) patterns. Nucleic Acids Res. *40*, 9392–9403.

Chen, X., Xu, H., Yuan, P., Fang, F., Huss, M., Vega, V.B., Wong, E., Orlov, Y.L., Zhang, W., Jiang, J., *et al.* (2008). Integration of external signaling pathways with the core transcriptional network in embryonic stem cells. Cell *133*, 1106–1117.

Crooks, G.E., Hon, G., Chandonia, J.-M., and Brenner, S.E. (2004). WebLogo: a sequence logo generator. Genome Res. *14*, 1188–1190.

Day, W.H., and McMorris, F.R. (1992). Critical comparison of consensus methods for molecular sequences. Nucleic Acids Res. *20*, 1093–1099.

Dunham, I., Kundaje, A., Aldred, S.F., Collins, P.J., Davis, C.A., Doyle, F., Epstein, C.B., Frietze, S., Harrow, J., Kaul, R., *et al.* (2012). An integrated encyclopedia of DNA elements in the human genome. Nature *489*, 57–74.

D'haeseleer, P. (2006). What are DNA sequence motifs? Nat. Biotechnol. *24*, 423–425.

Favorov, A.V., Gelfand, M.S., Gerasimova, A.V., Ravcheev, D.A., Mironov, A.A., and Makeev, V.J. (2005). A Gibbs sampler for identification of symmetrically structured, spaced DNA motifs with improved estimation of the signal length. Bioinformatics *21*, 2240–2245.

Fejes, A.P., Robertson, G., Bilenky, M., Varhol, R., Bainbridge, M., and Jones, S.J.M. (2008). FindPeaks 3.1: a tool for identifying areas of enrichment from massively parallel short-read sequencing technology. Bioinformatics *24*, 1729–1730.

Feng, J., Liu, T., Qin, B., Zhang, Y., and Liu, X.S. (2012). Identifying ChIP-seq enrichment using MACS. Nat. Protoc. *7*, 1728–1740.

Flicek, P., and Birney, E. (2009). Sense from sequence reads: methods for alignment and assembly. Nat. Methods *6*, S6–S12.

Galas, D.J., and Schmitz, A. (1978). DNAse footprinting: a simple method for the detection of protein-DNA binding specificity. Nucleic Acids Res. *5*, 3157–3170.

Gershenzon, N.I., Stormo, G.D., and Ioshikhes, I.P. (2005). Computational technique for improvement of the position-weight matrices for the DNA/protein binding sites. Nucleic Acids Res. *33*, 2290–2301.

Gotea, V., Visel, A., Westlund, J.M., Nobrega, M.A., Pennacchio, L.A., and Ovcharenko, I. (2010). Homotypic clusters of transcription factor binding sites are a key component of human promoters and enhancers. Genome Res. *20*, 565–577.

Guillon, N., Tirode, F., Boeva, V., Zynovyev, A., Barillot, E., and Delattre, O. (2009). The oncogenic EWS-FLI1 protein binds *in vivo* GGAA microsatellite sequences with potential transcriptional activation function. PLoS One *4*, e4932.

Guo, Y., Mahony, S., and Gifford, D.K. (2012). High resolution genome wide binding event finding and motif discovery reveals transcription factor spatial binding constraints. PLoS Comput. Biol. *8*, e1002638.

Hannenhalli, S., and Wang, L.-S. (2005). Enhanced position weight matrices using mixture models. Bioinformatics *21* (Suppl 1), i204–12.

Hooghe, B., Broos, S., Van Roy, F., and De Bleser, P. (2012). A flexible integrative approach based on random forest improves prediction of transcription factor binding sites. Nucleic Acids Res. *40*, e106.

Horak, C.E., and Snyder, M. (2002). ChIP-chip: a genomic approach for identifying transcription factor binding sites. Methods Enzymol. *350*, 469–483.

Hu, M., Yu, J., Taylor, J.M.G., Chinnaiyan, A.M., and Qin, Z.S. (2010). On the detection and refinement of transcription factor binding sites using ChIP-Seq data. Nucleic Acids Res. *38*, 2154–2167.

Impey, S., McCorkle, S.R., Cha-Molstad, H., Dwyer, J.M., Yochum, G.S., Boss, J.M., McWeeney, S., Dunn, J.J., Mandel, G., and Goodman, R.H. (2004). Defining the CREB regulon: a genome-wide analysis of transcription factor regulatory regions. Cell *119*, 1041–1054.

Ji, H., Jiang, H., Ma, W., Johnson, D.S., Myers, R.M., and Wong, W.H. (2008). An integrated software system for analyzing ChIP-chip and ChIP-seq data. Nat. Biotechnol. *26*, 1293–1300.

Johnson, D.S., Mortazavi, A., Myers, R.M., and Wold, B. (2007). Genome-wide mapping of *in vivo* protein–DNA interactions. Science *316*, 1497–1502.

Jothi, R., Cuddapah, S., Barski, A., Cui, K., and Zhao, K. (2008). Genome-wide identification of *in vivo* protein-DNA binding sites from ChIP-Seq data. Nucleic Acids Res. *36*, 5221–5231.

Kaplan, T., Li, X.-Y., Sabo, P.J., Thomas, S., Stamatoyannopoulos, J.A., Biggin, M.D., and Eisen, M.B. (2011). Quantitative models of the mechanisms that control genome-wide patterns of transcription factor binding during early Drosophila development. PLoS Genet. *7*, e1001290.

Kharchenko, P.V., Tolstorukov, M.Y., and Park, P.J. (2008). Design and analysis of ChIP-seq experiments for DNA-binding proteins. Nat. Biotechnol. *26*, 1351–1359.

Kim, H., Kim, J., Selby, H., Gao, D., Tong, T., Phang, T.L., and Tan, A.C. (2011). A short survey of computational analysis methods in analysing ChIP-seq data. Hum. Genomics *5*, 117–123.

Kulakovskiy, I., Levitsky, V., Oshchepkov, D., Bryzgalov, L., Vorontsov, I., and Makeev, V. (2013a). From binding motifs in ChIP-Seq data to improved models of transcription factor binding sites. J. Bioinform. Comput. Biol. *11*, 1340004.

Kulakovskiy, I.V., and Makeev, V.J. (2010). Discovery of DNA motifs recognized by transcription factors through integration of different experimental sources. Biophysics *54*, 667–674.

Kulakovskiy, I.V., Boeva, V.A., Favorov, A.V., and Makeev, V.J. (2010). Deep and wide digging for binding motifs in ChIP-Seq data. Bioinformatics *26*, 2622–2623.

Kulakovskiy, I.V., Medvedeva, Y.A., Schaefer, U., Kasianov, A.S., Vorontsov, I.E., Bajic, V.B., and Makeev, V.J. (2013b). HOCOMOCO: a comprehensive collection of human transcription factor binding sites models. Nucleic Acids Res. *41*, D195–D202.

Kundaje, A., Kyriazopoulou-Panagiotopoulou, S., Libbrecht, M., Smith, C.L., Raha, D., Winters, E.E., Johnson, S.M., Snyder, M., Batzoglou, S., and Sidow, A. (2012). Ubiquitous heterogeneity and asymmetry of the chromatin environment at regulatory elements. Genome Res. *22*, 1735–1747.

Kuttippurathu, L., Hsing, M., Liu, Y., Schmidt, B., Maskell, D.L., Lee, K., He, A., Pu, W.T., and Kong, S.W. (2011). CompleteMOTIFs: DNA motif discovery platform for transcription factor binding experiments. Bioinformatics *27*, 715–717.

Landt, S.G., Marinov, G.K., Kundaje, A., Kheradpour, P., Pauli, F., Batzoglou, S., Bernstein, B.E., Bickel, P., Brown, J.B., Cayting, P., *et al.* (2012). ChIP-seq guidelines and practices of the ENCODE and modENCODE consortia. Genome Res. *22*, 1813–1831.

Langmead, B., Trapnell, C., Pop, M., and Salzberg, S.L. (2009). Ultrafast and memory-efficient alignment of short DNA sequences to the human genome. Genome Biol. *10*, R25.

Leleu, M., Lefebvre, G., and Rougemont, J. (2010). Processing and analyzing ChIP-seq data: from short reads to regulatory interactions. Brief Funct. Genomics *9*, 466–476.

Lelli, K.M., Slattery, M., and Mann, R.S. (2012). Disentangling the many layers of eukaryotic transcriptional regulation. Annu. Rev. Genet. *46*, 43–68.

Levitsky, V.G., Ignatieva, E.V., Ananko, E.A., Turnaev, I.I., Merkulova, T.I., Kolchanov, N.A., and Hodgman, T.C. (2007). Effective transcription factor binding site prediction using a combination of optimization, a genetic algorithm and discriminant analysis to capture distant interactions. BMC Bioinformatics *8*, 481.

Li, H., and Durbin, R. (2009). Fast and accurate short read alignment with Burrows-Wheeler transform. Bioinformatics *25*, 1754–1760.

Li, H., and Homer, N. (2010). A survey of sequence alignment algorithms for next-generation sequencing. Brief. Bioinformatics *11*, 473–483.

Li, X.-Y., Thomas, S., Sabo, P.J., Eisen, M.B., Stamatoyannopoulos, J.A., and Biggin, M.D. (2011). The role of chromatin accessibility in directing the widespread, overlapping patterns of Drosophila transcription factor binding. Genome Biol. *12*, R34.

Lifanov, A.P., Makeev, V.J., Nazina, A.G., and Papatsenko, D.A. (2003). Homotypic regulatory clusters in Drosophila. Genome Res. *13*, 579–588.

Linhart, C., Halperin, Y., and Shamir, R. (2008). Transcription factor and microRNA motif discovery: the Amadeus platform and a compendium of metazoan target sets. Genome Res. *18*, 1180–1189.

Liu, C.-M., Wong, T., Wu, E., Luo, R., Yiu, S.-M., Li, Y., Wang, B., Yu, C., Chu, X., Zhao, K., *et al.* (2012). SOAP3: ultra-fast GPU-based parallel alignment tool for short reads. Bioinformatics *28*, 878–879.

Liu, X., Brutlag, D.L., and Liu, J.S. (2001). BioProspector: discovering conserved DNA motifs in upstream regulatory regions of co-expressed genes. Pac. Symp. Biocomput. 127–138.

Liu, X., Noll, D.M., Lieb, J.D., and Clarke, N.D. (2005). DIP-chip: rapid and accurate determination of DNA-binding specificity. Genome Res. *15*, 421–427.

Ma, X., Kulkarni, A., Zhang, Z., Xuan, Z., Serfling, R., and Zhang, M.Q. (2012). A highly efficient and effective motif discovery method for ChIP-seq/ChIP-chip data using positional information. Nucleic Acids Res. *40*, e50.

Machanick, P., and Bailey, T.L. (2011). MEME-ChIP: motif analysis of large DNA datasets. Bioinformatics *27*, 1696–1697.

Matys, V., Kel-Margoulis, O.V., Fricke, E., Liebich, I., Land, S., Barre-Dirrie, A., Reuter, I., Chekmenev, D., Krull, M., Hornischer, K., *et al.* (2006). TRANSFAC and its module TRANSCompel: transcriptional gene regulation in eukaryotes. Nucleic Acids Res. *34*, D108–10.

Mercier, E., Droit, A., Li, L., Robertson, G., Zhang, X., and Gottardo, R. (2011). An integrated pipeline for the genome-wide analysis of transcription factor binding sites from ChIP-Seq. PLoS One *6*, e16432.

Monteiro, P.T., Mendes, N.D., Teixeira, M.C., d'Orey, S., Tenreiro, S., Mira, N.P., Pais, H., Francisco, A.P., Carvalho, A.M., Lourenço, A.B., *et al.* (2008). YEASTRACT-DISCOVERER: new tools to improve the analysis of transcriptional regulatory associations in Saccharomyces cerevisiae. Nucleic Acids Res. *36*, D132–6.

Myers, E.W., and Miller, W. (1989). Approximate matching of regular expressions. Bull. Math. Biol. *51*, 5–37.

Myers, R.M., Stamatoyannopoulos, J., Snyder, M., Dunham, I., Hardison, R.C., Bernstein, B.E., Gingeras, T.R., Kent, W.J., Birney, E., Wold, B., *et al.* (2011). A user's guide to the encyclopedia of DNA elements (ENCODE). PLoS Biol. *9*, e1001046.

Ng, P., Tan, J.J.S., Ooi, H.S., Lee, Y.L., Chiu, K.P., Fullwood, M.J., Srinivasan, K.G., Perbost, C., Du, L., Sung, W.-K., *et al.* (2006). Multiplex sequencing of paired-end ditags (MS-PET): a strategy for the ultra-high-throughput analysis of transcriptomes and genomes. Nucleic Acids Res. *34*, e84.

Ni, Y., Weber Hall, A., Battenhouse, A., and Iyer, V.R. (2012). Simultaneous SNP identification and assessment of allele-specific bias from ChIP-seq data. BMC Genet. *13*, 46.

Oshchepkov, D.Y., Vityaev, E.E., Grigorovich, D.A., Ignatieva, E.V., and Khlebodarova, T.M. (2004). SITECON: a tool for detecting conservative conformational and physicochemical properties in transcription factor binding site alignments and for site recognition. Nucleic Acids Res. *32*, W208–12.

Park, P.J. (2009). ChIP-seq: advantages and challenges of a maturing technology. Nat. Rev. Genet. *10*, 669–680.

Pepke, S., Wold, B., and Mortazavi, A. (2009). Computation for ChIP-seq and RNA-seq studies. Nat Methods *6*, S22–32.

Portales-Casamar, E., Thongjuea, S., Kwon, A.T., Arenillas, D., Zhao, X., Valen, E., Yusuf, D., Lenhard, B., Wasserman, W.W., and Sandelin, A. (2010). JASPAR 2010: the greatly expanded open-access database of transcription factor binding profiles. Nucleic Acids Res. *38*, D105–10.

Rhee, H.S., and Pugh, B.F. (2011). Comprehensive genome-wide protein–DNA interactions detected at single-nucleotide resolution. Cell *147*, 1408–1419.

Ridinger-Saison, M., Boeva, V., Rimmele, P., Kulakovskiy, I., Gallais, I., Levavasseur, B., Paccard, C., Legoix-Ne, P., Morle, F., Nicolas, A., *et al.* (2012). Spi-1/PU.1 activates transcription through clustered DNA occupancy in erythroleukemia. Nucleic Acids Res..

Robertson, G., Hirst, M., Bainbridge, M., Bilenky, M., Zhao, Y., Zeng, T., Euskirchen, G., Bernier, B., Varhol, R., Delaney, A., *et al.* (2007). Genome-wide profiles of STAT1 DNA association using chromatin immunoprecipitation and massively parallel sequencing. Nat. Methods *4*, 651–657.

Rozowsky, J., Euskirchen, G., Auerbach, R.K., Zhang, Z.D., Gibson, T., Bjornson, R., Carriero, N., Snyder, M., and Gerstein, M.B. (2009). PeakSeq enables systematic scoring of ChIP-seq experiments relative to controls. Nat. Biotechnol. 27, 66–75.

Rye, M.B., Sætrom, P., and Drabløs, F. (2011). A manually curated ChIP-seq benchmark demonstrates room for improvement in current peak-finder programs. Nucleic Acids Res. *39*, e25.

Sandelin, A., and Wasserman, W.W. (2004). Constrained binding site diversity within families of transcription factors enhances pattern discovery bioinformatics. J. Mol. Biol. *338*, 207–215.

SantaLucia, J., and Hicks, D. (2004). The thermodynamics of DNA structural motifs. Annu. Rev. Biophys. Biomol. Struct. 33, 415–440.

Siddharthan, R. (2010). Dinucleotide weight matrices for predicting transcription factor binding sites: generalizing the position weight matrix. PLoS One *5*, e9722.

Sinha, S., and Tompa, M. (2002). Discovery of novel transcription factor binding sites by statistical overrepresentation. Nucleic Acids Res. *30*, 5549–5560.

Van Steensel, B., Delrow, J., and Henikoff, S. (2001). Chromatin profiling using targeted DNA adenine methyltransferase. Nat. Genet. *27*, 304–308.

Stormo, G.D. (2000). DNA binding sites: representation and discovery. Bioinformatics *16*, 16–23.

Thomas-Chollier, M., Herrmann, C., Defrance, M., Sand, O., Thieffry, D., and Van Helden, J. (2012). RSAT peak-motifs: motif analysis in full-size ChIP-seq datasets. Nucleic Acids Res. *40*, e31.

Tompa, M., Li, N., Bailey, T.L., Church, G.M., De Moor, B., Eskin, E., Favorov, A.V., Frith, M.C., Fu, Y., Kent, W.J., *et al.* (2005). Assessing computational tools for the discovery of transcription factor binding sites. Nat. Biotechnol. *23*, 137–144.

Tuerk, C., and Gold, L. (1990). Systematic evolution of ligands by exponential enrichment: RNA ligands to bacteriophage T4 DNA polymerase. Science *249*, 505–510.

Tuteja, G., White, P., Schug, J., and Kaestner, K.H. (2009). Extracting transcription factor targets from ChIP-Seq data. Nucleic Acids Res. *37*, e113.

Valouev, A., Johnson, D.S., Sundquist, A., Medina, C., Anton, E., Batzoglou, S., Myers, R.M., and Sidow, A. (2008). Genome-wide analysis of transcription factor binding sites based on ChIP-Seq data. Nat. Methods *5*, 829–834.

Wang, G., Yu, T., and Zhang, W. (2005). WordSpy: identifying transcription factor binding motifs by building a dictionary and learning a grammar. Nucleic Acids Res. *33*, W412–6.

Wang, J., Zhuang, J., Iyer, S., Lin, X., Whitfield, T.W., Greven, M.C., Pierce, B.G., Dong, X., Kundaje, A., Cheng, Y., *et al.* (2012). Sequence features and chromatin structure around the genomic regions bound by 119 human transcription factors. Genome Res. *22*, 1798–1812.

Warren, C.L., Kratochvil, N.C.S., Hauschild, K.E., Foister, S., Brezinski, M.L., Dervan, P.B., Phillips, G.N., and Ansari, A.Z. (2006). Defining the sequence-recognition profile of DNA-binding molecules. Proc. Natl. Acad. Sci. U.S.A. *103*, 867–872.

Wei, C.-L., Wu, Q., Vega, V.B., Chiu, K.P., Ng, P., Zhang, T., Shahab, A., Yong, H.C., Fu, Y., Weng, Z., *et al.* (2006). A global map of p53 transcription-factor binding sites in the human genome. Cell *124*, 207–219.

Wei, G., Hu, G., Cui, K., and Zhao, K. (2012). Genome-wide mapping of nucleosome occupancy, histone modifications, and gene expression using next-generation sequencing technology. Methods Enzymol. *513*, 297–313.

Weirauch, M.T., Cote, A., Norel, R., Annala, M., Zhao, Y., Riley, T.R., Saez-Rodriguez, J., Cokelaer, T., Uedenko, A., Talukdev, S., *et al.* (2013). Evaluation of methods for modelling transcription factor sequence specificity. Nat. Biotechnol. *31*, 126–134.

Whitington, T., Frith, M.C., Johnson, J., and Bailey, T.L. (2011). Inferring transcription factor complexes from ChIP-seq data. Nucleic Acids Res. *39*, e98.

Wilbanks, E.G., and Facciotti, M.T. (2010). Evaluation of algorithm performance in ChIP-seq peak detection. PLoS One *5*, e11471.

Xie, X., Rigor, P., and Baldi, P. (2009). MotifMap: a human genome-wide map of candidate regulatory motif sites. Bioinformatics *25*, 167–174.

Zambelli, F., Pesole, G., and Pavesi, G. (2012). Motif discovery and transcription factor binding sites before and after the next-generation sequencing era. Brief. Bioinformatics *14*, 225–237.

Zhao, Y., Ruan, S., Pandey, M., and Stormo, G.D. (2012). Improved models for transcription factor binding site identification using nonindependent interactions. Genetics *191*, 781–790.

Mammalian Enhancer Prediction

6

Dongwon Lee and Michael A. Beer

Abstract

We are still far from a complete understanding of regulatory elements in mammalian genomes, even though their central role in most biological processes is widely appreciated. Development of sequence-based models to predict the function and activity of regulatory elements is a fundamental step in being able to address many unsolved questions. Here we describe the current state of the art in computational methods to predict enhancers, especially recent developments using a support vector machine (SVM) framework, which can accurately identify tissue specific enhancers using only genomic sequence and an unbiased set of general sequence features. These models reveal both enriched and depleted predictive sequence features that are critical for specifying these enhancer activities, and can also be used to identify novel enhancers. Some of these predictions have been validated by several independent experiments both *in vitro* and *in vivo*. These methods can be applied to computationally predict the functional consequences of common sequence variants in regulatory regions. We believe that these efforts will significantly contribute to our understanding of mammalian regulatory systems and their role in common disease.

Introduction

Uncovering the function of regulatory DNA elements in the genome is an essential step in gaining a deeper understanding of complex biological processes. These elements, by regulating the expression of their associated genes, are widely believed to play a key role in human development and disease. As highlighted by a catalogue of published genome-wide association studies (GWAS) by The National Human Genome Research Institute (NHGRI) (Hindorff *et al.*, 2009), almost 90% of common human single nucleotide polymorphisms (SNPs), significantly associated with a phenotypic trait or a disease (*P*-value $<10^{-8}$), are located in intergenic or intronic regions. Thus, changes in regulatory elements are implicated in most human diseases and traits of clinical significance. Much effort in current genomic studies is, therefore, devoted to understanding regulatory systems on genome-wide levels, i.e. identifying the networks of regulatory factors and their interactions, and specifying how they modulate their target genes.

Why does a specific genomic locus act as an enhancer in one cell type, but not in another? What mutations in that locus would affect its function, and how? In mathematical terms, building a sequence-based discriminative model that can separate a set of regulatory regions with shared functions from others, such as cell-type-specific enhancers, is a critical step in being able to address these questions. In addition, it would be particularly useful for inferring the underlying molecular mechanisms by which enhancers function. Such models not only provide a means to detect additional regulatory regions that have similar sequence properties and, thereby, presumably similar biological functions, but also give clues about transcription factors (TFs) that may play crucial roles in specifying the regulatory regions by exploiting predictive sequence features. These approaches were successful especially in simple model organisms (Beer and Tavazoie,

2004). However, until recently, development of sequence-based predictive models for mammalian regulatory system has been limited by their extremely large genome sizes, making it difficult to identify even a small set of regulatory regions of a common function.

The advent of new massively parallel sequencing technologies has made the experimental identification of cell type specific regulatory regions a common feature of many investigations. Furthermore, large-scale collaborative efforts have generated comprehensive maps of regulatory regions associated with many key TFs, coactivators and epigenomic histone markers in a broad spectrum of human cell-lines (ENCODE Consortium, 2012) and mouse tissues (Mouse ENCODE Consortium, 2012). These publicly available large-scale genomic datasets provide us with unique opportunities to systematically study the underlying mechanisms of a wide range of regulatory elements.

In this chapter we review the biology of transcriptional enhancers, and describe new computational approaches for enhancer prediction. We highlight a recent computational method, referred to as 'kmer-SVM' (Lee *et al.*, 2011), to predict tissue specific mammalian enhancers using a complete set of oligomers or '*k*-mers' as features in a support vector machine (SVM) (Boser *et al.*, 1992; Vapnik, 1995). Some of these computationally predicted enhancers have been experimentally validated *in vitro* and *in vivo*, showing that these methods can be used to identify novel regulatory regions (Gorkin *et al.*, 2012). We describe in detail general issues affecting classifier evaluation and performance. We believe that these results and continuing efforts to improve computational approaches to enhancer identification will significantly contribute to our understanding of the human regulatory system.

Transcriptional enhancers

Transcriptional enhancers are a key component of the information encoded in the human genome. As a primary step in the regulation of protein product concentrations, transcriptional regulation contributes significantly to virtually all biological responses of the human organism, especially those of embryonic development and cell fate determination.

In this section, we first introduce a brief history of mammalian transcription studies followed by discussion of the role of enhancers as a key element in this essential biological process (see 'Mammalian transcription and enhancers' below). We then discuss efforts to identify disease-causing mutations, and introduce some representative examples of human disorders caused by mutations in enhancers (see 'Enhancers and human disease' below). Current methods to identify enhancers are then discussed in detail (see 'Experimental methods for genome-wide enhancer detection' below). Finally, we direct readers to an extensive body of earlier literature for reviews of a wide range of enhancer biology (see 'Additional reviews on enhancers' below).

Mammalian transcription and enhancers

Since the pioneering work of Jacob and Monod (1961), it was recognized that the rate of transcription was a regulated process which had the potential to impact all downstream processes. Early work on the lysis-lysogeny switch in phage lambda (Ptashne, 2004) identified DNA binding transcription factors as the key molecular mediators of this process. Subsequent studies on the lac-operon in *E. coli* and additional pathways in yeast showed how transcription factors could interact with small molecules or surface signalling receptors to modulate the expression of sets of genes in order to respond to external signals and conditions, by activating or repressing immediately downstream genes. These same basic molecular events are essential steps in the processing of information regarding the extracellular or developmental state of multicellular organisms via cell–cell or diffuse morphogen interactions with cell-surface receptors in embryonic development.

Early studies in mammalian systems uncovered a significantly different regulatory element structure. Mammalian transcription typically involves sets of regulatory elements, which act at a distance, in contrast to the local regulation in single cellular organisms. These elements can broadly be classified as promoters, enhancers, and insulators. Promoters are regulatory elements immediately upstream from transcriptional start sites and are locations of RNA polymerase II complex initiation. Enhancers are distal regulatory elements that can enhance transcriptional activities of specific genes roughly independent of their orientation and distance from the site of action (Banerji, 1981). Insulators block interactions between distal enhancers and promoters.

The scale of these DNA regulatory elements are on the order of several hundred base pairs (up to a few kilobases (kb)) of DNA sequence, comparable to the intergenic space in the prokaryotic or yeast genomes. In contrast, however, these elements can act over a much longer intervening distance in linear sequence; many examples of enhancer–promoter interactions are known to act over several hundred kb, and up to a megabase (Mb), via DNA looping (Blackwood and Kadonaga, 1998; Bulger and Groudine, 1999). While insulators block promoter–enhancer interactions in some cases and contribute to the establishment of specific promoter-enhancer connections, the general mechanisms which determine specific promoter–enhancer association are not yet clear.

It is likely that this modular structure of mammalian transcriptional regulation and the larger mammalian and vertebrate genome size have evolved concomitantly. It has been convincingly argued that intronic recombination can facilitate modular protein evolution by allowing exon shuffling (Patthy, 1999), and this process was associated with a burst of evolutionary diversity at the time of metazoan radiation. Similarly, the dissociation of enhancers and promoters may have allowed more modular regulatory evolution. Typically, a developmentally regulated gene will have several distal enhancers that can interact with the gene's promoter in different cell types or at different times during development. The increased genome size at a roughly fixed gene number greatly increases the available regulatory space relative to single-celled organisms, and would have allowed this modular structure to evolve more easily than in a very compact genome. Intergenic recombination events that swap enhancers to new locations can allow the rapid evolution of novel combinations of enhancers and regulatory circuits. It seems likely that this was crucial in the evolution of the increased cell diversity and developmental regulatory complexity of multicellularity.

Since the discovery of the interferon-beta enhancer, much progress has been made towards elucidating the fundamental molecular mechanisms by which enhancers regulate the transcription of their target genes. The transcriptional consequences of enhancers are mediated through direct binding of sequence specific transcription factors (TFs) and their interactions with complexes, which modulate chromatin accessibility. It is now fairly well established that all of three main classes of regulatory elements (promoters, enhancers, and insulators) are comprised of binding sites of several cooperating TFs. While all of these elements share some common features, recent studies indicate that a key difference between enhancers and promoters and insulators is that while most enhancers are activated in a temporal or cell-type specific manner, promoters and insulators tend to have a broader function, i.e. they are active across most cell types in human and mouse (Heintzman *et al.*, 2009; Shen *et al.*, 2012).

Enhancers and human disease

Most human traits have a familial hereditary component, but demonstrate complex patterns of inheritance. Genome-wide association studies (GWAS) have been widely used to identify complex trait loci, and now more than 7000 SNPs are known to be associated with variation in more than 700 traits and diseases (www.genome.gov/gwastudies) (Hindorff *et al.*, 2009). Although GWAS can still only explain a small fraction (5–10%) of the phenotypic variance (Manolio *et al.*, 2009), the list of regulatory mutations responsible for heritable susceptibility to specific disease is growing at a steady rate. The significant role of regulatory variation in complex trait heritability is underscored by the finding that the majority (>80%) of the trait associated SNPs are non-exonic (Manolio, 2010). When a regulatory variant is identified, it is often hypothesized that the variant disrupts a TF binding site, or creates a new binding site, or both.

Although it can be extremely difficult to identify causal regulatory variants associated with human disorders and their underlying molecular mechanisms, there are some well-studied cases. The classic example of how disruption of a regulatory element can directly cause human disease is in the β-globin locus, where removal of the locus control region can disrupt the high expression levels required in erythroid cells and lead to thalassemia (Grosveld *et al.*, 1987; van Assendelft *et al.*, 1989). The first vertebrate insulator to be systematically characterized was also in the β–globin locus (Chung *et al.*, 1993). Preaxial polydactyly (PPD), a limb malformation, is another dramatic example. PPD in humans was genetically mapped to a 450 kb chromosomal locus on 7q36, but it was not until a fortuitous mutant mouse was generated by a transgene insertion that these mutations were associated with mis-expression of *Sonic hedgehog* (*Shh*), approximately 1 Mb away from the gene (Lettice *et al.*, 2002). This transgene insertion disrupted a limb distal enhancer of *Shh*, in which several different single-nucleotide variations, segregating with the PPD phenotype in four unrelated human families, have been identified (Lettice *et al.*, 2003).

Other important examples include cases in which disruption of a PAX6 enhancer are associated with aniridia (Lauderdale *et al.*, 2000); deletion of sequences approximately 1 Mb upstream of *POU3F4* causing X-linked deafness (Kleinjan and Van Heyningen, 2005); and a common SNP in RET intron 1 associated with a high risk of Hirschsprung's disease (Emison *et al.*, 2005; Grice *et al.*, 2005). In a more recent example, two SNPs associated with prostate cancer are both found in an enhancer which regulates the 1 Mb distant *SOX9* gene, and these variants modulate the enhancer's activity by strengthening or weakening binding sites for FoxA1, AP1, and AR within the enhancer (Zhang *et al.*, 2012).

Since the early observation of the extreme similarity of proteins across different vertebrate species, both in sequence and biochemical activity, it has been postulated that much evolutionary diversity is generated by changes in gene regulation mediated by mutation of regulatory elements (Britten and Davidson, 1971; King and Wilson, 1975). It is thus reasonable to expect that human evolution is also operating largely at the level of regulatory variation, and that much of the existing variation in the human population is contributing to variable susceptibility to common disease.

Experimental methods for genome-wide enhancer detection

Early experimental exploration of the genomic regulatory landscape in *Drosophila* used a method known as enhancer trapping, employing a mobile GAL4 gene inserted randomly into the genome, driving GAL4 expression from flanking genomic enhancers (Brand and Perrimon, 1993). When crossed with lines carrying a GAL4 responsive reporter gene, the patterns of the flanking enhancer's activity can be observed. The genome sequencing projects ushered in a new generation of systematic approaches which aim to map all genomic regulatory elements. The initial sequencing of the human genome revealed that the gene number was surprisingly low, 20,000–25,000 genes, comparable to other model organisms (*C. elegans*, *Drosophila*). This was an initial indication that the differences between humans and these model organisms is not due to a dramatic increase in the repertoire of tissue-specific genes, but instead results from a dramatic evolution in the structure and repertoire of regulatory elements. Subsequent sequencing of the mouse genome (Waterston *et al.*, 2002) showed that roughly 5% of the human genome was under selection, and since only 1.5% of the human genome was accounted for by protein coding genes, it was strongly suggested that at least 3.5% of the genome encoded regulatory function.

Genome sequencing allowed the construction of the first generation of microarrays, which used hybridization to detect DNA bound by key components of regulatory complexes. This method, known as ChIP-chip, was used successfully to map TF–DNA interactions of hundreds of TFs in yeast (Harbison *et al.*, 2004), and similar techniques (ChIP-PET) were applied to detect genome wide binding of MYC in human cells (Zeller *et al.*, 2006).

Soon after the Human Genome Project was completed, the ENCODE pilot project was then initiated with the ambitious goal of cataloguing all possible functional elements in the human genome (ENCODE Project Consortium, 2007). Although only 1% of the human genome was initially evaluated with various techniques, several biological insights were already gained from this pilot project. One striking finding was that a surprisingly large fraction of potential regulatory DNA elements did not appear to be evolutionarily constrained (as assessed by sequence alignment). These initial findings encouraged a more complete approach applying similar analysis to the whole genome (ENCODE Consortium, 2012).

At about the same time, the parallel development of next-generation sequencing technologies and the accompanying dramatic drop in the cost of deep sequencing enabled completely different approaches to the identification of regulatory elements. For example, chromatin immunoprecipitation followed by sequencing (ChIP-seq) using antibodies specific to the protein of interest is now a routine process when genomic occupancies of a certain DNA interacting factor are in question. Specifically, enhancer-specific protein markers, such as EP300/CREBBP coactivators and covalent modifications of histone proteins (H3K4me1 and H3K27ac) have further facilitated the identifications of genome-wide enhancers in mammalian genomes (Heintzman *et al.*, 2007, 2009; Visel *et al.*, 2009a). DNase-seq is another example of using next-generation sequencing technologies to detect regulatory elements (Crawford *et al.*, 2006; Boyle *et al.*, 2008). The accessibility of the genome to DNA binding factors is not invariant, but rather extremely variable under different conditions, and it has long been known that highly accessible DNA is associated with various kinds of regulatory elements such as promoters, insulators, and enhancers. To experimentally detect such regions, biologists have taken advantage of the fact that deoxyribonuclease I (DNase I), an endonuclease, exhibits varying cleavage efficiency depending on the DNA accessibility. Now, combined with new sequencing technology, DNase-seq has become a standard technique to find genome-wide open chromatin regions.

The initial successful ENCODE pilot project has now been significantly expanded to the main human and mouse ENCODE projects fully equipped with new sequencing technologies. The ENCODE project (ENCODE Consortium, 2012) has produced maps of chromatin accessibility (Thurman *et al.*, 2012) (via Dnase I hypersensitivity (DHS)), genomic binding of many key TFs (Gerstein *et al.*, 2012) (via ChIP-seq), and specific chromatin marks in a broad array of human cell lines (Ernst *et al.*, 2011) and mouse tissues (Shen *et al.*, 2012). Together, these maps have identified a large set of previously undocumented regulatory regions and either directly or indirectly reflect cell-specific TF occupancy. Each of these experimental methods is subject to its own limitations and they are under continuing development. In particular, the set of genomic regions defined as being positive in any of these assays will depend sensitively on the signal threshold chosen. However, as we will demonstrate, these data provide a rich substrate which can be used to develop a predictive computational model of enhancer activity.

Additional reviews on enhancers

Several valuable and somewhat overlapping reviews of enhancer structure and function include those by Kleinjan and Lettice (2008), which focuses on examples of human disease associated with regulatory mutation; those by Maston *et al.* (2006), Visel *et al.* (2009b) and Noonan and McCallion (2010) which focus on molecular mechanisms, conservation, and comparative genomics; that by Zhou *et al.* (2011), which emphasizes chromatin structure and histone modifications; those by Bulger and Groudine (2011) and Spitz and Furlong (2012), which focus on the functional diversity and structure of enhancers; and, finally, that by Levine (2010), which emphasizes the central role of enhancers in vertebrate development and evolutionary diversity.

Computational prediction of enhancers

While consensus on the general mechanisms is rapidly emerging, we do not yet understand how enhancers work at the level of a predictive model of regulatory element activity, which can specify the set of cell types and environmental conditions under which the enhancer would stimulate the expression of its target gene(s). Further, a predictive enhancer model should describe how specific mutations to that enhancer sequence would affect its activity. Our philosophy here is akin to protein coding gene prediction: although direct experimental validation of each individual gene's transcription is ultimately required to verify the predictions, current gene prediction algorithms which integrate incomplete and perhaps noisy experimental data (e.g. ESTs) and genomic sequence features (e.g. ORFs and models of splice donors and acceptors) are able to provide a highly accurate picture of the set of protein coding genes in many organisms (Brent, 2008, and references therein). In the case of enhancer prediction, we know that the key features are transcription factor binding sites, but we have incomplete knowledge of their binding specificities and function, especially which TF binding events are able to modulate chromatin structure and vice versa.

In this section, we first review early computational approaches to enhancer prediction (see 'Early approaches based on TF binding sites (TFBS) clustering and conservation' below), then we introduce methods which use primary DNA sequence and frame the problem as a discriminative classification problem (see 'Early sequence based discriminative approaches' below), and present a recent successful SVM-based discriminative approach (see 'Enhancer prediction using support vector machines' below). Lastly, we discuss important general issues affecting classifier evaluation and performance, and use as an example prediction of mouse heart enhancers from DNA sequence (see 'Evaluation of classifier performance' below).

Early approaches based on TF binding sites (TFBS) clustering and conservation

Attempts to predict regulatory elements from DNA sequence in mammalian genomes still face major challenges in computational biology. As we have gained more knowledge about regulatory elements, various strategies to computationally identify regulatory regions have been developed. Until recently, however, none of the previous methods have shown success rates in predicting mammalian enhancers that would encourage their use as a general tool in biological or medical investigation, as assessed in a benchmark study (Su *et al.*, 2010).

Several early approaches took advantage of the observation that TFBSs tend to cluster together within relatively short DNA stretches ranging from several hundreds to thousands base pairs. These approaches showed some success, especially in the *Drosophila* genome (Berman *et al.*, 2002; Bailey and Noble, 2003; Johansson *et al.*, 2003; Sinha and Tompa, 2003; for review, see Wasserman and Sandelin, 2004), but application to mammalian genomes has been much less promising. These methods essentially identify regions that harbour TFBSs more than expected by chance within a given window of DNA sequence by using relatively simple counting methods or more sophisticated probabilistic methods such as hidden Markov models. However, this strategy always relies on prior knowledge about TF binding specificities, which is still thought to be far from complete. Also, some of these methods only identify regions where TFs are densely clustered without regard to the identity of the TFs in the combinations, a biologically implausible assumption that might lead to large number of false positives in their predictions.

There are other strategies that utilize sequence conservation information in combination with aforementioned methods. Since regulatory function is under evolutionary constraint, it is a widely accepted idea that significant fraction of conserved non-coding DNA is likely to function as regulatory elements (Waterston *et al.*, 2002), although the converse is not necessarily true (Fisher *et al.*, 2006; ENCODE Project Consortium, 2007; McGaughey *et al.*, 2008; Blow *et al.*, 2010). Sequence conservation information can be used to detect putative regulatory elements under purifying selection by comparing different species, as well as to detect individual TFBSs within regulatory elements, a technique known as phylogenetic footprinting. Several methods have been developed based on this idea, mostly focusing on the *Drosophila* genome (Sinha *et al.*, 2004; Sinha and He, 2007; He *et al.*, 2009), and some for mammalian genomes (Xie *et al.*, 2005; Hallikas *et al.*, 2006; Pennacchio *et al.*, 2007). However, subsequent experiments are always required since sequence conservation gives essentially no information about the element's specific biological function. Moreover, these validation experiments are typically labour-intensive and time-consuming. One notable study set out to systematically assay conserved non-coding regions in the human genome *in vivo* using a LacZ reporter system in transgenic mice to discover developmental tissue-specific enhancers. Among over 2000 regions tested so far, they discovered that at least 40–50% can act as tissue specific enhancers at a single developmental time in the early mouse embryo (Pennacchio *et al.*, 2006).

Unfortunately, all the efforts discussed so far have achieved only limited success and systematic computational approaches fell far short of desired predictive accuracy (Su *et al.*, 2010), especially in mammalian genomes. These results strongly suggest that current knowledge about TF binding specificities and overall sequence conservation information is not sufficient to describe the function of regulatory elements from primary DNA sequence.

Early sequence based discriminative approaches

The demonstrated limitations of sequence conservation and current knowledge about TF binding specificities as predictors of regulatory function have led many computational biologists to develop more sophisticated approaches. Models which integrate limited experimental evidence of chromatin state or cofactor binding and sequence features enriched in these regions can provide a more accurate description of the genomic enhancer landscape than either approach used in isolation. Such sequence-based models can be framed as classifiers which can discriminate between regulatory elements and non-regulatory DNA after training on a suitable but incomplete set of sequence regions with the function of interest.

One of the first successful studies using a sequence based discriminative method, also known as 'Regulatory Potential', showed that simple Markov models can distinguish regulatory regions from non-regulatory regions with reasonable accuracy (Elnitski *et al.*, 2003; Kolbe *et al.*, 2004; King *et al.*, 2005). In this approach, two second-order Markov models separately trained on a set of aligned known regulatory regions and a set of neutral DNA regions (ancestral repeats) were used to calculate the log odds ratio of a given DNA sequence. This remarkably simple method demonstrated for the first time that regulatory elements in mammalian genomes can be predicted from primary DNA sequence without prior knowledge of TF binding specificities, although the overall accuracy was not high enough to be useful for genome-wide prediction. More recently, several studies have achieved notable successes in predicting different classes of regulatory elements in mammalian genome using various techniques: transcription start site prediction using SVMs (Sonnenburg *et al.*, 2006); promoter prediction using logistic regression (Megraw *et al.*, 2009); enhancer prediction using LASSO regression (Narlikar *et al.*, 2010); and enhancer prediction using SVMs (Lee *et al.*, 2011; Gorkin *et al.*, 2012). These recent successes are mostly due to (1) state-of-the-art supervised machine learning algorithms and (2) appropriate experimental datasets for model training. Especially for enhancer prediction, one of the most accurate methods to date is kmer-SVM previously developed by our group (Lee *et al.*, 2011). Since the development of the original kmer-SVM, this model has been applied to other problems, and some of computationally predicted regions have been experimentally validated both *in vitro* and *in vivo* (Gorkin *et al.*, 2012). In the next section, we will discuss kmer-SVM methods in great detail.

Enhancer prediction using support vector machines

Since the development of support vector machines (SVMs) in the early 1990s (Boser *et al.*, 1992; Vapnik, 1995), the SVM has become one of the most popular machine learning techniques and has been successfully applied to almost every problem in computational biology [for reviews, see Schölkopf *et al.* (2004) and Ben-Hur *et al.* (2008)]. A SVM is a general binary classifier that learns a decision boundary, called a hyperplane, by maximizing margins between the two sets in the feature vector space, formalized as follows. Suppose we have N number of n-dimensional real-valued vectors $\mathbf{x}_i \in R^n$ with associated class labels $y_i \in \{+1, -1\}$ for $i = 1, \ldots, N$. Then, a hyperplane is found by minimizing $\|\mathbf{w}\|^2$ such that $y_i(\mathbf{x}_i \cdot \mathbf{w} + b) \geq 1$ for all i. In practice, however, the optimal solution is obtained by maximizing the Wolfe's dual form:

$$\sum_{i=1}^{N}\alpha_i - 1/2\sum_{i=1}^{N}\sum_{j=1}^{N}\alpha_i\alpha_j y_i y_j(\mathbf{x}_i \cdot \mathbf{x}_j)$$

subject to $\alpha_i \geq 0$ for any $i = 1, \ldots, N$, and

$$\sum_{i=1}^{N}\alpha_i y_i = 0.$$

In the dual form, the inner product $(\mathbf{x}_i \cdot \mathbf{x}_j)$ can be considered as a measure of the similarity between two data points *i* and *j* in the feature space. Moreover, since this is the only term that has feature vectors in the object function, it can be replaced by a more general function, called a kernel function $K(\mathbf{x}_i \cdot \mathbf{x}_j)$. This generalization makes SVMs very powerful methods because it relaxes the requirement of an explicit feature space as long as a kernel function between any two data points is defined. A very simple yet powerful measure of sequence similarity is the *k*-spectrum kernel (Leslie *et al.*, 2002), which calculates the inner product of frequencies of all possible *k*-mers of two sequences. This kernel was first introduced to classify functional domains from protein sequence, and have been successfully used in several different contexts such as nucleosome positioning (Peckham *et al.*, 2007), as well as enhancer predictions (Lee *et al.*, 2011).

In the initial study of enhancer prediction, our kmer-SVM method was originally applied to several different EP300/CBP bound genomic data in various tissue and cell types; mouse embryonic tissues (Visel *et al.*, 2009a), activated cultured neurons (Kim *et al.*, 2010), and embryonic stem cells (Chen *et al.*, 2008). In Visel's dataset, for example, several thousands of EP300-bound DNA elements were collected by ChIP-seq in micro-dissected forebrain, midbrain, and limb in embryonic day 11.5 mice. We formulated this as a simple discriminative classification problem: given a positive set of EP300 bound enhancers, and a set of unbound negative regions with matched sequence properties (length, GC and repeat fraction), we sought to determine the set of sequence features and a scoring function which can discern the positive and negative sequences. Because EP300 binding is facilitated through indirect binding with a set sequence specific TFs whose composition is tissue specific and generally unknown, we chose to use as sequence features the complete set of *k*-mers of a given length (e.g. $k=6$ or hexamer). We used these *k*-mer counts as a feature vector used to train an SVM classifier for each of the positive enhancer sets, as shown in Fig. 6.1. Parallel developments have also achieved similar success using a set of position weight matrix (PWM) scores as features in an SVM model (Busser *et al.*, 2012). As discussed earlier, however, PWM approaches may be limited by the possible incompleteness of the known PWM databases and can suffer from some unreliable

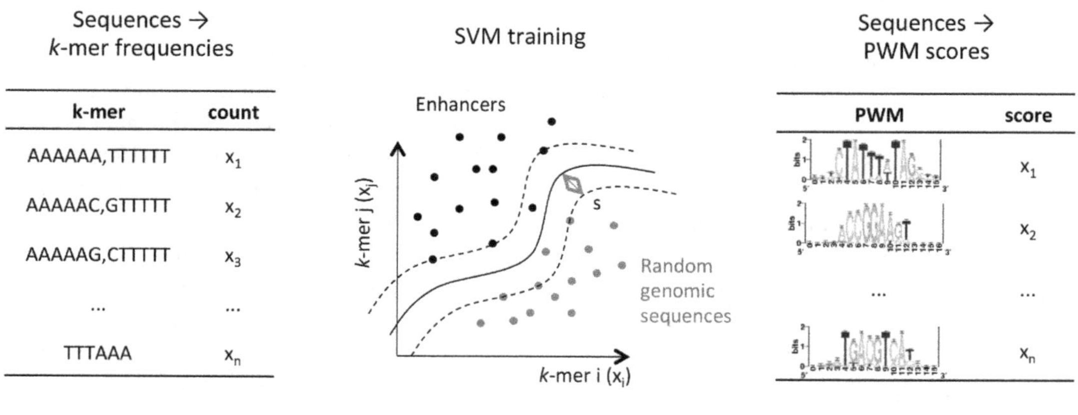

Figure 6.1 An SVM classifier to predict enhancers describes a boundary that maximally separates enhancers from non-enhancer sequences in an *n*-dimensional feature space. Both *k*-mers (Lee *et al.*, 2011) and PWMs (Busser *et al.*, 2012) have been successfully used as features.

PWM models in the databases. After a SVM is trained, a class label of each tested sequence is predicted based on the SVM output score, *s*, which describes the distance of a given element from the decision boundary in the *n* dimensional feature space.

The quality of each classifier was evaluated by calculating the test set AUC (area under the ROC curve) using five-fold cross validation, and surprisingly high accuracy (AUC > 0.9) was achieved in every case we tested. These results were the first strong indication that mammalian enhancers are primarily specified by their DNA sequence, through the binding sites of tissue specific regulators. We also demonstrated the ability to detect tissue-specific enhancers by comparing between the three EP300 ChIP-seq datasets, treating one tissue's enhancers as the positive set, and the other's as the negative set. While forebrain and midbrain enhancers can be discriminated from limb enhancers with a reasonable AUC (0.85–0.86), the SVM failed to separate forebrain and midbrain enhancers, suggesting that predictive sequence features in forebrain and midbrain enhancers are similar to each other, but sufficiently different from those in limb enhancers.

Although a SVM is generally considered as a complete black box method and, therefore, is thought to be very hard to interpret, we found that a weight of each *k*-mer in the primary feature space can be interpreted as its overall contribution to enhancer activity. Consistent with this idea, the majority of the most predictive *k*-mers (those with the largest positive weights) for forebrain enhancers were identified as binding sites for TFs critical in nervous system development. We also established additional supporting evidence by showing that these sequence features are evolutionarily conserved and spatially clustered. A notable distinguishing feature of the kmer-SVM approach is its ability to identify sequence features that are significantly depleted in EP300 enhancers, *k*-mers with large negative weights. This can be biologically interpreted as binding sites for TFs which can disrupt enhancer function in a specific tissue. In forebrain enhancers, for example, ZEB1-related *k*-mers were observed with the largest negative weights, suggesting that ZEB1 binding may play a repressive role, or block the activation of forebrain enhancers. It is worth noting that the SVM algorithm does not directly calculate the weights, but finds an optimal set of support vectors, data points on the positive and negative margins (dashed lines in Fig. 6.1). SVM weights can then be calculated from the set of support vectors in the case of a linear kernel function.

Besides identifying important sequence features within the training enhancer set, kmer-SVM can be directly used to detect additional enhancers. To validate our genome-wide predictions of novel forebrain enhancers with independent experiments, we used DNase I hypersensitivity in embryonic mouse whole brain (Mouse ENCODE Consortium, 2012). Although these experiments were not performed under exactly the same conditions, strong enrichment of DNase I hypersensitivity in the original EP300 forebrain regions is an independent experimental validation of our predictions. Similar to the EP300 bound forebrain regions, we observed a dramatic increase in DNase I hypersensitivity only for high scoring SVM regions, but we did not observe such enrichment of DNase I signal in these genomic regions in other tissues (i.e. we used mouse kidney as a negative control). From the Dnase I experiments we roughly estimate that our genome-wide predictions have ~50% precision.

After the successful development of kmer-SVM, this method was further validated by several independent experiments in different biological conditions (Gorkin *et al.*, 2012). Using the kmer-SVM trained on 2489 putative melanocyte enhancers (AUC = 0.912), a set of 7361 additional melanocyte enhancers were predicted with an SVM score threshold expected to give 50% precision. Eleven of these predicted enhancers were then tested and discovered that eight regions directed significant luciferase expression *in vitro* in melanocytes (>3× minimal promoter alone, 8/11 or 73%). Furthermore, two of three predicted enhancers assayed *in vivo* directed GFP expression in the melanocytes of mosaic transgenic zebrafish. These additional validation experiments both *in vitro* and *in vivo* provide strong evidence that our kmer-SVM method can be effectively used to predict novel regulatory regions.

In this section, we demonstrated three major features of kmer-SVM methods: (1) predicted sequence features reveal biologically relevant sequence elements enriched in the enhancers; (2) kmer-SVM can also identify other sequence features that are significantly depleted in enhancers; (3) kmer-SVM can predict putative enhancers with reasonable precision which were further supported by both *in vitro* and *in vivo* validation experiments.

Evaluation of classifier performance

There are several metrics used to evaluate the prediction performance of machine learning algorithms. The most common approach is to use cross-fold validation, where a subset of the training data is not used for training, but reserved for evaluating the model. These elements are known as the test set, and their known class membership can be compared to the predictions. Test set *accuracy*, defined to be the fraction of correct positive or negative predictions, is the perhaps the most obvious measure, but has two limitations. First, when class membership is very unbalanced, as in the case of genomic predictions of enhancers, *accuracy* can be quite high even for a completely uninformative predictor. For example, if we are predicting enhancers among random genomic intervals, and if 1% of these intervals are positive, we can get 99% *accuracy* with a classifier that simply predicts all elements to be negative, or non-enhancers. Second, accuracy does not take into account that the classifier is typically very confident about the predicted class for some elements and less confident for others. The investigator is usually most interested in those elements which are predicted to be positive with high confidence, in our case these predicted positive elements would be chosen for experimental tests of enhancer activity. Since validation experiments are usually fairly time consuming and difficult, clearly we wish to test those elements which are most likely to be successful positive predictions and impart useful biological knowledge. The most common way of separately representing the accuracy of the positive and negative predictions is a confusion matrix: a table showing the number of true positive, false positive, true negative, and false negative (TP, FP, TN, FN) test set predictions. The confusion matrix will depend on the particular threshold used to separate the positive and negative predictions.

To look at a specific example, we now consider enhancers bound by EP300 in adult mouse heart tissue as determined from ChIP-seq experiments (Mouse ENCODE Consortium, 2012). After processing this dataset using MACS (Zhang *et al.*, 2008) to detect peaks, we defined a set of 3000 high confidence EP300 bound 400 bp regions as the positive set. Improved techniques to detect which of these peaks are truly significant are under continuing development, and could improve classification accuracy by reducing noise in the training set (Ghandi and Beer, 2012).

We then generated a GC and repeat fraction matched set of random genomic negative regions, and varied the size of this negative set from 3000 to 300,000 negative elements, i.e. 1× to 100× the positive set. After generating the SVM model from the training set, the SVM score function can be used to score each element in the test set. The score distributions for training a $k = 7$ kmer-SVM for 1×, 5×, and 10× negative sets are shown in Fig. 6.2, where the positive enhancer elements have systematically higher scores. It can be seen that the kmer-SVM score produces a reasonable but imperfect separation between the enhancer and non-enhancer elements. In these plots, the score for each element is shown only when it is a member of the test set. When we retrain on the larger negative sets, the shape of the score distributions are largely unchanged, and remain approximately normal with mean Δ and standard deviation σ, $N(s,\Delta,\sigma)$, as shown in Fig. 6.2D. This motivates the use of the true positive rate = $TPR = TP/P$ and false positive rate = $FPR = FP/P$ as performance metrics, where P and N are the number of positive and negative elements in the test set, respectively.

The largest weight *k*-mers are those that distinguish the heart enhancers, and largely consist of binding sites for TFs known to play a role in the development and differentiation of heart tissue. Some of the largest weight 7-mers from training against the 10× negative set are shown in Fig. 6.3, and matched to PWMs for specific factors that are known to direct the cardiomyocyte transcriptional program: DNA-binding TFs MEF2A

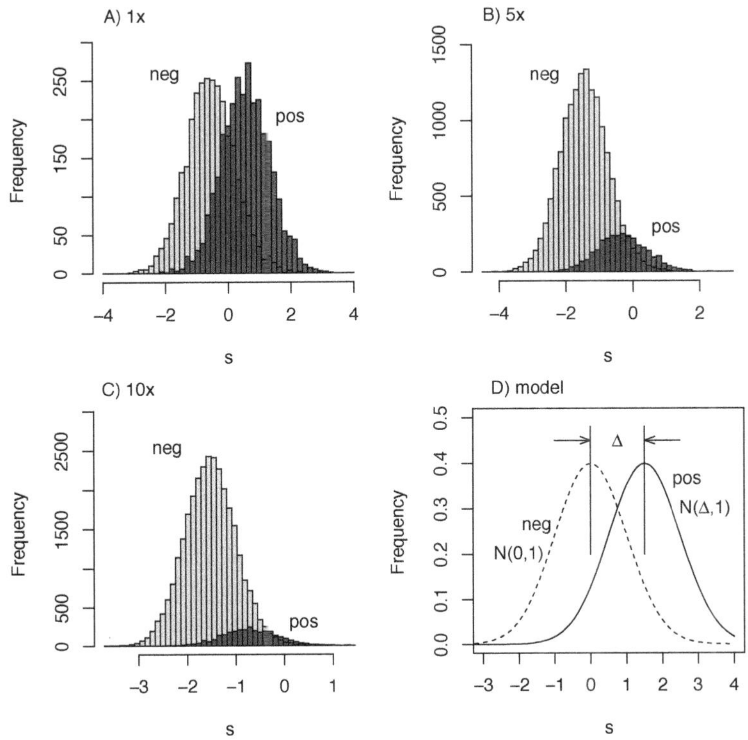

Figure 6.2 Score distributions for training a kmer-SVM on EP300 bound heart enhancers with a varying size of the negative set (A) 1×, (B) 5×, (C) 10×. As the negative set size varies the distributions remain close to normal, and the classifier can be modelled using shifted normal distributions (D).

(myocyte-specific enhancer factor 2A), GATA4 (GATA binding protein 4), NKX2.5 (NK2 homeobox 5) and SRF (serum response factor) are known to play important roles for cardiomyocyte differentiation and functional specificity (Lyons *et al.*, 1995; Kuo *et al.*, 1997; Molkentin *et al.*, 1997; Naya, 2002; Miano *et al.*, 2004; Niu *et al.*, 2005; Srivastava, 2006). Additionally, *Gata4*, *Mef2a*, *Nkx2.5* and *Srf* are also known to regulate each other's expression (Spencer and Misra, 1996; Searcy *et al.*, 1998; Balza and Misra, 2006; Karamboulas *et al.*, 2006), but little is known about the complex interplay of these factors in cardiomyocyte specification and homeostasis. Recent studies have begun to investigate the co-occurrence of binding sites for these TFs near heart expressed genes (Schlesinger *et al.*, 2011), but have focused on binding sites only within several kb of transcriptional start sites (TSS), when many can be located much further away. Many of the most predictive *k*-mers are identifiable as binding sites of known cardiac regulators, but also identify more general factors involved in enhancer specification, such as CTCF. Also shown in Fig. 6.3 is the enrichment of binding sites for all of the highlighted PWMs in the heart enhancers relative to the genomic average. Although several factors known to play key roles in cardiomyocyte regulation are only weakly enriched individually (GATA4, MITF, NFAT, Nkx2.5, SRF), the kmer-SVM finds that their binding sites are predictive in combination, and thus they receive high *k*-mer weights.

A standard metric to evaluate classifier performance is the ROC curve, shown in Fig. 6.4A. This curve is produced by plotting the parametric curve $y(s) = \text{TPR}(s)$ vs. $x(s) = \text{FPR}(s)$ where the true positive rate and false positive rate are calculated for elements above a varying threshold score, s. At very high s, the curve starts at (0,0). As s is decreased slowly, the score histogram in Fig. 6.2 shows that most of the elements are true

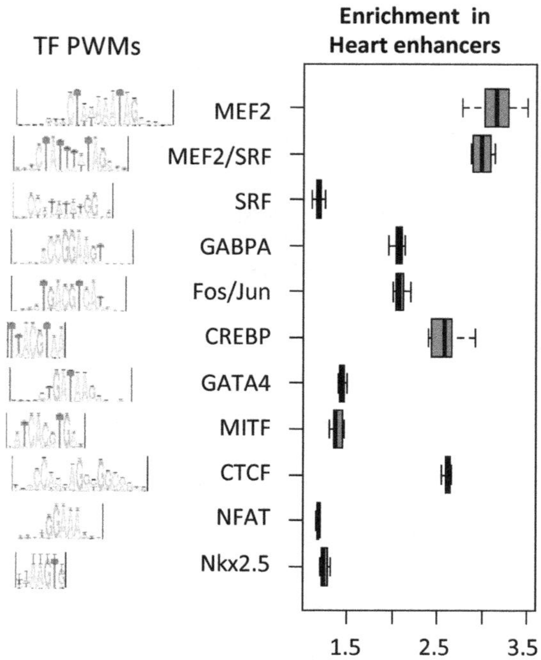

Associated k-mer SVM

TF	k-mer	wt
MEF2	AAAATAG	2.13
	AAATAGC	1.99
MEF2/SRF	TAAAAAT	1.92
	CTAAAAA	1.77
SRF	AAAAATA	1.66
GABPA	CGGAAGT	2.11
	ACCGGAA	1.75
	GCCGGAA	1.69
Fos/Jun CREBBP	TGACGTC	2.15
	GTGACGC	2.06
	TACGTAA	0.60
GATA4	AGATAAG	1.97
MITF	CATGTGA	1.99
CTCF	CCACTAG	1.30
	GGTGGCG	1.19
	AGAGGGC	1.17
NFAT	CCGGAAA	2.15
Nkx2.5	GGAAGTG	1.99

Figure 6.3 SVM weights reflect diversity of TF binding sites involved in heart enhancer specification. *K*-mers with large weights found after training on heart enhancers are mapped to known PWMs. Although the weights are found to be significant in describing the SVM boundary, many of these PWMs are not significantly enriched in the heart enhancers relative to genomic background levels.

positives, but as s decreases more, some negative elements exceed the threshold and false positives are predicted. Eventually, at negative s thresholds, the TPR and FPR both approach one, as all elements are predicted to be positive at the extreme negative value of s. The area under this ROC curve (AUC) is a rough aggregate measure of overall classifier performance. A random classifier has AUC = 0.5.

The ROC for the heart trained SVM with $k=7$ is shown in Fig. 6.4A for different negative set sizes, and the AUC is shown. The AUC of around 0.85 is clearly much better than random, but is on the low range of our experience modelling different datasets (Lee *et al.*, 2011). This is primarily due to the quality of the input dataset, but the sequence composition and complexity of the TFs involved, e.g. their length and GC content, may also play a role. EP300 may not be the best molecular target to identify heart enhancers with high precision. Two separate factors contribute to the initial increase in performance from 1× to 5×, and then the decline in performance from 10× to 100×. First, with more training data the estimation of the SVM classification boundary becomes more statistically robust, but then the imbalance of the positive and negative set sizes begins to modestly degrade performance. This can be corrected during SVM training with the class imbalance parameter, but for simplicity this was not done here.

AUC, however, is not the best practical measure of classification performance. The investigator is usually interested in performing experimental tests to validate the predictions, and the crucial measure is how many of these predictions will be correct. This measure is precision, the ratio of true positives to predicted positives (PP). In the ROC, the status of the test set elements are assumed to be known, but in validation experiments, this is not the case, so the relevant rate is TP/PP, not TP/P. The precision-recall, or P-R, curve is the standard measure in this case, and is shown in Fig. 6.4B. Recall is the same as the True positive rate, and the P-R curve is also a parametric curve as the score threshold s is varied. The P-R curves with

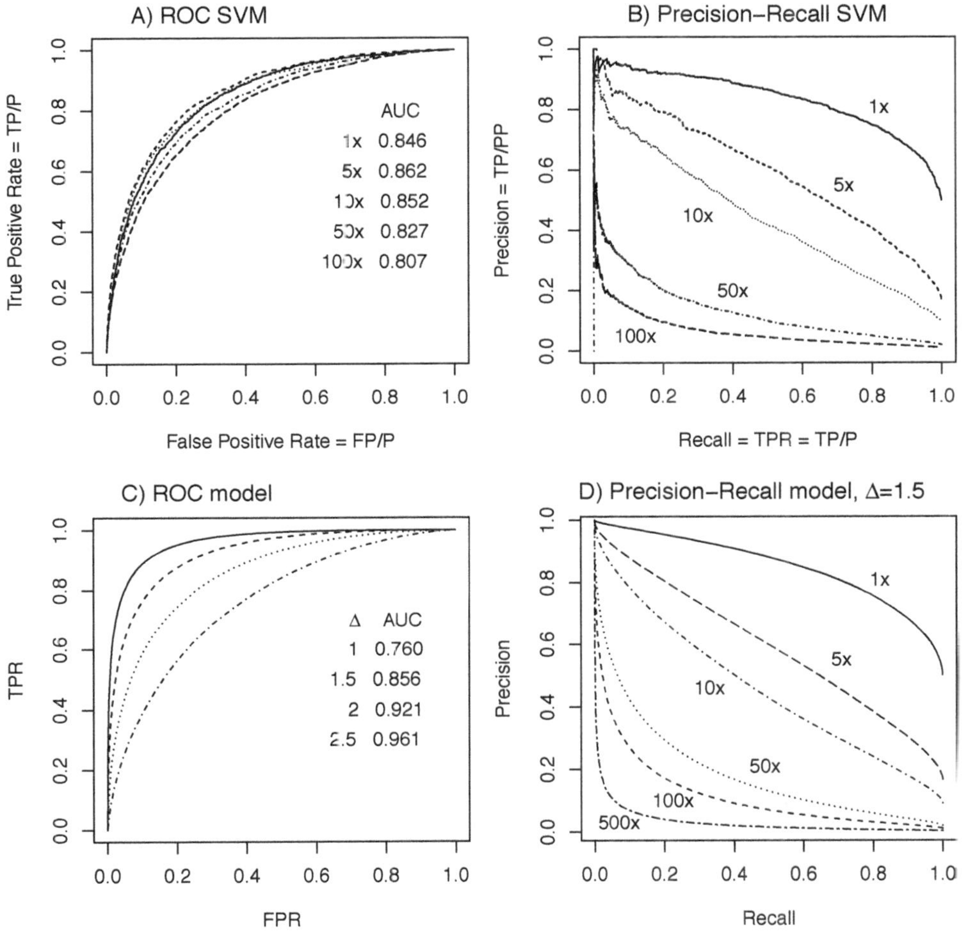

Figure 6.4 ROC analysis of SVM classification performance. (A) ROC and AUC are not very sensitive to the size of the negative set. (B) The Precision of the classifier, defined as the ratio of True Positives to Predicted Positives, is extremely sensitive to the size of the negative set, as shown on the P-R curve. (C) Modelling the positive and negative score distributions as shifted normal distributions with mean Δ as in Figure 6.2D, the shift Δ of the positive distribution relative to the negative distribution centred at 0 directly correlates with AUC. (D) The P-R curves for this model with Δ=1.5, to roughly match AUC, are in close agreement with the actual SVM performance, and allow us to extrapolate to larger negative set size (500×).

varying negative set sizes change dramatically. For a 1× negative set, the precision at recall = 0.5 is 0.866. For a 100× negative set, the precision at recall = 0.5 is 0.045. Recall = 0.5 is a reasonable point to consider, since at this stringent score threshold, half of our known true positive training set elements would be predicted to be positive.

This highlights a crucial difference between training and experimental validation of these predicted enhancers. While each of the models in Fig. 6.4A has a reasonable AUC of around 0.85, the precision of the predictions varies dramatically with the size of the negative set. In practice, we only have rough estimates of the number of true heart enhancers in the human genome, and it seems reasonable to use as a negative set the entire available regulatory genome. At this time, in principle all non-repetitive (50%) and non-coding regions of the 3×10^9bp genome are potentially regulatory, but introns should certainly be included in the potential regulatory space, and it is not inconceivable that regulatory elements could also be in exons. Repetitive regions could also in principle have regulatory function. In this heart enhancer example, the positive set covers 1.2×10^6 bp. If we had further information to accurately constrain the negative regions to just those which are thought to be regulatory, perhaps 2–10% of the genome, the negative genomic space is still $1.5\times10^{7-8}$ bp, or 24–120, but since this is not accurately known, we should really consider the whole non-repetitive genome as potential regulatory space, so the negative genomic space is 1200× our positive set. As shown in Fig. 6.1C, as the negative set gets this large, the high scoring tail of the negative distribution can become competitive with, and even exceed the positive set score distribution. Using the negative set size of 100×, the Precision of 0.045 means that approximately 1 of 20 predicted enhancers tested will actually be positive in experiments. This would clearly be a disappointing result for the experimenter.

Using the model introduced above, where the positive set score distribution is a normal distribution shifted from the negative set score distribution with mean Δ and standard deviation σ, $N(s, \Delta, \sigma)$, as shown in Fig. 6.2D, we can investigate ways to improve this situation. First considering both positive and negative distributions to have the same standard deviation, $\sigma=1$, which is consistent with the observed scores in Fig. 6.1, we can calculate AUC and P-R curves as shown in Fig. 6.4C. Increasing Δ leads to higher AUC as expected, and $\Delta=1.5$ has AUC = 0.856 in close agreement with the heart enhancer kmer-SVM. Since $\mathrm{TPR}(s)=\mathrm{N}(s, \Delta, 1.5)$ and $\mathrm{FPR}(s)=N(s, 0, 1)$ are given by the normal distributions, we can calculate the precision exactly, and $\text{precision}=\mathrm{TP}/(\mathrm{FP}+\mathrm{TP})= 1/(1+(n_-/n_+)(\mathrm{TPR}(s)/\mathrm{FPR}(s))$, where n_-/n_+ and Δ are the only parameters. n_-/n_+ is the estimated ratio of negatives to positives in the genome, and as discussed above 100–500× seems a reasonable estimate. The model P-R curves shown in Fig. 6.4D closely match the kmer-SVM performance in Fig. 6.4B, and in the model we can extrapolate to larger negative set sizes (500×). It is clear that making high Precision genomic predictions is quite challenging, because the potential negative regulatory genomic space is so large.

We can use the normal model distributions to address this challenge. Fig. 6.5 shows how the Precision of the model depends on AUC for various values of n_-/n_+. For each value of n_-/n_+, we vary Δ to simulate classifiers with a range of AUC. For each value of Δ, we calculate the AUC and the precision at recall = 0.5. This shows that Precision is a very steep function of AUC. For example, at $n_-/n_+=500\times$, there is a cliff in Precision above AUC = 0.96. While, our heart enhancer predictions had AUC = 0.85, any improvements in the quality of the training data or improvements in the details of the SVM kernel or classification accuracy could dramatically improve the Precision of the predictions, and drastically improve the success rate in the enhancer validation experiments.

Finally, we would also like to emphasize that classifiers with the same AUC can give very different Precision. Using the model introduced above, we can consider the effect of changing the shape of the positive and negative score distributions. In Fig. 6.6 we vary the shape of the positive set score distribution relative to the negative score distribution, $N(s, \Delta, \sigma)$, by changing Δ and σ at the same time. In Fig. 6.6A three different positive set score distributions – $N(s, 1.55, 0.8)$, $N(s, 1.5, 1)$ and $N(s, 1.28, 1.2)$ – are shown and they especially differ in the high scoring tail

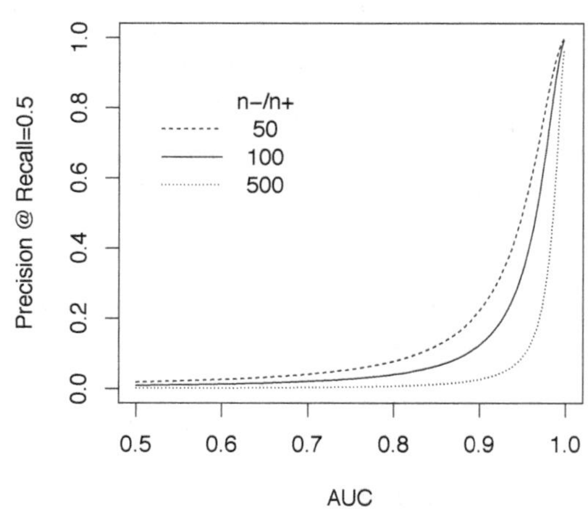

Figure 6.5 Precision is a very steep function of AUC. In the range AUC=0.85–1.0, any improvements in the kmer-SVM modelling or training data that increase AUC will have dramatic impact on the Precision of the enhancer predictions.

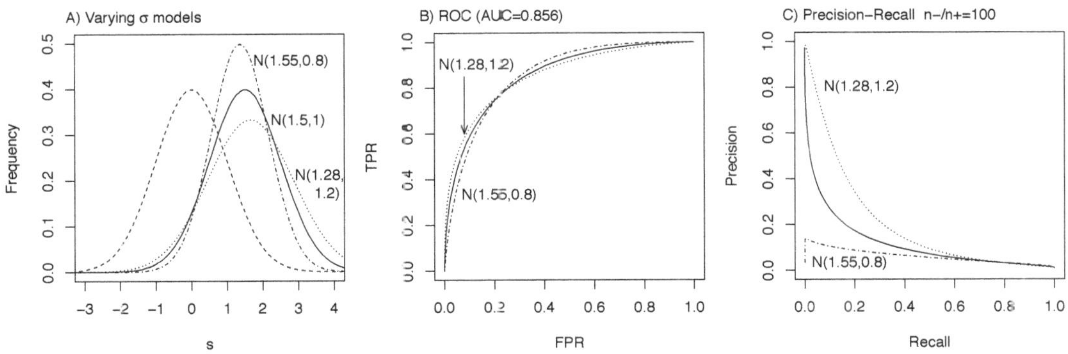

Figure 6.6 AUC is not a good indication of Precision if the shape of the score distribution varies. (A) Three different positive set score distributions are shown, with varying mean and s, and significantly different high scoring tails. (B) These three positive set score distributions all have the same AUC, but different ROC curves. (C) The three models have dramatically different Precision, the true positive rate at low FPR is the most important range.

of the distribution. These three parameter sets have slightly different ROC curves, as shown in Fig. 6.6B, but by construction all have the same AUC. However, as shown in Fig. 6.6C the precision-recall curves for these three distributions are dramatically different. With $n_-/n_+ = 100$, the Precision of the model with $N(s, 1.28, 1.2)$ is significantly greater than the model with N(s, 1.55, 0.8). This is because there is a high scoring tail of large SVM scores. This shows that the most important part of the ROC curve for genomic predictions with large n_-/n_+ is the low-FPR region, where $N(s, 1.28, 1.2)$ has significantly higher TPR than $N(s, 1.55, 0.8)$. Any improvements to enhancer prediction algorithms that effect these distributions will have dramatically improved precision.

Discussion and conclusions

The field of regulatory genomics is undergoing rapid and exciting growth, spurred by the combination of many factors. It seems almost unnecessary to mention that the human genome sequence has provided the framework for the development of several technologies which are coming together to revolutionize our understanding of the function and role of gene regulation. Key directions among these are (1) the development of new experimental technologies for producing genomic maps of chromatin accessibility, histone state, and DNA binding by regulator proteins in an ever-growing number of cell types, environmental conditions, and disease states; (2) the development of DNA sequence based machine learning approaches to detect regulatory elements; and (3) the assessment of common human sequence variation and associations with disease. However, this enterprise is still in a relatively early stage, and is sure to yield many surprises as we develop a more complete understanding of regulatory mechanisms and evolution.

Future trends

As discussed above, the accuracy of current enhancer prediction algorithms has recently improved dramatically, but there is significant room for improvement, and the development of higher precision classifiers would have a substantial impact on the rate at which biologically and medically relevant enhancers can be identified. Progress will likely come in the area of development of better kernel functions or distance measures used by the classifiers. A key ingredient that has not been fully exploited is detailed information about the spatial and configurational constraints between clusters of binding sites. While current approaches are improving in their ability to detect combinations of cofactor binding sites, most scoring functions are invariant to arbitrary reshuffling of these binding sites. It seems unlikely that this type of variation would completely preserve enhancer function. A challenge here will be that high order grammatical structures and more complex statistical learning models require more data for training, and there are is limited variation in the existing data (evolution tends to discard its mistakes). It could be that generation of large synthetic enhancer datasets is necessary to fully test the space of regulatory variation.

While predicting enhancers is interesting in its own right and aids the testing and validation of functional regulatory elements, we are also interested in predicting the fine scale features within the enhancer (the functional TF binding sites) at single base pair resolution for comparison with genomic footprinting experiments. Many statistical learning classification methods, such as SVM, succeed at overall classification by robustly describing the boundary between enhancers and non-enhancers, but are not very good at determining precisely how important each feature contributes to this boundary. This weakness arises because there may be many successful classifiers with significantly different hyperplane boundaries in the high-dimensional feature space. Robust feature detection will again require larger amounts of training data.

Ultimately, enhancer prediction tools should be able to predict the impact of DNA variants on cell type specific enhancer function. This would be a significant advance, but it is worth pointing out that even being able to precisely predict the strength of a mutated enhancer in isolation may not be sufficient to predict the phenotypic consequence of the mutation. Each enhancer has multiple inputs and operates within a highly connected regulatory network. A mutation, which strengthens an enhancer in one individual, may have a stronger or weaker effect in another individual because of nonlinear interactions with other variants. Nevertheless, our biological networks are extremely robust, so there may be simple design principles which help quantify these interactions. The critical mutations are those which most dramatically affect the overall output of the regulatory element in the context of its biological circuit.

As a final point we would like to emphasize a very important experiment whose conclusions impact the degree to which studies of gene regulation in model organisms are transferable to human biology. While studies of the genomic binding of several TFs important in liver differentiation showed that binding events were only conserved between human and mouse about a third of the time (Odom *et al.*, 2007), this evolution could have occurred at the DNA sequence level, or at the level of variation in the DNA binding specificities of regulatory factors. In a stunning follow up experiment, genomic binding of these TFs in chimeric mice carrying human chromosome 21 showed that the mouse proteins bound the human DNA in patterns almost identical to the binding of the human TFs to the human DNA in human cells (Wilson *et al.*, 2008). This indicates that the evolutionary changes between mice and humans are indeed occurring at the level of DNA sequence, and that the human and mouse enhancer binding proteins and transcriptional apparatus have very similar DNA binding specificities. We thus believe that studies of genomic binding and regulatory element structure in the mouse model system will have direct and relevant impact on the development of DNA sequence based enhancer prediction algorithms in humans.

Web resources

PWM databases

- JASPAR
  - http://jaspar.genereg.net/
  - The JASPAR database contains a manually curated, non-redundant collection of PWMs from publications of experimentally defined TFBSs for eukaryotes.
- TRANSFAC
  - http://www.gene-regulation.com/pub/databases.html
  - The TRANSFAC Public database provides PWMs from experimentally identified eukaryotic TFBSs. More comprehensive versions are available as commercial products.

- UniPROBE
  - http://the_brain.bwh.harvard.edu/uniprobe/
  - The Universal PBM Resource for Oligonucleotide Binding Evaluation (UniPROBE) database provides the *in vitro* DNA binding specificities of proteins for all possible *k*-mers, as well as PWMs, generated by protein binding microarray (PBM) technology.

Motif finding

- MEME Suite
  - http://meme.nbcr.net/meme/
  - The MEME Suite provides several tools for motif analysis, including *de novo* motif finding and motif comparison.

SVM

- SVM-light
  - http:/t/svmlight.joachims.org/
  - SVM-light is a C implementation of SVMs with very fast optimization algorithms. Both source codes and binaries are available.
- LIBSVM
  - http://www.csie.ntu.edu.tw/~cjlin/libsvm/
  - LIBSVM is software for SVM classification, regression, and distribution estimation. Both source codes and binaries are available.
- kmer-SVM
  - http://kmersvm.beerlab.org/
  - kmer-SVM is a Galaxy-based web-server that provides tools for regulatory analysis of next-generation sequencing (NGS) data (ChIP-seq, DNase-seq). Kmer-SVM identifies predictive combinations of short transcription factor binding sites within the larger regulatory elements.
- Rätsch Galaxy
  - http://galaxy.raetschlab.org/
  - This is a customized Galaxy-based web-server that provides general machine learning based tools for sequence and tiling array data analysis.

Genomic tools

- UCSC genome browser
  - http://genome.ucsc.edu/
  - The UCSC genome browser hosts a large collection of reference genomes, annotations and various related genomic data with convenient visualization tools.
- ENCODE datasets
  - http://genome.ucsc.edu/ENCODE
  - All ENCODE datasets can be accessed via this website.
- modENCODE
  - http://www.modencode.org
  - All modENCODE dataset can be accessed via this website.
- GWAS catalogue
  - http://www.genome.gov/gwastudies
  - The GWAS catalogue provides a comprehensive list of the most significantly trait-associated SNPs from previously published genome-wide association studies.

References

Van Assendelft, G.B., Hanscombe, O., Grosveld, F., and Greaves, D.R. (1989). The β-globin dominant control region activates homologous and heterologous promoters in a tissue-specific manner. Cell *56*, 969–977.

Bailey, T.L., and Noble, W.S. (2003). Searching for statistically significant regulatory modules. Bioinformatics *19*, ii16–ii25.

Balza, R.O., and Misra, R.P. (2006). Role of the serum response factor in regulating contractile apparatus gene expression and sarcomeric integrity in Cardiomyocytes. J. Biol. Chem. *281*, 6498–6510.

Banerji, J. (1981). Expression of a beta-globin gene is enhanced by remote SV40 DNA sequences. Cell *27*, 299–308.

Beer, M.A., and Tavazoie, S. (2004). Predicting gene expression from sequence. Cell *117*, 185–198.

Ben-Hur, A., Ong, C.S., Sonnenburg, S., Schölkopf, B., and Rätsch, G. (2008). Support vector machines and kernels for computational biology. PLoS Comput. Biol. *4*, e1000173.

Berman, B.P., Nibu, Y., Pfeiffer, B.D., Tomancak, P., Celniker, S.E., Levine, M., Rubin, G.M., and Eisen, M.B. (2002). Exploiting transcription factor binding site clustering to identify cis-regulatory modules involved in pattern formation in the Drosophila genome. Proc. Natl. Acad. Sci. *99*, 757–762.

Blackwood, E.M., and Kadonaga, J.T. (1998). Going the distance: a current view of enhancer action. Science *281*, 60–63.

Blow, M.J., McCulley, D.J., Li, Z., Zhang, T., Akiyama, J.A., Holt, A., Plajzer-Frick, I., Shoukry, M., Wright, C., Chen, F., *et al.* (2010). ChIP-Seq identification of weakly conserved heart enhancers. Nat. Genet. *42*, 806–810.

Boser, B.E., Guyon, I.M., and Vapnik, V.N. (1992). A training algorithm for optimal margin classifiers. (New York, NY: ACM), pp. 144–152.

Boyle, A.P., Davis, S., Shulha, H.P., Meltzer, P., Margulies, E.H., Weng, Z., Furey, T.S., and Crawford, G.E. (2008). High-resolution mapping and characterization of open chromatin across the genome. Cell *132*, 311–322.

Brand, A.H., and Perrimon, N. (1993). Targeted gene expression as a means of altering cell fates and generating dominant phenotypes. Development *118*, 401–415.

Brent, M.R. (2008). Steady progress and recent breakthroughs in the accuracy of automated genome annotation. Nat. Rev. Genet. *9*, 62–73.

Britten, R.J., and Davidson, E.H. (1971). Repetitive and non-repetitive DNA sequences and a speculation on the origins of evolutionary novelty. Q. Rev. Biol. 111.

Bulger, M., and Groudine, M. (1999). Looping versus linking: toward a model for long-distance gene activation. Genes Dev. *13*, 2465–2477.

Bulger, M., and Groudine, M. (2011). Functional and mechanistic diversity of distal transcription enhancers. Cell *144*, 327–339.

Busser, B.W., Taher, L., Kim, Y., Tansey, T., Bloom, M.J., Ovcharenko, I., and Michelson, A.M. (2012). A machine learning approach for identifying novel cell type–specific transcriptional regulators of myogenesis. PLoS Genet. *8*, e1002531.

Chen, X., Xu, H., Yuan, P., Fang, F., Huss, M., Vega, V.B., Wong, E., Orlov, Y.L., Zhang, W., Jiang, J., *et al.* (2008). Integration of external signaling pathways with the core transcriptional network in embryonic stem cells. Cell *133*, 1106–1117.

Chung, J.H., Whiteley, M., and Felsenfeld, G. (1993). A 5′ element of the chicken β-globin domain serves as an insulator in human erythroid cells and protects against position effect in Drosophila. Cell 74, 505–514.

Crawford, G.E., Holt, I.E., Whittle, J., Webb, B.D., Tai, D., Davis, S., Margulies, E.H., Chen, Y., Bernat, J.A., Ginsburg, D., *et al.* (2006). Genome-wide mapping of DNase hypersensitive sites using massively parallel signature sequencing (MPSS). Genome Res. *16*, 123–131.

Elnitski, L., Hardison, R.C., Li, J., Yang, S., Kolbe, D., Eswara, P., O'Connor, M.J., Schwartz, S., Miller, W., and Chiaromonte, F. (2003). Distinguishing Regulatory DNA From Neutral Sites. Genome Res. *13*, 64 –72.

Emison, E.S., McCallion, A.S., Kashuk, C.S., Bush, R.T., Grice, E., Lin, S., Portnoy, M.E., Cutler, D.J., Green, E.D., and Chakravarti, A. (2005). A common sex-dependent mutation in a RET enhancer underlies Hirschsprung disease risk. Nature *434*, 857–863.

E.N.C.O.D.E Consortium (2012). An integrated encyclopedia of DNA elements in the human genome. Nature *489*, 57–74.

E.N.C.O.D.E Project Consortium (2007). Identification and analysis of functional elements in 1% of the human genome by the ENCODE pilot project. Nature *447*, 799–816.

Ernst, J., Kheradpour, P., Mikkelsen, T.S., Shoresh, N., Ward, L.D., Epstein, C.B., Zhang, X., Wang, L., Issner, R., Coyne, M., *et al.* (2011). Mapping and analysis of chromatin state dynamics in nine human cell types. Nature *473*, 43–49.

Fisher, S., Grice, E.A., Vinton, R.M., Bessling, S.L., and McCallion, A.S. (2006). Conservation of RET regulatory function from human to zebrafish without sequence similarity. Science *312*, 276–279.

Gerstein, M.B., Kundaje, A., Hariharan, M., Landt, S.G., Yan, K.-K., Cheng, C., Mu, X.J., Khurana, E., Rozowsky, J., Alexander, R., *et al.* (2012). Architecture of the human regulatory network derived from ENCODE data. Nature *489*, 91–100.

Ghandi, M., and Beer, M.A. (2012). Group normalization for genomic data. PLoS One *7*, e38695.

Gorkin, D.U., Lee, D., Reed, X., Fletez-Brant, C., Bessling, S.L., Loftus, S.K., Beer, M.A., Pavan, W.J., and McCallion, A.S. (2012). Integration of ChIP-seq and machine learning reveals enhancers and a predictive regulatory sequence vocabulary in melanocytes. Genome Res. *22*, 2290–2301.

Grice, E.A., Rochelle, E.S., Green, E.D., Chakravarti, A., and McCallion, A.S. (2005). Evaluation of the RET regulatory landscape reveals the biological relevance of a HSCR-implicated enhancer. Hum. Mol. Genet. *14*, 3837–3845.

Grosveld, F., Van Assendelft, G.B., Greaves, D.R., and Kollias, G. (1987). Position-independent, high-level expression of the human β-globin gene in transgenic mice. Cell *51*, 975–985.

Hallikas, O., Palin, K., Sinjushina, N., Rautiainen, R., Partanen, J., Ukkonen, E., and Taipale, J. (2006). Genome-wide prediction of mammalian enhancers based on analysis of transcription-factor binding affinity. Cell *124*, 47–59.

Harbison, C.T., Gordon, D.B., Lee, T.I., Rinaldi, N.J., Macisaac, K.D., Danford, T.W., Hannett, N.M., Tagne, J.-B., Reynolds, D.B., Yoo, J., *et al.* (2004). Transcriptional regulatory code of a eukaryotic genome. Nature *431*, 99–104.

He, X., Ling, X., and Sinha, S. (2009). Alignment and Prediction of cis-Regulatory Modules Based on a Probabilistic Model of Evolution. PLoS Comput. Biol. *5*, e1000299.

Heintzman, N.D., Stuart, R.K., Hon, G., Fu, Y., Ching, C.W., Hawkins, R.D., Barrera, L.O., Van Calcar, S., Qu, C., Ching, K.A., *et al.* (2007). Distinct and predictive chromatin signatures of transcriptional promoters and enhancers in the human genome. Nat. Genet. *39*, 311–318.

Heintzman, N.D., Hon, G.C., Hawkins, R.D., Kheradpour, P., Stark, A., Harp, L.F., Ye, Z., Lee, L.K., Stuart, R.K., Ching, C.W., *et al.* (2009). Histone modifications at human enhancers reflect global cell-type-specific gene expression. Nature *459*, 108–112.

Hindorff, L.A., Sethupathy, P., Junkins, H.A., Ramos, E.M., Mehta, J.P., Collins, F.S., and Manolio, T.A. (2009). Potential etiologic and functional implications of genome-wide association loci for human diseases and traits. Proc. Natl. Acad. Sci. U.S.A. *106*, 9362–9367.

Jacob, F., and Monod, J. (1961). On the Regulation of Gene Activity. Cold Spring Harb. Symp. Quant. Biol. *26*, 193–211.

Johansson, O., Alkema, W., Wasserman, W.W., and Lagergren, J. (2003). Identification of functional clusters of transcription factor binding motifs in genome sequences: the MSCAN algorithm. Bioinformatics *19*, i169–i176.

Karamboulas, C., Dakubo, G.D., Liu, J., Repentigny, Y.D., Yutzey, K., Wallace, V.A., Kothary, R., and Skerjanc, I.S. (2006). Disruption of MEF2 activity in cardiomyoblasts inhibits cardiomyogenesis. J. Cell Sci. *119*, 4315–4321.

Kim, T.-K., Hemberg, M., Grey, J.M., Costa, A.M., Bear, D.M., Wu, J., Harmin, D.A., Laptewicz, M., Barbara-Haley, K., Kuersten, S., *et al.* (2010). Widespread transcription at neuronal activity-regulated enhancers. Nature *465*, 182–187.

King, D.C., Taylor, J., Elnitski, L., Chiaromonte, F., Miller, W., and Hardison, R.C. (2005). Evaluation of regulatory potential and conservation scores for detecting cis-regulatory modules in aligned mammalian genome sequences. Genome Res. *15*, 1051–1060.

King, M.C., and Wilson, A.C. (1975). Evolution at two levels in humans and chimpanzees. Science *188*, 107–116.

Kleinjan, D.A., and Van Heyningen, V. (2005). Long-range control of gene expression: emerging mechanisms and disruption in disease. Am. J. Hum. Genet. *76*, 8–32.

Kleinjan, D.A., and Lettice, L.A. (2008). Long-range gene control and genetic disease. In Advances in Genetics, van Heyningen, V., and Hill, R.E., eds. (Academic Press), pp. 339–388.

Kolbe, D., Taylor, J., Elnitski, L., Eswara, P., Li, J., Miller, W., Hardison, R., and Chiaromonte, F. (2004). Regulatory potential scores from genome-wide three-way alignments of human, mouse, and rat. Genome Res. *14*, 700–707.

Kuo, C.T., Morrisey, E.E., Anandappa, R., Sigrist, K., Lu, M.M., Parmacek, M.S., Soudais, C., and Leiden, J.M. (1997). GATA4 transcription factor is required for ventral morphogenesis and heart tube formation. Genes Dev. *11*, 1048–1060.

Lauderdale, J.D., Wilensky, J.S., Oliver, E.R., Walton, D.S., and Glaser, T. (2000). 3′ deletions cause aniridia by preventing PAX6 gene expression. PNAS *97*, 13755–13759.

Lee, D., Karchin, R., and Beer, M.A. (2011). Discriminative prediction of mammalian enhancers from DNA sequence. Genome Res. *21*, 2167–2180.

Leslie, C., Eskin, E., and Noble, W.S. (2002). The spectrum kernel: a string kernel for SVM protein classification. Pac. Symp. Biocomput. 564–575.

Lettice, L.A., Horikoshi, T., Heaney, S.J.H., Baren, M.J. van, Linde, H.C. van der, Breedveld, G.J., Joosse, M., Akarsu, N., Oostra, B.A., Endo, N., *et al.* (2002). Disruption of a long-range cis-acting regulator for Shh causes preaxial polydactyly. PNAS *99*, 7548–7553.

Lettice, L.A., Heaney, S.J.H., Purdie, L.A., Li, L., Beer, P. de, Oostra, B.A., Goode, D., Elgar, G., Hill, R.E., and Graaff, E. de (2003). A long-range Shh enhancer regulates expression in the developing limb and fin and is associated with preaxial polydactyly. Hum. Mol. Genet. *12*, 1725–1735.

Levine, M. (2010). Transcriptional enhancers in animal development and evolution. Curr. Biol. *20*, R754–R763.

Lyons, I., Parsons, L.M., Hartley, L., Li, R., Andrews, J.E., Robb, L., and Harvey, R.P. (1995). Myogenic and morphogenetic defects in the heart tubes of murine embryos lacking the homeo box gene Nkx2–5. Genes Dev. *9*, 1654–1666.

Manolio, T.A. (2010). Genomewide association studies and assessment of the risk of disease. N. Engl. J. Med. *363*, 166–176.

Manolio, T.A., Collins, F.S., Cox, N.J., Goldstein, D.B., Hindorff, L.A., Hunter, D.J., McCarthy, M.I., Ramos, E.M., Cardon, L.R., Chakravarti, A., *et al.* (2009). Finding the missing heritability of complex diseases. Nature *461*, 747–753.

Maston, G.A., Evans, S.K., and Green, M.R. (2006). Transcriptional regulatory elements in the human genome. Annu. Rev. Genomics Hum. Genet. *7*, 29–59.

McGaughey, D.M., Vinton, R.M., Huynh, J., Al-Saif, A., Beer, M.A., and McCallion, A.S. (2008). Metrics of sequence constraint overlook regulatory sequences in an exhaustive analysis at phox2b. Genome Res. *18*, 252–260.

Megraw, M., Pereira, F., Jensen, S.T., Ohler, U., and Hatzigeorgiou, A.G. (2009). A transcription factor affinity-based code for mammalian transcription initiation. Genome Res. *19*, 644–656.

Miano, J.M., Ramanan, N., Georger, M.A., Bentley, K.L. de, M., Emerson, R.L., Balza, R.O., Xiao, Q., Weiler, H., Ginty, D.D., and Misra, R.P. (2004). Restricted inactivation of serum response factor to the cardiovascular system. PNAS *101*, 17132–17137.

Molkentin, J.D., Lin, Q., Duncan, S.A., and Olson, E.N. (1997). Requirement of the transcription factor GATA4 for heart tube formation and ventral morphogenesis. Genes Dev. *11*, 1061–1072.

Mouse E.N.C.O.D.E Consortium (2012). An encyclopedia of mouse DNA elements (Mouse ENCODE). Genome Biol. *13*, 418.

Narlikar, L., Sakabe, N.J., Blanski, A.A., Arimura, F.E., Westlund, J.M., Nobrega, M.A., and Ovcharenko, I. (2010). Genome-wide discovery of human heart enhancers. Genome Res. *20*, 381–392.

Naya, F.J. (2002). Mitochondrial deficiency and cardiac sudden death in mice lacking the MEF2A transcription factor. Nat. Med. *8*, 1303.

Niu, Z., Yu, W., Zhang, S.X., Barron, M., Belaguli, N.S., Schneider, M.D., Parmacek, M., Nordheim, A., and Schwartz, R.J. (2005). Conditional mutagenesis of the murine serum response factor gene blocks cardiogenesis and the transcription of downstream gene targets. J. Biol. Chem. *280*, 32531–32538.

Noonan, J.P., and McCallion, A.S. (2010). Genomics of long-range regulatory elements. Annu. Rev. Genomics Hum. Genet. *11*, 1–23.

Odom, D.T., Dowell, R.D., Jacobsen, E.S., Gordon, W., Danford, T.W., MacIsaac, K.D., Rolfe, P.A., Conboy, C.M., Gifford, D.K., and Fraenkel, E. (2007). Tissue-specific transcriptional regulation has diverged significantly between human and mouse. Nat. Genet. *39*, 730–732.

Patthy, L. (1999). Genome evolution and the evolution of exon-shuffling – a review. Gene Combis. *238*, 103.

Peckham, H.E., Thurman, R.E., Fu, Y., Stamatoyannopoulos, J.A., Noble, W.S., Struhl, K., and Weng, Z. (2007). Nucleosome positioning signals in genomic DNA. Genome Res. *17*, 1170–1177.

Pennacchio, L.A., Ahituv, N., Moses, A.M., Prabhakar, S., Nobrega, M.A., Shoukry, M., Minovitsky, S., Dubchak, I., Holt, A., Lewis, K.D., *et al.* (2006). *In vivo* enhancer analysis of human conserved non-coding sequences. Nature *444*, 499–502.

Pennacchio, L.A., Loots, G.G., Nobrega, M.A., and Ovcharenko, I. (2007). Predicting tissue-specific enhancers in the human genome. Genome Res. *17*, 201–211.

Ptashne, M. (2004). A Genetic Switch: Phage Lambda Revisited (New York, Cold Spring Harbor Laboratory Press).

Schlesinger, J., Schueler, M., Grunert, M., Fischer, J.J., Zhang, Q., Krueger, T., Lange, M., Tönjes, M., Dunkel, I., and Sperling, S.R. (2011). The cardiac transcription network modulated by Gata4, Mef2a, Nkx2.5, Srf, histone modifications, and microRNAs. PLoS Genet. 7, e1001313.

Schölkopf, B., Tsuda, K., and Vert, J.P. (2004). Kernel Methods in Computational Biology (Cambridge, MA, The MIT Press).

Searcy, R.D., Vincent, E.B., Liberatore, C.M., and Yutzey, K.E. (1998). A GATA-dependent nkx-2.5 regulatory element activates early cardiac gene expression in transgenic mice. Development *125*, 4461–4470.

Shen, Y., Yue, F., McCleary, D.F., Ye, Z., Edsall, L., Kuan, S., Wagner, U., Dixon, J., Lee, L., Lobanenkov, V.V., *et al.* (2012). A map of the cis-regulatory sequences in the mouse genome. Nature *488*, 116–120.

Sinha, S., and He, X. (2007). MORPH: probabilistic alignment combined with hidden markov models of cis-regulatory modules. PLoS Comput. Biol. 3, e216.

Sinha, S., and Tompa, M. (2003). YMF: a program for discovery of novel transcription factor binding sites by statistical overrepresentation. Nucleic Acids Res. *31*, 3586–3588.

Sinha, S., Schroeder, M.D., Unnerstall, U., Gaul, U., and Siggia, E.D. (2004). Cross-species comparison significantly improves genome-wide prediction of cis-regulatory modules in Drosophila. BMC Bioinformatics 5, 129.

Sonnenburg, S., Rätsch, G., Schäfer, C., and Schölkopf, B. (2006). Large scale multiple kernel learning. J. Mach. Learn. Res. 7, 1531–1565.

Spencer, J.A., and Misra, R.P. (1996). Expression of the serum response factor gene is regulated by serum response factor binding sites. J. Biol. Chem. *271*, 16535–16543.

Spitz, F., and Furlong, E.E.M. (2012). Transcription factors: from enhancer binding to developmental control. Nat. Rev. Genet. *13*, 613–626.

Srivastava, D. (2006). Making or breaking the heart: from lineage determination to morphogenesis. Cell *126*, 1037–1048.

Su, J., Teichmann, S.A., and Down, T.A. (2010). Assessing computational methods of cis-regulatory module prediction. PLoS Comput. Biol. *6*, e1001020.

Thurman, R.E., Rynes, E., Humbert, R., Vierstra, J., Maurano, M.T., Haugen, E., Sheffield, N.C., Stergachis, A.B., Wang, H., Vernot, B., *et al.* (2012). The accessible chromatin landscape of the human genome. Nature *489*, 75–82.

Vapnik, V.N. (1995). The Nature of Statistical Learning Theory (New York, NY, Springer).

Visel, A., Blow, M.J., Zhang, T., Akiyama, J.A., Holt, A., Plajzer-Frick, I., Shoukry, M., Wright, C., Chen, F., Afzal, V., *et al.* (2009a). ChIP-seq accurately predicts tissue-specific activity of enhancers. Nature *457*, 854–858.

Visel, A., Rubin, E.M., and Pennacchio, L.A. (2009b). Genomic views of distant-acting enhancers. Nature *461*, 199–205.

Wasserman, W.W., and Sandelin, A. (2004). Applied bioinformatics for the identification of regulatory elements. Nat. Rev. Genet. *5*, 276–287.

Waterston, R.H., Lindblad-Toh, K., Birney, E., Rogers, J., Abril, J.F., Agarwal, P., Agarwala, R., Ainscough, R., Alexandersson, M., An, P., *et al.* (2002). Initial sequencing and comparative analysis of the mouse genome. Nature *420*, 520–562.

Wilson, M.D., Barbosa-Morais, N.L., Schmidt, D., Conboy, C.M., Vanes, L., Tybulewicz, V.L.J., Fisher, E.M.C., Tavare, S., and Odom, D.T. (2008). Species-specific transcription in mice carrying human chromosome 21. Science *322*, 434–438.

Xie, X., Lu, J., Kulbokas, E.J., Golub, T.R., Mootha, V., Lindblad-Toh, K., Lander, E.S., and Kellis, M. (2005). Systematic discovery of regulatory motifs in human promoters and 3[prime] UTRs by comparison of several mammals. Nature *434*, 338–345.

Zeller, K.I., Zhao, X., Lee, C.W.H., Chiu, K.P., Yao, F., Yustein, J.T., Ooi, H.S., Orlov, Y.L., Shahab, A., Yong, H.C., *et al.* (2006). Global mapping of c-Myc binding sites and target gene networks in human B cells. PNAS *103*, 17834–17839.

Zhang, X., Cowper-Sal Lari, R., Bailey, S.D., Moore, J.H., and Lupien, M. (2012). Integrative functional genomics identifies an enhancer looping to the SOX9 gene disrupted by the 17q24.3 prostate cancer risk locus. Genome Res. *22*, 1437–1446.

Zhang, Y., Liu, T., Meyer, C.A., Eeckhoute, J., Johnson, D.S., Bernstein, B.E., Nussbaum, C., Myers, R.M., Brown, M., Li, W., *et al.* (2008). Model-based Analysis of ChIP-Seq (MACS). Genome Biol. *9*, R137.

Zhou, V.W., Goren, A., and Bernstein, B.E. (2011). Charting histone modifications and the functional organization of mammalian genomes. Nat. Rev. Genet. *12*, 7–18.

DNA Patterns for Nucleosome Positioning

7

Ilya Ioshikhes

Abstract

There is an abundant experimental evidence for the role of specific nucleosome positioning in gene regulation. The nucleosome positioning is determined by DNA sequence and non-sequence factors such as ATP dependent remodelling factors. The nucleosome positions differ between various cell types for the same species as well as for similar genes of the different species. A nucleosome shift by just a few base pairs can alter the entire regulation of a gene. Knowing precise nucleosome location is critical for understanding how cis-regulatory elements control genetic information. Among different factors affecting nucleosome positioning on the DNA, the DNA sequence itself is the most important, and various sequence motifs have been described as guiding nucleosome positioning in a sequence specific manner. These motifs or patterns eventually were termed nucleosome positioning sequence (NPS) patterns although this term is not necessarily universal. In this chapter, we describe various classes of such NPS patterns known from literature and possible biological implications thereof. This chapter does not embark to provide an exhaustive review of all related points of view, neither its emphasis is on new results presented here for the first time. Rather it reflects the author's point of view on general tendencies in this area of science and tries to provide a possible answer to the most difficult question in the area: why does nucleosome positioning differ between different tissues while respective DNA sequence of the involved genes is essentially identical?

Nucleosome as the basic unit of chromatin

The role of DNA as the basic hereditary substance in all living cells does not require an extensive introduction, which is neither necessary for a classical structure of DNA double helix: every student knows at least the basic, structural and functional, features of the DNA since the famous work by Watson and Crick on deciphering the basic DNA structure and its genetic implications in the 1950s (Watson and Crick, 1953a,b). But of course, as the time went by, it quickly appeared that a classical picture of the double helical DNA with an almost straight axis, as it appears in almost every textbook, is only the initial (albeit very fundamental) model. Various estimates of the DNA length were given at an early onset of the development of the Human Genome project, but all of them concluded that an actual size of a human cell's DNA, if taken as a straight molecule, varies around 2 milion in total with particular estimates between 1.5 and 3 million (Leltninger, 1975; Zelenyi, 1996; Mitchell, 1997; Parker, 1997), i.e. comparable to a length of the human body. Provided that a genomic DNA of eukaryotic cells should fit in a much smaller compartment of a cell nucleus, of a size ~10 μm, it cannot fit there as a mere straight double helix. Instead, it should be very tightly packed on several levels and form an intricate protein–DNA complex named chromatin. There are several levels of such packaging, each facilitated by specific proteins. Chromosomes, chromatin units visible in a microscope, are in fact such DNA–protein complexes representing the highest level of the DNA packaging. (As one may remember from a history of science, the role of the hereditary substance was initially assigned to proteins, and it was the discovery of the double helix that dramatically changed the picture.) Yet, this highest level of the DNA packaging is based on several lower levels, with the basic unit named as a nucleosome: dyad symmetrical protein core consisting of specific proteins called histones and ~146 base pairs (bp) of DNA (rather between 144 and 147 bp by various estimates) wrapped around it.

The structure of the nucleosome core particle was originally solved at 7 Å resolution in the early 1980s, and refined in later studies. Among these, the crystallographic structure at 2.8 Å resolution (Luger *et al.*, 1997) has revealed how the histone protein octamer is assembled, and how the 146 bp of DNA are organized into a superhelix around it. The nucleosome consists of 146 bp of DNA wrapped in a left-handed superhelix around an octameric histone core formed by two copies of each of the histone molecules H2A, H2B, H3, and H4 (Richmond *et al.*, 1984; Luger *et al.*, 1997). The $(H3–H4)_2$ tetramer occupies a central position in the octameric core structure, flanked on both sides by the (H2A-H2B) dimers (Arents *et al.*, 1991), as can be viewed in Fig. 7.1A. The assembly of a stable nucleosome core depends on the initial heterodimerization of the H3 and H4 molecules, and their subsequent dimerization to form the $(H3–H4)_2$ tetramer (Eickbush and Moudrianakis, 1978), followed by the dimerization of the histones H2A and H2B that bind to both sides of the $(H3–H4)_2$ tetramer (Hayes *et al.* 1990, 1991). Changes in the accessibility of DNA to histones in response to environmental stimuli affect the mechanism of transcription and gene regulation. The core histones share a structurally conserved motif called the histone fold and mobile extended regions at the chain termini (also named histone tails). The histone fold consists of three α-helices (α1, α2, and α3) connected by short loops L1 and L2, respectively. During dimerization, loop L1 of one of the monomers (e.g. H3) aligns against loop L2 of the other monomer (e.g.

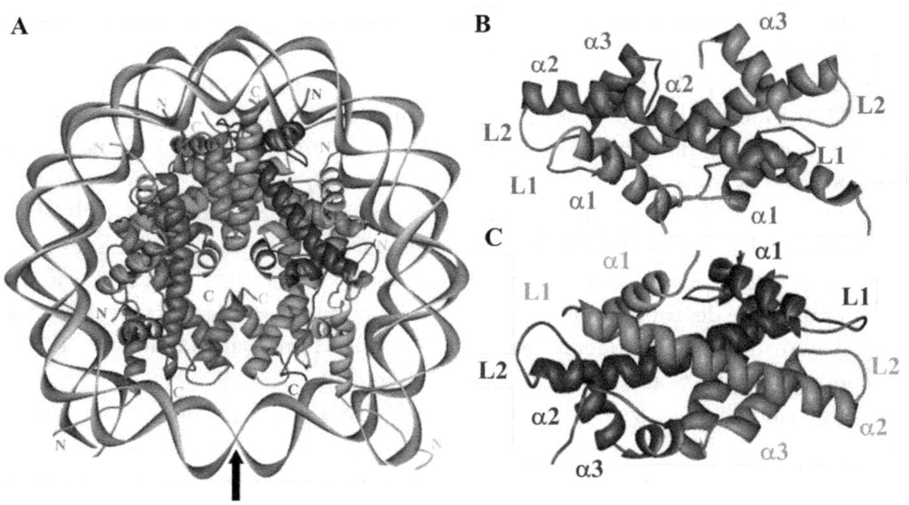

Figure 7.1 Architecture of the histone octamer core in the nucleosome. **(A)** Mutual positioning of the histone octamer and DNA in the nucleosome. The dyad axis is shown by an arrow at the middle of the nucleosome. H3 (pink), H4 (green), H2A.Z (blue) and H2B (orange), and DNA (grey) in the nucleosome. Dark and light colours distinguish each copy of the monomers. The N- and C-termini are indicated. The arrow indicates the direction of the dyad axis. The structure has been constructed using PDB file 1F66 for the nucleosome with the histone variant H2A.Z, deposited by Luger and coworkers (Suto *et al.*, 2000). **(B)** Handshake motifs formed by H3-H4 and H2A-H2B. The monomers H3, H4, H2Z, and H2B are collared pink, green, blue, and orange, respectively, consistent with **(A)**. Adopted from (Ramaswamy *et al.*, 2005). A colour version of this figure is located in the plate section at the back of the book.

H4) to form the so-called handshake motif (Fig. 7.1B) that interacts with DNA. The flexible tails of the core histones interact with DNA via the minor groove. The histone tails are the major targets for post-translational modifications such as acetylation, methylation, and phosphorylation, and the key modulators of chromatin function (Davie, 1998). The octameric histones and the DNA are connected by a network of hydrogen bonds and the major DNA–protein interaction sites of the dimers are the two pairs of adjoining loops L1 and L2, and the $\alpha 1$ helices of the monomers.

Role of nucleosome positioning in gene regulation

An immense evidence for the important role that chromatin plays in regulatory mechanisms of gene expression has been accumulated in multiple experimental studies (see, for example, Grunstein, 1990, 1997; Felsenfeld, 1992; Grunstein *et al.* 1992; Lu *et al.*, 1994; Wolffe, 1994a,b, 2001; Kornberg and Lorch, 1995, 2002; Kingston *et al.*, 1996; Peterson, 1996; Svaren and Horz, 1996; Gottesfeld and Forbes, 1997; Beato and Eisfeld, 1997; Li *et al.*, 1997, 1998; Narlikar *et al.*, 2002). It appears that nucleosomes serve not just as a mere DNA packaging device but also as a part of various regulatory schemes in which their precise positioning along the DNA molecule is important (Radman-Livaja and Rando, 2010). The subject was first broached in a 1982 *Nature* article by S. and B. Wittig titled 'Function of tRNA gene promoter depends on nucleosome position' (Wittig and Wittig, 1982). Since then, multiple additional evidences of the role of precise nucleosome positioning in eukaryotic promoters have been gathered. The works of M. Beato should be particularly mentioned as pioneering works on the subject. In the studies of Chavez *et al.* (1995), Spangenberg *et al.* (1998) and Di Croce *et al.* (1999) MMTV viral promoter was investigated in the context of nucleosome positioning and its influence onto activity of the TATA box and binding of various TFs. Recently, study of chromatin remodelling by SWI/SNF complex draw the most attention. This appears to be related to nucleosome translocations in 9- to 11- or approximately 50-bp increments by ISW2 and SWI/SNF complexes (Zofall *et al.*, 2006). Precise knowledge of nucleosome positioning would be very important for studying these processes. To name only few most relevant works, we would like to mention a study of Georgel (Georgel, 2005) on chromatin remodelling and events leading to transcription initiation of the *D. melanogaster* hsp70 gene by GAGA-factor. There, by matching the relative position of the GAGA-factor binding sites with the distribution of nucleosomes over the hsp70 promoter, the GAGA site 2 appeared to be the most accessible, i.e. located close to a nucleosomal edge or within the linker DNA. Komura and Ono (Komura and Ono, 2003) investigated nucleosome positioning in the human c-Fos promoter. The *in vivo* nucleosomal translational position in this promoter appears to be strongly influenced by the presence of transcription factors (TFs), which may function as boundaries, while the rotational position appears to be determined predominantly by the DNA sequence. General mechanism of influence of the nucleosome positioning on gene regulation is related to the chromatin compaction by phased nucleosomes, resulting in a lesser availability of such DNA regions for transcriptional machinery (Dlakic *et al.*, 2004). Assuming specific positions along the DNA, nucleosomes affect gene regulation. Positions of the nucleosomes relative to the locations of promoter cis-acting elements may be crucial for function of these elements. Overlap of the TF binding sites (TFBS) and nucleosome positions may lead to repression of the regulatory elements. Transcriptional activation is also related to specific nucleosome positioning (Martinez-Campa *et al.*, 2004). In principle, TFs interact with nucleosomal DNA by either binding to the exposed periodic DNA surface or by displacing the nucleosome. Knowing precise nucleosomes location is critical for understanding how cis-regulatory elements control genetic information. A nucleosome can inhibit as well as activate the gene expression (Ioshikhes *et al.*, 2006). Without proper nucleosomes, erroneous transcription, called cryptic transcription can occur (Lickwar *et al.*, 2009). Nucleosomes involving H2A.Z variant in place of a regular H2A histone may serve as a barrier for core promoter region so that TFs can work (Bargaje *et al.*, 2012). Together with negative elongation factor (NELF), nucleosomes may be related to polymerase pausing (Levine, 2011). By manipulating the nucleosome disfavouring sequences, gene expression could be controlled (Raveh-Sadka *et al.*, 2012). The position of nucleosomes is different between different cell types (Valouev *et al.*, 2011; Bargaje *et al.*, 2012). Not only the nucleosome translational position, but the rotational setting is also important to expose the binding motifs to other factors (Hebert and Roest Crollius, 2010). A nucleosome shift by just a few base pairs can alter the entire regulation of a gene (Lu *et al.*, 1994).

The question is still open, however, what are the determinants of nucleosome positioning. Many cis- and trans-acting factors that may direct positioning of a histone octamer have been suggested and studied (reviewed in Simpson, 1991; Thoma, 1992; Lu *et al.*, 1994; Wolffe, 1994a,b; Radman-Livaja and Rando, 2010). Different viewpoints on the problem were introduced quite early and modified with the development of the science:

1 The nucleosome positioning could be random or at least statistically defined.
2 Nucleosomes may have preferential affinity to certain DNA sequence motifs.
3 DNA-binding proteins (not histones) may define nucleosome positioning.
4 Neighbouring 'strong' nucleosomes may act as a barrier for adjacent nucleosomes.
5 Chromatin higher-order structure affects positioning of individual nucleosomes.

Although these points of view were first expressed more than 20 years ago, they are, to some extent, still in use today, being adjusted to new experimental data. Generally it is well accepted that the nucleosome positioning is determined by sequence factors (Valouev *et al.*, 2011) and non-sequence factors such as ATP-dependent remodelling factors (Gkikopoulos *et al.*, 2011).

With that said, studying the nucleosome interaction with other DNA-binding elements, in particular chromatin remodellers and TFs would be important both for more reliable prediction of nucleosome positioning and for understanding sophisticated processes of gene regulation at various levels. Involvement of nucleosomes in the promoter activity (e.g. Beato and Eisfeld, 1997; Gottesfeld and Forbes, 1997; Grunstein, 1997; Li *et al.*, 1997, 1998; Wolffe, 2001, Kornberg and Lorch, 2002) and regulation (Grunstein, 1990; Felsenfeld, 1992; Wolffe, 1994a,b, 2001; Chavez *et al.*, 1995; Svaren and Horz, 1996; Beato and Eisfeld, 1997; Li *et al.*, 1998; Spangenberg *et al.*, 1998; Kornberg and Lorch, 2002; Zofall *et al.*, 2006; Wang *et al.*, 2012) suggests that the nucleosomes would occupy certain positions in the vicinity of promoters to provide specific spatial environment for the recognition of the promoters and for interactions with various TFs. As mentioned in Komura and Ono (2003) (see also Wang *et al.*, 2012) 87% of TFBS are positioned in nucleosome-depleted areas. It is also possible that locally TFs may compete with nucleosomes for preferential positioning sites along DNA sequence, thus altering nucleosome positioning defined by DNA sequence properties. TFs interact with nucleosomal DNA either by binding to the exposed DNA surface or displacing the nucleosome. Our *PNAS* paper (Ioshikhes *et al.*, 1999) (see also Albert *et al.*, 2007) showed clear periodical distributions of particular TFs with the periodicity of ~10bp characteristic for NPS DNA pattern. It is related to phasing of specific TFBS relative to the surface of nucleosome histone core and to a typical spacing between respective TFBS. Provided a tissue specific nature of particular TFs, this may help in answering the most difficult question related to the nucleosome positioning: why it differs between different tissues of the same species, sharing the same DNA sequence? Still, the answer should be in line with a sequence-related model of the nucleosome positioning to have these two factors (TFs and DNA) to work cooperatively rather than antagonistically.

Different nucleosome sequence patterns

In any event, it is frequently assumed that the role of DNA sequence is important. However, the positioning signal in the sequence appears to be very weak and hard to detect. Several rather different nucleosome DNA sequence patterns have been suggested (Satchwell *et al.*, 1986; Ioshikhes *et al.*, 1996; Lowary and Widom, 1998; Segal et al., 2006; Kogan et al., 2006; Cohanim et al., 2006; Albert et al., 2007; Salih et al., 2007). The difference in the patterns was in part caused by the difference in the respective nucleosome datasets, particularly in the techniques of the nucleosome mapping and differences in the algorithms applied to extract the patterns. One should remember that prior to a ground-breaking paper by Yuan et al. (2005), which introduced high-throughput techniques to the nucleosome research, only few hundreds nucleosome positions were reliably mapped, often with a low precision, which made an extraction of the nucleosome sequence pattern a challenging task. It appeared to be generally accepted that the segment of DNA double helix, which is wrapped around the histone octamer, should possess sequence-dependent anisotropic deformability (bendability) properties that would provide a stabilizing contribution to a free energy of the nucleosome 3D structure. Since DNA deformability is dictated by the deformability of adjacent nucleotide pairs, the dinucleotide motifs should be those primarily responsible for the DNA bending. In particular it is believed that AA and TT dinucleotides play a special role in the nucleosome positioning, at least for some species such as yeast. The observation of periodical appearance of some dinucleotides, primarily AA and TT, along eukaryotic DNA sequences with the period close to DNA helical repeat was spelled out as a nucleotide sequence pattern that may facilitate anisotropic DNA bendability and nucleosome formation (Trifonov and Sussman, 1980). It was suggested that such a periodicity may reflect orientation of synonymous base pair stacks preferentially in the same direction relative to the surface of a histone octamer (Trifonov, 1980). The nucleosome DNA, thus, would have a preferred side to which a histone octamer would bind (rotational setting). Evidence in favour of such rotational preference was provided by experiments with DNase I digestion of DNA within the nucleosomes (Noll, 1974; Sollner-Webb et al., 1978; Drew and Travers, 1985). Nuclease digestion experiments (Sollner-Webb et al., 1978) and high resolution X-ray diffraction studies (Richmond et al., 1984; Luger et al., 1997) have demonstrated that at the exit point of the dyad axis of the nucleosome core the minor groove of DNA double helix is positioned in outward orientation. Therefore, both translational and rotational settings of a DNA double helix on the surface of the histone octamer are unambiguously defined by the position where the dyad axis passes through the midpoint of the nucleosomal DNA.

Trifonov and Sussman were the first who obtained the 10.2 bp periodicity of certain dinucleotides in nucleosome preferred DNA sequence (Trifonov and Sussman, 1980). Specific patterns and/or periodicity were found in variety of eukaryotic organisms (Ioshikhes et al., 1996), particularly in *S. cerevisiae* (Segal et al., 2006; Ioshikhes et al., 2011; Kaplan et al., 2009; Reynolds et al., 2010), *C. elegans* (Gabdank et al., 2010a), *H. sapiens* (Schones et al., 2008), and *D. Melanogaster* (Fitzgerald et al., 2006; Mavrich et al., 2008b). In addition to the periodicity, GC-content and the frequency of some AT-rich tetranucleotides explain nucleosome occupancy *in vitro* and *in vivo* as in Kaplan et al. (2009). GC-content is dominant, alone explaining ~50% of the variation in nucleosome occupancy *in vitro* (Tillo and Hughes, 2009). Besides the DNA sequence, transcription factors and RNA polymerase are involved in generating the nucleosome positioning *in vivo* (Weiner et al., 2009; Hughes et al., 2012). Genetic region 200 bp downstream the transcription start site (TSS) has periodicity of 10 bp that assists in chromatin remodelling by shifting nucleosome downstream by tens of bps (Hebert et al., 2010). Intrinsic DNA curvature determined by 10 bp periodicity is associated with nucleosomal DNA (Nair et al., 2010).

Most of the existing methods of nucleosome mapping *in silico* are based on comparison of genomic sequences with certain sequence features extracted from DNA of experimentally mapped nucleosomes. These features represent nucleosome sequence pattern, described with variety of details (sequence elements and their distribution along the nucleosomal DNA) and generalization. Despite the variety of nucleosomal DNA sequence patterns, they generally may be divided into two major distinct classes. In the first class, the AA and TT nucleotides have identical phases in their positional distributions along nucleosomal DNA sequence. Structurally that means that AA and TT are preferentially located on the same side of nucleosomal DNA with respect to the nucleosome surface. From structural considerations, such dinucleotide positioning should be related to intrinsically curved DNA segments. Internal DNA shape of such segments with dinucleotide periodicity of ~10 bp is suitable for nucleosome accommodation. Segal and colleagues (Segal *et al.*, 2006) presented pattern [basically consistent with earlier works by Satchwell *et al.* (1986) and Lowary and Widom (1998)] that places AA and TT (and also TA) dinucleotides inside the nucleosomal DNA structure, on the side of DNA facing histone octamer (Fig. 7.2, left). In this model, GC dinucleotides are positioned outside of the histone octamer. Albert *et al.* (2007) showed a pattern highly consistent with that presented in Segal *et al.* (2006), with addition of all SS dinucleotides to GC, although differing in WW/SS placement around the dyad (Travers *et al.*, 2010). Trifonov (2010) switched positions of WW and SS dinucleotides in his model of the canonical WW/SS pattern. For either model, such dinucleotide positioning should be related to intrinsically and persistently curved DNA segments (Drew and Travers 1985; Cohanim *et al.* 2006), and the periodicity of ~10 bp would further augment nucleosome stability.

In the second class of the dinucleotide patterns (including Ioshikhes *et al.*, 1996), AA and TT dinucleotides have opposite phases in their positional distributions. Maximum of AA distribution coincides with the minimum of TT and vice versa.

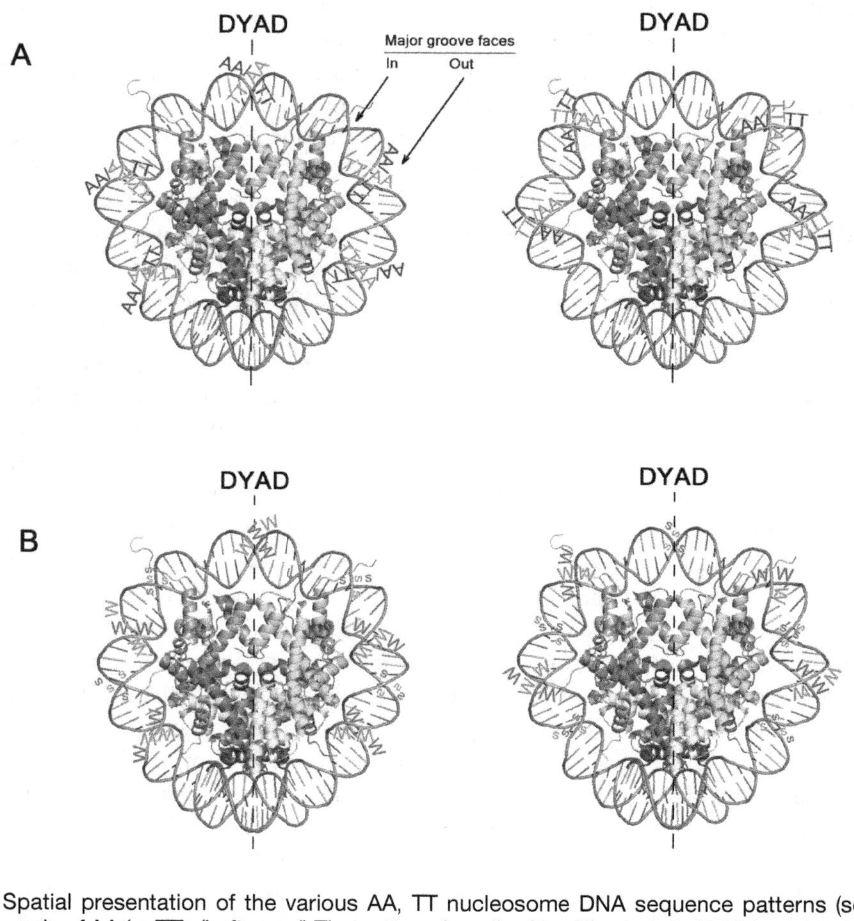

Figure 7.2 Spatial presentation of the various AA, TT nucleosome DNA sequence patterns (scheme). (A) Specific example of AA (or TT). (Left panel) The pattern described by Albert *et al.*, 2007 (consistent with Segal *et al.*, 2006). (Right panel) The 'anti' configuration. Note, AA or TT at each indicated position on each strand is allowable. (B) The general case from panel A is shown. Smaller letters for SS indicate a small contribution. The patterns are shown only in the area close to the dyad. Adapted from Ioshikhes *et al.* (2011).

Structurally this is related to preferential separation of these dinucleotides with respect to the surface of the histone octamer: AA dinucleotides are preferentially located on the octamer surface, while TT are positioned outside of it (Fig. 7.3, left). Such positioning is consistent with AA and TT geometrical properties (Salih *et al.*, 2007). Kogan *et al.* (2006) showed preferential role of CC and GG dinucleotides in human chromatin with GG preferentially positioned on the histone core surface and CC facing outside of it. Overall, the patterns of the second class have RR (PuPu) dinucleotides (AA, GG, AG, GA) grouped together on the side of the DNA facing histone octamer, and YY (PyPy) dinucleotides grouped outside. The second class of patterns is related to relatively flexible DNA with relatively moderate intrinsic curvature, but that is able to relatively easily wind around the histone octamer (Trifonov, 1985). In the following section we will try to provide a historic overview of discovery of the mentioned classes of the patterns, followed by the description of further development in the area.

Early history (pre-genomic and genomic era)

While it seems to be intuitively correct that the nucleosome sequence patterns should be derived from the experimentally obtained nucleosome DNA sequences, in reality some of the patterns may be derived from the analysis of the whole

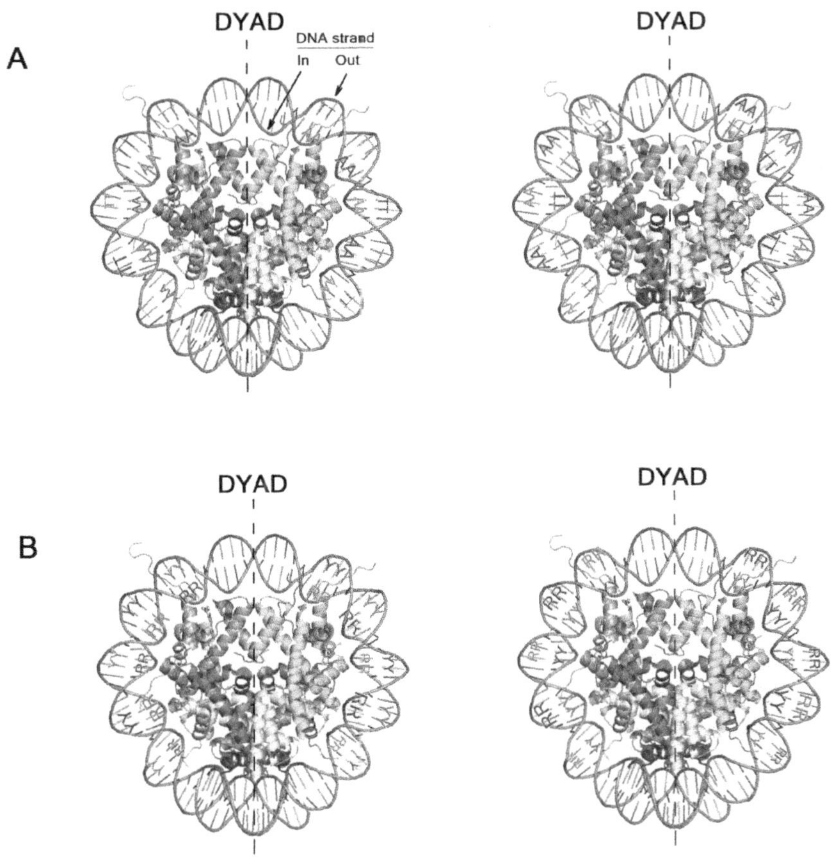

Figure 7.3 Spatial presentation of the various nucleosome DNA sequence patterns (scheme). (*A*) Counter-phase AA/TT (*left*) and anti-AA/TT (*right*) patterns. (*B*) RR/YY (*left*) and anti-RR/YY (*right*) patterns. The variable size of the letters reflects variable peak magnitude of respective dinucleotide distributions at subsequent figures (no precise scale was kept). The patterns are shown only in the area close to the dyad. Adopted from (Ioshikhes *et al.*, 2011).

genomic sequences where precise nucleosome positioning is not known. This method is based on the assumption that most of the eukaryotic genomes (~70% provided that almost entire genomes are covered by nucleosome binding DNA segments of ~145 bp length intermitted by linkers of a length ~50 bp) are covered by the nucleosomes, hence the nucleosomal sequence patterns most likely will be dominant along the genomic DNA. Another reason for that approach is that high-throughput chromatin digestion data obtained by next-generation sequencing techniques, which are presently in a wide use, still require significant additional analyses to obtain precise nucleosome positions. The paucity of precise nucleosome positioning data was yet more distinct in the earlier years, when only handful of nucleosomes were reliably mapped.

Historically, the Trifonov's group and his coworkers were the first to analyse both types of the patterns. In the 1980 paper in *Proceedings of the National Academy of Sciences of the USA* (Trifonov and Sussman, 1980) it was reported, based on the autocorrelation analysis of the SV40 genomic sequence (known to have been chromatin packed when in a host cell), that certain dinucleotides show clear periodicity of ~10.5 bp, which was previously associated with a pitch of the nucleosomal DNA (Sussman and Trifonov, 1978; Trifonov and Bettecken, 1979). Since then, such periodicity became associated with the nucleosome sequence patterns, and further works on the pattern search had to take this feature into account to be credible. In a subsequent work (Mengeritsky and Trifonov, 1983) similar analysis was applied to a larger variety of viral genomes and eukaryotic genome sequences on which nucleosome positioning was previously experimentally studied. At this time the periodicity of the dinucleotide distributions was supplemented by the notion of sequence complementarity as another major feature of the nucleosome pattern. Relative phase shift of various dinucleotide distributions was also estimated, and combination of the periodical distributions of particular dinucleotides with a specified phase shift inside DNA segment of 10–11 bp constituted the so-called bendability matrix of the nucleosomal DNA. The bendability matrix was generally introduced some time earlier (Trifionov, 1980, 1981) to characterize regular deformability of DNA required to wrap around the histone octamer. Combined with the sequence information, the bendability matrix represented the initial vision of the nucleosome sequence pattern, which could be further used for nucleosome mapping in silico. Several successful examples of the computational mapping of the experimental nucleosomes were also provided, which established a credibility of the pattern. With the basic ideas of this scientific field thus successfully established, further progress was significantly hindered by a limited number of available nucleosome sequences. This hurdle was to be overcome with a passage of time, when more significant experimental data became available. These were accumulated in a nucleosome sequence database (e.g. Ioshikhes and Trifonov, 1993) and served a subject of further analysis.

In the meantime, an alternative approach was applied to resolve the data limitation problem. Generally, it is possible to reconstitute chromatin *in vitro* by combining histones and DNA, either from different species or the same species (as mentioned above, the histones are highly conserved across the eukaryotes). The reconstituted chromatin may be then digested by variety of enzymes, with respective resolution varying from 1 to 50 bp depending on the enzymes and concentration used. To achieve the nucleosome mapping with 1 bp resolution, the easiest way is to apply chromatin DNA digestion with a very high concentration of the MNase. While the MNase generally cuts DNA in the linkers between the nucleosomes, it cuts it just on the linker–nucleosome border when the enzyme concentration is high. Such procedure produces segments of the nucleosome DNA which sequence may be identified afterwards, although such dataset may be biased by an excess of the nucleosomes surviving high concentrations of the MNase with others shifted or degraded during the process. Because of the MNase cutting preference to AT base pairs (Chung *et al.*, 2010), the sequence bias also exists and should be factored out to rely on respective nucleosome positioning data. This method was applied in Satchwell *et al.* (1986) who anaysed the sequences of 177 different DNA molecules from chicken erythrocyte core particle. Variety of dinucleotides, as well as some trinucleotides were shown to have a periodicity of ~10 bp. However, the presented distributions combined complementary sequence elements in one distribution (so that, e.g. AA were always combined with TT), which was rather questionable from a point of view presented by Trifonov. While no separate (non-combined) distributions of AA and TT, and so of other di- and trinucleotides, were published or otherwise made publicly available, the validity of such combination was questioned in the years to come (see discussion in Ioshikhes *et al.*, 1996). In the meantime, one more dataset was analysed along similar lines (Lowary and Widom, 1978) and resulted in generally quite similar patterns, yet only combined distributions were again shown for each dinucleotide and its complementary counterpart.

To resolve the controversy about how and whether dinucleotide motifs should be combined was only possible by analysing an alternative nucleosome dataset of a size similar to those of (Satchwell *et al.*, 1986), but obtained by different methods. This work culminated in a collection of the nucleosome sequence database (Ioshikhes and Trifonov, 1993) and its analysis (Ioshikhes *et al.* 1992, 1996). Our original algorithm of the multiple alignment designed for detecting dinucleotide pattern in nucleosome sequences indicated the presence of specific AA and TT positional preferences (Ioshikhes *et al.*, 1992) in the earlier version of the database. In that paper 118 nucleosome sequences compiled from published experimental data have been aligned. The alignment revealed that the distances between major maxima of the AA and TT positional distributions correspond to multiples of a period of 10.4 bases. The AA and TT positional patterns were found to have six base phase shift between them (see also (Mengeritsky and Trifonov, 1983)). Since the nucleosome DNA sequence pattern is very weak one has to be sure that the pattern derived is not an artefact of a chosen computational technique. Several different techniques should be tried, and only those features of the pattern, which are common for all outputs, can be considered as belonging to the pattern sought for. The results of this elaborate analysis were presented in our subsequent work (Ioshikhes *et al.*, 1996) for a larger set of nucleosomal sequences; and, by using several techniques of multiple alignment, we confirmed the earlier conclusions on the AA and TT dinucleotides periodicity. The periodicity was found to span a whole length of the nucleosomal DNA with the exception of the small region around the dyad axis.

Ioshikhes *et al.* (1996) searched for the DNA sequence pattern, specifically for the AA and TT dinucleotide profiles, characteristic for the large ensemble of the experimentally mapped nucleosome DNA sequences. The choice of the sequence elements was dictated by their over-representation in the nucleosome DNA sequences (~10–11% of each AA and TT instead of 6.25% randomly expected). Our analysis was free of any a priori limitations on the generated patterns, including bendability considerations, which were used only as a basis for interpretation of the generated patterns. The analysis did not account for other possible nucleosome positioning factors, or for internucleosomal interactions. The pattern found in this study was later proven to be helpful for prediction of nucleosome positions in DNA sequence. Only a few of the nucleosome DNA sequences of those determined in experiments and available in literature have been mapped with high accuracy (21 bp) with respect to nucleosome centre. For others the uncertainty in midpoint position was higher, up to 50 bp. Therefore, in order to obtain a common nucleosome pattern we had to extract it using a multiple alignment procedure. That is, for each sequence of the database the 'true' midpoint position had to be found, within the experimental error limits, such that it would fit best to a 'consensus' sequence pattern derived from the whole ensemble of similarly aligned sequences of the database. Since a DNA molecule is in continuous contact with the surface of the histone octamer (with no loops or bulges), the alignment should also be without gaps. Since the conventional algorithms of sequence alignment are usually based on evolutionary considerations and search for segments of high local sequence similarity thought to be the remnants of the ancestor DNA or proteins, which do not exist in nucleosomal DNA, we had to develop our own algorithms of the multiple sequence alignment for this particular problem or substantially modify some of the existing algorithms (such as Gibbs sampler).

Dinucleotide positional frequency distributions for AA and TT dinucleotides have been obtained from multiple alignments of experimentally mapped nucleosome sequences (Ioshikhes *et al.*, 1996). While the profiles corresponding to different alignment techniques were rather different, several common maxima appeared in three, four or all five of the distributions simultaneously. Direct summation (averaging) of five AA frequency distributions revealed also other maxima, which followed a rather regular pattern. The regular peaks in the averaged pattern corresponded to positions separated by 10.3 bases or multiples thereof. The TT profiles also showed regularly positioned peaks, following fairly well a periodic pattern, but in the positions different from those for the AA. Spectral analysis of the averaged profiles gave the values of the period around the average 10.3 (±0.2) bases [10.2 (±0.2) and 10.5 (±0.3) for AA and TT, respectively]. Although the periodically spaced peaks were present in both AA and TT positional frequency distributions, the periodicity of AA profile appeared to be more pronounced. In time the difference was attributed to the presence of scattered AA and TT dinucleotides that do not participate in nucleosome positioning signal and, thus, could be viewed as a random component that interferes with the multiple alignments and distorts the output profiles. As it was shown in a later work of (Bolshoy *et al.*, 1996), the magnitude of the random component was even much stronger than it was initially thought. It was obtained that only 3–5 AA and TT dinucleotides in total are situated in the 'correct' positions following the periodic pattern in an average nucleosome. With the proportion of ~10–11% of each AA and TT among all 16 dinucleotides in the analysed nucleosome dataset, and consequently 14–15 of each AA and TT per an average nucleosome, the vast majority of these dinucleotides constituted the random composite or noise. However the periodic component still remained apparent in the overall pattern.

Another fundamental feature of the pattern was that several of the TT-peaks of the periodic set have symmetrical counterparts in the periodic set of the AA-peaks, with the axis of symmetry passing through the midpoint position of the nucleosome. Since DNA is a complementary duplex the TT-pattern calculated for the given collection of nucleosome DNA sequences is identical to the AA-pattern for corresponding complementary strands, if the latter one is read in the complementary 3′ to 5′ direction. Reading this AA-pattern in the opposite, that is conventional, direction (5′ to 3′) would result in the mirror-symmetrical picture. In other words, the average AA and TT-patterns calculated for the same strand were mirror-symmetrical to one another, relative to the midpoint of the nucleosome DNA sequence, with minor differences attributed to the interference of the noise components. Our data disagreed with the suggestion made by Satchwell *et al.* (1986) that the AA and TT patterns are identical and mirror-symmetrical to themselves. Indeed, the preferred positions for AA and TT found in our work were far from being identical and neither AA nor TT alone showed the symmetry.

The mirror symmetry of AA-pattern to TT-pattern, as observed in our work, provided an additional opportunity to refine the AA and TT profiles and decrease the noise. The AA- and TT-distributions were combined together by the symmetry rules. That is, new AA- and TT-profiles were calculated: $AAsym = [AA(x) + TT(-x)]/2$, and $TTsym = [TT(x) + AA(-x)]/2$. The coordinate x here is counted from the middle of the nucleosome DNA sequence, resulting in the even more enhanced periodicity of the AA and TT distributions. The lack of mirror symmetry of the AA distributions relative to the midpoint is clearly seen in the Fig. 7.4.

In the two top panels of Fig. 7.4 the averaged distributions AAsym and TTsym are shown together. When these patterns are compared, it appears that almost every maximum of AA-profile corresponds to a minimum in the TT profile, and *vice versa*. This indicates that the AA and TT dinucleotides are major contributors to the overall nucleosome dinucleotide pattern, so that an excess of AA at some position automatically causes lack of TT at this position. According to our calculations, other dinucleotides, indeed, contributed insignificantly, as compared to AA and TT (data is not shown). Ideally, if AA and TT are sole contributors to the overall pattern the plots AAsym

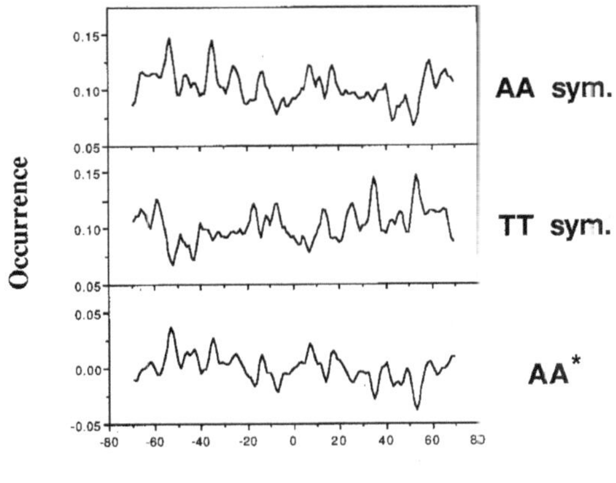

Figure 7.4 The dinucleotide distributions refined by averaging AAsym and TTsym patterns. AA*, distribution of AA dinucleotides after averaging AAsym and -TTsym patterns: AA*=[AAsym – TTsym]/2. Adapted from Ioshikhes *et al.* (1996).

and TTsym should have equal amplitudes but opposite signs. Considering observed deviations from such ideal case as largely due to still remaining noise, a further improvement of the signal can be obtained by subtracting the patterns from each other, that is, calculating AA* = [AAsym(x) – TTsym(x)]/2. The refined distribution (for AA dinucleotides) is shown in Fig. 7.4, bottom. This oscillating pattern has no peaks other than ones of the periodic family, which appears as an almost complete series of maxima separated by 9 to 12 bases. The deviations from the average distance 10.3 bases are, apparently, due to remaining noise. This AA*-pattern, together with the TT*-pattern mirror-symmetrical to it (not shown), represents only a first approximation to the eventually full description of the nucleosome sequence pattern that will include some other dinucleotides, as well as, perhaps, trinucleotide and higher oligonucleotide contributions. This AA (TT) dinucleotide approximation is remarkably close to the pattern derived in Mengeritsky and Trifonov (1983), when the number of available experimentally mapped nucleosomes was insufficient for the alignment analysis as above. Both AA (TT) periodicity (rounded to 10.5 bases in the earlier work) and the about half period phase shift between preferred positions for AA and TT were confirmed in this work. It was also shown that the periodicity is disrupted in the area around the dyad, which presentation in different nucleosome models varies more significantly than in the rest of the nucleosomal DNA (Travers *et al.*, 2010).

Another significant feature of the AA/TT pattern was its gradient: a magnitude of the AA peaks generally decreases if moving from the 5′ to 3′ direction of a given DNA strand whereas a magnitude of the TT peaks decreases in that direction and instead increases in the opposite direction. Although noticed in that time, this feature was not understood and was considered as being rather an artefact of either the dataset or alignment procedure. Obviously, it could not be observed in the (Satchwell *et al.*, 1986) since there AA and TT distributions were combined and resulting distributions were self-symmetrical.

Although other dinucleotides did not exhibit obvious periodic distributions at this work, CC and GG were likely to be the next largest contributors to the nucleosomal pattern. As it was found by two-dimensional autocorrelation analysis of the nucleosome sequences (Bolshoy, 1995), these dinucleotides also display significant periodicity, although in absolute terms their contribution is lower as compared to AA (TT) dinucleotides. For detection of possible contributions of tri- and tetranucleotides to the nucleosomal pattern the database of 204 nucleosomes was too small.

Hence collecting of larger nucleosome sequence datasets became the next issue to resolve. The database of Ioshikhes and Trifonov that was first published in 1993 (Ioshikhes and Trifonov, 1993) with the next release in 1996 (Ioshikhes *et al.*, 1996) contained 204 sequences from 18 eukaryotes and three viruses. The database of Widlund *et al.* (1997) contained only 87 nucleosomal DNA sequences of mouse genome. The next database which consisted of 1002 human dinucleosome DNA sequences has been established in Kato *et al.* (2003) based on *in vitro* MNase mapping technique. Since dinucleosome formation is the first step in the organization of the higher order chromatin structure, it was considered to be more stable than single nucleosome. It has been shown that different eukaryotes may vary in their nucleosome positioning pattern (Herzel *et al.*, 1999; Kato *et al.*, 2003; Cohanim *et al.*, 2005; Kogan and Trifonov, 2005). Therefore, next bigger nucleosome sequence database had to be either species-specific or consist of significant number of sequences originated from a broad spectrum of organisms.

The absence of large nucleosome datasets also impeded application of the available nucleosome patterns to the nucleosome mapping *in silico:* there was no available test dataset to verify the predictions by different patterns. Although single experiments of nucleosome mapping were still performed, they usually were not accompanied by computational analyses which typically require statistically significant data. In that situation, the nucleosome patterns could be improved and tested only indirectly, through the genome sequence analyses. For instance, basic features of the AA/TT nucleosome pattern have been confirmed by positional correlation analysis for the complete genome of *S. cerevisiae* (Cohanim *et al.*, 2005). Besides the strong periodicity with the period 10.4 bases for the AA and TT, the oscillations were observed also in the distributions of other dinucleotides. However, the respective amplitudes were small, consistent with secondary effects, due to dominant periodicity of AA and TT. The observations were in accord with earlier data on the chromatin sequence periodicities and nucleosome DNA sequence patterns. The autocorrelations of AA and TT dinucleotides in yeast included both the in-phase and counter-phase component. A tentative DNA sequence pattern for the yeast nucleosomes was suggested and verified by comparison of its autocorrelation plots with the respective natural autocorrelations. The nucleosome mapping guided by the pattern was in accord with experimental data on the linker length distribution in yeast. However, the suggested patterns were essentially limited by the dinucleotide distributions inside the canonical period of ~10 bp, and the whole-nucleosome pattern was considered as the periodical reproduction of that picture, with the exception of the area around the dyad.

Post-genomic era and high-throughput data

The situation changed dramatically only in 2005, when Rando's group introduced high-throughput techniques to the nucleosome research (Yuan *et al.*, 2005). They mapped 2278 nucleosomes in a single experiment – way more than anything available prior to their work. Another highlight of their work was establishment of a typical nucleosome positioning architecture of the yeast promoters: a stable nucleosome around TSS (named a +1 nucleosome) preceded by a nucleosome-free region (NFR) upstream and another stable nucleosome (the so-called −1 nucleosome) just upstream of the NFR, and regular nucleosome arrays both upstream and downstream of this area. With certain variations (mostly related to the positioning of the +1 and −1 nucleosomes) this model generally was also confirmed later for other species. Most importantly for the computational nucleosome research, this work finally provided an ample amount of data on which nucleosome sequence patterns (interpreted now as nucleosome POSITIONING sequence or NPS patterns (Ioshikhes *et al.*, 2006) could be tested, for their potential of the correct nucleosome mapping *in silico.*

Segal, Widom and their coworkers were among the first to stand up to the challenge. In the paper named 'Genomic code for nucleosome positioning' (Segal *et al.*, 2006) they applied a statistical-mechanic based algorithm for comparison of the genomic sequence and 16 dinucleotide distributions, derived from isolated nucleosome-bound sequences at high resolution from yeast, with the combined (and thus self dyad symmetrical) AA/TT and other complementary dinucleotide distributions. The results of Segal *et al.* (2006) of the nucleosome mapping *in silico* were generally in line with the findings of (Yuan *et al.*, 2005) and correctly mapped just over 50% of all nucleosomes with a precision of +–35 bp.

The paper by Ioshikhes *et al.* (2006) appeared just couple of months later. The Ioshikhes–Pugh group worked simultaneously but independently from the Segal–Widom group, although the first got a chance to compare the results just before their paper was finally accepted for publication. The Ioshikhes–Pugh group applied the AA/TT pattern from Ioshikhes *et al.* (1996) accompanied by an algorithm of the nucleosome mapping based on correlations of dinucleotide distributions in the pattern and genomic sequence. They also generally confirmed results by (Yuan *et al.*, 2005) but were able also to discover several additional effects. For instance, they were able to capture differential nucleosome distributions in the nucleosome-inhibited and nucleosome-stimulated promoters, promoters with and without TATA-box, establish several types of promoter nucleosome positioning in different promoter classes and relay it to gene function and particularly to a type of chromatin remodelling. On the individual nucleosome level, the technique applied in Ioshikhes *et al.* (1996) performed better than that in Segal *et al.* (2006), with a somewhat higher level of true positive predictions (putative positions predicted inside ±35 bp from the experimental positions) and significantly lower level of false positive ones (putative positions predicted outside of the ±35 bp from the experiment).

While some of the differences in the performance of the two approaches may be attributed to the difference in the algorithms of the mapping, the major difference is in the patterns used: Segal *et al.* (2006) used the pattern with AA, TT and TA grouped together (i.e. positioned in-phase) with GC positioned between the latter, whereas Ioshikhes *et al.* (2006) used the pattern with AA and TT separated and even mutually exclusive (and positioned in a counter-phase as explained above), both at the sequence and structural level. Therefore, it could seem at that point that the latter pattern found better confirmation on the experimental nucleosome dataset mapped on a genome-wide level.

But the situation appeared to be much more complex. In 2007, Pugh and coworkers (Albert *et al.*, 2007) mapped 322,000 nucleosomes in yeast on genomic scale. The patterns extracted from these nucleosomes were highly resembling those obtained in Segal *et al.* (2006) and could serve as further generalization thereof: AA, TT, TA and AT (or all WW dinucleotides, with W = A or T) grouped together closer to the histone octamer, with CC, GG, GC and CG (or all SS, SS = C or G) grouped together further away from the octamer. Hence it seemed that in fact the 'in-phase' pattern (as far as AA and TT or other complementary pairs concerned) found better confirmation at the genomic scale. And yet, in the following work of Mavrich *et al.* (2008a), arrays of the nucleosome positions were successfully mapped computationally using the nucleosome pattern derived from these very nucleosomes, both around the TSS and transcription termination site (TTS). But this result was not observed by the pattern published in Albert *et al.* (2007): the combined in-phase AA and TT did not fit well the mapping algorithm from Ioshikhes *et al.* (2006), which explored the individual dinucleotide distributions separately. Instead, a computational scheme of iterative adjustment of the pattern from Ioshikhes *et al.* (1996) to the new data was suggested, and although the resulting pattern could converge to the in-phase AA/TT, it did not: it generally still was following the trend in Ioshikhes *et al* (1996) but had more evident gradient of the AA and TT distributions. The gradient was even more evident than the AA and TT periodicity. Hence the paradox resurfaced: although the in-phase AA/TT pattern seemed to be more readily popping out of the sequence data, the counter-phase once again scored better in terms of the nucleosome mapping.

While Segal *et al.* and Ioshikhes *et al.* deduced nucleosome sequence patterns from a training set of experimentally defined nucleosome DNA sequences and used it for predictions of the tentative nucleosome positions on genomic DNA, subsequent works concerning the nucleosome identification *in silico* went in rather different direction. These works mostly focused on discrimination between the nucleosome and non-nucleosome sequences using ROC curves, and in that they achieved better performance than either in Ioshikhes *et al.* (2006) or Segal *et al.* (2006) (e.g. see Chung and Vingron, 2009). However, it is not clear how efficient such approach would be in terms of the mapping as performed by Segal *et al.* (2006) and Ioshikhes *et al.* (2006) and hence there is no evidence of the superiority of the new approaches in these terms. In addition, the new approaches also have some inherent conceptual problems. First, it is rather difficult to provide experimentally verified non-nucleosome negative dataset of DNA. Sequences with low nucleosome occupancy may be related to nucleosome free regions, but these have rather specific sequence content, e.g. over-representation of the poly-A motifs. Linkers, another class of the nucleosome-void sequences, are much shorter than the nucleosomes and generally mapped with a lower experimental reliability than the nucleosomes. Hence there is no negative dataset available which could be really consistent with the nucleosome DNA. Consequently, the results of the nucleosome/non-nucleosome DNA discrimination would strongly depend on the choice of the analysed negative dataset. The choice of the positive dataset is also not very straightforward, provided that various datasets may differ (Jiang and Pugh, 2009) by positioning many individual nucleosomes, although this hurdle may be overcome by the selection of the only consistent nucleosomes across various datasets. Use of computationally designed sequences for the negative dataset is obscured by the fact that nucleosome DNA correlation with the genomic DNA is comparable to those on shuffled or randomized sequences (Salih *et al.*, 2007). Attempt to achieve better performance of the ROC discrimination may therefore result in selection of skewed negative datasets: e.g. smaller specifically selected datasets of most reliable nucleosome sequences and most trustworthy non-nucleosome ones not necessarily would reflect the most common nucleosome and linker sequences genome-wide. Hence, although the ROC sequence discrimination approaches achieved better performance comparing to that in Segal *et al.* (2006) and Ioshikhes *et al.* (2006), it is still important to achieve the better nucleosome mapping efficiency in the same fashion as both groups did. The latest developments in that area are described further.

Positive versus negative, combining and splitting the patterns

While so far our attention was attracted mostly to the patterns derived from the nucleosomal DNA, another idea requiring exploration is that some motifs tend to be void of nucleosomes and thus could act as the negative nucleosome patterns. These patterns would be situated between the nucleosomes, i.e. in the linkers or NFR and unlike the previous patterns would position the adjacent nucleosomes by exclusion: the nucleosomes would tend to avoid the regions enriched with these motifs, and would be positioned between them. In particular, it was shown earlier that poly(dA:dT) motifs (longer stretches of A or T nucleotides) are detrimental to the nucleosomes because of their rigidity, which makes them hardly bendable around the DNA. Such motifs were described by Field *et al.* (2008) in the NFRs and possibly linkers (though it is rather difficult to reliably attribute their presence to the actual linkers because of a relatively short linker length and relatively low experimental precision of the mapping for most of the nucleosomes, comparable to the linker length itself). Mavrich *et al.* (2008a) focused on these motifs around the TTS where they often occur. They implemented masking of such motifs as part of their nucleosome mapping procedure to achieve the better results.

At the same time, Trifonov and his group tried to combine the in-phase and counter-phase patterns (in the sense described above) to achieve the 'ultimate' nucleosome positioning pattern. This idea was based on consideration of various patterns retrieved from analysis of genomic sequences of different species and the discovery of the fact that although the nucleosomes are highly conserved structures among all eukaryotes, this is mostly due to the high conservation of the histone proteins. However the nucleosome DNA sequence patterns as described here in fact may vary among the species. The reason for the latter is not completely clear. One possibility is that the patterns vary due to the variations in the genome sequence content for the various species. Another option is that this pattern variability is related to the variability of the mechanisms of gene regulation: e.g. yeast may have more periodical nucleosomes with a more stable shape formed by AA and TT dinucleotides because the main function of the nucleosomes there is just to tightly pack DNA. In the higher species however the gene regulatory role of the chromatin becomes more important, and this is related to looser and less periodical nucleosome patterns. In this situation, however the quality of the sequence data becomes crucially important: so far only nucleosomes in just a few species are mapped with a high resolution. However, the general fact of species specificity of the patterns is now well established, even though specific features of each pattern are subject to change as soon as new data or methods of their analysis become available.

It is commonly believed that the sequence pattern responsible for nucleosome positioning is periodical alternation of AA and TT dinucleotides. While it is true, indeed, for the nucleosomes of many organisms, e.g. for yeast (Cohanim *et al.*, 2005; Ioshikhes *et al.*, 2006), human nucleosomes display a very different pattern – periodicity of RR and YY dinucleotides (Kato *et al.*, 2003), especially of GG and CC dinucleotides (Kogan and Trifonov, 2005). Exploiting an early observation that the nucleosomes are preferentially centred at the gene splice junction sites encoded in DNA, a large database of the junctions has been analysed. Dinucleotide distributions around the splice junctions demonstrate that the main contributors to the nucleosome sequence periodicity are RR and YY dinucleotides. The periodical RR and YY usage is different in different species (Kogan and Trifonov, 2005). E.g., in human and mouse these are, primarily, periodical GG and CC, with smaller contributions by GA (TC) and AG (CT). In *Arabidopsis* and in *C. elegans*, on the other hand, these are periodical AA and TT, with weak contributions by other RR and YY dinucleotides. The special role of RR and YY dinucleotides in the deformation of DNA in the nucleosome can be explained by low resistance of outwardly positioned YY base stacks to bending in the nucleosome. As we can see from these findings of the Trifonov's group, however, the conclusions were made mostly based on the analysis of the genomic DNA or those associated with certain genome segments like splice junctions or Alu-repeats (Salih *et al.*, 2008) known to be associated with phased nucleosome positioning, rather than on the analysis of real nucleosome sequences. Would it be the same if the nucleosome sequences per se were analysed? Obviously, the answer only may be given when the nucleosomes are reliably mapped with a high resolution in variety of species. Yet the latter is not an easy task. While the chromatin may be digested with a variety of enzymes (of which the MNase overdigestion cutting DNA just on the border of the nucleosomes and linkers resulting in a high-resolution nucleosome mapping is the most widely used) it is still a challenge to identify nucleosome-associated DNA cuts among all others.

Drosophila melanogaster was the next species after the yeast in which the nucleosomes were mapped genome-wide with a high resolution, using the MNase overdigestion, and analysed computationally (Mavrich *et al.*, 2008b). To extract the NPS pattern, only nucleosomes mapped with a high resolution were used. Similarly to yeast (Albert *et al.*, 2007), the MNase digestion sites situated on a nucleosome size distance with 1 bp shift on either one or two strands were extracted. These segments could be most reliably associated with the nucleosomes, and such selection also allowed filtering out the sequences for the NPS extraction: while in our earlier work (Ioshikhes *et al.*, 2006) only handful of the nucleosomes were mapped with a high resolution and we had to invent original multiple sequence alignment algorithms for analysis of the whole dataset of ~200 nucleosome sequences. Such alignment would be hardly feasible for hundreds of thousands of nucleosomes obtained in the modern high-throughput experiments. The analysis of the high-resolution nucleosomes resulted in combined periodical WW and SS patterns very similar to those in yeast. But once again, these patterns were not the patterns, that define the nucleosome mapping. In this case again only pattern with distinct positioning of each of the complementary dinucleotides was able to achieve a high match between the computational prediction and an experiment. While AA/TT pattern extracted from the *Drosophila* nucleosomes showed some correlation with the experimental data, it was CC/GG pattern (with CC and GG clearly shifted from each other), which mapped the nucleosomes in this case. Thus species specificity of the pattern was shown on real nucleosome sequences.

Further developments in the area were dependent on availability of datasets for various species, which could provide with a high-resolution sequence data. Schones *et al.* (2008) provided significant data concerning chromatin digestion in humans. This dataset has approximately 40 million sequences and was generated from the normal CD4+ T lymphocytes by micrococcal nuclease overdigestion. The ends of the isolated mononucleosome-sized DNA segments from MNase-digested chromatin were sequenced using the Solexa sequencing technology. Unique non-paired nucleosomal sequences with the length of 24–25 bp were obtained as described previously (Barski *et al.*, 2007) from the normal CD4+ T lymphocytes. The raw data consisted of roughly 40 million non-paired reads of length 24–25 bp. Therefore, the data potentially represent genome-wide nucleosome positioning map in human. However, the exact nucleosome positioning with 1 bp resolution is not directly provided by this dataset. Instead, the nucleosome positions should be retrieved using a pre-processing algorithm, and veracity of that algorithm by itself may be debatable. Without the pre-processing, on the other hand, the resolution of tentative nucleosome positioning would be rather low (comparable to the length of the sequenced DNA segments) and thus hardly may serve for retrieval of the NPS sequence patterns comparable to those previously discussed. In the absence of high-resolution nucleosome positioning data one could resort to multiple sequence alignment to retrieve the patterns, but as it has been already said, the multiple sequence alignment is rather infeasible with the given number of sequences. In addition, even with a claim to map the nucleosomes with a high resolution by any of the high-throughput techniques, there would still remain a question about the extent, to which the given positions are biased by the given experimental approach. To resolve this question, alternative dataset obtained by an alternative experimental technique would be needed, which makes it even more difficult problem. While multiple datasets are presently available for yeast (see Jiang and Pugh, 2009) and *Drosophila* (Mavrich *et al.*, 2008b; Teves and Henikoff, 2011), there is still a paucity of such data for other species. In this situation, Trifonov's group continued further to retrieve the nucleosome patterns by whole-genome analyses of various species.

As already mentioned above, basic features of the AA/TT nucleosome pattern have been confirmed by positional correlation analysis for the complete genome of *S. cerevisiae* (Cohanim *et al.*, 2005). A strong periodicity with the period 10.4 bases was detected in the distance histograms for the dinucleotides AA and TT whose autocorrelations include both the in-phase and counter-phase component. This analysis was further developed in the subsequent work of (Cohanim *et al.*, 2006) by application to tree different eukaryotic genomes: *S. cerevisiae, C. elegans* and *D. melanogaster*. Once again, existence of two different AA/TT periodical patterns associated with the nucleosome positioning was confirmed: (1) the pattern with the counter-phase oscillation of AA and TT dinucleotides (named 'the nucleosome DNA pattern'); (2) the in-phase oscillation of the AA and TT dinucleotides with the same nucleosome DNA period, 10.4 bp, corresponding to the curved DNA, that also participates in the nucleosome formation. These two patterns, with the addition of the specific linker sizes (preferably 8 or 18 bp), dictated by the steric exclusion rules, constituted 'Three sequence rules for the chromatin', which was the title of the paper.

From our standpoint, the main message of this article was independent confirmation of simultaneous existence of the in-phase and the counter-phase nucleosome patterns in the nucleosome DNA. This was obtained from the genome-wide analysis, independently of any experimental techniques of the nucleosome mapping and thus could not be just attributed to some experimental or formal artefacts. It was still necessary however to explain differential behaviour of the in-phase and counter-phase patterns in the available nucleosome datasets, both in terms of relative representation of the patterns and their performance in the nucleosome mapping. For that, two opposing approaches were applied. First approach, developed by the Trifonov's group, strived to combine the different patterns, which eventually resulted in the 'ultimate nucleosome sequence pattern' as expected 'finale' of the nucleosome pattern story. Second approach was discussing comparative roles of the in-phase and counter-phase patterns and their biological implications. As the result of this second approach, additional patterns were discovered which looked the inverse to the previously described patterns, and hence neither could be reduced to the conventional patterns, nor obtained by combining the latter. Therefore, the expected finale may need to be postponed until the time when the role of new patterns is fully clarified. We will address the issue of the separation of the patterns and discovery of the new patterns later, and now will focus on combining the patterns as a more straightforward approach.

As already mentioned, the species specific nucleosome pattern features were observed with primarily variable roles of different dinucleotides in the sequence patterns: the AA/TT dinucleotides were apparently dominant sequence motifs in the yeast, yet primary role in nucleosome positioning in *Drosophila* and in human was attributed to CC and GG dinucleotides. To reveal a role of different sequence motifs in the nucleosome patterns, more comprehensive picture discussing the role of all dinucleotides and possibly the longer sequence motifs in a variety of species was needed. While such an attempt had been made in the earlier works, it was performed on rather limited nucleosome sequence data, and hence the conclusions had to be updated when new data became available. At that time a third genome-wide nucleosome positioning set (after yeast and *Drosophila*) of the *C. elegans* nucleosomes, mapped with a high resolution (Johnson *et al.*, 2006), was analysed (Gabdank *et al.*, 2009). Although very large database of the nucleosome DNA sequences of the length 146 bp was used, the positional preferences of various dinucleotides within the 10.4 bp nucleosome DNA repeat rather than in the entire sequences, as in Ioshikhes *et al.* (1996) or Segal *et al.* (2006), were calculated. First, the sequence structure of the 10.4 bp repeat was partially reconstructed by analysis of preferred distances between various dinucleotides in [R, Y] alphabet: $(YYYYYRRRRR)_n$, and in [A, T] alphabet: $(TTTYTARAAA)_n$ (Salih *et al.*, 2008). Afterwards, similar information for other sequence elements was extracted using iterative approach. Since clear distributions of the dinucleotides other than AA and TT was rather difficult to obtain from the sequence data alone due to the weakness of the signal, it was done step by step. First, the most periodical dinucleotides were identified and their distributions inside the 10.4 bp sequence segment retrieved; then nucleosome sequences were aligned to that pattern and the next most periodical dinucleotides were identified in the aligned sequences and incorporated to the pattern. The derived bendability matrix involved positional distributions of the particular dinucleotides in the segment of 10–11 bp long, comparable with a period of the nucleosomal DNA. The peaks of the distributions were related to the dinucleotide relative phasing in the bendability matrix and periodical nucleosome pattern, and hence the resulting pattern could be derived by combining all periodic dinucleotides, taking into account their mutual phase shift. As a result it was shown that the common positional preferences can be described by one-line consensus CGGAAATTTCCG, the same pattern for all six chromosomes, or, in more general form, YRRRRRYYYYYR (Gabdank *et al.*, 2009). The latter pattern as well as the other similar, however, should be properly understood. It cannot imply any possible combination of Y and R in respective positions because, e.g. AAAA or TTTT motifs are nucleosome detrimental. Rather the pattern should be *ab initio* strongly degenerated with only few nucleotides appearing in the specific defined positions. The highest selectivity to the positions within the period was displayed by CG, AT, GG/CC, GA/TC and AA/TT dinucleotides. The strong periodicity of the YY, CG and AT dinucleotides was in a good agreement with the results obtained, by physical considerations, from the analysis of nucleosome crystal structures (Cui and Zhurkin, 2010). The bendability matrix derived in Gabdank *et al.* (2009) spans only a length of 10 bp. The next step was to reconstitute the whole nucleosome length sequence pattern for *C. elegans* based on this bendability matrix. This was done in the subsequent work of Gabdank *et al.* (2010a) where the nucleosome sequences from the database were scanned using the bendability matrix obtained in Gabdank *et al.* (2009). Instead of a regular nucleosome length of ~147 bp, shorter sequences of 116 bp centred around the nucleosome dyad were analysed as exhibiting tighter binding to the histone octamer (Johnson *et al.*, 2006). The correlation of the nucleosome segments with the bendability matrix was calculated, resulting in rather noisy picture with uneven and asymmetric distribution of the correlations along the sequences, which in part could be attributed to the interference of tandem repeats of a length 35 bp. After removing the nucleosome sequences with highest magnitude of 35 and

70 bp oscillations, more regular distribution of motifs with periodicity of ~10 bp emerged in the remaining sequences. Combined with the dyad symmetry considerations, the bendability matrix was extrapolated to the 117 bp length. Technically the resulting pattern could be also directly derived from two known binary presentations of the nucleosome DNA sequence pattern – $(R_5Y_5)_n$ (Salih *et al.*, 2008), and $(S_5W_5)_n$ (Chung and Vingron, 2009). The resulting pattern featured the repeat of the motif (GGAAATTTCC), which was declared a universal nucleosome pattern. The obtained pattern was also successfully applied to mapping, with a high precision of 1 bp, of several nucleosomes whose crystal structures were available in the Protein Data Bank (PDB). However, since these nucleosomes involve artificially designed complementarily symmetrical sequences made of halves of the primate alpha-satellite repeats, the application of the universal pattern to genome-wide mapping of the nucleosomes majority, of which are both less symmetrical and less periodical, still remained a problem, although the proposed method (Gabdank *et al.*, 2010b) provided a mapping with efficiency comparable to other approaches (Kaminsky *et al.*, 2012). However, the correlation of the universal pattern with the nucleosomal sequences, found in the same study (Gabdank *et al.*, 2010b), was comparable to those in the shuffled sequences, and only ~20% of the nucleosomes showed significant correlation with the pattern (similarly to the percentage of the significantly correlating nucleosomes in the shuffled sequences). This estimate was in line with the estimated percentage of the nucleosomes mapped by sequence pattern simply by chance (Segal *et al.*, 2005).

While the universal pattern should be present in all the species, its particular implementation in various species would be defined by the sequence composition of the respective genomes and specific roles that nucleosomes may play in the gene regulation thereof. The species specific implementation of the patterns was shown in (Bettecken and Trifonov, 2009; Rapoport *et al.*, 2011) where periodicity of various dinucleotides differed among the considered species. Thus in (Bettecken and Trifonov, 2009) the distance analysis technique was applied for determining the dinucleotides that display the 10.4 base periodicity in thirteen diverse eukaryotic genomes. A total of 208 periodicity plots for the 13 genomes and all 16 dinucleotides were calculated, revealing that each of the 16 dinucleotides clearly shows the periodical positioning in at least one of the genomes analysed. Although this was interpreted as a difference in the dinucleotide repertoire, it certainly correlates with the DNA sequence composition: for instance CG dinucleotide was found to be highly periodical by 10.4 bp for honey bee, whose genome is known to be relatively CG-rich (Elango *et al.*, 2009).

The basic concept of these works is the existence of the general nucleosome sequence pattern, with the recurring motif CCGGRAATTYCCGG, identified as a universal nucleosome bendability sequence pattern in Trifonov (2010a,b). It hardly could be found in its entirety in any genome, because nucleosomes containing the entire pattern would be too strong to facilitate chromatin remodelling and gene regulation. Yet its particular components (e.g. trinucleotides) are statistically well represented in different genomes (Rapoport *et al.*, 2011) with particular motif subsets differing among the genomes and depending on their sequence composition and specific gene regulation mechanisms.

The latter pattern was further simplified to the repeat CGGAAATTTCCG and further to (GGAAATTTCC)*n* to take into account periodicity of 10.4 bp instead of 10 bp as in the previous models (Trifonov, 2010). The latter work explicitly mentioned the final pattern as a combination of the earlier commonly accepted patterns: YRRRRRYYYYYR in the purine/pyrimidine alphabet and SWWWWWSSSSSW in the strong/weak alphabet. It also mentioned the latest pattern as the 'apparent finale' of the long-standing problem of nucleosome positioning, which provides simple means for nucleosome mapping *in silico* with single-base resolution.

However, our own following works presented a somewhat surprising development to the nucleosome positioning problem and evidence that it may be a bit premature to talk about the finale. The study (Ioshikhes *et al.*, 2011) initially tried to address controversy of a visible preference of the WW/SS pattern in the collection of the yeast nucleosomes versus better performance of the RR/YY pattern in the nucleosome mapping. Careful analysis of the AA/TT distributions in the yeast nucleosomes showed that they are neither strictly in phase nor in the opposite phase but rather may represent a mixture of both patterns. While this idea is similar to those advanced by Trifonov in his latest studies, it in fact evolved rather independently even before those publications. In order to identify the nucleosomes following either of the two patterns, they were separated by their correlation to the WW/SS pattern, its AA/TT component as well as to the counter-phase AA/TT pattern from Ioshikhes *et al.* (1996).

To elucidate the WW/SS (including in-phase AA/TT) and RR/YY (including counter-phase AA/TT) patterns independently, we separated H2A.Z-containing nucleosomal sequences mapped with 1 bp resolution (Albert *et al.*, 2007) into 5718 sequences in which the distribution of AA dinucleotides had a positive correlation to the WW pattern, and 3422 that had a negative correlation. The rationale of this separation was that the WW/SS and RR/YY patterns would work not in a cooperative (as implied by Trifonov) but rather antagonistic manner obscuring each other. Indeed the cooperation of the WW/SS with RR/YY would result in mutual enhancement of the patterns instead of their mutual distortion as in these data. In that case, WW/SS or RR/YY patterns could be in rather opposite phases. The black traces in Fig. 7.5 display the WW dinucleotide distribution for both sets. RR distributions were also examined (grey traces in Fig. 7.5). As expected, the WW and RR patterns for the same subsets were indeed in opposite phases. Remarkably, separation of the nucleosomal sequences into the positively and negative correlating subsets lead to a clearer periodicity of ~10 bp for all considered dinucleotide combinations.

Much more surprising was the existence of well pronounced oppositely phasing counterparts for both conventional WW and RR patterns (compare WW+ with WW-, and compare RR+ with RR– in Fig. 7.5). We refer to the patterns attained from the negatively correlating sequences as anti-WW and anti-RR. As these nucleosomal sequences were not pre-selected to have correlations with any of the NPS patterns, it is quite surprising that a significant fraction showed the anti-correlation. We were therefore prompted to directly examine the positive versus negative correlations of individual nucleosome sequences with the in-phase and counter-phase AA/TT patterns.

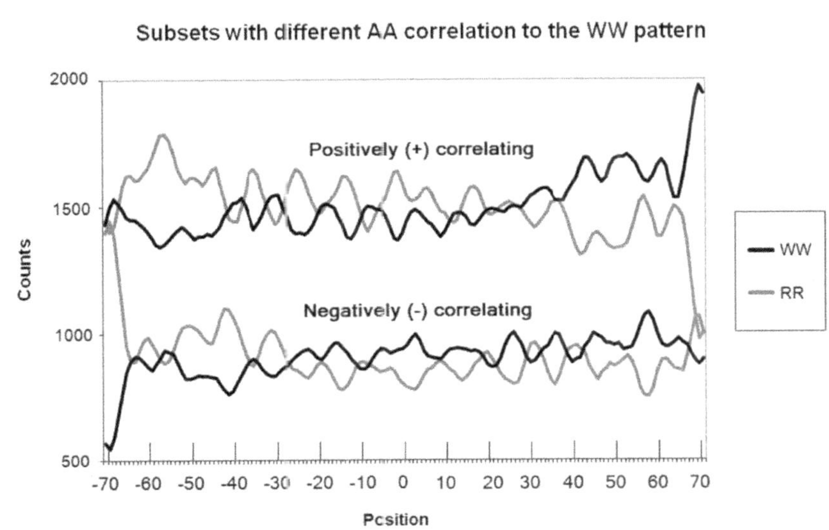

Figure 7.5 Combined dinucleotide distributions (smoothed by 3 points) for subsets with AAs positively correlating with the major WW pattern from Albert *et al.* (2007) (+, higher in the graph) and with AAs negatively correlating with the major WW pattern (–, lower in the graph). Adopted from Ioshikhes *et al.* (2011).

We divided the entire nucleosome set into two subsets based on dinucleotide correlation with each of the major patterns:

1. As shown in Fig. 7.6, those with positive (A) and negative (B) correlations to the counter-phase AA/TT pattern from (Ioshikhes *et al.*, 1996).
2. As shown in Fig. 7.7, those with positive (A) and negative (B) correlation to the WW/SS pattern from (Albert *et al.*, 2007).

In Fig. 7.6, both the regular and anti- RR and YY patterns were very well pronounced. The patterns for sequences with positive correlation to the counter-phase AA/TT are highly consistent with the counter-phase pattern, with RR and YY in opposite phases to each other (compare RR and YY in Fig. 7.6A), which is expected. Consistent with expectations, WW and SS patterns are relatively weak for this subset (Fig. 7.6A). The patterns for the sequences having a negative correlation to the counter-phase AA/TT

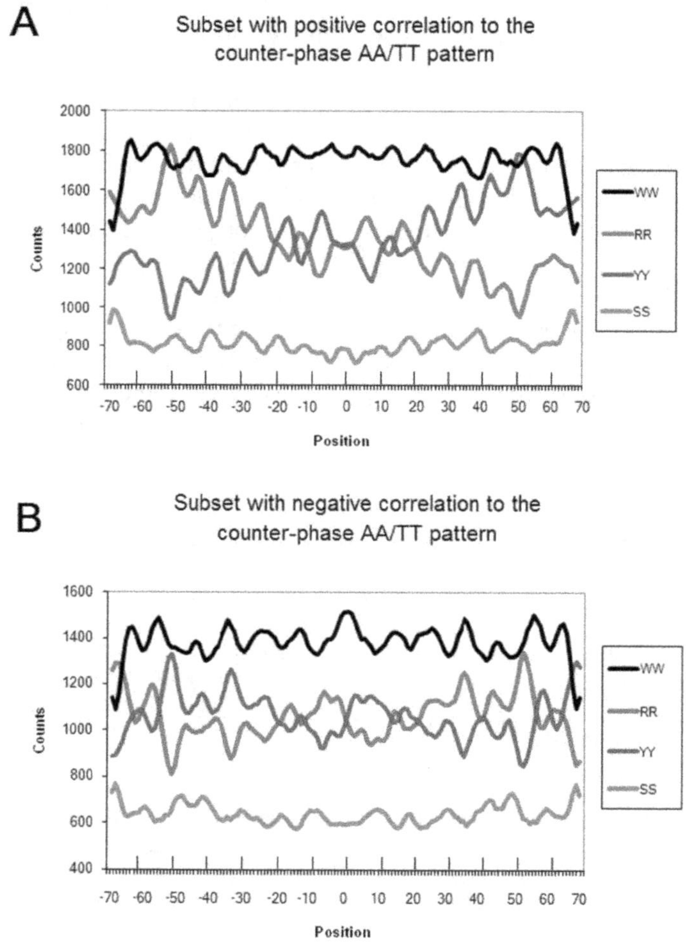

Figure 7.6 Combined dinucleotide distributions (smoothed by 3) for nucleosome subsets with positive (A) and negative (B) AA/TT correlation to the counter-phase AA/TT pattern from (Ioshikhes *et al.*, 1996). Notice opposite phases for RR patterns at A and B, for YY patterns at A and B, and steep gradients for RR and YY patterns at A (no obvious gradients at B). The opposite phases for the patterns in the A and B panels are related to inverse positioning of respective sequence elements in conventional (major) patterns and respective anti-patterns (presented in Figures 7.2 and 7.3, left and right sides, respectively). Adopted from Ioshikhes *et al.* (2011). A colour version of this figure is located in the plate section at the back of the book.

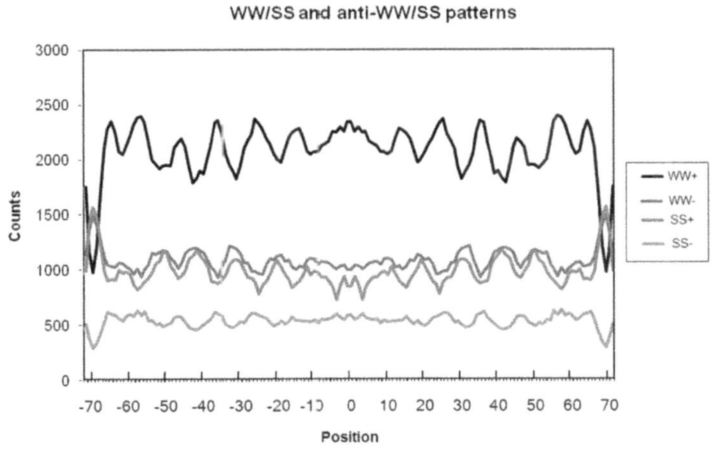

Figure 7.7 WW and SS patterns (smoothed by 3) for subsets with positive (+) and negative (–) WW/SS correlation to the major WW/SS patterns from Albert *et al.* (2007). Notice identical phases for the WW– and SS+ and for the WW+ and SS– patterns. Adapted from Ioshikhes *et al.* (2011). A colour version of this figure is located in the plate section at the back of the book.

pattern were also clearly visible (see RR and YY in Fig. 7.6B). The magnitude of the dinucleotide distributions in the latter sequences is clearly above the random level, which is rather surprising. The WW/SS pattern is better pronounced for these sequences. A subset of 5107 sequences from the entire sequence set of 9140 showed positive correlation to the counter-phase AA/TT pattern (Ioshikhes *et al.*, 1996), while 4033 sequences showed negative correlation. The respective ratio of the number of NPS/anti-NPS nucleosomes (i.e. those with positive and negative correlation to the counter-phase AA/TT pattern) is ~5/4 = 1.25. The statistical significance of the separation by the chi-square test is almost 8 standard deviations (SD) ($P < 0.0001$), i.e. extremely high if the number of members in each group is compared with those randomly expected (50% or 4570). Yet the most intriguing finding of our study is that the number of nucleosomes described by an anti-NPS pattern is almost as high as the number obtained by the standard NPS pattern, which almost doubles the number of predictable nucleosomes.

We next parsed the H2A.Z nucleosomal sequences according to their positive or negative correlation to the WW/SS pattern of (Albert *et al.*, 2007). This resulted in clear WW/SS and anti-WW/SS patterns (Fig. 7.7): WW(+), SS(+), and WW(–), SS(–) respectively. Of these, 6030 sequences were related to the WW/SS, and 3110 were related to the anti-WW/SS pattern, with the NPS/anti-NPS ratio being approximately 2:1 (22 SD; $P < 0.0001$). As seen from (Ioshikhes *et al.*, 2011), 50% of the nucleosomal sequences clearly conform to either the WW/SS or RR/YY patterns, whereas 14% conform to neither. Magnitudes of the positive and negative correlations are overall comparable. Hence the contribution of the anti-patterns to nucleosome formation is similar to those of the conventional patterns.

While the existence of the canonical WW/SS and RR/YY patterns in the H2A.Z nucleosomal sequence set is somewhat expected, the existence of the 'anti' patterns (both anti-WW/SS and anti-RR/YY) is a novel finding. Further examination of the WW/SS and anti-WW/SS patterns shows that, even though these patterns have the same periodicity (~10 bp) while being in opposite phases, they are genuinely different patterns that cannot be mutually converted by simply shifting one of the patterns by half of the period (5 bp). Thus, experimental mapping uncertainty of nucleosome positions, for example by 5 bp, would alter potential alignments but would not produce an altered pattern. Indeed the ~10 bp dinucleotide periodicity is disrupted in the central part of the nucleosome (around its dyad), so a shift of either pattern by 5 bp could not result in its anti-counterpart. The same is true for the RR/YY and anti-RR/YY patterns. The high amplitudes of the patterns mean that they both are relatively well pronounced on their respective nucleosome subsets. If the canonical WW/SS and RR/YY patterns are related to stable nucleosome structures, their 'anti' counterparts appear opposite to them and thus may be related to relatively unstable nucleosome structures (Figs. 7.2 and 7.3, right panels) and genomic regions harbouring unstable nucleosomes.

Structural and biological implications of WW/SS and RR/YY patterns

Conceivably, the four NPS patterns presented here (Figs. 7.2 and 7.3) might be related to nucleosomes of different stability. The WW/SS pattern (Fig. 7.2, left panel) defined by Segal *et al.* (2006) and Albert *et al.* (2007), or as depicted in the right panel of Fig. 7.2 according to Trifonov (2010), may be the most stable structure. This arrangement would involve intrinsically curved DNA that may promote nucleosome formation (Drew and Travers. 1985; Anselmi *et al.*, 2000). The WW/SS class of nucleosomal DNA is relatively stiff but intrinsically curved, and nucleosomes containing such DNA should be relatively stable. Its DNA is analogous to an aluminium wire wrapped around a pencil: its shape persists when the pencil is removed, and the pencil may fit back into it. However, rigid DNA of a different shape (e.g. those containing poly dA:dT tracts) would rather preclude nucleosome formation in that it would be energetically unfavourable to bend it around the histone octamer.

On the other hand, the RR/YY pattern (Fig. 7.3, left) relates to nucleosomes that have lower intrinsic curvature, but higher bendability. Its DNA rather resembles thread wrapped around the pencil: it may be easily wrapped around (high bendability) but its shape does not persist when the pencil is removed (no intrinsic curvature). The rotational positioning of the YY backbone further from the octamer surface compared to RR in such nucleosomes may be explained by the lower energy of stacking interaction between pyrimidines *vs.* purines (Salih *et al.* 2007; Trifonov, 1985) that favours DNA bending in the direction of the inward facing RR backbone (Trifonov, 1985). DNA in such nucleosomes may be easier to wind/unwind around a histone octamer, yet it would not preserve its curvature in nucleosome-free DNA. These nucleosomes are less stable and may be more easily displaced, which may be beneficial for the nucleosomes around transcription start sites, TATA boxes, and other promoter elements. Indeed, this was a reason that our computational mapping of nucleosome positioning (Ioshikhes *et al.*, 2006) by the counter-phase AA/TT pattern (AA/TT component of the RR/YY pattern discussed here) (Ioshikhes *et al.*, 1996) in combination with a comparative genomics approach (Ioshikhes *et al.* 2006) was significantly more successful for yeast promoters than mapping by WW/GC pattern (Segal *et al.*, 2006). Mapping by the updated AA/TT pattern shows an even better performance in the proximal promoter area (Mavrich *et al.*, 2008a).

The anti-RR/YY pattern represents a more unstable structure (Fig. 7.3, right), with unfavourable RR/YY positioning for DNA bending. Of the four patterns, anti-WW/SS may generate the most unstable nucleosomes, with DNA that would bend around a histone octamer in direction opposite to its intrinsic curvature. Intuitively, the latter nucleosomes should be quite unfavourable in their occurrence along the genome. Hence the anti-WW/SS pattern may be related to genomic regions that are relatively depleted of nucleosomes.

The anti-RR/YY pattern has the backbone of AAs and other RRs facing mostly outward from the histone surface and TTs, and other YYs facing inward, which is not a favourable structure according to the geometrical properties of dinucleotides (Salih *et al.*, 2007). It may be favourable, however, if the number of RRs and YYs following the anti-pattern is greater than the number of YYs and RRs following the regular pattern (see Fig. 7.3B). In addition, inspection of the regular RR/YY patterns (Ioshikhes *et al.*, 1996; Mavrich *et al.*, 2008a) shows a clear and steep gradient in RR distributions from the 5′ to 3′ end of the nucleosomal DNA (see Fig. 7.6A). However, the opposite (3′ to 5′) gradient is much less (if at all) pronounced in the anti-RR/YY pattern (Fig. 7.6B).

What are the biological implications of the different patterns? The RR/YY pattern may be related to nucleosomes that require an easy shifting from their positions upon chromatin remodelling. This may be relevant to nucleosomes situated in the proximal promoter area, in particular around the TSS, TATA box or interacting with specific TFs, which are especially susceptible to chromatin remodelling. To test this hypothesis directly, we counted the nucleosomal sequences that correlated positively or negatively with either the WW/SS or counter-phase AA/TT patterns in the promoter regions of various classes of genes. Generally for the promoter nucleosomes, more nucleosomes were associated with the counter-phase AA/TT (RR/YY) pattern compared to the WW/SS pattern, which is the opposite of what is observed genome-wide. In addition, unlike genome-wide, the number of promoter nucleosomes following the anti-WW/SS pattern was higher or close to the number of nucleosome following the conventional WW/SS pattern. Thus, promoter nucleosomes utilize RR/YY and anti-WW/SS patterns in a disproportionately higher number than the rest of the genome. They may be well suited for promoting chromatin remodelling at promoters.

Thus it becomes possible to explain the former controversy regarding comparative role of the WW/SS and RR/YY patterns in nucleosome positioning. While WW/SS pattern may be prevalent in a bulk chromatin, which may also be related to the use of the MNase over-digestion for the nucleosome mapping in respective experiments. As mentioned above, the outcome of such experiments would predominantly involve most stable nucleosomes, i.e. those which follow the WW/SS pattern. On the other hand, computational nucleosome mapping as performed in Segal *et al.* (2006), Ioshikhes *et al.* (2006) and Mavrich *et al.* (2008) was focused primarily on gene promoters. Since promoter nucleosomes (at least in the yeast) primarily follow RR/YY pattern, the latter should provide better mapping of those nucleosomes.

Role of alternative anti-NPS patterns

We examined whether the regular and anti-counter-phase AA/TT NPS patterns (both of the RR/YY class) could model nucleosome organization around the 5′ ends of genes, which is where DNA-encoding of nucleosome organization is most predominant. To facilitate the examination of a large number of genes and look for overarching patterns, we utilized groups of genes that were previously clustered by K-means (Zhang *et al.*, 2011), based upon their *in vivo*-determined nucleosome organization (Fig. 7.8).

Each cluster was defined by the combination of two attributes: nucleosome positions and occupancy levels. Cluster 1 is characterized with a low-occupancy and thus potentially unstable +1 nucleosome. This nucleosome was positioned upstream of the canonical location, which places the TSS in a more repressive location closer to the nucleosome dyad. Cluster 2 had high nucleosome occupancy at the −1 position. Clusters 3–5 had genic nucleosomal arrays starting at different distances from the TSS. Like cluster 1, cluster 5 also displayed nucleosome encroachment in the NFR, but near the −1 position instead of the +1 position.

Clusters 1 and 5 genes tend to be TATA-containing, SAGA-dominated, chromatin-regulated, and stress-induced. Genes of such character tend to have plasticity and stochastic fluctuations in gene expression (Newman *et al.*, 2006; Tirosh and Barkai, 2008). Clusters 3 and 4 are enriched with genes in the TATA-less, TFIID-dominated, housekeeping class, which represents the majority of all yeast genes and is consistent with their nucleosome organization being more canonical.

Fig. 7.8 (black trace) shows that for clusters 2–4, which comprise most genes, stronger counter phase AA/TT NPS correlations to a large extent matched with the measured positions of the −1 and +1 nucleosomes (and to a small extent +2 and +3) in native chromatin. Remarkably, the anti-counter-phase AA/TT pattern (grey trace) matched with NFRs. This relationship was also most evident in subsets of stress-related genes, but was also found generally at most genes. As reported in (Schones *et al.* 2008; Jin *et al.*, 2007; Henikoff, 2009; Weiner *et al.*, 2010), some NFRs in fact may be occupied by unstable nucleosomes. Such nucleosomes may be easily removed from their sites during gene expression, yet cannot be mapped well by regular NPS

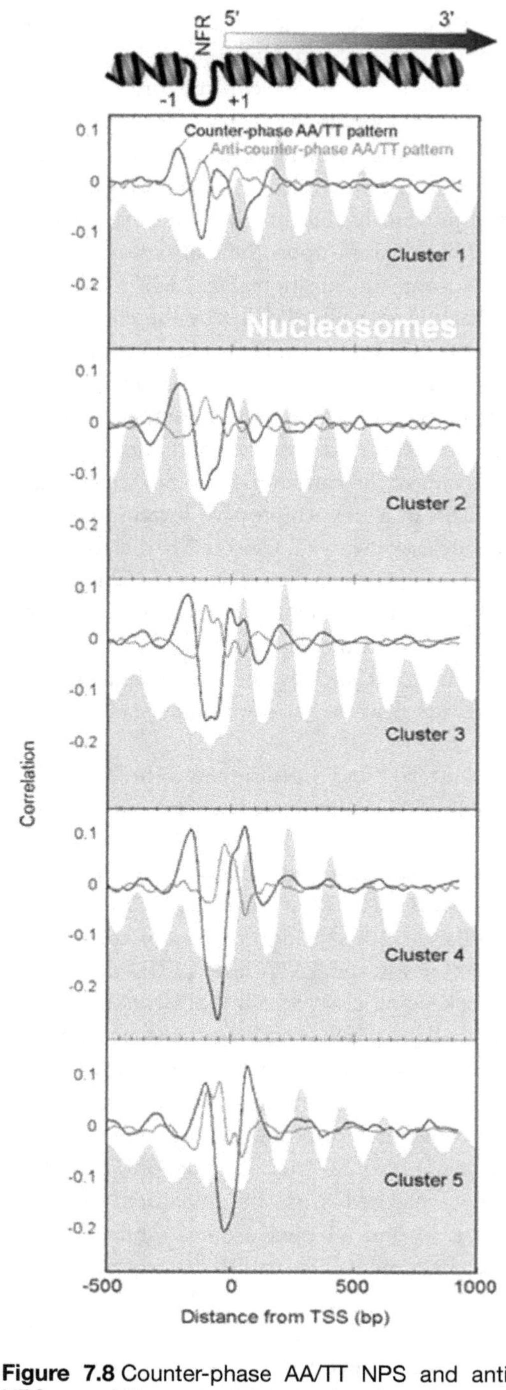

Figure 7.8 Counter-phase AA/TT NPS and anti-NPS correlations predict subsets of nucleosome locations. Regular (black trace) and anti- (grey trace) counter-phase AA/TT correlations are shown for composite subsets of nucleosomes defined by K–means clustering (from –500 to +1000bp) of *in vivo* patterns for genes aligned by their TSS (Zhang *et al.* 2011). The grey-filled background shows the composite nucleosome distribution for each cluster of genes.

patterns, which show strong negative correlation to the NFR.

For cluster 1 and to a lesser extent cluster 5, the regular counter-phase AA/TT NPS correlation reached a local minimum at or near the +1 nucleosome position (Fig. 7.8, top panel for cluster 1). In contrast, the anti-counter-phase AA/TT pattern reached a local maximum at the +1 position. This provides evidence for intrinsically sequence-encoded nucleosome destabilization at this position. Such destabilization may contribute to the expression plasticity that is characteristic of cluster 1 genes.

We used the conventional counter-phase AA/TT pattern published by us earlier (Mavrich *et al.* 2008a) and anti-counter-phase AA/TT pattern (AA/TT component of the anti-RR/YY pattern described here; Fig. 7.6B) for genome-wide nucleosome mapping of *in vivo* nucleosome positions defined by the consensus of six experimental high resolution nucleosome positioning datasets considered previously (Jiang and Pugh, 2009). Of the 55,387 identified nucleosomes, we chose for modelling a subset of 10,961 well-positioned nucleosomes. Dinucleotide frequency distributions for the anti-AA/TT pattern were double symmetrized as in Ioshikhes *et al.* (1996). The percentage of nucleosomes mapped within 35 bp of the *in vivo* determined position by the conventional counter-phase AA/TT pattern was ~45%. This percentage is consistent with the previous accuracy assessments obtained for much smaller experimental datasets (Ioshikhes *et al.*, 2006; Segal *et al.*, 2006). The percentage of the nucleosomes mapped by the anti- counter-phase AA/TT pattern alone was lower (38.5%), yet consistent with the proportions mentioned in the previous sections. However, involvement of both the regular and anti- counter-phase AA/TT patterns allowed successful mapping of up to ~75% of the nucleosomes compared to ~39% randomly expected (Z score 13 SD). Mapping by the counter-phase AA/TT pattern obtained here (AA/TT component of the RR/YY pattern represented in Fig. 7.6A) brought similar results (data is not shown). By comparison, the alternative approach (Kaplan *et al.* 2008, http://genie.weizmann.ac.il/software/nucleo_prediction.html) allows successful mapping of only 40–42%

of the nucleosomes considered here. Only 9% of the nucleosomes are detected by the both patterns simultaneously, much lower than 17.5% expected by chance (Z-score 17 SD), which means that the patterns essentially are mutually exclusive. Hence addition of the alternative anti-NPS pattern allows dramatic improvement of the nucleosome mapping.

Discussion and conclusions

As follows from the previous review, the existence of the nucleosome positioning sequence patterns is well established fact, even though an exact percentage of the nucleosomes positioned by the sequence rules versus those positioned by other factors is still subject to debate. It is notable, however that taking into account variety of the patterns significantly improves efficiency of the nucleosome mapping.

Existence of several alternative nucleosome positioning patterns may be considered an established fact. While it is possible to some extent to combine these patterns in a consistent non-contradictory manner, existence of the anti-NPS (Ioshikhes *et al.*, 2011) patterns shows that ultimately there are patterns that cannot be combined. One important thing to note here is that the nucleosome positioning pattern is periodical by variety of sequence motifs (primarily dinucleotides) but even though such periodicity is a major feature of any nucleosome pattern, it is not the only one. Different dinucleotide distributions may also have various gradients: simple 5′ to 3′ gradient as in Ioshikhes *et al.* (1996) and Mavrich *et al.* (2008a) or a convex curved gradient better seen for the dyad symmetrical WW or SS distributions. In the latter, dinucleotide frequency distributions around nucleosome dyad may generally have higher magnitude than those at the flanks or vice versa. The difference depends on the kind of the dinucleotides (WW and SS are typically mutually exclusive and hence would show opposite gradients, the same as the other mutually exclusive dinucleotides) and types of the nucleosomes. Our own unpublished data show that the gradients may differ for nucleosomes associated with different genes or gene regulatory processes. Nucleosome mapping by the smoothed patterns where 10 bp periodicity is eliminated but the preserved gradient still results in certain correlation with experimental data, which shows importance of the gradient as part of the nucleosome pattern. Furthermore, existence of the anti-NPS patterns, which cannot be obtained from the conventional patterns by a simple shift because of the periodicity disruption around the dyad and different dinucleotide gradients, shows importance of the dinucleotide phase relative the dyad. While different patterns could be combined so far at the 10–11 bp segment representation, combining them at the whole nucleosome length, taking into account their correct gradients and shifts, is virtually impossible. Our unpublished data further reinforce conclusions made in the work of Ioshikhes *et al.* (2011) that while different nucleosome subsets may follow different patterns, percentage of the nucleosomes where such patterns are combined is pretty low. Therefore, it is probably justified to say that the different patterns compete with each other, rather than define the nucleosome positioning in the co-operative manner. Indeed, while various patterns were identified in various studies, in no single case did the significant nucleosome subset show affinity to a pattern combined from some part of the RR/YY pattern and some part of the WW/SS pattern. Hence, more careful analysis of the different patterns is necessary, but this work is as yet far from completion.

As already mentioned above, existence of the different patterns may be related, at least in part, to the difference in the sequence composition of the respective DNA sequences, and to some extent to the difference in the underlying mechanisms of gene regulation (which perhaps may appear to work in concert with each other). This is well observed on the species level. However, it is possible that similar reasons are also responsible for differential nucleosome patterns and positioning in different gene classes or sequences of different function (e.g. promoters, 3′ ends of the genes, etc.) (Ioshikhes *et al.*, 2006; Mavrich *et al.*, 2008; Ioshikhes, unpublished data). This may also be true for various classes of the tissue-specific genes. The latter may be also related to various post-translational modifications of the core histone proteins, which may alter nucleosome pattern for the nucleosomes with modified histones. At the present time, there is not enough data to reliably identify nucleosome patterns for the various classes of the tissue-specific genes. However, our unpublished data provide evidence that rather specific nucleosome patterns exist, e.g. for the genes involved in the apoptosis or nucleosomes with H2A.Z histone variant versus those with regular H2A. In that case, difference in the nucleosome positioning for the various tissues may also find explanation along the lines discussed above. This hypothesis needs verification in future research.

Acknowledgements

The present work is based on the data published earlier. In particular, the manuscript (Ioshikhes *et al.*, 2011) was re-used here to a significant extent in the last sections of the chapter. The author also used materials from Ioshikhes *et al.* (1996) and Ramaswamy *et al.* (2005), all these in agreement with the relevant copyright policies. The credits should be shared with other authors of these manuscripts, particularly Frank Pugh, Ed Trifonov, Sergey Hosid, Amutha Ramaswamy, Alex Bolshoy, and others. The author is thankful to all of them and also to Doo Yang and Sergey Hosid for useful discussions and information, particularly in the literature review. Any interpretation of the data beyond the previous publications is responsibility of the author of this chapter.

References

Albert, I., Mavrich, T.N., Tomsho, L.P., Qi, J., Zanton, S.J., Schuster, S.C., and Pugh, B.F. (2007). Translational and rotational settings of H2A.Z nucleosomes across the Saccharomyces cerevisiae genome. Nature *446*, 572–576.

Anselmi, C., Bocchinfuso, G., De Santis, P., Savino, M., and Scipioni, A. (2000). A theoretical model for the prediction of sequence-dependent nucleosome thermodynamic stability. Biophys. J. *79*, 601–613.

Arents, G., Burlingame, R.W., Wang, B.C., Love, W.E., and Moudrianakis, E.N. (1991). The nucleosomal core histone octamer at 3.1 Å resolution: a tripartite protein assembly and a left-handed super helix. Proc. Natl. Acad. Sci. U.S.A. *88*, 10148–10152.

Bargaje, R., Alam, M.P., Patowary, A., Sarkar, M., Ali, T., Gupta, S., Garg, M., Singh, M., Purkanti, R., Scaria, V., Sivasubbu, S., Brahmachari, V., and Pillai, B. (2012). Nucleic Acids Res. *40*, 8965–8978.

Barski, A., Cuddapah, S., Cui, K., Roh, T.Y., Schones, D.E., Wang, Z., Wei, G., Chepelev, I., and Zhao, K. (2007). High-resolution profiling of histone methylations in the human genome. Cell *129*, 823–837.

Beato, M., and Eisfeld, K. (1997). Transcription factor access to chromatin. Nucleic Acids Res. *25*, 3559–3563.

Bettecken, T., and Trifonov, E.N. (2009). Repertoires of the nucleosome-positioning dinucleotides. PLoS One *4*, e7654.

Bolshoy, A. (1995). CC dinucleotides contribute to the bending of DNA in chromatin. Nat Struct Biol. *2*, 446–448.

Bolshoy, A., Ioshikhes, I., and Trifonov, E.N. (1996). Applicability of the multiple alignment algorithm for detection of weak patterns: periodically distributed DNA pattern as a study case. Comput Appl Biosci. *12*, 383–389.

Brogaard, K., Xi, L., Wang, J.P., and Widom, J. (2012). A map of nucleosome positions in yeast at base pair resolution. Nature *486*, 496–501.

Chavez, S., Candau, R., Truss, M., and Beato, M. (1995). Constitutive repression and nuclear factor I-dependent hormone activation of the mouse mammary tumor virus promoter in Saccharomyces cerevisiae. Mol. Cell. Biol. *15*, 6987–6998.

Chung, H.R., and Vingron, M. (2009). Sequence-dependent nucleosome positioning. J. Mol. Biol. *386*, 1411–1422.

Chung, H.R., Dunkel, I., Heise, F., Linke, C., Krobitsch, S., Ehrenhofer-Murray, A.E., Sperling, S.R., and Vingron, M. (2010). The effect of micrococcal nuclease digestion on nucleosome positioning data. PLoS One *5*, e15754.

Cohanim, A.B., Kashi, Y., and Trifonov, E.N. (2005). Yeast nucleosome DNA pattern: deconvolution from genome sequences of *S. cerevisiae. J. Biomol. Struct. Dyn.* *22*, 687–694.

Cohanim, A.B., Kashi, Y., and Trifonov, E.N. (2006). Three sequence rules for chromatin. J. Biomol. Struct. Dyn. *23*, 559–565.

Cui, F., and Zhurkin, V.B. (2010). Structure-based analysis of DNA sequence patterns guiding nucleosome positioning *in vitro*. J. Biomol. Struct. Dyn. *27*, 821–841.

Cui, F., Cole, H.A., Clark, D.J., and Zhurkin, V.B. (2012). Transcriptional activation of yeast genes disrupts intragenic nucleosome phasing. Nucleic Acids Res. *40*, 10753–10764.

Davie, J.R. (1998). Covalent modifications of histones: expression from chromatin templates. Curr. Opin. Genet. Dev. *8*, 173–178.

Di Croce, L., Koop, R., Venditti, P., Westphal, H.M., Nightingale, K.P., Corona, D.F., Becker, P.B., and Beato, M. (1999). Two-step synergism between the progesterone receptor and the DNA-binding domain of nuclear factor 1 on MMTV minichromosomes. Mol. Cell *4*, 45–54.

Dlakić, M., Ussery, D.W., and Brunak, S. (2004). DNA bendability and nucleosome positioning in transcriptional regulation. In DNA Conformation and Transcription, Ohyama, T., ed. (New York, Springer), pp. 189–202.

Drew, H.R., and Travers, A.A. (1985). DNA bending and its relation to nucleosome positioning. J. Mol. Biol. *186*, 773–790.

Eickbush, T.H., and Moudrianakis, E.N. (1978). The histone core complex: An octamer assembled by two sets of protein–protein interactions. Biochemistry *17*, 4955–4964.

Elango, N., Hunt, B.H., Goodisman, M.A.D., and Yi, S.V. (2009). DNA methylation is widespread and associated with differential gene expression in castes of the honeybee, Apis mellifera. Proc. Natl. Acad. Sci. U.S.A. *106*, 11206–11211.

Felsenfeld, G. (1992). Chromatin as an essential part of the transcriptional mechanism. Nature *355*, 219–224.

Field, Y., Kaplan, N., Fondufe-Mittendorf, Y., Moore, I.K., Sharon, E., Lubling, Y., Widom, J., and Segal, E. (2008). Distinct modes of regulation by chromatin encoded through nucleosome positioning signals. PLoS Comput. Biol. *4*, e1000216.

FitzGerald, P.C., Sturgill, D., Shyakhtenko, A., Oliver, B., and Vinson, C. (2006). Comparative genomics of Drosophila and human core promoters. Genome Biol. *7*, R53.

Gabdank, I., Barash, D., Trifonov, E.N. (2009). Nucleosome DNA bendability matrix (C. elegans). J. Biomol. Struct. Dyn. *26*, 403–411.

Gabdank, I., Barash, D., and Trifonov, E.N. (2010a). FineStr: a web server for single-base-resolution nucleosome positioning. Bioinformatics *26*, 845–846.

Gabdank, I., Barash, D., and Trifonov, E.N. (2010b). Single-base resolution nucleosome mapping on DNA sequences. J. Biomol. Struct. Dyn. *28*, 107–122.

Georgel, P.T. (2005). Chromatin potentiation of the hsp70 promoter is linked to GAGA-factor recruitment. Biochem. Cell Biol. *83*, 555–565.

Gkikopoulos, T., Schofield, P., Singh, V., Pinskaya, M., Mellor, J., Smolle, M., Workman, J.L., Barton, G.J., and Owen-Hughes, T. (2011). A role for Snf2-related nucleosome-spacing enzymes in genome-wide nucleosome organization. Science *333*, 1758–1760.

Gottesfeld, J.M., and Forbes, D.J. (1997). Mitotic repression of the transcriptional machinery. Trends. Biochem. Sci. *22*,197–202.

Grunstein, M. (1990). Histone function in transcription. Annu. Rev. Cell Biol. *6*, 643–678.

Grunstein, M. (1997). Histone acetylation in chromatin structure and transcription. Nature *389*, 349–352.

Grunstein, M., Durrin, L.K., Mann, R.K., Fisher-Adams, G., and Johnson, L.M. (1992). Histones: regulators of transcription in yeast. In Transcriptional *Regulation*, McKnight, S., and Yamamoto, K., eds. (New York, Cold Spring Harbor Laboratory Press), pp. 1295–1315.

Hayes, J.J., Tullius, T.D., and Wolffe, A.F. (1990). The structure of DNA in a nucleosome. Proc. Natl. Acad. Sci. U.S.A. *87*, 7405–7409.

Hayes, J.J., Clark, D.J., and Wolffe, A.P. (1991). Histone contributions to the structure of DNA in the nucleosome. Proc. Natl. Acad. Sci. U.S.A. *88*, 6829–6833.

Hebert, C., and Roest Crollius, H. (2010). Nucleosome rotational setting is associated with transcriptional regulation in promoters of tissue-specific human genes. Genome Biol. *11*, p.R51.

Henikoff, S. (2009). Labile H3.3+H2A.Z nucleosomes mark 'nucleosome-free regions'. Nat Genet. *41*, 865–866.

Herzel, H., Weiss, O., and Trifonov, E.N. (1999). 10–11 bp periodicities in complete genomes reflect protein structure and DNA folding. Bioinformatics *15*, 187–193.

Hughes, A.L., Jin, Y., Rando, O.J., and Struhl, K. (2012). A functional evolutionary approach to identify determinants of nucleosome positioning: a unifying model for establishing the genome-wide pattern. Mol. Cell. *48*, 5–15.

Ioshikhes, I., and Trifonov, E.N. (1993). Nucleosomal DNA sequence database. Nucleic Acids Res. *21*, 4857–4859.

Ioshikhes, I., Bolshoy, A., and Trifonov, E.N. (1992). Preferred positions of AA and TT dinucleotides in aligned nucleosome DNA sequences. J. Biomol. Struct. Dynam. *9*,1111–1117.

Ioshikhes, I., Bolshoy, A., Derenshteyn, K., Borodovsky, M., and Trifonov, E.N. (1996). Nucleosome DNA sequence pattern revealed by multiple alignment of experimentally mapped sequences. J. Mol. Biol. *262*, 129–139.

Ioshikhes, I., Trifonov, E.N., and Zhang, M.Q. (1999). Periodical distribution of transcription factor sites in promoter regions and connection with chromatin structure. Proc. Natl. Acad. Sci. U.S.A. *96*, 2891–2895.

Ioshikhes, I., Hosid, S., Pugh, B.F. Variety of genomic D.N.A patterns for nucleosome positioning. (2011). Genome Res. *21*, 1863–1871.

Ioshikhes, I.P., Albert, I., Zanton, S.J., and Pugh, B.F. (2006). Nucleosome positions predicted through comparative genomics. Nat. Genet. *38*, 1210–1215.

Jin, C., and Felsenfeld, G. (2007). Nucleosome stability mediated by histone variants H3.3 and H2A.Z. Genes Dev. *21*, 1519–1529.

Jiang, C., and Pugh, B.F. (2009). A compiled and systematic reference map of nucleosome positions across the Saccharomyces cerevisiae genome. Genome Biol. *10*, R109.

Johnson, S.M., Tan, F.J., McCullough, H.L., Riordan, D.P., and Fire, A.Z. (2006). Flexibility and constraint in the nucleosome core landscape of *Caenorhabditis elegans* chromatin. Genome Res. *16*, 1505–1516.

Kaplan, N., Moore, I.K., Fondufe-Mittendorf, Y., Gossett, A.J., Tillo, D., Field, Y., LeProust, E.M., Hughes, T.R., Lieb, J.D., Widom, J., and Segal, E. (2009). The DNA-encoded nucleosome organization of a eukaryotic genome. Nature *458*, 362–636.

Kato, M., Onishi, Y., Wada-Kiyama, Y., Abe, T., Ikemura, T., Kogan, S., Bolshoy, A., Trifonov, E.N., and Kiyama, R. (2003). Dinucleosome DNA of human K562 cells: experimental and computational characterizations. J. Mol. Biol. *332*, 111–125.

Kingston, R.E., Bunker, C.A., and Imbalzano, A.N. (1996). Repression and activation by multiprotein complexes that alter chromatin structure. Genes Dev. *10*, 905–920.

Kogan, S., and Trifonov, E.N. (2005). Gene splice sites correlate with nucleosome positions. *Gene 352*, 57–62.

Kogan, S.B., Kato, M., Kiyama, R., and Trifonov, E.N. (2006). Sequence structure of human nucleosome DNA. J. Biomol. Str. Dyn. *24*, 43–48.

Komura, J., and Ono, T. (2003). Nucleosome positioning in the Human c-FOS promoter analyzed by *in vivo* footprinting with Psoralen and ionizing radiation. Biochemistry *42*, 15084–15091.

Kornberg, R.D., and Lorch, Y. (1995). Interplay between chromatin structure and transcription. Curr. Opin. Cell. Biol. *7*, 371–375.

Kornberg, R.D., and Lorch, Y. (2002). Chromatin and transcription: where do we go from here. Curr. Opin. Genet. Dev. *12*, 249–251.

Leltninger, A.L. (1975). Biochemistry (New York: Worth).

Levine, M. (2011). Paused RNA Polymerase II as a developmental checkpoint. Cell *145*, 502–511.

Li, G., Chandler, S.P., Wolffe, A.P., and Hall, T.C. (1998). Architectural specificity in chromatin structure at the TATA box *in vivo:* nucleosome displacement upon beta-phaseolin gene activation. Proc. Natl. Acad. Sci. U.S.A. *95*, 4772–4777.

Li, Q., Wrange, O., and Eriksson, P. (1997). The role of chromatin in transcriptional regulation. Int. J. Biochem. Cell. Biol. *29*, 731–742.

Lickwar, C.R., Rao, B., Shabalin, A.A., Nobel, A.B., Strahl, B.D., and Lieb, J.D. (2009). The Set2/Rpd3S pathway suppresses cryptic transcription without regard to gene length or transcription frequency. PLoS One *4*, e4886.

Lowary, P.T., and Widom, J. (1998). New DNA sequence rules for high affinity binding to histone octamer and sequence-directed nucleosome positioning. J. Mol. Biol. *276*, 19–42.

Lu, Q., Wallrath, L.L., and Elgin, S.C.R. (1994). Nucleosome positioning and gene regulation. J. Cell. Biochem. *55*, 83–92.

Luger, K., Mader, A.W., Richmond, R.K., Sargent, D.F., and Richmond, T.J. (1997). Crystal structure of the nucleosome core particle at 2.8Å resolution. Nature *389*, 251–260.

Mavrich, T.N., Ioshikhes, I.P., Venters, B.J., Jiang, C., Tomsho, L.P., Qi, J., Schuster, S.C., Albert, I., and Pugh, B. (2008a). A barrier nucleosome model for statistical positioning of nucleosomes throughout the yeast genome. Genome Res. *18*, 1073–1083.

Mavrich, T.N., Jiang, C., Ioshikhes, I.P., Li, X., Venters, B.J., Zanton, S.J., Tomsho, L.P., Qi, J., Glaser, R.L., Schuster, S.C., Gilmour, D.S., Albert, I., and Pugh, B.F. (2008b). Nucleosome organization in the Drosophila genome. Nature *453*, 358–362.

Mengeritsky, G., and Trifonov, E.N. (1983). Nucleotide sequence-directed mapping of the nucleosomes. Nucleic Acids Res. *11*, 3833–3851.

Campbell, N.A., Mitchell, L.G., and Reece, J.B. (1997). Biology: Concepts and Connections (2nd edn. Menlo Park, California).

Nair, T.M. (2010). Sequence periodicity in nucleosomal DNA and intrinsic curvature. BMC Struct. Biol. *10* (Suppl 1), S8.

Narlikar, G.J., Fan, H.Y., and Kingston, R.E. (2002). Cooperation between complexes that regulate chromatin structure and transcription. Cell *108*, 475–487.

Newman, J.R., Ghaemmaghami, S., Ihmels, J., Breslow, D.K., Noble, M., DeRisi, J.L., and Weissman, J.S. (2006). Single-cell proteomic analysis of S. cerevisiae reveals the architecture of biological noise. Nature *441*, 840–846.

Noll, M. (1974). Internal structure of the chromatin subunit. Nucleic Acids Res. *1*, 1573–1578.

Parker, S.P. (ed.) (1997). McGraw Hill Encyclopedia of Science and Technology. (New York: McGraw Hill).

Peterson, C.L. (1996). Multiple SWItches to turn on chromatin? Curr. Opin. Genet. Dev. *6*, 171–175.

Radman-Livaja, M., and Rando, O.J. (2010). Nucleosome positioning: how is it established, and why does it matter? Dev Biol. *339*, 258–266.

Ramaswamy, A., Bahar, I., and Ioshikhes, I. (2005). Structural dynamics of nucleosome core particle: comparison with nucleosomes containing histone variants. Proteins *58*, 63–69.

Rapoport, A.E., Frenkel, Z.M., and Trifonov, E.N. (2011). Nucleosome positioning pattern derived from oligonucleotide compositions of genomic sequences. J. Biomol. Struct. Dyn. *28*, 567–574.

Raveh-Sadka, T., Levo, M., Shabi, U., Shany, B., Keren, L., Lotan-Pompan, M., Zeevi, D., Sharon, E., Weinberger, A., and Segal, E. (2012). Manipulating nucleosome disfavoring sequences allows fine-tune regulation of gene expression in yeast. Nat. Genet. *44*, 743–750.

Reynolds, S.M., Bilmes, J.A., and Noble, W.S. (2010). Learning a weighted sequence model of the nucleosome core and linker yields more accurate predictions in *Saccharomyces cerevisiae* and *Homo sapiens*. PLoS Comput. Biol. *6*, e1000834.

Richmond, T.J., Finch, J.T., Rushton, B., Rhodes, D., and Klug, A. (1984). Structure of the nucleosome core particle at 7Å resolution. Nature *311*, 532–537.

Salih, F., Salih, B., and Trifonov, E.N. (2007). Sequence-directed mapping of nucleosome positions. J. Biomol. Struct. Dyn. *24*, 489–494.

Salih, F., Salih, B., Kogan, S., Trifonov, E.N. (2008). Epigenetic nucleosomes: Alu sequences and CG as nucleosome positioning element. J. Biomol. Struct. Dyn. *26*, 9–16.

Satchwell, S.C., Drew, H.R., and Travers, A.A. (1986). Sequence periodicities in chicken nucleosome core DNA. J. Mol. Biol. *191*, 659–675.

Schones, D.E., Cui, K., Cuddapah, S., Roh, T.Y., Barski, A., Wang, Z., Wei, G., and Zhao, K. (2008). Dynamic regulation of nucleosome positioning in the human genome. Cell *132*, 887–898.

Segal, E., Fondufe-Mittendorf, Y., Chen, L., Thastrom, A., Field, Y., Moore, I.K., Wang, J.P., and Widom, J. (2006). A genomic code for nucleosome positioning. Nature *442*, 772–778.

Simpson, R.T. (1991). Nucleosome positioning: occurrence, mechanisms, and functional consequences. Prog. Nucl. Acid Res. *40*, 143–184.

Sollner-Webb, B., Melchior Jr, W., and Felsenfeld, G. (1978). DNAase I., DNAase II and staphylococcal nuclease cut at different, yet symmetrically located, sites in the nucleosome core. Cell *14*, 611–627.

Spangenberg, C., Eisfeld, K., Stunkel, W., Luger, K., Flaus, A., Richmond, T.J., Truss, M., and Beato, M. (1998). The mouse mammary tumour virus promoter positioned on a tetramer of histones H3 and H4 binds nuclear factor 1 and OTF1. J. Mol. Biol. *278*, 725–739.

Suto, R.K., Clarkson, M.J., Tremethick, D.J., and Luger, K. (2000). Crystal structure of a nucleosome core particle containing the variant histone H2A.Z. Nat. Struct. Biol. 7, 1121–1124.

Sussman, J.L., and Trifonov, E.N. (1978). Possibility of nonkinked packing of DNA in chromatin. Proc. Natl. Acad. Sci. U.S.A. *75*, 103–107.

Svaren, J., and Horz, W. (1996). Transcription factors vs nucleosomes: regulation of the PHO5 promoter in yeast. Curr. Opin. Genet. Dev. *6*, 164–170.

Teves, S.S., and Henikoff, S. (2011). Heat shock reduces stalled RNA polymerase II and nucleosome turnover genome-wide. Genes Dev. *25*, 2387–2397.

Thoma, F. (1992). Nucleosome positioning. Biochim. Biophys. Acta *1130*, 1–19.

Tillo, D., and Hughes, T.R. (2009). G+C content dominates intrinsic nucleosome occupancy. BMC Bioinformatics *10*, 442.

Tirosh, I., and Barkai, N. (2008). Two strategies for gene regulation by promoter nucleosomes. Genome Res. *18*, 1084–1091.

Travers, A., Hiriart, E., Churcher, M., Caserta, M., Di Mauro, E. (2010). The DNA sequence dependence of nucleosome positioning *in vivo* and vitro. *J. Biomol. Struct. Dyn.* 27, 713–724.

Trifonov, E.N. (1980). Sequence-dependent deformational anisotropy of chromatin DNA. Nucleic Acids Res. *8*, 4041–4053.

Trifonov, E.N. (1981). International Cell Biology 1980–1981, Schweiger, H., ed. (Springer-Verlag, New York), pp. 128–138.

Trifonov, E.N. (1985). Curved DNA. CRC Crit. Rev. Biochem. *19*, 89–106.

Trifonov, E.N. (2010a). Base pair stacking in nucleosome DNA and bendability sequence pattern. J. Theor. Biol. *263*, 337–339.

Trifonov, E.N. (2010b). Nucleosome positioning by sequence, state of the art and apparent finale. *J. Biomol. Struct. Dyn.* 27, 741–746.

Trifonov, E.N., and Bettecken, T. (1979). Noninteger pitch and nuclease sensitivity of chromatin DNA. Biochemistry. *18*, 454–456.

Trifonov, E.N., and Sussman, J.L. (1980). The pitch of chromatin DNA is reflected in its nucleotide sequence. Proc. Natl Acad. Sci. U.S.A. 77, 3816–3820.

Valouev, A., Johnson, S.M., Boyd, S.D., Smith, C.L., Fire, A.Z., and Sidow, A. (2011). Determinants of nucleosome organization in primary human cells. Nature *474*, 516–520.

Wang, J., Zhuang, J., Iyer, S., Lin, X., Whitfield, T.W., Greven, M.C., Pierce, B.G., Dong, X., Kundaje, A., Cheng, Y., *et al.* (2012). Sequence features and chromatin structure around the genomic regions bound by 119 human transcription factors. Genome Res. *22*, 1798–1812.

Watson, J.D., and Crick, F.H. (1953a). Genetical implications of the structure of deoxyribonucleic acid. Nature *171*, 964–967.

Watson, J.D., and Crick, F.H. (1953b). Molecular structure of nucleic acids; a structure for deoxyribose nucleic acid. Nature *171*, 737–738.

Weiner, A., Hughes, A., Yassour, M., Rando, O.J., and Friedman, N. (2010). High-resolution nucleosome mapping reveals transcription-dependent promoter packaging. Genome Res. *20*, 90–100.

Widlund, H.R., Cao, H., Simonsson, S., Magnusson, E., Simonsson, T., Nielsen, P.E., Kahn, J.D., Crothers, D.M., and Kubista, M. (1997). Identification and characterization of genomic nucleosome-positioning sequences. J. Mol. Biol. *267*, 807–817.

Wittig, S., and Wittig, B. (1982). Function of a tRNA gene promoter depends on nucleosome position Nature *297*, 31–38.

Wolffe, A.P. (1994a). Nucleosome positioning and modification: chromatin structures that potentiate transcription. Trends Biochem. Sci. *19*, 240–244.

Wolffe, A.P. (1994b). Transcription: in tune with the histones. Cell 77, 13–16.

Wolffe, A.P. (2001). Transcriptional regulation in the context of chromatin structure. Essays Biochem. 37, 45–57.

Yuan, G.C., Liu, Y.J., Dion, M.F., Slack, M.D., Wu, L.F., Altschuler, S.J., and Rando, O.J. (2005). Genome-scale identification of nucleosome positions in S. cerevisiae. Science *309*, 626–630.

Zelenyi, R.O. (ed.) (1996). 'Cell.' The World Book Encyclopedia (Chicago: Field Enterprises).

Zhang, Z., Wippo, C.J., Wal, M., Ward, E., Korber, P., and Pugh, B.F. (2011). A packing mechanism for nucleosome organization reconstituted across a eukaryotic genome. Science 332, 977–980.

Zofall, M., Persinger, J., Kassabov, S.R., and Bartolomew, B. (2006). Chromatin remodeling by ISW2 and SWI/SNF requires DNA translocation inside the nucleosome. Nat. Struct. Mol Biol. *13*, 339–346.

Hypermethylation in Cancer

8

Marta Sánchez-Carbayo

Abstract

A compelling body of evidences sustains the importance of epigenetic mechanisms in the development and progression of cancer. Assessing the epigenetic component of tumour samples is strongly improving our understanding of their biology and clinical behaviour. In terms of DNA methylation, cancer cells show genome-wide hypomethylation and site-specific CpG island promoter hypermethylation. In the context of other epigenetic alterations, this chapter will focus on the hypermethylation of CpG islands in promoter regions, as the most widely described epigenetic modification in cancer. CpG islands hypermethylation is believed to be critical in the transcriptional silencing and regulation of tumour suppressor and crucial cancer genes involved in the major molecular pathways controlling cancer development and progression. In particular, several biological pathways of frequently methylated genes include cell cycle, DNA repair, apoptosis and invasion, among others. Furthermore, translational aspects of tumour methylomes described to date will be discussed towards their potential application as cancer biomarkers. Several tissue methylation signatures and individual candidates have been evidenced, that could potentially stratify tumours histopathologically, and discriminate patients in terms of their clinical outcome. Tumour methylation profiles could also be detected in body fluid specimens showing a promising role as non-invasive markers for cancer diagnosis towards an early detection and potentially for the surveillance of cancer patients in a near future. However, the epigenomic exploration of cancer has only just begun. Genome-scale DNA methylation profiling studies will further highlight the relevance of the epigenetic component to gain knowledge of cancer biology, and identify those profiles and candidates better correlating with clinical behaviour.

Introduction: hypermethylation in the context of other epigenetic modifications

Disruption of the epigenetic landscape is considered a common hallmark of many widespread diseases and is a feature of all human neoplasias (Jones and Laird, 1999; Bird, 2002; Laird, 2003). Epigenetic alterations can be defined as stable molecular changes of the phenotype of a cell that are heritable during somatic cell divisions (and sometimes germ line transmissions) but do not involve changes of the DNA sequence itself. The major epigenetic phenomena are mediated by several molecular mechanisms comprising DNA methylation, histone modifications, nucleosome remodelling, and RNA-associated silencing. These different modifications are closely interconnected. All together they may manifest as global alterations in chromatin packaging and specific promoter changes that influence the transcription of associated genes (Feinberg, 2005; Esteller, 2007; Jones and Baylin, 2007).

Types of DNA methylation

CpG methylation

Cytosine methylation takes place along the whole genome, and its disruption is the most widely studied epigenetic modification in humans. DNA methylation occurs predominantly in the context of CpG dinucleotides, although non CpG methylation also occurs in the human genome. The CpG dinucleotides tend to cluster in regions called CpG islands, defined as regions of more than 200 bases with a G+C content of at least 50% and a ratio of observed to statistically expected CpG frequencies of at least 0.6. About 60% of human gene promoters are associated with CpG islands and are usually unmethylated in normal cells, although some of them (~6%) become methylated in a tissue-specific manner during early development or in differentiated tissues (Esteller, 2008; Straussman *et al.*, 2009).

In terms of CpG DNA methylation, cancer cells show genome-wide hypomethylation and site-specific CpG island promoter hypermethylation. These processes coexist in human neoplasias: specific hypermethylation of specific CpG islands can be located at promoters or at shores of critical genes, and a global hypomethylation of repetitive sequences (for example, centromeres and retrotransposon elements) is present in cancer cells (Bird 2002; Wang and Leung, 2004). Both hyper- and hypo- methylation patterns may differ among tumour development and characterize each tumour subtype (Eden *et al.*, 2003; Lister *et al.*, 2009).

DNA hypomethylation occurs at many genomic sequences, resulting in genomic instability (Esteller 2008). At repeat sequences, this is achieved by a higher rate of chromosomal rearrangements and, at retrotransposons, by a higher probability of translocation to other genomic regions (Eden *et al.*, 2003). Furthermore, aberrant DNA hypomethylation can also account for the activation of some proto-oncogenes and lead to loss of imprinting (Esteller, 2007). Thus, DNA methylation is needed to protect chromosomal integrity, which is achieved by preventing reactivation of endoparasitic sequences that cause chromosomal instability, translocations and gene disruption. Although most of these CpGs are methylated to prevent chromosome instability, the majority of CpG islands remain unmodified during development and in differentiated tissues (Suzuki and Bird, 2008). Naturally occurring CpG island methylation takes place during developmental phenomena such as X chromosome inactivation or genomic imprinting (Meissner *et al.*, 2008) (Fig. 8.1).

In general, CpG-island hypermethylation is associated with gene silencing. DNA methylation can inhibit gene expression by various mechanisms. Methylated DNA can promote the recruitment of methyl-CpG-binding domain (*MBD*) proteins. MBD family members in turn recruit histone-modifying and chromatin-remodelling complexes to methylated sites (Esteller 2007). DNA methylation can also directly inhibit transcription by precluding the recruitment of DNA binding proteins from their target sites (Kuroda *et al.*, 2009). In contrast, unmethylated CpG islands would generate a chromatin structure favourable for gene expression (Thomson *et al.*, 2010).

DNA methylation does not occur exclusively at CpG islands. The term CpG island shores, referring to regions of lower CpG density that lie in close proximity (~2 kb) of CpG islands, has recently been coined. Extensive DNA methylation changes caused by differentiation take place at these 'shores', regions of comparatively low CpG density and close to CpG islands (Meissner *et al.*, 2008; Doi *et al.*, 2009; Irizarry *et al.*, 2009). The methylation of these CpG island shores is closely associated with transcriptional inactivation. It is believed that most of the tissue-specific DNA methylation seems to occur not at CpG islands but at CpG island shore (Irizarry *et al.*, 2009). Differentially methylated CpG island shores have also been shown to differ between specific tissues and are conserved between human and mouse. Moreover, 70% of the differentially methylated regions in reprogramming are associated with CpG island shores (Doi *et al.*, 2009).

DNA methylation is less frequently coupled with transcriptional activation, when, for instance, it occurs at gene bodies (Fig. 8.1). Gene body methylation is common in ubiquitously expressed genes and is positively correlated with gene expression (Hellman and Chess, 2007). It has been proposed that it might be related to elongation efficiency and prevention of spurious initiation of transcription.

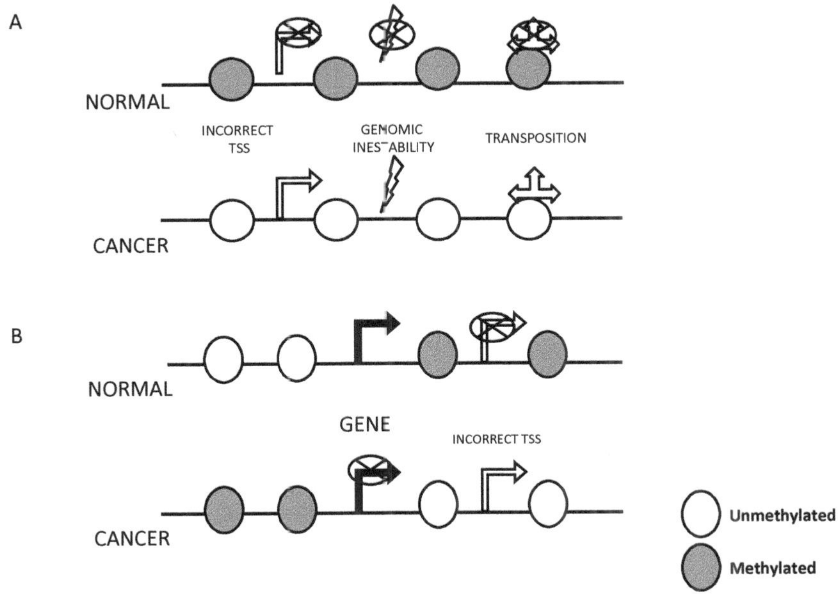

Figure 8.1 Major mechanisms of DNA methylation in cancer. (A) In normal cells (top), methylation of repetitive sequences prevents genomic instability and spurious transcription initiations. Moreover, transposable elements cannot be activated in a methylated environment. In cancer cells (bottom), global hypomethylation triggers genomic instability and aberrant transcription initiations. Concomitant activation of transposons may also lead to gene disruption. (B) In normal cells (top), CpG islands and CpG island shores usually remain unmethylated, allowing gene transcription. Additionally, DNA methylation within the gene bodies avoids spurious transcription initiations. In cancer cells (bottom), although both CpG islands and CpG island shores may be strongly methylated, gene bodies lack this modification. As a result, transcription of many genes gets blocked, and aberrant transcription may occur from incorrect transcription start sites (TSSs).

Non CpG methylation

It is important to be aware of the presence of other sites of methylation, which is less studied in cancer disease. Although DNA methylation mainly occurs in the CpG dinucleotide context in mammals, non-CG methylation has recently been described in humans at CHG and CHH sites (where H is A, C or T). CHG and CHH methylation has been found in stem cells and seems to be enriched in gene bodies directly correlated with gene expression. Almost one-quarter of all DNA methylation found in embryonic stem cells occurs in a non-CpG context (Lister *et al.*, 2009). However, mechanisms of non-CpG methylation remain unclear (Futscher *et al.*, 2002; Hattori *et al.*, 2004).

Furthermore, in addition to 5-methylcytosines, 5-hydroxymethyl-2′-deoxycytidine has also been described. 5-methylcytosine (5-mC) can be converted into 5-hydroxymethylcytosine (5-hmC) by the 2-oxoglutarate- and Fe(II)-dependent oxygenases such as the tet methylcytosine dioxygenases (*TET1*), (*TET2*) and (*TET3*) (Tahiliani *et al.*, 2009). The 5-hydroxymethyl-2′-deoxycytidine has been reported in Purkinje cells (constituting 0.6% of total nucleotides) and in granule cells (constituting 0.2% of total nucleotides), but has not been reported to be present in cancer cell lines (Esteller 2007; Tahiliani *et al.*, 2009; Ito *et al.*, 2010). These new DNA modifications need to be further studied to determine their implications for normal and cancer diseased epigenetic regulation. They have not been yet characterized in the context of many tumour types.

This chapter will focus on CpG islands hypermethylation, more specifically on the hypermethylation that occur mostly in the promoters of critical cancer related genes. The biological pathways affected by DNA hypermethylation in the regulation of crucial cancer genes and the potential clinical applications, such as tumour biomarkers, will be discussed in the following sections.

Hypermethylation machineries: the role of DNMTs

DNA methylation usually takes place at the 5′ position of the cytosine ring within CpG dinucleotides, and its consequence is the silencing of genes and noncoding genomic regions. DNA methylation is mediated by the cytosine DNA methyltransferase (*DNMT*) family of enzymes that catalyse the transfer of a methyl group from S-adenosyl methionine to DNA. Five members of the DNMT family have been reported: DNMT1, DNMT2, DNMT3a, DNMT3b and DNMT3L, but only DNMT1, DNMT3a and DNMT3b possess methyltransferase activity (Robertson *et al.*, 1999; Robert *et al.*, 2003).

The catalytic members of the DNMT family can be classified into maintenance *DNMTs* (*DNMT1*) and *de novo DNMTs* (*DNMT3A* and *DNMT3B*). *DNMT1* maintains the existing methylation patterns following DNA replication, and *DNMT3A* and *DNMT3B*, are *de novo* enzymes that target previously unmethylated CpGs (Bird, 2002). The maintenance *DNMT*, *DNMT1*, has a 30- to 40-fold preference for hemimethylated DNA, generated during semi-conservative DNA replication, and also has *de novo DNMT* activity. *DNMT1* is the most abundant *DNMT* in the cell and is transcribed mostly during the S phase of the cell cycle (Robertson *et al.*, 1999; Robertson *et al.*, 2000; Robert *et al.*, 2003). *DNMT3A* and *DNMT3B* are thought to be responsible for establishing the pattern of methylation during embryonic development (Bird, 2002). Disruption of the epigenetic machineries, either by mutation, deletion or the altered expression of any of their components, is known to provoke aberrant gene expression patterns that give rise to several cancer characteristics (Esteller, 2008). In fact, these 'epimutations' sometimes provide the second hit for cancer initiation postulated by the two-hit model, as they can silence the remaining active allele of previously mutated tumour suppressors (Jones and Laird, 1999). A few studies have explored the link between the methylation patterns observed in cancer with the expression of these proteins (Velicescu *et al.*, 2002; Nakagawa *et al.*, 2003), as compared to studies assessing candidate-driven hypermethylation. The disturbance of the DNA methylation landscape in transformed cells has been recently supported by the finding of somatic mutations in *DNMT3A* in acute myeloid leukaemia (Ley *et al.*, 2010). Homozygous null mutations and chromosomal deletions involving the *TET2* locus have been described in various myeloid malignancies (Delhommeau *et al.*, 2009), so that the impairment of the conversion of 5-mC into 5-hmC might also be related to cancer. It remains to be elucidated whether mutations affecting the *DNMTs* or *TETs* machineries are present or relevant in other tumour types. Furthermore, it remains to be determined which *DNMT* is primarily regulating the methylation patterns observed in each cancer and whether reversible methylation targeting this machinery could be therapeutically useful in each cancer.

It is important to be aware that the gene silencing subsequent to CpG hypermethylation involves the modification of both the genetic and histone code. More specifically, when a cytosine *DNMT* catalyses methylation of the 5th position of the cytosine ring within a CpG island located in the promoter region of a gene, using S-adenosylmethionine as the donor molecule, it is also accompanied by changes in chromatin composition around the island that denies access to regulatory proteins needed for transcription. The chromatin structure is modified during gene silencing by affecting acetylation, phosphorylation, methylation and/or ubiquitylation of histone tails (Esteller 2007).

Epigenetic factors contribute to tumourigenesis and cancer progression

Cancer is driven by the multistep accumulation of molecular alterations at the DNA, RNA and protein levels, among others. These molecular alterations result in uncontrolled cellular proliferation, cell cycle deregulation, decrease in cell death or apoptosis, blockage of differentiation, and invasion, potentially leading to metastatic spread (Sanchez-Carbayo, 2003; Sanchez-Carbayo and

Cordon-Cardo, 2003, 2007; Wolff *et al.*, 2005). It is becoming evident that the epigenomic component plays an important role in the regulation of the expression of many critical genes participating in these pathways. The particular alterations, including epigenetic changes, occurring as part of the crosstalk between these pathways, will in great part determine the biological behaviour of the tumour including its ability to grow, recur, progress and metastasize.

Likewise many other human diseases, cancer shows aberrant epigenetic regulation. In particular, the better characterized cancer epigenome alterations include global changes in DNA methylation, altered histone modification and miRNAs patterns (Esteller, 2007). Because epigenetic features such as global DNA hypomethylation and promoter-specific hypermethylation can be commonly observed in benign neoplasias and early-stage tumours, it is becoming apparent that epigenetic deregulation may precede preliminary transforming events such as mutations in tumour suppressors, proto-oncogenes, or both, and genomic instability (Feinberg, 2005). However, it has not been characterized yet when many of these epigenetic alterations are acquired along each step in cancer initiation and progression.

The heterogeneity and complexity of human tumours, described at the histological, transcriptional and protein levels, exist also at the epigenetic level. Genes frequently reported to be epigenetically inactivated by hypermethylation in the majority of solid tumours include tumour suppressors and critical cancer related genes of the main molecular pathways related to cancer whose expression is known to be lost by genetic and protein analyses, such as *CDKN2A*, E-cadherin, among others, and part of them will be described below. As could be expected, DNA methylation profiles not only confirm previous observations for critical cancer-related genes based on histological, transcriptomic and protein analyses, but also add new levels of complexity to the picture, requiring to integrate the epigenetic mechanisms of expression and unravelling novel candidates that contribute to interpretation of the biological phenotypes. However, there is still a need in the cancer research field to incorporate epigenetic alterations to the complexity of genetic and protein events to improve our understanding of the disease and the potential applications for therapies and biomarker development.

Biological pathways of frequently methylated genes in cancer

As commented above, the most common epigenetic event studied in cancer is the hypermethylation in CpG islands in promoter regions, frequently associated with inappropriate transcriptional silencing of tumour suppressor genes and other critical cancer related genes (Sanchez-Carbayo, 2012b), which will be summarized below. This transcriptional silencing by CpG-island hypermethylation affects genes involved in all aspects of normal cell function and may rival genetic changes that affect the coding sequence as a crucial trigger for neoplastic development and progression (Gonzalo *et al.*, 2007). This is an important aspect since for many genes showing a tumour suppressor role in cancer, together with DNA mutations, CpG hypermethylation represents one of the two events in the two-hit Hudson model of inactivation of tumour suppressor genes.

In the following section, we discuss several genes, inactivated by promoter hypermethylation in solid tumours. A great number of these genes exhibit a tumour suppressor function while the mechanisms may differ. Rather than describing all the genes reported to be inactivated by methylation, we group them focusing on crucial pathways and processes that, when disrupted as a result of gene methylation (alone or together with other epigenetic or genetic mechanism), are believed to influence the transformation of cancer cells. Additionally, for many of these genes, their epigenetic silencing has been proven *in vitro* by comparing their expression in different cancer cells with or without treatment by demethylating agents, as a proof of principle of the association of methylation with gene silencing (Aleman *et al.*, 2008a; Sanchez-Carbayo, 2012a). In this section, methylated genes have been grouped by major biological pathways. Part of the reports has evaluated many genes simultaneously. Several of these genes could have been annotated within more than one pathway and listed only in one of them, such as for *TP53*. The genes described below have been selected based on the tumour suppressor role shown for many of them by the studies *in vitro* and *in vivo* though the mechanisms of suppression can differ. It is important to be aware that many other genes have been shown to be methylated, but might not be mentioned within the selected annotation groups.

Cell cycle

The cyclin-dependent kinase inhibitor 2A (CDKN2A) gene is inactivated by hypermethylation at high prevalences in human cancer. *P16-INK4A* is an inhibitor of cyclin-dependent kinases 4 (*CDK4*) and 6 (*CDK6*), which bind cyclin D1 and phosphorylate the retinoblastoma (RB1) tumour-suppressor gene (Merlo *et al.*, 1995; Lukas *et al.*, 1995; Weinberg, 1995; Herman *et al.*, 1996). *P16-INK4A* contributes to the maintenance of RB in the unphosphorylated state, which inhibits cell-cycle progression. Interestingly, not only the CDKN2A is altered by hypermethylation but also RB is inactivated through mutation or deletion in cancer. *P16* (Dominguez *et al.*, 2003; Agundez *et al.*, 2011; Cabello *et al.*, 2011), p14 (Dominguez *et al.*, 2003) and Cyclin D2 (*CCND2*) (Agundez *et al.*, 2011) are among the most studied and frequently methylated genes in solid tumours. It has also been reported that *RB1* (Agundez *et al.*, 2011; Cabello *et al.*, 2011), and *P15* (Yurakh *et al.*, 2006) can be hypermethylated as well. This reciprocal relationship between RB1 and CDKN2A alterations underscores the importance of this pathway in cancer development and as a potential target (Weinberg, 1995).

Disruption of cell cycle progression in cancer could involve inactivating other genes through methylation. Cell cycle related genes, recently reported in multiplexing assays, include the cell-differentiation and embryonic–development paired box genes 5 (*PAX5*), and 6 (*PAX6*) (Salem *et al.*, 2000; Agundez *et al.*, 2011; Cabello *et al.*, 2011). The checkpoint with forkhead and ring fingers domain gene (CHFR) is a mitotic-stress checkpoint gene that mediates entry of cells into metaphase, thereby coordinating another phase of cell-cycle progression, mitotic prophase (Scolnick and Halazonetis, 2000), also reported to be hypermethylated in cancer (Agundez *et al.*, 2011; Cabello *et al.*, 2011). Thus, genes regulating both the G1 and G2 checkpoints are silenced through promoter hypermethylation. Several genes regulating cell cycle have been found methylated in solid tumours (Agundez *et al.*, 2011; Cabello *et al.*, 2011).

DNA repair

The O^6-methylguanine-DNA methyltransferase (*MGMT*) is a DNA-repair enzyme that protects cells from the carcinogenic effects of alkylating agents by removing adducts from the O6 position of guanine. MGMT is frequently inactivated by promoter methylation in human tumours (Agundez *et al.*, 2011; Cabello *et al.*, 2011). Failure to repair this DNA adduct can lead to transition mutations in genes such as TP53 and others. It remains to be evaluated whether tumour subtypes with a silenced MGMT gene might more often contain a mutation in the *TP53* tumour suppressor gene or others, indicating that loss of *MGMT* might cause the ensuing mutations. Other genes known to play a role in DNA repair that have been shown to be frequently hypermethylated in cancer include glutathione transferase 1 gene (*GSTP1*) (Agundez *et al.*, 2011; Cabello *et al.*, 2011), the adenomatous poliposis coli gene (*APC*) (Esteller, 2007), the ataxia telangiectasia mutated gene (*ATM*), the Mut S homologue 6 gene (*MSH6*), the breast cancer 1 and 2, early onset genes (*BRCA1* and *BRCA2*), or the tumour suppressor gene *TP53* (Agundez *et al.*, 2011; Cabello *et al.*, 2011).

Apoptosis

The Death-associated protein kinase gene (*DAPK*) is a serine/threonine, microfilament-bound kinase involved in the apoptosis induced by-interferon-, the tumour necrosis factor (TNF-α) or the apoptosis-mediating surface antigen (FAS) genes (Cohen *et al.*, 1999). Loss of function of this kinase could significantly affect cellular apoptosis. Methylation of the DAPK promoter is frequently observed in several cancers (Cohen *et al.*, 1999; Christoph *et al.*, 2008). DAPK can suppress oncogenic transformation of primary embryonic fibroblasts by activating p53 in an *ARF*-dependent manner, indicating an important role for this gene in eliminating premalignant cells during cancer development (Christoph *et al.*, 2008).

The death-inducing ligands [for example, FAS and the TNF-related apoptosis inducing ligand (TRAIL)] and their receptors [FASR, death receptors (DR4) and (DR5)] recruit the adaptor Fas-associating death domain-containing protein (FADD), resulting in the activation of caspase-8. Inactivation of the caspase-8 gene by promoter hypermethylation also occurs in human tumours (Christoph *et al.*, 2008). Although the FADD, DR4 and DR5 genes contain CpG islands in their promoters, they have not been found methylated in human tumours (Christoph *et al.*, 2008). These studies highlight how an apoptotic signalling network can be disrupted through promoter hypermethylation and provides some insight into why tumours could be highly resistant to apoptosis mediated by death receptors. Other apoptotic related genes reported to be methylated in human cancer include the antiapoptotic decoy receptors lacking death domains (*DRC1*) and (*DRC2*) (Shivapurkar *et al.*, 2004), the apoptotic protease activating factor-1 (*APAF1*) (Christoph *et al.*, 2008), B-cell lymphoma 2 (*BCL2*) (Agundez *et al.*, 2011; Cabello *et al.*, 2011), and the adaptor protein that is composed of two protein–protein interaction domains: a N-terminal PYRIN-PAAD-DAPIN domain (PYD) and a C-terminal caspase-recruitment domain (CARD) gene (*PYCARD*) (Agundez *et al.*, 2011; Cabello *et al.*, 2011), among others.

The *RAS* superfamily of GTP-binding proteins has an important role in signal transduction pathways that control cell proliferation, differentiation and cell death (Khokhlatchev *et al.*, 2002). Belonging to a family of genes encoding *RAS*-binding proteins, the *RAS* association domain protein family 1a (RASSF1A) has strongly been studied in solid tumours and was found frequently hypermethylated. Although the function of this gene is still not fully characterized, *RASSF1A* binds with the pro-apoptotic protein macrophage stimulating 1 (Khokhlatchev *et al.*, 2002). The binding of *RAS* to this complex might mediate *RAS*-dependent apoptosis. Therefore, silencing RASSF1A could effectively block the apoptosis mediated by *RAS* activation.

Invasion

A crucial step in the progression of an invasive tumour towards metastatic disease is its invasion through the lamina propria and stromal tissue (Takeichi, 1991). Several genes regulating these processes have been reported to be methylated in human cancer. The cadherins are a family of cell-surface glycoproteins responsible for cell recognition and adhesion (Takeichi, 1991). E-cadherin (*CDH1*) and other cadherins are inactivated in cancer. E-cadherin methylation has been frequently found in human tumours (Agundez *et al.*, 2011; Cabello *et al.*, 2011).The tissue inhibitor of metalloproteinases (*TIMPs*) inhibits the proteolytic activity of the matrix metalloproteinases, the activation of which can lead to degradation of the basement membrane (Chambers and Matrisian, 1997). Methylation of TIMP3 was seen in a high proportion of solid tumours. The laminins, code for polypeptide chains that interact with the integrins to maintain the structural integrity of the basement membrane (Colognato and Yurchenco, 2000). Laminin hypermethylation was also frequent in human tumours. Recent methylation affecting genes important on the invasive properties of cancer cells include stromal genes such as collagens or the Intercellular adhesion molecule 1 (*ICAM1*), or involved in the epithelial mesenchymal transition, or controlling metastasis signalling such as the cell surface glycoproteins *CD44* (Agundez *et al.*, 2011; Cabello *et al.*, 2011), or metastasis suppressor genes modifying the migration and invasion of cancer cells such as *KiSS-1* (Ruppen *et al.*, 2010; Cebrian *et al.*, 2011; Moya *et al.*, 2012), which represents the methylation study of a single candidate including the highest number of tumours ($n=804$) to date (Cebrian *et al.*, 2011). The silencing of these genes by methylation is believed to play a crucial event in promoting tumour-cell invasion and migration into surrounding tissue.

Miscellaneous

Among other networks reported to be methylated in cancer, the *WNT* signalling pathway has strongly been found methylated in human cancer. Several members of the *WNT*/beta-catenin pathway have been reported to be frequently methylated in solid tumours, including the secreted frizzled-related proteins (*SFRP*) or the Wnt inhibitory factor-1 (*WIF1*) (Costa *et al.*, 2010), highlighting the targetable role of *WNT* pathway in cancer. In the last few years several genes have been shown to be hypermethylated in human cancer and play a tumour suppressor role, which has not been fully characterized in this disease to date. These include fine analyses of genomic regions being hypermethylated, and individual tumour suppressor candidates such as the Runt-related transcription factor 3 gene (*RUNX3*) (Sanchez-Carbayo, 2012a), myopodin, a cytoskeleton actin binding protein whose methylation was found useful for diagnosis in bladder and colon cancer and to predict BCG response (Sanchez-Carbayo *et al.*, 2003; Cebrian *et al.*, 2008; Alvarez-Mugica *et al.*, 2010; Esteban *et al.*, 2012), the phosphatase and tensin homologue gene (*PTEN*), the von Hippel–Lindau tumour suppressor gene (*VHL*), the Wilms tumour 1 gene (*WT1*), or the immunoglobulin superfamily member 4 gene (*IGSF4*) (Agundez *et al.*, 2011; Cabello *et al.*, 2011). Differentiation related genes have also been found methylated with high prevalence in solid tumours such as the Sry-related-hmg box 9 gene (*SOX9*) (Aleman *et al.*, 2008a), the polyamine modulated factor 1 gene (*PMF1*) (Aleman *et al.*, 2008b; Alvarez-Mugica *et al.*, 2013), the oestrogen receptor gene (*ESR1*), or the retinoic acid receptor gene (*RARB*) (Agundez *et al.*, 2011; Cabello *et al.*, 2011). It is important to be aware that other candidates not described here have also been found methylated in human cancer in individual studies. Moreover, *in vitro* analyses using cell lines after exposure to demethylating agents have associated methylation with gene silencing revealing candidates, which might have not been fully evaluated in human clinical material. Further evaluations are required to define the tumour suppressor role and mechanisms of these genes in cancer using *in vitro* and/or *in vivo* functional analyses. Further analyses are also necessary using independent sets of human material to further define the step in cancer progression in which their role show the highest biological relevance.

The relevance of high-throughput technologies as accelerating discovery means of epigenetic events in cancer

It is important to be aware that the rapid advance in the study of gene-promoter methylation in cancer has been facilitated by two important discoveries. The first one was the demonstration that an important tumour-suppressor gene, CDKN2A – which encodes *INK4A* (also known as p16) and *ARF* (also known as p14) – was silenced in many cancers through aberrant promoter hypermethylation, including cancer (Merlo *et al.*, 1995). The second advance was the development of the methylation-specific PCR assays that allows for rapid detection of methylation in genes through the selective amplification of methylated alleles within a specific gene promoter (Herman *et al.*, 1996). It is important to mention that until the advent of high-throughput approaches, the discovery of the methylation of many of the genes reported to be hypermethylated in specific cancers was subsequent to the previous reported methylation of a given cancer related gene in other tumour types.

A major advantage of the analysis of DNA methylation is that it can be detected with high degree of specificity even in the presence of excessive unmethylated DNA. This allows detecting methylation with PCR techniques requiring just few numbers of methylated cells. The initial studies evaluating hypermethylated candidates in cancer utilized qualitative or quantitative methylation specific PCR reactions after bisulfite treatment of genomic DNA. From the methodological standpoint, the successful initial single gene candidate driven approach for describing the cancer methylome is being complemented by comprehensive epigenomic approaches, which allow for the interrogation of epigenomes to uncover novel genes and pathways in the search for translational applications in an unbiased manner. The advent of high-throughput technologies is accelerating the discovery process of such alterations, better describing those pathways that are more relevant from the biological or clinical points of views. Powerful techniques are constantly being implemented and can now detect minimal amounts of aberrant DNA methylation, especially for candidate-driven approaches (Bailey *et al.*, 2009; Li *et al.*, 2009; Malentacchi *et al.*, 2009). Furthermore, rapidly emerging high-throughput methods can comprehensively survey the epigenetic profiles, characteristic for distinct tumour types, and identify targets and pathways that may underlie a particular biological or clinical behaviour.

Several groups have used epigenomic profiling of cancer related samples using mainly different CpG array modalities (Aleman *et al.*, 2008a; Fernandez *et al.*, 2012; Sanchez-Carbayo 2012b). Methylation signatures are starting to be identified differentiating cancer cases from controls without the disease and showing promise to robustly distinguish cancer subclasses, and epigenetic candidates and pathways underlying tumourigenesis or cancer progression. Individual novel candidates are also being discovered using array identification strategies. In order to prove whether these candidates are epigenetically silenced by hypermethylation it is required to perform complementary validation analyses using *in vitro* strategies with exposure of cells to demethylating drugs (also applying high-throughput approaches), and using independent series of samples, such as described for *SOX9* in a CpG array study (Aleman *et al.*, 2008a). Several epidemiological studies are associating methylation patterns with etiologic variables, uncovering a potential additional utility of methylation patterns as risk factors (Sanchez-Carbayo 20012b). As a subsequent step after discovering methylation patterns, the multiplexing technologies are being developed to test the focused panels of identified genes in high number of samples for addressing specific clinical endpoints. The multiplexed profiles are showing promise by being associated with clinicopathological variables. In this regard, the methylation specific multiplex ligation-dependent probe amplification (MS-MLPA) technique has recently been applied to cancer tissues and urine samples as a simple and reliable method for the detection of changes in methylation status of 20 up to 40 selected genes per assay. Initial MS-MLPA studies have tested several series of tumour suppressor genes for tumour stratification, diagnosis, and even for predicting therapy response (Agundez *et al.*, 2011; Cabello *et al.*, 2011). However, further research is warranted in the field to expand the number of patients and the tumour subtypes in such initial analyses and to translate the identification of molecular epigenetic candidates into diagnostic, potential prognostic biomarkers of clinical behaviour, and predictive markers of therapeutic response. The challenge remains to optimize the measurement of these targets in non-invasive specimens with low DNA starting material. Efforts are also required for selecting and validating the most optimal combinations of methylation candidates for advancing diagnosis, tumour stratification, and outcome prediction. All these strategies will finally improve the overall clinical management of cancer patients, from early detection, staging and prognosis to surveillance purposes.

Translational applications of methylome analyses as a source of cancer biomarkers

The analyses of the methylation patterns and individual candidates in human tumours are relevant not only to unravel how frequent methylation is in human tissue clinical material, but also to reveal the critical steps along cancer progression where methylation takes place and the networks involved in these processes. Furthermore it is important to translate these pieces of information also to potentially aid as biomarkers in the clinical management of cancer patients. The initial clinical analyses in tissue specimens aimed to describe the prevalence of methylation in tumour specimens, so that they may aid for tumour stratification, or whether they may be helpful for clinical outcome prognosis. When the analytical methylation specific method shows reliability and sensitivity enough for low abundance of cancer cells, it can then be optimized and tested in the body fluids for diagnostic purposes.

Thus, the progress in the field of cancer epigenetics research includes not only the biological characterization of methylation alterations but moves towards translational applications including the development of potential new biomarkers. To date, three major areas seem to potentially benefit from DNA methylation–based biomarkers: cancer detection, tumour stratification and prognosis, and recent studies start to address prediction of treatment responses, a field known as pharmacoepigenetics.

Cancer detection

Important insights of the aberrant DNA methylation events occurring in tumours have been gained mostly by means of driven-candidate analyses, particularly for the hypermethylated CpG islands of tumour suppressor genes with their different specific mechanisms of tumour suppression described above (Wolff *et al.*, 2005; Sanchez-Carbayo and Cordon-Cardo, 2007; Gonzalgo *et al.*, 2007). Analyses from all tumour stages along cancer initiation and progression, including normal epithelium analyses have highlighted the existence of profiles of hypermethylated CpG islands that may potentially characterize tumours. The emphasis for defining potential diagnostic biomarkers for clinical routine practice relates to focusing on those aberrant methylation events that are shown to be absent in normal cells together with development of techniques that provide reliable, sensitive and fast results. Advances in technologies coupling bisulphite modification of DNA with PCR have been key in the development of reliable molecular methods. CpG island hypermethylation has been used as a tool to detect cancer cells not only for tissue specimens but also for non-invasive biological fluids, such as the serum (Dominguez *et al.*, 2003), and the urine (Cebrian *et al.*, 2008; Aleman *et al.*, 2008b, Sanchez-Carbayo, 2012b). Further studies will define whether urinary and serum methylomes may serve for the diagnosis of cancer diseases in clinical practice. The initial methylation studies were initially performed by MS-PCR in qualitative or quantitative fashion modalities and currently it is feasible by comprehensive multiplexed analyses such as by MS-MLPA, multiplexed quantitative RT-PCR strategies or homemade or commercial CpG arrays (Fernandez *et al.*, 2012; Sanchez-Carbayo, 2012b). Thus, detection of cancer-specific hypermethylation events in body fluids of individuals with cancer is feasible using this easily accessible biological material in the clinical setting. Powerful techniques detecting minimal amounts of aberrant DNA methylation will be critical to further expand the field and its applications (Bailey *et al.*, 2009; Li *et al.*, 2009; Malentacchi *et al.*, 2009) for clinical routine practice in the near future.

Several studies have shown that CpG island promoter hypermethylation of tumour suppressor genes occurs early in tumourigenesis (Wolff *et al.*, 2005; Sanchez-Carbayo and Cordon-Cardo, 2007; Gonzalgo *et al.*, 2007). Thus, the presence of aberrant CpG island methylation alone does not necessarily indicate an invasive cancer, since premalignant or precursor lesions can also carry these epigenetic signatures. This finding has implications for early detection of cancer, especially in people with inherited genetic risk, or exposed to carcinogenic environment. This represents an important potential application for tumours where environmental exposure is associated with carcinogenesis. This feature justifies the use of epigenetic profiling as a means to provide cancer risk biomarkers using germline DNA, and potentially in urine samples in the near future (Sanchez-Carbayo, 2012b).

Prognosis

A longstanding goal in biomarker research is the capacity to look at two morphologically and/or clinically identical tumours and distinguish which one will grow at a fast pace and which one will have a more indolent behaviour. An increasing number of hypermethylation individual events at single gene loci are indicating differential outcomes among cancer patients. Among the genes most frequently studied, for example, hypermethylation of the genes encoding such as *CDKN2A*, *DAPK*, *RASFF1A*, *CDH1*, *SFRPs*, *RUNX3*, or *myopodin*, among others, has strongly been linked to poor outcomes in tissue specimens (Wolff *et al.*, 2005; Sanchez-Carbayo and Cordon-Cardo, 2007; Gonzalgo *et al.*, 2007; Sanchez-Carbayo, 2012b). Combinations of tumour suppressor genes have also been utilized for this purpose (Agundez *et al.*, 2011; Cabello *et al.*, 2011). Importantly, individual candidates such as myopodin (C) (Sanchez-Carbayo *et al.*, 2003; Cebrian *et al.*, 2008; Alvarez-Mugica *et al.*, 2010; Esteban *et al.*, 2012), or a group of tumour suppressor genes

(Agundez *et al.*, 2011), served to predict recurrence, progression and survival.

An increasing number of global DNA methylation approaches is being devoted to accomplishing the goal of identifying aberrant methylation signatures. Epigenomics unveils prognostic dendrograms, where a combination of aberrant DNA methylation markers from CpG arrays is used (Aleman *et al.*, 2008a; Fernandez *et al.*, 2012). Importantly, these epigenomic profiles can also be developed with DNA extracted from archived material (Agundez *et al.*, 2011). Most important, single markers with great clinical potential, such as predictors of recurrence, progression or survival, can be identified and developed from these global DNA methylation screenings. The establishment of prognosis using aberrant DNA methylation signatures obtained using CpG arrays is beginning to emerge in cancer (Sanchez-Carbayo, 2012b). Larger multicentre and prospective studies are needed to further validate these candidates revealed by current array approaches and other single-gene DNA methylation markers.

Pharmacoepigenetics

Despite the lack of targeted therapies in several tumour types to date, the hypermethylation patterns of particular genes have been shown to enable to predict response to specific treatments, such as *myopodin* to predict BCG response (Alvarez-Mujica *et al.*, 2010). Moreover, the methylation profiles of a panel of tumour suppressor genes have shown ability to predict response to the intravesical immunotherapy with the BCG in non-muscle invasive tumours in paraffin embedded material (Agundez *et al.*, 2011). It is expected that in the near future other promising epigenetic candidates will be applied to predict pharmacoepigenetic cisplatin response in muscle-invasive disease.

Conclusions

Hypermethylation, as a hallmark of human cancer is a frequent event leading to gene silencing of critical cancer-related genes. Nowadays, the epigenetic silencing by hypermethylation for critical cancer related candidates, especially tumour suppressor genes, has been widely described in human cancer. These hypermethylation changes have a recognized role in the initiation and progression of cancer, by affecting critical molecular pathways participating in these processes. Hypermethylation changes could potentially aid for the detection and monitoring the clinical evolution of cancer patients. Methylation multiplexed profiles and individual candidates measured in tissue specimens could enable not only tumour stratification but also clinical outcome prognosis using recurrence, progression, disease-specific and overall survival as clinical endpoints. The use of body fluids allows non-invasive methylation assessment for early diagnosis and potentially for disease follow-up and risk assessment in the near future. However, the epigenomic exploration of cancer has only just begun. Genome-scale DNA methylation profiling studies will continue highlighting the relevance of the epigenetic component to gain better knowledge of cancer biology, and to identify those profiles and candidates better correlating with clinical behaviour.

Acknowledgements

We would like to thank all the members of the laboratory of M. Sánchez-Carbayo, especially Rodrigo Garcia-Baquero, Valle Montalvo, Noemi Pompas, and Patricia Moya, for technical support and constructive suggestions in the preparation of this chapter. This work is supported by grants from the Spanish Ministry of Science and Innovation grant SAF2009-13035, and Mutua Madrileña 2010 (to M. Sánchez-Carbayo).

References

Agundez, M., Grau, L., Palou, J., Algaba, F., Villavicencio, H., and Sanchez-Carbayo, M. (2011). Evaluation of the methylation status of tumour suppressor genes for predicting bacillus Calmette-Guérin response in patients with T1G3 high-risk bladder tumours. Eur. Urol. *60*, 131–140.

Aleman, A., Adrien, L., Lopez-Serra, L., Cordon-Cardo, C., Esteller, M., Belbin, T.J., and Sanchez-Carbayo, M. (2008a). Identification of DNA hypermethylation of SOX9 in association with bladder cancer progression using CpG microarrays. Br. J. Cancer. *98*, 466–473.

Aleman, A., Cebrian, V., Alvarez, M., Lopez, V., Orenes, E., Lopez-Serra, L., Algaba, F., Bellmunt, J., López-Beltrán, A., Gonzalez-Peramato, P., *et al.* (2008b).

Identification of PMF1 methylation in association with bladder cancer progression. Clin. Cancer Res. *14*, 8236–8243.

Alvarez-Mugica, M., Cebrian, V., Fernandez-Gomez, J.M., Fresno, F., Escaf, S., and Sanchez-Carbayo, M. (2010). Myopodin methylation is associated with clinical outcome in patients with T1G3 bladder cancer. J. Urol. *4*, 1507–1513.

Alvarez-Múgica, M., Fernández-Gómez, J.M., Cebrian, V., Fresno, F., Escaf, S., and Sánchez-Carbayo, M. (2013). PMF-1 methylation predicts BCG response in patients with high grade non-muscle invasive bladder carcinoma. Eur Urol. *63*, 364–70.

Bailey, V.J., Easwaran, H., Zhang, Y., Griffiths, E., Belinsky, S.A., Herman, J.G., Baylin, S.B., Carraway, H.E., and Wang, T.H. (2009). MS-qFRET: a quantum dot-based method for analysis of DNA methylation. Genome Res. *19*, 1455–1461.

Bird, A. (2002). DNA methylation patterns and epigenetic memory. Genes Dev. *16*, 6–21.

Cabello, M.J., Grau, L., Franco, N., Orenes, E., Alvarez, M., Blanca, A., Heredero, O., Palacios, A., Urrutia, M., Fernández, J.M., *et al.* (2011). Multiplexed methylation profiles of tumor suppressor genes in bladder cancer. J. Mol. Diagn. *13*, 29–40.

Cebrian, V., Alvarez, M., Aleman, A., Palou, J., Bellmunt, J., Gonzalez-Peramato, P., Cordón-Cardo, C., García, J., Piulats, J.M., and Sánchez-Carbayo, M. (2008). Discovery of myopodin methylation in bladder. J. Pathol. *216*, 111–119.

Cebrian, V., Fierro, M., Orenes-Piñero, E., Grau, L., Moya, P., Ecke, T., Alvarez, M., Gil, M., Algaba, F., Bellmunt, J., *et al.* (2011). KISS1 methylation and expression as tumor stratification biomarkers and clinical outcome prognosticators for bladder cancer patients. Am. J. Pathol. *179*, 540–6.

Chambers, A.F., and Matrisian, L.M. (1997). Changing views of the role of matrix metalloproteinases in metastasis. J. Natl. Cancer Inst. *89*, 1260–1270.

Christoph, F., Hinz, S., Weikert, S., Kempkensteffen, C., Schostak, M., Miller, K., and Schrader, M. (2008). Comparative promoter methylation analysis of p53 target genes in urogenital cancers. Urol. Int. *80*, 398–404.

Cohen, O., Inbal, B., Kissil, J.L., Raveh, T., Berissi, H., Spivak-Kroizaman, T., Feinstein, E., and Kimchi, A. (1999). DAP-kinase participates in TNF- and Fas-induced apoptosis and its function requires the death domain. J. Cell Biol. *146*, 141–148.

Colognato, H., and Yurchenco, P.D. (2000). Form and function: the laminin family of heterotrimers. Dev. Dyn. *218*, 213–234.

Costa, V.L., Henrique, R., Ribeiro, F.R., Carvalho, J.R., Oliveira, J., Lobo, F., Teixeira, M.R., and Jerónimo, C. (2010). Epigenetic regulation of Wnt signaling pathway in urological cancer. Epigenetics. *4*, 343–345.

Delhommeau, F., Dupont, S., DellaValle, V., James, C., Trannoy, S., Massé, A., Kosmider, O., Le Couedic, J.P., Robert, F., Alberdi, A., *et al.* (2009). Mutation in TET2 in myeloid cancers. N. Engl. J. Med. *360*, 2289–2301.

Doi, A., Park, I.H., Wen, B., Murakami, P., Aryee, M.J., Irizarry, R., Herb, B., Ladd-Acosta, C., Rho, J., Loewer, S., *et al.* (2009). Differential methylation of tissue- and cancer-specific CpG island shores distinguishes human induced pluripotent stem cells, embryonic stem cells and fibroblasts. Nat. Genet. *41*, 1350–1353.

Dominguez, G., Silva, J., Garcia, J.M., Silva, J.M., Rodriguez, R., Muñoz, C., Chacón, I., Sanchez, R., Carballido, J., Colás, A., *et al.* (2003). Prevalence of aberrant methylation of p14ARF over p16INK4a in some human primary tumors. Mutat. Res. *530*, 9–17.

Eden, A., Gaudet, F., Waghmare, A., and Jaenisch, R. (2003). Chromosomal instability and tumors promoted by DNA hypomethylation. Science. *300*, 455.

Esteban, S., Moya, P., Fernandez-Suarez, A., Vidaurreta, M., González-Peramato, P., and Sánchez-Carbayo, M. (2012). Diagnostic and prognostic utility of methylation and protein expression patterns of myopodin in colon cancer. Tumour Biol. *33*, 337–346.

Esteller, M. (2007a). Cancer epigenomics: DNA methylomes and histone-modification maps. Nat. Rev. Gene. *8*, 286–298.

Esteller, M. (2007b). Epigenetic gene silencing in cancer: the DNA hypermethylome. Hum. Mol. Genet. *16*, 50–59.

Esteller, M. (2008). Epigenetics in cancer. N. Engl. J. Med. *358*, 1148–1159.

Feinberg, A.P. (2005). Cancer epigenetics is no Mickey Mouse. Cancer Cell. *8*, 267–268.

Fernandez, A.F., Assenov, Y., Martin-Subero, J.I., Balint, B., Siebert, R., Taniguchi, H., Yamamoto, H., Hidalgo, M., Tan, A.C., Galm, O., *et al.* (2011). A DNA methylation fingerprint of 1628 human samples. Genome Res. *22*, 407–419.

Futscher, B.W., Oshiro, M.M., Wozniak, R.J., Holtan, N., Hanigan, C.L., Duan, H., and Domann, F.E. (2002). Role for DNA methylation in the control of cell type specific maspin expression. Nat. Genet. *31*, 175–179.

Gonzalgo, M.L., Datar, R.H., Schoenberg, M.P., and Cote, R.J. (2007). The role of deoxyribonucleic acid methylation in development, diagnosis, and prognosis of bladder cancer. Urol. Oncol. *25*, 228–235.

Hattori, N., Nishino, K., Ko, Y.G., Hattori, N., Ohgane, J., Tanaka, S., and Shiota, K. (2004). Epigenetic control of mouse Oct-4 gene expression in embryonic stem cells and trophoblast stem cells. J. Biol. Chem. *279*, 17063–17069.

Hellman, A., and Chess, A. (2007). Gene body-specific methylation on the active X chromosome. Science. *315*, 1141–1143.

Herman, J.G., Graff, J.R., Myöhönen, S., Nelkin, B.D., and Baylin, S.B. (1996). Methylation-specific PCR: a novel PCR assay for methylation status of CpG islands. Proc. Natl. Acad. Sci. U.S.A. *93*, 9821–9826.

Ito, S., D'Alessio, A.C., Taranova, O.V., Hong, K., Sowers, L.C., and Zhang, Y. (2010). Role of Tet proteins in 5mC to 5hmC conversion, ES-cell self-renewal and inner cell mass specification. Nature *466*, 1129–1133.

Irizarry, R.A., Ladd-Acosta, C., Wen, B., Wu, Z., Montano, C., Onyango, P., Cui, H., Gabo, K., Rongione, M., Webster, M., *et al.* (2009). The human colon cancer methylome shows similar hypo- and hypermethylation at conserved tissue-specific CpG island shores. Nat. Genet. *41*, 178–186.

Jones, P.A., and Baylin, S.B. (2007). The epigenomics of cancer. Cell. *128*, 683–692.

Jones, P.A., and Laird, P.W. (1999). Cancer epigenetics comes of age. Nat. Genet. *21*, 163–167.

Khokhlatchev, A., Rabizadeh, S., Xavier, R., Nedwidek, M., Chen, T., Zhang, X.F., Seed, B., and Avruch, J. (2002). Identification of a novel Ras-regulated proapoptotic pathway. Curr. Biol. *12*, 253–265.

Kuroda, A., Rauch, T.A., Todorov, I., Ku, H.T., Al-Abdullah, I.H., Kandeel, F., Mullen, Y., Pfeifer, G.P., and Ferreri, K. (2009). Insulin gene expression is regulated by DNA methylation. PLoS One *4*, e6953.

Laird, P.W. (2003). The power and the promise of DNA methylation markers. Nat. Rev. Cancer. *3*, 253–266.

Ley, T.J., Ding, L., Walter, M.J., McLellan, M.D., Lamprecht, T., Larson, D.E., Kandoth, C., Payton, J.E., Baty, J., Welch, J., *et al.* (2010). DNMT3A Mutations in acute myeloid leukaemia. N., Engl, J. [Year?] J.. *363*, 2424–2333.

Li, M., Chen, W.D., Papadopoulos, N., Goodman, S.N., Bjerregaard, N.C., Laurberg, S., Levin, B., Juhl, H., Arber, N., Moinova, H., *et al.* (2009). Sensitive digital quantification of DNA methylation in clinical samples. Nat. Biotechnol. *27*, 858–863.

Lister, R., Pelizzola, M., Dowen, R.H., Hawkins, R.D., Hon, G., Tonti-Filippini, J., Nery, J.R., Lee, L., Ye, Z., Ngo, Q.M., *et al.* (2009). Human DNA methylomes at base resolution show widespread epigenomic differences. Nature *462*, 315–322.

Lukas, J., Parry, D., Aagaard, L., Mann, D.J., Bartkova, J., Strauss, M., Peters, G., and Bartek, J. (1995). Retinoblastoma-protein-dependent cell-cycle inhibition by the tumour suppressor p16 Nature. *375*, 503–506.

Malentacchi, F., Forni, G., Vinci, S., and Orlando, C. (2009). Quantitative evaluation of DNA methylation by optimization of a differential-high resolution melt analysis protocol. Nucleic Acids Res. *37* e86.

Meissner, A., Mikkelsen, T.S., Gu, H., Wernig, M., Hanna, J., Sivachenko, A., Zhang, X., Bernstein, B.E., Nusbaum, C., Jaffe, D.B., *et al.* (2008). Genome-scale DNA methylation maps of pluripotent and differentiated cells. Nature *454*, 766–770.

Merlo, A., Herman, J.G., Mao, L., Lee, D.J., Gabrielson, E., Burger, P.C., Baylin, S.B., and Sidransky, D. (1995). 5′ CpG island methylation is associated with transcriptional silencing of the tumour suppressor p16/CDKN2/MTS1 in human cancers. Nat. Med *1*, 686–692.

Moya, P., Esteban, S., Fernandez-Suarez, A., Maestro, M., Morente, M., and Sánchez-Carbayo, M. (2013). KiSS-1 methylation and protein expression patterns contribute to diagnostic and prognostic assessments in tissue specimens for colorectal cancer. Tumor Biol. *34*, 471–479.

Nakagawa, T., Kanai, Y., Saito, Y., Kitamura, T., Kakizoe, T., and Hirohashi, S. (2003). Increased DNA methyltransferase 1 protein expression in human transitional cell carcinoma of the bladder. J. Urol. *170*, 2463–2466.

Robert, M.F., Morin, S., Beaulieu, N., Gauthier, F., Chute, I.C., Barsalou, A., and MacLeod, A.R. (2003). DNMT1 is required to maintain CpG methylation and aberrant gene silencing in human cancer cells. Nat. Genet. *33*, 61–65.

Robertson, K.D., Uzvolgyi, E., Liang, G., Talmadge, C., Sumegi, J., Gonzales, F.A., and Jones, P.A. (1999). The human DNA methyltransferases (DNMTs) 1, 3a and 3b: coordinate mRNA expression in normal tissues and overexpression in tumors. Nucleic Acids Res. *27*, 2291–2298.

Robertson, K.D., Keyomarsi, K., Gonzales, F.A., Velicescu, M., and Jones, P.A. (2000). Differential mRNA expression of the human DNA methyltransferases (DNMTs) 1, 3a and 3b during the G(0)/G(1) to S phase transition in normal and tumor cells. Nucleic Acids Res. *28*, 2108–2113.

Ruppen, I., Grau, L., Orenes-Piñero, E., Ashman, K., Gil, M., Algaba, F., Bellmunt, J., and Sánchez-Carbayo, M. (2010). Differential protein expression profiling by iTRAQ-two-dimensional LC-MS/MS of human bladder cancer EJ138 cells transfected with the metastasis suppressor KiSS-1 gene. Mol. Cell Proteomics. *10*, 2276–2291.

Salem, C.E., Markl, I.D., Bender, C.M., Gonzales, F.A., Jones, P.A., and Liang, G. (2000). PAX6 methylation and ectopic expression in human tumor cells. Int. J. Cancer *87*, 179–85.

Sánchez-Carbayo, M. (2003). Use of high-throughput DNA microarrays to identify biomarkers for bladder cancer. Clin. Chem. *49*, 23–31.

Sánchez-Carbayo, M. (2012a). Hypermethylation in bladder cancer: biological pathways and translational applications. Tumour Biol. *33*, 347–361.

Sánchez-Carbayo, M. (2012b). Urine epigenomics: a promising path for bladder cancer diagnostics. Exp. Rev. Mol. Diagn. *12*, 29–32.

Sánchez-Carbayo, M., and Cordon-Cardo, C. (2003). Applications of array technology: identification of molecular targets in bladder cancer. Br. J. Cancer. *89*, 2172–2177.

Sánchez-Carbayo, M., and Cordon-Cardó, C. (2007). Molecular alterations associated with bladder cancer progression. Semin. Oncol. *34*, 75–84.

Scolnick, D.M., and Halazonetis, T.D. (2000). Chfr defines a mitotic stress checkpoint that delays entry into metaphase. Nature. *406*, 430–5.

Sanchez-Carbayo, M., Schwarz, K., Charytonowicz, E., Cordon-Cardo, C., and Mundel, P. (2003). Tumor suppressor role for myopodin in bladder cancer: loss of nuclear expression of myopodin is cell-cycle dependent and predicts clinical outcome. Oncogene. *22*, 5298–5305.

Straussman, R., Nejman, D., Roberts, D., Steinfeld, I., Blum, B., Benvenisty, N., Simon, I., Yakhini, Z., and Cedar, H. (2009). Developmental programming of CpG island methylation profiles in the human genome. Nat. Struct. Mol. Biol. *16*, 564–571.

Shivapurkar, N., Toyooka, S., Toyooka, K.O., Reddy, J., Miyajima, K., Suzuki, M., Shigematsu, H., Takahashi, T., Parikh, G., Pass, H.I., *et al.* (2004). Aberrant methylation of trail decoy receptor genes is frequent in multiple tumor types. Int. J. Cancer. *109*, 786–789.

Suzuki, M.M., and Bird, A. (2008). DNA methylation landscapes: provocative insights from epigenomics. Nat. Rev. Genet. *9*, 465–476.

Tahiliani, M., Koh, K.P., Shen, Y., Pastor, W.A., Bandukwala, H., Brudno, Y., Agarwal, S., Iyer, L.M., Liu, D.R., Aravind, L., *et al.* (2009). Conversion of 5-methylcytosine to 5-hydroxymethylcytosine in mammalian DNA by MLL partner TET1. Science. *324*, 930–935.

Takeichi, M. (1991). Cadherin cell adhesion receptors as a morphogenetic regulator. Science. *251*, 1451–1455.

Thomson, J.P., Skene, P.J., Selfridge, J., Clouaire, T., Guy, J., Webb, S., Kerr, A.R., Deaton, A., Andrews, R., James, K.D., *et al.* (2012). CpG islands influence chromatin structure via the CpG-binding protein Cfp1. Nature. *464*, 1082–1086.

Velicescu, M., Weisenberger, D.J., Gonzales, F.A., Tsai, Y.C., Nguyen, C.T., and Jones, P.A. (2002). Cell division is required for *de novo* methylation of CpG islands in bladder cancer cells. Cancer Res. *62*, 2378–2384.

Wang, Y., and Leung, F.C. (2004). An evaluation of new criteria for CpG islands in the human genome as gene markers. Bioinformatics. *20*. 1170–1177.

Weinberg, R.A. (1995). The retinoblastoma protein and cell cycle control. Cell. *81*, 323–330.

Wolff, E.M., Liang, G., and Jones, P.A. (2005). Mechanisms of Disease: genetic and epigenetic alterations that drive bladder cancer. Nat. Clin. Pract. Urol. *2*, 502–510.

Yurakh, A.O., Ramos, D., Calabuig-Fariñas, S., López-Guerrero, J.A., Rubio, J., Solsona, E., Romanenko, A.M., Vozianov, A.F., Pellin, A., and Llombart-Bosch, A. (2006). Molecular and immunohistochemical analysis of the prognostic value of cell-cycle regulators in urothelial neoplasms of the bladder. Eur. Urol. *50*, 506–15.

Identification and Analysis of Transposable Elements in Genomic Sequences

9

Laurent Modolo and Emmanuelle Lerat

Abstract

Genome sequences are composed of different compartments, among which transposable elements (TEs) represent one of the most important. Not only do these elements correspond to a particularly large proportion of genomes, they are also involved in different mechanisms implicated in the evolution of genomes, such as chromosome rearrangement and gene innovation. Thus, the precise determination of TEs in genomes is of significant importance. This step is becoming more and more complex with the emergence of new types of sequence data coming from next-generation sequencing (NGS) technologies. In this chapter, we present the current status of bioinformatic developments made in the detection and analysis of TEs in genomic sequences. We first present the classic tools dedicated to the identification of TEs in classic genomic data, which originate from whole-genome sequences. Because these sequences are significantly different from the new types of sequences generated by NGS and because the problem of repeats in these data is not trivial, we then present how it is possible to handle TEs in NGS data. We also provide some examples of tools designed to answer particular questions about TEs using NGS data and how these types of data are particularly valuable for deepening our knowledge of the dynamics of TEs. Although this is a still a fast-growing field for which new developments are made every day, we hope to provide a broader view of what currently exists in this field and what allows for TE analyses in genomic sequences.

Introduction

Eukaryotic genomes are composed of different elements, which are classified according to their function in the organism. The protein-coding genes, the non-coding genes (specifically rRNA, tRNA, and small RNA genes), and the regulatory elements associated with these genes are typically considered the most important groups. The remaining elements, which include pseudogenes and repeated sequences, have long been considered trivial in terms of genome functioning. However, this limited vision has begun to change over the past several years. Measured genome size has been demonstrated to be highly variable. However, this metric has been found to have no relationship with the 'complexity' of the organism. When considering the expected number of genes, this paradox has had the implication that a given genome can contain more DNA than required. Early genome sequencing projects helped to answer this question with the discovery that the functional genome, which predominantly consists of protein-coding genes, represents only a small percentage of the genome, whereas the more variable regions of the genome, the 'non-coding regions', were demonstrated to represent a higher proportion of the complete genome. With regard to the human genome, this finding has been particularly striking. The first estimations of the gene number in our own genome were radically revised after the human genome sequencing project, which revealed a small number of genes (fewer than 25,000) and more than 98% 'non-functional' DNA (Lander *et al.*, 2001). The main source of size variation in genomes was then identified as the non-coding regions of genomes.

Transposable elements (TEs) are a part of these still underestimated genomic components, which can play important roles in the functioning and in the evolution of organisms.

TEs are dispersed repeat DNA sequences that have the ability to move from one position to another along chromosomes. These elements typically encode for all the proteins necessary for their movement and possess internal regulatory regions, allowing for their independent expression. Different categories of TEs have been identified, and several attempts were made to classify them (Wicker *et al.*, 2007; Kapitonov and Jurka, 2008). Globally, two main classes have been described according to their transposition intermediates (retrotransposons use an RNA intermediate and form class I, while transposons use a DNA intermediate and form class II). Within each class, subclasses have been created to group sequences with the same structural features (Fig. 9.1). According to the classification system of Wicker *et al.*, five orders containing 17 superfamilies were described as class I elements, and two subclasses, containing 12 superfamilies, were described as class II elements. As the number of the newly sequenced genomes increases, new elements and potential new types of TEs are being discovered, which increases and enriches the complexity of TE classification.

Since B. McClintock discovered and first described these elements in maize in the 1950s (McClintock, 1956), TEs have been searched for and discovered in almost all eukaryotic organisms. Depending on the organism, the proportion of TEs can be highly variable and at times large, for example, 3% in yeast (Kim *et al.*, 1998), 15%

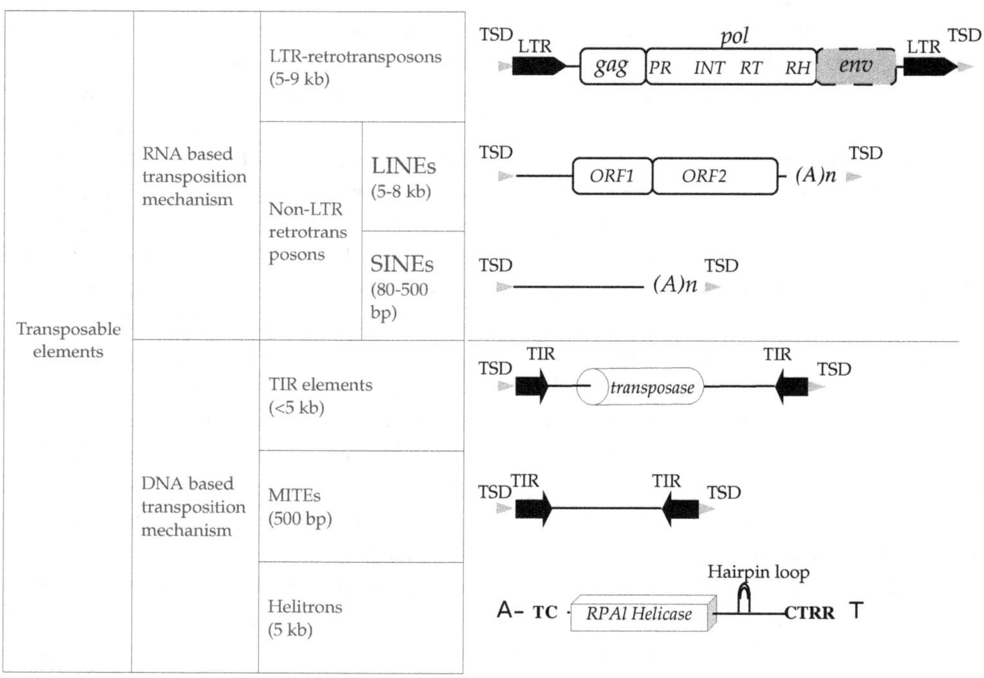

Figure 9.1 The different types of transposable elements (TEs) according to their transposition intermediates and their structural features. Almost all TEs possess two target site duplication (TSD) sequences at each extremity of the copy, corresponding to duplications of the insertion site. LTR-retrotransposons possess long terminal repeat (LTR) sequences at each extremity. Some LTR-retrotransposons encode for a third ORF specifying the ENV protein. The pol gene is formed by different domains, which encode for a protease (PR), an integrase (INT), a reverse transcriptase (RT) and an RNaseH (RH), respectively. The non-LTR retrotransposons contain a poly-A tail at their 3′ extremity. The LINEs possess two ORFs, whereas the SINEs have no coding capacity. Contrary to the MITEs, TIR elements contain an ORF encoding for a transposase. Both contain terminal inverted repeat (TIR) sequences at each extremity. The autonomous helitrons possess a coding capacity for a helicase and an RPA-like (RAPI) single-stranded DNA-binding protein.

in *Drosophila* (Dowsett and Young, 1982), 45% in human (Lander *et al.*, 2001) and more than 80% in maize (Schnable *et al.*, 2009). By proportion, TEs are directly linked to the host's genome size. Their proportion is typically very high in organisms with a very large genome, indicating their role in the expansion of the host's genome size. The variability of TE proportions in closely related species is linked to different parameters, such as the effective population size and reproduction mode of the host organism, the genetic drift, and the host's regulation of transposition activity. For example, in the sibling species *Drosophila melanogaster* and *D. simulans*, the proportion of TEs varies threefold (15% in *D. melanogaster* and 5% in *D. simulans*), although the two species have only been separated for 2.3 to 5.4 million years (Li *et al.*, 1999; Tamura *et al.*, 2004; Cutter, 2008). This discrepancy in TE content has been hypothesized to be linked to either a stronger selection against the deleterious effects of TE insertion in *D. simulans*, which could be due to a larger effective population size compared to *D. melanogaster*, or to a stronger resistance of *D. simulans* to an increase in TE copy number (Kimura and Kidwell, 1994; Vieira *et al.*, 1999). The self-incompatible plant *Arabidopsis lyrata* presents a larger genome (207 Mb) than does the self-compatible plant *A. thaliana* (125 Mb); however, these two species are only separated by 10 million years (Hu *et al.*, 2011). The size variation observed between these two plant species is predominantly due to TEs and appears, in part, to be related to differences in their mating systems (Lockton and Gaut, 2010).

Because of their presence in genomes, TEs have a significant impact on genome evolution and not only on the evolution of genome size (Lisch and Kidwell, 2000; Biémont and Vieira, 2006). TEs can promote mutations, which can be deleterious. For example, in humans, approximately 96 transposition events are directly linked to single-gene diseases (Hancks and Kazazian, 2012), and half of the spontaneous mutations observed in *Drosophila* are due to TEs (Eickbush and Furano, 2002). However, the effects of TEs can also increase the genetic diversity of an organism. The repetitive nature of TEs makes them responsible for chromosomal rearrangements via homologous recombination between copies, which, in some cases, can lead to the emergence of new species, such as has been hypothesized for *D. virilis*, for example (Evgen'ev *et al.*, 2000). When inserting in or near a gene, TEs can provide new regulatory elements, altering the expression of the gene, or they can contribute to gene innovation by providing a new coding region to a gene (Lisch and Kidwell, 2000). The implications of TEs in all epigenetic mechanisms have now been clearly established (Slotkin and Martienssen, 2007). All of these facts make TEs particularly important in the adaptation of organisms to environmental changes. During the years since their discovery, the status of TEs has moved from simple junk DNA to major players in genome evolution (for a historical review, see Biémont, 2010).

Thus, TEs are important components that cannot be neglected when analysing genome sequences. These elements can be quite numerous and, therefore, very important, and they should not be simply removed to ease gene annotations. The study of TEs is crucial for understanding their dynamics, which then allows us to better appreciate how genomes function and evolve. The question of identifying TEs in genomic sequences is a crucial point that has become more and more complex with the emergence of new types of sequencing data. In this chapter, we will first summarize the classic methods that exist to search and annotate TEs in assembled genome sequences. We will then identify the difficulties that can be encountered when using next-generation sequencing (NGS) data for this task and new methods that have been developed. Finally, we will provide examples concerning the specific analyses that can be performed on TEs using NGS data and how these new types of data constitute a significant advance in the field of TE dynamics.

Classic detection methods for TEs in genome sequences

Since the beginning of genome sequencing, significant efforts have been made to annotate the functional regions of genomes. The presence of transposable elements (TEs) and other repeats has made this task particularly difficult. Thus, to facilitate the annotation of genomes, methods to identify TEs and other repeats in genome sequences were developed (Tang, 2007). Indeed, the ability to recognize these types of sequences has been a suitable starting point to allow for the assembly of a genome and also to ease the prediction of genes. This task is particularly important for genomes containing very high proportions of TEs. Moreover, given the importance of TEs in genome evolution, the identification of these sequences has been considered crucial to allow the access to entire populations of TE copies present in a given organism. Having access to all copies of a given TE family is particularly interesting for studying the evolution and dynamics of a TE family. For example, an analysis of the TE copies from the majority of families integrated into the *D. melanogaster* genome surprisingly demonstrated that the majority of these TEs had recently moved because these TE copies were almost identical within families and very few ancient copies were present (Bowen and McDonald, 2001; Lerat *et al.*, 2003). These observations are in favour of either the hypothesis of recurrent and numerous horizontal transfers of these elements or the hypothesis of a very high turnover in this genome to remove ancient and inactive copies. The identification of all the TE copies present in the human genome has allowed us to obtain information concerning the waves of amplification of the non-LTR retrotransposons LINE and SINE and the formation of retro-processed pseudogenes in this genome (Lander *et al.*, 2001; Ohshima *et al.*, 2003). Numerous methods dedicated to the identification and classification of TEs have been developed over the past 15 years. Several reviews have described these methods exhaustively and provide lists of available programs in each category (Bergman and Quesneville, 2007; Saha *et al.*, 2008a; Lerat, 2010; Janicki *et al.*, 2011). In this section, we will mainly summarize the different categories of existing programs and those, which are currently more used and more successful in performing their tasks.

Similarity- or library-based methods

The principle of these methods is to compare genome sequences to a library of TE reference sequences to search for the occurrence of the TEs in a genome. The library used can be defined by the user or can be a public database. The most widely used public database employed in this type of work is REPBASE (Jurka *et al.*, 2005). This database contains the consensus sequences of different repeat sequences from a large set of eukaryotic organisms and is typically employed jointly with the program REPEATMASKER (Smit *et al.*, 1996–2010), which performs a similarity search using the library as a reference. The main advantage of this type of method is that it is fast and accurate. Although, it obviously cannot discover new TE families, this method is still a good starting point to explore a new genome, particularly if TE sequences from closely related species are described.

Signature-based methods

This type of method uses particular structural features (such as nucleotide or protein motifs) of known TE classes to determine their occurrence in a genome sequence. Thus, this approach can locate new elements from a given class but will fail to discover new classes of elements. Another drawback of this method is that it will only discover nearly complete and potentially active copies and miss degraded ones. Thus, such an approach can be complemented using a library-based method once complete reference elements have been discovered based on their structure. Moreover, such an approach will depend on the level of knowledge available for a given class and if it is possible to determine fixed and shared features among several families of the same class. Signature-based programs typically concentrate on a particular type of TE.

For example, it is possible to specifically search for LTR-retrotransposons given several shared characteristics between families, such as the presence of an LTR (long terminal repeat) at each end of the sequence, the fact that the two LTRs are almost identical for complete and potentially active copies, a particular distance between them, or the presence of particular protein motifs in the ORFs contained inside the element. Different programs have been designed to detect LTR-retrotransposons, of which the most successful to date is LTRHARVEST (Ellinghaus *et al.*, 2008; Lerat, 2010). However, the user needs to determine the perfect parameters for the analysed genome to avoid the occurrence of numerous false positives. Because of structural features such as the presence of a poly-A tail at the 3′ end of the sequence or target site duplications at each extremity of the copy, other programs have been designed to detect non-LTR retrotransposons (Szak *et al.*, 2002; Tu *et al.*, 2004; Lucier *et al.*, 2007). Particular DNA transposons known as MITEs have also been the subject of several programs because of their specific features, *i.e.* a short size (approximately 500 bp) and the presence of terminal inverted repeats (TIRs) at each end of a copy; and because the use of similarity-based methods to find these transposons is difficult due to their short size and a lack of coding capacity. Recently, the program MITE-HUNTER (Han and Wessler, 2010) was developed to decrease the number of false positives typically obtained using other programs to locate MITEs.

De novo methods

With these types of approaches, it is possible to search for new types of elements because there is no *a priori* knowledge of the sequence itself. Indeed, these programs take advantage of the repetitive nature of TEs. These methods are particularly interesting when sequencing the genomes of species for which no close relatives are currently annotated and for which nothing is known about their repeat content. However, these methods are particularly sensitive to genomic coverage and to the quality of the sequence assembly. Another drawback is that these approaches will find any type of repeated sequences, even tandem repeats, satellites or segmental duplications, in addition to identifying TE sequences, which implies a classification step for the results to identify TEs. Moreover, TE families containing very few copies will not be detected.

There are two main approaches that are considered *de novo* methods. In the first approach, the genome sequence is first compared against itself to locate all the repeated sequences. Several programs use BLAST (Altschul *et al.*, 1990) to perform this step. The repeated sequences that are located are then grouped into clusters of similar sequence. A consensus sequence is then built for each cluster, and all the consensus sequences are used in a library-based approach to retrieve all occurrences within the genome. Among the most utilized programs that are currently used in the annotation of genomes, we can cite RECON (Bao and Eddy, 2002), PILER (Edgar and Myers, 2005), and BLASTER (Quesneville, unpublished).

In the second approach, the occurrence of multiple small words known as k-mers is searched for within the genome sequence. The k-mer can then be extended to obtain longer sequences. Among the existing programs using this approach, some have been used to discover TEs in genome sequences, for example, REPUTER (Kurtz and Schleiermacher, 1999), REPEATSCOUT (Price *et al.*, 2005), and REAS (Li *et al.*, 2005). This last program has the peculiarity of running not on an assembled genome but on the unassembled reads of a whole genome shotgun sequence to avoid the problems related to a bad assembly.

TEs in the next-generation sequencing data era

The availability of next-generation DNA sequencing (NGS) technologies has revolutionized our approach to genomics (Margulies *et al.*, 2005). These technologies allow us to obtain huge amounts of data at a relatively low cost and with less bias than older technologies (Wicker *et al.*, 2006), thus opening new avenues to the study of TEs. These new types of data also imply that the classic methods described previously will no longer be adapted. To describe how to deal with TEs in NGS data, it is first important to understand why these data are different from older sequencing data and what methodologies are currently available to handle them before performing the TE studies.

With NGS technologies, not only has the volume of data generated dramatically increased, but the range of applications has broadened from methylation pattern detection (MeDIP-Seq) and the study of DNA–protein interactions (ChIP-Seq) to quantifying and detecting gene expression (RNA-Seq). Whatever the application, three steps can always be highlighted in sequencing methodologies. First, the DNA of interest is randomly fragmented and amplified. Second, the ends of each of these fragments are sequenced into reads. Finally, the original sequence of the

DNA is reconstructed from the reads. Currently, the first two steps are highly automated, and the only concern of the researcher is to determine the appropriate sequencing cost to balance the read length and the depth of coverage necessary for the study. Even if the read size has increased with the development of NGS technologies, the reconstruction of the original DNA sequences (assembly) is currently still the most challenging and time-consuming step.

NGS data are subject to particular artefacts that need to be taken into account before performing analyses. These artefacts are predominantly adaptor sequences originating from failed or short DNA insertions during library preparation or near identical reads originating from PCR error. This step of trimming and filtering can be performed using several tools, such as SeqTrim or Quake (Falgueras *et al.*, 2010; Kelley *et al.*, 2010). Another problem is the presence of sequencing errors. Variations between reads can be caused by real sequencing errors or by single-nucleotide polymorphisms (SNPs), accounting for polyploidy and pooled samples. Thus, it is important to perform error corrections that will be linked to the NGS technologies used to generate the data because the different sequencing methods produce different types of sequencing error.

Sequencing and analysing DNA using NGS

Once the reads are sequenced, an important step is to perform their assembly in order to reconstruct the complete genome. This step is complex but can be eased if a reference genome is available. In such a case, it is possible to map the reads directly onto the reference genome. As the number of reads can be very large, classic mapping programs such as the one from the BLAST suite (Altschul *et al.*, 1990) have become too computationally demanding. Thus, a number of alternative approaches have been developed over the past three years to handle NGS data. Two strategies exist for mapping reads onto a genome. The first uses a hash table of the reads, and the other uses a Burrows-Wheeler (BW) transform of the genome (Schbath *et al.*, 2012). Generally, a hash table can better address mismatches, whereas the more complex BW transform approach can easily handle repeats. There are different alternatives to deal with reads occurring at multiple positions, which are known as multi-reads. The program can either ignore them, keep the best matches, keep a specific number of them, or ignore the ones mapping to more than a specific number of locations (Treangen and Salzberg, 2012). Taking into account all of these considerations, it appears that the programs BWA (Li *et al.*, 2010) and Bowtie (Langmead *et al.*, 2009) can outperform the other programs on many criteria, such as computational time, the correct positions of the reads, the number of unmapped reads, and the multi-reads with no more than three mismatches (Schbath *et al.*, 2012). Other parameters that need to be taken into account in addition to mismatches are indels. For this question, BWA is currently the only mapping program using the BW transform that is able to handle indels. Mapping programs using the BW transform appear to be the most appropriate for the study of TEs because they can better handle genomic repeats. Moreover, even if this class of algorithms relies on heuristics to address mismatches, we expect to have a better error correction and less mismatch on TE sequences because they are sequenced with a better coverage than the rest of the genome. For example, with a coverage of 10×, we expect an average coverage of 50× for a given TE that is present with five copies in the genome.

Most of the time there is no reference genome, and the original DNA sequences have to be reconstructed *de novo*. The mapping approach can still be used with the LAST program, which can take into account the divergence between species to relax the mapping parameters and perform *xeno-mapping* (Frith *et al.*, 2010). Otherwise, two different approaches exist for assembling without a reference genome, the first using a seed approach and the other using a de Bruijn graph. In the first approach, the algorithm tries to elongate short sequences of *k* nucleotides (*k*-mers) using overlapping reads by computing an overlapping graph where all the paths in the graph consist of overlapping reads. Developed for Sanger technologies, the construction of such a graph is often computationally intractable in the case of NGS data (Pevzner *et al.*, 2001). Since the publication of the EULER program, most assemblers use the de Bruijn graph approach to assemble reads (Fig. 9.2). The first step of this approach is to build an index of all the possible sequences of size *k* (*k*-mers, often between 24 and 27 bp). The graph itself is built by adding the information from each read, and the sequence of a read is represented by a path between nodes. The nodes correspond to the *k*-mers and their reverse complements to handle more efficiently the two strands of DNA. The addition of each read will correspond to the addition of more edges between the nodes. The construction of a contig is a byproduct of the graph itself. Retrieving the original DNA sequence is a matter of linearizing the information contained in the graph by following the most supported edge, i.e. the one with the highest number of reads (Pevzner *et al.*, 2001; Zerbino and Birney, 2008).

In practice, even with a high coverage, there are always some parts of a DNA sequence that are more difficult to assemble, such as repetitive elements. If the repeat length is larger than the read length, which is often the case for TEs, then the coverage cannot help to reconstruct the original sequence. In this case, a read cannot be associated with one particular copy of the repeated element. Unassembled or uncovered regions are going to form gaps in the assembly, thus decreasing the connectivity between the resulting sequences. Paired-end sequencing technology can be very useful for solving some of the problems caused by repetitive content and short or very short reads (Treangen and Salzberg, 2012). This type of technology is a specific way of sequencing a DNA fragment at both ends. As the size of the fragments is known, the two resulting reads can be reliably positioned relative to each other. To obtain pairs of reads that are more than 500 bp apart (called the insert size), a specific library must be built (a mate-pair or long paired-end library) (Bentley *et al.*, 2008). With paired-end sequencing, a coverage of 10× is sufficient to have at least one mate-pair spanning every instance of each repeat in the genome and to be able to anchor this repeat if the paired read is uniquely mappable (Wetzel *et al.*, 2011). This mate pair information is very useful to position and order the contigs between each other, to fill the gaps, and to build scaffolds (Fig. 9.3). Contrary to the read length, longer inserts will not correspond to a better assembly (Wetzel *et al.*, 2011). It appears that the best strategy is to use different insert sizes to be able to resolve small repeats with short inserts and long repeats with large inserts. The size of the insert can be specifically tuned to optimally assemble the repeat content of the genome under study (Wetzel *et al.*, 2011). For example, it was possible to obtain the same assembly quality found using the Sanger-based approach by using insert sizes of 180 bp, 3 kb, 6 kb and 40 kb for the mouse and human genomes (Gnerre *et al.*, 2011). As with the classic Sanger sequencing methods, the mis-assembly of the repetitive parts of a genome can lead to numerous errors in the reconstructed sequences (Phillippy *et al.*, 2008). For example, a collapse is formed when the assembler incorrectly joins reads originating from distinct repeat copies. On the contrary, expansions are formed

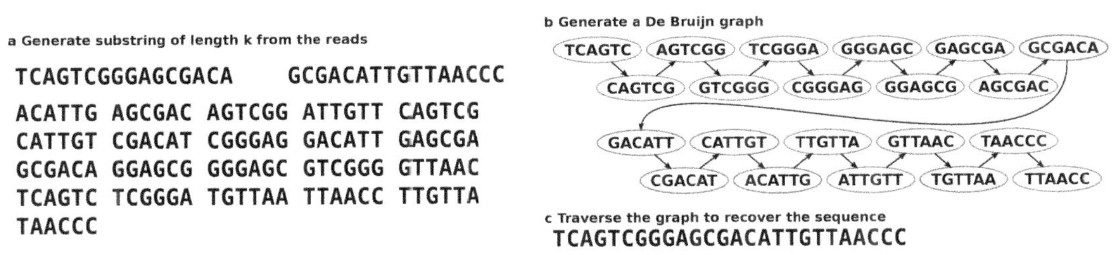

Figure 9.2 Overview of a sequence reconstruction using the de Bruijn graph-based strategy. (a) All substrings of length k (k-mers) were generated from the reads (here k=6). (b) Each unique k-mer represents a node in the de Bruijn graph. The read information is stored in the graph by connecting two nodes with an edge each time that two k-mers have a k – 1 overlap in the reads. Note that generally each node in the graph represents a k-mer and its complement. (c) Nodes from a chain of adjacent nodes link with each other by only one edge and are collapsed into a single node. The graph is then traversed to form contigs.

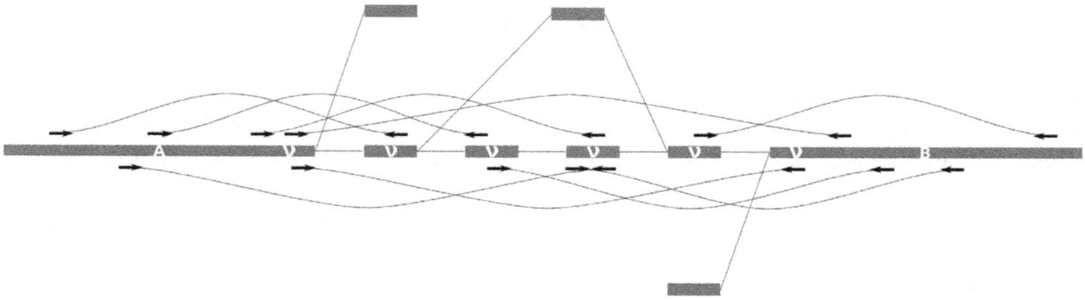

Figure 9.3 Utilization of paired-end information to resolve repeats and to fill gaps. A and B represent long contigs, and the small grey squares are other nodes in the graph. The dashed lines represent all the possible paths built from the read information. The reads are represented by black arrows, and the paired-end information is represented by grey curves linking the two arrowheads to each other. Finding the exact path in the graph from A to B is not a simple task. However, by using the paired-end information linking each pair of reads, we can explore a much simpler graph composed of the 'v' marked blocks.

when reads originating from different copies of a repeat are included in the assembly of one copy of this repeat. These cases are often associated with a greater or lesser density of reads than expected over the repeat(s), and the mate-pairs are stretched out or compressed. Sequence rearrangements are another type of assembly artefact and appear when blocks of DNA are separated by repeats. In this case, the order of these blocks can be wrongly recovered because they are anchored by similar repeats. A special case of rearrangements is inversion, where the two repeat copies are in opposite directions. Sometimes all the above cases of mis-assembly can be possible without violating any constrains on the paired-end information. The amosvalidate package is a collection of tools aimed at detecting mis-assemblies in an automated pipeline (Phillippy *et al.*, 2008).

All the DNA reconstruction problems that we have just described will appear in the de Bruijn graph by forming three types of structures (Zerbino and Birney, 2008). The first corresponds to 'tips', where a chain of nodes is disconnected from the rest of the graph at one end. The second corresponds to 'bubbles', where two paths start from the same node. The third corresponds to erroneous connections, which have no identifiable structure and correspond to chimeric reads. The 'bubble' structures of the graph can correspond to perfect circles, in the case of SNPs or can form densely connected 'tangles' in the case of repetitive sequences (Pevzner *et al.*, 2001). The shape of these 'tangles' can be used for the characterization of the TE, as described in Macas *et al.* (2007). There are many programs that can handle whole genome assemblies using the de Bruijn graph and that are capable of using paired-end information, such as the programs VELVET (Zerbino and Birney, 2008) and ABySS (Simpson *et al.*, 2009). More recently, SOAPDENOVO (Li *et al.*, 2010) was used for the *de novo* assembly of the giant panda genome with an average of 20× coverage using 52 bp reads and 37 paired-end libraries with insert sizes ranging from 150 bp to 10 kb. This genome contains 36.1% TEs. The most recent program, ALLPATHS-LG (Gnerre *et al.*, 2011), was created to account for paired-end information with mixed insert size libraries. This program produces a better assembly quality than SOAPDENOVO but it is slower (for mammalian-sized genomes, it requires three weeks when SOAPDENOVO requires only three days (Gnerre *et al.*, 2011)). Even if the development of assemblers allows for an ever improving assembly of genomes, including their repeat content, there will always remain some unassembled fractions in the data. The study of these leftover contigs or reads can be a source of information about the TE content. These analyses can be conducted using mapping approaches of the data to known TE databases of DNA or protein sequences (Sun *et al.*, 2012).

Sequencing and analysing RNAs using NGS

New sequencing technologies have opened up new vistas of knowledge at the various levels of an organism's biology. Having access to an entire genome is particularly valuable, but having access to the landscape of gene expression allows us to go deeper into the analysis of genomes. Thus, RNA-Seq technology has been developed, which predominantly consists of adding a reverse transcription step to transform RNA into DNA before performing the regular NGS steps (Marioni *et al.*, 2008). The goal of this approach is to obtain the sequence and the abundance of all the transcripts of a given organism in a given condition. Using these types of sequencing methods, it is important to handle particular artefacts. In addition to the errors generated by the reverse transcription step, the reconstruction of the RNA sequences accounts for the different levels of expression for the different transcripts and for the mechanism of alternative splicing (Martin and Wang, 2011). Because of the presence of shared exons in different transcripts, expression counting or quantification is not trivial for RNA-Seq data (Trapnell *et al.*, 2010). Moreover, there is a competition between the most abundant transcripts, which can be over-sequenced, and the less abundant ones, which can be missed. To sequence and quantify less abundant transcripts, hybridization-based depletion methods to remove the more abundant transcripts can be utilized (He *et al.*, 2010). However, these depletion methods induce biases for the quantification and for the assembly of the most expressed transcripts.

Once the main artefacts are corrected, there are two strategies to analyse the RNA-Seq data. The first approach, known as 'map first', consists of mapping the reads onto a reference genome. This approach can be confronted with many problems, ranging from the correctness of the alignment, the possibility of losing the splicing information and the completeness of the reference genome. There are many programs for mapping RNA-Seq data to a reference genome, but it appears that the most commonly utilized is a combination of the program TopHat, which is able to discover splice junctions, and the program Cufflinks, which can be used for the quantification of transcripts (Trapnell *et al.*, 2010). The second approach, 'assembly first' (*de novo* method), consists of directly assembling reads to reconstruct the transcripts. Using this approach, the resultant transcripts can subsequently be mapped to a reference genome. A number of transcriptome assembly programs have been developed, and most use the de Bruijn graph approach. This data structure naturally handles the high redundancy of the data because each repeat or transcript is only present once in the graph. The most commonly used transcriptome assemblers are Velvet (Zerbino and Birney, 2008) and Trans-ABySS (Robertson *et al.*, 2010). More recently, the Trinity program was proposed (Grabherr *et al.*, 2011). Other approaches, such as KISSPLICE (Sacomoto *et al.*, 2012), have been developed to identify and quantify *de novo* polymorphisms such as alternative splicing, SNP and tandem repeats in RNA-Seq data.

Even with a reference genome, the best results are obtained using mixed strategies of the two approaches (Surget-Groba and Montoya-Burgos, 2010). Based on the confidence we have in the reference genome, there are two possibilities. When the reference genome is of a very good quality, it is possible to first align the reads onto it and then to assemble the reads using the mapped reads as long contigs. The assembling step allows for resolving of the reads coming from the expressed regions not present in the reference genome. In a case where the reference genome is not of a very good quality, another strategy is used, consisting of first making the assembly of reads before mapping them onto the reference genome. In this case, the errors present in the reference genome have little impact because they are not present in the assembled contigs. The mapping step is then used to resolve scaffolds from the more fragmented contigs obtained using the *de novo* approaches. This last approach was used to successfully assemble the transcriptome of the mosquito *Anopheles funestus* (Crawford *et al.*, 2010)

What do NGS data bring to TE analyses?

With the new technologies of DNA and RNA sequencing, new opportunities to study TEs have appeared, once the difficulty for handling these repeated sequences taken into account when processing the data. In this section, we will exemplify different analyses that have allowed us to improve on our understanding of TEs.

TE identification in a genome survey

Even if the cost of NGS sequencing has continuously diminished it is still of interest to perform genomic surveys to characterize large genomes. The historic approach of using the end sequences of bacterial artificial chromosome (BAC) vectors is biased towards sequences that can be successfully loaded into BACs. NGS technologies allow us to perform more representative surveys of large genomes by sequencing at a low depth using a whole genome shotgun (WGS) approach. This approach was first used to further characterize the soybean genome, with particular attention paid to its repeat content (Swaminathan *et al.*, 2007). The first step of the data analysis is to characterize a maximum number of reads using databases of annotated sequences. For this step, classical programs such as BLASTX and BLASTN may be used. For example, in the survey analysis of the barley genome using 454 sequencing at 0.1× coverage, it was possible to determine 7.4% of the barley genes using BLASTX on a database of predicted rice proteins and to characterize the presence of many TEs using BLASTN on a TE plant database (Wicker *et al.*, 2009).

Another characteristic resulting from genome sampling with WGS and NGS technologies is to expect an increased coverage of the repetitive content. For example, with a genomic coverage of 0.01×, we can expect a coverage of 10× for each repeat occurring in the genome with 1000 copies (Macas *et al.*, 2007). With this characteristic, the identification of repeat content using a NGS survey is not limited to a homology search. The first program developed to use the expected increase in coverage for TE sequences is REAS, which has already been referenced in section 1 of this chapter and allows us to assemble consensus TE sequences (Li *et al.*, 2005). The TEs must exist at a sufficient copy number to be recognized by their read number and must not be too degraded to have sufficient sequence similarities between their copies to be able to build a consensus sequence. The REAS program starts by building a k-mer index of the reads. The high copy number k-mers are then picked out to retrieve the reads containing them. The reads are then assembled and expanded to recover the consensus sequences of the different elements. When possible, the contigs formed are linked using the paired-end information. The main drawback of this method is the fact that it is not designed for short or very short reads and cannot process reads less than 104 bp (Macas *et al.*, 2007). The AAARF (Assisted Automated Assembler of Repeat Families) program was designed to overcome this problem and can process short or very short reads (DeBarry *et al.*, 2008). This program uses one read as a query sequence to obtain its nucleotide coverage against the rest of the dataset using BLASTN. This nucleotide coverage is then used to select the overlapping reads, which are then aligned using CLUSTALW to build a new query sequence. This program iteratively elongates each query sequence and assembles a set of TE contigs.

To recover repetitive sequences, it is also possible to cluster overlapping unannotated DNA sequences. This method was used to assemble 41% of the reads of the soybean genome into contigs using the PHRAP program (Swaminathan *et al.*, 2007). A similar approach was used to assemble 31.6% of the unannotated reads into contigs for the barley genome survey (Wicker *et al.*, 2009). However, using this approach and because of the low depth of the data, some links are absent in the overlapping graphs, leading to contig fragmentation (Novák *et al.*, 2010). A slightly different approach was used to reconstruct the repeat sequences from a survey of the pea genome (Macas *et al.*, 2007). In this study, the program TCLUST from the TGICL package was used to cluster the reads based on a mutual similarity and to assemble each cluster into contigs. With this method, each cluster contains related repeat sequences, which can be used to better characterize the variability in the TE content of a genome. This method was successfully used to characterize the repeat content of the banana genome (Hribová *et al.*, 2010). One drawback of this more sensible approach is the formation of chimeric clusters, which are caused by the presence of reads spanning two TE sequences and form bridges between two clusters (Macas *et al.*, 2007). More recently, a novel cluster-based approach was proposed with the program SeqGrapheR (Novák *et al.*, 2010). This program uses a hierarchical agglomeration algorithm to cluster the reads and to characterize the TE sequences. This graph-based clustering allows for a better segregation of groups of unrelated sequences than TCLUST. Moreover, it allows for a better characterization of the cluster structure by computing various graph metrics to discriminate between different types of repeats. The assembly of TE sequences from a cluster results in the formation of consensus sequences. With this type of sequence representation, a significant amount of information about TE sequence variability is lost. SeqGraphR can provide an alternative representation of TEs by direct graph visualization. This approach can be very useful for deciphering contig assembly or for distinguishing between two closely related TEs. SeqGraphR was successfully used to characterize the TE content of three species of *Nicotiana tabacum* in a genome survey using 454 sequencing at 0.1× (Renny-Byfield *et al.*, 2011).

All previous methods work for genome survey data with relatively small dataset sizes, ranging from 33 Mb to 90 Mb, but may not be suitable for larger datasets. Moreover, approaches using read overlap information will only work for sparse genome survey data, where the only sequences that can be assembled come from repeat sequences. If the coverage reaches or exceeds 1×, most of the genome can be assembled, and the described approaches will lose their specificity for TE sequences. However, a genome survey dataset can always be generated from a deep genome sequencing output by randomly selecting a subset of reads to easily study the most abundant repeats. LTR retrotransposons can also be *de novo* identified using mapping approaches. For this class of TEs, the reads will pileup on the two LTRs, and a 'batman ears'-like structure will appear when using programs such as Tallymer or JellyFish if we allow for multi-read mapping (Kurtz *et al.*, 2008; Marçais and Kingsford, 2011).

Structural variant detection

One of the most interesting characteristics of TEs is their ability to replicate and to colonize a genome. This transposition activity can be studied between species or populations. NGS technologies allow us to study copy number variation (CNV) by sequencing pooled DNA from different individuals (or pool-Seq). The main advantages of these approaches are that the approaches are fast and not copy specific and offer a higher sensitivity compared to other technologies (Alkan *et al.*, 2011). To perform these types of studies, one needs an assembled reference genome, a database of the TE sequences (which can be built from the sequencing data) and, optionally, paired-end technology to better resolve the TE information.

For this purpose, the T-lex program was developed to compute the population frequencies of individual TE insertions (Fiston-Lavier *et al.*, 2011). This program is a pipeline using four modules. The first module uses RepeatMasker to identify TEs and their flanking regions in the reference genome. The second module uses MAQ (Li *et al.*, 2008) to determine the presence of TEs by mapping reads across the sequences formed by an identified TE and its flanking regions in the reference genome. The third module uses SHRiMP (Rumble *et al.*, 2009), which can align sequences with long gaps, to identify the absence of a TE insertion in the analysed populations by mapping reads spanning only the two flanking regions of the TE sequence in the reference genome. Finally, the last module combines the information of the previous modules to obtain the frequencies of each TE family in the populations. The second version of the T-lex program is able to automatically use paired-end information to detect novel TE insertions. By using a similar approach to T-lex, it was possible to successfully analyse the activity of TEs using pooled DNA samples from 114 isofemale lines of *D. melanogaster* (Kofler *et al.*, 2012). For this study, 80 million paired-end fragments were produced with the Illumina Genome Analyser IIx. These reads were mapped onto the reference genome, where all the repeats were first masked using RepeatMasker. The mapping step was then performed using the BWA-SW program (Li *et al.*, 2010). The authors were then able to identify novel insertions with at least three paired couples of reads, with, for each couple, one read mapping to a genomic locus and the others mapping to a TE sequence. In this study, novel TE insertions were detected if they were present in at least 7% of the populations.

Without a reference genome, comparative studies of the TE content of different species can also be achieved using genome surveys. For example, an analysis of genomic gigantism in plethodontid salamanders was performed using a genome survey of six species sequenced with a coverage of 0.1× using 454 technology to obtain reads of a maximum length of 400 bp (Sun *et al.*, 2012). In this study, the REPEATMODELER program (Smit and Hubley 2008–2010) was used to identify those repeats covered by a minimum of four reads. The REPCLASS program (Feschotte *et al.*, 2009) was then used to further classify the unknown repeats. The results demonstrated that the analysed salamander species accumulate large amounts of LTR-retrotransposons compared to other vertebrates.

Analysis of TE regulation by the host

With the prevalence of TEs and their capacity to invade a genome, it is crucial for the host to be able to regulate their activity to avoid too many deleterious effects. During the past few years, the links between TEs and the epigenetic systems of regulation, such as DNA methylation, histone modifications, and RNA interference, have been shown to be linked to the repeat content of a genome (Siomi and Siomi, 2008; Rebollo *et al.*, 2010). In particular, diverse RNA-mediated defences have been discovered in different eukaryotic organisms (Slotkin and Martienssen, 2007; Blumenstiel, 2011). These discoveries have allowed for the development of new models for TE dynamics in natural populations in which four phases have been described: an initial phase of TE invasion; a second phase of TE proliferation, leading to the appearance of TE insertion alleles initiating the production of small RNAs; and finally a quiescent state, leading to the stabilization of TE copy number (see (Blumenstiel, 2011) for a review).

With the development of NGS technologies, it has become easier to analyse the epigenetic control of TEs. Particular modifications can regulate TE activity, such as DNA methylation. This type of modification can be determined using a MeDIP-seq approach and has already been used in several organisms. For example, in black cottonwood, the TEs possessed variable methylation according to their family, with LTR retrotransposons being globally more methylated than other classes (Vining *et al.*, 2012).

RNA-Seq is also a reliable way to study active TEs because the complete RNA sequences of the TEs can be found in the data output. The control of TEs can also be studied when considering small RNAs such as piRNAs or siRNAs, which are expected to be copy specific to the TEs that they control. This type of analysis can also be used to validate *de novo* annotations of TEs. The results of these approaches are naturally linked with the condition or the stage where the different TEs are expressed. Due to piRNA regulation, the quantification aspect of RNA-Seq is of a lesser interest for the study of TEs because the number of transcripts is not directly correlated with the activity of a TE (Brennecke *et al.*, 2007). However, these studies allow us to obtain information on the potentially complete and active copies that are inserted into the genome. The analysis of small RNAs is typically performed by mapping the reads on to a reference genome to help identify clusters of small RNAs and to determine which copy in the genome is associated with a particular small RNA. This approach has been used in the analysis of the control of particular TEs in *D. melanogaster* (Brennecke *et al.*, 2007; Brennecke *et al.*, 2008; Grentzinger *et al.*, 2012) and in plant species (Hollister *et al.*, 2011). Of course, these genome-mapping approaches have limitations due to the differences that exist between individuals. For example, in the case of an analysis of *P*-elements, it was not possible to map them onto the *D. melanogaster* reference genome because this particular element is absent in the sequenced strain (Brennecke *et al.*, 2008). Thus, it was necessary to find another strategy, in this case, using those reads that did not map onto the reference genome.

The regulation of TEs can also be linked to the histone modifications of the DNA. Using ChIP-seq data, it is possible to determine what types of modifications are associated with TEs or with genes given their TE neighbourhood. For example, in mouse embryonic stem cells, an analysis of ChIP-seq data revealed that the majority of gene promoters surrounded by numerous TEs were depleted of the bivalent marks H3K27me3 + H3K4me3 compared to genes surrounded by few or no TEs. This bivalent mark has been demonstrated to be specific to a 'poised state' of developmental genes that are temporarily repressed in embryonic stem cells but that will be activated later during development (Zhang and Mager, 2012). This previous analysis indirectly observed modifications associated with TEs. Generally, during the mapping step of reads sequenced using ChIP-seq analysis, the reads cannot be associated with TE copies because only uniquely mappable reads are conserved (those having a unique location on the genome). Thus, an alternative mapping approach was proposed by Huda *et al.* to allow for the direct examination of ChIP-seq reads associated with TEs (Huda *et al.* 2010). In that analysis, the authors took advantage of several ChIP-seq experiments, which allowed them access to a genome-wide map of 38 histone modifications in human CD4^{+} T cells (Barski *et al.*, 2007; Wang *et al.*, 2008). They used the MAQ program to align the reads, allowing for redundant genomic locations. It was then possible to characterize what type of TE was present in the mapped reads. Their results demonstrated a high variation in TE histone modifications according to the TE family, with the older TE families and the TEs close to genes carrying more modifications than the younger TE families and those TEs distal to genes.

These different examples demonstrate how valuable NGS data are in the analysis of TEs, and also identify the need for specific tools to handle these data for studying TEs.

Conclusions

Transposable elements are important components of genomes that cannot simply be put aside when analysing genomes. It is important to understand how TEs function and evolve to better understand all of the impacts of TEs on genome evolution. Given the large amount of genomic data that has been continuously generated, this task becomes more and more difficult. However, these data allow us an access to new information that we did not have several years ago.

Since the first sequencing projects were undertaken, efforts have been made to develop programs that allow for the detection and analysis of TE sequences. Various programs have been developed that use different or complementary approaches to detect TE sequences. These tools have very different performances, but even the best ones cannot discover all the TE sequences in a genome because each has its own drawback(s) that prevents it from finding each and every TE (Saha *et al.*, 2008b; Lerat, 2010). Thus, the best approach to exhaustively describe the landscape of TEs in a genome is to use several of these different programs and to cross-reference the results. Similarly, the best approach to locate TE sequences in complete genomes appears to reside in the use of pipelines of programs. For example, the RepeatModeler pipeline includes different programs to build, refine and classify consensus sequences of putative interspersed repeats (Smit and Hubley, 2008–2010). The REPET pipeline has been built to integrate the findings of similarity- and *de novo*-based programs (Quesneville *et al.*, 2005). This pipeline was recently updated to retain those programs that provide the best results after the authors tested different *de novo* programs (Flutre *et al.*, 2011). Other pipelines have generally been developed to answer specific questions (see Lerat, 2010).

In all cases, another important step after the identification of putative TE sequences is the classification of the repeats into families. This is a difficult step because it must take into account the biological aspect of TEs, such as the fact that some copies can be fragmented and thus not only full-length elements exist in a genome and that TEs often insert inside each other, producing what are known as nested TEs. Some programs have been developed to integrate the classification step, such as the TEclass program, which tries to determine the main classes of unknown elements using machine-learning algorithms (Abrusán *et al.*, 2009), or the REPCLASS program, which uses different approaches to annotate TEs (Feschotte *et al.*, 2009).

With the new type of genomics data generated by next-generation sequencing technologies, it is necessary to develop new approaches to detect and analyse TEs. Indeed, most of the programs that have been designed for classic genomic data cannot handle these new types of data. This is predominantly because NGSs produce small sequence fragments, which increase the difficulties involved in assembling the repeat content of a genome, but also because the amounts of these data are too large to be handled with the existing tools. However, new programs have been designed to take these problems into account. Even if all these programs are not specific to TEs, specific programs are now available to answer particular questions on TEs with regards to population genomics, and we can hope for new developments in more specific areas. Questions about the epigenetic regulation of TEs are particularly important at different levels, such as the impact of TEs in cancer development in humans. NGS data now offer the possibility of having access to this information, which will necessitate the development of particular tools specific for TEs.

Future trends

With the development of NGS data, access to individuals' sequence data should provide us with valuable information on the specific content of TEs, allowing us to be more precise in terms of the insertion profiles of TEs. Currently, the only available possibility is to compare data to reference genomes. However, this shortcut is not precise enough when delving deeper into our understanding of the mechanisms of TE dynamics.

Web resources

A list of existing TE detection tools is available at:

- http://bergmanlab.smith.man.ac.uk/?page_id=295
- Quesneville. *BLASTER suite* <http://urgi.versailles.inra.fr/Tools/Blaster>.
- Smit, AFA, Hubley, R and Green, P. *RepeatMasker Open-3.0.* 1996–2010 <http://www.repeatmasker.org>.
- Smit, AFA, Hubley, R. *RepeatModeler Open-1.0.* 2008–2010 <http://www.repeatmasker.org>.

References

Abrusán, G., Grundmann, N., DeMester, L., and Makalowski, W. (2009). Teclass – a tool for automated classification of unknown eukaryotic transposable elements. Bioinformatics *25*, 1329–1330.

Alkan, C., Coe, B.P., and Eichler, E.E. (2011). Genome structural variation discovery and genotyping. Nat. Rev. Genet. *12*, 363–376.

Altschul, S.F., Gish, W., Miller, W., Myers, E.W., and Lipman, D.J. (1990). Basic local alignment search tool. J. Mol. Biol. *215*, 403–410.

Bao, Z., and Eddy, S.R. (2002). Automated *de novo* identification of repeat sequence families in sequenced genomes. Genome Res. *12*, 1269–1276.

Barski, A., Cuddapah, S., Cui, K., Roh, T.-Y., Schones, D.E., Wang, Z., Wei, G., Chepelev, I., and Zhao, K. (2007). High-resolution profiling of histone methylations in the human genome. Cell *129*, 823–837.

Bentley, D.R., Balasubramanian, S., Swerdlow, H.P., Smith, G.P., Milton, J., Brown, C.G., Hall, K.P., Evers, D.J., Barnes, C.L., Bignell, H.R., *et al.* (2008). Accurate whole human genome sequencing using reversible terminator chemistry. Nature *456*, 53–59.

Bergman, C.M., and Quesneville, H. (2007). Discovering and detecting transposable elements in genome sequences. Brief. Bioinform. *8*, 382–392.

Biémont, C. (2010). A brief history of the status of transposable elements: from junk DNA to major players in evolution. Genetics *186*, 1085–1093.

Biémont, C., and Vieira, C. (2006). Genetics: junk DNA as an evolutionary force. Nature *443*, 521–524.

Blumenstiel, J.P. (2011). Evolutionary dynamics of transposable elements in a small RNA world. Trends Genet. *27*, 23–31.

Bowen, N.J., and McDonald, J.F. (2001). Drosophila euchromatic LTR retrotransposons are much younger than the host species in which they reside. Genome Res. *11*, 1527–1540.

Brennecke, J., Aravin, A.A., Stark, A., Dus, M., Kellis, M., Sachidanandam, R., and Hannon, G.J. (2007). Discrete small RNA-generating loci as master regulators of transposon activity in Drosophila. Cell *128*, 1089–1103.

Brennecke, J., Malone, C.D., Aravin, A.A., Sachidanandam, R., Stark, A., and Hannon, G.J. (2008). An epigenetic role for maternally inherited piRNAs in transposon silencing. Science *322*, 1387–1392.

Crawford, J.E., Guelbeogo, W.M., Sanou, A., Traoré, A., Vernick, K.D., Sagnon, N., and Lazzaro, B.P. (2010). *De novo* transcriptome sequencing in *Anopheles funestus* using Illumina RNA-seq technology. PLoS One *5*, e14202.

Cutter, A.D. (2008). Divergence times in Caenorhabditis and Drosophila inferred from direct estimates of the neutral mutation rate. Mol. Biol. Evol. *25*, 778–786.

DeBarry, J.D., Liu, R., and Bennetzen, J.L. (2008). Discovery and assembly of repeat family pseudomolecules from sparse genomic sequence data using the assisted automated assembler of repeat families (AAARF) algorithm. BMC Bioinformatics *9*, 235.

Dowsett, A.P., and Young, M.W. (1982). Differing levels of dispersed repetitive DNA among closely related species of Drosophila. Proc. Natl. Acad. Sci. U.S.A. *79*, 4570–4574.

Edgar, R.C., and Myers, E.W. (2005). PILER: identification and classification of genomic repeats. Bioinformatics *21* (Suppl 1), i152–i158.

Eickbush, T.H., and Furano, A.V. (2002). Fruit flies and humans respond differently to retrotransposons. Curr. Opin. Genet. Dev. *12*, 669–674.

Ellinghaus, D., Kurtz, S., and Willhoeft, U. (2008). LTRharvest, an efficient and flexible software for *de novo* detection of LTR retrotransposons. BMC Bioinformatics *9*, 18.

Evgen'ev, M.B., Zelentsova, H., Poluectova, H., Lyozin, G.T., Veleikodvorskaja, V., Pyatkov, K.I., Zhivotovsky, L.A., and Kidwell, M.G. (2000). Mobile elements and chromosomal evolution in the virilis group of Drosophila. Proc. Natl. Acad. Sci. U.S.A. *97*, 11337–11342.

Falgueras, J., Lara, A.J., Fernández-Pozo, N., Cantón, F.R., Pérez-Trabado, G., and Claros, M.G. (2010). SeqTrim: a high-throughput pipeline for pre-processing any type of sequence read. BMC Bioinformatics *11*, 38.

Feschotte, C., Keswani, U., Ranganathan, N., Guibotsy, M.L., and Levine, D. (2009). Exploring repetitive DNA landscapes using REPCLASS, a tool that automates the classification of transposable elements in eukaryotic genomes. Genome Biol. Evol. *1*, 205–220.

Fiston-Lavier, A.-S., Carrigan, M., Petrov, D.A., and González, J. (2011). T-lex: a program for fast and accurate assessment of transposable element presence using next-generation sequencing data. Nucleic Acids Res. *39*, e36.

Flutre, T., Duprat, E., Feuillet, C., and Quesneville, H. (2011). Considering transposable element diversification in *de novo* annotation approaches. PLoS One *6*, e16526.

Frith, M.C., Wan, R., and Horton, P. (2010). Incorporating sequence quality data into alignment improves DNA read mapping. Nucleic Acids Res. *38*, e100.

Gnerre, S., Maccallum, I., Przybylski, D., Ribeiro, F.J., Burton, J.N., Walker, B.J., Sharpe, T., Hall, G., Shea, T.P., Sykes, S., *et al.* (2011). High-quality draft assemblies of mammalian genomes from massively parallel sequence data. Proc. Natl. Acad. Sci. U.S.A. *108*, 1513–1518.

Grabherr, M.G., Haas, B.J., Yassour, M., Levin, J.Z., Thompson, D.A., Amit, I., Adiconis, X., Fan, L., Raychowdhury, R., Zeng, Q., *et al.* (2011). Full-length transcriptome assembly from RNA-Seq data without a reference genome. Nat. Biotechnol. *29*, 644–652.

Grentzinger, T., Armenise, C., Brun, C., Mugat, B., Serrano, V., Pelisson, A., and Chambeyron, S. (2012). piRNA-mediated transgenerational inheritance of an acquired trait. Genome Res. *22*, 1877–1888.

Han, Y., and Wessler, S.R. (2010). MITE-Hunter: a program for discovering miniature inverted-repeat transposable elements from genomic sequences. Nucleic Acids Res. *38*, e199.

Hancks, D.C., and Kazazian, H.H. (2012). Active human retrotransposons: variation and disease. Curr. Opin. Genet. Dev. *22*, 191–203.

He, S., Wurtzel, O., Singh, K., Froula, J.L., Yilmaz, S., Tringe, S.G., Wang, Z., Chen, F., Lindquist, E.A., Sorek, R., *et al.* (2010). Validation of two ribosomal RNA removal methods for microbial metatranscriptomics. Nat. Methods *7*, 807–812.

Hollister, J.D., Smith, L.M., Guo, Y.-L., Ott, F., Weigel, D., and Gaut, B.S. (2011). Transposable elements and small RNAs contribute to gene expression divergence between *Arabidopsis thaliana* and *Arabidopsis lyrata*. Proc. Natl. Acad. Sci. U.S.A. *108*, 2322–2327.

Hribová, E., Neumann, P., Matsumoto, T., Roux, N., Macas, J., and Dolezel, J. (2010). Repetitive part of the banana (*Musa acuminata*) genome investigated by low-depth 454 sequencing. BMC Plant Biol. *10*, 204.

Hu, T.T., Pattyn, P., Bakker, E.G., Cao, J., Cheng, J.-F., Clark, R.M., Fahlgren, N., Fawcett, J.A., Grimwood, J., Gundlach, H., *et al.* (2011). The *Arabidopsis lyrata* genome sequence and the basis of rapid genome size change. Nat. Genet. *43*, 476–481.

Huda, A., Mariño-Ramírez, L., and Jordan, I.K. (2010). Epigenetic histone modifications of human transposable elements: genome defense versus exaptation. Mob. DNA *1*, 2.

Janicki, M., Rooke, R., and Yang, G. (2011). Bioinformatics and genomic analysis of transposable elements in eukaryotic genomes. Chromosome Res. *19*, 787–808.

Jurka, J., Kapitonov, V.V., Pavlicek, A., Klonowski, P., Kohany, O., and Walichiewicz, J. (2005). Repbase Update, a database of eukaryotic repetitive elements. Cytogenet. Genome Res. *110*, 462–467.

Kapitonov, V.V., and Jurka, J. (2008). A universal classification of eukaryotic transposable elements implemented in Repbase. Nat. Rev. Genet. *9*, 411–412.

Kelley, D.R., Schatz, M.C., and Salzberg, S.L. (2010). Quake: quality-aware detection and correction of sequencing errors. Genome Biol. *11*, R116.

Kim, J.M., Vanguri, S., Boeke, J.D., Gabriel, A., and Voytas, D.F. (1998). Transposable elements and genome organization: a comprehensive survey of retrotransposons revealed by the complete *Saccharomyces cerevisiae* genome sequence. Genome Res. *8*, 464–478.

Kimura, K., and Kidwell, M.G. (1994). Differences in *P* element population dynamics between the sibling species *Drosophila melanogaster* and *Drosophila simulans*. Genet. Res. *63*, 27–38.

Kofler, R., Betancourt, A.J., and Schlötterer, C. (2012). Sequencing of pooled DNA samples (Pool-Seq) uncovers complex dynamics of transposable element insertions in *Drosophila melanogaster*. PLoS Genet. *8*, e1002487.

Kurtz, S., and Schleiermacher, C. (1999). REPuter: fast computation of maximal repeats in complete genomes. Bioinformatics *15*, 426–427.

Kurtz, S., Narechania, A., Stein, J.C., and Ware, D. (2008). A new method to compute K-mer frequencies and its application to annotate large repetitive plant genomes. BMC Genomics *9*, 517.

Lander, E.S., Linton, L.M., Birren, B., Nusbaum, C., Zody, M.C., Baldwin, J., Devon, K., Dewar, K., Doyle, M., FitzHugh, W., *et al.* (2001). Initial sequencing and analysis of the human genome. Nature *409*, 860–921.

Langmead, B., Trapnell, C., Pop, M., and Salzberg, S.L. (2009). Ultrafast and memory-efficient alignment of short DNA sequences to the human genome. Genome Biol. *10*, R25.

Lerat, E. (2010). Identifying repeats and transposable elements in sequenced genomes: how to find your way through the dense forest of programs. Heredity *104*, 520–533.

Lerat, E., Rizzon, C., and Biémont, C. (2003). Sequence divergence within transposable element families in the *Drosophila melanogaster* genome. Genome Res. *13*, 1889–1896.

Li, H., Ruan, J., and Durbin, R. (2008). Mapping short DNA sequencing reads and calling variants using mapping quality scores. Genome Res. *18*, 1851–1858.

Li, R., Ye, J., Li, S., Wang, J., Han, Y., Ye, C., Wang, J., Yang, H., Yu, J., Wong, G.K.-S., *et al.* (2005). ReAS: Recovery of ancestral sequences for transposable elements from the unassembled reads of a whole genome shotgun. PLoS Comput. Biol. *1*, e43.

Li, R., Fan, W., Tian, G., Zhu, H., He, L., Cai, J., Huang, Q., Cai, Q., Li, B., Bai, Y., *et al.* (2010). The sequence and *de novo* assembly of the giant panda genome. Nature *463*, 311–317.

Li, Y.J., Satta, Y., and Takahata, N. (1999). Paleo-demography of the *Drosophila melanogaster* subgroup: application of the maximum likelihood method. Genes Genet. Syst. *74*, 117–127.

Lisch, D.R., and Kidwell, M.G. (2000). Transposable elements and host genome evolution. Trends Ecol. Evol. *15*, 95–99.

Lockton, S., and Gaut, B.S. (2010). The evolution of transposable elements in natural populations of self-fertilizing *Arabidopsis thaliana* and its outcrossing relative *Arabidopsis lyrata*. BMC Evol. Biol. *10*, 10.

Lucier, J.-F., Perreault, J., Noël, J.-F., Boire, G., and Perreault, J.-P. (2007). RTAnalyzer: a web application for finding new retrotransposons and detecting *L1* retrotransposition signatures. Nucleic Acids Res. *35*, W269–W274.

Macas, J., Neumann, P., and Navrátilová, A. (2007). Repetitive DNA in the pea (*Pisum sativum* L.) genome: comprehensive characterization using 454 sequencing and comparison to soybean and *Medicago truncatula*. BMC Genomics *8*, 427.

Margulies, M., Egholm, M., Altman, W.E., Attiya, S., Bader, J.S., Bemben, L.A., Berka, J., Braverman, M.S., Chen, Y.-J., Chen, Z., *et al.* (2005). Genome sequencing in microfabricated high-density picolitre reactors. Nature *437*, 376–380.

Marioni, J.C., Mason, C.E., Mane, S.M., Stephens, M., and Gilad, Y. (2008). RNA-seq: an assessment of technical reproducibility and comparison with gene expression arrays. Genome Res. *18*, 1509–1517.

Martin, J.A., and Wang, Z. (2011). Next-generation transcriptome assembly. Nat. Rev. Genet. *12*, 671–682.

Marçais, G., and Kingsford, C. (2011). A fast, lock-free approach for efficient parallel counting of occurrences of k-mers. Bioinformatics *27*, 764–770.

McClintock, B. (1956). Controlling elements and the gene. Cold Spring Harb. Symp. Quant. Biol. *21*, 197–216.

Novák, P., Neumann, P., and Macas, J. (2010). Graph-based clustering and characterization of repetitive sequences in next-generation sequencing data. BMC Bioinformatics *11*, 378.

Ohshima, K., Hattori, M., Yada, T., Gojobori, T., Sakaki, Y., and Okada, N. (2003). Whole-genome screening indicates a possible burst of formation of processed pseudogenes and Alu repeats by particular L1 subfamilies in ancestral primates. Genome Biol. *4*, R74.

Pevzner, P.A., Tang, H., and Waterman, M.S. (2001). An Eulerian path approach to DNA fragment assembly. Proc. Natl. Acad. Sci. U.S.A. *98*, 9748–9753.

Phillippy, A.M., Schatz, M.C., and Pop, M. (2008). Genome assembly forensics: finding the elusive mis-assembly. Genome Biol. *9*, R55.

Price, A.L., Jones, N.C., and Pevzner, P.A. (2005). *De novo* identification of repeat families in large genomes. Bioinformatics *21* (Suppl 1), i351–i358.

Quesneville, H., Bergman, C.M., Andrieu, O., Autard, D., Nouaud, D., Ashburner, M., and Anxolabehere, D. (2005). Combined evidence annotation of transposable elements in genome sequences. PLoS Comput. Biol. *1*, 166–175.

Rebollo, R., Horard, B., Hubert, B., and Vieira, C. (2010). Jumping genes and epigenetics: towards new species. Gene *454*, 1–7.

Renny-Byfield, S., Chester, M., Kovařík, A., Comber, S.C.L., Grandbastien, M.-A., Deloger, M., Nichols, R.A., Macas, J., Novák, P., Chase, M.W., *et al.* (2011). Next generation sequencing reveals genome downsizing in allotetraploid *Nicotiana tabacum*, predominantly through the elimination of paternally derived repetitive DNAs. Mol. Biol. Evol. *28*, 2843–2854.

Robertson, G., Schein, J., Chiu, R., Corbett, R., Field, M., Jackman, S.D., Mungall, K., Lee, S., Okada, H.M., Qian, J.Q., *et al.* (2010). De novo assembly and analysis of RNA-seq data. Nat. Methods *7*, 909–912.

Rumble, S.M., Lacroute, P., Dalca, A.V., Fiume, M., Sidow, A., and Brudno, M. (2009). SHRiMP: accurate mapping of short color-space reads. PLoS Comput. Biol. *5*, e1000386.

Sacomoto, G.A.T., Kielbassa, J., Chikhi, R., Uricaru, R., Antoniou, P., Sagot, M.-F., Peterlongo, P., and Lacroix, V. (2012). KISSPLICE: *de-novo* calling alternative splicing events from RNA-seq data. BMC Bioinformatics *13* (Suppl 6), S5.

Saha, S., Bridges, S., Magbanua, Z., and Peterson, D. (2008a) Computational approaches and tools used in identification of dispersed repetitive DNA sequences. Trop. Plant Biol. *1*, 85–96.

Saha, S., Bridges, S., Magbanua, Z.V., and Peterson, D.G. (2008b) Empirical comparison of *ab initio* repeat finding programs. Nucleic Acids Res. *36*, 2284–2294.

Schbath, S., Martin, V., Zytnicki, M., Fayolle, J., Loux, V., and Gibrat, J.-F. (2012). Mapping reads on a genomic sequence: an algorithmic overwiew and a practical comparative analysis. J. Comput. Biol. *19*, 796–813.

Schnable, P.S., Ware, D., Fulton, R.S., Stein, J.C., Wei, F., Pasternak, S., Liang, C., Zhang, J., Fulton, L., Graves, T.A., *et al.* (2009). The B73 maize genome: complexity, diversity, and dynamics. Science *326*, 1112–1115.

Simpson, J.T., Wong, K., Jackman, S.D., Schein, J.E., Jones, S.J.M., and Birol, I. (2009). ABySS: a parallel assembler for short read sequence data. Genome Res. *19*, 1117–1123.

Siomi, H., and Siomi, M.C. (2008). Interactions between transposable elements and Argonautes have (probably) been shaping the Drosophila genome throughout evolution. Curr. Opin. Genet. Dev. *18*, 181–187.

Slotkin, R.K., and Martienssen, R. (2007). Transposable elements and the epigenetic regulation of the genome. Nat. Rev. Genet. *8*, 272–285.

Sun, C., Shepard, D.B., Chong, R.A., López Arriaza, J., Hall, K., Castoe, T.A., Feschotte, C., Pollock, D.D., and Mueller, R.L. (2012). LTR retrotransposons contribute to genomic gigantism in plethodontid salamanders. Genome Biol. Evol. *4*, 168–183.

Surget-Groba, Y., and Montoya-Burgos, J.I. (2010). Optimization of *de novo* transcriptome assembly from next-generation sequencing data. Genome Res. *20*, 1432–1440.

Swaminathan, K., Varala, K., and Hudson, M.E. (2007). Global repeat discovery and estimation of genomic copy number in a large, complex genome using a high-throughput 454 sequence survey. BMC Genomics *8*, 132.

Szak, S.T., Pickeral, O.K., Makalowski, W., Boguski, M.S., Landsman, D., and Boeke, J.D. (2002). Molecular archeology of L1 insertions in the human genome. Genome Biol. *3*, research0052.

Tamura, K., Subramanian, S., and Kumar, S. (2004). Temporal patterns of fruit fly (*Drosophila*) evolution revealed by mutation clocks. Mol. Biol. Evol. *21*, 36–44.

Tang, H. (2007). Genome assembly, rearrangement, and repeats. Chem. Rev. *107*, 3391–3406.

Trapnell, C., Williams, B.A., Pertea, G., Mortazavi, A., Kwan, G., van Baren, M.J., Salzberg, S.L., Wold, B.J., and Pachter, L. (2010). Transcript assembly and quantification by RNA-Seq reveals unannotated transcripts and isoform switching during cell differentiation. Nat. Biotechnol. *28*, 511–515.

Treangen, T.J., and Salzberg, S.L. (2012). Repetitive DNA and next-generation sequencing: computational challenges and solutions. Nat. Rev. Genet. *13*, 36–46.

Tu, Z., Li, S., and Mao, C. (2004). The changing tails of a novel short interspersed element in *Aedes aegypti*: genomic evidence for slippage retrotransposition and the relationship between 3′ tandem repeats and the poly(dA) tail. Genetics *168*, 2037–2047.

Vieira, C., Lepetit, D., Dumont, S., and Biémont, C (1999). Wake up of transposable elements following *Drosophila simulans* worldwide colonization. Mol. Biol Evol. *16*, 1251–1255.

Wang, Z., Zang, C., Rosenfeld, J.A., Schones, D.E., Barski, A., Cuddapah, S., Cui, K., Roh, T.-Y., Peng, W., Zhang, M.Q., *et al.* (2008). Combinatorial patterns of histone acetylations and methylations in the human genome. Nat. Genet. *40*, 897–903.

Wetzel, J., Kingsford, C., and Pop, M. (2011). Assessing the benefits of using mate-pairs to resolve repeats in *de novo* short-read prokaryotic assemblies. BMC Bioinformatics *12*, 95.

Wicker, T., Schlagenhauf, E., Graner, A., Close, T.J., Keller, B., and Stein, N. (2006). 454 sequencing put to the test using the complex genome of barley. BMC Genomics *7*, 275.

Wicker, T., Sabot, F., Hua-Van, A., Bennetzen, J.L., Capy, P., Chalhoub, B., Flavell, A., Leroy, P., Morgante, M., Panaud, O., *et al.* (2007). A unified classification system for eukaryotic transposable elements. Nat. Rev. Genet. *8*, 973–982.

Wicker, T., Taudien, S., Houben, A., Keller, B., Graner, A., Platzer, M., and Stein, N. (2009). A whole-genome snapshot of 454 sequences exposes the composition of the barley genome and provides evidence for parallel evolution of genome size in wheat and barley. Plant J. *59*, 712–722.

Zerbino, D.R., and Birney, E. (2008). Velvet: algorithms for *de novo* short read assembly using de Bruijn graphs. Genome Res. *18*, 821–829.

Zhang, Y., and Mager, D.L. (2012). Gene properties and chromatin state influence the accumulation of transposable elements in genes. PLoS One *7*, e30158.

The Current State of Metagenomic Analysis 10

Pieter De Maayer, Angel Valverde and Don A. Cowan

Abstract

The Earth harbours an enormous diversity of organisms, small and large, which produce a wealth of proteins and enzymes that we can potentially utilize to address many of our current challenges, including food safety, sustainable energy sources and human health. However, the majority of these organisms have been of limited value because they cannot be grown in culture, which has long been a prerequisite in order to study or exploit them. A new field of research that can provide access to the 'unculturable majority' has recently emerged. The field is termed 'metagenomics'. With the advent of genome sequencing and the development of next-generation sequencing technologies, metagenomics has provided access to the genomes of unculturable organisms, allowing us to study their diversity and to access the potential in these organisms and their cellular constituents. Other technologies, such as metatranscriptomics, metaproteomics and metabolomics, have been incorporated into the metagenomics toolkit, all enhancing the power of this field of research. Here, we address the state of metagenomics, its current methodologies and pitfalls, recent developments and future perspectives, and show how this technology is transforming modern biological science.

Introduction

Biology, or the study of life, has developed very rapidly in the past six decades, with the discovery of DNA (Watson and Crick, 1953), the advent of DNA amplification by PCR (Mullis and Faloona, 1987), DNA sequencing (Sanger *et al.*, 1977), the cataloguing of organisms in the three domains of life on the basis of universal molecular markers (Woese *et al.*, 1990) and the revolutionary development of whole genome sequencing. Since the complete genome of *Haemophilus influenzae* (Fleischmann *et al.*, 1995) and the first human genome sequence (Venter *et al.*, 2001), the field of genome sequencing has 'exploded'. This has been facilitated by the development of novel high-throughput sequencing technologies, which allow the cost-effective and rapid sequencing of whole genomes. At the time of writing, 3701 genomes have been completed, and a further 13,819 genome projects are under way (www.genomesonline.org). The majority of complete genomes are those of prokaryotes, with 3359 and 159 complete genomes of diverse representative bacterial and archaeal taxa sequenced, respectively (www.genomesonline.org). These microbial genomes have given in depth insights into their biology, ecology, diversity and phylogeny and have facilitated massive advances in all branches of scientific research, including the medical, agricultural, industrial and biotechnological fields.

It has been estimated that only 0.1–1% of prokaryotes on Earth can currently be grown in pure cultured (Amann *et al.*, 1995). Culturing has long been a prerequisite for the biological characterization of prokaryotes, even in the genomic era (Schloss and Handelsman, 2005). The requirement of pure cultures of microorganisms for genome sequencing has resulted in an inability to analyse more than half of the known bacterial and most archaeal phyla, where no cultivable members are known (Schloss and Handelsman, 2005). However, this limitation can now be resolved through the use of a new technology termed 'metagenomics'. The term metagenomics was coined by Jo Handelsman (Handelsman *et al.*, 1998) and was defined as 'the genomic analysis of microorganisms by direct extraction and cloning of DNA from an assemblage of microorganisms.' By this means, the 'unculturable majority' of environmental bacteria has become accessible (Schmieder and Edwards, 2012).

Early metagenomic analyses involved the following steps: (1) extraction of DNA from an environmental sample, (2) cloning of the DNA fragments in a vector, (3) transformation of the clones into a culturable bacterium, and (4) screening of the transformants (Handelsman, 2004). The metagenomics field has also, more recently, grown to incorporate sequence-based analyses, or metagenome sequencing. Here we use the term metagenome sequencing to reflect approaches used to determine the DNA sequences of the entire metagenome. More targeted approaches, which make use of primers to target specific genes, amplification of these genes and sequencing of the resulting amplicons, have also been developed (Tamaki *et al.*, 2011; Logares *et al.*, 2012). These targeted approaches have been of particular value in determining the diversity of organisms within a particular ecosystem. As is the case for whole genome sequencing, sequence-based and gene-targeted metagenomic studies have been greatly enhanced with the application of next-generation sequencing technologies (Shokralla *et al.*, 2012; Thomas *et al.*, 2012). Similarly, the field of functional metagenomics has evolved rapidly, and a number of new 'omics' technologies have allowed the incorporation of new fields of research into the metagenomics sphere (Simon and Daniel, 2011). These include metatranscriptomics (characterization of the RNA-based regulation and the expression of target genes within complex ecosystems), metaproteomics (characterization of the expressed proteins within an environmental sample at a given time) (Wilmes and Bond, 2004; Siggins *et al.*, 2012) and metametabolomics (characterization of the metabolites produced through the cellular metabolic processes of the microbiota within a given environmental sample) (Siggins *et al.*, 2012). The data from these fields can then be combined to study the functional activities of communities, including the metabolic pathways, cellular processes and interaction occurring within a given ecosystem (Siggins *et al.*, 2012), in a new field now known as 'systems biology'.

Here we give a comprehensive overview of the current state of metagenomics. We include an overview of the principles and practices of metagenomic studies, the advances made within the field and potential pitfalls in current metagenomic protocols, advances in both functional and sequence-based metagenomic analysis, applications of metagenomics and future perspectives in the field of metagenomics.

Metagenome sequencing

Early metagenome sequencing studies, such as those performed by Venter *et al.* (2004) on samples obtained from the Sargasso Sea, and Tyson *et al.* (2004) on acid mine drainage, made use of random sequencing of bacterial clones (Handelsman, 2004). A major drawback of this classical sequence-based methodology is that the preparation of clone libraries and subsequent sequencing of these libraries are expensive, laborious and time-consuming (Eyers *et al.*, 2004; Handelsman, 2004). The random sequencing also represented a blind 'hit-and-miss' approach, where the identification of genes encoding novel and desirable products relied on manageable sequence datasets and was limited by the capacities of the laboratories undertaking the studies, making the analysis of more complex environmental samples unfeasible (Shokralla *et al.*, 2012). With the advent of next-generation sequencing (NGS) technologies, it has become possible to sequence metagenomic DNA directly from the environment without the required cloning step and its associated problems, and the massively parallel high-throughputs achieved with these technologies has allowed for the cost-effective and rapid sequencing of thousands of metagenomes (Shokralla *et al.*, 2012; Thomas *et al.*, 2012).

Metagenome sequencing protocols

Current metagenome sequencing protocols (Fig. 10.1) make use of the following steps: (i) DNA extraction and purification, (ii) DNA

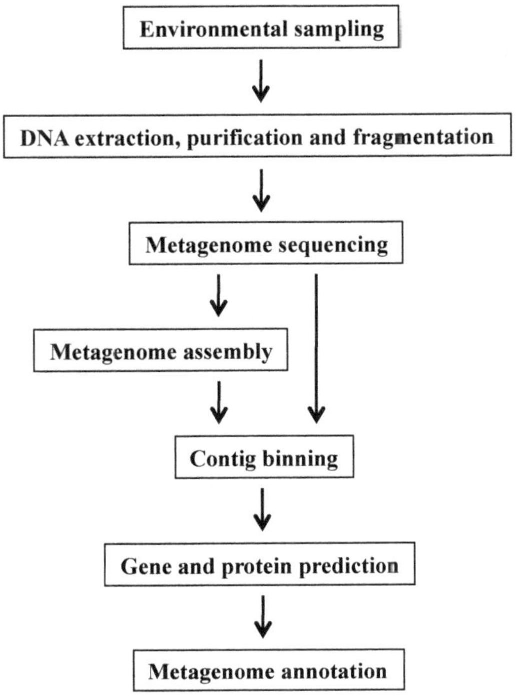

Figure 10.1 Schematic diagram of current metagenome sequencing protocols.

fragmentation and metagenome sequencing, (iii) sequence assembly, (iv) contigs binning and (v) gene and protein annotation. (Thomas *et al.*, 2012).While we discuss the current metagenome sequencing protocol in this order, some of the steps (e.g. binning before assembly) may be interchangeable or may be omitted (e.g. assembly for complex metagenome samples).

DNA extraction and purification

Metagenome sequencing requires access to high-quality DNA (Vogel *et al.*, 2009). However, because each DNA extraction method exhibits its own specific biases, a different picture of the community in a sample can be obtained according to the DNA extraction method used (Inceoglu *et al.*, 2010). It may therefore be important to determine which DNA extraction method provides access to the most representative picture of the extant community. This is not an easy task as numerous DNA extraction methods have been developed for different environments (Zhou *et al.*, 1996; Reigstad *et al.*, 2011) and numerous commercial kits are available (e.g. SoilMaster DNA Extraction Kit (Epicentre), PowerSoil DNA Isolation Kit (MoBio), Qiagen Plant DNA Extraction Kit (Qiagen), FastDNA SPIN Kit (MP)). It is therefore recommended that when undertaking a comparison of communities, a single DNA extraction method should be chosen. In contrast, when a robust picture of the community in one environment is the aim of the study, several DNA extraction methods should be combined in order to capture the greatest diversity. The application of different methods to the same samples has a two-fold purpose (Delmont *et al.*, 2012): (i) it can enable the detection of different portions of the metagenome, and (ii) it may increase the number of functions and species detected.

Here, we focus on soil because this environment is extremely complex and especially challenging due to the heterogeneous distribution of cells inside soil aggregates. Furthermore, similar principles are used in a large majority of commercial kits as well as in custom-made protocols. Two main approaches to bacterial DNA extraction from soil are currently available: (i) direct extraction, which relies on direct cell lysis inside the soil matrix and (ii) indirect extraction, in which bacterial cells are first extracted from the soil matrix, and then lysed for subsequent DNA extraction, purification and analysis.

Direct DNA extraction

The majority of disruptive methods for lysing cells and breaking down soil aggregates include a physical treatment of soil slurries with, for instance, glass beads in a cell disruptor. For higher efficiency, physical cell lysis is often combined with enzymatic lysis using lysozyme and proteinase K (Zhou *et al.*, 1996). Detergents such as SDS, which dissolve the hydrophobic compounds of cell membranes, are also used (Herrera and Cockell, 2007). According to some studies, up to 90% of the bacterial cells in a given soil can be disrupted using the combined method (Howeler *et al.*, 2003). DNA extracted from soil is usually highly contaminated by humic and fulvic acids that have chemical properties similar to those of DNA. Such organic compounds are strong inhibitors of enzymatic reactions (Wilson, 1997), interfering negatively with, for instance, subsequent PCR-based DNA amplification or cloning procedures. Precipitation/purification with polyethylene glycol, cetyltrimethyl ammonium bromide and/or polyvinylpolypyrrolidone minimizes the concentration of contaminants, but further purification is often necessary when the soil is particularly rich in humic substances.

The separation of soil DNA from humic substances (and other contaminants) can be achieved using methods that are based either on charge or on molecular weight, such as ion exchange chromatography (hydroxyapatite gel) or size exclusion chromatography (Sephadex/Sepharose gels) (Desai and Madamwar, 2007). Gel electrophoresis of the crude DNA extracts is another way to separate DNA from soil contaminants. Whereas both DNA and negatively charged humic substances run towards the cathode, positively charged contaminants migrate in the opposite direction (Kallmeyer and Smith, 2009).

Soil prewashes with chelating agents (EDTA) or surfactants (Triton X-100) are also beneficial to DNA purification by promoting the desorption of contaminants before *in situ* cell lysis (Inceoglu *et al.*, 2010). However, most purification procedures incur losses of DNA, thus decreasing the overall DNA extraction efficiency and leading to a compromise between purity and quantity. According to various studies, direct DNA extraction methods rarely yield more than 60% of the total microbial DNA contained in soil (Howeler *et al.*, 2003). Moreover, 30% of bacterial DNA may be lost during the subsequent purification procedure performed to remove humic acids (Howeler *et al.*, 2003).

Indirect DNA extraction

To avoid adsorption of DNA onto soil components and to limit humic acid contamination, bacterial cells may first be extracted from soil samples before further DNA extraction and purification. However, the majority of cells in soil samples are attached to soil particles (Nannipieri *et al.*, 2003). Such cells have physical and chemical properties that affect the efficiency with which they can be detached and extracted. For instance, bacterial cell surfaces contain positive and negative charges, as well as global hydrophobic and hydrophilic characteristics, which all affect their adhesion to soil particles (Doyle, 2000).

After disaggregation and desorption, cells can be separated from the soil particles by differential centrifugation to remove the densest soil particles. This may be followed by density gradient centrifugation to remove other particles and contaminants such as humic acids (Bakken, 1985). The resulting cell suspension can be subjected to classical DNA extraction techniques (enzymatic and/or chemical treatments), which may be less drastic than the ones used in direct DNA extraction, but sufficient to lyse 90% of the cells (Steffan *et al.*, 1988). The resulting DNA solution is often of good quality, but fragments may be somewhat sheared (range 10–50 kb), with moderate humic acid levels (Steffan *et al.*, 1988). On the other hand, DNA fragments ranging in size from 150 kb to 1000 kb are routinely obtained when cells are lysed in an agar plug to avoid DNA shearing (Bertrand *et al.*, 2005). However, persistent interactions of cells with soil particles often prevent the recovery of 20–50% of the cells in soil (Robe *et al.*, 2003), making this method less suitable for representative analysis of soil organisms.

Metagenome sequencing

Pioneering sequence-based metagenomic studies made use of a shotgun sequencing approach (Tyson *et al.*, 2004; Venter *et al.*, 2004). These studies relied on the shearing of DNA into 2–10 kb fragments which were cloned into a plasmid or fosmid vector for subsequent transformation into a bacterial host. The fragments were consequently amplified by bacterial culturing and sequenced using the Sanger dye-termination method (Mardis, 2008; Wooley *et al.*, 2010). Although this technique is still frequently used in functional metagenomic studies, it suffers from several drawbacks when employed for large-scale sequence-based metagenomic studies. The preparation of clone libraries is laborious and time-consuming (Thomas *et al.*, 2012). Furthermore, some DNA sequences are more amenable to cloning than others, resulting in cloning bias. One of the main factors influencing clonal bias is the G+C composition, with G+C rich sequences frequently forming secondary structures, while A+T-rich regions can be toxic to cloning vectors such as *Escherichia coli* (Sorek *et al.*, 2007; Thomas *et al.*, 2012). Another major drawback of dye-termination-based sequencing is the low throughput achieved with this technology, with current automated Sanger capillary sequencers capable of sequencing only 96 samples, with maximum read lengths of 900 bp, giving ~86.4 kb per run (Table 10.1), while metagenome samples often consist of Gigabases of DNA sequence. There is also the inherently high cost of sequencing with this technology, at US $400,000 per gigabase (Shokralla *et al.*, 2012; Thomas *et al.*, 2012). This makes cloning/Sanger sequencing-based metagenomic analyses unfeasible to all but the biggest laboratories.

Next-generation sequencing technologies

The development and commercial introduction of revolutionary next-generation sequencing technologies has dramatically changed the field of genome sequencing and sequence-based metagenomic studies (Mardis, 2008). These NGS technologies allow massively parallel sequencing, currently allowing up to 600 Gigabases of sequence to be generated per run, at a fraction of the cost associated with Sanger capillary-based sequencing (Shokralla *et al.*, 2012), both of which are limiting factors to large-scale metagenomic analyses. Two distinct classes of NGS platforms are currently available, which differ in the template which is sequenced (Mardis, 2008; Loman *et al.*, 2012; Shokralla *et al.*, 2012). The 'second-generation' platforms rely on the preparation and sequencing of clonally amplified template libraries, while more recently, 'third-generation' platforms have become available which can determine the sequences of single DNA molecules without the need of amplification (Loman *et al.*, 2012).

Second-generation NGS technologies

Second-generation NGS technologies are currently the most widely used, with a number of commercially introduced platforms available. These platform, make use of different enzymologies, chemistries, visualization technologies and hardware, giving rise to wide variations in their throughput, read lengths, run-time and costs. Such variations give the end-user a wide range of options in selecting the appropriate sequencing strategy to address their particular research questions and samples (Loman *et al.*, 2012). There are currently four available platforms, namely the 454 (Roche), Illumina (Life Technologies), AB SOLiD (Life Technologies) and Ion Torrent (Life Technologies) sequencers.

In general, second-generation NGS sequencing involves three steps: library preparation, clonal amplification of the template and sequencing (Mardis, 2008; Loman *et al.*, 2012). As no cloning into bacterial vectors is required, this technology is not affected by the clonal bias

associated with traditional shotgun-sequencing methods (Mardis, 2008). Following DNA or RNA extraction and quantity/quality control of the templates, they are sheared into fragments, using mechanical (e.g. nebulization or ultrasonication) or enzymatic (commercial nuclease-based kits such as the New England Biolabs NEBnext fragmentase) means (Knierim *et al.*, 2011). This is followed by size selection of DNA fragments using paramagnetic beads or agarose gel separation and ligation of the blunt-ended fragments with platform-specific adaptors which serve as primer-binding sites for template amplification (Loman *et al.*, 2012). Subsequently, adaptor-bound DNA fragments are immobilized onto different solid surfaces, such as flow cells (Illumina sequencers), beads (454 and SOLiD sequencers) or ionized spheres (Ion Torrent sequencers) (Loman *et al.*, 2012). Template amplification occurs within the sequencing instrument (Illumina Miseq and Hiseq sequencers) or by emulsion PCR in separate microreactors (SOLiD, 454 and Ion Torrent) (Loman *et al.*, 2012).

Major differences between the various commercially available second generation platforms are the different chemistries, outputs and sequencing approaches used. Some platforms use sequencing-by-synthesis (454, Ion Torrent and Illumina platforms) while the SOLiD platform makes use of a sequencing-by-ligation approach (Shokralla *et al.*, 2012). The Roche 454 and Ion Torrent sequencing platforms make use of pyrosequencing technology, where the incorporation of each nucleotide by DNA polymerase is accompanied with the release of a pyrophosphate and

Table 10.1 Comparison of first, second and third generation platforms which can be applied for metagenome sequencing, based on the field guide to next-generation DNA-sequencers (Glenn, 2011; http://www.molecularecologist.com/next-gen-fieldguide/)

Platform	Sequencing chemistry	Read length (bp)	Reads/run (million)	Accuracy	Run time	Mb/run	Cost/run (US $)[a]	Major advantage	Major disadvantage
3730xl (Life Technol.)	Capillary	650	0.000096	99–99.9%	2 hours	0.06	144	Long reads	Small datasets
Second generation platforms									
454 GS FLX+ (Roche)	Pyrosequencing	600–800	1	99–99.9%	20 hours	650	6,200	Long reads	High cost
Hiseq 2000/2500 (Illumina)	Reversible terminator	150–300	6000	99.90%	11 days	600,000	23,500	High throughput	Short reads
Genome Analyser IIx (Illumina)	Reversible terminator	150–300	300	99.90%	14 days	96,000	17,500	High throughput	Short reads
5500xl SOLiD (Life Technol.)	Ligation	35–75	6000	>99.90%	8 days	155,000	10,500	High throughput	Short reads
Benchtop platforms									
454 GS Junior (Roche)	Pyrosequencing	400–450	0.1	99–99.9%	10 hours	50	1,100	Long reads	High cost
Ion Torrent PGM (Life Technol.)	Pyrosequencing	100–200	11	99–99.9%	4 hours	800[b]	750[b]	Short runs	Short reads
Ion Proton I (Life Technol.)	Pyrosequencing	100–200	11	99–99.9%	4 hours	40,000	1,000	Short runs	Short reads
Miseq (Illumina)	Reversible terminator	150–300	7	99.90%	26 hours	1200	1,160	Short runs	Short reads
SMS technologies									
Heliscope (Helicos Biosci.)[c]	Reversible terminator	30–35	1000[c]	99.90%	1 day	28[c]	Unknown	Short runs	No longer sold[c]
PacBio RS SMRT (Pacific Biosci.)	Real-time sequencing	1500	0.01	>85.00%	2 hours	5–10	1700	Long reads	High error rate
MinION (Oxford Nanopore)	Real-time Nanopore	6000[d]	0.1[d]	99.60%	6 hours	1000[d]	900	Long reads	High error rate
GridION (Oxford Nanopore)	Real-time Nanopore	10,000[d]	4[d]	99.60%	Unknown	40,000[d]	Unknown	Long reads	High error rate

[a]Reagent cost per run.
[b]Calculation for Ion Torrent Personal Genome Machine (PGM) is on the basis of 318 chip.
[c]Heliscope information as per Http://www.lgtc.nl/services/next_generation_sequencing/Helicos_SalesSpec.pdf.
[d]Values as project by Oxford Nanopore at Advances in Genome Biology and Technology Conference (AGBT), USA, Http://www.nanoporetech.com/news/press-releases/view/39

hydrogen ions as by-products (Margulies *et al.*, 2005; Loman *et al.*, 2012; Shokralla *et al.*, 2012). During sequencing, repetitive cycles of nucleotides (dNTPs) (G, A, T and C) are passed over the beads/sequencing templates. In the 454 platform, when a complementary nucleotide is incorporated, the released pyrophosphate activates an enzymatic cascade, resulting in the emission of light by the enzyme luciferase (Margulies *et al.*, 2005; Shokralla *et al.*, 2012). This emitted light from each template-containing bead is detected by a CCD (Charge Coupled Device) camera. The intensity of emitted light is proportional to the number of nucleotides incorporated (i.e. single nucleotides or homopolymers) (Margulies *et al.*, 2005; Shokralla *et al.*, 2012). The Ion Torrent, by contrast, makes use of a modified silicon chip that detects the hydrogen ion pyrosequencing by-product, and thus represents the first 'post-light' platform that does not rely on imaging for sequence determination (Rothberg *et al.*, 2011). The Illumina platforms make use of a flow cell, containing the library of template fragments. The flow cell is cyclically flooded with polymerase and the four differentially labelled fluorescent nucleotides (i.e. A, T, G and C) with inactivated 3′ OH groups. These terminator nucleotides are incorporated where complementary nucleotides are present on the template and a concomitant excitation and imaging step are used to determine the sequence at each position in the template fragments. A chemical de-blocking step ensures the signal is quenched and the 3′ OH group is reactivated, prior to the next sequencing step, or flow cycle, where terminator fluorescent nucleotides are incorporated at the next position in the template (Loman *et al.*, 2012; Shokralla *et al.*, 2012). The Applied Biosystems SOLiD platform makes use of sequencing-by-ligation technology, where four semi-degenerate fluorescent 8-mer oligonucleotides (corresponding to each of the A, T, G and C nucleotides) and DNA ligase are washed over the bead-immobilized DNA templates with a universal sequencing primary ligated at the end (Shokralla *et al.*, 2012). When a matching 8-mer is hybridized and ligated by the DNA ligase at the 3′ end of the universal primer on the template fragment, the phosphate backbone is sealed, and a fluorescent read-out identifies the specific 8-mer oligonucleotide that is ligated, corresponding to complementary nucleotide at each site (Loman *et al.*, 2012; Shokralla *et al.*, 2012). The fluorescent group is removed, allowing the next ligation round to commence (Loman *et al.*, 2012; Shokralla *et al.*, 2012).

Third generation, single-molecule sequencing

One of the major short-falls of all current commercially available second generation NGS platforms is the introduction of PCR artefacts or biases during the amplification of the template DNA fragments, which can negatively impact the sequencing outcomes. This is particularly problematic in metagenomic diversity studies, where certain templates may be preferentially amplified, resulting in an unbalanced picture of the true diversity within a given sample (Polz and Cavanaugh, 1998). This is discussed in more detail in the section *Current pitfalls in metagenomic diversity studies* below. A promising new technology, termed third generation, or 'single-molecule' sequencing (SMS), by-passes the amplification step, thereby overcoming the PCR artefacts and biases associated with second generation sequencing (Gupta, 2008; Treffer and Deckert, 2010; Thompson and Milos, 2011; Loman *et al.*, 2012; Shokralla *et al.*, 2012).

The first SMS technology on the market, the Heliscope platform (Helicos Bioscience) makes use of sequence-by-synthesis on single DNA molecules, where repetitive cycles of DNA polymerase and each of the four individual fluorescently labelled nucleotides are passed over the single stranded DNA templates, resulting in template-dependent extension and capturing of the incorporated fluorescence by means of a CCD camera attached to a fluorescent microscope (Harris *et al.*, 2008; Pushkarev *et al.*, 2009; Shokralla *et al.*, 2012). This technology can produce up to 1 billion reads, but its use has been limited by the high costs of the platform (Loman *et al.*, 2012). The Pacific BioScience (PacBio) SMRT (single molecule real time) platform also uses sequencing-by-synthesis, but single stranded DNA templates are attached to a chip containing thousands of zero-mode waveguides (ZMWs) which allow real-time observation of DNA polymerization or sequence synthesis (Gupta, 2008; Eid *et al.*, 2009; Thompson and Milos, 2011). These ZMWs are 10-nm-diameter cavities, which have a highly processive, strand-displacing DNA polymerase tethered at the bottom. The sequence is then determined in real-time as the complementary strand is formed in each ZMW, with the incorporation of differentially labelled fluorescent nucleotides (Gupta, 2008; Eid *et al.*, 2009). Major drawbacks of the PacBio SMRT system are the rather limited number of sequencing reads (100,000) compared to the Helicos system, and the high error rates associated with this platform (Thompson and Milos, 2011). Starlight, a platform under development by Life Technologies, relies on fluorescence resonance energy transfer (FRET)-based SMS technology, which makes use of quantum dots, semiconducting nanoparticles to label the DNA polymerase, rather than fluorescent dyes (Thompson and Milos, 2011; Pennisi, 2010). Real-time sequencing occurs in a similar fashion to the PacBio SMRT system, but the DNA polymerase can be replaced once it becomes inactive, allowing sequencing along the entire length of single DNA molecule template (Gupta, 2008; Thompson and Milos, 2011). This provides a major advantage to this technology, but its release date, and platform and sequencing costs, are not known (Gupta, 2008).

Another technology that is finding use in novel next-generation sequencing platforms is nanopore technology. Nanopore sequencing relies on a thin membrane that contains 1.5 to 2 nm 'nanopores' which can consist of carbon nanotubes, or bacterial membrane pore proteins such as α-haemolysin or the mycobacterial porin protein MspA (Treffer and Deckert, 2010; Gupta, 2011; Thompson and Milos, 2011). The negatively charged single stranded DNA template is electrophoretically threaded through the nanopore across which a current is applied. This results in nucleotides blocking the nanopore, resulting in a change in the electrical conductance across the membrane, which is dependent on the nucleotide blocking the pore, thereby allowing the nucleotide at each position in the template DNA strand to be discriminated (Treffer and Deckert, 2010; Gupta, 2011). This technique has the advantage of the Ion Torrent second-generation sequencer, in that no labels or expensive imaging equipment are required and the sequence of very long templates could be determined (Thomson and Milos, 2011). One pitfall that is holding back the commercial application of nanopore sequencing technologies is the required sensitivity of detection, as the differences in conductivity between the different nucleotides is extremely small. Furthermore, the DNA template may traverse the nanopore at too high a speed to resolve single nucleotides. Several different methods have been used to overcome these limitations to nanopore sequencing technologies and Oxford Nanopore have recently announced the imminent release of their MinION and GridION platforms (late 2012) (Eisenstein, 2012; Loman *et al.*, 2012). The former will represent the first single-use sequencing unit, where a USB cable can stream sequence data directly onto a computer. This sequencer will consist of 512 nanopores and will be capable of 120–1000 base pairs per hour per pore for up to 6 hours, yielding up to 900 Mb of sequence, at a projected price of US$900 (Loman *et al.*, 2012). The GridION rack-mountable sequencer will consist of 2000 nanopores and 20 GridION sequencers run in parallel and will potentially be capable of sequencing a human genome in 15 minutes (Eisenstein, 2012; Maitra *et al.*, 2012).

Choosing NGS platforms for metagenome sequencing

The availability of a number of different commercial sequencing technologies means that end-users have a wide and often difficult choice of which NGS platform to apply to their particular experiment (Shokralla *et al.*, 2012). Such choice should take several factors into account. One factor is the complexity of the environmental sample being studied and the scale of the metagenomic analysis to be undertaken, which together will dictate the throughput required for a representative coverage of the DNA within the sample (Shokralla *et al.*, 2012). Another important factor to be considered is the read length required. Increased throughput and reduced read length represent a major trade-off with current NGS platforms (Logares *et al.*, 2012). Long reads are particularly important in diversity and taxonomic metagenomic studies, where these sequences must provide a large enough taxonomic signal to differentiate between different taxa (Wommack *et al.*, 2008). The error rates of the various NGS platforms must also be considered (Loman *et al.*, 2012). Furthermore, when choosing a NGS platform, practical factors such as cost and running-time are important determinants (Loman *et al.*, 2012). The different platforms currently available all have their associated advantages and disadvantages (Table 10.1).

A number of benchtop devices for the end-user laboratory (Table 10.1), which have smaller throughputs and are suitable for small-scale microbial applications, are also available (Loman *et al.*, 2012). The availability of several competing platforms from different companies drives continuous technical and performance improvements to the sequence systems, ensuring more cost- and time-effective systems with longer read lengths and higher throughputs are available to the user (Loman *et al.*, 2012; Shokralla *et al.*, 2012). The revolutionary developments in NGS technologies are such that at present sequencing capacities are doubling every 6–9 months (Loman *et al.*, 2012). One of the major drawbacks to this rapid advancement is that increases in sequencing capacity greatly exceed the rate of development of the computing power required to handle the sequencing data, as predicted by Moore's law. This law states that computing power, as a measure of the number of transistors on integrated circuits, doubles approximately every five years (Moore, 1965). This can result in computing costs surpassing those associated with the actual sequencing. A run using the Illumina HiSeq 2500 can yield up to 600 Gb of sequence data, requiring 0.6 Tb of short-term storage and long-term archiving computer disk space (Glenn, 2011; Logares *et al.*, 2012; Scholz *et al.*, 2012). The analysis of this data will also require a suitably powerful processor with sufficient RAM and efficient software algorithms (Logares *et al.*, 2012). Given the large amounts of sequence generated in sequence-based metagenomic analyses, the informatics capacities must thus also be taken into consideration when choosing a NGS sequencing platform.

NGS enhancements and their application in metagenomic studies

Aside from the continuously longer read lengths and higher throughputs available for the various NGS platforms, there are several other enhancements to these sequencing technologies which have a direct bearing on sequence-based metagenomic analyses. One such enhancement, available for all second generation platforms, is paired-end sequencing, where DNA segments derived through fragmentation are sequenced from both ends, thereby nearly doubling the read length (Loman *et al.*, 2012; Magi *et al.*, 2010). Another strategy, mate-pair sequencing, involves fragmentation of the genomic DNA into segments of a pre-determined size (typically 3, 6, 8 or 20 kb), circularization of these segments, refragmentation and paired-end sequencing of the two ends. This allows the determination of sequence reads a known distance from each other on the genome, simplifying the assembly process (Loman *et al.*, 2012; Magi *et al.*, 2010). These two technologies are of particular use when assembling whole genomes, but can also be applied in metagenomic studies that make use of platforms yielding shorter sequence reads (Rodrigue *et al.*, 2010).

The ability of NGS platforms to sequence DNA fragments in a massively parallel fashion also allows the incorporation of multiplex sequencing. By constructing tagged or bar-coded DNA fragments it becomes possible to target different genes within a sample, or to bulk multiple samples and sequence them within a single sequencing run (Binladen *et al.*, 2007; Shokralla *et al.*, 2012). This parallel tagged sequencing relies on short unique oligomers or multiplex identifiers (MIDs) which are ligated to the template DNA fragments and can be used to distinguish distinct samples after sequencing (Meyer *et al.*, 2007). Several commercial kits consisting of tens to hundreds of MIDs are currently available for the different second-generation platforms. This technology is very useful in metagenomic diversity studies, and has been used, for instance, in an analysis of the human gut microbiome (Wu *et al.*, 2010b), comparison of the microbial diversity between freshwater, intertidal wetland, and marine sediments sequenced in a single run (Wang *et al.*, 2012), and the high-throughput multilocus profiling of pathogenic bacteria in the genera *Streptococcus*, *Staphylococcus*, *Pseudomonas* and *Legionella* (Boers *et al.*, 2012).

Metagenome assembly

Metagenome sequencing on an NGS platform results in millions of short reads, where the read lengths are a function of the NGS platform used and incorporation of paired-end and mating-pair sequencing technologies. The further analysis of these sequences, the reconstruction of complete or draft genome sequences and of full-length genes, depends on assembly of these short read fragments (Wooley and Ye, 2010; Thomas *et al.*, 2012). This assembly involves the construction of multiple alignments on the basis of overlaps between the reads, thereby generating consensus sequences which are termed contigs (Logares *et al.*, 2012). Two means of gene and genome assembly from metagenomic samples are currently employed, namely reference-based and *de novo* assembly (Thomas *et al.*, 2012). The former is dependent on the metagenomic sample containing sequences for which reference genomes, that are sufficiently closely related and similar, are available. However, large insertions, deletions or polymorphisms in the metagenomic sequence can result in fragmented assemblies (Thomas *et al.*, 2012).

Several software packages which were designed for the assembly of genomes from single, cultured organisms, can be used for reference-based assembly from metagenome datasets (Thomas *et al.*, 2012). These include Newbler (Roche) and MIRA (Chevreux *et al.*, 1999), as well as a number of assemblers for short and ultra-short reads, such as Velvet (Hernandez *et al.*, 2008), Abyss (Simpson *et al.*, 2009) and ALLPATHS (Butler *et al.*, 2008). The application of these 'single organism' assemblers to metagenome sequence datasets suffers from two complicating factors, namely that there is an uneven representation of the organisms within a metagenomic sample, and that highly conserved regions between closely and distantly related members of an environment may exist (Pop, 2009). Due to uneven coverage, the dominant members of a community could be considered by the assembler as repeats, resulting in their exclusion by the assembler in order to avoid mis-assemblies (Pop, 2009). The presence of highly conserved genomic regions, such as bacteriophage insertions or genomic islands, in phylogenetically distant organisms can result in contigs being built from reads belonging to different genomes (Logares *et al.*, 2012).

Several metagenome-specific assemblers have been developed that take these factors into account. Most of these make use of de Bruijn graphs (DBG) which are built based on short pieces of all the reads. These graphs are constructed without the calculation of overlaps between reads, but are relatively memory intensive (Logares *et al.*, 2012). Overlap-layout-consensus (OLC) based algorithms do incorporate calculation of the overlaps between reads, laying out of the reads in a graph and derivation of consensus sequences (Logares *et al.*, 2012). Commonly used *de novo* metagenomic DBG assemblers include MetaVelvet (Namiki *et al.*, 2011), Genovo (Laserson *et al.*, 2011) and Meta-IDBA (Peng *et al.*, 2011). OLC-based assemblers include MAP (Lai *et al.*, 2012) and SSAKE (Warren *et al.*, 2007). Bambus2 is a metagenomic scaffolder, which takes contigs from various assemblers and scaffolds them to assemble larger contigs (Koren *et al.*, 2011). Another means of assembling metagenomes is by detecting open reading frames from the reads, clustering and assembling these ORFs. This annotation-before-assembly approach is applied in MetaORFA (Ye and Tang, 2009) and allows for the generation of long peptides even when sequence coverage is low and if they have synonymous mutations (Wooley and Ye, 2010). It must be noted that accurate metagenomic assembly is dependent on sufficient sequencing depth and coverage, as multiple reads covering a particular DNA segment will provide better accuracy in assembly than single reads. As metagenomes of complex communities will generally consist of data with a lower sequencing depth and coverage, assembly in these cases may be of limited value (Thomas *et al.*, 2012). Complex genetic elements, including repeat regions such as those observed in CRISPRs, cannot be analysed without assembly. While sequencing strategies such as mate pair can partly solve this problem, there is an urgent need for the development of metagenomic assembly strategies to tackle such DNA regions (Logares *et al.*, 2012; Thomas *et al.*, 2012).

Contig binning

Following assembly of the reads into larger contigs, these contigs can be subjected to binning. Binning refers to the sorting of contigs into groups or 'bins' representing individual or closely related genomes (Thomas *et al.*, 2012). Two different approaches of binning have been developed: compositional and similarity-based binning (Simon and Daniel, 2011). Compositional-based binning algorithms group contigs on the basis of sequence characteristics, such as G+C contents, codon usage or the abundance distributions of oligonucleotides (*k*-mers) and their co-occurrence in reference genomes of known taxonomic origins (Simon and Daniel, 2011; Thomas *et al.*, 2012). Per-contig read depth can also be used for binning purposes, as those genomes that are most abundant in a metagenomic sample will contribute the majority of reads (Logares *et al.*, 2012). Several software packages for compositional-based binning are available, such as Phylopythia (McHardy *et al.*, 2007), S-GSOM (Chan *et al.*, 2008), PCAHIER (Zheng and Wu, 2011) and TETRA (Teeling *et al.*, 2004). One problem with compositional-based approaches is that they are unreliable for the binning of short reads (e.g. 100 bp reads at best contain less than half of all possible 4-mer combinations, insufficient to determine the k-mer distribution). The use of training datasets for the compositional algorithms can, however, improve their assignment accuracy (Thomas *et al.*, 2012). On the other hand, similarity-based binning classifies reads on the basis of sequence homology to reference sequences of known taxonomy, contained in a database, using tools like BLAST (Logares *et al.*, 2012). Software packages based on this approach include Sort-ITEMS (Monzoorul Haque *et al.*, 2009), MEGAN (Huson *et al.*, 2007), CARMA (Krause *et al.*, 2008) and Metaphyler (Liu *et al.*, 2011). Similarity-based algorithms also rely on availability of a representative database. Given the bias of most databases towards model or culturable organisms the assembly and binning of many of the metagenomic genome sequences may be precluded, whereas compositional binning, which relies on intrinsic sequence features, may prove more useful (Simon and Daniel, 2011). Several software programs that make use of both composition and similarity-based binning are also available, including PhymmBL (Brady and Salzberg, 2009) and MetaCluster (Leung *et al.*, 2011). Assembled and binned contigs can then be further used to construct partial genomes of uncultured organisms, which can subsequently be used as datasets for the similarity-based binning of other metagenomic samples. Binning can also be applied prior to assembly to reduce the computational requirements of subsequent assemblies (Thomas *et al.*, 2012). Importantly, care must be taken to ensure uncontaminated genomic bins for effective binning of other metagenomic datasets (Thomas *et al.*, 2012).

Metagenome annotation

Assembled and binned metagenomes can be submitted for annotation. This annotation step can be defined as the addition of biological information to raw DNA sequences and is vital for subsequent comparative metagenomic analyses and bioprospecting (Logares *et al.*, 2012). Current metagenomic annotation approaches depend on homology searches against functionally and/or taxonomically annotated data. For smaller, simpler datasets, automated annotation can be supplemented with manual curation. However, as metagenomic datasets are generally very large, manual annotation is not plausible. Fast, accurate and computationally robust approaches are required (Thomas *et al.*, 2012). The annotation process consists of two sequential steps, namely structural and functional annotation. The former involves the identification of open reading frames (ORFs) and genomic elements in the DNA fragments, while the latter involves the assignment of functions to the predicted ORFs (Thomas *et al.*, 2012). Several structural annotation software packages designed for ORF prediction in whole genomes are freely available, including AMIGene (Bocs *et al.*, 2003), GeneMark (Besemer and Borodovsky, 2005) and Glimmer (Delcher *et al.*, 1999). These programs are useful for the prediction of ORFs from large contigs; for metagenomic datasets consisting of shorter reads, a number of metagenome-specific tools, such as FragGeneScan

(Rho *et al.*, 2010), MetaGeneMark (McHardy *et al.*, 2007), MetaGeneAnnotator (Noguchi *et al.*, 2008) and Orphelia (Hoff *et al.*, 2009), are available. These software packages routinely fail to identify a significant number of ORFs, which must be manually identified using BLAST algorithms when less complex metagenomic samples are sequenced (Thomas *et al.*, 2012). It must be noted that ORFs are generally larger (e.g. an average prokaryote ORFs is 900–1000 nt in size) than the reads or assembled contigs derived from metagenomic size, making ORF prediction and functional annotation a difficult task (Logares *et al.*, 2012). More accurate gene prediction will thus be facilitated with the increased read lengths achieved with newer NGS platforms and chemistries.

Functions are assigned to predicted ORFs based on similarity searches (e.g. BLAST) against function databases. These include KEGG (Kanehisa *et al.*, 2004) which currently contains over 9 million functionally annotated genes, NCBI (Sayers *et al.*, 2012), Gene Ontology (GO) databases (Ashburner *et al.*, 2000), Clusters of Orthologous Groups (COG) databases (e.g. eggNOG and COG/KOG) (Tatusov *et al.*, 2003; Muller *et al.*, 2010) and protein family databases (e.g. PFAM and TIGRFAM) (Selengut *et al.*, 2007; Finn *et al.*, 2010). Programs for the functional annotation of non-protein coding genes such as tRNAs (tRNAscan-SE) (Lowe *et al.*, 1997), non-coding RNAs (RFAM) (Gardner *et al.*, 1997) and CRISPRs (CRT) (Bland *et al.*, 2007) are also available. BLAST-based searches against these databases are computationally expensive and no single database covers all possible biological functions (Thomas *et al.*, 2012). There is thus a need for the development of pipelines which can handle the rapid and accurate functional annotation of large data volumes such as those presented by metagenomic samples.

Several automated pipelines have been developed which incorporate a number of the tools for the functional annotation of metagenomic datasets. These include CAMERA (Community cyberinfrastructure for Advanced Microbial Ecology Research and Analysis – Sun *et al.*, 2010), IGM/M (Markowitz *et al.*, 2008) and MG-RAST (Meyer *et al.*, 2008). These automated pipelines incorporate some of the metagenome-specific ORF prediction algorithms (see above) and enable functional annotation by comparing predicted protein coding sequences against a range of functional databases and merging them within a single framework (Thomas *et al.*, 2012). Furthermore, IMG/M and MG-RAST are capable of binning sequences on the basis of taxonomic assignment. These pipelines also serve as important repositories for metagenomic data, providing databases for the publication, storage and sharing of this data (Thomas *et al.*, 2012). Currently, there are over 8000 registered users on the MG-RAST server, with 62,234 metagenome datasets in the database, of which over 11,000 are publicly available and over 17 Terabase pairs analysed (http://metagenomics.anl.gov/). One of the major future challenges will be to standardize metagenome outputs, thereby allowing interchange of data between different metagenomic analysis platform. The outputs from different pipelines such as those provided in MG-RAST, IMG/M and CAMERA could be combined to provide a multifaceted and standardized output for end-users. The Genomic Standards Consortium (GSC) M5 project is driving the development of such a standard for the effective exchange of metagenomic information (Thomas *et al.*, 2012).

One of the major challenges with the current approaches is that only between 20% and 50% of all sequences in a metagenomic sample can currently be annotated (Gilbert *et al.*, 2010a), due to that fact that they show no homology to known sequences (Thomas *et al.*, 2012). There is thus still an extensive requirement for the functional, structural and biochemical characterization of these genes and their protein products.

Reconstruction of complete genomes of 'unculturable' organisms from metagenomes

One of the major objectives of metagenomic studies has been the reconstitution of whole genomes of individual organisms within a metagenomic population, particularly those that are uncultured (Luo *et al.*, 2011). These genomes can then serve as repositories for the discovery of novel peptides or enzymes (see *functional metagenomics and bioprospecting* section). One of the specific objectives of this approach is the identification of predicted physiological functions which may aid the development of strategies for culturing the 'unculturables' (Vartoukian *et al.*, 2010). The first reconstruction of a genome of an 'unculturable' microorganism was for the aphid obligate endosymbiont *Buchnera aphidicola* (Abbot *et al.*, 2001).

Reconstruction of genomes is particularly difficult, given the complexities of the binning and assembly processes when applied to more complex metagenomic sequence datasets. However, with sufficient sequence coverage, long read lengths and powerful computational resources, genome reconstruction from complex samples can still be achieved (Tyson *et al.*, 2004; Venter *et al.*, 2004; Hess *et al.*, 2011; Iverson *et al.*, 2012). It is therefore likely that new NGS platforms, with their longer sequence reads, higher volumes, decreasing sequencing costs and greater computing power and resources, will lead to many more complete genomes to be derived from metagenomic sources in the near future.

Comparative metagenomics

The advances in NGS technologies and reduced cost of sequencing has resulted in the sequencing of numerous metagenomes from a wide range of environments: over 62,000 metagenome datasets have been deposited in the MG-RAST server (http://metagenomics.anl.gov/). The availability of multiple metagenomes has stimulated the rapid development of a new research field: comparative metagenomics. Comparative studies allow individual metagenomes to be placed into environmental and physiological context, giving insights into both the functioning of communities in different environments and their responses to biotic and abiotic factors (Raes *et al.*, 2007). Comparative metagenomic studies have, for example, shown the metabolic and functional differences of communities in response to light and pressure conditions at different depths in the sea (DeLong *et al.*, 2006), and have provided an explanation for the differences in G+C contents between open ocean and littoral communities (Rusch *et al.*, 2007).

A number of different analytical and statistical software packages and tools have been developed for, or are used in, comparative metagenomic studies. Multivariate statistics, such as Principal Component Analysis (PCA), Principal Coordinate Analysis (PCoA), Non-metric multidimensional scaling (NMDS) and hierarchical cluster analysis are frequently used to identify factors which significantly affect the metagenomic data, and for visualizing metagenome comparisons (Wooley *et al.*, 2010; Logares *et al.*, 2012). Software tools to elucidate differences between metagenomes include Xipe (Rodriguez-Brito *et al.*, 2006), STAMP (Parks and Beiko, 2010), ShotgunFunctionalizeR (Kristiansson *et al.*, 2009), LefSe (Segata *et al.*, 2011) and MG-DOTUR (Schloss and Handelsman, 2008). Newer versions of MEGAN (version 2.0 –4.0) also allow functional comparative analyses (Huson *et al.*, 2011). Comparative tools are also incorporated into some of the major pipelines and repositories hosted on servers on the internet, such as CAMERA (Sun *et al.*, 2011), JCVI's METAREP (Goll *et al.*, 2010) and IMG/M (Markowitz *et al.*, 2008). Similarly, MG-RAST allows comparison of metagenome datasets to other metagenomes or complete genomes and incorporates PCA as a statistical analysis tool (Meyer *et al.*, 2008).

There are several factors that need to be taken into consideration when comparative metagenomic studies are undertaken. Caution must be exercised in the comparison of two or more metagenomes, as biological and sampling factors, such as differences in average genome size, sample size, phylogenetic complexity, sequencing technology and methodology, and the approaches employed for assembly and annotation can hamper direct comparison (Raes *et al.*, 2007). The metadata associated with a metagenome sequence, that is, where and when the samples were taken, under which conditions, what sampling protocols were used, as well as the sequencing technologies, assembly and annotation programs used, must also be taken into consideration (Wooley *et al.*, 2010). For a meaningful comparative analysis it is imperative that the correct types of datasets are compared. There is thus an urgent need for a standardized description of metagenome studies so that they can be exchanged and compared. A minimum information of genomic and metagenomic (MIGS/MIMS) metadata has been proposed (Field *et al.*, 2008; Kottmann *et al.*,

2008) and major metagenomics servers including IMG/M, CAMERA and MG-RAST are already MIMS-compliant.

Metagenomic diversity studies

One of the most widely used applications of metagenomics is to assess the diversity of organisms within a particular environment. Initially, the metagenomic study of diversity within a given environment, such as determined by Venter *et al.* (2004) for the Sargasso Sea and Tyson *et al.* (2004) in acid mine drainage, relied on whole genome sequencing and assembly. This methodology is, however, very time-consuming, costly and impractical for complex environmental samples with a high diversity. These problems have been overcome using the current widely applied technology of pyrotagging (Sogin *et al.*, 2006; Degnan and Ochman, 2011). This technology involves the selective amplification of a universal phylogenetic marker or pyrotag and making use of the massively parallel capacities of NGS platforms to create a profile of the diversity and richness of an environmental sample (Sogin *et al.*, 2006; Degnan and Ochman, 2011). Other revolutionary technologies such as microarrays containing universal tags to which phylogenetic markers can hybridize, known as phylochips, have also become available.

Current metagenomic diversity analysis methodology

The current pyrotagging-based strategy consist of several steps (Fig. 10.2), including selection of a phylogenetic marker, primer design and amplification of these markers, amplicon sequencing, selection of quality reads, binning and phylogenetic species richness and biodiversity analyses.

Phylogenetic marker selection, clonal amplification and sequencing

In the preliminary steps of amplicon sequencing, a suitable gene or stretch of DNA representing a phylogenetic marker or anchor is selected. The selection will depend on the type of organism, (Bacteria, Archaea or Eukarya) being studied, and the marker should be universal for the selected group. Markers can be chosen to target any

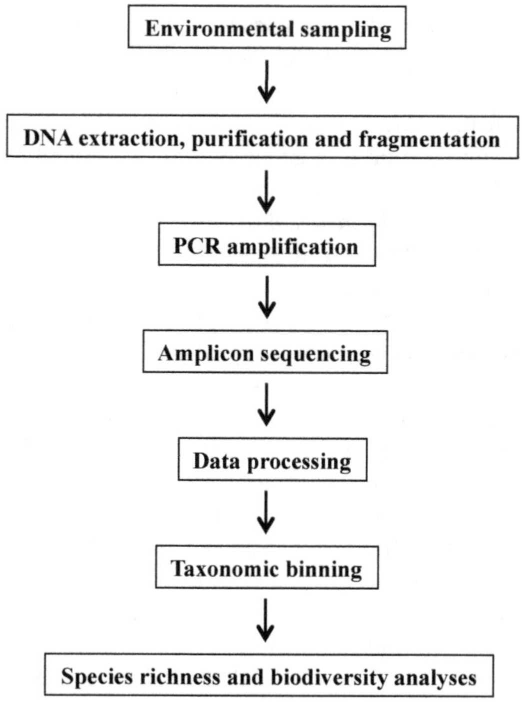

Figure 10.2 Schematic diagram of current metagenome diversity analysis protocols.

taxonomic tier, ranging from Kingdom to Genus, or to target organisms with a particular phenotype, such as methanogenesis or sulphate-reduction.

16S rRNA, as a universal phylogenetic marker among the prokaryotes, has been used to pyrotag and analyse a very wide range of bacterial and archaeal communities (Sogin *et al.*, 2006; Degnan and Ochman, 2011). Other phylogenetic markers, such as the house-keeping genes *recA* and *rpoB*, have also been utilized (Wooley *et al.*, 2010; Vos *et al.*, 2012). Pyrotagging within the eukaryal domain makes use of the 18S ribosomal small subunit (Huber *et al.*, 2007; Wooley *et al.*, 2010). For more restricted diversity studies, pyrotagging of the internal transcribed spacer (ITS) of the nuclear ribosomal locus has been used for fungi (Nilsson *et al.*, 2008), while plant and animal studies have used the maturase K (*matK*)/ribulose-bisphosphate carboxylase (*rbcI*) and cytochrome *c* oxidase subunit I (*cox1*) genes, respectively (Meusnier *et al.*, 2008; Burgess *et al.*, 2011). Several factors relating to the fragment to be amplified and sequenced must be taken into consideration. Currently, a maximum read length of ~750 bp can be achieved with the 454 GS FLX Titanium + platform, while much shorter reads lengths are obtained using other second generation sequencing platforms. Such read lengths cover only a portion of the ~1550 bp 16S rRNA gene. Furthermore, the sequenced reads should be sufficiently variable to discriminate between microorganisms in an environmental sample at the deeper taxonomic levels (Vos *et al.*, 2012). The 16S rRNA gene contains nine hypervariable regions interspersed between conserved regions (Neefs *et al.*, 1993). These conserved regions can be used to design PCR primers that are nearly universal to all bacterial and archaeal taxa which can be used to amplify the variable regions (Claesson *et al.* 2010). Various hypervariable regions have been proposed and used in different pyrotag sequencing-based studies, including V2 (Costello *et al.*, 2009), V3 (Dethlefsen *et al.*, 2008) and V4 (Wang *et al.*, 2007) regions. Similarly, different variable regions of the 18S rRNA gene have been used to study eukaryotic diversity in different environmental samples (Bråte *et al.*, 2010; Cheung *et al.*, 2010; Monchy *et al.*, 2011). The choice of hypervariable region was based on recommendations and familiarity within different research groups (Claesson *et al.*, 2010). It has been noted that some hypervariable regions are better able to assign reads to different taxa, e.g. to the genus level, than others (Claesson *et al.*, 2009). It is recommended that a combination of several hypervariable regions should be used in order to generate a more representative sampling of the population. At present, there are still discrepancies in the recommended combinations used for metagenomic phylotyping and biodiversity studies (e.g. V1/V2/V4 regions were favoured by Sundquist *et al.* (2007), while V2/V3/V4 or V2/V3 combinations of the 16S rRNA gene have been suggested by Liu *et al.* (2008) and Chakravorty *et al.* (2007).

After choosing the variable regions to be amplified and sequenced, primers are designed on the basis of the conserved regions flanking the hypervariable target. As the 454 GS-FLX pyrosequencer platform is the most commonly used for amplicon-based sequencing, the primer design based on this technology is described here. The 454 amplicon sequencing primers consist of the target (amplicon) specific primers which are linked to standardized 19-mer adapter sequencing primers; these anneals the amplicons to the capture beads used for emulsion PCR and carry a sequencing key. Two different adapter primers, A and B, are linked to the target primers which amplify in opposite directions of the phylogenetic anchor (Jarvie, 2007; Tamaki *et al.*, 2011). These adapter primers contain a sequencing key and may further be adapted by adding a multiplex identifier (MID), discussed in the section '*NGS enhancements and their application in metagenomic studies*'), which will allow multiple amplicon sets to be amplified in parallel (Meyer *et al.*, 2007). Using these adapter/MID-linked primers targeted PCR amplification, PCR from the environmental DNA sample is performed, and the resulting amplicon library can then be sequenced using an NGS platform. While the 454 Titanium/GS-FLX systems are most commonly used for amplicon sequencing, similar studies using the Illumina (Degnan and Ochman, 2011) and Ion Torrent (Jünemann *et al.*, 2012) platforms have also been undertaken.

NGS sequence data processing

As some NGS technologies are prone to sequencing errors, the first step in a typical workflow is to subject the data to rigorous quality control checks. This step is of benefit in whole metagenome sequencing projects, but is imperative for amplicon sequences. Quality control measures usually include the removal of reads that have any ambiguous base calls, have read lengths significantly longer or shorter than the typical distribution of sequence lengths, or that have inexact matches to the initial primers used for amplicon sequencing (Gilles *et al.*, 2011; Glenn, 2011). Sequences that pass the quality checks have to be trimmed to remove primers. When MID tags are incorporated, these must first be used to bin sequences prior to trimming (Degnan and Ochman, 2011).

Several software packages for NGS data processing have been developed to facilitate the process for users not familiar with bioinformatics. For example, AmpliconNoise (Quince *et al.*, 2011) is a collection of programs for the removal of errors from 454 sequenced PCR amplicons.

PRINSEQ (Schmieder and Edwards, 2011) is a web-based application for easy and rapid quality control and data preprocessing, while MOTHUR (Schloss *et al.*, 2009) facilitates a standard operating procedure (SOP) to process 16S rRNA gene sequences.

Taxonomic binning

As is the case for whole metagenome sequencing, phylogenetic marker amplicons are binned. This separation into different groups is imperative for the measurement of how many different taxa are present in a given sample. A major problem with the binning of bacterial taxa is derived from the fact that the species concept (i.e. a species is a group of organisms that remain phenotypically similar because of recombination between them (Mayr 1944)) is not well suited to prokaryotes. Bacteria generally reproduce asexually and recombination events can occur between distantly related organisms through horizontal gene transfer (HGT) (Doolittle and Zhaxybayeva, 2009). This has given rise to the concept of operational taxonomic units (OTUs), where related sequences with a similarity above a pre-determined level can be clustered into taxonomic groups (Schloss and Handelsman, 2004). Here, the most common thresholds are 3% to differentiate at the species level, 5% at the genus level, 10% at the family/class level, and 20% at the phylum level (Schloss and Handelsman, 2004).

A number of different tools, such as the Ribosomal Database Project (RDP) classifier (Cole *et al.*, 2009), Greengenes Classifier (DeSantis *et al.*, 2006) and SINA classifier associated with SILVA (Pruesse *et al.*, 2007), can be used to assign amplicon sequences to OTUs. These tools are linked to extremely large databases of aligned ribosomal DNA sequences, both complete and incomplete, and amplicon sequences can therefore be mapped against reference OTUs to generate a higher-order taxonomic context (Logares *et al.*, 2012). The latest updated version of the RDP database contains over 2.5 million aligned and annotated 16S rRNA bacterial and archaeal gene sequences (http://rdp.cme.msu.edu/). These tools can provide information on which class, order, family, genus and sometimes even species, a particular taxon, or phylotype belongs to. Other taxonomic binning tools include PPLACER (Matsen *et al.*, 2010), ARB (Ludwig *et al.*, 2004) and MEGAN (Huson *et al.*, 2009).

Phylogenetic species richness and biodiversity analyses

A number of parametric and nonparametric methods can be applied to study OTU composition and abundance within a given environmental sample (Sogin *et al.*, 2006). Diversity can be calculated in many ways (Magurran, 2004; Lozupone and Knight, 2008), but three main distinctions are possible. The first is a focus on the measurement of the diversity within a community (alpha-diversity) as compared to the diversity between two or more communities (beta-diversity) (Whittaker *et al.*, 2001). Secondly, the analysis can be either 'qualitative,' examining only presence-absence data, or 'quantitative,' when relative abundances are taken into account (Lozupone and Knight, 2008). Finally, OTUs can be treated as equally related to one another (taxon-diversity), or can consider phylogenetic distance (phylogenetic-diversity) or lifestyles and ecological roles (functional-diversity) (Magurran, 2004).

Another important issue is the identification of factors that contribute to the clustering of compositionally similar communities. Several approaches exist for elucidating diversity relationships among samples, including cluster analyses (where samples are assigned to discrete groups), ordination methods (where samples are arranged in low-dimensional space), and explicit hypothesis testing methods (e.g. ANOVA and Mantel tests) (Ramette, 2007).

A number of software tools to study taxon diversity are available, including Species Prediction and Diversity Estimation (SPADE) package (Chao and Shen, 2005), EstimateS (Colwell, 2005), and various packages for R, including SPECIES (Wang, 2011). Simple beta-diversity comparisons between two sample datasets can be done using the RDPs LIBCOMPARE tool, where a statistical test estimates the likelihood that taxon membership is the same between the two samples (Cole *et al.*, 2009). Phylogenetic diversity can be assessed using Phylocom (Webb *et al.*, 2008) and UniFrac (Lozupone *et al.*, 2006). Functional diversity can be calculated with FD package for R (Laliberté and Legendre, 2010) or FDiversity (Casanoves et al., 2011). Cluster analysis, ordination and hypothesis testing can be performed using R software (R Development Core Team, 2008), PAST (Hammer et al., 2001) or PRIMER v6 (Clarke and Gorley, 2006). MOTHUR (Schloss et al., 2009), QIIME (Caporaso et al. 2010) and VAMPS (http://vamps.mbl.edu/index.php) can be used for visualization and production of publication-quality graphics.

Pitfalls with metagenomic diversity study protocols

Several factors relating to the processes of DNA amplification, sequencing and assembly can result in problems in amplicon-based metagenome diversity studies, leading to the under- or over-estimation of the abundance of some species (Acinas et al., 2005). The first problem is the introduction of PCR bias during amplification and library preparation (Shokralla et al., 2012). Factors contributing to PCR bias include differences in the amplification efficiency of templates; e.g. influenced by the copy number of target genes, low G+C contents and secondary structures of rRNA (Qui et al., 2001), or the inhibited amplification of less abundant DNA fragments by more abundant templates (Suzuki and Giovanoni, 1996). A number of solutions to amplification bias have been developed, including adjusting the annealing temperature, increasing the denaturation ramping rate, increasing template concentrations and lowering the number of amplification cycles (Qui et al., 2001; Shokralla et al., 2012).

Another technical problem is the inability of many NGS platforms to read homopolymer regions, resulting in insertion/deletion errors. This may result in sequencing errors being interpreted as novel phylotypes (Sogin et al., 2006; Shokralla et al., 2012). Several software tools are available for the filtering and removal of these erroneous sequences (Shokralla et al., 2012). Similarly, PCR chimeras result from incomplete extension of a template sequence, where two templates anneal to form a recombinant molecule which is amplified in subsequent PCR cycles. While chimeras usually occur between highly similar sequences, they can, in rare cases, also occur between phylogenetically distant sequences, and thereby produce novel sequences which are substantially different to both parental sequences (Wooley and Ye, 2010; Logares et al., 2012). When present at high frequencies, these chimeras can also lead to inaccurate diversity measurements (Logares et al., 2012). Novel protocols, such as emulsion PCR, have been developed to minimize chimera generation (Williams et al., 2006). A number of software packages, including Bellerophon (Huber et al. 2004), Pintail (Ashelford et al., 2005), Perseus (Quince et al., 2011) and ChimeraSlayer (Haas et al., 2011), have been developed for the filtering and removal of chimeras from sequence datasets prior to analysis.

Applications of metagenomic diversity studies

Thousands of studies using amplicon-based metagenomic approaches to study diversity have been published. These have mostly looked at the diversity of microorganisms associated with a wide range of environments (the 'microbiome'), including fresh water (Telias et al., 2011), marine water (Herlemann et al., 2011), soil (Daniel, 2005), extreme environments such as Antarctic and hot deserts (Smith et al., 2006; Andrew et al., 2012), deep sea hydrothermal vents (Xie et al., 2011) and many more. The microbiomes associated with various host species, including plants (Lundberg et al., 2012), insects (Suen et al., 2010) and different parts of the human body (Turnbaugh et al., 2007; Zhao, 2010) have also been analysed. One important finding, derived from a reanalysis of the data from the Sargasso Sea metagenome, has been the identification of a potential fourth domain of life (Wu et al., 2011). The comparison of metagenome amplicon datasets has also provides extensive insights into the differences in diversity between environmental samples over space and time and in response to biotic and abiotic stimuli. For example, amplicon dataset comparisons have shown fundamental differences in the microorganisms associated with the gut of patients with inflammatory bowel disease and healthy individuals (Willing et al., 2010), and the changes in environmental microbial communities in response to faecal pollution (Wu et al., 2010a).

Many of the projects focusing on the cataloguing of microorganisms in the human body have fallen under the consortium of the Human Microbiome Project (Turnbaugh *et al.*, 2007), which provides funding for metagenomes sequenced from this host to answer specific biological questions. Similarly, large consortia have been created for the study of the microbiomes in all of the earth's biospheres. These include the Earth Microbiome Project (Gilbert *et al.*, 2010b), the Microbial Earth Project (Gilbert *et al.*, 2010b) and the Terragenome project focusing on soil (Vogel *et al.*, 2009). These consortia will aim to sequence terabases of metagenomes, thereby providing a means of tapping into the wealth of microorganisms found on earth, including both culture and unculturable bacteria, archaea, viruses, fungi and other eukaryotes.

Diversity studies using phylogenetic microarrays

Another means by which bacterial diversity in environmental samples has been studied is through the use of microarrays of phylogenetic markers. This methodology was first employed by Guschin and co-workers, who developed oligonucleotide probes based on the 16S rRNA gene sequences of several nitrifying bacterial genera and immobilized these on a glass slide (Guschin *et al.*, 1997). Palmer *et al.* (2007) used 16S probes targeting over 1,600 bacterial and archaeal species to study the temporal and interpersonal variation in faecal microbiota among children. These microarray-based technologies typically target selected taxa, based on the sample being studied. However, a commercially available Affymetrix-based microarray, Phylochip, was designed with the intention to cover all available 16S rRNA genes (DeSantis *et al.*, 2007), including OTUs representing thousands of families and subfamilies (Paliy and Agans, 2012). Phylochips have been used to study bacterial diversity in a number of environments, including the human stomach (Cox *et al.*, 2010), soil (Deangelis *et al.*, 2011) and plant rhizosphere communities (Mendes *et al.*, 2011). Advantages of these phylogenetic microarrays include their moderate cost and the rapid processing and analysis of samples (Tringe and Hugenholtz, 2008). This is facilitated by dedicated software, PhyloTrac, which can be utilized to identify the phylotypes in a given sample (Tringe and Hugenholtz, 2008). Furthermore, these phylogenetic microarrays can provide quantitative and abundance information by monitoring the intensity of hybridization (Paliy and Agans, 2012). However, there are several drawbacks with the use of phylogenetic microarrays such as the PhyloChip. One is that these arrays are capable of identifying only those phylotypes which are targeted on the array (Tringe and Hugenholtz, 2008). The *de novo* design of these phyloarrays is also not straight-forward, requiring extensive design and validation prior to use (Paliy and Agans, 2012).

Functional metagenomics and bioprospecting

One of the key purposes of metagenomic analysis is for the exploitation of a metagenome as a genetic resource, and a basis for the discovery of novel proteins and enzymes. The mining of metagenome libraries, also known as bioprospecting, enables access to the genes and protein products of communities of organisms, both the culturable, and the uncultured majority (Schloss and Handelsman, 2003; Handelsman, 2004). This approach has been successfully applied for discovery of a range of novel proteins including novel antibiotics, lipases, esterases, cellulases, DNA polymerases, proteases and antibiotics (Tuffin *et al.*, 2009; Simon and Daniel, 2011).

Current bioprospecting methodology

Bioprospecting protocols generally include the following steps: (i) cloning of environmental metagenomic DNA into a vector for the construction of clone libraries; (ii) transformation into a suitable host; and (iii) phenotypic and genotypic screening to identify novel proteins and enzymes with a desired activity (Simon and Daniel, 2011; Iqbal *et al.*, 2012). Screening protocols are generally categorized as function-based and sequence-based approaches.

Construction of clone libraries

Clone library preparation commences with the generation of DNA fragments of a pre-selected average size, either by mechanical shearing or restriction enzyme treatment. This is followed by the insertion of the DNA fragments into a suitable vector system (Daniel, 2005; Simon and Daniel, 2011; Ekkers *et al.*, 2012). The choice of vector is dependent on several factors, including DNA quality, the target genes, insert size and screening strategy (Daniel, 2005). Plasmids can carry inserts of up to 15 kb, and because of their high copy numbers can result in stronger expression of target genes (Dionisi *et al.* 2012). Fosmid and cosmid vectors are capable of holding inserts up to 40 kb in size, while bacterial artificial chromosomes (BACs) can host inserts up to 350 kb. These are maintained in the bacterial host in low copy number, but can be used for the cloning of large gene clusters coding for complete pathways (Handelsman *et al.*, 1998; Simon and Daniel, 2011). The resulting clones will each contain a DNA fragment from the metagenome community (Dionisi *et al.*, 2012). It must be stressed that while the preparation of a clone library fully representing the selected metagenome is the desired outcome of the cloning process, in practice this may not be possible. Biases introduced by the sampling and lysis protocols, by selection of a particular DNA fragment size range, and by the dominance of certain organisms within a community will all result in a library which is not fully representative (Ekkers *et al.*, 2012). It is therefore important to take the abundance of organisms within the environment from which the metagenomic DNA was obtained into account when selecting the vector system, as well as the host into which the clone libraries are subsequently transformed (Ekkers *et al.*, 2012).

Transformation and heterologous expression in a suitable host

Vector libraries are subsequently transformed into a suitable heterologous host, in which the genes or pathways present in the inserts are expressed. The most frequently used host is *E. coli* for which, because of its status as a model host, a number of effective gene expression strategies have been developed (Ekkers *et al.*, 2012). However, the heterologous expression in *E. coli* is not always feasible, as is described in further detail in the section '*Pitfalls of current functional metagenomics protocols*'.

Sequence-based screening

Sequence-based screening makes use of hybridization or PCR amplification to detect and identify genes encoding target proteins and enzymes in the metagenomic clone library (Schloss and Handelsman, 2003; Simon and Daniel, 2011). This process relies on the design of DNA probes or primers on the basis of conserved regions of genes encoding known proteins or protein families. In consequence, one of the major drawbacks of this approach is that only novel variants of previously identified functional protein families can be identified (Simon and Daniel, 2011; Dionisi *et al.*, 2012). Regardless of this, however, a large number of novel enzymes in characterized protein classes have been discovered using sequence-based metagenome screening. These include alkane monooxygenases (Xu *et al.*, 2008), nitrite reductases (Bartossek *et al.*, 2010), [NiFe] hydrogenases (Maróti *et al.*, 2009) and chitinases (Hjort *et al.*, 2010). An approach by which complete gene sequences can be determined, termed gene-targeted (GT)-metagenomics, makes use of PCR-based screening in combination with massively parallel NGS of the amplicons to design probes for the recovery of the entire full-length genes (Iwai *et al.*, 2010). The sequencing of clones which carry phylogenetic anchors (e.g. bacterial 16S rRNA and archaeal DNA repair gene *radA*) has provided insights into the organisms with which a certain gene or function is associated. For example, a clone carrying a bacterial 16S rRNA gene showed that bacteriorhodopsin genes are not only associated with Archaea, as previously assumed, while heterologous expression in *E. coli* showed that this protein is functional in bacteria (Beja *et al.*, 2000). This anchor-guided approach is of particular use in providing information on the physiology and function of taxa in complex environmental samples, where reassembly of genomes from the metagenome is not possible (Handelsman, 2004). With the increasing availability of complete genome sequences, more phylogenetic markers are becoming available. While less robust, codon usage and G+C content can also be used to link a function to a particular taxon (Schloss and Handelsman, 2003).

An alternative to the marker-driven approaches is the mass random sequencing of clones in the metagenomic library (Handelsman, 2004). The introduction of NGS technologies has also made this approach much more accessible. The massive throughput from these technologies and the use of associated bioinformatic tools to detect and identify conserved functional motifs in proteins and enzymes allows hybridization and PCR methods to be by-passed and enables the use of more flexible homology searching (Iqbal *et al.*, 2012). A relatively novel field, 'synthetic metagenomics', has recently been developed. This approach relies on the chemical synthesis and cloning- and expression-optimization of genes of interest, which are then cloned and expressed in a heterologous host (Bayer *et al.*, 2009; Iqbal *et al.*, 2012)

Function-based screening

Function-based screening approaches involve the detection or selection of desired functions from a metagenomic clone library expressed in a heterologous host (Simon and Daniel, 2011). As discussed above, the heterologous expression is dependent on the correct transcription, translation, expression and, in some cases, secretion of the genes/gene products of interest (Handelsman, 2004). This methodology, unlike sequence-based approaches, can be used to identify entirely novel genes and/or functions (Simon and Daniel, 2011). However, the number of clones in a metagenomic library expressing any given function is generally very low. For example, only 1 in 730,000 clones from a metagenome derived from soil showed lipolytic activity (Henne *et al.*, 2000). There is thus a need for high sensitivity screening methods in order to identify rare functions in a metagenomic library. Three major function-based approaches are currently utilized for screening: (i) phenotypic insert detection (PID), (ii) modulated detection (MD) by heterologous complementation of hosts or mutants and (iii) substrate-induced gene expression (SIGEX).

PID relies on subjecting the metagenomic clone library to phenotypic assays for the detection of specific functions (Dionisi *et al.*, 2012). Direct detection is facilitated by directly observing alterations in colony morphology, pigmentation or halo formation which may result from the expression of the heterologous gene (Ekkers *et al.*, 2011). More commonly, in the absence of directly detectable phenotypes, metagenomic clones are grown in a medium containing a chemical dye or chromophore-linked substrate in order to identify those clones with the desirable phenotype, which can break down the substrate to release the chromophore (Simon and Daniel, 2011). The direct detection and indicator medium approaches can also be adapted to use microfluidic approaches in nanolitre volumes and automated using colony pickers and microplate readers in order to develop high-throughput screening protocols (Ekkers *et al.*, 2012). These approaches have been used for the detection and identification of a number of different phenotypes including the production of small molecules, metabolism and catabolism of particular substrates as well as antibiotic production (Ekkers *et al.*, 2012). Another PID approach relies on the co-expression of the target gene and a reporter gene, e.g. *lacZ*, and has been used for the identification of enzymes that inhibit quorum sensing-based intracellular communication (Schipper *et al.*, 2009; Ekkers *et al.*, 2012).

MD, through heterologous complementation, depends on the presence of the target gene being a prerequisite for growth of the clone host or mutant (Simon and Daniel, 2011). Phenotypes and genotypes thus tested include factors for antibiotic resistance (Donato *et al.*, 2010) and the production of DNA polymerases (Simon *et al.*, 2009). This method allows for the highly specific and fast detection of clones with desirable traits from very large clone libraries, with very few false positives (Simon and Daniel, 2011; Dionisi *et al.*, 2012). It must be noted, however, that only a limited number of phenotypes can be selected for, which limits the usefulness of this approach for many function-based screening studies (Handelsman, 2005).

SIGEX and related methodologies make use of specific expression vectors that produce fluorescence linked to the activity related to the protein of interest (Dionisi *et al.*, 2012; Ekkers *et al.*, 2012). SIGEX was designed specifically for the detection and recovery of catabolic genes whose expression is induced by specific substrates. A promoter- or operon-trap green fluorescent protein (GFP) expression vector is used, where induction of a target gene in response to the substrate leads to co-expression of the GFP protein, thereby resulting in a fluorescent signal (Uchiyama *et al.*, 2005). Fluorescence-assisted cell sorting (FACS) then allow the separation of the fluorescent clones (Dionisi *et al.*, 2012; Ekkers *et al.*, 2012). As the proximity of regulatory elements to the target gene is a prerequisite for cloning and expression, those targets where transcriptional regulatory are located some distance from the target gene or are constitutively expressed are often not be detected (Dionisi *et al.*, 2012). Another drawback of SIGEX is that non-specific substrates can activate the transcriptional regulators, resulting in false positives (Simon and Daniel, 2011). Several variants of SIGEX have been developed. METabolite-Regulated EXpression (METREX) can be used for the detection of small biologically active signal molecules (Williamson *et al.*, 2005). When threshold concentrations of the small molecule encoded by the metagenomic fragments are reached, a quorum sensing promoter linked to the gfp gene results in the production of GFP which can be detected using fluorescent microscopy or FACS (Williamson *et al.*, 2005; Simon and Daniel, 2011). In another variant, Product-Induced Gene Expression (PIGEX), *gfp* gene expression is triggered by product formation. This has been used, for example, to screen for amidases, where the transcriptional regulator BenR which responds to benzoate incorporated into the clone construct controlled the expression of *gfp*. Upon benzoate production, those clones containing amidases fluoresce (Uchiyama and Miyazaki, 2010). Although these different methodologies have been treated independently, they can also be combined (e.g. SIGEX can be combined with phenotypic insert detection assays; Ekkers *et al.*, 2012) to develop optimized screening protocols.

Applications of metagenomic approaches for bioprospecting

The application of sequence- and function-based screening of metagenomic libraries from a wide range of environments has resulted in the identification and isolation of thousands of novel proteins and vastly expanded our knowledge of the protein universe. For example, the Sorcerer II Global Ocean Sampling (GOS) resulted in a doubling of the known protein families (Yooseph *et al.*, 2007).

Within the range of metagenomic 'protein products' are a number of novel biocatalysts and enzymes which have potential or actual biotechnological applications. Here, we present a few examples of industries that have benefited from molecules identified using metagenomic approaches. One of the key areas where metagenomic bioprospecting has been of extensive value is industrial biotechnology (Dionisi *et al.*, 2012). It has been predicted that biotechnological means will be used for the production of 20% of all chemical entities by 2020 (Meyer and Turner, 2009).

A sequence-based metagenomic bioprospecting study, targeting the cow rumen microbiome, yielded a number of biomass-degrading enzymes, including endoglucanases, glucosidases and cellobiohydrolases and other activities of potential application in biological fuel production (Hess, 2011). Other examples of industrially important enzymes isolated from metagenomes include glycerol and diol dehydratases, used for the production of propanediol (Knietsch *et al.*, 2003), and a high performance amylase which can be used, for example, in textile de-sizing, baking and paper recycling (Richardson *et al.*, 2002). Proteases extracted from metagenomic libraries from the Gobi and Death Valley desert soils have application in various industrial processes including stain removal, leather preparation and biofilm removal (Neveu *et al.*, 2011). Synthetic metagenomics has also been employed to produce industrially relevant enzymes, such as methyl halide transferases used in synthesizing agricultural fumigants (Bayer *et al.*, 2009) and glycoside hydrolases (Allgaier *et al.*, 2010). In these studies, genes identified from metagenomes were chemically synthesized and heterologously expressed in a host in order to analyse their function.

Metagenomic bioprospecting has also been applied in the agricultural and environmental biotechnology fields, where enzymes and proteins mined from metagenome libraries can potentially be used to protect environmental resources, including water, soil and energy, as well as to develop sustainable crop production and agricultural practices (Dionisi *et al.*, 2012). A tannase capable of hydrolysing various tannin derivatives, isolated from a cotton field metagenome, could be applied in the degradation of tannin-polluting industrial and agricultural effluents, or as a clarifying agent in the tea, fruit juice or beer manufacturing process (Yao *et al.*, 2011). Copper ATPases and xenobiotic dioxygenases recovered from soil metagenomes can be used for the removal of copper and herbicide contaminants, respectively (De la Iglesia *et al.*, 2010; Zaprasis *et al.*, 2010).

Another field on which metagenomic bioprospecting has great potential is pharmaceutical biotechnology. Metagenomics approaches have the capacity to yield completely novel antimicrobial peptides, and reveal their biosynthetic pathways. Soil metagenomic bioprospecting has already yielded novel antibiotics such as turbomycins (Gillespie *et al.*, 2002), terragenines (Wang *et al.*, 2000) and novel *N*-acyltyrosine variants (Brady *et al.*, 2002). The novel antibiotic patellamide was isolated from the marine ascidian, *Prochloron* (Long *et al.*, 2005). Metagenomic screening methods have also provided insights into antimicrobial resistance mechanisms in environmental communities (Schmieder and Edwards, 2012). For example, a novel tetracycline resistance determinant, *tet37*, was identified in metagenomic libraries constructed from the human oral cavity microbiome (Diaz-Torres *et al.*, 2003).

One of the major concerns with the surge of metagenomic bioprospecting is that many of these novel enzymes and other products will be subject to patent protection, their metagenomic origins will not be released by the manufacturers, and the novel products will be unavailable to the public (Ekkers *et al.*, 2012). However, the publicly available database MetaBioMe provides access to over 510 'commercially useful enzymes' (CUEs) with potential applications in the agricultural, biosensor, biotechnology, energy, environmental, food and nutrition, and medical industries (Sharma *et al.*, 2010). This database can be used for the sequence-based identification of CUE homologues in metagenomic libraries.

With the rapid technological advancement of molecular biotechnology and the increasing automatization of screening approaches, it is envisaged that many novel screening technologies will be developed in the near future. Some promising technologies that can be used in functional screening approaches include chip- or microarray-based technologies, microfluidic devices, flow cytometry and fluorescence-based assays (Ekkers *et al.*, 2012). It is also likely that more sophisticated approaches combining both sequence- and function-based screening will become core bioprospecting technologies, in particular where genes identified directly from metagenomic sequencing will be used to develop highly targeted functional screens (Eyers *et al.*, 2004). This approach has already been used successfully for the isolation and characterization of xylose isomerases (Parachin and Gorwa-Grauslund, 2011) and lipases (Wang *et al.*, 2010) from soil metagenomic libraries.

To date, soil (Ekkers *et al.*, 2012), marine (Dionisi *et al.*, 2012) and freshwater (Ghai *et al.*, 2011) habitats have been most frequently targeted for metagenomic bioprospecting. However, there is increasing interest in specialist environments, particularly those exhibiting 'extreme' characteristics (such as hot and cold environments, deserts, sites with low pH, high salinity or alkalinity, high pressure, etc.). Such environments not only provide a wide range of novel genetic elements that have evolved as adaptive strategies, but many have highly unique microbial communities.

Pitfalls of current functional metagenomics protocols

Functional metagenomics has become a well-entrenched tool for the mining of microbial communities for novel products and pathways. However, several major hurdles limit the effectiveness of these technologies. One, in particular, relates to the effective expression of target genes in any heterologous host. Due to factors such as different codon usage, G+C content, failure to recognize promoters, improper protein folding, absence of required co-factors, toxicity of the expressed protein to the host or failure to secrete the protein, expression of genes from the metagenomic clone library usually has relatively low efficiency (Dionisi *et al.*, 2012; Ekkers *et al.*, 2012). An experiment using *E. coli* as heterologous host for the expression of various enzymatic activities was able to express only 40% of the total enzymatic activities (Gabor *et al.*, 2004).

One means to counteract this inefficiency in heterologous expression is through the genetic modification of *E. coli* strains. For example, a strain with an altered S1 ribosomal protein was able to express high G+C content genes 12-fold more efficiently than an unmodified *E. coli* host (Bernstein *et al.*, 2007). An alternative solution is through the use of a different heterologous host, such as *Streptomyces*, *Pseudomonas*, *Bacillus* and *Burkholderia* (Ekkers *et al.*, 2012). Multiple hosts, used either sequentially or in parallel, can substantially improve the chances of successful gene expression (Ekkers *et al.*, 2012).

Metagenomic libraries typically provide poor genomic coverage of organisms that are present in low numbers in a complex environment (Eyers *et al.*, 2004). In order to effectively access genomes occurring at low frequencies, very large libraries (10^5–10^7 clones, depending on the insert size) are required. At these scales, the use of automated equipment for high-throughput colony-picking, inoculation and the use of microfluidic and microtitre plates technologies is essential (Eyers *et al.*, 2004; Ekkers *et al.*, 2012). Alternatively, enrichment strategies may be used to reduce the subset of metagenomic clones to be screened (Eyers *et al.*, 2004). Several effective enrichment strategies have been developed, including the selection of high G+C content DNA by differential centrifugation (Schleper *et al.*, 1998), and bromodeoxyuridine (BrdU) enrichment and immunocapture of metabolically active cells (Urbach *et al.*, 1999). Stable isotope probing (SIP) uses a heavy isotope-labelled substrate where enriched DNA is recovered by density gradient centrifugation (Radajewski *et al.*, 2000). However, it is argued that enrichment protocols result in preferential selection of faster growing organisms and loss of the microbial diversity which is key to the discovery of novel genes and proteins (Daniel, 2004).

Metagenomics and 'meta-omics' approaches for ecosystems biology studies

With the growth of genomics and metagenomics, an emerging field of research in ecology has developed, termed 'ecosystems biology'. This approach aims to study the functional interactions occurring within an ecosystem (i.e. the metabolic pathways, cellular processes and molecular interactions taking place within a complex community) (Siggins *et al.*, 2012). However, it is widely recognized that metagenomic DNA sequences alone cannot give a complete picture of the functional processes within an ecosystem. Metagenomic DNA sequences reveal the genes that are present in the ecosystem, but not the expression levels of these genes, the protein titres produced, or the roles played by metabolites in the metabolic processes within the ecosystem (Raes and Bork, 2008; Röling *et al.*, 2010). This has led to the integration of a number of novel technologies and methodologies, termed 'omics' which can be used in combination (systems biology). These 'omic' technologies include metatranscriptomics, metaproteomics and metametabolomics (Kandpal *et al.*, 2009; Zhang *et al.*, 2011). Collectively, these approaches can be used to assess functional interactions in mixed communities within an environment (Siggins *et al.*, 2012). A range of other 'omics' technologies have also been developed, including microRNomics (the study of micro-RNAs), interferomics (RNA interference) and interactomics (protein–protein interactions) (Kandpal *et al.*, 2009).

Metatranscriptomics

The term transcriptomics refers to the study of transcribed and expressed genes, by determining the mRNA transcriptional profile within a sample (McGrath *et al.*, 2008; Warnecke and Hess, 2009). Transcriptomics can give insights into the biological or functional activity within an organism at a specific time, as well as enabling the study of RNA-based gene regulation (Sorek and Cossart, 2010). This technology has been extensively used to study the expression of genes and associated phenotypes in response to different biotic and abiotic stimuli, for example the differential gene expression in response to diel and circadian rhythms in the malaria vector *Anopheles gambiae* (Rund *et al.*, 2011), gene expression in human skin during cutaneous wound healing (Nuutila *et al.*, 2012), gene expression linked to fluoroquinolone resistance (Yamane *et al.*, 2012) and zinc response (Lee *et al.*, 2005) in *E. coli*. Its usefulness in studying gene expression in mono-isolates has also led to its incorporation in the metagenomics field, in what is termed metatranscriptomics, whereby the functional activity and gene regulation can be studied at the community level. Traditionally, transcriptome analysis consisted of the following steps: (i) total RNA extraction, (ii) RNA enrichment, (iii) cDNA synthesis, (iv) an optional RNA amplification step, followed by (v) microarray hybridization and profiling or sequencing of cDNA clone libraries (Warnecke and Hess, 2009; Desai *et al.*, 2010). DNA microarrays are of limited value to metatranscriptomics because genes or complete genome sequences are prerequisites for the design of microarray chips and hence cannot cover the unculturable majority in a metagenome sample. Furthermore, they are limited by their low detection and quantitative sensitivity (Darby and Hall, 2008). Similarly, cloning biases resulting from the targeting of abundant transcripts represent a major restriction in the use of cDNA libraries in metatranscriptomic studies. However, metatranscriptomics has greatly benefited from the advent of NGS-based transcriptome sequencing technologies. NGS transcriptomics, referred to as 'RNA-seq' technology, enables the high-throughput sequencing of transcriptomes without the requirement of a cloning step or access to the genome sequences from which the transcripts were derived (Wang *et al.*, 2009).

Effective protocols for the extraction of RNA from various environments have been established (Poretsky *et al.*, 2005; Bailly *et al.*, 2007). One factor that must be taken into consideration is that RNA is far less stable than DNA, with mRNA transcripts of *E. coli* and *Bacillus subtilis* having half-lives of ~5 minutes (Moran *et al.*, 2012). Therefore, samples must be snap-frozen immediately or stored in a stabilizing buffer (e.g. RNAlater, Ambion, USA) that protect RNA fragments from nuclease digestion (Warnecke and Hess, 2009). Following extraction, it is advisable (though not essential) to enrich the messenger RNA (mRNA) transcripts. In a typical environmental sample, mRNA contributes only 1–5% of the total RNA, with the larger fraction made up of rRNA and tRNA elements (McGrath *et al.*, 2008). Several methods for removal of the rRNAs and tRNAs have been developed, including subtractive hybridization, exonuclease treatment, selective polyadenylation of mRNAs and antibody capture (Sorek and Cossart, 2010; Filiatrault, 2011). The MICROBExpress kit (Ambion, USA) makes use of probes for conserved regions in bacterial 16S and 23S rRNAs. These probes are attached to magnetic beads, which hybridize and remove the rRNA fraction from the mRNA (Warnecke and Hess, 2009; Sorek and Cossart, 2010). Another method, NSR-SEQ (Not-So-Random) enriches mRNA transcripts through hexamer priming during cDNA synthesis (Armour *et al.*, 2009). A number of different protocols for the preparation of cDNA for RNA-seq have been developed. Typically these involve the conversion of RNA into cDNA by means of hexamer-primed reverse transcription (Croucher and Thomson, 2010). This was initially followed by second strand synthesis, but this resulted in a loss of information on the direction of transcription, as both sense and antisense strands are available in equal proportions. Various methods for maintaining the directional signals of transcripts have been developed, including the use of single strand cDNA, ligation of adapters in an orientation-specific manner, or the addition of sequence tags into the primers (Croucher and Thomson, 2010).

Where insufficient RNA is available for library preparation for RNA-seq, RNA amplification can be incorporated into the protocol (Filiatrault, 2011). Several different approaches have been developed. These include amplification by random hexamer priming and T7 RNA polymerase (Frias-Lopez *et al.*, 2008), multiple strand displacement amplification (MDA) of cDNA (Gilbert *et al.*, 2008) and cDNA amplification through emulsion PCR (Blow *et al.*, 2008). Subsequently, the various NGS platforms can be applied to sequence the cDNA fragments. As is the case for the construction of genomes or larger fragments from metagenomic samples, longer mRNA transcript read lengths are desirable. As only the transcripts are targets for sequencing, the required sequence throughput is also much lower (Warnecke and Hess, 2009).

The cDNA synthesis step presents several problems to the RNA-seq workflow, including difficulties with the preparation of strand-specific libraries, template switching where the synthesizing cDNA strand dissociates from template RNA, anneals to a different RNA transcript and generates chimeric cDNAs, as well as the introduction of biases (Ozsolak and Milos, 2011). Novel methods have been developed which bypass the conversion of RNA transcripts to cDNA templates (Ozsolak *et al.*, 2009; Ozsolak and Milos, 2011). The direct RNA sequencing (DRS) approach captures the RNAs poly(A) tails on a surface coated with poly(dT) oligonucleotides prior to sequencing, thereby by-passing the cDNA synthesis step (Ozsolak *et al.*, 2009).

Numerous studies over the past few years using cDNA-based RNA-seq and direct RNA sequencing approaches have yielded hundreds of metatranscriptomes from a wide range of environments, including marine (Gilbert *et al.*, 2008; Poretsky *et al.*, 2010; Shi *et al.*, 2012) and soil (Urich *et al.*, 2008) communities.

Metaproteomics

Proteomics is the characterization of the protein complement, or proteome, of an expressed genome (Keller and Hettich, 2009). Similarly, metaproteomics can be defined as the identification of all proteins expressed within an environment at a given time (Wilmes and Bond, 2004). The analysis of these proteins can give vital information on the function of proteins within an environment, their role in metabolic pathways, as well as providing insights into responses to external changes, protein–protein interactions, post-translational modification and protein stability (Keller and Hettich, 2009). A typical workflow for metaproteomic studies involves the following steps: (i) protein extraction, (ii) protein separation and fractionation, (iii) mass spectrometric analysis, (iv) database comparison and (v) interpretation (Siggins *et al.*, 2012). Protein extraction is dependent on the environment, and the type of sample being analysed, and whether intra- or extracellular proteins are being studied. As proteins are present in much lower copy number than DNA and RNA, and cannot be amplified, protein extraction from environmental samples can be challenging (Siggins *et al.*, 2012). Humic acids in soils are commonly co-extracted and interfere with protein extraction, quantification and identification (Bastida *et al.*, 2009). To gain access to intracellular proteins, a cellular lysis step, typically sonication or bead beating, is incorporated. Proteins can then be separated into cytoplasmic, membrane and extracellular fractions by centrifugation (VerBerkmoes *et al.*, 2009). Proteins subfractions are further fractionated, either by two-dimensional gel electrophoresis (2D-GE) of high resolution liquid chromatography. In 2D-GE, the proteins are first separated according to their isoelectric point (pI) by isoelectric focusing (IEF), followed by separation according to their molecular weights with SDS-PAGE (Desai *et al.*, 2010). Individual protein spots are extracted from the gel and fragmented to peptides by proteolysis (e.g. using trypsin) (VerBerkmoes *et al.*, 2009). Alternatively, proteins can be separated by 2D-liquid chromatography (2D-LC), on the basis of charge and hydrophobicity, followed by proteolytic cleavage of individual protein fractions (VerBerkmoes *et al.*, 2009). The peptide mixtures are then subjected to mass spectrometry (MS) analysis. Mass spectrometers consist of three component parts: an ionizer which converts the peptide mixture into gas phase ions; a mass analyser which separates the ions based on their mass/charge ratio (m/z); and an ion detector. The two most commonly used MS approaches are the electrospray ionization (ESI)-MS and matrix-assisted laser desorption ionization (MALDI). These are also known as soft ionization technologies, as they can be used for the fingerprinting of whole proteins or peptide fragments, without extensive degradation of the peptide substrates (Bakthiar and Nelson, 2000). Both these MS are now mostly used in combination with a Time of Flight (TOF) analyser, which measures the time taken by ions to travel down a tube to the detector to determine the m/z ratio (Bakthiar and Nelson, 2000). The MS spectra obtained for peptide fragments and proteins are then compared to extensive databases of known proteins and peptide fragments, and those predicted from genomic sequence data (VerBerkmoes *et al.*, 2009). Several proteomics databases are available, including the PRoteomics IDEntification database (PRIDE) and the Open Proteomics Database (OPD) (Prince *et al.*, 2004; Vizcaíno *et al.*, 2010). A number of algorithms, such SEQUEST (Eng *et al.*, 1994), MASCOT (Perkins *et al.*, 1999) and X-tandem (Craig and

Beavis, 2004) have been developed to interpret the peptide amino acid sequence from MS spectra, which can then be used to search protein databases (VerBerkmoes *et al.*, 2009). In some cases peptide fragment and protein MS fingerprints cannot be linked to any known proteins in the databases. In this event, peptide fragments can be further degraded into short peptides, which could then be matched to known proteins or peptide regions. This can be done using two sequential MS analysers, separated by a Collision-Induced Dissociation (CID) cell where the ions are excited, resulting in fragmentation of peptides, thereby improving the sensitivity and detection limit of MS analysis (VerBerkmoes *et al.*, 2009).

Metaproteomics forms an integral part of systems biology. In combination with metagenomic, metatranscriptomic and metabolomic approaches, it can be used to gain key insights into microbial metabolic, cellular and community activities. It can also serve as a powerful tool for novel protein discovery (Simon and Daniels, 2011; Siggins *et al.*, 2012). While metaproteomics studies focussing on microorganisms inhabiting various environments, including soil and fresh water (Benndorf *et al.*, 2007), marine water (Morris *et al.*, 2010), the human gut and saliva (Rooijers *et al.*, 2011; Rudney *et al.*, 2010) and other natural and man-made environments have been reported, the integrated use of 'omic' technologies is still in its infancy, and is predicted to become one of the major growth areas of environmental sciences in the next decades (Siggins *et al.*, 2012).

Metabolomics

The term metabolomics refers to the study of the cellular metabolites (i.e. low molecular weight compounds) generated by an organism or group of organisms within a given environment or under specific conditions (Turnbaugh and Gordon, 2008). The impetus for studying these metabolites is that they play a critical role in metabolic pathways and the biochemical network within an ecosystem. For example, it has been noted that the regulation of the bacterial glycolysis pathway is 30% by gene expression and 70% by metabolism and metabolites (ter Kuile and Westerhoff, 2001). The clear implication is that mapping the roles of DNA, mRNA transcripts and proteins is insufficient in order to understand global biochemical networks, and that the metabolome is a key determinant of network function. A metabolome is generally considered to consist of intracellular (endometabolome) and extracellular (exometabolome) metabolites, which are analysed using metabolome fingerprinting and footprinting, respectively (Mapelli *et al.*, 2008). These approaches are complemented by metabolite profiling, which involves the identification and quantification of groups of metabolites with similar physical and chemical properties or those that participate in the same metabolic pathway, and metabolite target analysis, where one or a few targeted metabolites are quantified (Mapelli *et al.*, 2008; Desai *et al.*, 2010). Various analytical techniques are employed in metabolomic studies, including mass spectrometry, nuclear magnetic resonance (NMR), high performance liquid chromatography (HPLC) and gas chromatography (Mapelli *et al.*, 2009; Desai *et al.*, 2010).

A growing number of studies have used metabolome analyses. The Human Metabolome Project used data mining for the identification and quantification of human metabolites (Wishart, 2007). Other examples include analysis of the effects of genetics and the environment on the metabolome of commercial maize hybrids (Asiago *et al.*, 2012), the discovery of a novel xylose metabolism pathway in the fungus *Aspergillus nidulans* (Panagiotou *et al.*, 2007), a survey of stress responses and optimum growth conditions for the yeast *Saccharomyces cerevisiae* during ethanol fermentation (Devantier *et al.*, 2005) and the identification of interactions between gut microbiota and human hosts (Ibrahim and Anishetty, 2012). A relatively new research field in metabolomics, fluxomics, looks at the real-time flux of metabolites within a cell (Kohlstedt *et al.*, 2010).

Given the wide field of applications, it can be expected that metabolomics will become well entrenched in the scientific and metagenomic arena. It must be noted, however, that metabolomic studies are currently hampered by the inability to identify many, previously undescribed, metabolites. There is thus an urgent need for the further characterization and classification of cellular metabolites (Kell, 2004).

Conclusions

Metagenomics has yielded a wealth of information of organisms, their genes, proteins and enzymes, in almost every environment on earth, regardless of whether these organisms are cultured or not (Handelsman, 2004). Functional metagenomics approaches have been used to 'prospect' the environment for many novel proteins and enzymes, including novel antibiotics, enzymes for use in bioethanol production, enzymes for the degradation of toxic wastes, and much more (Ekkers *et al.*, 2012). With the advent of NGS technologies, which are rapidly improving in terms of cost-effectiveness, throughput and read lengths, there has been a dramatic shift towards metagenome sequencing (Shokralla *et al.*, 2012). This has resulted in the sequencing of thousands of metagenomes, which are now available for comparison. These sequenced metagenomes can also be used for whole-genome assembly and the design of medium and growth conditions to culture 'the unculturables' (Vartoukian *et al.*, 2010). The availability of enormous databases of gene sequences can be used to streamline the screening strategies for functional metagenomic approaches. The integration of '-omics' technologies, such as metatranscriptomics, metaproteomics and metametabolomics in the metagenomics sphere has provided sophisticated strategies to enhance our understanding of the complex processes and pathways occurring in different ecosystems. Their combination with functional and sequence-based metagenomics has also proven effective for the identification and characterization of novel bioprospected proteins (Kandpal *et al.*, 2009).

Future perspectives

It seems reasonable to predict that the field of metagenomics will continue to grow as an increasingly powerful technology, helping us to understand the wealth of organisms and communities that inhabit the Earth, and giving us access to these diverse genomes as a resource for biotechnology (Dionisi *et al.*, 2012). Recent trends may be a guide to the future of environmental research. Large sequencing consortia, such as the Terragenome project focussing on soil metagenomes (Vogel *et al.* 2009), the Earth Microbiome Project, the Microbial Earth Project (Gilbert *et al.*, 2010b) and the Human Microbiome Project (Turnbaugh *et al.*, 2007) may represent a new model for metagenomics. At a stage in technology development where the generation of sequence data is almost a facile process, and the real challenge is the manipulation and exploitation of the sequence data, such large collaborative trans-disciplinary consortia may become the future vehicles for research.

In the foreseeable future, the largest advances can be expected in the sequence-based metagenomic approaches. In just the past 5 years the development of NGS technologies have brought about massive increases in sequencing throughput, cost-effectiveness and read length (Glenn, 2011). With the promise of single-molecule sequencing technologies within the next 1–2 years, we can expect even cheaper DNA and RNA sequencing, with unparalleled read lengths which will (we hope) bypass current metagenome assembly problems.

A number of basic conceptual and technical questions must still be addressed. With whole genome and metagenome sequencing projects currently outpacing the available computing power, how will we deal with the massive volume of sequence data? How do we reliably gain access to rare taxa among more abundant phylotypes? How will we classify the organisms we discover through metagenomics, and how will a shift from single locus to multilocus phylogenetics affect the currently fragile species concept?

Amazingly, the field of metagenomics was only born at the end of the 1990s and, despite its apparent sophistication, must be considered to still be in a state of 'infancy'. With the current (frightening) rate of technology development, we must assume that metagenomics will lead us into new, unexplored and unimagined worlds. The future of molecular ecology is very bright, but the fundamental tenet of science remains: the hardest task is always to ask the right questions.

Web resources

Bay Paul Center, MBL. (2009). Visualization and Analysis of Microbial Population Structures (VAMPs). Http://vamps.mbl.edu/index.php.

Chao, A., and Shen, T-J. (2010). SPADE: Species

Prediction And Diversity Estimation. Http://chao.stat.nthu.edu.tw.

Clarke, K.R., and Gorley, R.N. (2006). PRIMER v6: User Manual/Tutorial. PRIMER-E, Plymouth. Http://www.primer-e.com.

Colwell, R.K. (2005). EstimateS: Statistical estimation of species richness and shared species from samples. Version 7.5. Http://purl.oclc.org/estimates.

Genome OnLine Database (GOLD). (2012). Http://www.genomesonline.org

Helicos BioSciences Corporation ®. (2008). Helicos™ genetic analysis system specification sheet. Http://www.lgtc.nl/services/next_generation_sequencing/Helicos_SalesSpec.pdf.

Jarvie, T. (2007). Amplicon sequencing, Application Note No. 5, Roche Applied Science. Http://www.454.com/downloads/protocols/5_AmpliconSequencing.pdf

NGS Field Guide. (2012). Http://www.molecularecologist.com/next-gen-fieldguide/

Meyer F, Paarmann D, D'Souza M, Olson R, Glass EM, Kubal M, Paczian, T., Rodriguez, A., Steverns, R., Wilke, A., *et al.* (2012). MG-RAST metagenomic analysis server. Http://metagenomics.anl.gov/

Oxford Nanopore Technologies ®. (2012). Press release. Http://www.nanoporetech.com/news/press-releases/view/39

R Development Core Team. (2008). R: a language and environment for statistical computing. R Foundation for Statistical Computing. Vienna, Austria. Http://www.R-project.org.

Ribosomal Database Project. (2012). RDP Release 10, Update 30. Http://rdp.cme.msu.edu/

References

Abbot, P., Withgott, J.H., and Moran, N.A.(2001). Genetic conflict and conditional altruism in social aphid colonies. Proc. Natl. Acad. Sci. U.S.A. *98*, 12068–12071.

Acinas, S.G., Sarma-Rupvtarm, R., Klepac-Ceraj, V., and Polz, M.F. (2005). PCR-induced sequence artefacts and bias: insights from comparison of two 16S rRNA clone libraries constructed from the same sample. Appl. Environ. Microbiol. *71*, 8966–8969.

Allgaier, M., Reddy, A., Park, J.I., Ivanova, N.D.P., Sapra, R., Hazen, T.C., Simmons, B.A., and Vandergheynst, J.S. (2010). Targeted discovery of glycoside hydrolases from a switchgrass-adapted compost community. PLoS One 5, e8812.

Amann, R.I., Ludwig, W., and Schleifer, K.H. (1995). Phylogenetic identification and in situ detection of individual microbial cells without cultivation. Microb. Rev. *59*, 143–169.

Andrew, D.R., Fitak, R.R., Manguia-Vega, M., Racolta, A., Martinson, V.G., and Dontsova, K. (2012). Abiotic factors shape microbial diversity in sonoran desert soils. Appl. Environ. Microbiol. *78*, 7527–7537.

Armour, C.D., Castle, J.C., Chen, R., Babak, T., Loerch, P., Jackson, S., Shah, J.K., Dey, J., Rohl, C.A., Johnson, J.M., *et al.* (2009). Digital transcriptome profiling using selective hexamer priming for cDNA synthesis. Nat. Methods *6*, 647–649.

Ashburner, M., Ball, C.A., Blake, J.A., Botstein, D., Butler, H., Cherry, J.M., Davis, A.P., Dolinski, K., Dwight, S.S., Eppig, J.T., *et al.* (2000). Gene ontology: Tool for the unification of biology. The Gene Ontology Consortium. Nat. Genet. *25*, 25–29.

Ashelford, K.E., Chuzhanova, N.A., Fry, J.C., Jones, A.J., and Weightman, A.J. (2005). At least 1 in 20 16S rRNAsequence records currently held in public repositories is estimated to contain substantial anomalies. Appl. Environ. Microbiol. *71*, 7724- 7736.

Asiago, V.M., Hazebroek, J., Harp, T., and Zhong, C. (2012). Effects of genetics and environment on the metabolome of commercial maize hybrids: A multisite study. J. Agric. Food Chem. *60*, 11195–11212.

Bailly, J., Fraissinet-Tachet, L., Verner, M.-C., Debaud, J.-C., Lemaire, M., Wesolowski- Louvel, M., and Marmeisse, R. (2007). Soil eukaryotic functional diversity, a metatranscriptomic approach. ISME J. *1*, 632–642.

Bakhtiar, R., and Nelson, R.W. (2000). Electrospray ionization and matrix-assisted laser desorption ionization mass spectrometry. Biochem. Pharmacol. *59*, 891–905.

Bakken, L.R. (1985). Separation and purification of bacteria from soil. Appl. Environ. Microbiol. *49*, 1482–1487.

Bartossek, R., Nicol, G.W., Lanzen, A., Klenk, H.P., and Schleper, C.(2010). Homologues of nitrite reductases in ammonia-oxidizing archaea: diversity and genomic context. Environ. Microbiol. *12*, 1075–1088.

Bastida, F., Moreno, J.L., Nicolás, C., Hernández, T., and García, C. (2009). Soil metaproteomics: a review of an emerging environmental science. Significance, methodology and perspectives. J. Soil Sci. *60*, 845–859.

Bayer, T.S., Widmaier, D.M., Temme, K., Mirsky, E.A., Santi, D.V., and Voigt, C.A. (2009). Synthesis of methyl halides from biomass using engineered microbes. J. Am. Chem. Soc. *131*, 6508–6515.

Beja, O., Aravind, L., Koonin, E.V., Suzuki, M.T., Hadd, A., Nguyen, L.P., Jovanovich, S.B., Gates, C.M., Feldman, R.A., Spudich, J.L., *et al.* (2000). Bacterial rhodopsin: evidence for a new type of phototrophy in the sea. Science *289*, 1902–1906.

Benndorf, D., Balcke, G.U., Harms, H., and Von Bergen, M. (2007). Functional metaproteome analysis of protein extracts from contaminated soil and groundwater. ISME J. *1*, 224–234.

Bernstein, J.R., Bulter, T., Shen, C.R., and Liao, J.C. (2007). Directed evolution of ribosomal protein

S1 for enhanced translational efficiency of high GC Rhodopseudomonas palustris DNA in *Escherichia coli*. J. Biol. Chem. *282*, 18929–36.

Bertrand, H., Poly, F., Van, V.T., Lombard, N., Nalin, R., Vogel, T.M., and Simonet, P. (2005). High molecular weight DNA recovery from soils prerequisite for biotechnological metagenomic library construction. J. Microbiol. Methods 62,1–11.

Besemer, J., and Borodovsky, M. (2005). GeneMark: Web software for gene finding in prokaryotes, eukaryotes and viruses. Nucleic Acids Res. *33*, 451–454.

Binladen, J., Gilbert, M.T., Bollback, J.P., Panitz, F., Bendixen, C., Nielsen, R., and Willerslev, E. (2007). The use of coded PCR primers enables high-throughput sequencing of multiple homolog amplification products by 454 parallel sequencing. PLoS One *2*, e197.

Bland, C., Ramsey, T.L., Sabree, F., Lowe, M., Brown, K., Kyrpides, N.C., and Hugenholtz, P. (2007). CRISPR recognition tool (CRT): A tool for automatic detection of clustered regularly interspaced palindromic repeats. BMC Bioinformatics *8*, 209.

Blow, M.J., Zhang, T., Woyke, T., Speller, C.F., Krivoshapkin, A., Yang, D.Y., Derevianko, A., and Rubin, E.M. (2008). Identification of ancient remains through genomic sequencing. Genome Res. *18*, 1347–1353.

Bocs, S., Cruveiller, S., Vallenet, D., Nuel, G., and Médigue, C. (2003). AMIGene: Annotation of MIcrobial Genes. Nucleic Acids Res. *31*, 3723–3726.

Boers, S.A., Reijden, W.A.V.D., and Jansen, R. (2012). High-throughput multilocus sequence typing: bringing molecular typing to the next level. PLoS One 7, e39630.

Brady, A., and Salzberg, S.L.(2009). Phymm and PhymmBL: metagenomic phylogenetic classification with interpolated Markov models. Nat. Methods *6*, 673–676.

Brady, S.F., Chao, C.J., and Clardy, J.(2002). New natural product families from an environmental DNA (eDNA) gene cluster. Am. Chem. Soc. *124*, 9968–9969.

Bråte, J., Logares, R., Berney, C., Ree, D.K., Klaveness, D., Jakobsen, K.S., and Shalchian-Tabrizi, K. (2010). Freshwater Perkinsea and marine-freshwater colonizations revealed by pyrosequencing and phylogeny of environmental rDNA. ISME J. *4*, 1144–1153.

Burgess, K.S., Fazekas, A.J., Kesanakurti, P.R., Graham, S.W., Husband, B.C., Newmaster, S.G., Percy, D.M., Hajibabaei, M., and Barrett, S.C.H. (2011). Discriminating plant species in a local temperate flora using the rbcL+matK DNA barcode. Methods Ecol. Evol. *2*, 333–340.

Butler, J., MacCallum, I., Kleber, M., Shlyakhter, I.A., Belmonte, M.K.,Lander, E.S., Nusbaum, C., and Jaffe, D.B. (2008). ALLPATHS: De novo assembly of whole-genome shotgun microreads. Genome Res. *18*, 810–820.

Caporaso, J.G., Kuczynski, J., Stombaugh, J., Bittinger, K., Bushman, F.D., Costello, E.K., Fierer, N., Peña, A.G., Goodrich, J.K., Gordon, J.I., *et al.* (2010). QIIME allows analysis of high-throughput community sequencing data. Nat. Methods *7*, 335–336.

Casanoves, F., Pla, L., Di Rienzo, J.A., and Díaz, S. (2011). FDiversity: a software package for integrated analysis of functional diversity. Methods Ecol. Evol. *2*, 233–237.

Chakravorty, S., Helb, D., Burday, M., Connell, N., and Alland, D. (2007). A detailed analysis of 16S ribosomal RNA gene segments for the diagnosis of pathogenic bacteria. J. Microbiol. Methods. *69*, 330–339.

Chan, C.K., Hsu, A.L., Halgamuge, S.K., and Tang, S.L. (2008). Binning sequences using very sparse labels within a metagenome. BMC Bioinformatics *9*, 215.

Cheung, M.K., Au, C.H., Chu, K.H., Kwan, H.S., and Wong, C.K. (2010). Composition and genetic diversity of picoeukaryotes in subtropical coastal waters as revealed by 454 pyrosequencing. ISME J. *4*, 1053–1059.

Chevreux, B., Wetter, T., and Suhai, S. (1999). Genome sequence assembly using trace signals and additional sequence information Computer Sci. Biol. *99*, 45–56.

Claesson, M.J., O'Sullivan, O., Wang, Q., Nikkilä, J., Marchesi, J.R., Smidt, H., de Vos, W.M., Ross, R.P., and O'Toole, P.W. (2009). Comparative analysis of pyrosequencing and a phylogenetic microarray for exploring microbial community structures in the human distal intestine. PLoS One *4*, e6669.

Claesson, M.J., Wang, Q., Sullivan, O.O., Greene-Diniz, R., Cole, J.R., Ross, R.P., and O'Toole, P.W. (2010). Comparison of two next-generation sequencing technologies for resolving highly complex microbiota composition using tandem variable 16S rRNA gene regions. Nucleic Acids Res. *38*, e200.

Cole, J.R., Wang, Q., Cardenas, E., Fish, J., Chai, B., Farris, R.J., Kulam-Syed-Mohideen, A.S., McGarrell, D.M., Marsh, T., Garrity, G.M., *et al.* (2009). The Ribosomal Database Project: improved alignments and new tools for rRNA analysis. Nucleic Acids Res. 37, D141–145.

Costello, E.K., Lauber, C.L., Hamady, M., Fierer, N., Gordon, J.I., and Knight, R. (2009). Bacterial community variation in human body habitats across space and time. Science *326*, 1694–1697.

Cox, M.J., Huang, Y.J., Fujimura, K.E., Liu, J.T., McKean, M., Boushey, H.A., Segal, M.R., Brodie, E.L., Cabana, M.D., and Lynch, S.V. (2010). Lactobacillus casei abundance is associated with profound shifts in the infant gut microbiome. PLoS One 5, e8745.

Craig, R., and Beavis, R. (2004). TANDEM: matching proteins with mass spectra. Bioinformatics *20*, 1466–1467.

Croucher, N.J., and Thomson, N.R. (2012). Studying bacterial transcriptomes using RNA-seq. Curr. Opin. Microbiol. *13*, 619–624.

Daniel, R. (2005). The metagenomics of soil. Nat. Rev. Microbiol. 3, 470–478.

Darby, A.C., and Hall, N. (2008). Fast forward genetics. Nat. Biotech. *26*, 1248–1249.

De la Iglesia, R., Valenzuela-Heredia, D., Pavissich, J.P., Freyhoffer, S., Andrade, S., Correa, J.A., and González, B. (2010). Novel polymerase chain reaction primers for the specific detection of bacterial copper P-type ATPases gene sequences in environmental isolates and metagenomic DNA. Lett. Appl. Microbiol. *50*, 552–562.

Deangelis, K.M., Allgaier, M., Chavarria, Y., Fortney, J.L., Hugenholtz, P., Simmons, B., Sublette, K., Silver, W.L., and Hazen, T.C. (2011). Characterization of trapped lignin-degrading microbes in tropical forest soil. PLoS One *6*, e19306.

Degnan, P.H., and Ochman, H. (2011). Illumina-based analysis of microbial community diversity. ISME J. *6*, 183–194.

Delcher, A.L., Harmon, D., Kasif, S., White, O., and Salzberg, S.L. (1999). Improved microbial gene identifi cation with GLIMMER. Nucleic Acids Res. *27*, 4636–4641.

Delmont, T.O., Simonet, P., and Vogel, T.M. (2012). Describing microbial communities and performing global comparisons in the 'omic era. ISME J. 6,1625–1628.

DeLong, E.F., Preston, C.M., Mincer, T., Rich, V., Hallam, S.J., Frigaard, N.U.,Martinez, A., Sullivan, M.B., Edwards, R., Brito, B.R., *et al.* (2006). Community genomics among stratified microbial assemblages in theocean's interior. Science *311*, 496–503.

Desai, C., and Madamwar, D. (2007). Extraction of inhibitor-free metagenomic DNA from polluted sediments, compatible with molecular diversity analysis using adsorption and ion- exchange treatments. Biores. Technol. 98,761–768.

Desai, C., Pathak, H., and Madamwar, D. (2010). Advances in molecular and '-omics' technologies to gauge microbial communities and bioremediation at xenobiotic/anthropogen contaminated sites. Biores. Technol. *101*, 1558–1569.

DeSantis, T.Z., Hugenholtz, P., Larsen, N., Rojas, M., Brodie, E.L., Keller, K., Huber, T., Dalevi, D., Hu, P., and Andersen, G.L. (2006). Greengenes, a chimera-checked 16S rRNA gene database and workbench compatible with ARB. Appl. Environ. Microbiol. *72*, 5069–5072.

DeSantis, T.Z., Brodie, E.L., Moberg, J.P., Zubieta, I.X., Piceno, Y.M., and Andersen, G.L. (2007). High-density universal 16S rRNA microarray analysis reveals broader diversity than typical clone library when sampling the environment. Microb. Ecol. *53*, 371–383.

Dethlefsen, L., Huse, S., Sogin, M.L., and Relman, D.A. (2008). The pervasive effects of an antibiotic on the human microbiota, as revealed by deep 16S rRNA sequencing. PLoS Biol. *6*, e280.

Devantier, R., Scheithauer, B., Villas-Bôas, S.G., Pedersen, S., and Olsson, L. (2005). Metabolite profiling for analysis of yeast stress response during very high gravity ethanol fermentations. Biotechnol. Bioeng. *90*, 703–714.

Diaz-Torres, M.L., McNab, R., Spratt, D.A., Villedieu, A., Hunt, N., Wilson, M., and Mullany, P. (2003). Novel tetracycline resistance determinant from the oral metagenome. Antimicrob. Agents Chemother. *47*, 1430–1432.

Dionisi, H.M., Lozada, M., and Olivera, N.L. (2012). Bioprospection of marine microorganisms: Biotechnological applications and methods. Rev. Arg. Microbiol. *44*, 49–60.

Donato, J.J., Moe, L.A., Converse, B.J., Smart, K.D., Berklein, F.C., McManus, P.S., and Handelsman, J. (2010). Metagenomic analysis of apple orchard soil reveals antibiotic resistance genes encoding predicted bifunctional proteins. Appl. Environ. Microbiol. *76*, 4396–4401.

Doolittle, W.F., and Zhaxybayeva, O. (2009). On the origin of prokaryotic species. Genome Res. *19*, 744–756.

Doyle, R.J. (2000). Contribution of the hydrophobic effect to microbial infection. Microbes Infect. *2*, 391–400.

Eid, J., Fehr, A., Gray, J., Luong, K., Lyle, J., Otto, G., Peluso, P., Rank, D., Baybayan, P., Bettman, B., *et al.* (2009). Real-time DNA sequencing from single polymerase molecules. Science *323*, 133–138.

Eisenstein, M. (2012). Oxford Nanopore announcement sets sequencing sector abuzz. Nat. Biotechnol. *30*, 295–296.

Ekkers, D.M., Cretoiu, M.S., Kielak, A.M., and Van Elsas, J.D. (2012). The great screen anomaly – a new frontier in product discovery through functional metagenomics. Appl. Microbiol. Biotechnol. *93*, 1005–1020.

Eng, J.K., McCormack, A.L., Yates, I.I.I., J.R. (1994). An approach to correlate tandem mass spectral data of peptides with amino acid sequences in a protein database. J. Am Soc. Mass Spectrom. *5*, 976–989.

Eyers, L., George, I., Schuler, L., Stenuit, B., Agathos, S., and El Fantroussi, S. (2004). Environmental genomics: exploring the unmined richness of microbes to degrade xenobiotics. Appl. Microbiol. Biotechnol. *66*, 123–130.

Field, D., Garrity, G., Gray, T., Morrison, N., Selengut, J., Sterk, P., Tatusova, T., Thomson, N., Allen, M.J., Anguioli, S.V., *et al.* (2008). The minimum information about a genome sequence (MIGS) specification. Nat. Biotech. *26*, 541–547.

Filiatrault, M.J. (2011). Progress in prokaryotic transcriptomics. Curr. Opin. Microbiol. *14*, 579–586.

Finn, R.D., Mistry, J., Tate, J., Coggill, P., Heger, A., Pollington, J.E., Gavin, O.L., Gunasekaran, P., Ceric, G., Forslund, K., *et al.* (2010). The Pfam protein families database. Nucleic Acids Res. *38*, 211–222.

Fleischmann, R., Adams, M., White, O., Clayton, R., Kirkness, E., Kerlavage, A., Bult, C., Tomb, J., Dougherty, B., and Merrick, J. (1995). Whole-genome random sequencing and assembly of Haemophilus influenza Rd. Science *269*, 496–512.

Frias-Lopez, J., Shi, Y., Tyson, G.W., Coleman, M.L., Schuster, S.C., Chisholm, S.W., and Delong, E.F. (2008). Microbial community gene expression in ocean surface waters. Proc. Natl. Acad. Sci. U.S.A. *105*, 3805–3810.

Gabor, E.M., Alkema, W.B., and Janssen, D.B. (2004). Quantifying the accessibility of the metagenome by random expression cloning techniques. Environ. Microbiol. *6*, 879–886.

Gardner, P.P., Daub, J., Tate, J.G., Nawrocki, E.P., Kolbe, D.L., Lindgreen, S., Wilkinson, A.C., Finn, R.D., Griffiths-Jones, S., Eddy, S.R., *et al.* (2009). Rfam: Updates to the RNA families database. Nucleic Acids Res. *37*, 136–140.

Ghai, R., Rodriguez-Valera, F., McMahon, K.D., Toyama, D., Rinke, R., Cristina Souze-de Oliviera, T., Wagner Garcia, J., Pellon de Miranda, F., and Henrique-Silva, F. (2011). Metagenomics of the water column in the pristine upper course of the Amazon river. PLoS One *6*, e23785.

Gilbert, J.A., Field, D., Huang, Y., Edwards, R., Li, W., Gilna, P., and Joint, I. (2008). Detection of large numbers of novel sequences in the metatranscriptomes of complex marine microbial communities. PLoS One *3*, e3042.

Gilbert, J.A., Field, D., Swift, P., Thomas, S., Cummings, D., Temperton, B., Weynberg, K., Huse. S., Hughes, M., Joint, I., *et al.* (2010a). The taxonomic and functional diversity of microbes at a temperate coastal site: a 'multi-omic' study of seasonal and diel temporal variation. PLoS One *5*, e15545.

Gilbert, J.A., Meyer, F., Jansson, J., Gordon, J., Pace, N., Tiedje, J., Ley, R., Fierer, N., Field, D., Kyrpides, N., *et al.* (2010b). The Earth Microbiome Project: meeting report of the '1 EMP meeting on sample selection nd acquisition' at Argonne National Laboratory October 6 2010. Stand. Genomics Sci. *3*, 249–253.

Gilles, A., Meglécz, E., Pech, N., Ferreira, S., Malausa, T., and Martin, J.F. (2011). Accuracy and quality assessment of 454 GS-FLX Titanium pyrosequencing. BMC Genomics *12*, 245.

Gillespie, D.E., Brady, S.F., Bettermann, A.D., Cianciotto, N.P., Liles, M.R., Rondon, M.R., Clardy, J., Goodman, R.M., and Handelsman, J. (2002). Isolation of antibiotics turbomycin a and B from a metagenomic library of soil microbial DNA. Appl. Environ. Microbiol. *68*, 4301–4306.

Glenn, T.C.(2011). Field guide to next-generation DNA sequencers. Mol. Ecol. Resour. *11*, 759–769.

Goll, J., Tanenbaum, D.M., Thiagarajan, M., Li, K., Methé, B.A., and Yooseph, S. (2010). METAREP: JCVI metagenomics reports – an open source tool for high-performance comparative metagenomics. Bioinformatics 26, 2631–2632.

Gupta, P.K. (2008). Single-molecule DNA sequencing technologies for future genomics. Trends Biotechnol. *26*, 602–611.

Guschin, D.Y., Mobarry, B.K., Proudnikov, D., Stahl, D.A., Rittmann, B.E., and Mirzabekov, A.D. (1997). Oligonucleotide microchips as genosensors for determinative and environmental studies in microbiology. Appl. Environ. Microbiol. *63*, 2397–2402.

Haas, B.J., Gevers, D., Earl, A.M., Feldgarden, M., Ward, D.V., Giannoukos, G., Ciulla, D., Tabbaa, D., Highlander, S.K., Sodergren, E., *et al.*(2011). Chimeric 16S rRNA sequence formation and detection in Sanger and 454- pyrosequenced PCR amplicons. Genome Res. *21*, 494–504.

Hammer, Ø., Harper, D.A.T., and Ryan, P.D.(2001). PAST: paleontological statistics software packagefor education and data analysis. Palaeontol. Electron. 4,1–9.

Handelsman, J. (2004). Metagenomics: Application of genomics to uncultured microorganisms. Microbiol. Mol. Biol. Rev. *68*, 669–685.

Handelsman, J. (2005). Sorting out metagenomes. Nat. Biotechnol. *23*, 38–39.

Handelsman, J., Rondon, M.R., Brady, S.F., Clardy, J., and Goodman, R. (1998). Molecular biological access to the chemistry of unknown soil microbes A new frontier for natural products. Chem. Biol. *5*, 245–249.

Harris, T., Buzby, P., Babcock, H., Beer, E., Bowers, J., Braslavsky, I., Causey, M., Colonell, J., Dimeo, J., and Efcavitch, J.W. (2008). Single-molecule DNA sequencing of a viral genome. Science *320*, 106–109.

Henne, A., Schmitz, R.A., Bomeke, M., Gottschalk, G., and Daniel, R. (2000). Screening of environmental DNA libraries for the presence of genes conferring lipolytic activity on *Escherichia coli*. Appl. Environ. Microbiol. *66*, 3113–3116.

Herlemann, D.P., Labrenz, M., Jürgens, K., Bertilsson, S., Waniek, J.J., and Andersson, A.F. (2011). Transitions in bacterial communities along the 2000 km salinity gradient of the Baltic Sea. ISME J. *5*, 1571–1579.

Hernandez, D., Francois, P., Farinelli, L., Osteras, M., and Schrenzel, J. (2008). De novo bacterial genome sequencing: Millions of very short reads assembled on a desktop computer. Genome Res. *18*, 802–809.

Herrera, A., and Cockell, C.S. (2007). Exploring microbial diversity in volcanic environments: a review of methods in DNA extraction. J. Microbiol. Methods *70*, 1–12.

Hess, M., Sczyrba, A., Egan, R., Kim, T.W., Chokhawala, H., Schroth, G., Luo, S., Clark, D.S., Chen, F., Zhang, T., *et al.* (2011). Metagenomic discovery of biomass-degrading genes and genomes from cow rumen. Science *331*, 463–467.

Hjort, K., Bergström, M., Adesina, M.F., Jansson, J.K., Smalla, K., and Sjöling, S. (2010). Chitinase genes revealed and compared in bacterial isolates, DNA extracts and a metagenomic library from a phytopathogen-suppressive soil. FEMS Microbiol. Ecol. *71*, 197–207.

Hoff, K.J., Lingner, T., Meinicke, P., and Tech, M. (2009). Orphelia: Predicting genes in metagenomic sequencing reads. Nucleic Acids Res. *37*, 101–105.

Howeler, M., Ghiorse, W.C., and Walker, L.P. (2003). A quantitative analysis of DNA extraction and purification from compost. J. Microbiol. Methods *54*, 37–45.

Huber, F., da Silva, S., Bomfim, T.C., Teixeira, K.R., and Bello, A.R. (2007). Genotypic characterization and phylogenetic analysis of Cryptosporidium sp. from domestic animals in Brazil. Vet. Parasitol. *150*, 65–74.

Huber, T., Faulkner, G., and Hugenholz, P. (2004). Bellerophon: A program to detect chimeric sequences in multiple sequence alignments. Bioinformatics *20*, 2317–2319.

Huson, D.H., Auch, A.F., Qi, J., and Schuster, S.C. (2007). MEGAN analysis of metagenomic data. Genome Res. *17*, 377–386.

Huson, D.H., Mitra, S., Weber, N., Ruscheweyh, H., and Schuster, S.C. (2011). Integrative analysis of environmental sequences using MEGAN4. Genome Res. *21*, 1552–1560.

Huson, D.H., Richter, D.C., Mitra, S., Auch, A.F., and Schuster, S.C. (2009). Methods for comparative metagenomics. BMC Bioinformatics *10*, S12.

Ibrahim, M., and Anishetty, S. (2012). A meta-metabolome network of carbohydrate metabolism: interactions between gut microbiota and host. Biochem. Biophys. Res. Commun. *428*, 278–284.

Inceoglu, O., Hoogwout, E.F., Hill, P., and van Elsas, J.D. (2010). Effect of DNA extraction method on the apparent microbial diversity of soil. Appl. Environ. Microbiol. *76*, 3378–3382.

Iqbal, H.A., Feng, Z., and Brady, S.F. (2012). Biocatalysts and small molecule products from metagenomic studies. Curr. Opin. Chem. Biol. *16*, 109–116.

Ishii, K., and Fukui, M. (2001). Optimization of annealing temperature to reduce bias caused by a primer mismatch in multitemplate PCR. Appl. Environ. Microbiol. *67*, 3753–3755.

Iverson, V., Morris, R.M., Frazar, C.D., Berthiaume, C.T., Morale, R.L., and Armburst, V. (2012). Untangling genomes from metagenomes: revealing an uncultured class of marine Euryarchaeota. Science *335*, 587–590.

Iwai, S., Chai, B., Sul, W.J., Cole, J.R., Hashsham, S.A., and Tiedje, J.M. (2010). Gene-targeted-metagenomics reveals extensive diver- sity of aromatic dioxygenase genes in the environment. ISME J. *4*, 279–285.

Jünemann, S., Prior, K., Szczepanowski, R., Harks, I., Ehmke, B., Goesmann, A., Stoye, J., and Harmsen, D. (2012). Bacterial community shift in treated periodontitis patients revealed by ion torrent 16S rRNA gene amplicon sequencing. PLoS One *7*, e41606.

Kallmeyer, J., and Smith, D.C. (2009). An improved electroelution method for separation of DNA from humic substances in marine sediment DNA extracts. FEMS Microbiol. Ecol. 69,125–131.

Kandpal, R.P., Saviola, B., and Felton, J. (2009). The era of 'omics unlimited. Biotechniques *46*, 351–355.

Kanehisa, M., Goto, S., Kawashima, S., Okuno, Y., and Hattori, M. (2004). The KEGG resource for deciphering the genome. Nucleic Acids Res. *32*, 277–280.

Kell, D.B. (2004). Metabolomics and systems biology: making sense of the soup. Curr. Opin. Microbiol. *7*, 296–307.

Keller, M., and Hettich, R. (2009). Environmental proteomics: a paradigm shift in characterizingmicrobial activities at the molecular level. Microbiol. Mol. Biol. Rev. *73*, 62–70.

Knierim, E., Lucke, B., Schwarz, J.M., Schuelke, M., and Seelow, D. (2011). Systematic comparison of three methods for fragmentation of long-range PCR products for next generation sequencing. PLoS One *6*, e28240.

Knietsch, A., Bowien, S., Whited, G., Gottschalk, G., and Daniel, R. (2003). Identification and characterization of genes encoding coenzyme B12-dependent glycerol and diol dehydratases from metagenomic DNA libraries derived from enrichment cultures_ Appl. Environ. Microbiol. *69*, 3048–3060.

Kohlstedt, M., Becker, J., and Wittmann, C. (2010). Metabolic fluxes and beyond-systems biology understanding and engineering of microbial metabolism. Appl. Microbiol. Biotechnol. *88*, 1065–1075.

Koren, S., Treangen, T.J., and Pop, M. (2011). Bambus2: scaffolding metagenomes. Bioinformatics *27*, 2964–2971.

Kottmann, R., Gray, T., Murphy, S., Kagan, L., Kravitz, S., Lombardot, T., Field, D., and Glöckner, F.O., Genomic Standards Consortium. (2008). A standard MIGS/MIMS compliant XML schema: toward the development of the Genomic Contextual Data Markup Language (GCDML). Omics *12*, 115–121.

Krause, L., Diaz, N.N., Goesmann, A., Kelley, S., Nattkemper, T.W., Rohwer, F., Edwards, R.A., and Stoye, J. (2008). Phylogenetic classification of short environmental DNA fragments. Nucleic Acids Res. *36*, 2230–2239.

Kristiansson, E., Hugenholtz, P., and Dalevi, D. (2009). ShotgunFunctionalizeR: an R-package for functional comparison of metagenomes. Bioinformatics *25*, 2737–2738.

Kurata, S., Kanagawa, T., Magariyama, Y., Takatsu, K., Yamada, K., Yokomaku, T., and Kamagata, Y. (2004). Reevaluation and reduction of a PCR bias caused by reannealing of templates. Appl. Environ. Microbiol. *70*, 7545–7549.

Lai, B., Ding, R., Li, Y., Duan, L., and Zhu, H. (2012). A de novo metagenomic assembly program for shotgun DNA reads. Bioinformatics *28*, 1455–1462.

Laliberté, E., and Legendre, P. (2011). A distance-based framework for measuring function diversity from multiple traits. Ecology *91*, 299–305..

Laserson, J., Jojic, V., and Koller, D. (2011). Genovo: de novo assembly for metagenomes. J. Comput. Biol. *18*, 429–443.

Lee, L.J., Barrett, J.A., and Poole, R.K. (2005). Genome-wide transcriptional response of chemostat-cultured *Escherichia coli* to zinc. J. Bacteriol. *187*, 1124–1134.

Leung, H.C., Yiu, S.M., Yang, B., Peng, Y., Wang, Y., Liu, Z., Chen, J., Qin, J., Li, R., and Chin, F.Y. (2011). A robust and accurate binning algorithm for metagenomic sequences with arbitrary species abundance ratio. Bioinformatics *27*, 1489–1495.

Liu, B., Gibbons, T., Ghodsi, M., Treangen, T., and Pop, M. (2011). Accurate and fast estimation of taxonomic profiles from metagenomic shotgun sequences. BMC Genomics *12*, S4.

Liu, Z., DeSantis, T.Z., Andersen, G.L., and Knight, R. (2008). Accurate taxonomy assignments from 16S rRNA sequences produced by highly parallel pyrosequencers. Nucleic Acids Res. *36*, e120.

Logares, R., Haverkamp, T.H.A., Kumar, S., Lanzén, A., Nederbragt, A.J., Quince, C., and Kauserud, H. (2012). Environmental microbiology through the lens of high-throughput DNA sequencing: synopsis of current platforms and bioinformatics approaches. J. Microbiol. Methods *91*, 106–113.

Loman, N.J., Constantinidou, C., Chan, J.Z., Halachev, M., Sergeant, M., Penn, C.W., Robinson, E.R., and Pallen, M.J. (2012). High-throughput bacterial genome sequencing: an embarrassment of choice, a world of opportunity. Nat. Rev. Microbiol. *10*, 599–606.

Long, P.F., Dunlap, W.C., Battershill, C.N., and Jaspars, M. (2005). Shotgun cloning and heterologous expression of the patellamide gene cluster as a strategy to achieving sustained metabolite production. Chembiochem. *6*, 1760–1765.

Lowe, T.M., and Eddy, S.R. (1997). tRNAscan-SE: a program for improved detection of transfer RNA genes in genomic sequence. Nucleic Acids Res. *25*, 955–964.

Lozupone, C., Hamady, M., and Knight, R. (2006). UniFrac – an online tool for comparing microbial community diversity in a phylogenetic context. BMC Bioinformatics *7*, 371.

Lozupone, C.A., and Knight, R. (2008). Species divergence and the measurement of microbial diversity. Fems Microbiol. Rev. *32*, 557–578.

Ludwig, W., Strunk, O., Westram, R., Richter, L., Meier, H., Yadhukumar, Buchner, A., Lai, T., Steppi, S., Jobb, G., *et al.*(2004). ARB: a software environment for sequence data. Nucleic Acids Res. *32*, 1363–1371.

Lundberg, D.S., Lebeis, S.L., Parades, S.H., Yourstone, S., Gehring, J., Malfatti, S., Tremblay, J., Engelbrektson, A., Kunun, V., del Rio, T.G., *et al.* (2012). Defining the core Arabidopsis thaliana root microbiome. Nature *488*, 86–90.

Luo, C., Tsementzi, D., Kyrpides, N.C., and Konstantinidis, K.T. (2011). Individual genome assembly from complex community short-read metagenomic datasets. ISME J. *77*, 6000–6011.

Magi, A., Benelli, M., Gozzini, A., Girolami, F., Torricelli, F., and Brandi, M.L. (2010). Bioinformatics for next generation sequencing data. Genes *1*, 294–307.

Magurran, A.E. (2004). Measuring Biological Diversity. (Oxford, UK, Blackwell Science).

Maitra, R.D., Kim, J., and Dunbar, W.B. (2012). Recent advances in nanopore sequencing. Electrophoresis *33*, 1–11.

Mapelli, V., Olsson, L., and Nielsen, J. (2008). Metabolic footprinting in microbiology: methods and applications in functional genomics and biotechnology. Trends Biotechnol. *26*, 490–497.

Mardisk, E.R. (2008). Next-generation DNA sequencing methods. Ann. Rev. Genomics Hum. Genet. *9*, 387–402.

Margulies, M., Egholm, M., Altman, W.E., Attiya, S., Bader, J.S., Bemblen, L.A., Berka, J., Braverman, M.S., Chen, Y.J., Chen, Z., *et al.* (2005). Genome sequencing in microfabricated high-density picolitre reactors. Nature *437*, 376–380.

Markowitz, V.M., Ivanova, N.N., Szeto, E., Palaniappan, K., Chu, K., Dalevi, D., Chen, I.M., Grechkin, Y., Dubchak, I., Anderson, I., *et al.* (2008). IMG/M: a data management and analysis system for metagenomes. Nucleic Acids Res. *36*, D534–8.

Maróti, G., Tong, Y., Yooseph, S., Baden-Tillson, H., Smith, H.O., Kovács, K.L., Frazier, M., Venter, J.C., and Xu, Q. (2009). Discovery of [NiFe] hydrogenase genes in metagenomic DNA: Cloning and heterologous expression in Thiocapsa roseopersicina. Appl. Environ. Microbiol. *75*, 5821–5830.

Matsen, F.A.,Kodner, R.B., and Armbrust, E.V.(2010). PPLACER: linear timemaximum-likelihood and Bayesian phylogenetic placement of sequences onto a fixed reference tree. BMC Bioinformatics *11*, 538.

Mayr, E. (1944). Systematics and the Origin Of Species from the Viewpoint of a Zoologist. (New York, USA, Columbia University Press).

Mcgrath, K.C., Thomas-Hall, S.R., Cheng, C.T, Leo, L., Alexa, A., Schmidt, S., and Schenk, P.M., (2008). Isolation and analysis of mRNA from environmental microbial communities. J. Microbiol. Methods *75*, 172.

McHardy, A.C., Martín, H.G., Tsirigos, A., Hugenholtz, P., and Rigoutsos, I. (2007). Accurate phylogenetic classification of variable-length DNA fragments. Nat. Methods *4*, 63–72.

Mendes, R., Kruijt, M., de Bruijn, I., Dekkers, E., van der Voort, M., Schneider, J.H., Piceno, Y.M., DeSantis, T.Z., Andersen, G.L., Bakker, P.A., *et al.* (2011). Deciphering the rhizosphere microbiome for disease-suppressive bacteria. Science *332*, 1097–1100.

Meusnier, I., Singer, G.A.C., Landry, J.-F., Hickey, D.A., Hebert, P.D., and Hajibabaei, M. (2008). A universal DNA mini-barcode for biodiversity analysis. BMC Genomics *9*, 214.

Meyer, F., Paarmann, D., D'Souza, M., Olson, R., Glass, E.M., Kubal, M., Paczian, T., Rodriguez, A., Stevens, R., Wilke, A., *et al.* (2008). The metagenomics RAST server – a public resource for the automatic phylogenetic and functional analysis of metagenomes. BMC Bioinformatics *9*, 386.

Meyer, H.-P., and Turner, N.J. (2009). Biotechnological manufacturing options for organic chemistry. Min. Rev. Org. Chem. *6*, 300–306.

Meyer, M., Stenzel, U., Myles, S., Pru, K., and Hofreiter, M. (2007). Targeted high-throughput sequencing of tagged nucleic acid samples. Nucleic Acids Res. *35*, 1–5.

Monchy, S., Sanciu, G., Jobard, M., Rasconi, S., Gerphagnon, M., Chabé, M., Cian, A., Meloni, D., Niquil, N., Christaki, U., *et al.* (2011). Exploring and quantifying fungal diversity in freshwater lake ecosystem using rDNA cloning/sequencing and SSU tag pyrosequencing. Environ. Microbiol. *13*, 1344–1453.

Monzoorul Haque, M., Ghosh, T.S., Komanduri, D., and Mande, S.S. (2009). SOrt-ITEMS: sequence orthology based approach for improved taxonomic estimation of metagenomic sequences. Bioinformatics *25*, 1722–1730.

Moore, G. (1965). Cramming more components onto integrated circuits. Electronics *38*, 114–117.

Moran, M.A., Satinsky, B., Gifford, S.M., Luo, H., Rivers, A., Chan, L.-K., Jun, M., Durham, B.P., Chen, S., Varaljay, V.A., *et al.* (2012). Sizing up metatranscriptomics. ISME J. *7*, 1–7.

Morris, R.M., Nunn, B.L., Frazar, C., Goodlett, D.R., Ting, Y.S., and Rocap, G. (2010). Comparative metaproteomics reveals ocean-scale shifts in microbial nutrient utilization and energy transduction. ISME J. *4*, 673–685.

Muller, J., Szklarczyk, D., Julien, P., Letunic, I., Roth, A., Kuhn, M., Powell, S., von Mering, C., Doerks, T., Jensen, L.J., *et al.* (2010). eggNOG v2.0: extending the evolutionary genealogy of genes with enhanced non-supervised orthologous groups, species and functional annotations. Nucleic Acids Res. *38*, 190–195.

Mullis, K., and Faloona, F. (1987). Specific synthesis of DNA in vitro via a polymerase-catalyzed chain reaction. Methods Enzymol. *155*, 335–350.

Namiki, T., Hachiya, T., Tanaka, H., and Sakakibara, Y. (2012). MetaVelvet: an extension of Velvet assembler to de novo metagenome assembly from short sequence reads. Nucleic Acids Res. *40*, e155.

Nannipieri, P., Ascher, J., Ceccherinim, M.T., Landi, L., Pietramellara, G., and Renella, G. (2003). Microbial diversity and soil functions. Eur. J. Soil Sci. *54*, 655–670.

Neefs, J.M., Van de Peer, Y., De Rijk, P., Chapelle, S., and De Wachter, R. (1993). Compilation of small ribosomal subunit RNA structures. Nucleic Acids Res. *21*, 3025–3049.

Neveu, J., Regeard, C., and Dubow, M.S. (2011). Isolation and characterization of two serine proteases from metagenomic libraries of the Gobi and Death Valley deserts. Appl. Environ. Microbiol. *91*, 635–644.

Nilsson, R.H., Kristiansson, E., Ryberg, M., Hallenberg, N., and Larsson, K.H. (2008). Intraspecific ITS variability in the kingdom fungi as expressed in the international sequence databases and its implications for molecular species identification. Evol. Bioinform. *4*, 193–201.

Noguchi, H., Taniguchi, T., and Itoh, T. (2008). MetaGeneAnnotator: detecting species-specific patterns of ribosomal binding site for precise gene prediction in anonymous prokaryotic and phage genomes. DNA Res. *15*, 387–396.

Nuutila, K., Siltanen, A., Peura, M., Bizik, J., Kaartinen, I., Kuokkanen, H., Nieminen, T., Harjula, A., Aarnio, P., Vuola, J., *et al.*(2012). Human skin transcriptome during superficial cutaneous wound healing. Wound Repair Regen. *20*, 830–839.

Ozsolak, F., and Milos, P.M. (2011). RNA sequencing: advances, challenges and opportunities. Nat. Rev. Genet. *12*, 87–98.

Ozsolak, F., Platt, A.R., Jones, D.R., Reifenberger, J.G., Sass, L.E., McInerney, P., Thompson, J.F., Jarosz, M., and Milos, P.N. (2009). Direct RNA sequencing. Nature *461*, 814–818.

Paliy, O., and Agans, R. (2012). Application of phylogenetic microarrays to interrogation of human microbiota. FEMS Microbiol. Ecol. *79*, 2–11.

Palmer, C., Bik, E.M., Digiulio, D.B., Relman, D.A., and Brown, P.O. (2007). Development of the human infant intestinal microbiota. PLoS Biol. *5*, e177.

Panagiotou, G., Kouskoumvekaki, I., Jónsdottir, S.O., and Olsson, L. (2007). Monitoring novel metabolic pathways using metabolomics and machine learning: induction of the phosphoketolase pathway in Asperigillus nidulans cultivations. Metabolomics 3, 503–516.

Pang, H., Zhang, P., Duan, C.-J., Mo, X.-C., Tang, J.-L., Feng, J.-X. (2009). Identification of cellulase genes from the metagenomes of compost soils and functional characterization of one novel endoglucanase. Curr. Microbiol. *58*, 404–408.

Parachin, N.S., and Gorwa-Grauslund, M.F. (2011). Isolation of xylose isomerases by sequence- and function-based screening from a soil metagenomic library. Biotechnol. Biofuels 4, 9.

Parks, D.H., and Beiko, R.G. (2010). Identifying biologically relevant differences between metagenomic communities. Bioinformatics *26*, 715–721.

Peng, Y., Leung, H.C., Yiu, S.M., and Chin, F.Y., (2011). Meta-IDBA: A de novo assembler for metagenomic data. Bioinformatics *27*, 94–101.

Pennisi, E. (2010). Genomics. Semiconductors inspire new sequencing technologies. Science *327*, 1190.

Perkins, D.N., Pappin, D.J., Creasy, D.M., and Cottrell, J.S. (1999). Probability-based protein identification by searching sequence databases using mass spectrometry data. Electrophoresis *20*, 3551–3567.

Polz, M.F., and Cavanaugh, C.M. (1998). Bias in template-to-product ratios in multitemplate PCR. Appl. Environ. Microbiol. *64*, 3724–3730.

Pop, M. (2009). Genome assembly reborn: recent computation challenges. Brief. Bioinform. *10*, 354–366.

Poretsky, R.S., Bano, N., Buchan, A., Lecleir, G., Kleikemper, J., Pickering, M., Pate, W.M., Moran, M.A., and Hollibaugh, J.T. (2005). Analysis of microbial gene transcripts in environmental samples. Appl. Environ. Microbiol. *71*, 4121–4126.

Poretsky, R.S., Sun, S., Mou, X., and Moran, M. (2010). Transporter genes expressed by coastal bacterioplankton in response to dissolved organic carbon. Environ. Microbiol *12*, 616–627.

Prince, J.T., Carlson, M.W., Wang, R., Lu, P., and Marcotte, E.M. (2004). The need for a public proteomics repository. Nat. Biotech. *22*, 471–472.

Pruesse, E., Quast, C., Knittel, K., Fuchs, B.M., Ludwig, W., Peplies, J., and Glöckner, F.O. (2007). SILVA: a comprehensive online resource for quality checked and aligned ribosomal RNA sequence data compatible with ARB. Nucleic Acids Res. *35*, 7188–7196.

Pushkarev, D., Neff, N.F., and Quake, S.R. (2009). Single-molecule sequencing of an individual human genome. Nat. Biotechnol. *9*, 847–850.

Qin, J., Li, R., Raes, J., Arumugam, M., Burgdorf, K.S., Manichanh, C., Nielsen, T., Pons, N., Levenez, F., Yamada, T., *et al.* (2010). A human gut microbial gene catalogue established by metagenomic sequencing. Nature *464*, 59–65.

Qiu, X., Wu, L., Huang, H., McDonel, P.E., Palumbo, A.V., Tiedje, J.M., and Zhou, J.(2001). Evaluation of PCR-generated chimeras, mutations, and heteroduplexes with 16S rRNA gene-based cloning. Appl. Environ. Microbiol. *67*, 880–887.

Quince, C., Lanzen, A., Davenport, R.J., and Turnbaugh, P.J. (2011). Removing noise from pyrosequencing amplicons. BMC Bioinformatics *12*, 38.

Radajewski, S., Ineson, P., Parekh, N.R., and Murrell, J.C. (2000). Stable-isotope probing as a tool in microbial ecology. Nature *403*, 646–649.

Raes, J., and Bork, P. (2008). Molecular eco-systems biology: towards an understanding of community function. Nat. Rev. Microbiol. *6*, 639–696.

Raes, J., Foerstner, K.U., and Bork, P. (2007). Get the most out of your metagenome: computational analysis of environmental sequence data. Curr. Opin. Microbiol. *10*, 490–498.

Ramette, A. (2007). Multivariate analyses in microbial ecology. FEMS Microbiol. Ecol. *62*, 142–160.

Reigstad, L.J., Bartossek, R., and Schleper, C. (2011). Preparation of high-molecular weight DNA and metagenomic libraries from soils and hotsprings. Methods Enzymol. *496*, 319–344.

Rho, M., Tang, H., and Ye, Y. (2010). FragGeneScan: predicting genes in short and error- prone reads. Nucleic Acids Res. *38*, e191.

Richardson, T.H., Tan, X., Frey, G., Callen, W., Cabell, M., Lam, D., Macomber, J., Short, J.M., Robertson, D.E., and Miller, C. (2002). A Novel, high Performance enzyme for starch liquefaction. Discovery and optimization of a low pH, thermostable α-amylase. J. Biol. Chem. *277*, 26501–26507.

Robe, P., Nalin, R., Capellano, C., Vogel, T.M., and Simonet, P. (2003). Extraction of DNA from soil. Eur. J. Soil Biol. *39*, 183–190.

Rodrigue, S., Materna, A.C., Timberlake, S.C., Blackburn, M.C., Malmstrom, R.R., Alm, E.J., and Chishclm, S.W. (2010). Unlocking short read sequencing for metagenomics. PLoS One *5*, e11840.

Rodriguez-Brito, B., Rohwer, F., and Edwards, R.A.(2006). An application of statistics to comparative metagenomics. BMC Bioinformatics *7*, 162.

Röling, W.F.M., Ferrer, M., and Golyshin, P.N. (2010). Systems approaches to microbial communities and their functioning. Curr. Opin. Biotechnol. *21*, 532–538.

Rooijers, K., Kolmeder, C., Juste, C., Dore, J., de Been, M., Boeren, S., Galan, P., Beauvallet, C., de Vos, W., and Schaap, P. (2011). An iterative workflow for mining the human intestinal metaproteome. BMC Genomics *12*, 6.

Rothberg, J., Hinz, W., Rearick, T., Schultz, J., Mileski, W., Davey, M., Leamon, J.H., Johnson, K., Milgrew, M.J., Edwards, M., *et al.* (2011). An integrated semiconductor device enabling non-optical genome sequencing. Nature *475*, 348–352.

Rudney, J.D., Xie, H., Rhodus, N.L., Ondrey, F.G., and Griffin, T.J. (2010). A metaproteomic analysis of the human salivary microbiota by three-dimensional peptide fractionation and tandem mass spectrometry. Mol. Oral Microbiol. *25*, 38–49.

Rund, S.S.C., Hou, T.Y., Ward, S.M., Collins, F.H., and Duffield, G.E. (2011). Genome-wide profiling of diel and circadian gene expression in the malaria vector Anopheles gambiae. Proc. Natl. Acad. Sci. U.S.A. *108*, 421–430.

Rusch, D.B., Halpern, A.L., Sutton, G., Heidelberg, K.B., Williamson, S., Yooseph, S., Wu, D., Eisen, J.A., Hoffman, J.M., Remington, K. *et al.* (2007). The Sorcerer II Global Ocean Sampling expedition: northwest Atlantic through eastern tropical Pacific. PLoS Biol. *5*, e77.

Sanger, F., Nicklen, S., and Coulson, A.R. (1977). DNA sequencing with chainterminating inhibitors. Proc. Natl. Acad. Sci. U.S.A. *74*, 5463–5467.

Sayers, E.W., Barrett, T., Benson, D.A., Bolton, E., Bryant, S.H., Canese, K., Chetvernin, V., Church, D.M., Dicuccio, M., Federhen, S., *et al.* (2012). Database resources of the National Center for Biotechnology Information. Nucleic Acids Res. 40,13–25.

Schipper, C., Hornung, C., Bijtenhoorn, P., Quitschau, M., Grond, S., and Streit, W. (2009). Metagenome-derived clones encoding two novel lactonase family proteins involved in biofilm inhibition in Pseudomonas aeruginosa. Appl. Environ. Microbiol. 75, 224–233.

Schleper, C., DeLong, E.F., Preston, C.M., Feldman, R.A., Wu, K.Y., and Swanson, R.V. (1998). Genomicanalysisrevealschromosomalvariationin natural populations of the uncultured psychrophilic archaeon Cenarchaeum symbiosum. J. Bacteriol. *180*, 5003–5009.

Schloss, P.D., and Handelsman, J. (2003). Biotechnological prospects from metagenomics. Curr. Opin. Biotechnol. *14*, 303–310.

Schloss, P.D., and Handelsman, J. (2004). Status of the microbial census. Microbiol. Mol. Biol. Rev. *68*, 686–691.

Schloss, P.D., and Handelsman, J. (2005). Metagenomics for studying unculturable microorganisms: cutting the Gordian knot. Genome Biol. *6*, 229.

Schloss, P.D., and Handelsman, J. (2008). A statistical toolbox for metagenomics: assessing functional diversity in microbial communities. BMC Bioinformatics *9*, 34.

Schloss, P.D., Westcott, S.L., Ryabin, T., Hall, J.R., Hartmann, M., Hollister, E.B., Lesniewski, R.A., Oakley, B.B., Parks, D.H., Robinson, C.J., *et al.* (2009). Introducting MOTHUR: Open-source, platform-independent, community-supported software for describing and comparing microbial communities. Appl. Environ. Microbiol. *75*, 7537–7541.

Schmieder, R., and Edwards, R. (2011). Quality control and preprocessing of metagenomic datasets. Bioinformatics *27*, 863–864.

Schmieder, R., and Edwards, R. (2012). Insights into antibiotic resistance through metagenomic approaches. Future Microbiol. *7*, 73–89.

Scholz, M.B., Lo, C.C., and Chain, P.S.(2012). Next generation sequencing and bioinformatic bottlenecks: the current state of metagenomic data analysis. Curr. Opin. Biotechnol. *23*, 9–15.

Segata, N., Izard, J., Waldron, L., Gevers, D., Miropolsky, L., Garrett, W., and Huttenhower, C. (2011). Metagenomic biomarker discovery and explanation. Genome Biol. *12*, R60.

Selengut, J.D., Haft, D.H., Davidsen, T., Ganapathy, A., Gwinn-Giglio, M., Nelson, W.C., Richter, A.R., and White, O. (2007). TIGRFAMs and genome properties: tools for the assignment of molecular function and biological process in prokaryotic genomes. Nucleic Acids Res. *35*, 260–264.

Sharma, V.K., Kumar, N., Prakash, T., and Taylor, T.D. (2010). MetaBioME: a database to explore commercially useful enzymes in metagenomic datasets. Nucleic Acids Res. *38*, 468–472.

Shi, Y., McCarren, J., and DeLong, E.F. (2012). Transcriptional responses of surface water marine microbial assemblages to deep-sea water amendment. Environ. Microbiol. *14*, 191–206.

Shokralla, S., Spall, J., Gibson, J., and Hajibabaei, M. (2012). Next-generation sequencing technologies for environmental DNA research. Mol. Ecol. *21*, 1794–1805.

Siggins, A., Gunnigle, E., and Abram, F. (2012). Exploring mixed microbial community functioning: recent advances in metaproteomics. FEMS Microbiol. Ecol. *80*, 265–280.

Simon, C., and Daniel, R. (2011). Metagenomic Analyses: Past and future trends. Appl. Environ. Microbiol. *77*, 1153–1161.

Simon, C., Herath, J., Rockstroh, S., and Daniel, R. (2009). Rapid identification of genes encoding DNA polymerases by function-based screening of metagenomic libraries derived from glacial ice. Appl. Environ. Microbiol. *75*, 2964–2968.

Simpson, J.T., Wong, K., Jackman, S.D., Schein, J.E., Jones, S.J.M., and Birol, I. (2009). ABYSS: a parallel assembler for short read sequence data. Genome Res. *19*, 1117–1123.

Smith, J.J., Tow, L.A., Stafford, W., Cary, C., and Cowan, D.A. (2006). Bacterial diversity in three different Antarctic cold desert mineral soils. Microb. Ecol. *51*, 413–421.

Sogin, M.L., Morrison, H.G., Huber, J.A., Welch, D.M., Huse, S.M., Neal, P.R., Arrieta, J.M., and Herndl, G.J. (2006). Microbial diversity in the deep sea and the underexplored 'rare biosphere'. Proc. Natl. Acad. Sci. U.S.A. *103*, 12115–12120.

Sorek, R., and Cossart, P. (2010). Prokaryotic transcriptomics: a new view on regulation, physiology and pathogenicity. Nat. Rev. Genet. *11*, 9–16.

Sorek, R., Zhu, Y., Creevey, C.J., Francino, M.P., Bork, P., and Rubin, E.M. (2007). Genome-wide experiment determination of barriers to horizontal gene transfer. Science *318*, 1449–1452.

Steffan, R.J., Goksoyr, J., Bej, A.K., and Atlas, R.M. (1988). Recovery of DNA from soils and sediments. Appl. Environ. Microbiol. *54*, 2908–2915.

Suen, G., Scott, J.J., Aylward, F.O., Adams, S.M., Tringe, S.G., Pinto-Tomás, A.A., Foster, C.E., Pauly, M., Weimer, P.J., Barry, K.W., *et al.* (2010). An insect herbivore microbiome with high plant biomass-degrading capacity. PLoS Genet. *6*, e1001129.

Sun, S., Chen, J., Li, W., Altintas, I., Lin, A., Peltier, S., Stocks, K., Allen, E.E., Ellisman, M., Grethe, J., *et al.* (2011). Community cyberinfrastructure for Advanced Microbial Ecology Research and Analysis: the CAMERA resource. Nucleic Acids Res. 39, D546–551.

Sundquist, A., Bigdeli, S., Jalili, R., Druzin, M.L., Waller, S., Pullen, K.M., El-Sayed, Y.Y., Taslimi, M.M., Batzoglou, S., and Ronaghi, M. (2007). Bacterial flora-typing with targeted, chip-based pyrosequencing. BMC Microbiol. *30*, 108.

Suzuki, M.T., and Giovannoni, S.J. (1996). Bias caused by template annealing in the amplification of mixtures of 16S rRNA genes by PCR. Appl. Environ. Microbiol. *62*, 625–630.

Tamaki, H., Wright, C.L., Li, X., Lin, Q., Hwang, C., Wang, S., Thimmapuram, J., Kamagate, Y., and Liu, W.-T. (2011). Analysis of 16S rRNA amplicon sequencing options on the Roche/454 next-generation Titanium sequencing platform. PLoS One *6*, e25236.

Tatusov, R.L., Fedorova, N.D., Jackson, J.D., Jacobs, A.R., Kiryutin, B., Koonin, E.V., Krylov, D.M., Mazumder, R., Mekhedov, S.L., Nikolskaya, A.N., *et al.* (2003). The COG database: an updated version includes eukaryotes. BMC Bioinformatics *4*, 41.

Teeling, H.,Waldmann, J., Lombardot, T., Bauer, M., and Glockner, F.O. (2004). TETRA: a web-service and a stand-alone program for the analysis and comparison of tetranucleotide usage patterns in DNA sequences. BMC Bioinformatics *5*, 163.

Telias, A., White, J.R., Pahl, D.M., Ottesen, A.R., and Walsh, C.S. (2011). Bacterial community diversity and variation in spray water sources and the tomato fruit surface. BMC Microbiol. *11*, 81.

ter Kuile, B.H., and Westerhoff, H.V. (2001). Transcriptome meets metabolome: hierarchical and metabolic regulation of the glycolytic pathway. FEBS Lett. *500*, 169–171.

Thomas, T., Gilbert, J., and Meyer, F. (2012). Metagenomics – a guide from sampling to data analysis. Microb. Inform. Exp. *2*, 3.

Thompson, J.F., and Milos, P.M. (2011). The properties and applications of single-molecule DNA sequencing. Genome Biol. *12*, 217.

Torsvik, V.L.,and Goksoyr, J. (1978). Determination of bacterial DNA in soil. Soil Biol. Biochem. *10*, 7–12.

Treffer, R., and Deckert, V. (2010). Recent advances in single-molecule sequencing. Curr. Opin. Biotechnol. *21*, 4–11.

Tringe, S.G., and Hugenholtz, P. (2008). A renaissance for the pioneering 16S rRNA gene. Curr. Opin. Microbiol. *11*, 442–446.

Tuffin, I.M., Anderson, D., Heath, C., and Cowan, D.A. (2009). Metagenomic gene discovery: how far have we moved in sequence space? Biotechnol. J. *4*, 1671–1683.

Turnbaugh, P.J., and Gordon, J.I. (2008). An invitation to the marriage of metagenomics and metabolomics. Cell *134*, 708–713.

Turnbaugh, P.J., Ley, R.E., Mahowald, M.A., Magrini, V., Mardis, E.R., and Gordon, J.I. (2006). An obesity-associated gut microbiome withincreased capacity for energy harvest. Nature *444*, 1027–1031.

Turnbaugh, P.J., Ley, L.E., Micah, H., Fraser-Liggett, C.M., Knight, R., and Gordon, J.I. (2007). The human genome project. Nature *449*, 804–810.

Tyson, G.W., Chapman, J., Hugenholtz, P., Allen, E.E., Ram, R.J., Richardson, P.M., Solovyev, V.V., Rubin, E.M., Rokhsar, D.S., and Banfield, J.F. (2004). Community structure and metabolism through reconstruction of microbial genomes from the environment. Nature *428*, 37–43.

Uchiyama, T., Abe, T., Ikemura, T., and Watanabe, K. (2005). Substrate-induced gene-expression screening of environmental metagenome libraries for isolation of catabolic genes. Nat. Biotech. *23*, 88–93.

Uchiyama, T., and Miyazaki, K. (2010). Product-induced gene expression, a product-responsive reporter assay used to screen meta- genomic libraries for enzyme-encoding genes. Appl. Environ. Microbiol. 76, 7029–35.

Urbach, E., Vergin, K.L., and Giovannoni, S.J. (1999). Immunochemical detection and isolation of DNA from metabolically active bacteria. Appl. Environ. Microbiol. 65, 1207–1213.

Urich, T., Lanzen, A., Qi, J., Huson, D.H., Schleper, C., and Schuster, S.C. (2008). Simultaneous assessment of soil microbial community structure and function through analysis of the meta-transcriptome. PLoS One 3, e2527.

Vartoukian, S.R., Palmer, R.M., and Wade, W.G. (2010). Strategies for culture of 'unculturable' bacteria. FEMS Micrbiol. Lett. 309, 1–7.

Venter, J.C., Adams, M.D., Myers, E.W., Li, P.W., Mural, R.J., Sutton, G.G., Smith, H.O., Yandell, M., Evans, C.A., Holt, R.A., *et al.* (2001). The sequence of the human genome. Science *291*, 1304–1351.

Venter, J.C., Remington, K., Heidelberg, J.F., Halpern, A.L., Rusch, D., Eisen, J.A., Wu, D., Paulsen, I., Nelson, K.E., Nelson, W., *et al.*(2004). Environmental genome shotgun sequencing of the Sargasso Sea. Science *304*, 66–74.

VerBerkmoes, N., Denef, V., Hettich, R., and Banfield, J.F. (2009). Functional analysis of natural microbial consortia using community proteomics. Nat. Rev. Microbiol. 7, 196–205.

Vizcaíno, J.A., Côté, R., Reisinger, F., Barsnes, H., Foster, J.M., Rameseder, J., Hermjakob, H., and Martens, L. (2010). The proteomics identifications database: 2010 update. Nucleic Acids Res. 38, 736–742.

Vogel, T.M., Simonet, P., Jansson, J.K., Hirsch, P.R., Tiedje, J.M., Van Elsas, J.D., Bailey, M.J., Nalin, R., and Philippot, L. (2009). TerraGenome: a consortium for the sequencing of a soil metagenome. Nature Rev. Microbiol. 7, 252.

Vos, M., Quince, C., Pijl, A.S., Hollander, M.D., and Kowalchuk, G.A. (2012). A comparison of rpoB and 16S rRNA as markers in pyrosequencing studies of bacterial diversity. PLoS One 7, e30600.

Wang, G.Y., Graziani, E., Waters, B., Pan, W., Li, X., McDermott, J., Meurer, G., Saxena, G., Andersen, R.J., and Davies, J. (2000). Novel natural products from soil DNA libraries in a streptomycete host. Org. Lett. 2, 2401–2404.

Wang, J.-P. (2011). SPECIES: An R package for species richness estimation. J. Stat. Soft. *40*, 1–15.

Wang, Q., Garrity, G.M., Tiedje, J.M., and Cole, J.R. (2007). Naive Bayesian classifier for rapid assignment of rRNA sequences into the new bacterial taxonomy. Appl. Environ. Microbiol. 73, 5261–5267.

Wang, Q., Wu, H., Wang, A., Du, P., Pei, X., Li, H., Yin, X., Huang, L., and Xiong, X. (2010). Prospecting metagenomic enzyme subfamily genes for DNA family shuffling by a novel PCR-based approach. J. Biol. Chem. 285, 41509–41516.

Wang, Y., Sheng, H.F., Wu, J.Y., Jiang, Y.X., Tam, N.F., and Zhou, H.W. (2012). Comparison of the levels of bacterial diversity in freshwater, intertidal wetland, and marine sediments by using millions of illumina tags. Appl. Environ. Microbiol. 78, 8264–8271.

Wang, Z., Gerstein, M.,and Snyder, M. (2009). RNA-Seq: a revolutionary tool for transcriptomics. Nature Rev. Genet. *10*, 57–63.

Warnecke, F., and Hess, M. (2009). A perspective: metatranscriptomics as a tool for the discovery of novel biocatalysts. J. Biotechnol. *142*, 91–95.

Warren, R.L., Sutton, G.G., Jones, S.J., and Holt, R.A. (2007). Assembling millions of short DNA sequences using SSAKE. Bioinformatics 23, 500–501.

Watson, J.D., and Crick, F.H.C. (1953). A structure for deoxyribose nucleic acid. Nature *171*, 737–738.

Webb, C.O., Ackerly, D.D., and Kembel, S.W. (2008). Phylocom: software for the analysis of phylogenetic community structure and trait evolution. Bioinformatics *24*, 2098–2100.

Whittaker, R.J., Willis, K.J., and Field, R. (2001). Scale and species richness: towards a general, hierarchical theory of species diversity. J. Biogeog. *28*, 453–470.

Williams, R., Peisajovich, S., Miller, O., Magdassi, S., Tawfik, D., and Griffiths, A. (2006). Amplification of complex gene libraries by emulsion PCR. Nat. Methods 3, 545–550.

Williamson, L.L., Borlee, B.R., Schloss, P.D., Guan, C., Allen, H.K., and Handelsman, J. (2005). Intracellular screen to identify metagenomic clones that induce or inhibit a quorum-sensing biosensor. Appl. Environ. Microbiol. *71*, 6335–6344.

Willing, B.P., Dicksved, J., Halfvarson, J., Andersson, A.F., Lucio, M., Zongli, Z., Järnerot, G., Tysk, C., Jansson, J.K., and Engstrand, L. (2010). A pyrosequencing study in twins shows that gastrointestinal microbial profiles vary with inflammatory bowel disease phenotypes. Gastroenterology *139*, 1844–1854.

Wilmes, P., and Bond, P.L. (2004). The application of two-dimensional polyacrylamide gel electrophoresis and downstream analyses to a mixed community of prokaryotic microorganisms. Environ. Microbiol. *6*, 911–920.

Wilson, I.G. (1997). Inhibition and facilitation of nucleic acid amplification. Appl. Environ. Microbiol. *63*, 3741–3751.

Wishart, D.S. (2007). Human Metabolome Database: Completing the 'human parts' list. Pharacogenomics 8, 683–686.

Woese, C., Kandler, O., and Wheelis, M. (1990). Towards a natural system of organisms: proposal for the domains Archaea, Bacteria and Eucarya. Proc. Natl. Acad. Sci. U.S.A. *87*, 4576–4579.

Wommack, K.E., Bhavsar, J., and Ravel, J. (2008). Metagenomics: Read length matters. Appl. Environ. Microbiol. *74*, 1453–1463.

Wooley, J.C., and Ye, Y. (2010). Metagenomics: facts and artifacts, and computational challenges. J. Computer Sci. *25*, 71–81.

Wooley, J.C., Godzik, A., and Friedberg, I. (2010). A primer on metagenomics. PLoS Comput. Biol. *6*, e1000667.

Wu, C.H., Sercu, B., Van de Werfhorst, L.C., Wong, J., DeSantis, T.Z., Brodie, E.L., Haezen, T.C., Holden, P.A., and Andersen, G.L. (2010a) Characterization of coastal urban watershed bacterial communities leads to alternative community-based indicators. PLoS One *5*, e11285.

Wu, D., Wu, M., Halpern, A., Rusch, D.B., Yooseph, S., Frazier, M., Venter, J.C., and Eisen, J. (2011). Stalking the fourth domain in metagenomic data: Searching for, discovering, and interpreting novel, deep branches in marker gene phylogenetic trees. PLoS One *6*, e18011.

Wu, G.D., Lewis, J.D., Hoffmann, C., Chen, Y.Y., Knight, R., Bittinger, K., Hwang, J., Chen, J., Berkowsky, R., Nessel, L., *et al.* (2010b). Sampling and pyrosequencing methods for characterizing bacterial communities in the human gut using 16S sequence tags. BMC Microbiol. *10*, 206.

Xie, W., Wang, F., Guo, L., Chen, Z., Sievert, S.M., Meng, J., Huang, G., Li, Y., Yan, Q., Wu, S., *et al.* (2011). Comparative metagenomics of microbial communities inhabiting deep-sea hydrothermal vent chimneys with contrasting chemistries. ISME J. *5*, 414–426.

Xu, M., Xiao, X., and Wang, F. (2008). Isolation and characterization of alkane hydroxylases from a metagenomic library of Pacific deep-sea sediment. Extremophiles *12*, 255–62.

Yamane, T., Enokida, H., Hayami, H., Kawahara, M., and Nakagawa, M. (2012). Genome-wide transcriptome analysis of fluoroquinolone resistance in clinical isolates of *Escherichia coli*. Int. J. Urol. *19*, 360–368.

Yao, J., Fan, X.J., Lu, Y., and Liu, Y.H. (2011). Isolation and characterization of a novel tannase from a metagenomic library. J. Agric. Food. Chem. *59*, 3812–3818.

Ye, Y., and Tang, H.(2009). An ORFome assembly approach to metagenomics sequences analysis. J. Bioinform. Comput. Biol. *7*, 455–471.

Yooseph, S., Sutton, G., Rusch, D.B., Halpern, A.L., Williamson, S.J., Remington, K., Eisen, J.A., Heidelberg, K.B., Manning, G., Li, W., *et al.* (2007). The Sorcerer II Global Ocean Sampling expedition: expanding the universe of protein families. PLoS Biol. *5*, e16.

Zaprasis, A., Liu, Y.-J., Liu, S.-J., Horn, M.A., and Drake, H.L. (2010). Abundance of novel and diverse tfdA-like genes, encoding putative phenoxyalkanoic acid herbicide-degrading dioxygenases, in soil. Appl. Environ. Microbiol. *76*, 119–128.

Zhang, K., Liu, Y., Yang, J.Y., Arabnia, H.R., Niemierko, A., Ghafoor, A., Li, W., and Deng, Y. (2011). From genes to networks: in systematic points of view. BMC Syst. Biol. *5*, I1.

Zhao, L. (2010). Genomics: the tale of our other genome. Nature *465*, 879–880.

Zheng, H., and Wu, H. (2010). Short prokaryotic DNA fragment binning using a hierarchical classifier based on linear discriminant analysis and principal component analysis. J. Bioinform. Comput. Biol. *8*, 995–1011.

Zhou, J., Bruns, M.A., and Tiedje, J.M. (1996). DNA recovery from soils of diverse composition. Appl. Environ. Microbiol. *62*, 316–322.

Metatranscriptomics

11

Atsushi Ogura

Abstract

Metagenomic studies, accelerated by the evolution of sequencing technologies and the rapid development of genomic analysis methods, can reveal genetic diversity and biodiversity in various samples including those of uncultured or unknown species. This approach, however, cannot be used to identify active functional genes under actual environmental conditions. Metatranscriptomics, which is similar in approach to metagenomics except that it utilizes RNA samples, is a powerful tool for the transcriptomic study of environmental samples. Unlike metagenomic studies, metatranscriptomic studies have not been popular to date due to problems with reliability, repeatability, redundancy and cost performance.

Introduction

Metatranscriptome encompasses genes and transcripts obtained directly from environmental samples. Metatranscriptome concept was proposed not only for seeking novel genes but also for studying dynamics of gene regulatory networks in various environmental conditions in order to understand complex system of life and environment. It is difficult to point out when the term 'metatranscriptome' appeared being a combination of vogue 'meta-' or '-ome'. However, the passage from the 'meta-genome' to the 'meta-transcriptome' was already predicted in 2005 (Cases and de Lorenzo, 2005), and the first attempt was performed by Leininger *et al.* in their study of ammonia oxidation in soils (Leininger *et al.*, 2006). Metatranscriptomics is tightly connected with metagenomics in terms of experimental methods and analytical pipelines, although applications and purposes are slightly distinct.

Metagenomics has been developed as a cutting-edge approach for the genomic study of environmental samples and complex microorganisms without the need for cultivation and isolation (Patil *et al.*, 2011). As of 2012, more than 1200 research articles related to metagenomic studies were identified using a PubMed title/abstract search under the keywords 'metagenome' or 'metagenomics', but the real number of studies related to metagenomics will be even higher. Most of these studies were published within the last 5 years, indicating that this field of research has grown rapidly (Fig. 11.1). This rapid growth was driven by recent developments in next-generation sequencing (NGS) technologies and high-throughput methods for genomic analysis (Petrosino *et al.*, 2009; Creer, 2010). In general, metagenome study has been carried out to understand taxonomic distribution of microorganisms in a particular environmental sample. A metagenomic approach has been applied to many samples, including those from seawater, soil, internal organs of animal species and has revealed the species and genetic diversity in various environment (Bailly *et al.*, 2007). For instance, marine and soil microbiome studies revealed complex genetic interactions among various species, and human microbiome studies uncovered that there are almost 10 times as many microorganisms that affect human health and disease as there exist in the human body (Table 11.1).

Metagenomics offers a valuable approach to study the functional and genetic diversity of microbial communities; however, this approach cannot

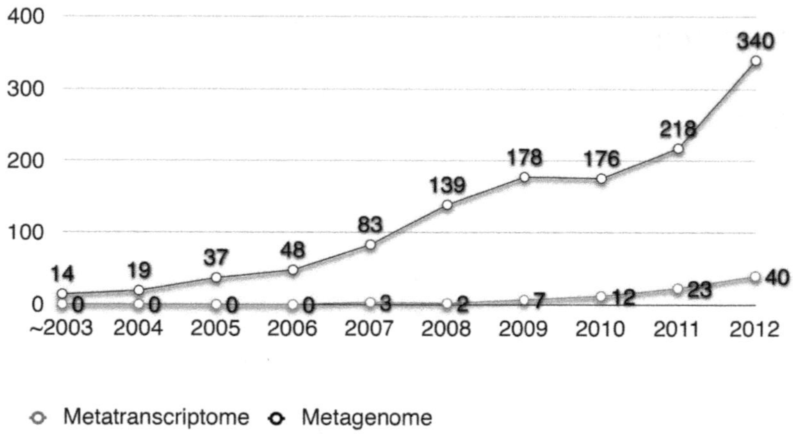

Figure 11.1 Publications.

provide information on active functional genes and gene regulatory networks in the environmental communities. Changes in the environment lead to variations in gene expression patterns in organisms, and the interactions of genes across species might change their environment. Therefore, comparative studies of metatranscriptome under various conditions or in various samples are essential to our understanding of genetic interactions under actual environmental conditions (McGrath *et al.*, 2008; Tartar *et al.*, 2009; Wu *et al.*, 2011). However, only 87 metatranscriptomic studies have been published as of 2012 according to the same search procedure as for 'metagenome' (Gilbert *et al.*, 2008; Bomar *et al.*, 2011; Gosalbes *et al.*, 2011; Hollibaugh *et al.*, 2011) (Fig. 11.1). This might be due to the difficulties to perform a metatranscriptomic study. Unlike genomic studies, for example, transcriptomic data vary according to environmental conditions, and a small number of highly expressed genes can disrupt the identification of other infrequently expressed genes (Nolte *et al.*, 2010). Furthermore, the metatranscriptome is composed of the transcriptomes of many organisms so that, unlike single transcriptomic studies, large-scale sequencing efforts are required.

Metatranscriptomic approaches

As it has been difficult to conduct metatranscriptome experiments, relatively few metatranscriptomic studies of microbial communities have been published to date (Fig. 11.1). First, collecting intact RNA from environmental samples is problematic because mRNA has short half-life, which is influenced by a contamination and enrichment bias at the collection phase and the extraction phase. Second, Sanger sequencer was too expensive to complete metatranscriptome analysis with sufficient coverage for gene expression profiles. Unlike an EST analysis of one sample, a metatranscriptome analysis screens for target genes from numerous organisms. Third, there was no suitable method to analyse large-scale metatranscriptomic data. Therefore, early metatranscriptomic studies utilized relatively small-scale samples and microarray technologies.

A microarray is a collection of DNA probes attached to a slide glass, and can be utilize to simultaneously measure the expression levels of large numbers of genes. Microarray-based approach requires pre-designed probe sequence information to detect gene expression profiles of metatranscriptome samples. Therefore, it is not possible to measure expression intensity of a certain gene if there is no probe designed for the gene. While originally limited to the microarray technology, metatranscriptomics studies have made use of direct high-throughput cDNA sequencing to provide whole-genome expression and quantification of a microbial community, as first employed by Leininger in their analysis of ammonia oxidation in soils (Leininger *et al.*,

Table 11.1 Metatranscriptome study identified by Pubmed search with 'Metatranscriptome [Title]'

Target, approach	Journal year	Authors	PMID
Deep-sea	ISME J, 2012	Lesniewski RA, *PubMed.*	22695860
Carbon-flux, marine	ISME J, 2012	Burow LC, *et al.*	23190731
Phosphorus limited lake	Environ Microbio ., 2012	Vila-Costa M, *et al.*	23176588
Soil	J. Microbiol Methods, 2012	Carvalhais LC, *et al.*	22963791
Aerobic, phylogeny	ISME J., 2012	Liu Z, *et al.*	22456447
Phenanthrene degradation, soil	Environ Microbio ., 2012	de Menezes A, *et al.*	22625871
Response to oil spill, marine	ISME J., 2012	Mason OU, *et al.*	22717885
Viral and microbial	J. Cyst. Fibros., 2012	Lim YW, *et al.*	22951208
Marine, phylogeny	Environ Microbio ., 2012	Radax R, *et al.*	22364353
Biogas plant	J. Biotechnol., 2012	Zakrzewski M, *et al.*	22342600
Protistan community	Genomics, 2012	Xie L, *et al.*	22326742
Microbial community	ISME J., 2012	Helbling DE, *et al.*	22094344
Periodontal pathogen	J Bacteriol, 2012	Frias-Lopez J, and Duran-Pinedo A.	22328675
Phytoplankton	Proc Natl Acad Sci U S A, 2012	Marchetti A, *et al.*	22308424
Suboxic water	ISME J., 2012	Feike J, *et al.*	21776032
Oxygen minimum zone, marine	Environ Microbio ., 2012	Stewart FJ, *et al.*	21210935
Mouse microbiome, network	PLoS One, 2012	Xiong X, *et al.*	22558305
Viral and ant	PLoS One, 2012	Valles SM, *et al.*	22384082
Forest soil	PLoS One, 2012	Damon C, *et al.*	22238585
Soil eukaryotes	ISME J., 2011	Damon C, *et al.*	21654847
Marine, NGS and array hybrid	BMC Genomics, 2011	Ogura A, *et al.*	22369384
Directed culturing	MBio, 2011	Bomar L, *et al.*	21467263
Marine	ISME J., 2011	Gifford SM, *et al.*	20844569
Sourdough fermentations	Appl. Environ. Microbiol., 2011	Weckx S, *et al.*	21097589
Method	Methods Mol. Biol., 2011	Gilbert JA, and Hughes M.	21431772
rRNA removal method	Nat. Methods, 2010	He S, *et al.*	20852648
Sourdough fermentations	Appl. Environ. Microbiol., 2010	Weckx S, *et al.*	20581179
Human faecal microbiome	Appl. Environ. Microbiol., 2010	Booijink CC, *et al.*	20562280
rRNA-subtraction	ISME J., 2010	Stewart FJ, *et al.*	20220791
Gut microbiome	Biotechnol. Biofuels, 2009	Tartar A, *et al.*	19832970
Review	Int. J. Food Microbiol., 2009	Weckx S, and De Vuyst L.	19446906
Review	J. Biotechnol., 2009	Warnecke F, and Hess M.	19480952
Marine	Nature, 2009	Shi Y, *et al.*	19444216
Lactic acid bacteria	Int. J. Food Microbiol., 2009	Nam YD, *et al.*	19215993

2006). This transition has been achieved not only by high-throughput sequencers but also by establishment of RNA extraction from environmental samples and development of bioinformatics tools brought by metagenome studies. However, in the case of metatranscriptome studies, the microarray approach still has an advantage over high-throughput sequencing method because microarray can limit the number of target genes to measure expression intensities in a cost-effective manner and reproducibility. Therefore, for the time being both high-throughput cDNA sequencing methods and the microarray approach could be utilized for metatranscriptomic studies (Ogura *et al.*, 2011).

NGS method for metatranscriptome

Development of the next- or second-generation sequencers have been triggered by a pyrosequencing method reported by 454 life sciences in 2005 (Margulies *et al.*, 2005). In comparison with the first-generation sequencers (Sanger method), the second-generation sequencers utilize slides or plates with millions of wells, in which nucleotide incorporation occurs, and imaging is used to identify nucleotide sequence. As an important characteristic, NGS technology can produce shorter but numerous sequences, whereas Sanger method can produce relatively longer pieces. One of the latest sequencer, Hiseq 2500 sequencer by Illumina, Inc., could generate more than 600 gigabases (Gb) per experiment that can cover 200-fold sequence of human genome, whereas Sanger method sequencer could generate less than 1 megabase (Mb) per experiment. Running costs of NGS machines are getting lower and lower, and draft genome project of an animal having 1-Gb genome can be performed within a few thousand dollars as of 2012. The cost of human genome sequencing was in the hundreds of thousands of dollars, but with latest techniques it might get down to a $10 genome project in the near future.

By utilizing these emerging technologies, it is possible to compare metatranscriptome samples by following the typical RNA-seq approach. In this approach, a population of RNA from an environmental sample is used to construct a library of cDNA fragments and then it is sequenced by NGS machine to obtain short sequences (reads) from one or both ends. In metatranscriptomic studies, the resulting reads are assembled *de novo* without the genomic sequence to produce a gene model. From the estimated gene model, gene expression frequencies can be calculated by the number of reads mapped to the assembled genes (contigs) with a statistical evaluation. Obtained gene expression profile data can be analysed with similar protocol as in the microarray approach.

Comparison of approaches

The microarray approach requires sequences of target genes for a probe design, and it is not possible to measure gene expression levels if the corresponding probe was not included. Therefore, the microarray approach is not suitable for seeking novel genes from environmental samples. However, if the target gene regulatory network is fixed, a microarray provide a cheap and reliable approach for metatranscriptome studies. Typical cost per one sample is several hundred dollars (Table 11.2).

NGS approach does not require sequence information as the microarray approach, but it requires *de novo* assembly of genes with no sequence information in advance. Typical cost per one sample with quality similar to the above microarray approach is several thousand dollars. In addition, samples can be contaminated by rRNAs, and normalization between samples is relatively difficult (Table 11.2).

The predominance of rRNAs in metatranscriptome samples is one of the major technical problems in NGS analysis of RNAs from microbiomes. Based on the fact that many rRNAs genes were present in mRNA samples where they limit the opportunity to sequence infrequently

Table 11.2 Comparison of Metagenome and metatranscriptome

Metagenome	Versus	Metatranscriptome
gDNA	Samples	Total RNA, mRNA
Genetic diversity, taxonomy	Targets	Gene interaction, network, functional genes
100~300Gb	Typical coverage	1~20Gb/sample
mapping, *de novo* assembly	Gene identification	Microarray, *de novo* assembly
$2000-$10,000/sample	Cost	$500~$5,000/sample

expressed genes, undertaking of metatranscriptomic studies using intact samples appears to be an inefficient and expensive strategy. Several methods developed to remove redundant genes or sequences that reduce usable sequences in sequence reads. Several approaches or methods, such as Ribo-Zero™ and RiboMinus™, could be applied to deplete rRNAs from metatranscriptomes, but rRNA depletion approaches introduce gene expression biases. Furthermore, these methods can deplete rRNAs only of known species such as human, mouse, yeast, and bacteria, so that they cannot work efficiently for metatranscriptome samples.

One of the reasonable approaches for metatranscriptomic study is a hybrid of the NGS and microarray approaches, in which gene contents can be collected by NGS technique, and then, comparison of metatranscriptome samples can be performed by a microarray designed from a gene collection obtained via NGS. Prior to the comparison of metatranscriptomic samples, creation of a comprehensive gene collection, even in the absence of information regarding expression frequency, would be useful in gaining a better understanding of active functional genes in the samples and would contribute to database construction and microarray design for the cost-effective monitoring of changes in gene expression in various samples. To improve efficiency of gene collection method, we propose normalization strategy of metatranscriptome samples. A typical flow chart for metatranscriptomic analysis is illustrated in Fig. 11.2. Normalization, in this case, is used to reduce the interference between highly expressed genes through the use of duplex-specific nuclease (Zhulidov *et al.*, 2004).

To measure gene expression levels of orthologous genes from metatranscriptome samples by microarray experiment, an algorithm presented in (Fukuzaki *et al.*, 2012) is used in which orthologous genes are measured with a single probe. In this algorithm, the effect of probe sequence mutations on the expression levels using systematically mutated probes is assessed, and mutations on the 3′-end of the probes have very little effect on the expression values whereas those on the 5′-end decreased the observed expression levels. This allows generating a microarray that can be used for orthologous gene expression comparison of metatranscriptome samples.

Bioinformatics for metatranscriptome: RNA assembly

Assembling metatranscriptome data is one of the most difficult problems in metatranscriptome analysis. Metatranscriptome samples consist of a huge number of microorganisms, and many genes with tiny differences, heterozygosity, orthology, SNPs and mutations interfere with a concise gene assembly and produce chimeric genes.

For metatranscriptomic samples, it is not possible to measure gene expression frequencies by mapping-based methods as the reference genome does not exist. Therefore, de novo assembly of metatranscriptome data is required prior to comparison of gene expression profiles of the samples. Gene expression frequencies are calculated as reads per kilobase per million reads (RPKM). Unlike genome sequence assembly, transcriptome data include different types of transcripts including transcripts with alternative splicing. Therefore, an assembly algorithm is essential for a transcriptome, but not for a genome. Unfortunately, a specialized assembly algorithm for metatranscriptome data has not been developed, and instead assembly algorithms for metagenome or transcriptome are used in metatranscriptomic analysis. Below are the examples of assembly programs:

Metagenome-based tools

MetaIDBA

This algorithm can efficiently reconstruct scaffolds from metagenome sequences by partitioning de Bruijn graph into isolated components of different species, then by capturing the slight variants of the genomes of subspecies from the same species using a consensus sequence (Peng *et al.*, 2011).

MetaVelvet

This algorithm can reconstruct scaffolds from metagenome sequences consisting of multiple species by decomposing de Bruijn graph constructed from mixed short reads into individual subgraphs, then by building scaffolds based on every decomposed de Bruijn subgraph as isolate species genome (Namiki *et al.*, 2012).

Transcriptome-based tools

Trinity

This program partitions the sequence data into many individual de Bruijn graphs, each representing the transcriptional complexity at a given gene or locus, and then processes each graph independently to extract full-length splicing isoforms and to tease apart transcripts derived from paralogous genes (Grabherr *et al.*, 2011).

SOAPdenovo-Trans

This is a *de novo* transcriptome assembler based on the SOAPdenovo framework that is adapted to alternative splicing and different expression level among transcripts (Luo *et al.*, 2012).

Any assembler can create fragmented or chimeric genes due to heterozygosity and complexity of the samples.

Bioinformatics for metatranscriptome

Gene estimation is the next step for metatranscriptome analysis. The annotation of gene model has two steps: estimation of gene models, in which open reading frames can be called using assembled scaffolds, and annotation of these gene models, performed using standardized terminology and ontology describing functions, expression patterns, and network involvement. This estimation process is often trained using known gene models for better quality. Either approach relies on collecting gene products found in metatranscriptome samples and annotating their molecular functions, biological roles, involvement in gene regulatory networks. Although there are many bioinformatic approaches, one of the basic methods is BLAST search that can show representatives from gene families of interest in the metatranscriptome data to determine their expression frequencies. Metatranscriptome data guarantee a transcribed sequence, in principle, so that usually gene model prediction is not required.

In metatranscriptome analyses, it is also important to distinguish orthologues and paralogues. Homologous genes are orthologues if they were

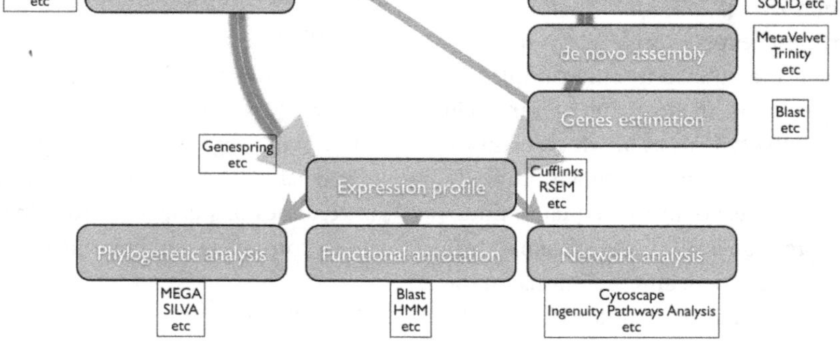

Figure 11.2 Flow chart.

separated by a speciation event, whereas they are paralogues if they were separated by a duplication event. Orthologues are useful not only for phylogenetic analysis but also for functional annotation because orthologous genes tend to have the same function. On the other hand, paralogues tend to have different functions as duplicated genes are often redundant in species and neo-functionalization or subfunctionalization occur to gain novel or substitute existing functions. One of the main motivations of metatranscriptome studies is to clarify dynamics of gene regulation and interaction, and in this respect discrimination between orthologues and paralogues and concise annotations of genes are essential.

Bioinformatics for metatranscriptome: network analysis

The next step is to examine functional representation of genetic and metabolic pathways within metatranscriptome data. By adopting a network-based approach to identify neighbourhoods of functionally related genes, well-represented pathways can be highlighted that are functional in the environment. Adopting a network analysis could identify node genes mediating essential roles within the network. Betweenness centrality is a metric that assesses the importance of the node in the network through determining the proportion of shortest path lengths that pass through that node. Focusing on the nodes of high betweenness centrality, we observe differences between samples. Network analysis of this type can be performed using tools and algorithms designed for microarray or RNA-seq studies, such as Cytoscape (Shannon *et al.*, 2003) and Ingenuity Pathways Analysis (Ingenuity® Systems).

Case study: metatranscriptome of marine microbiome

Marine microorganisms represent a major target for genetic resources and environmental monitoring (25). There remain, however, many uncultured organisms so that comprehensive studies at the molecular level have long been ignored. As for the marine microorganisms samples, we focused on plankton samples taken from the Inland Sea of Japan. Prefectural research institutes connected with Japan Fisheries have been conducting sampling of organisms for environmental monitoring in this area since the early 1970s, and have accumulated data on the appearance of phytoplankton and zooplankton. Phytoplankton monitoring has shown that diatoms have been the dominant phytoplankton group (>90%) over a 35-year period, and that there was a drastic shift from Skeletonema (~70%) to Chaetoceros dominance in the mid 1980s. While the monitoring of the dominant species has been conducted and reported, there is no information available on rare species and/or smaller-sized plankton species, such as Cryptophyceae, Haptophyceae and Prasinophyceae. Very recently, a new method for plankton metagenomic analysis was developed and this technique allows all-encompassing analyses of almost all plankton components, including zooplankton and protozoa, in coastal waters. Therefore, integrated metagenomic and metatranscriptomic analyses will allow us to obtain detailed information on all plankton species existing in coastal waters as well as on the gene expression in each plankton component, resulting in a more complete understanding of coastal ecosystems. For instance, metatranscriptomic analyses before and after red tides (abnormal growth of phytoplankton) may lead to the identification of the mechanisms behind red tides and the associated harmful microalgae. It may also be possible to develop a new environmental assessment technique for fishing grounds and give more scientific input to the healthy management of fishing grounds through the comparison of highly polluted and non-polluted areas.

To assess normalization effect of metatranscriptome study, a plankton sample in Hiroshima Bay (34.16′N; 132.16′E), in the Inland Sea of Japan was collected, and NGS libraries with and without normalization have been constructed. Transcriptome data does not reflect proportionally species diversity or gene functions, but it is thought that the frequencies of expressed genes in a sample reflect the activities of functional genes in seawater. For comparison of the two libraries, a reciprocal homology search using BLAT software has been performed and, as a result, 56.1% of genes in the non-normalized library were found to have identical or highly conserved homologues in the normalized library, whereas only 21.6% of genes in the normalized library had identical or highly conserved genes in the non-normalized library (Fig. 11.3A). In other words, 43.9% and 78.4% of genes were unique in the non-normalized and normalized libraries, respectively. Normalization, therefore, can reduce redundancy among the expressed genes and is suitable for the collection of various genes from marine transcriptomic samples. The taxonomic distribution of marine microorganisms is a typical focus of metagenomic studies, in which we examine the species diversity of samples (Yarza *et al.*, 2008). In the case of metatranscriptomic studies, the distribution of genes does not imply the distribution of species. However, it remains of interest to understand activity of marine microorganisms. For this purpose, taxonomic distribution analysis using rDNA database maintained at ARB (Ludwig *et al.*, 2004) has been conducted, which contains all known rRNA genes with taxonomic annotation. From this analysis, the majority of the species, at least at the level of rRNA activity, belonged to the Eukaryota domain, occupying more than 95% of the sample. This result is consistent with the fact that diatoms and dinoflagellates, which belong to the Eukaryota domain, are known to be the dominant species in the area where sample has been collected. In fact, Stramenopiles, which include many kinds of diatoms, is the major group. As the normalization protocol reduces highly expressed gene redundancy, it is much more difficult to understand the taxonomic distribution from the data obtained. However, a comparison with the non-normalized library indicates that the reduction in the number of species in the normalized library might be due to the fact that most were major species groups without genetic diversity. A comparison of the two libraries further suggested that those species are often members of the Archaea or Glaucocystophyceae. On the other hand, groups in which the proportions were increased in the non-normalized library, such as Metazoa, might contain various genetically diversified species. The reason why the taxonomic distribution of sequences is slightly changed following normalization is not evident from our results, but one possible explanation is that compression of taxonomic distribution could not be achieved due to insufficient depletion of rRNA variation (Fig. 11.3B).

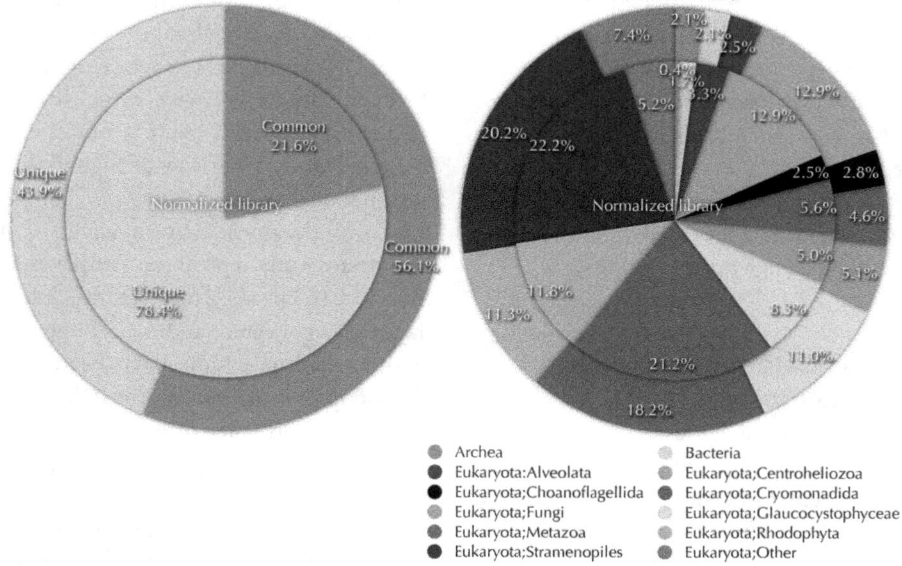

Figure 11.3 Normalization effect.

Future trends

Metatranscriptome studies shall be fundamental tool to understand complexity of life from inner universe of human body to global-scale environment. Metatranscriptome-oriented assembly tools will be developed by tweaking the algorithms of metagenome assembler or RNA assembler. Metatranscriptome-oriented network analysis tools will also be developed to include genes, species and data from an environmental monitoring. Metatranscriptome specified microarray for particular purpose will be available if the target genes are fixed.

Conclusion

Despite the technical and financial difficulties to conduct metatranscriptome studies, the field of metatranscriptome research has grown rapidly accelerated by the development of next-generation sequencers. An integrated metagenomic and metatranscriptomic analysis will allow us to obtain detailed information on marine, soil and any microbiome samples, for understanding our environment.

Web resources

- Ribo-Zero™: http://www.epibio.com/products/rna-sequencing/rrna-removal/
- RiboMinus™: http://goo.gl/F99N0
- Meta-IDBA: http://i.cs.hku.hk/~alse/hkubrg/projects/metaidba/
- Meta-velvet: http://metavelvet.dna.bio.keio.ac.jp
- Trinity: http://trinityrnaseq.sourceforge.net
- SOAPdenovo-Trans: http://soap.genomics.org.cn/SOAPdenovo-Trans.html
- Cytoscape: http://www.cytoscape.org
- Ingenuity Pathways Analysis: http://www.ingenuity.com

References

Bailly, J., Fraissinet-Tachet, L., Verner, M.-C., Debaud, J.-C., Lemaire, M., Wésolowski-Louvel, M., and Marmeisse, R. (2007). Soil eukaryotic functional diversity, a metatranscriptomic approach. ISME J. *1*, 632–642.

Bomar, L., Maltz, M., Colston, S., and Graf, J. (2011). Directed culturing of microorganisms using metatranscriptomics. MBio. *2*, e00012–11.

Cases, I., and de Lorenzo, V. (2005). Promoters in the environment: transcriptional regulation in its natural context. Nat. Rev. Microbiol. *3*, 105–118.

Creer, S. (2010). Second-generation sequencing derived insights into the temporal biodiversity dynamics of freshwater protists. Mol Ecol. *19*, 2829–2831.

Fukuzaki, M., Yoshida, M.-A., Ogura, A., and Sese, J. (2012). Systematic measurement of missmatch effect for designing inter-species microarray. Paper presented at: IEEE International Conference on Bioinformatics and Biomedicine *0*, 1–4.

Gilbert, J.A., Field, D., Huang, Y., Edwards, R., Li, W., Gilna, P., and Joint, I. (2008). Detection of large numbers of novel sequences in the metatranscriptomes of complex marine microbial communities. PLoS One *3*, e3042.

Gosalbes, M.J., Durbán, A., Pignatelli, M., Abellan, J.J., Jiménez-Hernández, N., Pérez-Cobas, A.E., Latorre, A., and Moya, A. (2011). Metatranscriptomic approach to analyze the functional human gut microbiota. PLoS One *6*, e17447.

Grabherr, M.G., Haas, B.J., Yassour, M., Levin, J.Z., Thompson, D.A., Amit, I., Adiconis, X., Fan, L., Raychowdhury, R., Zeng, Q., *et al.* (2011). Full-length transcriptome assembly from RNA-Seq data without a reference genome. Nat. Biotechnol. *29*, 644–52.

Hollibaugh, J.T., Gifford, S., Sharma, S., Bano, N., and Moran, M.A. (2011). Metatranscriptomic analysis of ammonia-oxidizing organisms in an estuarine bacterioplankton assemblage. ISME J. *5*, 866–878.

Leininger, S., Urich, T., Schloter, M., Schwark, L., Qi, J., Nicol, G.W., Prosser, J.I., Schuster, S.C., and Schleper, C. (2006). Archaea predominate among ammonia-oxidizing prokaryotes in soils. Nature *442*, 806–809.

Ludwig, W., Strunk, O., Westram, R., Richter, L., Meier, H., Yadhukumar, Buchner, A., Lai, T., Steppi, S., Jobb, G., *et al.* (2004). ARB: a software environment for sequence data. Nucleic Acids Res. *32*, 1363–1371.

Luo, R., Liu, B., Xie, Y., Li, Z., Huang, W., Yuan, J., and He, G. (2012). SOAPdenovo2: an empirically improved memory-efficient short-read de novo assembler. GigaScience *1*, 18.

Margulies, M., Egholm, M., Altman, W.E., Attiya, S., Bader, J.S., Bemben, L.A., Berka, J., Braverman, M S., Chen, Y.-J., Chen, Z., *et al.* (2005). Genome sequencing in microfabricated high-density picolitre reactors. Nature *437*, 376–380.

McGrath, K.C., Thomas-Hall, S.R., Cheng, C.T., Leo, L., Alexa, A., Schmidt, S., and Schenk, P.M. (2008). Isolation and analysis of mRNA from environmental microbial communities. J. Microbiol. Methods *75*, 172–176.

Namiki, T., Hachiya, T., Tanaka, H., and Sakakibara, Y. (2012). MetaVelvet: an extension of Velvet assembler to de novo metagenome assembly from short sequence reads. Nucleic Acids Res. *40*, e155.

Nolte, V., Pandey, R.V., Jost, S., Medinger, R., Ottenwälder, B., Boenigk, J., and Schlötterer, C. (2010). Contrasting seasonal niche separation between rare and abundant taxa conceals the extent of protist diversity. Mol Ecol. *19*, 2908–2915.

Ogura, A., Lin, M., Shigenobu, Y., Fujiwara, A., Ikeo, K., and Nagai, S. (2011). Effective gene collection from the metatranscriptome of marine microorganisms. BMC Genomics *12*, S15.

Patil, K.R., Haider, P., Pope, P.B., Turnbaugh, P.J., Morrison, M., Scheffer, T., and Mchardy, A.C. (2011). Taxonomic metagenome sequence assignment with structured output models. Nat. Methods *8*, 191–192.

Peng, Y., Leung, H.C.M., Yiu, S.M., and Chin, F.Y.L. (2011). Meta-IDBA: a de Novo assembler for metagenomic data. Bioinformatics *27*, i94–i101.

Petrosino, J.F., Highlander, S., Luna, R.A., Gibbs, R.A., and Versalovic, J. (2009). Metagenomic pyrosequencing and microbial identification. Clin. Chem. *55*, 856–866.

Shannon, P., Markiel, A., Ozier, O., Baliga, N.S., Wang, J.T., Ramage, D., Amin, N., Schwikowski, B., and Ideker, T. (2003). Cytoscape: a software environment for integrated models of biomolecular interaction networks. Genome Res. *13*, 2498–2504.

Tartar, A., Wheeler, M.M., Zhou, X., Coy, M.R., Boucias, D.G., and Scharf, M.E. (2009). Parallel metatranscriptome analyses of host and symbiont gene expression in the gut of the termite Reticulitermes flavipes. Biotechnol. Biofuels *2*, 25.

Wu, J., Gao, W., Zhang, W., and Meldrum, D.R. (2011). Optimization of whole-transcriptome amplification from low cell density deep-sea microbial samples for metatranscriptomic analysis. J. Microbiol. Methods *84*, 88–93.

Yarza, P., Richter, M., Peplies, J., Euzeby, J., Amann, R., Schleifer, K.-H., Ludwig, W., Glöckner, F.O., and Rosselló-Móra, R. (2008). The All-Species Living Tree project: a 16S rRNA-based phylogenetic tree of all sequenced type strains. Syst. Appl. Microbiol. *31*, 241–250.

Zhulidov, P.A., Bogdanova, E.A., Shcheglov, A.S., Vagner, L.L., Khaspekov, G.L., Kozhemyako, V.B., Matz, M.V., Meleshkevitch, E., Moroz, L.L., Lukyanov, S.A., *et al.* (2004). Simple cDNA normalization using kamchatka crab duplex-specific nuclease. Nucleic Acids Res. *32*, e37.

Inferring Viral Quasispecies Spectra from Shotgun and Amplicon Next-generation Sequencing Reads

12

Irina Astrovskaya, Nicholas Mancuso, Bassam Tork, Serghei Mangul, Alex Artyomenko, Pavel Skums, Lilia Ganova-Raeva, Ion Măndoiu and Alex Zelikovsky

Abstract

Many clinically relevant viruses, including hepatitis C virus (HCV) and human immunodeficiency virus (HIV), exhibit high genomic diversity within infected hosts which may explain the failure of vaccines and resistance to existing antiviral therapies. Characterizing the viral population infecting a host requires reconstructing all co-existing (related, but non-identical) viral variants, referred to as quasispecies, and inferring their relative abundances. Next-generation sequencing is a promising approach for characterizing viral diversity due to its ability to generate a large number of reads at a low cost. However, standard assembly software was originally designed for a single genome assembly and cannot be used to assemble multiple closely related quasispecies sequences and estimate their abundances. In this chapter, we focus on the problem of reconstructing viral quasispecies populations from next-generation sequencing reads produced by two most commonly used strategies: the shotgun sequencing and the sequencing of partially overlapping PCR amplicons. We discuss computational challenges associated with each strategy and review existing approaches to quasispecies reconstruction with focus on two state-of-the-art software tools – Viral Spectrum Assembler (ViSpA), designed for the shotgun reads, and Viral Assembler (VirA), which handles the amplicon reads. Both tools have been tested on simulated and real read data from HCV, HIV (ViSpA) and HBV (VirA) quasispecies, and shown to compare favourably with other existing methods.

Introduction

Viral quasispecies

Many medically important viruses, including influenza viruses, hepatitis C virus (HCV), and human immunodeficiency virus (HIV), exhibit high genomic diversity within and between their hosts. In RNA viruses, absence of proofreading mechanisms results in the inability to detect and repair mistakes during replication (Duarte *et al.*, 1994). As a result, the mutation rate may be as high as one mutation per each thousand bases copied per replication cycle (Drake *et al.*, 1999). Besides replication errors that result in substitutions, small insertions, and small deletions, many viruses also undergo recombination and genome segment reassortment. Much of this sequence variation is well tolerated and passed down to descendants, producing in each infected host a family of co-existing variants of the original viral genome referred to as mutant clouds, or *viral quasispecies*, a concept that originally described a mutation-selection balance (Domingo *et al.*, 1985; Steinhauer *et al.*, 1987; Eigen *et al.*, 1989; Martell *et al.*, 1992; Domingo and Holland, 1997).

Short replication cycles, small size of viral genomes, large population sizes, and strong selective pressure exerted by the immune response of an infected patient additionally contribute to the evolution of the quasispecies, for instance, resulting in 10^{10} to 10^{12} new variants per day in case of HIV-1, HBV, or HCV infections (Domingo *et al.*, 1998, Neumann *et al.*, 1998). These quasispecies variants differ in biological properties such as their ability to cause disease (virulence), ability to escape host immune responses, resistance to antiviral therapies, and the type of host cells and tissues infected by the virus (tissue tropism). As a consequence, the best adapted variants survive different environmental changes and selective pressures, resulting in various viral populations in infected hosts.

The diversity of viral sequences and frequent generation of new variants in an infected patient can cause vaccines failures and virus resistance to existing therapies. Drug-resistant viral quasispecies, even present as a minor portion of population, result in rapid viral adaptation in case of HIV-1 and HCV and failure of antiretroviral therapy (Metzner *et al.*, 2009, Li *et al.*, 2011; Skums *et al.*, 2011, 2012). As a result, very few RNA viruses are effectively controlled by vaccination or antiviral therapies (Holland *et al.*, 1992; Domingo *et al.*, 1998). Furthermore, since live viruses mutate rapidly, they can become virulent for a host at any replication cycle, posing another challenge for vaccine development. Since population dynamics cannot be explained just by the evolution of the most frequent viral variant, there is a great interest in reconstructing genomic diversity of all viral quasispecies in a host. Knowing sequences of virulent variants and their abundances in infected patients can improve our understanding of viral evolutionary dynamics, mechanisms of viral persistence and drug resistance, and, as result, may help to design an effective drugs (Beerenwinkel *et al.*, 2005; Rhee *et al.*, 2007) and vaccines (Gaschen *et al.*, 2002; Douek *et al.*, 2006; Luciani *et al.*, 2012), targeting particular viral variants *in vivo*.

Next-generation sequencing technologies

Evolution of the next-generation sequencing (NGS) technologies has significantly changed experimental analysis of viral communities (Barzon *et al.*, 2011) since their massive throughput and decreasing cost allow deep sampling of an entire viral population in a single experiment. Although existing NGS technologies have similar workflows, they differ in underlying biochemistries and sequencing protocols as well as their throughput and average sequence length. Applied Biosystem's SOLiD technology uses ligation to sequence genomes, whereas the sequencing technology developed by Illumina incorporates reversible dye-terminators (Mardis 2008, 2009). Both technologies produce high accuracy reads, enabling detection of local low-frequency variants; however, the shorter read length is not as efficient and effective for reconstruction of full-length quasispecies sequences (Zagordi *et al.*, 2012a). Ion Torrent's NGS technology uses integrated circuits that detect the pH change caused by the release of ions during template-directed DNA polymerase synthesis. Currently, Ion Torrent reads have average length around 200–400 bases (Barzon *et al.*, 2011), and future systems are expected to generate even longer reads (Rothberg *et al.*, 2011).

Since average read length strongly influences global viral quasispecies sequence reconstruction (Zagordi *et al.*, 2012a), to date, the 454/Roche pyrosequencing technology has been the most commonly used NGS technology for viral quasispecies analysis (Margulies *et al.*, 2005). In brief, the 454 pyrosequencing system shears the source genetic material into the fragments and amplifies them on beads using emulsion PCR. Millions of amplified template fragments are then sequenced by synthesizing their complementary strands. Repeatedly, the nucleotide reagents are flown over the fragments, one nucleotide (A, C, T or G) at a time. Light is emitted at a fragment location when the flown nucleotide base complements the first unpaired base of the fragment (Fakhrai-Rad *et al.*, 2002; Margulies *et al.*, 2005). Multiple identical nucleotides may be incorporated in a single cycle, and the light intensity would correspond to the number of the incorporated bases. However, in practice, this number (referred to as a homopolymer length) cannot be accurately estimated for the long homopolymers, resulting in a relatively high percentage of insertion and deletion sequencing errors, which respectively represent 65–75% and 20–30% of all sequencing errors (Brockman *et al.*, 2008; Quinlan *et al.*, 2008).

Shotgun versus amplicon sequencing reads

Contiguous NGS reads can be generated in two essentially different ways – as shotgun or as amplicon reads. In the first case, the fragments (the shotgun reads) are randomly produced from a given DNA segment in such way that their starting positions are uniformly distributed across the genome. In contrast, the amplicon reads are generated by using PCR from the overlapping 'windows' of the viral genome so the beginnings and the endings of reads are essentially fixed (see Fig. 12.1). Currently, GS FLX Titanium XL+ can generate up to 1 million shotgun and 700,000 amplicon reads with average read length around 700 and 400 bases, respectively.

Quasispecies spectrum reconstruction

The software provided by the instrument manufacturers was originally designed to assemble all reads into a single (consensus) genome sequence, rather than multiple similar but non-identical sequences. Thus, new software must be developed to solve the following problem.

Quasispecies spectrum reconstruction (QSR) problem. *Given a collection of shotgun or amplicon NGS reads generated from a viral sample, reconstruct the quasispecies spectrum, i.e. the set of sequences and the relative frequency of each sequence in the sample population.*

A major challenge in solving the QSR problem is that the quasispecies sequences are only slightly different from each other. The amount of differences between quasispecies and their distribution along the genome vary significantly across known viruses due to variation in their mutation rates and genomic architectures. A limited read length and a relatively high error rate of current high-throughput sequencing data also add to the complexity of the QSR problem.

Related work

The QSR problem is related to several well-studied problems: *de novo* genome assembly (Myers, 2005; Sundquist *et al.*, 2007; Chaisson *et al.*, 2008), haplotype assembly (Lippert *et al.*, 2002; Bansal *et al.*, 2008), population phasing (Brinza *et al.*, 2006) and metagenomics (Venter *et al.*, 2004). *De novo* assembly methods are designed to build a single genome sequence and are not well-suited for reconstructing a large number of closely related quasispecies sequences. Haplotype assembly does seek to infer two closely related haplotype sequences, but existing methods do not easily extend to the reconstruction of a large (and a priori unknown) number of sequences. Computational methods developed for population phasing deal with a large number of haplotypes, but rely on the availability of genotype data that conflate information about pairs of haplotypes. Metagenomic samples do consist of sequencing reads generated from the genomes of a large number of species. However, differences between the genomes of these species are considerably larger than those between viral quasispecies. Furthermore, existing tools for metagenomic data analysis focus on species identification since reconstruction of complete genomic sequences would require much higher sequencing depth than that typically provided by current metagenomic datasets.

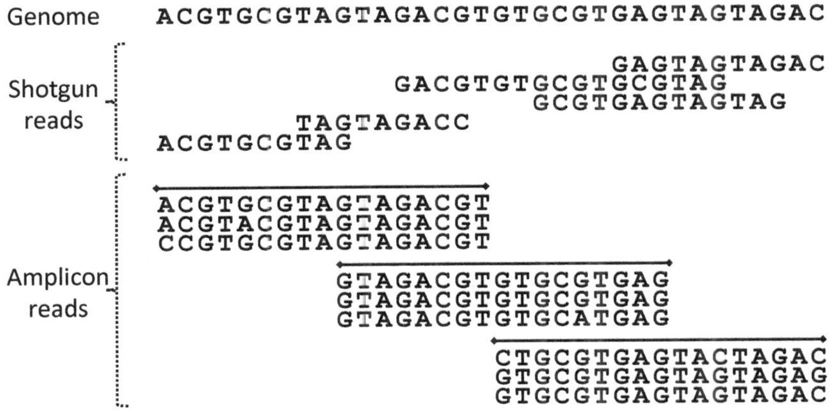

Figure 12.1 Shotgun versus amplicon reads.

In contrast, achieving high sequencing depth for the viral samples is very inexpensive, owing to the short length of the viral genomes. Mapping based approaches to the QSR problem are naturally preferred to *de novo* assembly since the reference genomes are available (or easy to obtain) for the viruses of interest, and the viral genomes do not contain repeats. Thus, it is not surprising that such approaches were adopted in two pioneering works on the QSR problem (Eriksson *et al.*, 2008; Westbrooks *et al.*, 2008). Both works independently introduced the concept of a read graph in which nodes represent possibly pre-processed reads. Two nodes are connected with a directed edge if reads sequences agree on their overlap. Eriksson *et al.* (2008) proposed a multi-step approach with focus on a local haplotype reconstruction. The method consists of sequencing error correction via probabilistic clustering, haplotype reconstruction via chain decomposition, and haplotype frequency estimation via expectation maximization (EM). This method was implemented in the software tool ShoRAH (Zagordi *et al.*, 2010a, 2011) and successfully applied to HIV data (Zagordi *et al.*, 2010b). In contrast to the previous approach, Westbrooks *et al.* (2008) globally reconstructed viral haplotypes via transitive reduction, overlap probability estimation and network flows, with application to simulated error-free HCV data. Astrovskaya *et al.* (2011) further extended this approach by allowing imperfect overlaps (i.e. overlaps with certain number of disagreements) during a read graph construction and by inferring quasispecies via max-bandwidth paths through the graph with adjusted overlaps probabilities estimations. This pipeline (referred to as ViSpA) was successfully applied to both simulated and real HCV and HIV data. Experimental results show that ShoRAH tends to overcorrect reads and that ViSpA outperforms ShoRAH in assembling quasispecies sequences. The idea of imperfect read overlaps is also exploited in the QColors's method (Huang *et al.*, 2011). In contrast to ViSpA, QColors uses an additional conflict graph to represent disagreement within read overlaps. The haplotypes are reconstructed by finding a partition of the reads into the minimal number of non-conflicting subsets. Although the QColors's method gives guidance on how to use short and non-contiguous NGS reads, it may be too sensitive to sequencing errors. Hapler (O'Neil and Emrich, 2012) reformulates the problem of finding a minimum number of paths needed to explain observed reads as a weighted bipartite graph matching problem. A randomization step with sampling from the space of possible haplotypes reduces the probability of reconstructing chimeric haplotypes. The local haplotype reconstruction was successful on a low read coverage sample from COI genes of the butterfly *Melitaea cinxia*.

In general, all mentioned methods pre-process the reads to correct sequencing errors before constructing a read graph. So all undetected/unfixed errors and mis-corrections are further treated as the true values and cannot be later altered by most of the tools, except ViSpA (Astrovskaya *et al.*, 2011). ViSpA deals with sequencing errors at several steps and may fix an incorrect base at the end if the position is covered with sufficient number of the reads. Alternatively, instead of correcting sequencing errors, one can model the stochastic process of the NGS reads generation and estimate which set of genome variants has larger likelihood to produce the observed reads if those reads are indeed generated under the proposed model (Jojic *et al.*, 2008). The model uses several hidden variables and parameters such as viral genome and relative concentration of viral variants, starting offset and depth of the coverage for different positions in the genome, error transformation parameters and uncertainty probability for a particular allele value at a particular position in the genome. The main drawback of the method is that the number of the inferred quasispecies has to be set in advance. PredictHaplo (Prabhakaran *et al.*, 2010) avoids this problem by using a truncated approximation of an infinite mixture model (Ewens, 1972; Ferguson, 1973; Rasmussen, 2000) to automatically choose the number of the reconstructed haplotypes. Under this model, a haplotype is represented as a mixture of the probability tables over four nucleotides and alignment gap. The experimental results on HIV quasispecies show that PredictHaplo is able to infer viral quasispecies if their relative proportion in a population is greater than 0.5%. V-Phaser (Macalalad *et al.*, 2012) also focuses on detecting rare (<1%) viral variants. The method estimates phase thresholds to distinguish between sequencing errors and variants that frequently occur on the same reads. Since sequencing error rate varies across a genome, V-Phaser further uses the thresholds in composite Bernoulli model that iteratively estimates error rate per each base with an EM algorithm. The model helps to identify HIV-1 variants with frequencies down to 0.2%. The probabilistic approach is further extended in QuasiRecomb (Zagordi *et al.*, 2012b) that explicitly targets recombination as an important mechanism of viral genetic diversity. The tool models point mutation and recombination via hidden Markov model and estimates the distribution of a whole viral population rather than individually reconstructs each viral variant. The model was validated on HIV-1 quasispecies and compared to ShoRAH (Zagordi *et al.*, 2010a, 2011) to illustrate the absence of chimeric viral reconstructions when recombination is explicitly taken into account.

Recently, several tools were developed for assembling viral quasispecies from the amplicon reads. QuRe (Prosperi, Prosperi *et al.*, 2011; Prosperi and Salemi, 2012) reconstructs quasispecies by coupling reads that have consistent overlaps and considering reads with similar frequencies across the various amplicons. Starting with the most frequent read in a 'guide' amplicon distribution, QuRe repeatedly selects the neighbour with closest frequency that has a consistent overlap until it reaches the first end. In the same way, it expands the sub-quasispecies to the other end. The frequency of the guide read is then subtracted from the selected reads' frequencies. This process is repeated until the last read from the guide distribution has been evaluated, or until all reads corresponding to an amplicon have zero frequencies. The method was applied to HIV and HBV data, and the results were similar to those of ShoRAH. Instead of minimizing the number of haplotypes, VirA (Mancuso *et al.*, 2012) solves the QSR problem as a problem of finding a minimum entropy distribution of the quasispecies that collectively explain the observed reads. Quasispecies are reconstructed by balancing haplotype frequencies between neighbouring amplicons with either a greedy or maximum bandwidth approaches. Experimental results on simulated error-free reads show that VirA outperforms ViSpA on error-free shotgun pyrosequencing reads. Table 12.1 summarizes available software for inferring viral spectrum from NGS data.

In the remainder of the chapter, we focus on two existing software tools – Viral Spectrum Assembler (ViSpA) for quasispecies reconstruction from the shotgun reads, and Viral Assembler (VirA) for reconstruction from the amplicon reads. Both methods were tested on simulated and experimental data from HCV, HIV (ViSpA) and HBV (VirA) samples and show better assembling quality compared to the other existing software tools.

Table 12.1 Available software for viral quasispecies spectrum reconstruction

Software	NGS reads	Approach	Haplotype reconstruction	Applications	References
Hapler	Shotgun	Read graph	Local/global	COI gene of *Melitaea cinxia*	O'Neil and Emrich, 2012
PredictHaplo	Shotgun	Probabilistic	Global	HIV	Prabhakaran *et al.*, 2010
QuasiRecomb	Shotgun	Probabilistic	Local/global	HIV	Zagordi *et al.*, 2012b
QuRe	Amplicons	Read graph	Global	HIV, HBV	Prosperi and Salemi, 2012
ShoRAH	Shotgun	Read graph	Local	HIV	Zagordi *et al.*, 2011
ViSpA	Shotgun	Read graph	Global	HCV, HIV	Astrovskaya *et al.*, 2011
VirA	Amplicons	Read graph	Global	HCV, HBV	Mancuso *et al.*, 2012
V-Phaser	Shotgun	Probabilistic	Global	HIV	Macalalad *et al.*, 2012

Viral Spectrum Assembler (ViSpA) (Astrovskaya *et al.*, 2011)

The ViSpA's method (Astrovskaya *et al.*, 2011) for inferring the quasispecies spectrum from 454 pyrosequencing shotgun reads consists of the following steps (see Fig. 12.2):

- constructing a viral consensus genome sequence for a given sample and aligning the reads onto this consensus;
- preprocessing the aligned reads to correct sequencing errors;
- constructing a (transitively reduced) read graph with vertices representing reads and edges representing overlaps between them;
- selecting paths in the read graph that correspond to the most probable quasispecies sequences and assembling candidate sequences from those paths by weighted consensus of the reads;
- estimating frequencies of candidate sequences by EM.

Read alignment and consensus construction

A viral quasispecies assembling starts with a read alignment wherein starting position of each read is determined relatively to either a reference or constructed consensus sequences. Usually, a reference genome sequence of a particular viral strain is publicly available [e.g. from NCBI (National Center for Biotechnology Information)]. Since viral genomes do not have sizable repeats and quasispecies sequences are usually close enough to a reference sequence, in general, majority of the reads can be uniquely aligned onto the reference genome. However, a significant number of the reads may remain unaligned due to the differences between the reference genome and sequences in the viral sample. These differences are caused by high mutation rate and viral evolution guided by an individual response of a host's immune system. In order to recover as many of these reads as possible, a consensus genome sequence is iteratively constructed from the already aligned reads.

In particular, 454 pyrosequencing shotgun reads are aligned to a reference sequence using available alignment software (in Astrovskaya *et al.*, 2011), we used SEGEMEHL (Hoffman *et al.*, 2009) with default parameters; we recently re-obtained results using Mosaik aligner (Mosaik Aligner)). The alignment score should satisfy two conditions. First, the lower score reflects the smaller number of differences between an aligned read and the reference sequence. And secondly, the alignment score penalizes mismatches less

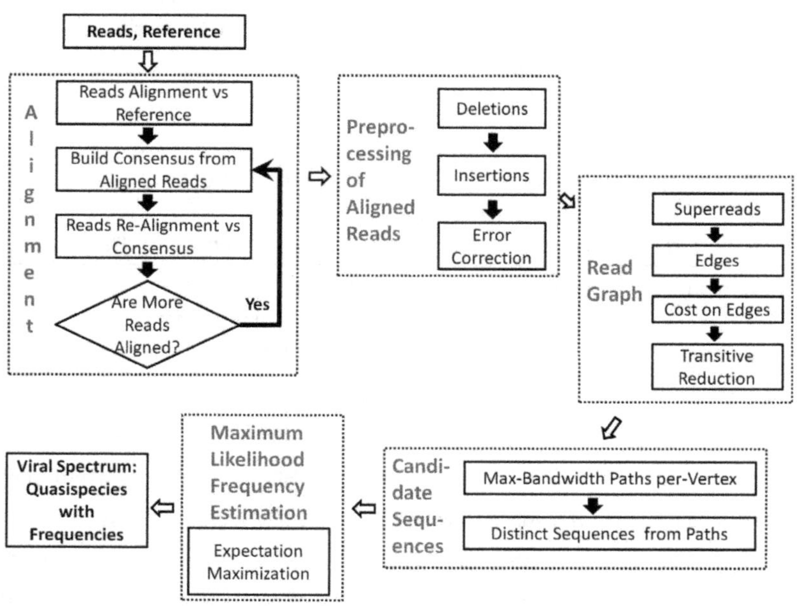

Figure 12.2 ViSpA.

than insertions into a read and deletions from a read with respect to the reference sequence. Then ViSpA extends the reference sequence with a placeholder I for each nucleotide inserted by at least one uniquely aligned read. Similarly, a placeholder D is added to a read sequence for each reference nucleotide missing from the aligned read. Next sequential multiple alignment of the previously aligned reads is performed against this extended reference sequence. Finally, the consensus sequence is obtained by (1) replacing each nucleotide in the extended reference with the nucleotide or placeholder in the majority of the aligned reads and (2) removing all I and D placeholders, which respectively correspond to rare insertions and deletions found in a majority of the reads. Reads may contain a small percentage of unidentified nucleotides denoted by N. In ViSpA, N is treated as a special allele value that matches to any of nucleotides A, C, T and G as well as placeholders I and D.

Iteratively, the reference is replaced with the consensus and the previously unaligned reads are re-aligned. In (Astrovskaya *et al.*, 2011), experiments on a dataset, consisting of approximately 31,000 454 pyrosequencing shotgun reads generated from 5.2 kb-long HCV fragment, showed that 85% of the reads were uniquely aligned onto the reference sequence and an additional 9% of the reads were aligned onto the final consensus sequences. Reads that cannot be aligned onto the final consensus were removed from further consideration.

Aligned reads preprocessing

Since aligned reads contain insertions and deletions, rationale of placeholders I and D is to simplify position referencing among the reads. All placeholders are treated as additional allele values, but they are removed from the final assembled sequences. First, each deletion in the aligned reads is substituted with the placeholder D (see Fig. 12.3). A deletion supported by a single read is replaced either with the allele value, which is present in all other reads overlapping this position, or with N, signifying an unknown value, otherwise. Next, the placeholders I fill each gap in a read corresponding to the insertions in the other reads. All insertions supported by a single read are removed from consideration.

Read graph

A read graph $G=(V,E)$ (Eriksson *et al.*, 2008; Westbrooks *et al.*, 2008) is a directed graph with vertices corresponding to the reads aligned with the consensus sequence. For a read u, we denote by $b(u)$ (respectively $e(u)$) the genomic coordinate at which the first (respectively the last) base of u gets aligned. A directed edge (u, v) connects read u to read v if a suffix of u overlaps with a prefix of v and they coincide across the overlap. Two auxiliary vertices – a source s and a sink t – are added such that s has edges into all reads with zero in-degree and t has edges from all reads with zero out-degree. Then each $s \to t$-path corresponds to a possible candidate quasispecies sequence. The read graph is transitively reduced (Westbrooks *et al.*, 2008), i.e. each edge $e=(u, v)$ is removed if there is a $u \to v$-path not including edge e. Currently, ViSpA does not require transitive reduction of the read graph since redundant edges do not influence candidate path construction (Astrovskaya 2011). Let a superread refer to a read that is not contained in any other read, and let the rest

```
Reference:  ...AGCGT---GAAGCT--T...
    Read1:  ...AG-GTG--GAAGCT
    Read2:  ...AGA-TGGAGAAA-T--T...
    Read3:  ...AGCGA---GTCG-TAAT...
    Read4:  ...AGCGA---GTCACT--T...

Reference:  ...AGCGTIIIGAAGCTIIT...
    Read1:  ...AGDGTGIIGAAGCT
    Read2:  ...AGADTGGAGAAADTIIT...
    Read3:  ...AGCGAIIIGTCGDTAAT...
    Read4:  ...AGCGAIIIGTCACTIIT...

Reference:  ...AGCGTIGAAGCTT...
    Read1:  ...AGNGTGGAAGCT
    Read2:  ...AGAGTGGAAADTT...
    Read3:  ...AGCGAIGTCGDTT...
    Read4:  ...AGCGAIGTCACTT...
```

Figure 12.3 (adapted from (Astrovskaya *et al.*, 2011)) The aligned reads preprocessing step. At the top, a multiple reads alignment is given. In the middle, *I* and *D* placeholders are added. At the bottom, deletions, supported by a single read, are corrected with either specific allele value or unknown value *N*; insertions, confirmed by a single read, are removed.

of the reads be called subreads. Subreads are not used in the construction of the read graph. In contrast to (Eriksson *et al.*, 2008; Westbrooks *et al.*, 2008), ViSpA takes into account all reads for the final assembly of candidate sequences and for estimation of their frequencies.

Since the number of different $s \rightarrow t$-paths is exponential, one wants to generate a set of paths that have high probability to correspond to real quasispecies sequences. In order to estimate a path probability, the probability $p(e)$ that edge e connects two reads from the same quasispecies is estimated for all edges in the path, and then those probabilities are multiplied. Under the assumption of independence between the edges, if we assign to each edge e a cost equal to

$$-\log(p(e)) = log\left(\frac{1}{p(e)}\right),$$

then the minimum-cost $s \rightarrow t$-path will have the maximum probability to represent a quasispecies sequence. In (Westbrooks *et al.*, 2008), the probability that two reads u and v, connected by the edge (u, v), belong to the same quasispecies is estimated as

$$p_\Delta \approx exp\left(\frac{\Delta N}{Lq}\right) = \Theta\left(e^{-\Delta}\right) \qquad (12.1)$$

where $\Delta = b(v) - b(u)$ is the *overhang* (shift) between reads u and v, N is the number of the reads, q is the number of the quasispecies in a viral population, and L is the number of the starting positions. Thus, in this case the cost of an edge with the overhang Δ can be approximated by

$$\Delta \propto log\left(1/p_\Delta\right).$$

To account for sequencing errors (Astrovskaya *et al.*, 2011), the read graph construction is altered as follows. We use three parameters: (1) n is the number of mismatches allowed between a read and a superread (see Fig. 12.4a), (2) m is the number of mismatches allowed in the overlap between two adjacent reads (see Fig. 12.4b), and (3) t is the number of mismatches expected between a read and a random quasispecies.

If imperfect overlaps are allowed in the graph then the probability of an edge (Eqn. 12.1) should be adjusted. First, we consider a simple model of the uniformly distributed quasispecies. After a quasispecies sequence Q is randomly chosen, a beginning position is generated from the uniform distribution, and a read r is produced from the quasispecies Q with j sequencing errors. Let A be an event that two reads u and v from the same quasispecies Q are connected with an edge (u, v) in the transitively reduced graph, and j mismatches between u and v occur in the overlap due to the sequencing errors. In the other words, the event A consists of the following independent events: (1) read u exists, starting at the position $b(u)$, (2) read v exists, starting at the position $b(v) = b(u) + \Delta$, (3) no read w from the same quasispecies Q satisfies $b(u) < b(w) < b(v)$, and (4) there are j sequencing errors in the overlap $o = e(u) - b(v)$ of the reads u, v. Given N reads, originated from q quasispecies of a length L, the probability that read u starts at the position $b(u)$ is

$$\frac{N}{Lq}.$$

The probability of the event $\Delta > k$ is the probability that there is no read from the quasispecies Q, starting at any position in $\{b(u)+1,\ b(u)+2, \ldots, b(u)+k\}$, that is

$$p_k = \left(1 - \frac{N}{Lq}\right)^k \approx exp\left(\frac{-kN}{Lq}\right).$$

Then

$$Pr(\Delta = k) = \left(\frac{N}{Lq}\right)^2 p_{k-1}.$$

Next we calculate probability of having j sequencing errors in the overlap o, which is

$$\Pr(j) = \binom{o}{j} \varepsilon^j (1-\varepsilon)^{o-j},$$

where ε is a sequencing error rate. Finally, the probability of the event A is a product of probabilities Pr $(\Delta = \mathrm{k})$ and Pr (j), or

$$\Pr(A) = \left(\frac{N}{Lq}\right)^2 exp\left(\frac{-\Delta N}{Lq}\right)\binom{o}{j}\varepsilon^j(1-\varepsilon)^{o-j}$$

$$\approx exp(-\Delta)\binom{o}{j}\varepsilon^j(1-\varepsilon)^{o-j}.$$

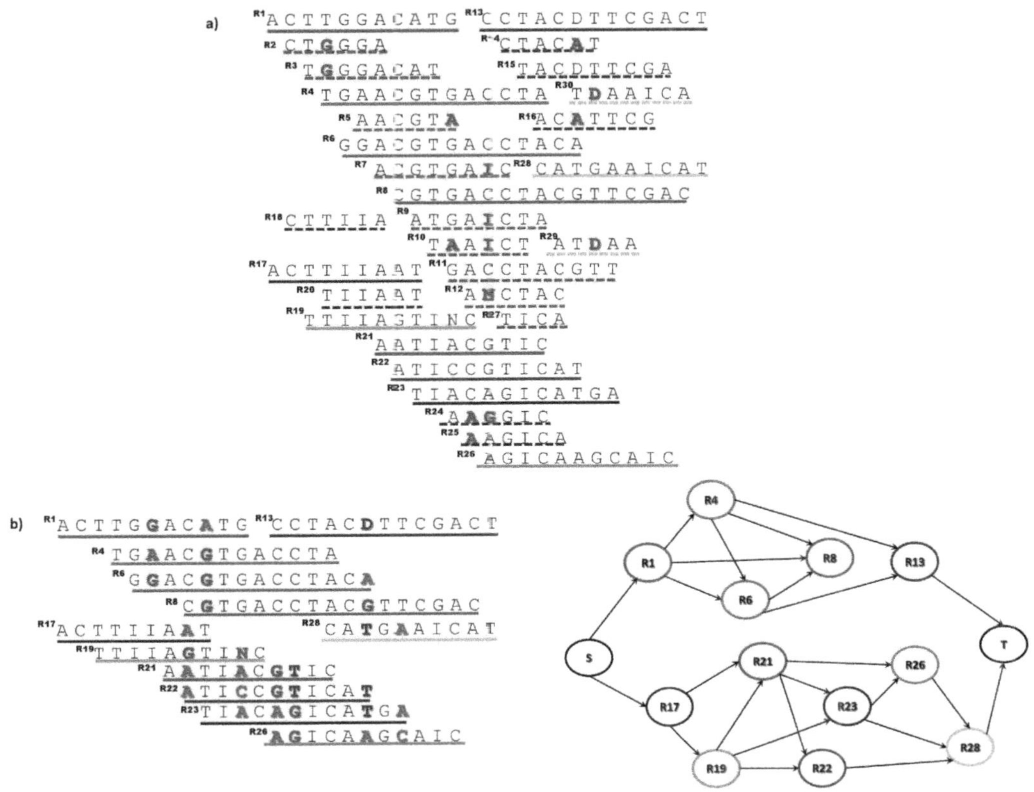

Figure 12.4 (adapted from (Astrovskaya *et al.*, 2011)) (a) Finding superreads from the aligned reads (*n*=2). The reads are divided into the superreads (underlined by a bold line) and the subreads (underlined by a dashed line of the corresponding colour). A mismatch between subread and its superread is in the red colour. For example, *R2* and *R3* are the subreads of *R1*, both subreads have one difference with the superread *R1*. (b) A read graph example. At the left, the aligned superreads (*n*=2) are given. The positions with several allelic values in overlap are highlighted (red (*G*), orange (*A*), green (*C*), blue (*T*), grey (*N*) and purple (*D*)). For simplicity of the presentation, differences are marked separately across reads *R1-R13* and *R17-R28*. At the right, a read graph (*n*=2, *m*=2) is drawn (the overlaps that are less than 3 bases are not considered). A colour version of this figure is located in the plate section at the back of the book.

As in the case of error-free reads, defining the edge costs as

$$\Delta\log\left(\binom{o}{j}^{-1}\varepsilon^{-j}(1-\varepsilon)^{j-o}\right)\propto log\left(1/\Pr(A)\right) \qquad (12.2)$$

ensures that $s \to t$-paths with low costs correspond to the most likely quasispecies sequences.

In practice, reads' starting positions indeed tend to follow uniform distribution with small amount of outliers. However, if it is not a case for a particular shotgun data, cost formula (Eqn. 12.2) can be adjusted by taking into account the number of reads, starting at any position in $\{b(u), b(u)+1, b(u)+2, \ldots, b(v)\}$. Since ViSpA constructs assembly by choosing at each step among edges, spanning almost the same positions, such adjustment is nonessential for ViSpA's haplotype reconstruction. The assumption of uniformly distributed quasispecies is most likely violated in real viral populations. Since there is no reliable way to approximate their unknown distribution, ViSpA treats all identical copies of the same quasispecies sequence as different entities during assembling and estimate their frequencies via maximum likelihood.

Candidate path selection

To generate a set of high-probability (low-cost) paths that are rich enough to explain observed reads, we compute for each vertex in the read graph the minimum cost $s \to t$-path passing through it. Finding these paths is computationally fast since two shortest-paths trees in G should be computed: one outgoing from s and one incoming into t; the shortest $s \to t$-path passing through a vertex v is the concatenation of the shortest $s \to v$ and $v \to t$-paths.

Preliminary simulation experiments show that better candidate sets are generated when the edge costs c defined by (Eqn. 12.1) and (Eqn. 12.2) are replaced by e^c. In fact, even faster dependency on c allows obtaining better candidate sets. The fastest growing cost effectively changes the shortest path into so called *max-bandwidth paths*, i.e. paths that minimize maximum edge cost for the entire path and for each subpath. So, ViSpA generates the max-bandwidth path per each vertex.

Candidate sequence assembly

When no mismatches are allowed in the construction of the read graph, finding a candidate sequence corresponding to an $s \to t$-path is trivial: by definition, adjacent superreads coincide across their overlap. When mismatches are allowed, ViSpA first assembles a consensus sequence from the superreads used by the $s \to t$-path. It may be not the best choice, especially, when the coverage with superreads is low. Hence, each initial candidate sequence is replaced with a weighted consensus sequence obtained on both superreads and subreads of the path, as described below.

For each read r, ViSpA computes the probability that it belongs to a particular initial candidate sequence cs as:

$$\mathrm{p}(cs,r)=\binom{l}{k}\left(1-\frac{t}{L}\right)^{l-k}\left(\frac{t}{L}\right)^{k}, \qquad (12.3)$$

where l and L denote the lengths of the read r and the initial candidate sequence cs, respectively, k is the number of the disagreements between the read r and the initial candidate sequence cs, and t/L is the estimated mutation rate. Then a final candidate sequence is computed as the weighted consensus over all reads, where the weight of a read is the probability (Eqn. 12.3) that it belongs to the sequence. Note that, unlike the case without mismatches, the same candidate sequence can be obtained from different candidate paths, so duplicates are removed.

Estimation of candidate quasispecies sequence frequencies

ViSpA assumes that reads R with observed frequencies $\{o_r\}_{r=1}^{|R|}$ were generated from a quasispecies population Q as follows. First, a quasispecies sequence $q \in Q$ is randomly chosen accordingly to its unknown frequency f_q. A read starting position is generated from the uniform distribution, and a read r is produced from the quasispecies q with j sequencing errors. The probability of this event has been estimated as

$$h_{q,r}=\binom{l}{j}(1-\varepsilon)^{l-j}\varepsilon^{j},$$

where l is the read length and ε is the sequencing error rate. Note that in the next release of ViSpA this probability will be calculated following (Nicolae *et al.*, 2011) as

$$h_{q,r}=\binom{l}{j}(1-\varepsilon)^{l-j}\left(\frac{\varepsilon}{3}\right)^{j}.$$

Thus, the probability of observing the read r under this model is

$$\Pr(r)=\sum_{q\in Q} f_q h_{q,r}$$

Quasispecies frequencies $\{f_q\}_{q=1}^{|Q|}$ are estimated by maximizing the log-likelihood function:

$$\ell\left(f_1,\ldots f_{|Q|}\right)=\sum_{r\in R} o_r \log\Pr(r)$$

by the EM algorithm (Laird *et al.*, 1977). Currently, convergence of the EM algorithm is determined at the tolerance level of 0.001.

VSEM enhancement of ViSpA (Mangul *et al.*, 2011)

Virtual string Expectation Maximization

The input to the EM algorithm (Laird *et al.*, 1977) can be viewed as a bipartite graph with two sets of vertices: one set corresponds to sequences (or strings), and the other one represents reads. If a read can be generated from a particular string, an edge connects corresponding vertices and probability of this event is associated with the edge. In practice, we miss certain information in a sample, making corresponding bipartite graph (further called panel) incomplete. VSEM method (Mangul *et al.*, 2011) estimates the incompleteness of such panel by adding a virtual string to the input dataset and by identifying reads that are more likely to be emitted by missing sequences. Accounting for missing sequences allows more accurately estimate frequencies for the known (reconstructed) strings.

Combining ViSpA with VSEM

Since VSEM can identify reads in a sample that are likely to be generated from unknown (not reconstructed) quasispecies sequences, ViSpA can extend its candidate set by proposing another explanation for those reads. Additionally, replacing ViSpA's EM with VSEM improves quasispecies frequencies estimates by taking into account incompleteness of the panel.

In brief, ViSpA's version, combined with VSEM (ViSpA-VSEM), works as follows (see Fig. 12.5). First, ViSpA gets a set of the aligned reads. Each read has a binary weight that corresponds to VSEM's probability of this read to be generated from a missing (so far unreconstructed) quasispecies sequence. Initially, all reads have weights equalled to 1, and ViSpA works as previously described, except frequencies for candidate quasispecies sequences are calculated by VSEM. At any other iteration, ViSpA separately finds superreads for covered (so called 0-weight) and uncovered (so called 1-weight) reads and removes those 0-weight superreads that share a subread with 1-weight superread. To ensure usage of the previously uncovered reads, modified version of ViSpA applies additional cost to the edges that have at least one endpoint among covered reads as follows:

$$cost'(u,v) = cost(u,v) + 0.5 \cdot L \cdot (2 - w(u) - w(v)),$$

where *cost* (u,v) is the original ViSpA's cost of the edge (u,v) (defined in Eqn. 12.2), L is the average

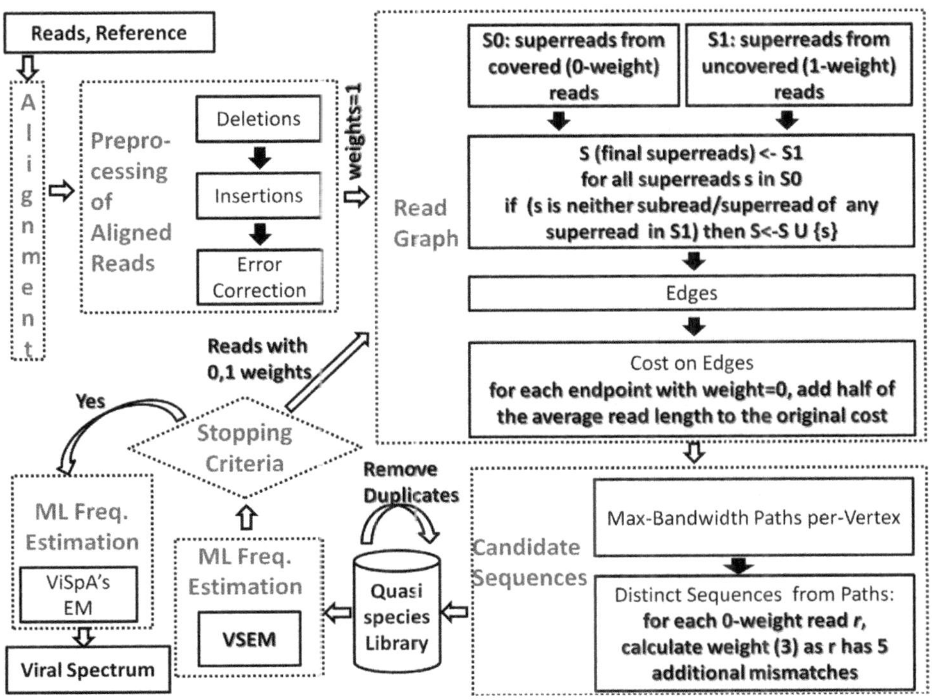

Figure 12.5 ViSpA-VSEM.

read length, and $w(u)$, $w(v)$ are the binary VSEM weights assigned to the reads u and v, respectively. The adjusted cost on the edges leads to different candidate sets across iterations. Next, VSEM reassigns new weights to the reads based on the cumulative set of the final assemblies, that is, on newly assembled sequences and all candidate sequences stored from the previous iterations. This cumulative set is called a quasispecies library. Once frequencies distribution is obtained, VSEM assigns 1 to the reads that are not adequately covered by already assembled sequences, and 0 to the reads that are explained by those assemblies. Very rare candidate sequences are removed from the quasispecies library. If the virtual string still has a high frequency, and the quasispecies library was expanded with respect to the previous iteration, then the reads and their weights are fed back to ViSpA, and the process is repeated. Otherwise, ViSpA's original EM is applied to all candidate sequences in the quasispecies library to obtain final maximum-likelihood estimates of their frequencies. Then the modified version of ViSpA with VSEM reports the reconstructed viral quasispecies spectrum.

Viral Assembler (VirA) (Mancuso *et al.*, 2012)*

In the amplicon-based assembly, the viral genome is covered with the set of K overlapping segments (*amplicons*) $A_1, \ldots, A_K$, started at the predefined positions within the genome. Each segment is bordered by primers and sequenced separately as a whole, producing the set of reads with equal lengths.

In the case of amplicon pyrosequencing reads, the QSR problem is usually formulated as the problem of finding the *Most Parsimonious Spectrum*, which asks to reconstruct the minimum number of distinct quasispecies that explains the observed reads. This formulation is NP-hard since there is a reduction from the SUBSET SUM problem: given a (multi) set S of integers and count c, is there a subset $S' \subseteq S$ such that its sum is equalled to c? (Garey and Johnson, 1979).

Claim 1. The QSR problem for amplicons is NP-hard.

Proof. Given an instance of the SUBSET SUM problem, a set $S = \{x_1, \cdots, x_k\}$ such that $\sum_{i=1}^{k} x_i = n$ and count c, let $A_1 = \{r_1, \cdots, r_k\}$ be reads with counts $\{x_1, \cdots, x_k\}$ and $A_2 = \{r_{k+1}, r_{k+2}\}$ be reads with counts $\{c, n-c\}$ with consistent overlap between any two reads $r_i \in A_1$, $r_j \in A_2$. A minimal solution to the QSR problem will cover exactly k reads in A_1. Taking $q \le k$ reads and respective counts from A_1 that correspond to quasispecies containing read r_{k+1}, we find the solution to the SUBSET SUM problem. If any read r_i, $0 \le i \le k$ is split between r_{k+1}, r_{k+2}, then at least $k+1$ quasispecies must exist (if all reads are covered), and it will not be a minimal solution.

In practice, each amplicon is expected to have a small number of different reads (small number of quasispecies), so the problem can be solved in practical time. If the sets of adjacent amplicons are partitioned into the subsets with the same parts in the intersection of these amplicons then the number of distinct reads in the overlap is at most the minimum of the number of the reads in the left and the right amplicons.

Further, we describe several other optimization formulations for the QSR problem, depending on the amplicons generation process. In (Mancuso *et al.*, 2012), we explored two sequencing error modes and focused on two specific frequencies models. Sequencing errors are taken into account as follows:

- In the error-free model, all sequencing errors are assumed to be previously corrected.
- In the error-prone model, the reads still contain sequencing errors that are expected to be corrected later.

Additionally, the following frequencies models were explored:

- In the ideal-frequency model, the observed reads distribution for each amplicon is assumed to represent the true distribution of the quasispecies.

* Reprinted from In Silico Biology 11, Mancuso, N., Tork, B., Skums, P., Ganova-Raeva, L., Măndoiu, I., and Zelikovsky, A. Reconstructing viral quasispecies from NGS amplicon reads, 237–249.

- In practice, distribution of the quasispecies on each amplicon differs from the true distribution in a sample. This model is called the skewed-frequency model.

Quasispecies assembly in the error-free, ideal-frequency model problem

The input data to the QSR problem can be represented as a K-staged read graph $G=(V=V_1\bigcup\ldots\bigcup V_k, E)$, where

- each vertex $v\in V_i$ corresponds to a distinct read in the i-th amplicon A_i and has a count $c(v)$ (i.e. the read is repeated $c(v)$ times in a sample) : $\sum_{v\in V_i} c(v)=D$;
- each edge (u,v) connects two reads from the consecutive amplicons A_i and A_{i+1} that agree in the overlap region.

Then solution can be represented by the set $Q=\{q_i\}$ of $u\rightarrow v$-paths, $u\in V_1, v\in V_k$. A path q_i has frequency f_i such that for each vertex $v\in V$,

$$\sum_{v\in q_i} f_i = c(v). \qquad (12.4)$$

In (Mancuso *et al.*, 2012), we first focused on the two-staged case and then used its solution to combine together all K stages. So let assume that there are only two stages, V_1 and V_2, meaning the read graph is bipartite. Now we check if there is a feasible solution for the problem.

Let f_e be the frequency of the quasispecies e that corresponds to the edge $e=(u,v)$. For each vertex v, we re-write the constraint (Eqn. 12.4) as the system of linear equations:

$$\forall v\in V_1\bigcup V_2, \sum_{e \text{ incident to } v} f_e = c(v). \qquad (12.5)$$

The system (Eqn. 12.5) is consistent if and only if the corresponding two-stage assembly problem is feasible. We denote by $G(c)$ the bipartite graph obtained from the read graph G by splitting each vertex v into $c(v)$ copies. It is clear that the following claim is true.

Claim 2. The system (Eqn. 12.5) is consistent if and only if G(c) has a perfect matching.

The feasibility of the two-stage assembly problem could be verified using any classical perfect matching algorithm. We assume that the system (Eqn. 12.5) is consistent.

Theorem 1. Any solution of the system (Eqn. 12.5) with the minimal number of edges is a spanning forest of the read graph G.

Proof. Let assume that the read graph G is connected. The system (Eqn. 12.5) can be re-written as $If=c$, or $I_1f_1+\cdots+I_mf_m=c$, where I is the incidence matrix of the graph G, I_j is the jth column of the matrix I.

Since the read graph G is bipartite, $rank(I)=n-1$ (Graham *et al.*, 1996). Let the first $n-1$ columns and rows of the matrix I be the basis columns and rows. The columns of the incidence matrix that correspond to the even cycle are linearly dependent. Since the graph G has no odd cycles, the edges $e_1,\cdots,e_{n-1}$ contain no cycles, and, therefore, form the spanning tree of the graph G. Set free variables $f_n,\cdots,f_m$ equal to 0.

Then the system (Eqn. 12.5) is equivalent to

$$I'f'=c', \qquad (12.6)$$

where I' is the submatrix of the incidence matrix I restricted to the basis columns and rows, $f'=(f_1,\cdots,f_{n-1})$, $c'=(c_1,\cdots,c_{n-1})$. Since the system (Eqn. 12.5) is consistent, the system (Eqn. 12.6) is also consistent. Moreover, since the graph G is bipartite, the incidence matrix I is totally unimodular (Graham *et al.*, 1996). Then the submatrix I' is also totally unimodular with $rank(I')=n-1$. Therefore, the system (Eqn. 12.6) has a positive integer solution. The non-zero edges of this solution form the spanning forest of the read graph G. This proves that the system (Eqn. 12.5) has a spanning forest solution. Let forest f, containing cycles, be a solution of the system (Eqn. 12.5). Consider a connected component of the forest f, which has cycles, and find the acyclic solution of the system (Eqn. 12.5) restricted to that component. This solution has a lesser number of edges. If we repeat this process for all connected components with cycles, we obtain the acyclic solution of the system (Eqn. 12.5).

Theorem 2. If there is a bipartite cycle between consecutive amplicons, then there is a 4-gamete rule violation, i.e. there should exist an additional (to existing in amplicons) recombination.

Proof. Let B be a bi-cycle (a bipartite cycle) with $2k$ vertices (with k vertices in each the left amplicon L and the right amplicon R). For the sake of simplicity, assume that the alphabet has two characters. Then k different haplotypes in the left amplicon L as well as in the right amplicon R can be distinguished in k-1 positions. There are exactly $2k$ different edges in the bi-cycle B. If there is no 4-gamete rule violation, we obtain $2k$ haplotypes, distinguished in $2k$–2 positions. This implies a contradiction.

A simple maximum likelihood approach assigns non-zero frequencies to all possible quasispecies since any edge (i.e. quasispecies) is assumed to be equally probable. A more meaningful parsimonious approach requires that most likely solutions should contain the minimum number of quasispecies, satisfying the linear system (Eqn. 12.5). Let us consider two distinct reads in each amplicon. In the other words, $|V_1|=|V_2|=2$, A and B are distinct reads in the first amplicon, and C and D are unique reads in the second amplicon (see Fig. 12.6). Also, let all four possible combinations of the reads be consistent in their overlap. Without loss of generality, assume that $d \le b \le a \le c$. If $a=c$, then $b=d$, and the minimum possible number (two) of non-zero edge frequencies is achieved. If $a \ne c$, then the four constraints have rank three. So it should be three edges with non-zero frequencies, meaning either of the following cases is possible:

$$AC=a,\ AD=0,\ BC=c-a,\ BD=d;$$

$$AC=a-d,\ AD=d,\ BC=b,\ BD=0.$$

If $a>b$, then the first case is more probable; in the case of the equality $(a=b)$, both cases are equally probable. Given the assumptions, a simple maximum likelihood approach seems to be inefficient.

Let us consider a more complex maximum likelihood model. We define a 'fork' $F=(L,R)$ to be a bi-clique with partitions L and R, where $|L|=l, |R|=r$. Since there are overlaps between the reads in both partitions, no single quasispecies can generate both its read in the partition L and its read in the partition R. With this assumption, $c(L)+c(R)$ is the total number of the quasispecies that can produce all observed reads.

For simplicity, let each quasispecies with overlap in the fork F generates exactly one read. So if a quasispecies does not produce a read, it is considered non-existent. Any possible resolution X_F of the fork F consists of $c(L)+c(R)$ instances of l_r distinct quasispecies $q_{l_1},\cdots,q_{l_r}$ with counts $x_{l_1},\cdots,x_{l_r}$. A resolution X_F is considered feasible if $X_F=\{x_{ij} : x_{ij}=x_{ij}^L+x_{ij}^R\}$ such that

$$\sum\nolimits_j^r x_{ij}^L = c_i,\ \forall i \in L$$

$$\sum\nolimits_i^l x_{ij}^R = d_i,\ \forall j \in R \qquad (12.7)$$

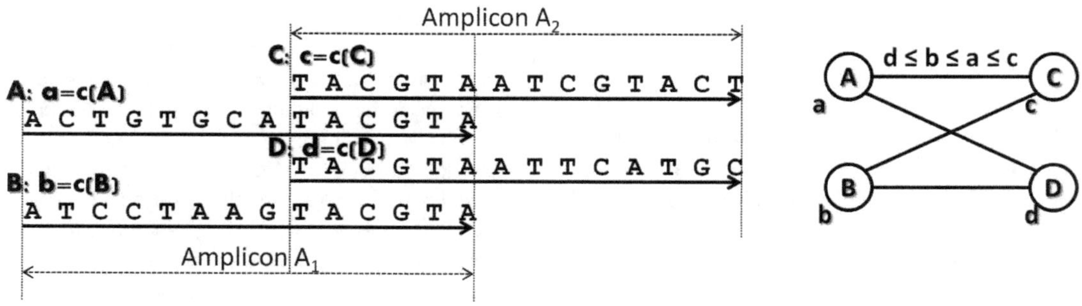

Figure 12.6 The case of two distinct reads in each out of two amplicons: *A* and *B* are distinct reads in the first amplicon, *C* and *D* are unique reads in the second amplicon. *a*, *b*, *c* and *d* are counts of the reads *A*, *B*, *C* and *D*, respectively. At the left, four reads are shown. At the right, the corresponding read graph is drawn.

Following the model (Eqn. 12.7), we estimate the probability $P(X_F)$ that X_F generates the observed reads in the partitions L and R as follows. In total, there are $T_F = 2^{c(L)+c(R)}$ ways to produce reads from $c(L) + c(R)$ quasispecies in the two-amplicons assembly. Then

$$P(X_F) = {C(X_F)}/{T_F},$$

where $C(X_F)$ is the number of the cases when such read counts can be observed.

Based on Theorem 1 (Mancuso *et al.*, 2012), we can assume that the resolution X_F is a forest. For a given leaf i and count c_i, there exists a single quasispecies q_{ij}, such that count $x_{ij} \geq c_i$, that can emit all copies of the ith read. There are

$$\binom{x_{ij}}{c_i}$$

possible quasispecies assignments to generate the read. Let F' be the fork obtained from the fork F after removing quasispecies q_{ij} and decreasing the count of the jth *stem*'s read to $x_{ij} - c_i$. Then

$$C(X_F) = C(X_{F'}) \cdot \binom{x_{ij}}{x_{ij}^L} = C(X_{F'}) \cdot \frac{x_{ij}!}{x_{ij}^L! \cdot x_{ij}^R!} \tag{12.8}$$

Eqn. 12.8 defines the recursive computation of the likelihood, given a forest resolution X_F for the fork $F = (L,R)$. If $k = k(X_F)$ is the number of connected components of the resolution X_F then the computation can be done in a leaf-removal order, that is $v_1, \cdots, v_{l+r-k}$, as follows. Let some leaf v_i has the remaining count c_i, and let x_{ij} and c_j be the counts for a covering quasispecies q_{ij} and the jth stem's read, respectively. Then

$$Likelihood(X_F \mid F = (L,R))$$

$$= P(X_F) = \frac{\prod_{i=1}^{l+r-k} \binom{x_i}{c_i}}{2^{c(L)+c(R)}}.$$

Therefore, maximizing the log-likelihood is equivalent to maximizing

$$logC(X_F) = \sum_{i=1}^{l+r-k} \binom{x_i}{c_i},$$

which, in turn, is equivalent to minimizing Shannon entropy (Halperin and Karp, 2005):

$$-\sum_{i=1}^{n} p(x_i) \log p(x_i).$$

The more quasispecies are in a resolution, the higher entropy is. Thus, a solution with the minimum amount of the quasispecies has the lower possible entropy and is the most parsimonious solution under the definition in (Mancuso *et al.*, 2012) (see Fig. 12.7).

Minimum entropy quasispecies assembly problem. *Given a read graph with frequencies on the reads, find the set of quasispecies (paths) and respective frequencies that explains all reads and frequencies, such that the resulting set has the minimum entropy.*

The read graph for the two-amplicon assembly consists of a set of the disjoint bi-cliques, each of which corresponds to a distinct overlap-part. The number of quasispecies cannot exceed the minimum number of distinct reads in the first and in the second amplicons.

Reduction of the skewed-frequency model to the ideal-frequency model

A fork $F = (L, R)$, where L and R are the sets of reads from the left and the right amplicons, respectively, is balanced if the total frequencies of L and R are equal to each other. In the previously explored model, an amplicon-read spectrum has all its forks balanced; and vice-versa, any amplicon-read spectrum with balanced forks has a feasible quasispecies spectrum with ideal frequencies.

Fork balancing problem. *Given a spectrum of the amplicon reads $\{r_i\}$ with the observed frequencies $\{o_i\}$ and forks $f_i = (L_f, R_f)$, find ideal frequencies $F = \{f_i\}$ such that all forks are balanced.*

Mancuso *et al.* (2012) solve the fork balancing problem via a quadratic program (QP) with linear constraints (Eqn. 12.9). The QP looks for the minimum squared adjustment of the observed read frequencies $X = \{x_i\}$. In general, weights can be added to the QP objective to normalize adjustment values: a more frequent read is expected to be adjusted to a greater extent than a rare one.

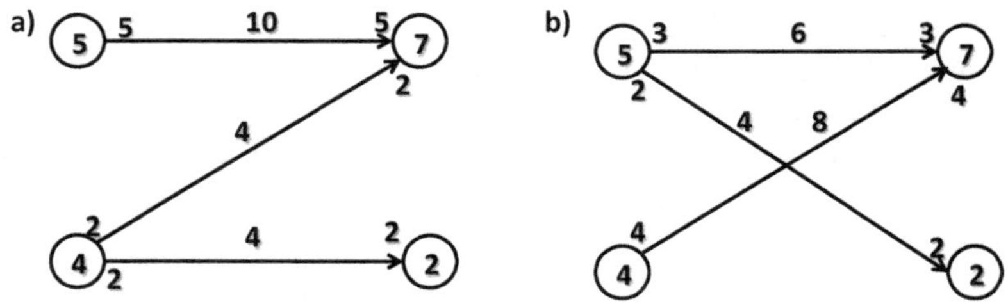

Figure 12.7 Two possible resolutions for the read graph. Likelihood of (a): $\binom{4}{2}\cdot\binom{4}{2}\cdot\binom{10}{5}\cdot 2^{-18} = 9072/2^{18} \approx 3.46\%$. Likelihood of (b): $\binom{8}{4}\cdot\binom{6}{3}\cdot\binom{4}{2}\cdot 2^{-18} = 8400/2^{18} \approx 3.2\%$. Although both resolutions produce 18 instances of three distinct quasispecies, the first resolution (a) has higher likelihood than the second resolution (b) as well as lower entropy $-\left[\frac{4}{18}\log\frac{4}{18}+\frac{4}{18}\log\frac{4}{18}+\frac{10}{18}\log\frac{10}{18}\right]\approx 0.43$ compared to (b) $-\left[\frac{8}{18}\log\frac{8}{18}+\frac{6}{18}\log\frac{6}{18}+\frac{4}{18}\log\frac{4}{18}\right]\approx 0.46$. Hence, resolution (a) is the more likely explanation of the read graph.

Typical weight values are inversely proportional to the frequency c_i of the read i, that is, $w_i = 1/c_i$.

Minimize:

$$\sum w_i x_i^2$$

Subject to:

$$\sum_{i\in L_f}(x_i+o_i)=\sum_{i\in R_f}(x_i+o_i), \forall f \in Forks$$
$$x_i + o_i \geq 0 \tag{12.9}$$

Viral quasispecies reconstruction algorithms

Graph data structure

A K-staged read graph from K amplicons is built as follows. For each unique read in the amplicon A_i, find consistent unique reads in the amplicon A_{i+1}. If there is a consistent read, add vertices per each read to the graph and join them with a directed edge. Repeat this process for each amplicon $A_1, \ldots, A_k$. Finally, add a source S and connect it to all reads in the first amplicon as well as add a sink T and connect all reads in the last amplicon to it.

Then the frequencies in the graph G may be adjusted by either solving the QP (Eqn. 12.9) or by simple scaling. In the latter case, for each complete bipartite subgraph in the directed graph G, separately sum the frequencies in the first and in the second partitions. If they do not equal one another, scale the values in the right partition to the frequencies in the left partition. The process is done for the entire graph in a breadth-first manner from the source to the sink.

The graph G consists of bipartite cliques (bi-cliques), and a clique corresponds to a consistent overlap in the reads. Next we transform the graph by adding a fork vertex for each bi-clique as follows. For each bi-clique $K_{n,m}$, connect n vertices in the first partition to the fork vertex and the fork vertex to the m vertices in the second partition. This modification results in the reduced number of edges in the final subgraph: indeed, $n \times m$ edges in each original bi-clique is replaced with $n+m$ edges in the 'forked' subgraph (Fig. 12.8).

In (Mancuso *et al.*, 2012), two general strategies were used to solve the QSR problem. In the first approach, an algorithm searches for $S \rightarrow T$ paths in the graph and partially resolves forks at each stage. In the second approach, the algorithm completely resolves each fork.

Maximum bandwidth path

The QSR problem can be solved by repeatedly constructing maximum bandwidth paths through the read graph. At each iteration, an $S \rightarrow T$ max-bandwidth path P is found. Its bandwidth m is subtracted from all edges in the path P, exhausting at least one edge from the read graph. The process is repeated until the source S and the sink T are

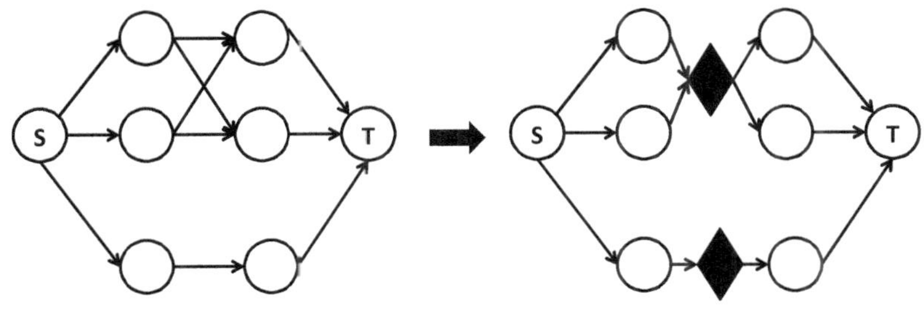

Figure 12.8 Adding forks to the original $S \to T$ connected read graph.

disconnected (edge does not exist if its frequency/count is equalled to zero or less than some positive number for the real valued edges). Finally, the frequencies of the quasispecies are normalized by dividing a quasispecies count (the maximum bandwidth) over the sum of all counts.

Maximum frequency path

A *maximum frequency path* in a graph G is a path from the source S to the sink T, such that the total frequency of the path is maximized. If the frequencies are integers, a maximum frequency path is equivalent to the longest path of the graph G. In general, the LONGEST PATH is NP-hard problem, but it can be solved on acyclic graphs in polynomial time using dynamic programming (Garey and Johnson, 1979). Furthermore, if the graph G is a directed acyclic graph (as is our case), the LONGEST PATH is solved in linear time via topological ordering of vertices and dynamic programming for computing path lengths. Then the QSR problem is resolved in a similar manner as in the previous case. In the other words, find the maximum frequency path P with the total frequency f until there exists no non-zero frequency path from the source S to the sink T. Then set the frequencies to zero for each edge in the path P. Finally, normalize the paths' frequencies. The concept is inspired by the greedy approximation algorithm for the MIN ENTROPY SET COVER problem (Halperin and Karp, 2005). The algorithm achieves the OPT + 3 approximation by the greedy selection of a set that covers the most elements from the given universe. The set is further removed, and the algorithm repeats on the remaining sets until every element is covered. Similarly, our algorithm covers the largest possible amount of the reads in the graph, removes them, and then repeats the process.

Greedy algorithm for resolving forks

Each fork vertex v (a vertex with at least two in-edges and two out-edges) should be replaced with an acyclic graph connecting inputs with outputs (a resolution). In (Mancuso *et al.*, 2012), the greedy approach was proposed to resolve the forks. First, the left and the right vertices are placed into the priority queues L and R, and the vertex with the maximum frequency is extracted from each of these queues to create a new edge (max_l, max_r). The frequency of this new edge is set to $min(max_l, max_r)$. The keys (frequencies) of both vertices are decreased; the larger maximum frequency is re-inserted; otherwise, both vertices are exhausted and not considered for further resolution. When either of the queues L or R is empty, the fork vertex is resolved. The process is repeated until no forks are left in the graph. It may be the case that the read vertices (not just overlap vertices) become forks and are eligible for the resolution.

General framework

The general framework (see Fig. 12.9) (Mancuso *et al.*, 2012) for solving the QSR problem is as follows. First, we construct the read graph. Then vertices frequencies are offset by solving the fork balancing problem, if necessary, by utilizing the quadratic program (Eqn. 12.9). Once the graph has been balanced, VirA solves the QSR problem by running any of the previously mentioned algorithms.

Results and discussions

Data

The following datasets were used for experimental validation of the previously discussed tools.

The error-free reads simulated from the real HCV quasispecies (von Hahn *et al.*, 2007)

Error-free reads were generated from 1739-bp-long fragment from the E1E2 region of 44 HCV sequences (von Hahn *et al.*, 2007). The quasispecies frequencies were generated according to some specific distribution. Each population was created by randomly selecting either 10, 20, 30 or 40 sequences among these HCV variants and assigning frequencies following either (1) the uniform (all sequences have the same frequency), or (2) the skewed uniform (a single sequence has high frequency; all other sequences have uniformly low frequency), or (3) the geometric (the ith sequence is a constant percentage more frequent that the $(i+1)$th sequence) distributions.

For ViSpA validation, the length of a read follows the normal distribution with a particular mean value and the variance of 400, and a starting position follows the uniform distribution. This simplified model of reads generation has two parameters: the number of the reads that varies from 20,000 up to 100,000 and the average read length that varies from 100 bp up to 600 bp.

In the case of VirA, we used the mean value of 320 bp and the variance of 10 bp. The total amount of reads varied from 5000 to 100,000.

454 reads simulated from the real HCV quasispecies (von Hahn *et al.*, 2007)

Additionally, reads with sequencing errors were produced by FlowSim (Balser *et al.*, 2010) from 10 quasispecies sequences (following the geometric distribution of the frequencies). The quasispecies were randomly chosen out of 44 HCV sequences (von Hahn *et al.*, 2007). FlowSim generated 30,000 reads with the average length of 350 bp, and almost 40,000 with the average length of 322 bp. 99.96% of the aligned reads has at least one insertion/deletion with respect to the reference: 99.97% of the deletions and 99.6% of the insertions are 1 bp long. Only 1.1% of the aligned reads have unknown base(s) (Astrovskaya, 2011).

454 simulated amplicons from the real HCV quasispecies (von Hahn *et al.*, 2007)

Ten different datasets were used as initial 'seeds' to generate viral populations following the uniform, geometric, and skewed distributions.

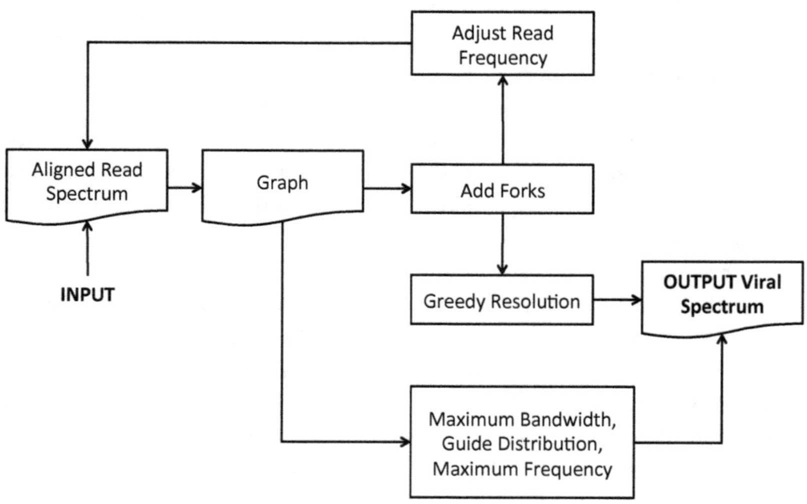

Figure 12.9 VirA. Reprinted from In Silico Biology 11, Mancuso, N., Tork, B., Skums, P., Ganova-Raeva, L., Măndoiu, I., and Zelikovsky, A. Reconstructing viral quasispecies from NGS amplicon reads, 237–249, © Mancuso *et al.* (2012), with permission from IOS Press.

Additionally, a shift size (an overlap between amplicons) of the reads was adjusted, along with the total read count.

The real 454 pyrosequencing reads from the HCV-1a samples

The dataset Data 1 was received from HCV Research Group in Institute of Biomedical Research, at University of Birmingham. Data 1 contains 30,927 reads obtained from the 5.2 kb-long fragment of HCV-1a genome (which is more than a half of the entire HCV genome). The average (aligned) read length is 292 bp. The depth of the reads coverage variability (see (Astrovskaya, 2011) for details) is due to a strong bias in the sequence start points, reflecting the secondary structure of the template DNA or RNA used to generate the initial PCR products. As a result, shorter reads are produced by GC-rich sequences.

The real 454 pyrosequencing reads from the HIV samples (Zagordi *et al.*, 2010b)

The HIV dataset (Zagordi *et al.*, 2010b) contains 55,611 reads from a mixture of 10 different 1.5-kb-long regions of HIV-1 quasispecies, including *pol* protease and part of the *pol* reverse transcriptase. The aligned reads length varies from 35 bp to 584 bp with the average of 345 bp. In contrast to (Zagordi *et al.*, 2010b), we did not filter out reads with low-quality scores.

The real 454 pyrosequencing amplicon data from the HBV samples

A total of 81,282 raw pyrosequencing reads (Ronaghi *et al.*, 1996) were produced by GS FLX LR25 from HBV quasispecies. Of these, 48,155 reads were longer than 150 bp. No reads were generated from the fragment 25. All reads were first aligned to a reference genome (MosaikAligner (Mosaik-aligner)), and then cleaned by cutting the remainders of the primers from both ends. The reads were separately corrected (Skums *et al.*, 2012) for each amplicon. The corrected reads were then realigned to the reference genome to facilitate identification of the distinct reads within the amplicons. Since several amplicons lack sufficient number of reads (see (Mancuso *et al.*, 2012) for details), we focus on a subset of a continuous region, in particular, the highly variable region in the quasispecies from Patient 1 (positions 1109 through 1642) and a conserved region in the quasispecies from Patient 2 (positions 2309 through 2748) (see Table 12.2).

Metrics

A viral spectrum assembler is judged by the quality of viral quasispecies sequences assembling and its ability to estimate the frequencies of a given viral population. In the ideal case of error-free reads, a quasispecies sequence is captured if one of the candidate sequences exactly matches it. If reads contain errors, a quasispecies sequence is considered to be reconstructed if a matched candidate sequence has at most k mismatches where k is some small integer value. The quality of an assembling is measured by portion of the real quasispecies sequences being captured by the candidate sequences

$$Sensitivity = \frac{|TruePositives|}{|TruePositives| + |FalseNegatives|}$$

and its portion among the candidate sequences (positive predictive value)

$$PPV = \frac{|TruePositives|}{|TruePositives| + |FalsePositives|}$$

in cross-validation tests.

Following (Eriksson *et al.*, 2008), the prediction quality of a frequencies distribution is usually measured with Kullback–Leibler divergence, or relative entropy. Given two probability distributions, the relative entropy measures the 'distance' (divergence) between them, or, in the other words, the quality of approximation of one probability distribution by the other distribution. Formally, the relative entropy between a true distribution P and an approximation distribution Q is given as:

$$D_{KL}(P||Q) = \sum_{i \in I} P(i) log \frac{P(i)}{Q(i)},$$

where sum is calculated over all reconstructed original sequences $I = \{i | P(i) > 0 \; and \; Q(i) > 0\}$, i.e. over all original sequences that have a match (exact or with at most k mismatches) among the assembled sequences.

Table 12.2 The number of all and distinct reads (after error correction) in the amplicons under consideration for each patient. An amplicon name contains its starting and ending positions. The data is taken from (Mancuso *et al.*, 2012)

	Patient 1		Patient 2	
Amplicon	No. reads	No. distinct reads	No. reads	No. distinct reads
Amp1109–1273	319	2	384	1
Amp1207–1403	1099	2	1552	1
Amp1263–1463	647	2	2370	1
Amp1445–1642	200	2	1260	1
Amp2309–2508	30	29	21	3
Amp2424–2629	270	1	5033	26
Amp2576–2748	1081	1	147	30

Alternatively, the similarity between two statistical distributions can be measured by Jensen–Shannon divergence, defined as

$$JSD(P||Q) = \frac{1}{2}D_{KL}(P||M) + \frac{1}{2}D_{KL}(Q||M),$$

where $M = \frac{1}{2}(P+Q)$. Since Jensen-Shannon metrics uses midpoint, this quasi-metric is continuous over all values of P and Q. As the size of the reconstructed spectrum may differ from the size of the 'true' spectrum, it still allows to measure how close the two distributions are.

In rare cases, Pearson's correlation coefficient and average prediction error

$$\overline{err} = \frac{\sum_{i\in I}|P_i - Q_i|}{|I|}$$

are reported to compare real and estimated frequencies distributions.

Validation on a real pyrosequencing data with unknown true viral spectrum

When a true viral spectrum is unknown in a given sample, Astrovskaya *et al.* (2011) proposed to validate a reconstructed viral spectrum (in particular, top ten most frequent inferred quasispecies) based on their reproducibility. The rationale is if a candidate sequence is indeed the true viral quasispecies rather than some randomly generated sequence, a viral assembler should be able to reconstruct it from any read sample, sequenced from the given viral population. Since the top 10 frequent inferred quasispecies usually have sufficient coverage by the reads, we used bootstrap strategy to generate different read sets from the given data. Repeatedly (100 times), we deleted randomly chosen 10% of the reads, run an assembler on each reduced read instance to reconstruct quasispecies spectrum, and compare this spectrum to the top ten frequent quasispecies, inferred from the whole read data. For the ith ($i = 1, .., 10$) most frequent sequence assembled from the whole data, we record its *reproducibility*, i.e. percentage of the runs when there is a match (with at most k mismatches) among 10 most frequent sequences found on the reduced data. The reproducibility can be used as additional validation of a theoretical spectrum before its verification in a laboratory. The biological experiments are the only way to fully validate predicted viral spectrum.

ViSpA results (Astrovskaya *et al.*, 2011)

Experimental validation on the simulated HCV data

In (Astrovskaya *et al.*, 2011), ViSpA was compared with ShoRAH (Zagordi *et al.*, 2011) and with ViSpA on the ShoRAH-corrected reads (ShoRAHreads+ViSpA). First, we analysed the sensitivity and PPV dependence on the number of the quasispecies in underlying viral population (see Fig. 12.10, left). Clearly, ViSpA outperforms ShoRAH here: it can correctly assemble all sequences out of 10 quasispecies and 29 sequences out of 40 quasispecies if the average

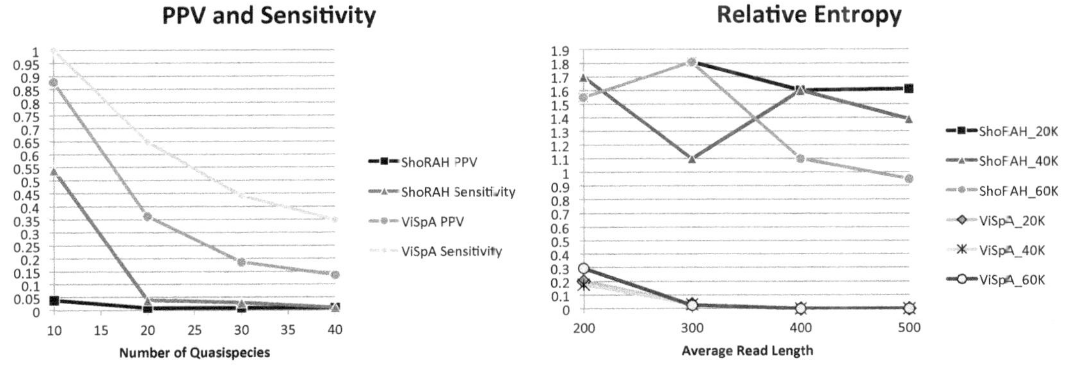

Figure 12.10 (adapted from (Astrovskaya *et al.*, 2011)) The statistical validation on the error-free reads from the known HCV quasispecies. Left: the PPV and sensitivity as a function of the number of the quasispecies in the original population (40,000 reads with the average read length of 300 bp, the geometric distribution). Right: the relative entropy as a function of the average read length (40,000 reads from 10 quasispecies under the geometric distribution).

read length is at least 300 bp. If the average read length is smaller (for example, in the range from 250 bp to 299 bp), the method can assemble at least 8 out of 10 sequences and 20 out of 40 sequences. The relative entropy of ViSpA's spectra is decreasing with an increase of the average read length (see Fig. 12.10, right). It is expected since the increase in sensitivity allows EM more accurately predict an underlying distribution. However, it is not the case for ShoRAH due to overcorrection of the reads and local haplotype reconstruction strategy.

On the other hand, ShoRAH has a significant advantage over ViSpA on the read data simulated by FlowSim both in prediction power and in robustness of the results (see Fig. 12.11). Indeed, ShoRAH correctly infers three out of 10 real quasispecies sequences whereas ViSpA reconstructs only one sequence. Additionally, 10 most frequent assemblies inferred by ShoRAH are more robust with repeating up to 45% of times on 10% reduced data versus 1% of times for ViSpA's assemblies. This advantage can be explained by the read correction algorithm in ShoRAH. If ViSpA is applied to the ShoRAH-corrected reads, the results drastically improve: five quasispecies sequences are inferred; they are reconstructed exactly 95% of times on the 10%-reduced read data. The latter confirms that ViSpA's global reconstruction of viral sequence is better than local haplotype reconstruction in ShoRAH.

Experimental validation on 454 pyrosequencing reads from the HCV samples

ViSpA's assembling depends on how many mismatches are allowed during clustering reads into the superreads (n) and during creation of the edges (m) in the read graph. For the dataset Data 1, the constructed read graph has no isolated vertices if up to six mismatches is used to determine superreads and up to 15 mismatches is allowed to create edges (see Astrovskaya *et al.*, 2011). To refine assembled candidate sequences, all reads were used, and mutation rate was varied from 1.75% up to 8% per sequence (which is in the range observed in (Fishman *et al.*, 2009)). Out of 3207 max-bandwidth paths, as much as 938 distinct sequences ($t=80$) and as low as 755 sequences ($t=350$) were obtained for different values of $t \in [80; 350]$.

The neighbour-joining tree for 10 most frequent candidate sequences obtained by each tool (ViSpA and ShoRAH) (see Fig. 12.12) reminds a neighbour-joining tree for the HCV quasispecies evolution. Additionally, the most frequent candidate sequence found by ViSpA is 99% identical to one of the actual ORFs obtained by cloning the quasispecies.

Viral sequences with internal stop codons after the starting codon are not viable since the entire HCV genome consists of a single coding region for a large polyprotein. Thus, the number

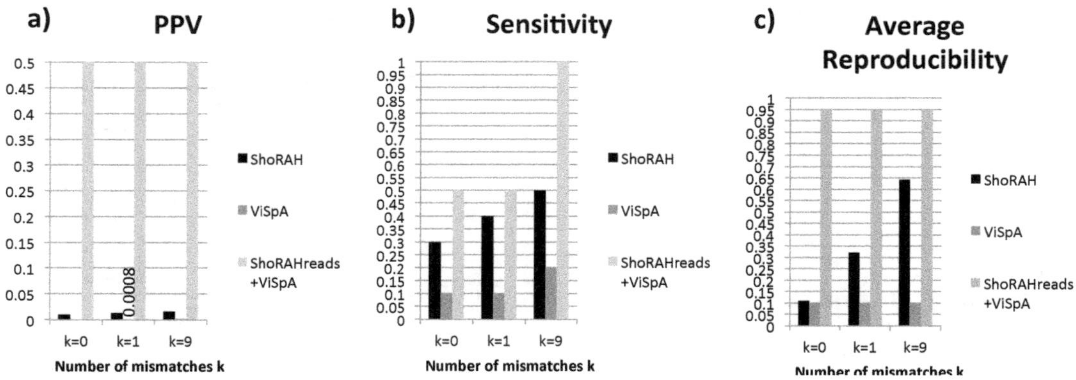

Figure 12.11 A comparison of three methods – ViSpA, ShoRAH, and ShoRAHreads+ViSpA – on the read data simulated by FlowSim. (a) and (b) report the PPV and sensitivity, respectively. (c) shows the average reproducibility across ten most frequent candidate sequences assembled on the whole read data.

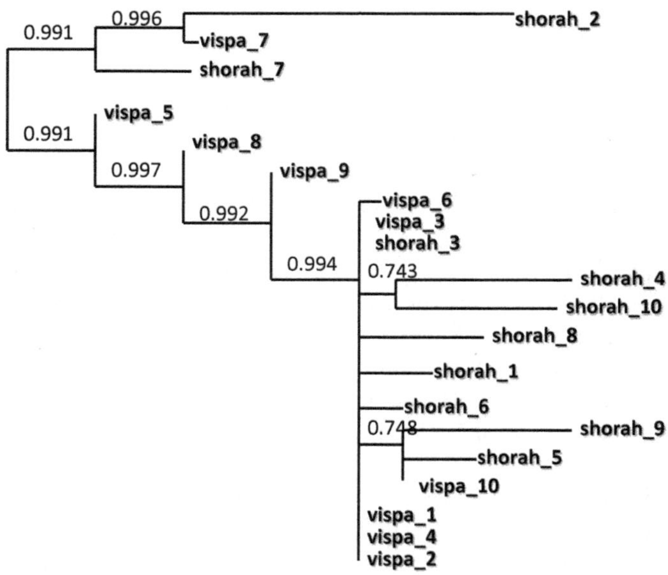

Figure 12.12 (adapted from (Astrovskaya *et al.*, 2011)) The neighbour-joining phylogenetic tree for 10 most frequent HCV quasispecies variants for the 5,205 bp-long fragment obtained by each tool – ViSpA and ShoRAH. Sequences are labelled with a software name and its rank among 10 most frequent assembled sequences.

of 'viable' (without stop codons after the starting codon) reconstructed sequences can serve as an additional accuracy measure for the quasispecies assembly. Out of 10 most frequent sequences reconstructed by ViSpA, only three are not viable while ShoRAH is able to reconstruct only one 'viable' sequence. This sequence has 99.94% similarity with the ViSpA's fourth most frequent theoretical assembly. Both methods returned similar frequency estimations for this sequence: 0.017% (ShoRAH) and 0.019% (ViSpA). Additionally, the stop codons in ViSpA's three non-viable predicted sequences occurred at the very last positions of the sequences and can be easily corrected after manual inspection. Unfortunately, it is not the case for ShoRAH's non-viable theoretical sequences.

ViSpA's and ShoRAH's viral spectra inferred from the dataset Data 1 were further validated by the bootstrap experiments. Two sequences are considered a match to each other if one can be transformed to the another with at most k ($k=0$, 1, 2, 5) substitutions. Fig. 12.13 shows the reproducibility of ViSpA's 10 most frequent sequences. For example, if we request the exact match ($k=0$), ShoRAH repeatedly (35% of times) reconstructs only the third most frequent sequence while ViSpA reconstructs seven sequences in at least 15% of times, and the most frequent sequence is reconstructed 40% of times.

Experimental validation on 454 pyrosequencing reads from the HIV samples

ViSpA and ShoRAH were also compared on the HIV dataset, used in the first experiment in (Zagordi *et al.*, 2010b). The low-quality reads were not filtered out that can explain poorer performance of ShoRAH. Indeed, ShoRAH correctly infers only two quasispecies sequences with at most four mismatches: one assembly has three mismatches with real quasispecies sequence, and the other has four mismatches.

ViSpA correctly reconstructs five quasispecies with at most two mismatches (three of them are among 10 most frequent assemblies): two sequences are inferred without any mismatches (one of them is among 10 most frequent assemblies), one assembly has a mismatch with real quasispecies sequence (and it is among 10 most frequent assemblies), and the rest have two mismatches (one of them is among 10 most frequent assemblies). The assemblies correspond to the 'viable' protein sequences.

If ViSpA is applied to the ShoRAH-corrected reads, it successfully infers three real quasispecies without any mismatches.

VSEM results (Mangul *et al.*, 2011)

In the first set of experiments (Mangul *et al.*, 2011), VSEM was validated on a completely known panel. For each simulated frequency distribution, repeatedly, a randomly chosen string was deleted from the panel, and the frequency of the missing quasispecies was calculated. If a panel is complete, the virtual string usually has the frequency less than 10^{-6}. And if the frequency of the deleted quasispecies is at least 1%, then the resulted virtual string frequency will grow to at least 0.5%. Hence, VSEM can reliably detect an incomplete panel in this case.

Next, ViSpA-VSEM performance was compared to the original ViSpA. Fig. 12.14 shows improvement over ViSpA's frequencies prediction

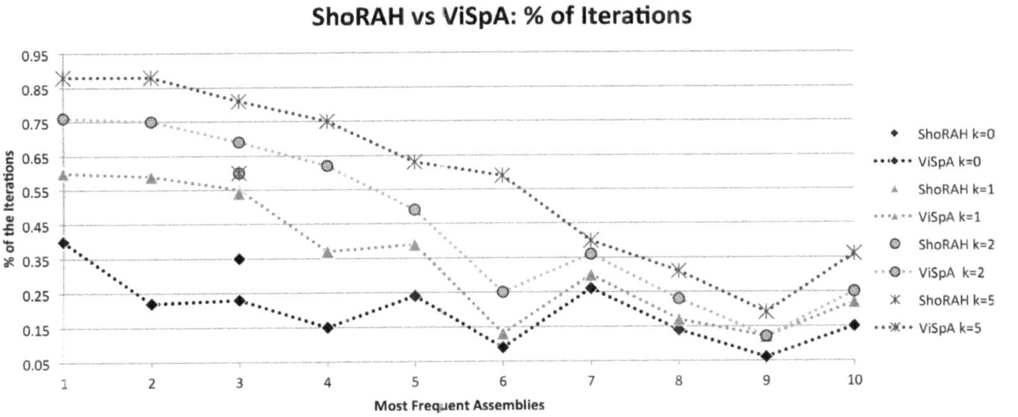

Figure 12.13 (adapted from (Astrovskaya *et al.*, 2011)) The reproducibility of ShoRAH's and ViSpA's (*n*=6, *m*=15, and *t*=120) viral spectra (10 most frequent assemblies) inferred from Data 1. The sequences are considered a match to each other if one can be transformed to another with at most *k* (*k*=0, 1, 2, 5) substitutions. The ith point at the *x*-axis corresponds to the ith most frequent sequence assembled on the 100% of the reads. The *y*-axis reports a percentage of the runs when the *i*th most frequent sequence assembled on all data is reproduced among 10 most frequent quasispecies assembled on the 10%-reduced set of the reads. No data are shown for the sequences that are reproduced less than 5% of the runs.

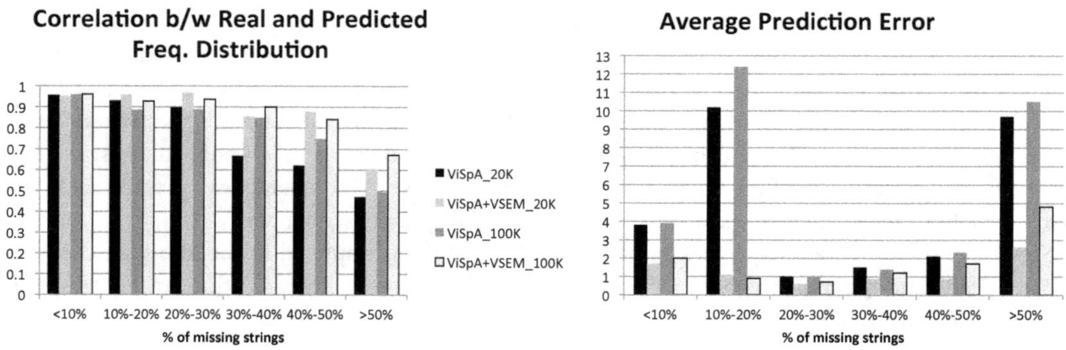

Figure 12.14 A comparison of ViSpA's and ViSpA+VSEM's frequencies distributions. At the left, the correlation between the real quasispecies frequencies and the estimated quasispecies frequencies is reported for each method if 20,000 or 100,000 error-free reads were generated with the average read length of 300bp from 40 quasispecies under the geometric distribution. At the right, the average prediction error is plotted for each method on the above-mentioned data.

if a missing string is taken into account in the EM. The correlation is slightly improved for the cases when the portion of missing strings is small and significantly increases (up to 15%) when a larger portion of sequences is missing. The results on the reads with sequencing errors, generated by Flowsim (Balser *et al.*, 2010) and corrected by ShoRAH (Zagordi *et al.*, 2010a) are similar to those on Fig. 12.14.

ViSpA-VSEM also improves predictive power of ViSpA. If there is no error in the reads (Fig. 12.15), on average, it reconstructs additional two (in the case of the geometric distribution of the quasispecies) to four sequences (in the case of the uniform and skewed uniform distributions) (Mangul *et al.*, 2011). Furthermore, frequencies estimations are closer to the real values with the decrease in the average error by 2.5 times for the geometric distribution and by more than 5 times for the skewed uniform and uniform distributions. In the case of the reads with errors, ViSpA and ViSpA-VSEM performance was compared on the ShoRAH-corrected reads (Fig. 12.16). ViSpA-VSEM additionally reconstructs exact sequence for one real quasispecies. As expected, ViSpA-VSEM has a drop in the average prediction error.

VirA results (Mancuso *et al.*, 2012)*

Experiments on the simulated amplicon reads from the HCV samples

The first set of experiments compared different strategies (see Fig. 12.17) to infer quasispecies sequences from the error-free amplicons. If quasispecies frequencies follow the uniform distribution, the maximum bandwidth strategy outperforms (in terms of sensitivity) the greedy fork resolution method, the maximum frequency approach and the guide distribution method (Prosperi, Prosperi *et al.*, 2011). VirA (with the maximum bandwidth path strategy) correctly assembles seven out of 10 quasispecies on average from 100,000 reads and the window length of 300 bp. The sensitivity of all four methods decreases with an increase in a shift position since a shorter overlap results in larger bi-cliques in a subgraph, thus making the problem more complex. Under the skewed distribution of the quasispecies frequencies, both the maximum bandwidth strategy and the greedy method assemble 8–9 out of 10 real viral sequences. The maximum bandwidth approach and the greedy method are also better if the quasispecies

* Reprinted from In Silico Biology 11, Mancuso, N., Tork, B., Skums, P., Ganova-Raeva, L., Măndoiu, I., and Zelikovsky, A. Reconstructing viral quasispecies from NGS amplicon reads, 237–249, © 2012 Mancuso *et al.* (2012), with permission from IOS Press.

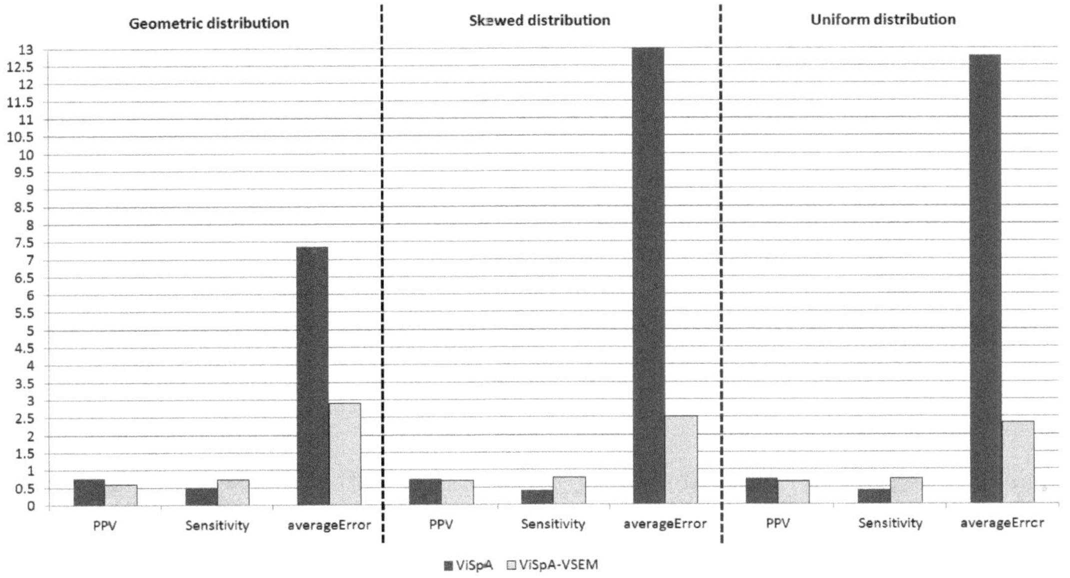

Figure 12.15 A comparison between ViSpA and ViSpA-VSEM. Experiments are run on 100,000 reads with the average read length of 300bp from 10 quasispecies under the geometric, skewed and uniform distributions.

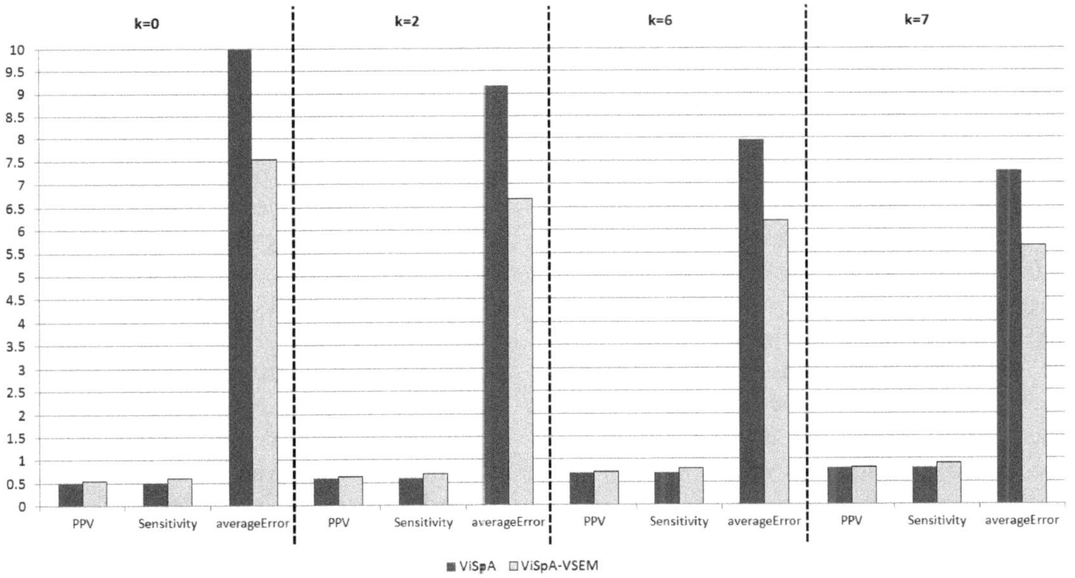

Figure 12.16 A comparison between 10 most frequent quasispecies assembled by ViSpA and by ViSpA-VSEM. 100,000 reads with sequencing errors were generated with the average read length of 300bp from 10 quasispecies under the geometric distribution. A quasispecies sequence is considered found if one of the candidate sequences matches it exactly (*k=0*) or with at most *k* (2, 6 or 7) mismatches.

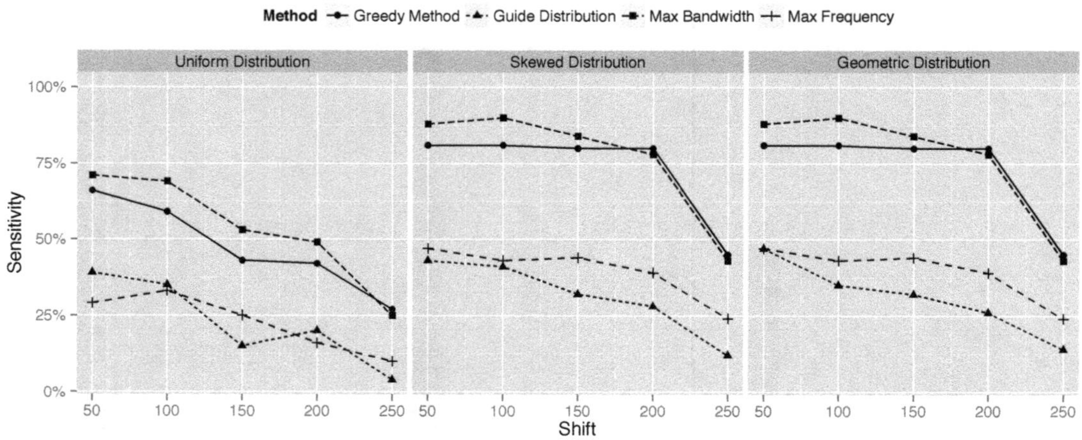

Figure 12.17 Sensitivity as a function of an amplicon window shift. 100,000 reads were generated from 10 quasispecies with the average read length of 300bp. Quasispecies are under the uniform (at the left), skewed uniform (in the middle), and geometric (at the right) distributions.

frequencies are followed the geometric distribution: the maximum bandwidth and the greedy resolution method correctly produce approximately 9 and 8 out of 10 quasispecies, respectively. Despite its typically lower sensitivity, the maximum frequency path approach tends to be more competitive in terms of the positive predicted values (Fig. 12.18). This can be explained by the rapid covering of many reads in the graph per iteration. Similarly, the maximum bandwidth approach exhausts the read graph much faster than the greedy approach, thus, has higher PPVs.

In terms of the quality of the produced viral spectra, the maximum bandwidth approach and the greedy resolution method, on average, outperform the other methods (Fig. 12.19). All four methods benefit from shorter shifts (that is, longer overlaps) since the divergence increases with the increase of the shift length.

All four methods also improve their sensitivity if fork balancing was used (Fig. 12.20). The QP approach has much less deviation from the ideal, comparing to both the original observed read frequencies and scaled read frequencies for each considered quasispecies frequencies distributions. The maximum bandwidth, the maximum frequency, and the guide distribution methods seem to benefit less from the initial data adjustment while the greedy fork resolution has a higher sensitivity, on average (Fig. 12.20). The correlation studies (Fig. 12.21) show that the weighted QP adjusting tends to perform the best, and simple scaling tends to hurt correlation in the lower read-count datasets.

The sensitivity and PPV of the VirA (with the best performing among the proposed methods *n* maximum bandwidth reconstruction) on the error-free amplicon reads is also compared to those of ViSpA (Astrovskaya *et al.*, 2011) on the error-free shotgun reads (Table 12.3). The VirA outperforms ViSpA under all generated quasispecies frequencies distributions and different numbers of the generated reads, except for the case of 20 quasispecies followed the uniform distribution. While ViSpA beats the maximum bandwidth on 100,000 reads, its performance degrades quickly as read count drops. This highlights the ability of the amplicon-based maximum bandwidth reconstruction to accurately assemble reads even at a low read coverage. The performance difference is more dramatic for the skewed and geometric distributions.

Experiments on the real amplicon data from the HBV samples

In Patient 1, both the maximum bandwidth and the greedy fork resolution reconstructed three quasispecies while the maximum frequency strategies inferred only one. In Patient 2, the maximum bandwidth, the maximum frequency, and the greedy fork resolution methods produced three, two and 11 quasispecies, respectively.

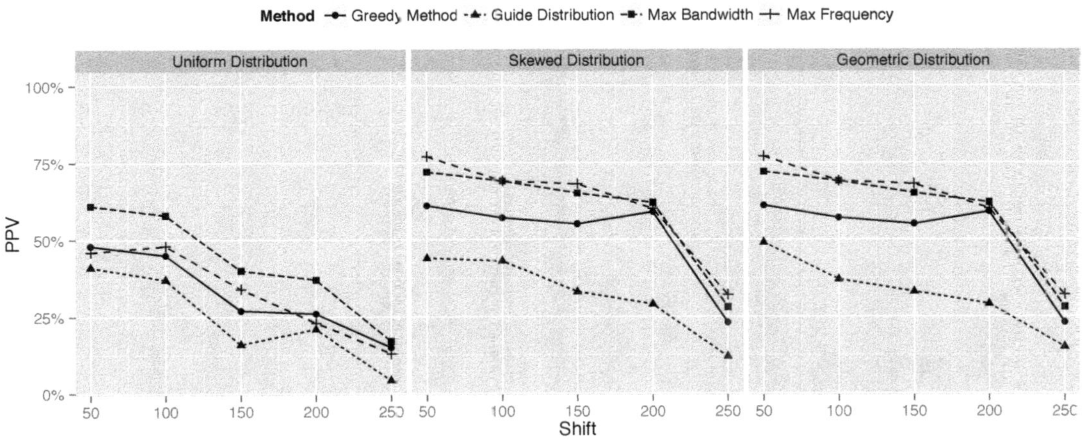

Figure 12.18 PPV as a function of an amplicon window shift. 100,000 reads were generated from 10 quasispecies with the average read length of 300bp. Quasispecies are under the uniform (at the left), skewed uniform (in the middle), and geometric (at the right) distributions.

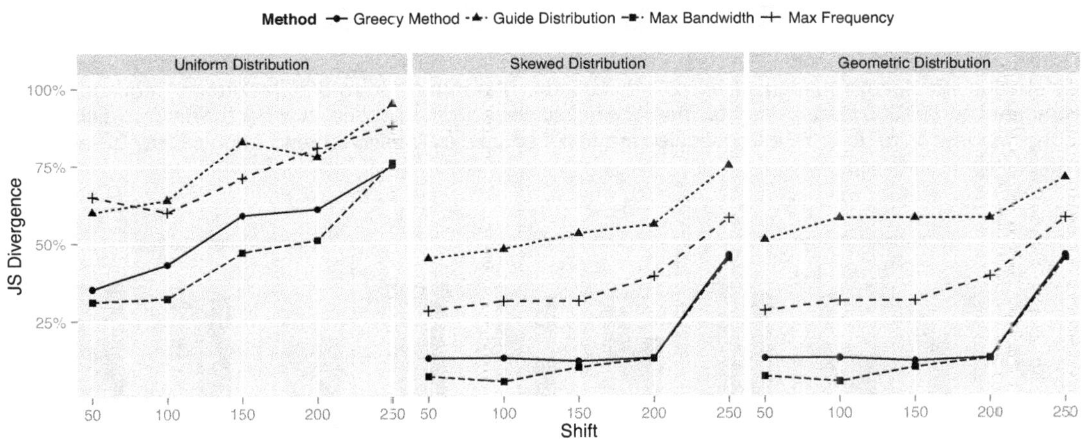

Figure 12.19 Jensen-Shannon divergence as a function of an amplicon window shift. 100,000 reads were generated from 10 quasispecies with the average read length of 300bp. Quasispecies are under the uniform (at the left), skewed uniform (in the middle), and geometric (at the right) distributions.

Future trends

Improvements in quasispecies spectrum reconstruction accuracy are expected from advances in both computational methods and sequencing technology. Computational advances are expected in algorithms for (1) correcting both systematic and random genotyping errors, (2) improved statistical models, and (3) improved reconstruction accuracy of full length viral haplotypes by using pair-end reads. Note that current EM-based methods described in this chapter and flow-based methods (Skums *et al.*, 2013) seem already sufficiently accurate in estimating frequencies of quasispecies.

Technological advances are likely to continue to increase a read length and an accuracy (Rothberg *et al.*, 2011), both of which are beneficial to accurate reconstruction of quasispecies. Improvements in sequencing accuracy are particularly important for detection of rare variants. A promising approach is to independently extend each viral fragment with a random tag during library preparation, then amplify and sequence multiple copies of each tagged fragment, thus virtually eliminating

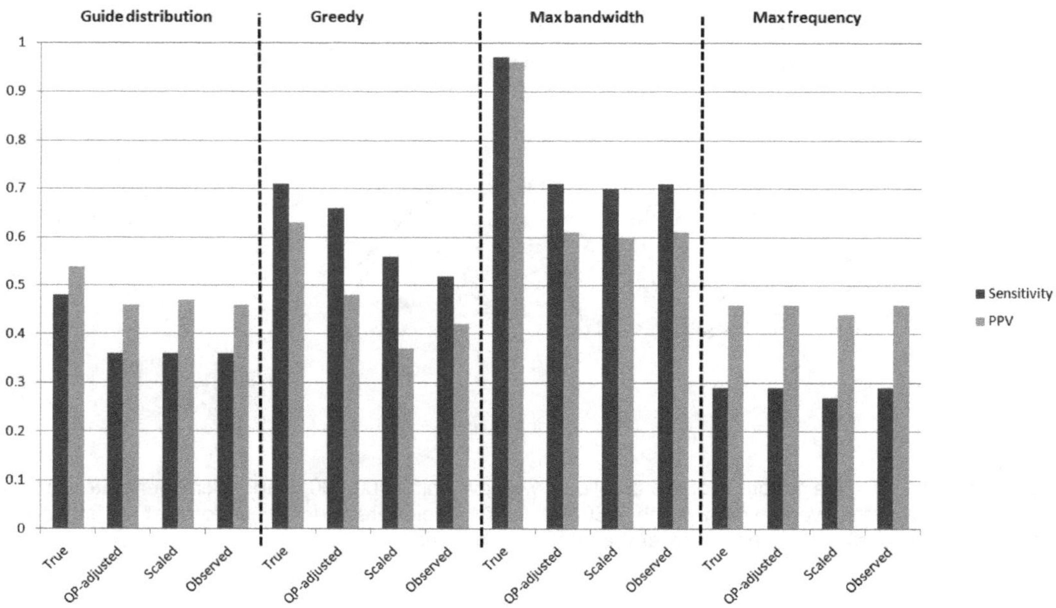

Figure 12.20 Sensitivity and PPV versus different fork frequencies. Four assembly methods (the Guide distribution (Prosperi, Prosperi *et al.*, 2011), Greedy, Maximum bandwidth, Maximum frequency) were compared on 100,000 reads generated from the quasispecies with frequencies under the uniform distribution (50bp window shift). Four fork frequencies are true frequencies, observed read frequencies, QP-adjusted and scaled frequencies.

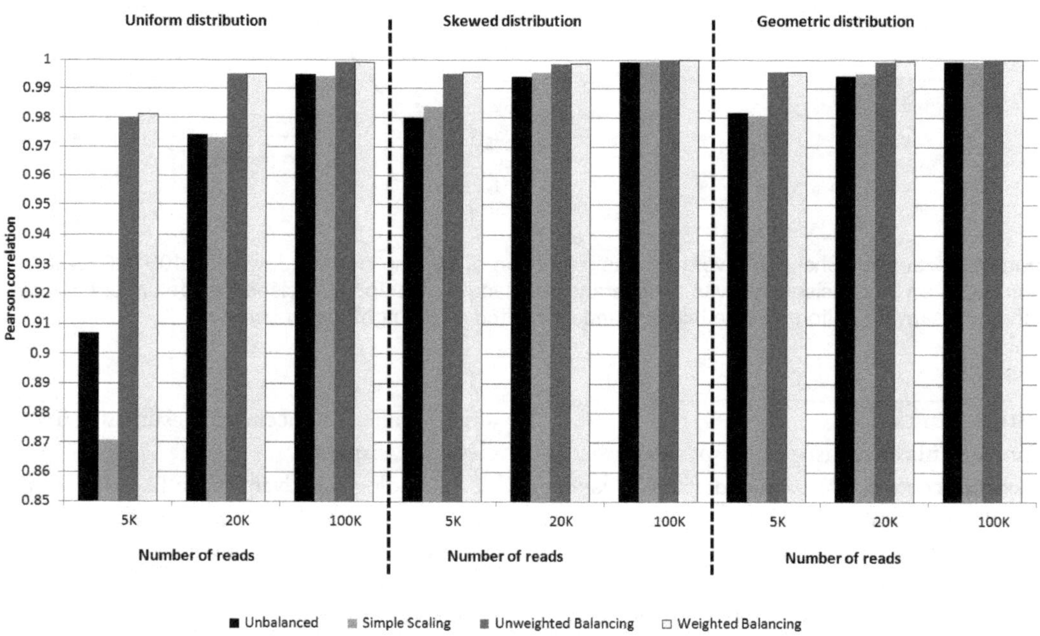

Figure 12.21 Pearson correlations for different balancing schemes – unbalanced, simple scaling, unweighted balancing and weighted balancing.

Table 12.3 The average sensitivity/PPV of ViSpA and VirA with the maximum bandwidth algorithm (MB) for respectively the shotgun and the amplicon error-free read datasets simulated from 10 or 20 quasispecies.

		Uniform distribution		Skewed distribution		Geometric distribution	
No. quasispecies	No. of reads	ViSpA	VirA(MB)	ViSpA	VirA (MB)	ViSpA	VirA (MB)
10	5K	0.17/0.45	0.42/0.36	0.18/0.13	0.63/0.55	0.18/0.13	0.56/0.53
	20K	0.48/0.42	0.55/0.48	0.46/0.33	0.86/0.72	0.41/0.30	0.87/0.72
	100K	0.65/0.56	0.71/0.61	0.62/0.47	0.88/0.72	0.68/0.52	0.95/0.79
20	5K	0.10/0.06	0.24/0.23	0.09/0.05	0.12/0.13	0.12/0.12	0.35/0.50
	20K	0.30/0.20	0.26/0.25	0.24/0.16	0.44/0.45	0.24/0.25	0.44/0.68
	100K	0.53/0.36	0.43/0.43	0.40/0.25	0.60/0.63	0.35/0.37	0.54/0.91

Reprinted from In Silico Biology 11, Mancuso, N., Tork, B., Skums, P., Ganova-Raeva, L., Măndoiu, I., and Zelikovsky, A. Reconstructing viral quasispecies from NGS amplicon reads, 237–249, © Mancuso *et al.* (2012), with permission from IOS Press.

sequencing errors (see, for example, (Schmitt *et al.*, 2012)).

Alternative technologies such as the i-Seq assay based on mass spectrometry (Stanssens *et al.*, 2004) are expected to significantly lower the costs for estimation of viral genetic heterogeneity as well as detection and surveillance of drug-resistant variants in chronically infected hosts. Viral quasispecies reconstruction from mass spectrometry data may also be feasible, but the more ambiguous nature of this type of data in comparison to the sequencing reads will require the development of novel computational methods.

Conclusions

In this chapter, we have discussed the problem of reconstruction quasispecies sequences and inferring their frequencies from NGS reads. Computational challenges and existing approaches have been reviewed, with a detailed description of two software tools – Viral Spectrum Assembler (ViSpA) and Viral Assembler (VirA) – for analysing the shotgun and the amplicon NGS reads, respectively. Both methods have been tested on simulated and real read data from HCV, HIV (ViSpA) and HBV (VirA) quasispecies and compared to several other software tools. The results of these comparisons show that NGS reads analysed with existing software can give a reasonable representation of viral quasispecies population, but better computational and technological tools are still required for further advancing applications in molecular epidemiology.

Web resources

Links to download different tools:

- Hapler (O'Neil and Emrich, 2012): http://nd.edu/~biocmp/hapler/
- Mosaic Aligner: http://bioinformatics.bc.edu/marthlab/Mosaik
- PredictHaplo (Prabhakaran *et al.*, 2010): http://bmda.cs.unibas.ch/HivHaploTyper/
- QuasiRecomb(Zagordi *et al.*, 2012b): http://www.bsse.ethz.ch/cbg/software/quasirecomb
- QuRe (Prosperi and Salemi, 2012): http://sourceforge.net/projects/qure/
- SEGEMEHL (Hoffman *et al.*, 2009): http://www.bioinf.uni-leipzig.de/Software/segemehl/
- ShoRAH (Zagordi *et al.*, 2011): http://www.bsse.ethz.ch/cbg/software/shorah
- ViSpA (Astrovskaya *et al.*, 2011): http://alla.cs.gsu.edu/~software/VISPA/vispa.html
- VirA (Mancuso *et al.*, 2011): https://bitbucket.org/nmancuso/bioa/
- V-Phaser (Macalalad *et al.*, 2012): http://www.broadinstitute.org/scientific-community/science/projects/viral-genomics/v-phaser

References

Astrovskaya, I.A. (2011). Inferring Genomic Sequences. Computer Science Dissertations, paper 59. Available at: http://digitalarchive.gsu.edu/cs_diss/59.

Astrovskaya, I., Tork, B., Mangul, S., Westbrooks, K., Măndoiu, I., Balfe, P., and Zelikovsky, A. (2011). Inferring viral quasispecies spectra from 454 pyrosequencing reads. BMC Bioinformatics *12* (Suppl. 6), S1.

Balser, S., Malde, K., Lanzen, A., Sharma, A., and Jonassen, I. (2010). Characteristics of 454 pyrosequencing data-enabling realistic simulation with FlowSim. Bioinformatics *26*, 420–425.

Bansal, V., and Bafna, V. (2008). HapCUT: an efficient and accurate algorithm for the haplotype assembly problem. Bioinformatics *24*, 153–159.

Barzon, L., Lavezzo, E., Militello, V., Toppo, S., and Palù, G. (2011). Applications of next-generation sequencing technologies to diagnostic virology. Int. J. Mol. Sci. *12*, 7861–7884.

Beerenwinkel, N., Sing, T., Lengauer, T., Rahnenführer, J., Roomp, K., Savenkov, I., Fischer, R., Hoffmann, D., Selbig, J., Korn, K., *et al.* (2005). Computational methods for the design of effective therapies against drug resistant HIV strains. Bioinformatics *21*, 3943–3950.

Brinza, D., and Zelikovsky, A. (2006). 2SNP: scalable phasing based on 2-SNP haplotypes. Bioinformatics *22*, 371–373.

Brockman, W., Alvarez, P., Young, S., Garber, M., Giannoukos, G., Lee, W.L., Russ, C., Lander, E.S., Nusbaum, C., and Jaffe, D.B. (2008). Quality scores and SNP detection in sequencing-by-synthesis systems. Genome Res. *18*, 763–770.

Chaisson, M.J., and Pevzner, P.A. (2008). Short read fragment assembly of bacterial genomes. Genome Res. *18*, 324–330.

Domingo, E., and Holland, J.J. (1997). RNA virus mutations and fitness for survival. Annu. Rev. Microbiol. *51*, 151–178.

Domingo, E., Martinez-Salas, E., Sobrino, F., de la Torre, J.C., Portela, A., Ortín, J., López-Galindez, C., Pérez-Breña, P., Villanueva, N., Nájera, R., *et al.* (1985). The quasispecies (extremely heterogeneous) nature of viral RNA genome populations: biological relevance – a review. Gene *40*, 1–8.

Domingo, E., Baranowski, E., Ruiz-Jarabo, C.M., Martín-Hernández, A.M., Sáiz, J.C., and Escarmís, C. (1998). Quasispecies structure and persistence of RNA viruses. Emerg. Infect. Dis. *4*, 521 – 527.

Douek, D.C., Kwong, P.D., and Nabel, G.J. (2006). The rational design of an AIDS vaccine. Cell *124*, 677–681.

Drake, J.W., and Holland, J.J. (1999). Mutation rates among RNA viruses. Proc. Natl. Acad. Sci. U.S.A. *96*, 13910–13913.

Duarte, E.A., Novella, I.S., Weaver, S.C., Domingo, E., Wain-Hobson, S., Clarke, D.K., Moya, A., Elena, S.F., de la Torre, J.C., and Holland, J.J. (1994). RNA virus quasispecies: significance for viral disease and epidemiology. Infect. Agents Dis. *3*, 201–214.

Eigen, M., Mccaskill, J., and Schuster, P. (1989). The molecular quasi-species. Adv. Chem. Phys. *75*, 149–263.

Eriksson, N., Pachter, L., Mitsuya, Y., Rhee, S.Y., Wang, C., Gharizadeh, B., Ronaghi, M., Shafer, R.W., and Beerenwinkel, N. (2008). Viral population estimation using pyrosequencing. PLoS Comput. Biol. *4*, e1000074.

Ewens, W.J. (1972). The sampling theory of selectively neutral alleles. Theor. Popul. Biol. *3*, 87–112.

Fakhrai-Rad, H., Pourmand, N., and Ronaghi, M. (2002). Pyrosequencing™: An accurate detection platform for single nucleotide polymorphisms. Hum. Mutat. *19*, 479–485.

Ferguson, T.S. (1973). A bayesian analysis of some nonparametric problems. Ann. Stat. *1*, 209–230.

Fishman, S.L., and Branch, A.D. (2009). The quasispecies nature and biological implications of the hepatitis C virus. Infect. Genet. Evol. *9*, 1158–1167.

Garey, M.R., and Johnson, D.S. (1979). Computers and Intractability: A Guide to the Theory of NP-completeness. (San Francisco, W.H. Freeman and Co.).

Gaschen, B., Taylor, J., Yusim, K., Foley, B., Gao, F., Lang, D., Novitsky, V., Haynes, B., Hahn, B.H., Bhattacharya, T., *et al.* (2002). Diversity considerations in HIV-1 vaccine selection. Science *296*, 2354–2360.

Graham, R., Grötschel, M., and Lovasz, L. (eds.) (1996). Handbook of Combinatorics (Cambridge, MA, The MIT Press).

von Hahn, T., Yoon, J.C., Alter, H., Rice, C.M., Rehermann, B., Balfe, P., and McKeating, J.A. (2007). Hepatitis C virus continuously escapes from neutralizing antibody and T-cell responses during chronic infection *in vivo*. Gastroenterology *132*, 667–678.

Halperin, E., and Karp, R. (2005). The minimum-entropy set cover problem. Theoret. Comput. Sci *348*, 733–744.

Hoffmann, S., Otto, C., Kurtz, S., Sharma, C.M., Khaitovich, P., Vogel, J., Stadler, P.F., and Hackermüller, J. (2009). Fast mapping of short sequences with mismatches, insertions and deletions using index structures. PLoS Comput. Biol. *5*, e1000502.

Holland, J.J., de la Torre, J.C., and Steinhauer, D.A. (1992). RNA virus populations as quasispecies. Curr. Top. Microbiol. Immunol. *176*, 1–20.

Huang, A., Kantor, R., Delong, A., Schreier, L., and Istrail, S. (2011). QColors: an algorithm for conservative viral quasispecies reconstruction from short and non-contiguous next generation sequencing reads. Paper presented at: IEEE International Conference on Bioinformatics and Biomedicine Workshops. (IEEE), pp. 130–136.

Jojic, V., Hertz, T., and Jojic, N. (2008). Population Sequencing Using Short Reads: HIV as a Case Study. In Pacific Symposium on Biocomputing, Altman, R.B., Dunker, A.K., Hunter, L., Murray, T., Klein, T.E., eds. (Bukit Timah, Singapore, World Scientific), 114–125.

Laird, N.M., Rubin, D.B., and Dempster, A.P. (1977). Maximum likelihood from incomplete data via the EM algorithm (with discussions). J. R. Stat. Soc. Series B Stat. Methodol. *39*, 1–38.

Lippert, R., Schwartz, R., Lancia, G., and Istrail, S. (2002). Algorithmic strategies for the single nucleotide polymorphism haplotype assembly problem. Brief. Bioinform. 3, 23–31.

Li, J.Z., Paredes, R., Ribaudo, H.J., Svarovskaia, E.S., Metzner, K.J., Kozal, M.J., Hullsiek, K.H., Balduin, M., Jakobsen, M.R., Geretti, A.M., *et al.* (2011). Low-frequency HIV-1 drug resistance mutations and risk of NNRTI-based antiretroviral treatment failure: a systematic review and pooled analysis. JAMA *305*, 1327–1335.

Luciani, F., Bull, R.A., and Lloyd, A.R. (2012). Next generation deep sequencing and vaccine design: today and tomorrow. Trends Biotechnol. *30*, 443–52.

Macalalad, A.R., Zody, M.C., Charlebois, P., Lennon, N.J., Newman, R.M., *et al.* (2012). Highly Sensitive and specific detection of rare variants in mixed viral populations from massively parallel sequence data. PLoS Comput. Biol. *8*, e1002417.

Mancuso, N., Tork, B., Skums, P., Ganova–Raeva, L., Măndoiu, I., and Zelikovsky, A. (2012). Reconstructing viral quasispecies from NGS amplican reads. In Silico Biol. *11*, 237–249.

Mangul, S., Astrovskaya, I., Nicolae, M., Tork, B., Măndoiu, I., and Zelikovsky, A. (2011). Maximum likelihood estimation of incomplete genomic spectrum from HTS data. In WABI'll Proceedings of the 11th International Conference on Algorithms in bioinformatics, Przytycka, T.M., and Sagot, M.-F., eds. (Springer-Verlag, Berlin Heidelburg), pp. 213–224.

Mardis, E.R. (2008). Next-generation DNA sequencing methods. Annu. Rev. Genomics Hum. Genet. *9*, 387–402.

Mardis, E.R. (2009). New strategies and emerging technologies for massively parallel sequencing: applications in medical research. Genome Med. *1*, 40.

Margulies, M., Egholm, M., Altman, W.E., Attiya, S., Bader, J.S., Bemben, L.A., Berka, J., Braverman, M.S., Chen, Y.J., Chen, Z., *et al.* (2005). Genome sequencing in microfabricated high-density picolitre reactors. Nature *437*, 376–380.

Martell, M., Esteban, J.I., Quer, J., Genesca, J., Weiner, A., Esteban, R., Guardia, J., and Gomez, J. (1992). Hepatitis C virus (HCV) circulates as a population of different but closely related genomes: quasispecies nature of HCV genome distribution. J. Virol. *66*, 3225–3229.

Metzner, K.J., Giulieri, S.G., Knoepfel, S.A., Rauch, P., Burgisser, P., Yerly, S., Gunthard, H.F., and Cavassini, M. (2009). Minority quasispecies of drug-resistant HIV-1 that lead to early therapy failure in treatment-naive and -adherent patients. Clin. Infect. Dis. *48*, 239–247.

Mosaik Aligner http://bioinformatics.bc.edu/marthlab/Mosaik.

Myers, G. (2005). Building fragment assembly string graphs. Proc. ECCB *2005*, Bioinformatics 21 Suppl2, 79–85.

National Centre for Biotechnology Information [http://www.ncbi.nlm.nih.gov]

Neumann, A.U., Lam, N.P., Dahari, H., Gretch, D.R., Wiley, T.E., Layden, T.J., and Perelson, A.S. (1998). Hepatitis C viral dynamics *in vivo* and the antiviral efficacy of interferon alpha therapy. Science *282*, 103–107.

Nicolae, M., Mangul, S., Mandoiu, I., and Zelikovsky, A. (2011). Estimation of alternative splicing isoform frequencies from RNA-Seq data. Algorithms Mol. Biol.: AMB *6*, 9.

O'Neil, S., and Emrich, S.J. (2012). Haplotype and minimum-chimerism consensus determination using short sequence data. BMC Genomics 13(Suppl 2), S4.

Prabhakaran, S., Rey, M., Zagordi, O., Beerenwinkel, N., and Roth, V. (2010). HIV haplotype inference using a constraint-based Dirichlet process mixture model. In NIPS Workshop on Machine Learning in Computational Biology.

Prosperi, M.C., Prosperi, L., Bruselles, A., Abbate, I., Rozera, G., Vincenti, D., Solmone, M.C., Capobianchi, M.R., and Ulivi, G. (2011). Combinatorial analysis and algorithms for quasispecies reconstruction using next-generation sequencing. BMC Bioinformatics *12*, 5.

Prosperi, M.C.F., and Salemi, M. (2012). QuRe: software for viral quasispecies reconstruction from next-generation sequencing data. Bioinformatics *28*, 132–133.

Quinlan, A.R., Stewart, D.A., Strömberg, M.P., and Marth, G.T. (2008). PyroBayes: an improved base caller for SNP discovery in pyrosequences. Nat. Methods *5*, 179–181.

Rasmussen, C.E. (2000). The infinite gaussian mixture model. In NIPS, Solla, S.A., Leen, T.K., Müller, K.-R., eds. (Cambridge, MA, The MIT Press), pp. 554–560.

Rhee, S.-Y., Liu, T.F., Holmes, S.P., and Shafer, R.W. (2007). HIV-1 subtype B protease and reverse transcriptase amino acid covariation. PLoS Comput. Biol. *3*, e87.

Ronaghi, M., Karamohamed, S., Pettersson, B., Uhlen, M., and Nyren, P. (1996). Real-time DNA sequencing using detection of pyrophosphate release. Anal. Biochem. *242*, 84–89.

Rothberg, J.M., Hinz, W., Rearick, T.M., Schultz, J., Mileski, W., Davey, M., Leamon, J.H., Johnson, K., Milgrew, M.J., Edwards, M., *et al.* (2011). An integrated semiconductor device enabling non optical genome sequencing. Nature *475*, 7356.

Schmitt, M.W., Kennedy, S.R., Salk, J.J., Fox, E.J., Hiatt, J.B., and Loeb, L.A. (2012). Detection of ultra-rare mutations by next-generation sequencing. Proc. Natl. Acad. Sci. U.S.A. *109*, 14508–14513.

Skums, P., Campo, D., Dimitrova, Z., Vaughan, G., Lau, D., and Khudyakov, Y. (2011–2012). Numerical detection, measuring and analysis of differential interferon resistance for individual HCV intra-host variants and its influence on the therapy response. In Silico Biol. *11*, 1–7.

Skums, P., Dimitrova, Z., Campo, D.S., Vaughan, G., Rossi, L., Forbi, J.C., Yokosawa, J., Zelikovsky, A., and Khudyakov, Y. (2012). Efficient error correction for next-generation sequencing of viral amplicons. BMC Bioinformatics *13* (Suppl. 10), S6.

Skums, P., Mancuso, N., Artyomenko, A., Tork, B., Mandoiu, I., Khudyakov, Y., and Zelikovsky, A. (2013). Reconstruction of viral population structure from next-generation sequencing data using multicommodity flows. BMC Bioinformatics *14* (Suppl. 9), S2.

Stanssens, P., Zabeau, M., Meersseman, G., Remes, G., Gansemans, Y., Storm, N., Hartmer, R., Honisch, C., Rodi, C.P., Böcker, S., *et al.* (2004). High-throughput MALDI-TOF discovery of genomic sequence polymorphisms. Genome Res. *14*, 126–133.

Steinhauer, D.A., and Holland, J.J. (1987). Rapid evolution of RNA viruses. Annu. Rev. Microbiol. *41*, 409–433.

Sundquist, A., Ronaghi, M., Tang, H., Pevzner, P., and Batzoglou, S. (2007). Whole-genome sequencing and assembly with high-throughput, short-read technologies. PLoS One *2*, e484.

Venter, J.C., Remington, K., Heidelberg, J.F., Halpern, A.L., Rusch, D., Eisen, J.A., Wu, D., Paulsen, I., Nelson, K.E., Nelson, W., *et al.* (2004). Environmental genome shotgun sequencing of the Sargasso sea. Science *304*, 66–74.

Westbrooks, K., Astrovskaya, I., Campo, D., Khudyakov, Y., Berman, P., and Zelikovsky, A. (2008). HCV quasispecies assembly using network flows. In Proceedings of the 4th International Conference on Bioinformatics Research and Applications (ISBRA'08), Măndoiu, I., Sunderraman, R., and Zelikovsky, A., eds. (Berlin Heidelberg, Springer-Verlag), pp. 159–170.

Zagordi, O., Geyrhofer, L., Roth, V., and Beerenwinkel, N. (2010a). Deep sequencing of a genetically heterogeneous sample: local haplotype reconstruction and read error correction. J. Comput. Biol. *17*, 417–428.

Zagordi, O., Klein, R., Daumer, M., and Beerenwinkel, N. (2010b). Error correction of next-generation sequencing data and reliable estimation of HIV quasispecies. Nucleic Acids Res. *38*, 7400–7409.

Zagordi, O., Bhattacharya, A., Eriksson, N., and Beerenwinkel, N. (2011). ShoRAH: estimating the genetic diversity of a mixed sample from next-generation sequencing data. BMC Bioinformatics *12*, 119.

Zagordi, O., Däumer, M., Beisel, C., and Beerenwinkel, N. (2012a). Read length versus depth of coverage for viral quasispecies reconstruction. PLoS One *7*, e47046.

Zagordi, O., Töpfer, A., Prabhakaran, S., Roth, V., Halperin, E., and Beerenwinkel, N. (2012b). Probabilistic inference of viral quasispecies subject to recombination. In RECOMB 2012, LNBI 7262, Chor, B., ed. (Berlin Heidelberg, Springer-Verlag), pp. 342–354.

DNA Instability in Bacterial Genomes: Causes and Consequences

13

Pedro H. Oliveira, Duarte M.F. Prazeres and Gabriel A. Monteiro

Abstract

DNA is a structurally dynamic molecule that is central to cellular processes such as replication, transcription and recombination. In order to maintain genomic integrity, bacteria have developed a finely tuned and interwoven network of mechanisms that operate at multiple levels, including damage recognition, signalling pathways, and DNA repair. On the other hand, without the capacity to accommodate genotypic variation up to a certain extent, bacteria would not be able to modify their fitness when faced with constantly changing environments. Herein we review our current knowledge on bacterial genome instability, with particular emphasis on findings gained from the often-studied Gram-negative model organism *Escherichia coli*. We will address topics such as spontaneous and stress-induced mutagenesis, major DNA repair pathways, and the design of more stable genomes. Major questions and future challenges will also be discussed.

Introduction

Bacterial DNA is recurrently exposed to insults of both endogenous and exogenous sources. Endogenous factors known to damage DNA in a spontaneous fashion, i.e. in actively dividing cells and in the absence of exogenous stressors, typically include DNA replication errors (Bzymek and Lovett, 2001; Hardin *et al.*, 2007), movement of mobile elements (Naas *et al.*, 1995), and lesions such as depurination, depyrimidation, deamination, alkylation and oxidation (Loeb and Preston, 1986; Cooke *et al.*, 2003; Drablos *et al.*, 2004; Gates, 2009). Many of these events are context specific, in the sense that they are highly dependent on the local DNA sequence and topology, as well as on the extent of replication, transcription and recombination activities. Typically, the overall rate of spontaneous mutations in wild-type *Escherichia coli* is maintained at low levels (around 10^{-3} per genome per generation) (Lee *et al.*, 2012), mostly due to the replication fidelity of the DNA polymerase III (Pol III) holoenzyme, as well as to the methyl-directed mismatch repair (MMR), base excision repair (BER) and nucleotide excision repair (NER) systems. Exogenous stresses, on the other hand, consist of alkylating agents (Drablos *et al.*, 2004), antimicrobials (Lopez and Blazquez, 2009), nutrient deprivation (Mandel and Silhavy, 2005), oxidative stress (Cooke *et al.*, 2003), ionizing (e.g. X-rays, gamma rays and atomic particles) (Tindall *et al.*, 1988; Bentchikou *et al.*, 2010) or ultraviolet (UV) radiation (Eichenbaum and Livneh, 1998), extreme temperature, pH and osmolarity shifts (Jolivet-Gougeon *et al.*, 2000; Ohtsubo *et al.*, 2005), among others. Under such stressful and adverse environments, the need to generate diversity, fuels maladapted cells to transiently up-regulate their mutation rates. This process of stress-induced mutagenesis varies according to the organism and persistence/intensity of the stress condition, and generally involves relaxing the consistency and fidelity of the DNA repair machinery, the use of error-prone polymerases, a shift to a transient hypermutable state, induced transposition of mobile elements and other rearrangements. For example, in the Gram-negative *E. coli* bacterium, a considerable number of mutations appearing in stressed cells results from error-prone repair of DNA double strand breaks (DSBs), in a process controlled by the SOS DNA-damage response and the general stress response sigma factor RpoS (Galhardo *et al.*, 2009; Rosenberg *et al.*, 2012).

Making sense of the biological significance and utility of mutational events and associated mechanisms is therefore fundamental to answer long-standing questions in multiple fields such as those dealing with evolution and species diversity, cancer progression, synthetic biology, among many others. This book chapter is designed to present a brief overview of the causes and consequences of spontaneous and stress-induced genomic instability in bacteria. In addition, we will address topics that are of utmost relevance to those in the biopharmaceutical sector engaged in the development of streamlined genomes and DNA-based molecules for research and therapeutic applications.

Spontaneous DNA lesions, replication and transcription as sources of instability

Spontaneous DNA lesions

Several types of DNA damage can be generated endogenously as a result of intrinsic DNA instability or the presence of by-products of the cellular metabolism. Examples of this are depurination, depyrimidation, deamination, alkylation or oxidation. Depurination and depyrimidation consist in the hydrolysis of the *N*-glycosidic bond between the deoxyribose moiety and guanine or adenine (depurination) or cytosine and thymine (depyrimidation). This leads to the formation of the so-called apurinic or apyrimidinic (AP or abasic) sites, the latter being formed at roughly 5% the rate of the former (Lindahl, 1993). AP sites are highly mutagenic and constitute a roadblock to replication and transcription (Loeb and Preston, 1986). Additional base modifications may occur through the spontaneous hydrolytic deamination of adenine, guanine, cytosine, and the modified base 5-methylcytosine (5meC), respectively yielding hypoxanthine, xanthine, uracil and thymine (Coulondre *et al.*, 1978; Duncan and Miller, 1980) (Fig. 13.1). Deamination of cytosine in double-stranded DNA (dsDNA) proceeds at a rate of about 7×10^{-13}/s (Frederico *et al.*, 1990), while 5meC deaminates 2–3 times faster (Shen *et al.*, 1994). Both are capable of generating G:C>A:T transitions. On the other hand, deamination of adenine and guanine occur at much slower rates (approximately 2–3%) than those observed for cytosine (Lindahl, 1979). Adenine deamination may result in A:T>G:C transitions, whereas deamination of guanine to xanthine may lead to DNA replication arrest and G:C>A:T transitions due to its ability to mispair with thymine. Hypoxanthine may also pair with cytosine leading to A:T>G:C transitions. Interestingly, deamination occurs at faster rates in single-stranded DNA (ssDNA) than in dsDNA, meaning that cellular processes such as replication, transcription or recombination have the potential to exacerbate those events (Frederico *et al.*, 1990; Beletskii and Bhagwat, 1996).

Nucleophilic sites in DNA represent potential targets to both endogenous and exogenous electrophilic alkylating agents. Examples of endogenous alkylating agents include the ubiquitous methyl donor S-adenosylmethionine (SAM) (Rydberg and Lindahl, 1982; Taverna and Sedgwick, 1996) as well as nitrosated amines or amides (Taverna and Sedgwick, 1996), that are capable of generating lesions in dsDNA such as O6-methylguanine (O6meG), 3-methyladenine (3meA) and 7-methylguanine (7meG) (Fig. 13.1). O6meG is known to cause G:C>A:T transitions (Loechler *et al.*, 1984), whereas 3meA has the ability to block replication and generate AP sites (Boiteux *et al.*, 1984; Larson *et al.*, 1985). 7meG is *per se* regarded as non-cytotoxic and non-mutagenic, but is capable of destabilizing the *N*-glycosidic bond and generate AP sites (Shrivastav *et al.*, 2010). In *E. coli*, O6meG lesions can be repaired by two DNA methyltransferases, the inducible Ada protein and the constitutive Ogt protein, whereas 3meA and 7meG are repaired by the Tag and AlkA glycosylases of the BER pathway (see below).

Reactive oxygen species (ROS) such as hydroxyl radicals (•OH), singlet oxygen ($^1O_2$), superoxide (O_2•$^-$), and hydrogen peroxide (H_2O_2) are by-products of the aerobic metabolism that are also capable of exerting a genotoxic effect on nucleic acids. The most intensively studied modified base is probably 8-oxo-7,8-dihydroguanine

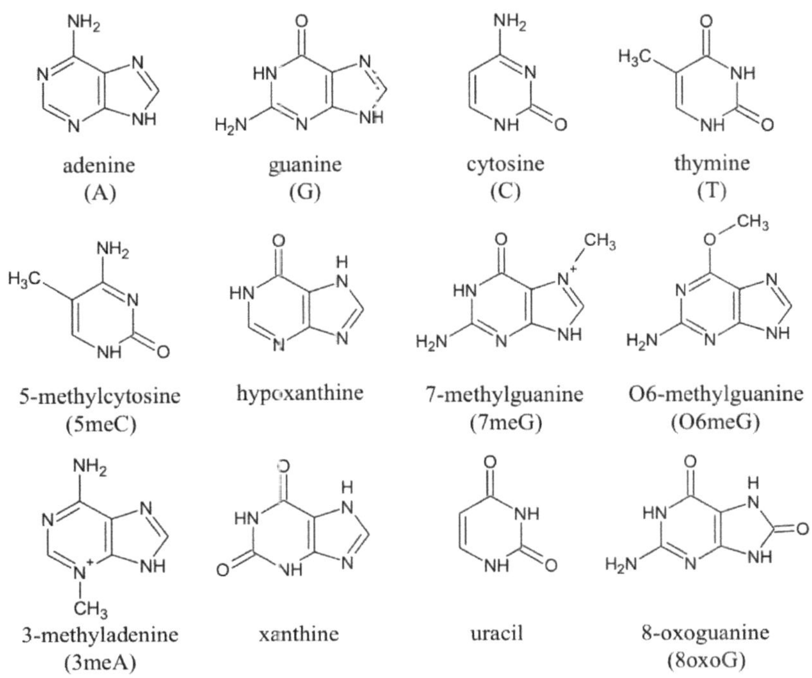

Figure 13.1 Structures of the DNA bases and of common endogenous base lesions.

(8-oxoguanine, 8oxoG), the product of guanine oxidation (Fig. 13.1). 8oxoG efficiently pairs with both cytosine and adenine mainly resulting in G:C>T:A transversions (Shibutani *et al.*, 1991). Many additional base lesions have been identified as the result of oxidative damage, and have been properly reviewed elsewhere (Cooke *et al.*, 2003).

The BER pathway is generally involved in the repair of structurally non-bulky damaged bases that do not cause major distortions to the double helix (e.g. after base excision, oxidation, alkylation) (reviewed by (Stivers and Jiang, 2003). In BER, mono- or bifunctional DNA glycosylases recognize and excise damaged bases by cleavage of the *N*-glycosidic bond, producing AP sites that are then processed by AP endonucleases, DNA polymerase and ligase to restore the original sequence (see Fig. 13.2 for further details). In order to mitigate the detrimental effect of 8oxoG, several bacteria make use of the oxidized guanine (GO) system, which typically comprises three proteins: MutT, MutM and MutY. MutT is a hydrolase that converts 8-oxoGTP to 8-oxoGMP and pyrophosphate in order to prevent the incorporation of the oxidized dGTP into the DNA (Maki and Sekiguchi, 1992). Conversely, MutM and MutY are glycosylases pertaining to the BER pathway. MutM excises 8oxoG when paired to cytosine, whereas MutY catalyses the excision of adenine when mismatched with 8oxoG (Michaels *et al.*, 1992; Tajiri *et al.*, 1995). The MMR system (see Fig. 13.3) corrects postreplicative errors but has also been shown to be involved in the direct excision of 8oxoG, mispaired adenines opposite 8oxoG, or both (Wyrzykowski and Volkert, 2003). Further details on these and other base lesions have been previously reviewed (Gates, 2009; Shrivastav *et al.*, 2010).

Replication errors occurring during DNA biosynthesis

Some of the most common errors generated during DNA biosynthesis are point mutations, which can appear either in the form of base substitutions or insertions and deletions (indels). Among the latter, those that cause a change in the reading frame are termed frameshift mutations. Frameshift mutations can be found in

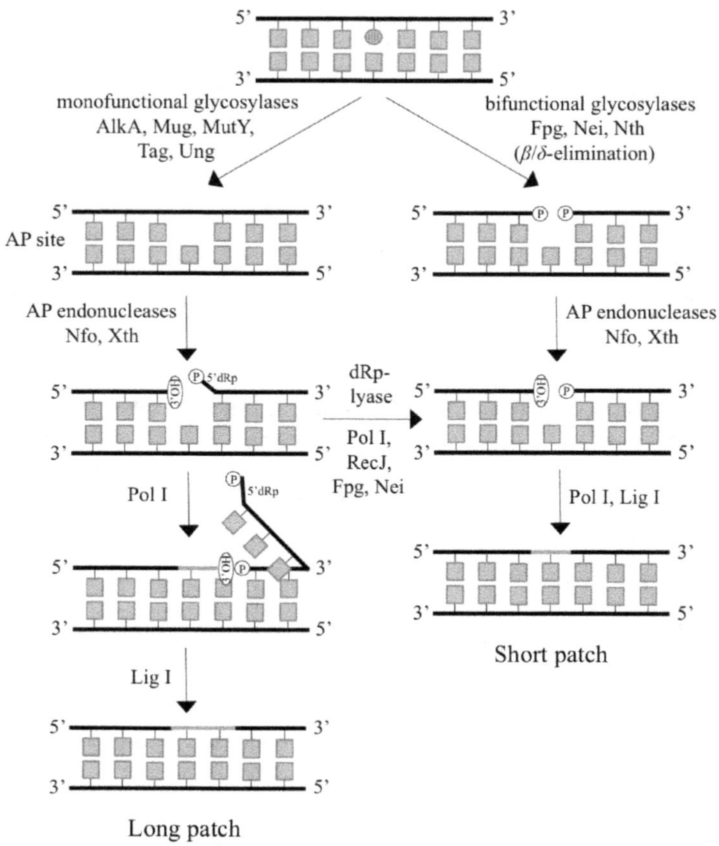

Figure 13.2 Scheme for the long-patch and small-patch BER pathways. The first step in BER involves the recognition and removal of base lesions (stippled circle) either by monofunctional glycosylases (for deaminated, alkylated or mismatched bases) or bifunctional glycosylases (for oxidized, or ring-saturated bases). Monofunctional glycosylases generate AP sites, which are subsequently cleaved at their 5′-side by an AP endonuclease leaving a 3′-hydroxyl (3′-OH) group and a 5′-deoxyribose phosphate (5′-dRp) terminus. Bifunctional glycosylases, on the other hand, possess AP lyase activity, with an associated β- or β/δ elimination activity (being the latter shown in the figure). In this situation, a single nucleotide gap is generated containing 5′- and 3′-phosphate (P) groups at the termini. Two distinct pathways are followed according to the type of glycosylases used in the process: if the repair is initiated by monofunctional glycosylases, it may follow both the long and short-patches, whereas repair initiated by bifunctional glycosylases may only follow the short-patch. In the long-patch pathway, the 5′-dRp and about 2–10 nucleotide patches are replaced during strand displacement by Pol I and subsequently ligated. On the other hand, in the short-patch pathway mediated by monofunctional glycosylases, the 5′-dRp terminus is removed by Fpg, Nei, RecJ or by the exonuclease activity of Pol I, generating a 5′-phosphate terminus. In the short-patch pathway mediated by bifunctional glycosylases, the 3′-phosphatase/phosphodiesterase activity of AP endonucleases will remove the corresponding blocking phosphate. Gap-filling and ligation steps then take place.

runs of repetitive DNA, and in this case, can be explained on the basis of misalignment between a nascent strand and its template, as proposed in the seminal work of Streisinger and colleagues (Streisinger *et al.*, 1966). Such misaligned structures allow the formation of loops which, according to their size, can be recognized and resolved by the MMR system (Lovett and Feschenko, 1996). In order to accommodate those situations in which frameshift mutations do not occur within homopolymeric DNA sequences, alternative models have been proposed. According to the so-called dislocation model, frameshift mutations could be initiated by direct misincorporation of a nucleotide (Kunkel and Soni, 1988) (Fig. 13.4a). It

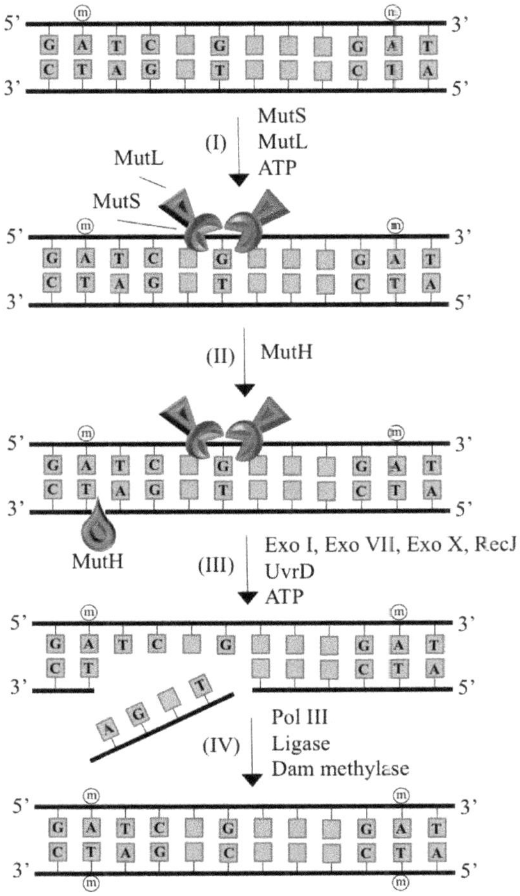

Figure 13.3 Scheme of the methyl-directed mismatch repair (MMR) pathway in *E. coli*. A base mismatch is recognized by the MutS homodimer, followed by recruiting of the MutL homodimer to the site of the lesion in a MutS- and ATP-dependent manner (I). Next, MutH is recruited to the nearest hemi-methylated (m) d(GATC) site and nicks the corresponding newly synthesized strand (non-methylated) (II). Because the nearest hemi-methylated d(GATC) site may be located in either sides of the lesion, the following step of strand degradation will involve UvrD (DNA helicase II) and different ssDNA exonucleases (III). Exo I, Exo VII and Exo X are involved in 3′→5′ degradation, whereas Exo VII and RecJ are involved in 5′→3′ degradation. DNA Pol III and ligase will fill the gap, and a Dam methylase will methylate the newly synthesized DNA strand (IV).

then follows that the misincorporated base may realign and pair with an adjacent complementary template, ultimately leading to an indel. Alternatively, base substitutions may originate if strand polymerization occurs in the absence of realignment (Fig. 13.4a). Many examples of spontaneous small-scale mutations have been reported to occur in a wide range of bacteria, such as in the *trpA* (Hardin *et al.*, 2007) and *lacI* (Schaaper and Dunn, 1991) genes of *E. coli*, in the *pyrE/pyrF loci* of *Thermus thermophilus* (Mackwan *et al.*, 2008), or in the *hisD3052* allele of *Salmonella typhimurium* (DeMarini *et al.*, 1998).

By a process resembling that of frameshift mutagenesis, slipped misalignment may take place over longer distances between interspersed DNA repeats. Such events can result in deletions or amplifications (deletion formation) of both repeat and spacer (intervening) sequence, in a process that is independent of the homologous recombination (HR) protein RecA, but highly dependent on repeat and spacer lengths (Dianov *et al.*, 1991; Mazin *et al.*, 1991; Bi and Liu, 1994; Chedin *et al.*, 1994) (Fig. 13.4b). Misalignment between direct repeats may also take place across a replication fork, in a process involving reciprocal sister-chromosome exchange (Lovett *et al.*, 1993; Bi and Liu, 1996; Feschenko and Lovett, 1998). Cross-fork slippage accompanied by sister-chromosome exchange involves an erroneous annealing between repeats from both nascent strands (see Fig. 13.4c for more details), and has been used to explain the formation of heterodimeric products in plasmids harbouring direct repeats (Morag *et al.*, 1999). In this regard, our group has previously reported the contamination of purified batches of a therapeutic non-viral plasmid DNA (pDNA) vector with deleted (monomeric, M) and head-to-tail heterodimeric (1+2, 1+3) forms (Ribeiro *et al.*, 2008; Oliveira *et al.*, 2010). The latter represent a safety concern, since there is the risk to administer small populations of unwanted plasmid molecules with unknown biological properties. These events may also lead to a loss of productivity during the expression of proteins of interest (Kawe *et al.*, 2009; Kim *et al.*, 2011). The frequency of such repeat-mediated recombination events can be estimated by using predictive mathematical tools (Oliveira *et al.*, 2008), and experimentally detected and/or quantified using techniques such as direct sequencing, PCR-based enrichment steps, temperature- or

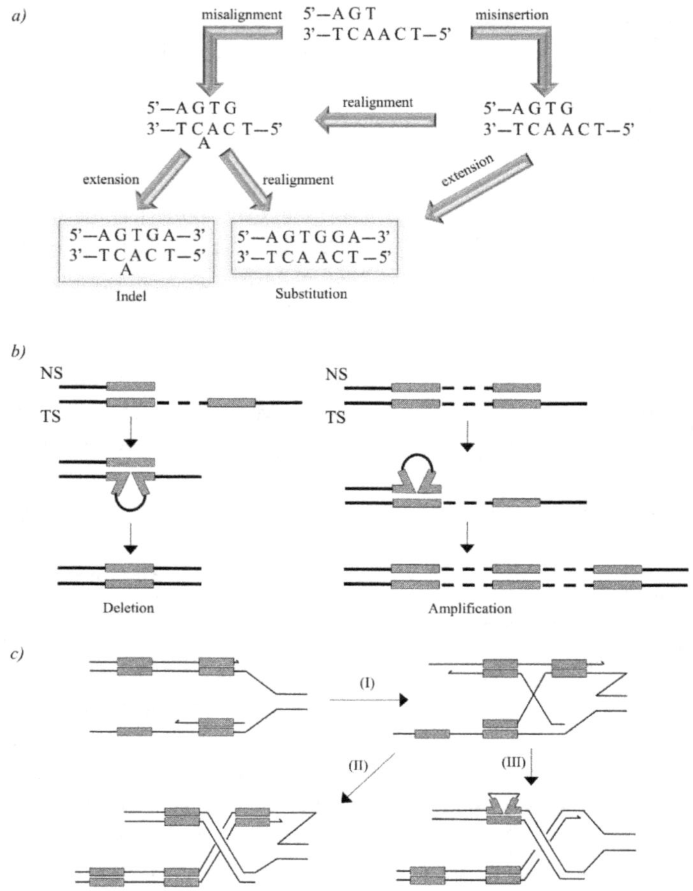

Figure 13.4 Slipped-misalignment events in DNA. (*a*) Indels and substitutions can be generated either by strand misalignment as initially proposed by Streisinger and co-workers (Streisinger *et al.*, 1966), or misinsertion as proposed in the dislocation model (Kunkel and Soni, 1988). (*b*) Slipped misalignment between interspersed direct repeats (rectangles) may take place when the nascent strand (NS) disengages from the template strand (TS) and re-anneals at a new downstream or upstream position, respectively generating deletion or amplification of genetic material. The spacer (or intervening) sequence between repeats is shown as an intermittent line. (*c*) Model of slipped misalignment accompanied by sister chromosome exchange. In a stalled replication fork, unequal pairing between nascent 3′-ends may take place involving repeated sequences (I). The intermediate form yielding a Holliday junction can be further processed in the absence of realignment (II) or with realignment of parental strands (III). This respectively leads to the formation of one deletion/triplication product or a deletion/duplication product. In circular molecules, the latter are termed 1+3 and 1+2 respectively. (b) and (c) were adapted from Bzymek and Lovett (2001).

denaturing-gradient gel electrophoresis, and real-time PCR (reviewed in (Oliveira *et al.*, 2011). In a very recent study using 454 pyrosequencing technology, large-scale spontaneous inversions, deletions and duplications were found at high frequency (20–40%) in growing cultures of *S. enterica* (Sun *et al.*, 2012), which highlights the importance of performing these types of analysis on a more frequent basis.

Instability mediated by non-B structures

Depending on its sequence and surrounding conditions, a DNA molecule has the ability to occasionally adopt several conformations beside the canonical right-handed double stranded B-form. These commonly include bent-inducing sequences, slipped structures, cruciforms, G-quadruplexes, triplexes and Z-DNA (Fig. 13.5).

Such non-canonical or non-B structures arise from separation of duplex DNA and negative supercoiling induced by several biological processes such as replication, transcription or protein binding. In addition, they have also been considered as 'fragile regions' prone to rearrangements.

Slipped-strand structures originate whenever a displaced nascent strand misaligns and pairs with an alternative complementary region in the template strand (previously shown in Fig. 13.4b). Cruciforms are typically generated by two main mechanisms (Lilley, 1985). The most common one, which is observed under physiological ionic conditions (S-type), involves a pre-nucleation unwinding step of 8 to 10 central base pairs, intrastrand hydrogen bonding between cognate bases flanking the symmetry centre (nucleation), and a final extrusion and complete intrastrand annealing between inverted repeats (Fig. 13.5a). Alternatively, the C-type mechanism has been observed under low-salt concentrations, and requires a long AT-rich DNA region flanking the inverted repeats, that allows the formation of a large bubble where nucleation and extrusion may take place. If ssDNA becomes available (e.g. during replication, DNA repair, or during uptake of exogenous DNA), single hairpins may be formed instead (Fig. 13.5b), with a stability that is proportional to its length and sequence context (Gacy *et al.*, 1995). Long inverted repeats have been shown to be replication stall sites both in prokaryotes and eukaryotes (Voineagu *et al.*, 2008). The authors demonstrate that hairpins, rather than cruciforms are preferentially formed, and thus should be regarded as the main agents of instability. Notwithstanding, hairpin DNA structures can be recognized and cleaved by the SbcCD nuclease (Connelly *et al.*, 1998), leaving a DSB for subsequent repair by the RecBCD pathway of HR (Eykelenboom *et al.*, 2008).

Concerning triplex structures, they can form at homopurine–homopyrimidine stretches, where a single strand portion can bind to the purine-rich strand of the duplex via non-canonical hydrogen bonding (Fig. 13.5c). Different triplexes can thus be formed according to the source of the triplex-forming strand: intermolecular if it originates from a second DNA molecule (e.g.

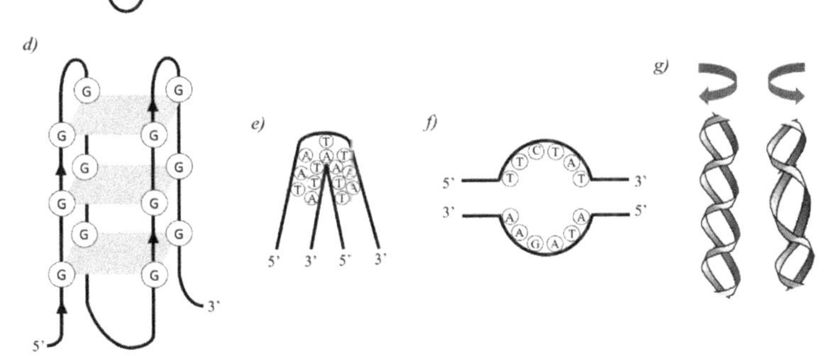

Figure 13.5 Examples of non-B (non-canonical) DNA structures. (a) Cruciform; (b) hairpin; (c) triplex or H-DNA (shown is the Y*R:Y structure formed by Hoogsteen hydrogen bonding); (d) intramolecular tetraplex, G4-DNA or G-quadruplex with diagonal loop (e) bent DNA; (*f*) unwound DNA; (g) comparison between B-DNA (left, clockwise) and Z- or left-handed DNA (right, anticlockwise). Figure was partially adapted from (Oliveira *et al.*, 2012).

triplex-forming oligonucleotides (TFOs)), or intramolecular if it involves regions with mirror symmetry (H-DNA) that originate from the same duplex strand. Irrespectively of the scenario, the third strand can be either pyrimidine-rich and bind with a parallel orientation to the purine strand in the duplex (Y*R:Y) via Hoogsteen base pairing, or purine-rich and bind with an anti-parallel orientation to the purine strand in the duplex (R*R:Y) via reverse Hoogsteen base pairs. It has been documented that H-DNA induces a yet unclear cascade of events encompassing the formation of DSBs, processing via non-homologous end-joining (NHEJ), and eventual formation of large-scale deletions or translocations (Jain *et al.*, 2008). Triple-helical structures have also been recognized as hot spots for the integration of mobile elements such as Tn7 (Rao *et al.*, 2000).

Tetraplex, G4-DNA or G-quadruplex structures (Fig. 13.5d) are stable stacks of coplanar G-quartets stabilized by monovalent cations, originating from transient single-stranded DNA bearing a signature motif of the type $G \geq_3 N_{1-7} G \geq_3 N_{1-7} G \geq_3 N_{1-7} G \geq_3$ or $C \geq_3 N_{1-7} C \geq_3 N_{1-7} C \geq_3 N_{1-7} C \geq_3$. Based on the number of DNA strands involved in quadruplex formation, these structures may be classified as intermolecular (two or four strands) or intramolecular (one strand), whereas loops may be classified as edgewise, diagonal or double-chain reversal. G-quadruplexes have been shown to interfere with important biological processes like replication, transcription, recombination and telomere protection, thus being regarded as potential agents of genetic instability (Han and Hurley, 2000; Belotserkovskii *et al.*, 2010; Boan and Gomez-Marquez, 2010). In addition, several genes involved in secondary metabolite biosynthesis and signal transduction, seem to be under the control of G4 motifs in *E. coli* (Rawal *et al.*, 2006).

The presence of properly phased A_nT_m ($n + m \geq 4$) tracts often leads to intrinsically curved regions (Fig. 13.5e), which play an important role in modulating replication and transcription initiation (Linial and Shlomai, 1988; Kahn and Crothers, 1993), and serve as hot targets for the insertion of mobile elements (Kobori *et al.*, 2009; Lewis *et al.*, 2012). Conversely, extremely AT-rich sequences are able to promote unwinding of DNA with increasing negative superhelical stress (Fig. 13.5f), and are highly hypersensitive to attack by nucleases (Kowalski *et al.*, 1988).

Alternating purine-pyrimidine repeat tracts such as $(CA:TG)_n$ and $(CG:CG)_n$ may fold into a left-handed structure with about 12 bp per turn called Z-DNA (Fig. 13.5g). Such structures have been shown to block transcription (Ditlevson *et al.*, 2008), and induce the formation of DSBs, resulting in gross deletions and complex rearrangements in mammalian cells, but only small deletions or expansions within the repetitive sequences in *E. coli*. Further processing either by NHEJ or HR is thought to be behind the different outcomes (Kha *et al.*, 2010).

Triplexes, G-quadruplexes and three- or four-way junctions have been described to be resolved by RecQ helicases, a ubiquitous subgroup of helicases highly conserved from prokaryotes to higher eukaryotes that ensures the faithful progression of the replication fork (Sharma *et al.*, 2006; Sharma, 2011).

A number of *ab initio* and consensus-based bioinformatics resources have been developed aiming at providing information about the presence of non-B elements. Several *k*-mer and seed extension approaches have been developed to detect repeated regions. Some examples include Reputer (Kurtz and Schleiermacher, 1999), Repeatoire (Treangen *et al.*, 2009) and Repseek (Achaz *et al.*, 2007) for the detection of exact or degenerate repeats, or Tandem Repeats Finder (Benson, 1999) for tandem repeats. Others focus on identifying elements such as triplex DNA, Z-DNA and quadruplexes (e.g. TTS (Jenjaroenpun and Kuznetsov, 2009), QuadParser (Huppert and Balasubramanian, 2005)). More recently, a database named Non-B DB has been developed, which takes into consideration all classes of non-B DNA-forming motifs, and provides the user with a search option for any input sequence as well as access to data in its genomic context (Cer *et al.*, 2011). Also, several experimental techniques such as two-dimensional gel electrophoresis, circular dichroism, nuclease cleavage, and others, have been used to detect non-B structures (Wang *et al.*, 2009).

Replication–transcription conflicts

In bacteria, the rate of replication is about 12–30 times higher than that of transcription (respectively 600 nucleotides/s versus 20–50 nucleotides/s) (Mirkin and Mirkin, 2007; Rudolph *et al.*, 2007), which leads to frequent physical conflicts between the replication and transcription machineries as well as occasional genomic rearrangements. While replisome and RNA polymerase (RNAP) collisions may occur either in a head-on or co-directional fashion, a greater impairment and stronger pausing of the replication fork is often seen in head-on collisions rather than in co-directional conflicts (Mirkin and Mirkin, 2005). To this end, bacterial genomes have evolved towards an over-representation of long, essential, and highly transcribed genes in the leading strand, presumably in an effort to limit head-on collisions (Brewer, 1988; Rocha and Danchin, 2003; Omont and Kepes, 2004; Price *et al.*, 2005). Studies performed *in vitro* show that co-directional collisions are often resolved by rapidly dislodging the RNAP from the template and resuming elongation using mRNA as a primer (Pomerantz and O'Donnell, 2008) (Fig. 13.6). However, *in vivo* studies performed in *E. coli*, revealed a less benign face of co-directional encounters, by showing that such thermodynamically stable DNA:RNA hybrids (R-loops) may cause replication fork stalling (Gan *et al.*, 2011) and favour the accumulation of DSBs mutations, particularly as a result of collisions with backtracked elongation complexes (Dutta *et al.*, 2011). RNAP backtracking allows the formation of longer and more stable R-loops, which provide accessible 3′-OH termini that can be used as primers for DNA synthesis (Dutta *et al.*, 2011). This leads to the formation of a single-strand break

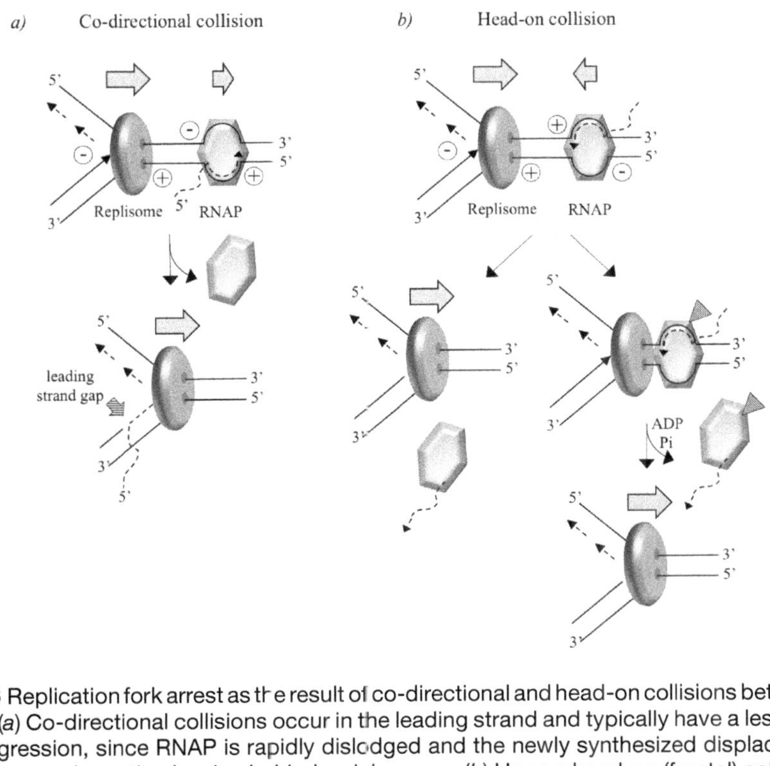

Figure 13.6 Replication fork arrest as the result of co-directional and head-on collisions between the replisome and RNAP. (*a*) Co-directional collisions occur in the leading strand and typically have a less detrimental effect on fork progression, since RNAP is rapidly dislodged and the newly synthesized displaced RNA is used as primer to resume elongation leaving behind a nick or gap. (*b*) Upon a head-on (frontal) collision in the lagging strand, the replisome stalls for longer periods of time because of the physical interaction with RNAP and the excess of positive supercoiling between both complexes (shown as plus signals). The replisome may then resume elongation after slowly displacing RNAP (left), or alternatively, the transcription-repair coupling factor Mfd (shown as a triangle) may bind to the halted transcription complex and force its dissociation in a process catalysed by ATP hydrolysis.

(SSB) that can be converted into a DSB after a second round of replication. To prevent excessive instability, the cells make use of several anti-backtracking mechanisms, namely termination factors and anti-termination complexes, among others (Dutta *et al.*, 2011). *In vivo* studies performed with *B. subtilis* also describe replication disruption and restart upon co-directional collisions at highly expressed ribosomal genes (Merrikh *et al.*, 2011). *In vivo* head-on collisions, on the other hand, induce a profound fork arrest eventually followed by fork collapse and recombination (Vilette *et al.*, 1996; Prado and Aguilera, 2005). This fact results from the strong physical contact and the positively supercoiled constraining domain generated between both machineries, being the former more detrimental than the latter. Further evidence on the fate of both replisome and RNAP after head-on collisions came from a study performed *in vitro* in *E. coli* (Pomerantz and O'Donnell, 2010). In this situation the replisome pauses but slowly resumes elongation after displacing RNAP from DNA (Fig. 13.6). This process of RNAP displacement and fork progression may be aided by the transcription-repair coupling factor Mfd in an ATP-driven reaction (Fig. 13.6). Recently it was shown that blocked forks arising from head-on collisions, may be reversed and further processed by HR or DNA degradation, allowing the reloading of a new replisome further away from the RNAP (De Septenville *et al.*, 2012).

Stress-induced instability

The concept of adaptive or stationary-phase mutation

Cells trying to cope with a changing environment find themselves in the need to effectively fine-tune the balance between genomic integrity and the chance of acquiring beneficial changes. According to the more classical view of DNA mutation, newly arising mutations occur spontaneously in growing cells, in a constant and stochastic manner, and in the absence of an environment in which such mutations might prove advantageous. This view has been challenged by the finding that mutations may also appear in non-growing or slowly growing cells under conditions that favour such mutant subpopulation (Shapiro, 1984; Cairns *et al.*, 1988). Nevertheless, these so-called adaptive or stationary-phase mutations accumulate at the same time that unselected mutations, and therefore, should not be regarded as directed to a particular advantageous gene (Foster, 1997; Torkelson *et al.*, 1997; Rosenberg, 2001). While several mechanisms of adaptive genetic changes have been described in bacteria, those occurring at the lactose catabolism (*lac*) operon in starving *E. coli* (adaptive point mutation and gene amplification) are among the best studied (Foster and Trimarchi, 1994; Hastings *et al.*, 2000; Powell and Wartell, 2001; Slack *et al.*, 2006; Gonzalez *et al.*, 2008). In this assay, cells carrying deleted chromosomal *lac* genes and an episomal *lac* +1 frameshift allele, respond to starvation on a lactose medium either by generating a compensatory frameshift (point mutation) Lac$^+$ allele (Foster and Trimarchi, 1994), or by allowing the amplification of the leaky *lac* allele to 20–100 copies (Hastings *et al.*, 2000). The mechanism underlying the appearance of adaptive point mutations presumes the differentiation towards a hypermutable subpopulation of cells (i.e. mutagenesis becomes up-regulated), during the repair of DNA lesions such as DSBs or double stranded ends (DSEs) via HR (Ponder *et al.*, 2005; Gonzalez *et al.*, 2008). The mutations are permanent and not directed preferentially to *lac* genes, as many unselected mutations also appear (Foster, 1997; Torkelson *et al.*, 1997). In this scenario there is an activation of the SOS and RpoS (σ^S)-mediated general stress responses, which leads the cell to shift to a mutagenic mode that makes use of the error-prone DNA Pols II (*polB*), IV (*dinB*) and V (*umuDC*) instead of the high-fidelity DNA Pol III (Lombardo *et al.*, 2004; Ponder *et al.*, 2005) (Fig. 13.7a). In this regard, it was recently shown that DSBs concentrate stress-inducible mutations within vicinal hotspots: the so called 'strong ones' are located at approximately 60 kb and arise via RecD-dependent resection of DSBs followed by gap-filling synthesis, and the 'weak ones' are located at roughly 1 Mb distance from a DSB and are formed via break-induced replication (BIR) (Shee *et al.*, 2012). To this matter, the authors also added that single-base differences observed preferentially at poorly expressed genes of the *E. coli* genome may result from the fact that

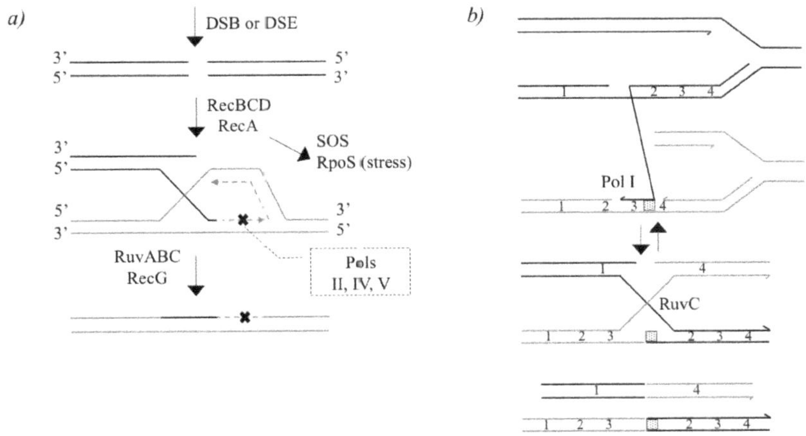

Figure 13.7 Models for DSB-dependent stress-induced point mutations (*a*) and gene amplification (*b*). In the initial stages involved in the repair of a DSB or DSE by homologous recombination (HR), RecBCD produces ssDNA and the RecA protein is loaded. The presence of DSB or DSE induces the SOS response, and the presence of a second stress (e.g. starvation) induces the general stress response mediated by the RpoS activator. This concatenation of events leads to a switch to a mutagenic mode that makes use of the error-prone DNA Pols II, IV and IV, instead of the high-fidelity DNA Pol III. Strand invasion then takes place with a sister molecule or duplicated region, and replication resumes, allowing for the accumulation of mutations (shown as X). On the other hand, the long-distance template-switching model has been proposed as a possible mechanism for adaptive gene amplification. According to this model, stalled replication forks may originate under starvation and in the presence of DSBs. In this situation, a 3′-end from a lagging strand might invade a template strand from a nearby fork, in a process aided by the flap-endonuclease function of Pol I and by a region of microhomology (stippled square). Such invasion is further stabilized by ligation to the 5′-extremity of the Okazaki fragment located ahead. The switch-strand intermediate then forms a Holliday junction that can be resolved by RuvC, yielding a duplicated and a deleted molecule. Schemes (a) and (b) were respectively adapted from Rosenberg *et al.* (2012) and Slack *et al.* (2006).

the latter are more DSB-prone, and thus more mutagenic (Shee *et al.*, 2012). In the mechanism of DSB-dependent adaptive gene amplification, the cells acquire an iterated duplication of the *lac* region. Unlike the *lac* point mutation mechanism, the cell does not enter a hypermutable state, does not require the activation of the SOS response nor Pol IV, and does not make use of HR to generate tandem duplications (Hastings *et al.*, 2000; Lombardo *et al.*, 2004; Slack *et al.*, 2006) (Fig. 13.7b). Amplifications appear to be the result of non-HR and are temporary, in the sense that they can revert to a single copy status by recombination.

Alternative mutagenic pathways have been proposed to take place in both growing and stationary-phase cells. In resting organisms in structured environments (ROSE), *E. coli* cells growing in colonies were found to acquire a rifampicin-resistant phenotype, in a process dependent of cAMP and the SOS regulon (Taddei *et al.*, 1997). In the growth advantage in stationary-phase (GASP) phenotype, a minor population of *E. coli* cells is able to outcompete their counterparts, presumably through an increased ability to catabolize some amino acids. Mutations in the GASP phenotype were found to require Pols II, IV and V, although not requiring the induction of the SOS system and DNA damage (Yeiser *et al.*, 2002).

Environmentally stressed bacteria have also evolved systems allowing for a local increase of mutagenesis at specific sites of the genome. In the next sections we will briefly comment on these 'contingency *loci*', and enumerate several environmental stressors, which have been described to induce mutagenesis.

Contingency loci

Contingency *loci* refer to specific hypermutable sites in a genome that are implicated in stochastic and reversible on–off phase variation events. Simple sequence repeats (SSRs) are the bacterial equivalents to eukaryotic microsatellites, being composed of tandem repeats of motifs sizing 1–6 bp. SSR-mediated indels arise by slipped-strand mispairing during replication (see Fig. 13.4a and b), and have the potential to, under environmental stress, favourably modulate gene expression by altering reading frames or promoter strength (Li *et al.*, 2004; Martin *et al.*, 2005; Vinces *et al.*, 2009). In the recent decades, whole-genome sequencing of several prokaryotes highlighted curious asymmetries concerning the length, abundance and distribution of SSRs. Particularly, it was observed that (i) longer SSRs are much less frequent in prokaryotes than eukaryotes and tend to be excluded from coding regions; (ii) smaller and AT-rich genomes tend to have an overrepresentation of SSRs when compared to larger and more GC-rich genomes; (iii) tetranucleotide SSRs are typically overrepresented in pathogenic bacteria (Coenye and Vandamme, 2005; Mrazek *et al.*, 2007). As an example, the small 1.8-Mb genome of the commensal bacterium *Haemophilus influenzae* has revealed a surprisingly high density of tetranucleotide SSRs, particularly associated with genes coding for virulence-like factors (Fleischmann *et al.*, 1995; Hood *et al.*, 1996).

Certain SSRs described as variable number of tandem repeats (VNTR) have been used as genetic markers owing to their inter-individual polymorphism. In particular, multilocus VNTR analysis (MLVA) typing assays are particularly useful in epidemiological and evolutionary studies, and essentially consist of a PCR amplification of a predefined set of VNTRs, for which the length and copy number can be estimated. To aid in this process, several bioinformatic resources are currently available that focus not only on the search for repetitive DNA (e.g. Tandem Repeats Finder; Benson, 1999), but also on assisting with the comparison between strains (Denoeud and Vergnaud, 2004; Chang *et al.*, 2007).

Use of antibiotics

Antibiotics are commonly used selectors of resistant individuals. Yet, compelling evidence has built up showing that they are also agents of bacterial evolvability. Antibiotic-induced mutagenesis seems to proceed through the induction of the SOS response, increased transposition of mobile elements, recombination and lateral gene transfer. As an example, studies involving exposure of *E. coli* cells to the broad-spectrum fluoroquinolone antibiotic ciprofloxacin have described the onset of adaptive mutations (Riesenfeld *et al.*, 1997) and provided insight into the mechanisms governing the emergence of resistance-conferring mutations. This response is mediated by cleavage of the LexA repressor, followed by transcription of genes encoding the SOS-regulated error-prone Pols II, IV and V (Cirz *et al.*, 2005). It has also been shown that ciprofloxacin is able to stimulate intrachromosomal recombination in a SOS-independent manner via either RecBCD or RecFOR (Lopez *et al.*, 2007). In another study that involved incubation of *Pseudomonas aeruginosa* for four days in the presence of the bacteriostatic antibiotic tetracycline, the authors found that the mutation rate per cell per day had increased by 10^5 times (Alonso *et al.*, 1999).

Different concentrations of selector also influence mutation rate and the spectra of mutations formed. We have previously shown that increasing concentrations of the aminoglycoside antibiotic kanamycin are able to shape the relative frequencies of deletions versus amplifications of genetic material in a bacterial plasmid model (Oliveira *et al.*, 2009). A shift towards the amplification of genetic material (heterodimeric forms) preferentially occurred at higher selector concentrations, in what constitutes an evidence of the advantage of gene amplification under stressful environments (Oliveira *et al.*, 2009). Increased transposition frequency of the mobile element IS2 was additionally observed to take place in a bacterial plasmid in the presence of kanamycin. The latter event was found to be region specific and capable of generating a novel hybrid promoter upstream of the kan^R gene (Oliveira *et al.*, 2009). Subinhibitory concentrations of antibiotics can also be the cause of increased genetic variability. A recent example describes the deletion of a haemolysin operon flanked by IS*91* sequences from an *E. coli* plasmid (Pedro *et al.*, 2011). Also, exposure to sublethal levels of a representative group of bactericidal antibiotics was found to increase the minimum inhibitory concentrations and to induce mutagenesis and resistance to antibiotics other than those used to treat cells, via formation of ROS species (Kohanski *et al.*, 2010).

From the standpoint of preventing the emergence of multiresistant bacteria in clinical settings, it becomes necessary to consider the antibiotic-induced increase in horizontal gene transfer and mutation rates, prior to devising the best strategy for treatment. It was recently shown that the combination of antibiotics is ideal to fight an infection, but detrimental to avoid a scenario of double resistance, in which case the use of cycling and mixing strategies proved to be more effective (Obolski and Hadany, 2012).

Starvation-induced mutations

Upon experiencing periods of nutrient limitation (e.g. when entering stationary phase), bacteria undergo transcriptional (e.g. RpoS-dependent general stress response), metabolic (e.g. stringent response), morphological and translational changes (reviewed in Navarro Llorens *et al.*, 2010). Mutation rate is also increased, in a SOS and RpoS-dependent manner. In this regard, we have already mentioned earlier in this chapter the example of *lac* point mutations and amplification events that take place in situations of cell starvation. Yet, other not so thoroughly studied examples merit our attention. For example, *E. coli* cells grown under carbon starvation on a lactose minimal medium were shown to acquire β-lactam resistance through the accumulation of *ampD* base substitutions and frameshift mutations. Similarly to what is known from ciprofloxacin-induced mutagenesis, loss of function of *ampD* also occurs during DSB repair and requires Pols IV and V (Petrosino *et al.*, 2009). An increasing body of evidence suggests that transposition of mobile elements is favoured at later growth stages under nutrient deprivation. Transposition of Tn*552* and IS*903* was found to occur at later stages of colony growth and to require the presence of GTP, despite the underlying mechanism remaining unclear (Coros *et al.*, 2005). Twiss and co-workers (Twiss *et al.*, 2005) generated random Tn5 insertion mutant libraries in *E. coli*, and selected several mutations capable of affecting transposition of other mobile elements. Those mutations disrupting the *aspA* gene of the TCA cycle were found to be of particular relevance, as they were able to promote transposition of IS*903* at earlier stages of growth. In another study in *P. putida*, the transposition of the toluene degradation transposon Tn*4652* also increased under starvation (Ilves *et al.*, 2001).

Temperature-induced mutations

For many microbes, changes in temperature can profoundly shape their growth, development, virulence profiles, and alter genomic integrity. A genome-wide study performed in several *E. coli* cell lines adapted for 2,000 generations to a temperature of 41.5°C, describes the occurrence of complex rearrangements involving deletions, duplications and insertion of mobile elements (Riehle *et al.*, 2001). Among the regions affected, the authors highlight duplications affecting *rpoS* as well as other genes involved in stress survival and adaptation to high temperature (Riehle *et al.*, 2001). According to a recent study performed in *E. coli* O157:H7 cell lines, the number of mutations involving tandem repeats more than triplicated when temperature increased from 10°C to 43°C (Cooley *et al.*, 2010). Only single repeat changes confined to a single locus were observed at lower temperatures, contrasting to double- and triple-step mutations in multiple *loci* at higher temperatures. In another study, growing *B. multivorans* at a high temperature (42°C) instead of a normal temperature (30°C), led to a sevenfold increase in the transposition activity of seven IS elements belonging to four different families (e.g. IS*401*, IS*402*, IS*406*, among others) (Ohtsubo *et al.*, 2005). Since transposase activity depends on temperature, the authors suggest that changes in the conformation of the donor or target DNA may take place at higher temperatures, favouring a more effective transposition. Also, consecutive submission of several *E. coli* cell lines to freeze–thaw–growth cycles for 1000 generations, led to the accumulation of small deletions and IS*150* insertions in the *cls* and *uspA-uspB* genes, which were found to increase membrane fluidity and enhance survival and recovery (Sleight *et al.*, 2008).

Oxidative stress

We have mentioned before the role of ROS from endogenous origin as a catalyst of spontaneous DNA lesions, in particular involving the modified base 8-oxoG. However, several studies have focused on the adaptation of bacteria to oxidative stress resulting from exogenous sources such as ionizing and UV radiation, or chemical compounds. For example, exposure of *Helicobacter pylori* to ROS and reactive nitrogen species, which are usually found in its natural environment, induced a hypermutable state characterized by the onset of point mutations, intergenomic recombination, and direct repeat-mediated rearrangements (Kang *et al.*, 2006). Exposure of *Burkholderia cenocepacia* to H_2O_2 also stimulated IS-mediated genomic rearrangements, which may consist in an adaptation mechanism to the ROS-rich environment typically found in the lungs of a cystic fibrosis patient (Drevinek *et al.*, 2010). Another interesting example involves the resistance of some enteric bacteria such as *E. coli* or *S. enterica* to the DNA damaging salts contained in bile, an antibacterial secretion produced by the liver and stored in the gall bladder. It has been shown that such salts cause oxidative damage (G:C>A:T transitions), activating a cascade of repair functions such as BER, the SOS response and DinB-mediated translesion synthesis (Prieto *et al.*, 2004, 2006).

UV radiation-induced mutations

UV light is typically classified according to the wavelength of its components: UVA (320–400 nm), UVB (290–320 nm) and UVC (<290 nm). Cell exposure to UV light (particularly to more energetic wavelengths) frequently results in the formation of dimers between adjacent pyrimidine bases, such as *cis-syn* cyclobutadipyrimidines (CPDs), pyrimidine 6–4 pyrimidone photoproducts (6–4PPs), and corresponding Dewar isomers (Fig. 13.8a) (Tyrrell, 1973). In this regard it has been found that T-T and T-C sequences are typically more photoreactive than C-C and C-T (Douki and Cadet, 2001). UV radiation is also known to cause base substitutions, frameshifts and deletions in bacterial DNA (Schaaper *et al.*, 1987). It has also been reported the selection of advantageous induced mutations in genes related to repair and replication of DNA upon submitting *E. coli* to cyclic UV light conditions (Alcantara-Diaz *et al.*, 2004). Such lesions may lead to replication fork arrest if allowed to persist in the genome, and for this reason bacteria make use of several mechanisms to correct them. Several bacteria, archaea, virus, fungi and others, have the ability to correct UV-induced pyrimidine dimers by means of a process named photoreactivation, which uses the light-dependent (300–500 nm) enzyme photolyase encoded by the *phr* gene. Alternatively, pyrimidine dimers as well as other bulky lesions can be corrected in a 'dark mode' by means of excision repair pathways such as BER (described before in this chapter) and NER. NER was first identified in *E. coli* (Pettijohn and Hanawalt, 1964), but it is now known to be highly ubiquitous among the entire tree of life. Recognition and repair of a lesion typically involve the proteins UvrA, UvrB, UvrC and UvrD as follows: first the ternary complex $UvrA_2UvrB$ recognizes and binds the DNA lesion; UvrA proteins dissociate from lesion site and UvrB recruits UvrC that cleaves the DNA at the 4th–5th nucleotide upstream the lesion and at the 7th–8th nucleotide downstream the lesion; finally UvrD (helicase II) removes the single stranded fragment carrying the lesion, and Pol I and DNA ligase complete the repair (Fig. 13.8b).

Exposure to UV light also seems to induce the transposition of mobile elements. In *E. coli* cells, a 28-fold increase in transposition of IS*10* was observed after irradiation with UV light, in a process dependent of the SOS response (Eichenbaum and Livneh, 1998). The authors of this study suggest that the cut-and-paste mechanism of IS*10* transposition may be favoured by the presence of ssDNA gaps resulting from stalled replication forks or DNA repair enzymes.

Cell storage and shipment

Studies on the influence of cell storage and shipment as potential sources of stress are scarce. Among the few studies available, it was found that long starvation periods in resting *E. coli* cells stored in agar stabs promoted increased genetic variation by transposition of mobile elements, namely IS2, IS3, IS5 and IS*30* (Naas *et al.*, 1994, 1995). In a more recent work, it was found that

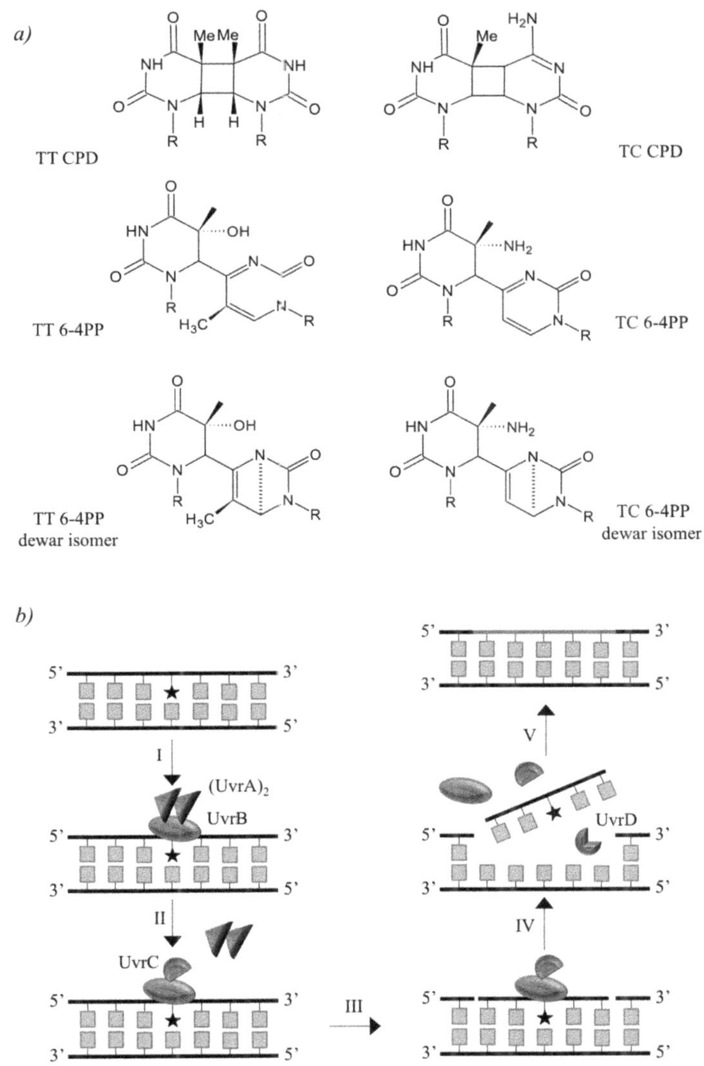

Figure 13.8 Main UV-induced pyrimidine photoproducts and nucleotide excision repair (NER) in eubacteria. (*a*) Three major classes of DNA lesions are formed upon exposure to UV radiation: cyclobutane pyrimidine dimers (CPDs), pyrimidine 6–4 pyrimidone photoproducts (6–4PPs), and corresponding Dewar isomers. (*b*) In NER, a complex formed by two UvrA molecules and one UvrB molecule is involved in the recognition of bulky lesions in the DNA such as pyrimidine dimers (black star) (I). After recognition, both UvrA molecules dissociate, and UvrC binds UvrB (II). In the next step, UvrC cuts 5′ and 3′ to the lesion site (III), and the helicase action of UvrD allows removal of the fragment and bound UvrB and UvrC (IV). Extension by DNA polymerase I and ligation restore the integrity of the fragment (V).

maintaining *E. coli* K12 cells for periods of time as short as a few days in LB-stabs is enough to confer mutations in the *rpoS* locus similar to the ones found in the GASP phenotype (Spira *et al.*, 2011). The authors of this study suggest resuspending bacteria in 15% glycerol (v/v) and applying it to a filter disk, as this strategy appears to maintain a high RpoS phenotype in all cells for a three-week period.

The problem of acquired bacterial resistance upon exposure to relatively mild stressors or certain growth-permissive environments gains particular relevance in the food industry. In an effort to adopt a more cost-effective and environmental-friendly approach, as well as to retain the sensory characteristics of certain food products, many companies are employing mild preservation approaches in a synergistic way, instead of using harsh preservation techniques such as heat sterilization. The former include washing steps, irradiation with UV light, controlled atmosphere packaging, among others. It turns out that exposure to these stresses at a non-detrimental level, confers the cells what is known as 'cross protection', i.e. an enhanced resistance against other stresses. Concomitantly, this has obvious implications in the assessment of food safety. Readers are referred to Capozzi *et al.* (2009) for a recent review on this topic.

More stable bacterial genomes and DNA-based therapeutic molecules

Severely reduced genomes can be found in nature (e.g. the human parasite *Mycoplasma genitalium* with a genome size of 0.58 Mb), being usually associated with unique and highly specific and stable ecological niches. Such minimal genomes are typically compositionally biased, highly compact, with few intergenic and non-functional regions, and preserve genes involved in core processes of replication, transcription and translation (reviewed in McCutcheon and Moran, 2012)). These naturally occurring examples of genome reduction and the notion that certain genome subsets may be dispensable under a particular environment immediately raised the question on what would be the minimal genomic content necessary to support life. Moreover, the wealth of sequencing data currently available has fuelled the development of projects dealing with minimal or streamlined chassis, particularly from industrial and academic parties. The practical advantages seem obvious: increased genomic stability, construction of microbial power-horses with streamlined pathways for production of high-value bioproducts, reduced production of contaminants by knocking out competing pathways, among many others.

Genome deconstruction by top-down approaches has been successfully achieved through the combinatorial removal of nonessential elements, namely mobile elements or transporter and cryptic genes (Posfai *et al.*, 2006; Ara *et al.*, 2007; Kato and Hashimoto, 2007; Lee *et al.*, 2009). In order to overcome IS-mediated structural instability, IS-free cellular chassis have been devised (Umenhoffer *et al.*, 2010). Such *E. coli* MDS42 cell lines allow the propagation of otherwise unstable clones, show reduced evolutionary potential, and represent a safer option to be used for amplification of DNA-based therapeutics. More recently, it was reported an *E. coli* MDS42-derived and stabilized strain lacking mobile elements and harbouring a triple deletion in the three error prone DNA Pols II, IV and V (Csorgo *et al.*, 2012).

Alternatively, bottom-up engineering attempts to specifically incorporate modular parts to form larger and more complex systems. In this regard, promising tailor-made DNA assembly strategies have been unveiled. For example, a successive hybridization assembly (SHA) relies on a simple *in vitro* hybridization of partially overlapping DNA fragments without the need of restriction enzymes, DNA ligases or recombinases (Jiang *et al.*, 2012). However, the prospect of constructing novel reduced genomes raises important issues that are worth some reflection. Namely, how will the current legislation on intellectual property cope with each novel assembled genome? How do these findings affect our view of life and the way we interact with it?

Within the biopharmaceutical industry, a similar tendency for the minimization of DNA-based therapeutics has been observed in the last couple of years. These strategies try to tackle or at least reduce structural instability and immunogenicity to a minimum, by simply removing those portions that are not needed for the therapeutic application. One possibility relies on the elimination of antibiotic resistance markers, since they pose a metabolic burden to the cells and may constitute a threat if disseminated (Cranenburgh *et al.*, 2001; Pfaffenzeller *et al.*, 2006; Goh and Good, 2008; Marie *et al.*, 2010; Peubez *et al.*, 2010; Vandermeulen *et al.*, 2011). Other strategies based on further plasmid minimization are also gaining wide acceptance mainly due to their small size, low content of unmethylated CpGs, increased transfection efficiency and higher levels of transgene expression in several tissues. This minimal vector design concept includes both the Minimalistic Immunogenically Defined Gene Expression (MIDGE) vectors and minicircles. MIDGE vectors are linear covalently closed dumbbell-shaped DNA fragments comprising the eukaryotic expression unit and terminal loop structures that protect against exonuclease degradation. In a similar way, minicircles are also devoid of bacterial elements, but are generated by site-specific recombination in *E. coli*, in a process that separates them from miniplasmids (which contain all bacterial elements). These constructs have demonstrated a great clinical significance when used as vectors for gene therapy (Chang *et al.*, 2008) and DNA vaccination (Lopez-Fuertes *et al.*, 2002; Moreno *et al.*, 2004), when used to control cancer growth (Schakowski *et al.*, 2001; Xu *et al.*, 2011), and as a non-viral carriers for reprogramming of stem cells (Jia *et al.*, 2010).

We can reasonably anticipate that in the next decades, the rational design of improved DNA-based biopharmaceuticals, whole synthetic pathways and even genomes will largely profit from the use of standardized biological parts (Lee *et al.*, 2011), as well as computational tools that must take into account important structural aspects that go beyond the simple functionality and assembly.

Concluding remarks

The purpose of this chapter was to provide the reader with a brief view on aspects dealing with genome instability in bacteria, with a focus on spontaneous and stress-induced mutations. Genomes have evolved mechanisms that limit the accumulation of potentially detrimental lethal lesions, while still harnessing beneficial genetic changes to their own profit. Genomes are now known to be highly responsive to the surrounding environment, and capable of using a *potpourri* of strategies to cope with adversity. These include the use of error-prone polymerases, activation of stress-response pathways, downregulation of DNA repair systems, increasing the rate of random mutagenesis, and decreasing the barriers of inter-species gene transfer. In the years to come, a deeper understanding of such mechanisms and the use of improved methods for analysing genetic variation, will pave the way to a more unified view on evolution, as well as to the development of streamlined genomes.

Acknowledgements

This work was supported by Fundação para a Ciência e a Tecnologia (FCT) (project PTDC/EBB-EBI/113650/2009 and Ph.D. grant BD/22320/2005 to P.H.O.) and by the MIT-Portugal Program.

References

Achaz, G., Boyer, F., Rocha, E.P., Viari, A., and Coissac, E. (2007). Repseek, a tool to retrieve approximate repeats from large DNA sequences. Bioinformatics *23*, 119–121.

Alcantara-Diaz, D., Brena-Valle, M., and Serment-Guerrero, J. (2004). Divergent adaptation of *Escherichia coli* to cyclic ultraviolet light exposures. Mutagenesis *19*, 349–354.

Alonso, A., Campanario, E., and Martinez, J.L. (1999). Emergence of multidrug-resistant mutants is increased under antibiotic selective pressure in Pseudomonas aeruginosa. Microbiology *145*, 2857–2862.

Ara, K., Ozaki, K., Nakamura, K., Yamane, K., Sekiguchi, J., and Ogasawara, N. (2007). Bacillus minimum genome factory: effective utilization of microbial genome information. Biotechnol. Appl. Biochem. *46*, 169–178.

Beletskii, A., and Bhagwat, A.S. (1996). Transcription-induced mutations: increase in C to T mutations in the nontranscribed strand during transcription in *Escherichia coli*. Proc. Natl. Acad. Sci. U.S.A. *93*, 13919–13924.

Belotserkovskii, B.P., Liu, R., Tornaletti, S., Krasilnikova, M.M., Mirkin, S.M., and Hanawalt, P.C. (2010). Mechanisms and implications of transcription blockage by guanine-rich DNA sequences. Proc. Natl. Acad. Sci. U.S.A. *107*, 12816–12821.

Benson, G. (1999). Tandem repeats finder: a program to analyze DNA sequences. Nucleic Acids Res. *27*, 573–580.

Bentchikou, E., Servant, P., Coste, G., and Sommer, S. (2010). A major role of the RecFOR pathway in DNA double-strand-break repair through ESDSA in Deinococcus radiodurans. PLoS Genet. *6*, e1000774.

Bi, X., and Liu, L.F. (1994). recA-independent and recA-dependent intramolecular plasmid recombination. Differential homology requirement and distance effect. J. Mol. Biol. *235*, 414–423.

Bi, X., and Liu, L.F. (1996). A replicational model for DNA recombination between direct repeats. J. Mol. Biol. *256*, 849–858.

Boan, F., and Gomez-Marquez, J. (2010). In vitro recombination mediated by G-quadruplexes. Chembiochem *11*, 331–334.

Boiteux, S., Huisman, O., and Laval, J. (1984). 3-Methyladenine residues in DNA induce the SOS function sfiA in *Escherichia coli*. EMBO J. *3*, 2569–2573.

Brewer, B.J. (1988). When polymerases collide: replication and the transcriptional organization of the E. coli chromosome. Cell *53*, 679–686.

Bzymek, M., and Lovett, S.T. (2001). Instability of repetitive DNA sequences: the role of replication in multiple mechanisms. Proc. Natl. Acad. Sci. U.S.A. *98*, 8319–8325.

Cairns, J., Overbaugh, J., and Miller, S. (1988). The origin of mutants. Nature *335*, 142–145.

Capozzi, V., Fiocco, D., Amodio, M.L., Gallone, A., and Spano, G. (2009). Bacterial stressors in minimally processed food. Int. J. Mol. Sci. *10*, 3076–3105.

Cer, R.Z., Bruce, K.H., Mudunuri, U.S., Yi, M., Volfovsky, N., Luke, B.T., Bacolla, A., Collins, J.R., and Stephens, R.M. (2011). Non-B DB: a database of predicted non-B DNA-forming motifs in mammalian genomes. Nucleic Acids Res. *39*, D383–391.

Chang, C.H., Chang, Y.C., Underwood, A., Chiou, C.S., and Kao, C.Y. (2007). VNTRDB: a bacterial variable number tandem repeat locus database. Nucleic Acids Res. *35*, D416–421.

Chang, C.W., Christensen, L.V., Lee, M., and Kim, S.W. (2008). Efficient expression of vascular endothelial growth factor using minicircle DNA for angiogenic gene therapy. J. Control Release *125*, 155–163.

Chedin, F., Dervyn, E., Dervyn, R., Ehrlich, S.D., and Noirot, P. (1994). Frequency of deletion formation decreases exponentially with distance between short direct repeats. Mol. Microbiol. *12*, 561–569.

Cirz, R.T., Chin, J.K., Andes, D.R., de Crecy-Lagard, V., Craig, W.A., and Romesberg, F.E. (2005). Inhibition of mutation and combating the evolution of antibiotic resistance. PLoS Biol. *3*, e176.

Coenye, T., and Vandamme, P. (2005). Characterization of mononucleotide repeats in sequenced prokaryotic genomes. DNA Res. *12*, 221–233.

Connelly, J.C., Kirkham, L.A., and Leach, D.R. (1998). The SbcCD nuclease of *Escherichia coli* is a structural maintenance of chromosomes (SMC) family protein that cleaves hairpin DNA. Proc. Natl. Acad. Sci. U.S.A. *95*, 7969–7974.

Cooke, M.S., Evans, M.D., Dizdaroglu, M., and Lunec, J. (2003). Oxidative DNA damage: mechanisms, mutation, and disease. FASEB J. *17*, 1195–1214.

Cooley, M.B., Carychao, D., Nguyen, K., Whitehand, L., and Mandrell, R. (2010). Effects of environmental stress on stability of tandem repeats in *Escherichia coli* O157:H7. Appl. Environ. Microbiol. *76*, 3398–3400.

Coros, A.M., Twiss, E., Tavakoli, N.P., and Derbyshire, K.M. (2005). Genetic evidence that GTP is required for transposition of IS903 and Tn552 in *Escherichia coli*. J. Bacteriol. *187*, 4598–4606.

Coulondre, C., Miller, J.H., Farabaugh, P.J., and Gilbert, W. (1978). Molecular basis of base substitution hotspots in *Escherichia coli*. Nature *274*, 775–780.

Cranenburgh, R.M., Hanak, J.A., Williams, S.G., and Sherratt, D.J. (2001). *Escherichia coli* strains that allow antibiotic-free plasmid selection and maintenance by repressor titration. Nucleic Acids Res. *29*, E26.

Csorgo, B., Feher, T., Timar, E., Blattner, F.R., and Posfai, G. (2012). Low-mutation-rate, reduced-genome *Escherichia coli*: an improved host for faithful maintenance of engineered genetic constructs. Microb. Cell Fact. *11*, 11.

De Septenville, A.L., Duigou, S., Boubakri, H., and Michel, B. (2012). Replication fork reversal after replication-transcription collision. PLoS Genet. *8*, e1002622.

DeMarini, D.M., Shelton, M.L., Abu-Shakra, A., Szakmary, A., and Levine, J.G. (1998). Spectra of spontaneous frameshift mutations at the hisD3052 allele of Salmonella typhimurium in four DNA repair backgrounds. Genetics *149*, 17–36.

Denoeud, F., and Vergnaud, G. (2004). Identification of polymorphic tandem repeats by direct comparison of genome sequence from different bacterial strains: a web-based resource. BMC Bioinformatics *5*, 4.

Dianov, G.L., Kuzminov, A.V., Mazin, A.V., and Salganik, R.I. (1991). Molecular mechanisms of deletion formation in *Escherichia coli* plasmids. I. Deletion formation mediated by long direct repeats. Mol. Gen. Genet. *228*, 153–159.

Ditlevson, J.V., Tornaletti, S., Belotserkovskii, B.P., Teijeiro, V., Wang, G., Vasquez, K.M., and Hanawalt, P.C. (2008). Inhibitory effect of a short Z-DNA forming sequence on transcription elongation by T7 RNA polymerase. Nucleic Acids Res. *36*, 3163–3170.

Douki, T., and Cadet, J. (2001). Individual determination of the yield of the main UV-induced dimeric pyrimidine photoproducts in DNA suggests a high mutagenicity of CC photolesions. Biochemistry *40*, 2495–2501.

Drablos, F., Feyzi, E., Aas, P.A., Vaagbo, C.B., Kavli, B., Bratlie, M.S., Pena-Diaz, J., Otterlei, M., Slupphaug, G., and Krokan, H.E. (2004). Alkylation damage in DNA and RNA – repair mechanisms and medical significance. DNA Repair (Amst) *3*, 1389–1407.

Drevinek, P., Baldwin, A., Lindenburg, L., Joshi, L.T., Marchbank, A., Vosahlikova, S., Dowson, C.G., and Mahenthiralingam, E. (2010). Oxidative stress of Burkholderia cenocepacia induces insertion sequence-mediated genomic rearrangements that interfere with macrorestriction-based genotyping. J. Clin. Microbiol. *48*, 34–40.

Duncan, B.K., and Miller, J.H. (1980). Mutagenic deamination of cytosine residues in DNA. Nature *287*, 560–561.

Dutta, D., Shatalin, K., Epshtein, V., Gottesman, M.E., and Nudler, E. (2011). Linking RNA polymerase backtracking to genome instability in E. coli. Cell *146*, 533–543.

Eichenbaum, Z., and Livneh, Z. (1998). UV light induces IS10 transposition in *Escherichia coli*. Genetics *149*, 1173–1181.

Eykelenboom, J.K., Blackwood, J.K., Okely, E., and Leach, D.R. (2008). SbcCD causes a double-strand break at a DNA palindrome in the *Escherichia coli* chromosome. Mol. Cell. *29*, 644–651.

Feschenko, V.V., and Lovett, S.T. (1998). Slipped misalignment mechanisms of deletion formation: analysis of deletion endpoints. J. Mol. Biol. *276*, 559–569.

Fleischmann, R.D., Adams, M.D., White, O., Clayton, R.A., Kirkness, E.F., Kerlavage, A.R., Bult, C.J., Tomb, J.F., Dougherty, B.A., Merrick, J.M., *et al.* (1995). Whole-genome random sequencing and assembly of Haemophilus influenzae Rd. Science 269, 496–512.

Foster, P.L. (1997). Nonadaptive mutations occur on the F' episome during adaptive mutation conditions in *Escherichia coli*. J. Bacteriol. *179*, 1550–1554.

Foster, P.L., and Trimarchi, J.M. (1994). Adaptive reversion of a frameshift mutation in *Escherichia coli* by simple base deletions in homopolymeric runs. Science *265*, 407–409.

Frederico, L.A., Kunkel, T.A., and Shaw, B.R. (1990). A sensitive genetic assay for the detection of cytosine deamination: determination of rate constants and the activation energy. Biochemistry *29*, 2532–2537.

Gacy, A.M., Goellner, G., Juranic, N., Macura, S., and McMurray, C.T. (1995). Trinucleotide repeats that expand in human disease form hairpin structures *in vitro*. Cell *81*, 533–540.

Galhardo, R.S., Do, R., Yamada, M., Friedberg, E.C., Hastings, P.J., Nohmi, T., and Rosenberg, S.M. (2009). DinB upregulation is the sole role of the SOS response in stress-induced mutagenesis in *Escherichia coli*. Genetics *182*, 55–68.

Gan, W., Guan, Z., Liu, J., Gui, T., Shen, K., Manley, J.L., and Li, X. (2011). R-loop-mediated genomic instability is caused by impairment of replication fork progression. Genes Dev. *25*, 2041–2056.

Gates, K.S. (2009). An overview of chemical processes that damage cellular DNA: spontaneous hydrolysis, alkylation, and reactions with radicals. Chem. Res. Toxicol. *22*, 1747–1760.

Goh, S., and Good, L. (2008). Plasmid selection in *Escherichia coli* using an endogenous essential gene marker. BMC Biotechnol. *8*, 61.

Gonzalez, C., Hadany, L., Ponder, R.G., Price, M., Hastings, P.J., and Rosenberg, S.M. (2008). Mutability and importance of a hypermutable cell subpopulation that produces stress-induced mutants in *Escherichia coli*. PLoS Genet. *4*, e1000208.

Han, H., and Hurley, L.H. (2000). G-quadruplex DNA: a potential target for anti-cancer drug design. Trends Pharmacol. Sci. *21*, 136–142.

Hardin, A., Villalta, C.F., Doan, M., Jabri, M., Chockalingham, V., White, S.J., and Fowler, R.G. (2007). A molecular characterization of spontaneous frameshift mutagenesis within the trpA gene of *Escherichia coli*. DNA Repair (Amst) *6*, 177–189.

Hastings, P.J., Bull, H.J., Klump, J.R., and Rosenberg, S.M. (2000). Adaptive amplification: an inducible chromosomal instability mechanism. Cell *103*, 723–731.

Hood, D.W., Deadman, M.E., Jennings, M.P., Bisercic, M., Fleischmann, R.D., Venter, J.C., and Moxon, E.R. (1996). DNA repeats identify novel virulence genes in Haemophilus influenzae. Proc. Natl. Acad. Sci. U.S.A. *93*, 11121–11125.

Huppert, J.L., and Balasubramanian, S. (2005). Prevalence of quadruplexes in the human genome Nucleic Acids Res. *33*, 2908–2916.

Ilves, H., Horak, R., and Kivisaar, M. (2001). Involvement of sigma(S) in starvation-induced transposition of Pseudomonas putida transposon Tn4652. J. Bacteriol. *183*, 5445–5448.

Jain, A., Wang, G., and Vasquez, K.M. (2008). DNA triple helices: biological consequences and therapeutic potential. Biochimie *90*, 1117–1130.

Jenjaroenpun, P., and Kuznetsov, V.A. (2009). TTS mapping: integrative WEB tool for analysis of triplex formation target DNA sequences, G-quadruplets and non-protein coding regulatory DNA elements in the human genome. BMC Genomics *10* (Suppl 3), S9.

Jia, F., Wilson, K.D., Sun, N., Gupta, D.M., Huang, M., Li, Z., Panetta, N.J., Chen, Z.Y., Robbins, R.C., Kay, M.A., *et al.* (2010). A nonviral minicircle vector for deriving human iPS cells. Nat. Methods 7, 197–199.

Jiang, X., Yang, J., Zhang, H., Zou, H., Wang, C., and Xian, M. (2012). In vitro assembly of multiple DNA fragments using successive hybridization. PLoS One 7, e30267.

Jolivet-Gougeon, A., David-Jobert, S., Tamanai-Shacoori, Z., Menard, C., and Cormier, M. (2000). Osmotic stress-induced genetic rearrangements in *Escherichia coli* H10407 detected by randomly amplified polymorphic DNA analysis. Appl. Environ. Microbiol. *66*, 5484–5487.

Kahn, J.D., and Crothers, D.M. (1993). DNA bending in transcription initiation. Cold Spring Harb. Symp. Quant. Biol. *58*, 115–122.

Kang, J.M., Iovine, N.M., and Blaser, M.J. (2006). A paradigm for direct stress-induced mutation in prokaryotes. FASEB J. *20*, 2476–2485.

Kato, J., and Hashimoto, M. (2007). Construction of consecutive deletions of the *Escherichia coli* chromosome. Mol. Syst. Biol. *3*, 132.

Kawe, M., Horn, U., and Pluckthun, A. (2009). Facile promoter deletion in *Escherichia coli* in response to leaky expression of very robust and benign proteins from common expression vectors. Microb. Cell Fact. *8*, 8.

Kha, D.T., Wang, G., Natrajan, N., Harrison, L., and Vasquez, K.M. (2010). Pathways for double-strand break repair in genetically unstable Z-DNA-forming sequences. J. Mol. Biol. *398*, 471–480.

Kim, M., O'Callaghan, P.M., Droms, K.A., and James, D.C. (2011). A mechanistic understanding of production instability in CHO cell lines expressing recombinant monoclonal antibodies. Biotechnol. Bioeng. [Epub ahead of print]

Kobori, S., Ko, Y., and Kato, M. (2009). A target site for spontaneous insertion of IS10 element in pUC19 DNA located within intrinsically bent DNA. Open Microbiol. J. 3, 146–150.

Kohanski, M.A., DePristo, M.A., and Collins, J.J. (2010). Sublethal antibiotic treatment leads to multidrug resistance via radical-induced mutagenesis. Mol. Cell. *37*, 311–320.

Kowalski, D., Natale, D.A., and Eddy, M.J. (1988). Stable DNA unwinding, not 'breathing,' accounts for single-strand-specific nuclease hypersensitivity of specific A+T-rich sequences. Proc. Natl. Acad. Sci. U.S.A. *85*, 9464–9468.

Kunkel, T.A., and Soni, A. (1988). Mutagenesis by transient misalignment. J. Biol. Chem. *263*, 14784–14789.

Kurtz, S., and Schleiermacher, C. (1999). REPuter: fast computation of maximal repeats in complete genomes. Bioinformatics *15*, 426–427.

Larson, K., Sahm, J., Shenkar, R., and Strauss, B. (1985). Methylation-induced blocks to *in vitro* DNA replication. Mutat. Res. *150*, 77–84.

Lee, H., Popodi, E., Tang, H., and Foster, P.L. (2012). Rate and molecular spectrum of spontaneous mutations in the bacterium *Escherichia coli* as determined by whole-genome sequencing. Proc. Natl. Acad. Sci. U.S.A. *109*, E2774–2783.

Lee, J.H., Sung, B.H., Kim, M.S., Blattner, F.R., Yoon, B.H., Kim, J.H., and Kim, S.C. (2009). Metabolic engineering of a reduced-genome strain of *Escherichia coli* for L-threonine production. Microb. Cell Fact. *8*, 2.

Lee, T.S., Krupa, R.A., Zhang, F., Hajimorad, M., Holtz, W.J., Prasad, N., Lee, S.K., and Keasling, J.D. (2011). BglBrick vectors and datasheets: A synthetic biology platform for gene expression. J. Biol. Eng. *5*, 12.

Lewis, L.A., Astatke, M., Umekubo, P.T., Alvi, S., Saby, R., Afrose, J., Oliveira, P.H., Monteiro, G.A., and Prazeres, D.M. (2012). Protein–DNA interactions define the mechanistic aspects of circle formation and insertion reactions in IS2 transposition. Mob. DNA 3, 1.

Li, Y.C., Korol, A.B., Fahima, T., and Nevo, E. (2004). Microsatellites within genes: structure, function, and evolution. Mol. Biol. Evol. *21*, 991–1007.

Lilley, D.M. (1985). The kinetic properties of cruciform extrusion are determined by DNA base-sequence. Nucleic Acids Res. *13*, 1443–1465.

Lindahl, T. (1979). DNA glycosylases, endonucleases for apurinic/apyrimidinic sites, and base excision-repair. Prog. Nucleic Acid Res. Mol. Biol. *22*, 135–192.

Lindahl, T. (1993). Instability and decay of the primary structure of DNA. Nature *362*, 709–715.

Linial, M., and Shlomai, J. (1988). Bent DNA structures associated with several origins of replication are recognized by a unique enzyme from trypanosomatids. Nucleic Acids Res. *16*, 6477–6492.

Loeb, L.A., and Preston, B.D. (1986). Mutagenesis by apurinic/apyrimidinic sites. Annu. Rev. Genet. *20*, 201–230.

Loechler, E.L., Green, C.L., and Essigmann, J.M. (1984). In vivo mutagenesis by O6-methylguanine built into a unique site in a viral genome. Proc. Natl. Acad. Sci. U.S.A. *81*, 6271–6275.

Lombardo, M.J., Aponyi, I., and Rosenberg, S.M. (2004). General stress response regulator RpoS in adaptive mutation and amplification in *Escherichia coli*. Genetics *166*, 669–680.

Lopez, E., and Blazquez, J. (2009). Effect of subinhibitory concentrations of antibiotics on intrachromosomal homologous recombination in *Escherichia coli*. Antimicrob. Agents Chemother. *53*, 3411–3415.

Lopez, E., Elez, M., Matic, I., and Blazquez, J. (2007). Antibiotic-mediated recombination: ciprofloxacin stimulates SOS-independent recombination of divergent sequences in *Escherichia coli*. Mol. Microbiol. *64*, 83–93.

Lopez-Fuertes, L., Perez-Jimenez, E., Vila-Coro, A.J., Sack, F., Moreno, S., Konig, S.A., Junghans, C., Wittig, B., Timon, M., and Esteban, M. (2002). DNA vaccination with linear minimalistic (MIDGE) vectors confers protection against Leishmania major infection in mice. Vaccine *21*, 247–257.

Lovett, S.T., Drapkin, P.T., Sutera, V.A., Jr., and Gluckman-Peskind, T.J. (1993). A sister-strand exchange mechanism for recA-independent deletion of repeated DNA sequences in *Escherichia coli*. Genetics *135*, 631–642.

Lovett, S.T., and Feschenko, V.V. (1996). Stabilization of diverged tandem repeats by mismatch repair: evidence for deletion formation via a misaligned replication intermediate. Proc. Natl. Acad. Sci. U.S.A. *93*, 7120–7124.

Mackwan, R.R., Carver, G.T., Kissling, G.E., Drake, J.W., and Grogan, D.W. (2008). The rate and character of spontaneous mutation in Thermus thermophilus. Genetics *180*, 17–25.

Maki, H., and Sekiguchi, M. (1992). MutT protein specifically hydrolyses a potent mutagenic substrate for DNA synthesis. Nature *355*, 273–275.

Mandel, M.J., and Silhavy, T.J. (2005). Starvation for different nutrients in *Escherichia coli* results in differential modulation of RpoS levels and stability. J. Bacteriol. *187*, 434–442.

Marie, C., Vandermeulen, G., Quiviger, M., Richard, M., Preat, V., and Scherman, D. (2010). pFARs, plasmids free of antibiotic resistance markers, display high-level transgene expression in muscle, skin and tumour cells. J. Gene Med. *12*, 323–332.

Martin, P., Makepeace, K., Hill, S.A., Hood, D.W., and Moxon, E.R. (2005). Microsatellite instability regulates transcription factor binding and gene expression. Proc. Natl. Acad. Sci. U.S.A. *102*, 3800–3804.

Mazin, A.V., Kuzminov, A.V., Dianov, G.L., and Salganik, R.I. (1991). Mechanisms of deletion formation in *Escherichia coli* plasmids. II. Deletions mediated by short direct repeats. Mol. Gen. Genet. *228*, 209–214.

McCutcheon, J.P., and Moran, N.A. (2012). Extreme genome reduction in symbiotic bacteria. Nat. Rev. Microbiol. *10*, 13–26.

Merrikh, H., Machon, C., Grainger, W.H., Grossman, A.D., and Soultanas, P. (2011). Co-directional replication-transcription conflicts lead to replication restart. Nature *470*, 554–557.

Michaels, M.L., Cruz, C., Grollman, A.P., and Miller, J.H. (1992). Evidence that MutY and MutM combine to prevent mutations by an oxidatively damaged form of guanine in DNA. Proc. Natl. Acad. Sci. U.S.A. *89*, 7022–7025.

Mirkin, E.V., and Mirkin, S.M. (2005). Mechanisms of transcription-replication collisions in bacteria. Mol. Cell. Biol *25*, 888–895.

Mirkin, E.V., and Mirkin, S.M. (2007). Replication fork stalling at natural impediments. Microbiol. Mol. Biol. Rev. *71*, 13–35.

Morag, A.S., Saveson, C.J., and Lovett, S.T. (1999). Expansion of DNA repeats in *Escherichia coli*: effects of recombination and replication functions. J. Mol. Biol. *289*, 21–27.

Moreno, S., Lopez-Fuertes, L., Vila-Coro, A.J., Sack, F., Smith, C.A., Konig, S.A., Wittig, B., Schroff, M., Juhls, C., Junghans, C., *et al.* (2004). DNA immunization with minimalistic expression constructs. Vaccine *22*, 1709–1716.

Mrazek, J., Guo, X., and Shah, A. (2007). Simple sequence repeats in prokaryotic genomes. Proc. Natl. Acad. Sci. U.S.A. *104*, 8472–8477.

Naas, T., Blot, M., Fitch, W.M., and Arber, W. (1994). Insertion sequence-related genetic variation in resting *Escherichia coli* K-12. Genetics *136*, 721–730.

Naas, T., Blot, M., Fitch, W.M., and Arber, W. (1995). Dynamics of IS-related genetic rearrangements in resting *Escherichia coli* K-12. Mol. Biol. Evol. *12*, 198–207.

Navarro Llorens, J.M., Tormo, A., and Martinez-Garcia, E. (2010). Stationary phase in Gram-negative bacteria. FEMS Microbiol. Rev. *34*, 476–495.

Obolski, U., and Hadany, L. (2012). Implications of stress-induced genetic variation for minimizing multidrug resistance in bacteria. BMC Med. *10*, 89.

Ohtsubo, Y., Genka, H., Komatsu, H., Nagata, Y., and Tsuda, M. (2005). High-temperature-induced transposition of insertion elements in burkholderia multivorans ATCC 17616. Appl. Environ. Microbiol. *71*, 1822–1828.

Oliveira, P.H., Lemos, F., Monteiro, G.A., and Prazeres, D.M.F. (2008). Recombination frequency in plasmid DNA containing direct repeats – predictive correlation with repeat and intervening sequence length. Plasmid *60*, 159–165.

Oliveira, P.H., Prather, K.J., Prazeres, D.M.F., and Monteiro, G.A. (2010). Analysis of DNA repeats in bacterial plasmids reveals the potential for recurrent instability events. Appl. Microbiol. Biotechnol. *87*, 2157–2167.

Oliveira, P.H., Prather, K.L., Prazeres, D.M.F., and Monteiro, G.A. (2011). Mutation detection in plasmid-based biopharmaceuticals. Biotechnol. J. *6*, 378–391.

Oliveira, P.H., Prazeres, D.M.F., and Monteiro, G.A. (2009). Deletion formation mutations in plasmid expression vectors are unfavored by runaway amplification conditions and differentially selected under kanamycin stress. J. Biotechnol. *143*, 231–238.

Oliveira, P.H., Prazeres, D.M.F., and Monteiro, G.A. (2012). Structural and segregational instability in plasmid biology. In Plasmids: Genetics, Applications and Health, Gonzalez, F.E.R., ed. (New York, Nova Science Publishers), pp. 79–99.

Omont, N., and Kepes, F. (2004). Transcription/replication collisions cause bacterial transcription units to be longer on the leading strand of replication. Bioinformatics *20*, 2719–2725.

Pedro, L., Banos, R.C., Aznar, S., Madrid, C., Balsalobre, C., and Juarez, A. (2011). Antibiotics shaping bacterial genome: deletion of an IS91 flanked virulence determinant upon exposure to subinhibitory antibiotic concentrations. PLoS One *6*, e27606.

Petrosino, J.F., Galhardo, R.S., Morales, L.D., and Rosenberg, S.M. (2009). Stress-induced beta-lactam antibiotic resistance mutation and sequences of stationary-phase mutations in the *Escherichia coli* chromosome. J. Bacteriol. *191*, 5881–5889.

Pettijohn, D., and Hanawalt, P. (1964). Evidence for repair-replication of ultraviolet damaged DNA in bacteria. J. Mol. Biol. *9*, 395–410.

Peubez, I., Chaudet, N., Mignon, C., Hild, G., Husson, S., Courtois, V., De Luca, K., Speck, D., and Sodoyer, R. (2010). Antibiotic-free selection in E. coli: new considerations for optimal design and improved production. Microb. Cell Fact. *9*, 65.

Pfaffenzeller, I., Mairhofer, J., Striedner, G., Bayer, K., and Grabherr, R. (2006). Using ColE1-derived RNA I for suppression of a bacterially encoded gene: implication for a novel plasmid addiction system. Biotechnol. J. *1*, 675–681.

Pomerantz, R.T., and O'Donnell, M. (2008). The replisome uses mRNA as a primer after colliding with RNA polymerase. Nature *456*, 762–766.

Pomerantz, R.T., and O'Donnell, M. (2010). Direct restart of a replication fork stalled by a head-on RNA polymerase. Science *327*, 590–592.

Ponder, R.G., Fonville, N.C., and Rosenberg, S.M. (2005). A switch from high-fidelity to error-prone DNA double-strand break repair underlies stress-induced mutation. Mol. Cell. *19*, 791–804.

Posfai, G., Plunkett, G. 3rd, Feher, T., Frisch, D., Keil, G.M., Umenhoffer, K., Kolisnychenko, V., Stahl, B., Sharma, S.S., de Arruda, M., *et al.* (2006). Emergent properties of reduced-genome *Escherichia coli*. Science *312*, 1044–1046.

Powell, S.C., and Wartell, R.M. (2001). Different characteristics distinguish early versus late arising adaptive mutations in *Escherichia coli* FC40. Mutat. Res. *473*, 219–228.

Prado, F., and Aguilera, A. (2005). Impairment of replication fork progression mediates RNA polII transcription-associated recombination. EMBO J. *24*, 1267–1276.

Price, M.N., Alm, E.J., and Arkin, A.P. (2005). Interruptions in gene expression drive highly expressed operons to the leading strand of DNA replication. Nucleic Acids Res. *33*, 3224–3234.

Prieto, A.I., Ramos-Morales, F., and Casadesus, J. (2004). Bile-induced DNA damage in Salmonella enterica. Genetics *168*, 1787–1794.

Prieto, A.I., Ramos-Morales, F., and Casadesus, J. (2006). Repair of DNA damage induced by bile salts in Salmonella enterica. Genetics *174*, 575–584.

Rao, J.E., Miller, P.S., and Craig, N.L. (2000). Recognition of triple-helical DNA structures by transposon Tn7. Proc. Natl. Acad. Sci. U.S.A. *97*, 3936–3941.

Rawal, P., Kummarasetti, V.B., Ravindran, J., Kumar, N., Halder, K., Sharma, R., Mukerji, M., Das, S.K., and Chowdhury, S. (2006). Genome-wide prediction of G4 DNA as regulatory motifs: role in *Escherichia coli* global regulation. Genome Res. *16*, 644–655.

Ribeiro, S.C., Oliveira, P.H., Prazeres, D.M., and Monteiro, G.A. (2008). High frequency plasmid recombination mediated by 28 bp direct repeats. Mol. Biotechnol. *40*, 252–260.

Riehle, M.M., Bennett, A.F., and Long, A.D. (2001). Genetic architecture of thermal adaptation in *Escherichia coli.* Proc. Natl. Acad. Sci. U.S.A. *98,* 525–530.

Riesenfeld, C., Everett, M., Piddock, L.J., and Hall, B.G. (1997). Adaptive mutations produce resistance to ciprofloxacin. Antimicrob. Agents Chemother. *41,* 2059–2060.

Rocha, E.P., and Danchin, A. (2003). Essentiality, not expressiveness, drives gene-strand bias in bacteria. Nat. Genet. *34,* 377–378.

Rosenberg, S.M. (2001). Evolving responsively: adaptive mutation. Nat. Rev. Genet. *2,* 504–515.

Rosenberg, S.M., Shee, C., Frisch, R.L., and Hastings, P.J. (2012). Stress-induced mutation via DNA breaks in *Escherichia coli*: A molecular mechanism with implications for evolution and medicine. Bioessays *34,* 885–892.

Rudolph, C.J., Dhillon, P., Moore, T., and Lloyd, R.G. (2007). Avoiding and resolving conflicts between DNA replication and transcription. DNA Repair (Amst) *6,* 981–993.

Rydberg, B., and Lindahl, T. (1982). Nonenzymatic methylation of DNA by the intracellular methyl group donor S-adenosyl-L-methionine is a potentially mutagenic reaction. EMBO J. *1,* 211–216.

Schaaper, R.M., and Dunn, R.L. (1991). Spontaneous mutation in the *Escherichia coli* lacI gene. Genetics *129,* 317–326.

Schaaper, R.M., Dunn, R.L., and Glickman, B.W. (1987). Mechanisms of ultraviolet-induced mutation. Mutational spectra in the *Escherichia coli* lacI gene for a wild-type and an excision-repair-deficient strain. J. Mol. Biol. *198,* 187–202.

Schakowski, F., Gorschluter, M., Junghans, C., Schroff, M., Buttgereit, P., Ziske, C., Schottker, B., Konig-Merediz, S.A., Sauerbruch, T., Wittig, B., *et al.* (2001). A novel minimal-size vector (MIDGE) improves transgene expression in colon carcinoma cells and avoids transfection of undesired DNA. Mol. Ther. *3,* 793–800.

Shapiro, J.A. (1984). Observations on the formation of clones containing araB-lacZ cistron fusions. Mol. Gen. Genet. *194,* 79–90.

Sharma, S. (2011). Non-B DNA Secondary Structures and Their Resolution by RecQ Helicases. J. Nucleic Acids *2011,* 724215.

Sharma, S., Doherty, K.M., and Brosh, R.M., Jr. (2006). Mechanisms of RecQ helicases in pathways of DNA metabolism and maintenance of genomic stability. Biochem. J. *398,* 319–337.

Shee, C., Gibson, J.L., and Rosenberg, S.M. (2012). Two mechanisms produce mutation hotspots at DNA breaks in *Escherichia coli.* Cell Rep. *2,* 714–721.

Shen, J.C., Rideout, W.M. 3rd, and Jones, P.A. (1994). The rate of hydrolytic deamination of 5-methylcytosine in double-stranded DNA. Nucleic Acids Res. *22,* 972–976.

Shibutani, S., Takeshita, M., and Grollman, A.P. (1991). Insertion of specific bases during DNA synthesis past the oxidation-damaged base 8-oxodG. Nature *349,* 431–434.

Shrivastav, N., Li, D., and Essigmann, J.M. (2010). Chemical biology of mutagenesis and DNA repair: cellular responses to DNA alkylation. Carcinogenesis *31,* 59–70.

Slack, A., Thornton, P.C., Magner, D.B., Rosenberg, S.M., and Hastings, P.J. (2006). On the mechanism of gene amplification induced under stress in *Escherichia coli.* PLoS Genet. *2,* e48.

Sleight, S.C., Orlic, C., Schneider, D., and Lenski, R.E. (2008). Genetic basis of evolutionary adaptation by *Escherichia coli* to stressful cycles of freezing, thawing and growth. Genetics *180,* 431–443.

Spira, B., de Almeida Toledo, R., Maharjan, R.P., and Ferenci, T. (2011). The uncertain consequences of transferring bacterial strains between laboratories – rpoS instability as an example. BMC Microbiol. *11,* 248.

Stivers, J.T., and Jiang, Y.L. (2003). A mechanistic perspective on the chemistry of DNA repair glycosylases. Chem. Rev. *103,* 2729–2759.

Streisinger, G., Okada, Y., Emrich, J., Newton, J., Tsugita, A., Terzaghi, E., and Inouye, M. (1966). Frameshift mutations and the genetic code. This paper is dedicated to Professor Theodosius Dobzhansky on the occasion of his 66th birthday. Cold Spring Harb. Symp. Quant. Biol. *31,* 77–84.

Sun, S., Ke, R., Hughes, D., Nilsson, M., and Andersson, D.I. (2012). Genome-wide detection of spontaneous chromosomal rearrangements in bacteria. PLoS One *7,* e42639.

Taddei, F., Halliday, J.A., Matic, I., and Radman, M. (1997). Genetic analysis of mutagenesis in aging *Escherichia coli* colonies. Mol. Gen. Genet. *256,* 277–281.

Tajiri, T., Maki, H., and Sekiguchi, M. (1995). Functional cooperation of MutT, MutM and MutY proteins in preventing mutations caused by spontaneous oxidation of guanine nucleotide in *Escherichia coli.* Mutat. Res. *336,* 257–267.

Taverna, P., and Sedgwick, B. (1996). Generation of an endogenous DNA-methylating agent by nitrosation in *Escherichia coli.* J. Bacteriol. *178,* 5105–5111.

Tindall, K.R., Stein, J., and Hutchinson, F. (1988). Changes in DNA base sequence induced by gamma-ray mutagenesis of lambda phage and prophage. Genetics *118,* 551–560.

Torkelson, J., Harris, R.S., Lombardo, M.J., Nagendran, J., Thulin, C., and Rosenberg, S.M. (1997). Genome-wide hypermutation in a subpopulation of stationary-phase cells underlies recombination-dependent adaptive mutation. EMBO J. *16,* 3303–3311.

Treangen, T.J., Darling, A.E., Achaz, G., Ragan, M.A., Messeguer, X., and Rocha, E.P. (2009). A novel heuristic for local multiple alignment of interspersed DNA repeats. IEEE/ACM Trans. Comput. Biol. Bioinform. *6,* 180–189.

Twiss, E., Coros, A.M., Tavakoli, N.P., and Derbyshire, K.M. (2005). Transposition is modulated by a diverse set of host factors in *Escherichia coli* and is stimulated by nutritional stress. Mol. Microbiol. *57,* 1593–1607.

Tyrrell, R.M. (1973). Induction of pyrimidine dimers in bacterial DNA by 365 nm radiation. Photochem. Photobiol. *17,* 69–73.

Umenhoffer, K., Feher, T., Baliko, G., Ayaydin, F., Posfai, J., Blattner, F.R., and Posfai, G. (2010). Reduced evolvability of *Escherichia coli* MDS42, an IS-less cellular chassis for molecular and synthetic biology applications. Microb. Cell Fact. *9*, 38.

Vandermeulen, G., Marie, C., Scherman, D., and Preat, V. (2011). New generation of plasmid backbones devoid of antibiotic resistance marker for gene therapy trials. Mol. Ther. *19*, 1942–1949.

Vilette, D., Ehrlich, S.D., and Michel, B. (1996). Transcription-induced deletions in plasmid vectors: M13 DNA replication as a source of instability. Mol. Gen. Genet. *252*, 398–403.

Vinces, M.D., Legendre, M., Caldara, M., Hagihara, M., and Verstrepen, K.J. (2009). Unstable tandem repeats in promoters confer transcriptional evolvability. Science *324*, 1213–1216.

Voineagu, I., Narayanan, V., Lobachev, K.S., and Mirkin, S.M. (2008). Replication stalling at unstable inverted repeats: interplay between DNA hairpins and fork stabilizing proteins. Proc. Natl. Acad. Sci. U.S.A. *105*, 9936–9941.

Wang, G., Zhao, J., and Vasquez, K.M. (2009). Methods to determine DNA structural alterations and genetic instability. Methods *48*, 54–62.

Wyrzykowski, J., and Volkert, M.R. (2003). The *Escherichia coli* methyl-directed mismatch repair system repairs base pairs containing oxidative lesions. J. Bacteriol. *185*, 1701–1704.

Xu, B.L., Yuan, L., Wu, J.X., Xu, N., Fang, W.J., Zhao, P., and Huang, W.L. (2011). Mc-hES, a novel plasmid carrying human endostatin gene, inhibits nasopharyngeal carcinoma growth. Cancer Gene Ther. *19*, 110–117.

Yeiser, B., Pepper, E.D., Goodman, M.F., and Finkel, S.E. (2002). SOS-induced DNA polymerases enhance long-term survival and evolutionary fitness. Proc. Natl. Acad. Sci. U.S.A. *99*, 8737–8741.

Comparative Methods for RNA Structure Prediction

14

Eckart Bindewald and Bruce A. Shapiro

Abstract

The appreciation of the pervasiveness of RNA biology continues to increase. The vastness of available sequence information calls for computational tools that aid in a variety of prediction problems, such as RNA structure prediction, RNA–RNA interaction prediction and genomic scans for conserved RNA structural elements. Comparative methods for RNA structure prediction employ a set of homologous RNA sequences; this additional information can then be used to not only estimate energy contributions, but also information in the form of compensatory base changes. Different software tools make use of this information in a fascinating variety of ways. This paper surveys current comparative approaches for RNA structure predictions and discusses a variety of future trends.

Introduction

In the last two decades it has become increasingly clear, that the central dogma of molecular biology, that DNA is transcribed to RNA, and RNA is translated to protein, is only part of cellular information flow. It is now increasingly appreciated, that RNA molecules can assume many more biomolecular functions in addition to acting (in the form of messenger RNAs) as an intermediate information-carrying step in between genomic material and protein production. After the early discovery of the specialized roles of functional RNAs in translation (transfer RNAs and ribosomal RNAs), it was the discovery of other types of RNAs, that made it increasingly clear, that RNAs play a role in virtually every cellular process such as transcription, transport, splicing, epigenetics and gene regulation. Examples of such classes of RNAs are long non-coding RNAs, microRNAs, small nuclear RNAs, small nucleolar RNAs (Lozada-Chavez *et al.*, 2012). Recently, more than 9000 human long noncoding genes were annotated as part of the ENCODE project (Derrien *et al.*, 2012; Dunham *et al.*, 2012). In hindsight it appears that the initially under-appreciated role seen in RNAs has been one of the great oversights of the field of molecular biology.

The RNA folding problem

RNAs are strands that are (in its unfolded state) a linear chain of ribonucleotides. The four types of ribonucleotides adenine (A), cytosine (C), guanine (G) and uracil (U) can form base pairs that are stabilized by hydrogen bonds. The standard Watson–Crick type pairings are A–U and G–C; also frequently found are G–U or 'wobble' type base pairs. The availability of structural data made clear, however, that there are many additional 'non-canonical' base pair types possible; Leontis and Westhof describe a systematic classification of 12 different base pair classes (Leontis and Westhof, 2001). Additionally, RNA strands can fold and form stable three-dimensional structures. The sequence information (defining the atoms and covalent bonds of an RNA) corresponds to the primary structure, the base pairing information corresponds to the secondary structure, the 3-dimensinoal folding (the x, y, z coordinates of all atoms) corresponds to the tertiary structure of an RNA. For a variety of structured RNA, it has been shown that the precise formation and orientation of core helices is critical for their function (Michel and Westhof, 1990; Montange and Batey, 2008; Gao *et al.*, 2012). It should, however, also be appreciated that, in some cases, it can be the existence of unstructured, single-stranded regions that facilitate function: for example, it has been reported that certain types of internal ribosome entry sites (IRES) that facilitate cap-independent translation of mRNAs are associated with unstructured regions upstream of the initiation codon (Xia and Holcik, 2009).

The RNA folding problem in its general sense is to computationally predict the secondary or tertiary structure of an RNA molecule using the available information (typically the RNA sequence and possibly additional experimental constraints). It has been argued, that the folding of an RNA from its single-stranded unfolded state into a folded structure can, as a reasonable approximation, be viewed as hierarchical: the single-stranded unfolded RNA forms base pairs that give rise to the secondary structure; next, the RNA with already formed base pairs folds in the presence of magnesium into a three-dimensional structure giving rise to a stable tertiary structure (Tinoco and Bustamante, 1999). Recently, evidence has been reported, that structural motifs and a network of tertiary interactions can guide the formation of helices and promote folding into the native structure (Behrouzi *et al.*, 2012). In other words, three-dimensional folding properties may influence the formation of base pairing. Nonetheless, the hierarchical view, although clearly an oversimplification, allows for a straightforward formulation of the single-sequence RNA secondary-structure folding problem:

Given an RNA sequence, what is the base pairing between the nucleotides that the RNA assumes as a result of its folding process?

Most approaches for RNA secondary structure prediction programs only consider the three types of canonical base pairs (AU, GC, GU), however, approaches for predicting secondary structures that possibly contain all 12 Leontis–Westhof classes of RNA base pairings have been described (Parisien and Major, 2008). In other words, there are variants of the RNA secondary structure prediction problem that researchers attempt to solve.

A schematic depiction of the secondary structure of an RNA is shown in Fig. 14.1. An important feature of RNA structure is the stacking of adjacent base pairs. This effect leads, in three dimensions, to the formation of RNA double helices. One can see in Fig. 14.1 additional structural elements that

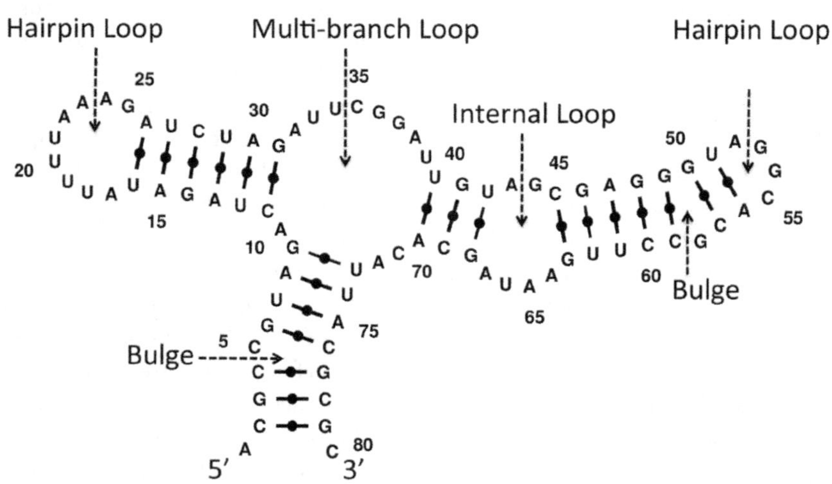

Figure 14.1 A schematic depiction of an RNA secondary structure containing structural elements such as a multi-branch loop, two hairpin loops, one internal loop and two bulges. Also shown are the 5′ and 3′ ends of the RNA sequence as well as nucleotide indices. The shown RNA corresponds to a secondary structure prediction performed by the RNAfold program (Hofacker *et al.*, 1994) applied to a randomized RNA nucleotide sequence that was generated for the purposes of this example.

have been identified such as hairpin loops (H), bulges (B), internal loops (I) and multi-branched loops (M).

The *consensus* RNA structure prediction problem for a set of *aligned* homologous sequences can be thought of as attempting to find the secondary structure (the set of base pairs) such that each base pair of the consensus secondary structure is present in at least half of the secondary structures of the individual RNA sequences. Different programs may, however, use different definitions of what constitutes a consensus, thus solving slightly different problems and complicating their comparison. The *consensus* RNA structure prediction problem for a set of *unaligned* homologous sequences has the additional non-trivial problem of also determining the correct RNA sequence alignment.

Conceptual building blocks for computational RNA structure prediction

There are several concepts and innovations that can be considered as seminal for RNA structure prediction approaches. Because these concepts or algorithms are being reused in a variety of ways in different software implementations, they will be briefly highlighted individually.

Energy models

The estimation of the free energy of folding of an RNA has a long history (Tinoco *et al.*, 1971, 1973). Such energy rule sets are based on experimentally determined thermodynamic parameters of various oligoribonucleotide structures. State of the art energy rule sets estimate the free energy of RNA folding using nearest neighbour base pair stacking parameters, as well as parameters corresponding to the structural elements shown in Fig. 14.1: bulge loops, internal loops, hairpin loops (Xia *et al.*, 1998; Mathews *et al.*, 1999, 2004; Mathews and Turner, 2006).

Pseudoknots versus nestedness

The base pairing of many RNAs has the property of 'nestedness'. This concept is made more clear by drawing RNA base pairing not in its standard way, but as a circular diagram as shown in Fig. 14.2. RNA structures with only nested base pairings are called 'pseudoknot-free' or 'non-pseudoknotted'. This means that if a certain base pair between two nucleotides *i,j* is given, there will be (for non-pseudoknotted structures) no base pairs corresponding to 'crossing arcs'. Structures corresponding to non-nested base pairings are called pseudoknotted.

The importance of the pseudoknot concept is, in addition to the biological importance,

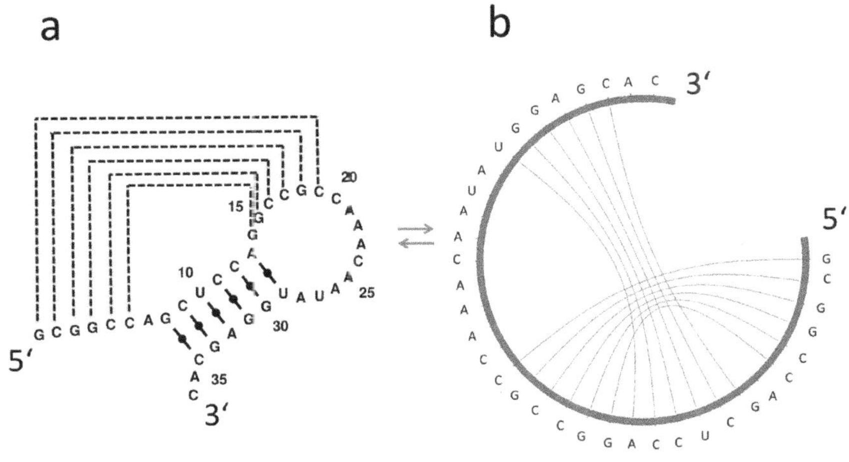

Figure 14.2 Two representations of an RNA secondary structure: (a) conventional secondary structure representation; (b) circular representation. The pseudoknot interaction is shown in (a) as dashed lines, and in the circular representation as 'crossing arcs'.

increased by the situation that there are efficient computational approaches for determining the minimum free energy secondary structure of pseudoknot-free RNAs. This in turn is the case because the nestedness of RNA structures makes possible the application of a special case of the divide-and-conquer paradigm, the so-called dynamic programming algorithm. Conversely, it has been shown, that the general case of determining the minimum free energy secondary structure allowing all possible pseudoknot interactions is equivalent to a class of computational problems for which no algorithm with polynomial complexity has yet been identified (Akutsu, 2000; Lyngso and Pedersen, 2000).

Dynamic programming (DP) applied to single-sequence RNA folding

Dynamic programming algorithms for non-pseudoknotted RNA structure prediction are a class of algorithms that exploit the 'nestedness' of RNA base pairing. It also means, that if a base pair between two nucleotides i,j is given, the structure prediction problem is vastly simplified, because now the optimal solution can be found by independently determining the optimal solution corresponding to the subsequence $(i+1,j-1)$ as well as the optimal solution corresponding to both sequence regions outside the (i,j) interval. The algorithm accomplishes this by determining scores of all subsequences of length one and working outwards according to a set of recursion rules (Dowell and Eddy, 2004).

First descriptions of this approach to maximize base pairing appeared in 1978 (Nussinov *et al.*, 1978; Waterman and Smith, 1978). The mfold program uses a dynamic programming algorithm to find the structure corresponding to the minimum free energy according to a detailed secondary structure energy model (Zuker and Stiegler, 1981). Nowadays, different variants of dynamic programming approaches are used in a wide variety of RNA structure prediction programs. Current research involves extending this basic idea of dynamic programming to RNA structures that can contain certain subclasses of pseudoknots (Rivas and Eddy, 1999; Dirks and Pierce, 2003, 2004; Reeder *et al.*, 2007).

Dynamic programming for simultaneous alignment and folding of RNA sequences

Sankoff described a model that allows for the simultaneous folding and aligning of a pair of RNA sequences (Sankoff, 1985). It is considered to be the most general solution to the folding and aligning problem, however the computational complexity of $O(L^{3n})$ in time and $O(L^{2n})$ in space for n sequences of length L [thus ($O(L^6)$ in time and $O(L^4)$ space for a pair of sequences] limits its usage in practical applications. We can view the list of comparative methods for consensus RNA structure prediction that start from unaligned sequences as a variety of attempts of providing more efficient approaches by using different computational shortcuts, approximations or limitations of the solution space.

RNA folding probability matrices and the McCaskill algorithm

McCaskill (1990) described an algorithm that generates from a dynamic programming matrix that contains the scores of all subsequences a matrix of base pair probabilities $p(i,j)$, where i and j are indices, each representing a position in an RNA sequence. These base pair probabilities should be interpreted as follows: assuming that every possible conformation of an RNA is assigned a Gibbs free energy, and the probability of a particular RNA molecule being chosen from an ensemble has this particular conformation with a probability that is given by a Gibbs-Boltzmann distribution, then, if we pick a large number of RNA structures, the probability of finding two nucleotides with indices i and j that participate in a base pairing is $p(i,j)$.

Compensatory base changes

A key guiding principle for the prediction of the consensus RNA structure of a set of homologous RNA sequences is the existence of compensatory base changes. In other words, if two positions i,j in a set of aligned RNA sequences corresponds to a conserved base pair, one should ideally only observe the 6 nucleotide combinations AU,CG, GC, GU, UA, UG (Gutell *et al.*, 1992). A more realistic view is that a certain amount of non-complementary bases should be tolerated, because there may be sequencing errors, alignment errors or non-complementary base pairs (Gutell *et al.*, 1992). This bias towards complementary bases in alignment position pairs that correspond to a conserved base pair is called compensatory base changes or compensatory information or simply covariation. A spectacular success that demonstrates the utility of the information that is inherent in compensatory base changes was the highly accurate prediction of the secondary structures of 16S and 23S ribosomal RNA (Gutell *et al.*, 2002). This has been accomplished largely by curation efforts of human experts who focused on maximizing compensatory base changes in homologous RNA sequences. In a later section different computational approaches will be mentioned that quantify the compensatory base change concept.

Measures for covariation

A key guiding principle for the prediction of the consensus RNA structure of a set of homologous RNA sequences is the existence of compensatory base changes. When designing a computational approach for the prediction of consensus RNA structures, one certainly wants to make use of this source of information when possible. Several approaches for incorporating covariation in an explicit and implicit manner exist (Bindewald *et al.*, 2006).

Mutual information is a concept based on Shannon's information theory. This approach essentially asks if one can detect additional patterns if one analyses a pair of alignment columns in combination as opposed to those two alignment columns individually (Bindewald *et al.*, 2006). The strength of this approach is its theoretical purity; a weakness is that the approach is ignorant about the properties of RNA base pairing (a bias towards complementarity). In other words, this approach can detect correlations in alignments, but is not able to distinguish a correlation that is due to conserved RNA base pairing from, for example, a correlation due to a RNA–protein interaction. That notwithstanding, mutual information is one of the most commonly used numerical measures for covariation, because its weakness can be compensated for by using either bioinformatic filters or scores (such as interaction energies) (Knight *et al.*, 2004; Bindewald and Shapiro, 2006).

Several approaches that are either extensions or alternatives to the mutual information concept have been described in order to improve upon this measure's ignorance of RNA base pairing preferences (Akmaev *et al.*, 2000; Hofacker *et al.*, 2002).

Covariance models

The concept of a covariance model (CM) has been described for the case of RNA by Eddy and Durbin (Eddy and Durbin, 1994; Durbin and Eddy, 1998). A covariance model can be thought of as a model that can generate the sequences that belong to an RNA family in a probabilistic fashion. *Training* the model means finding that covariance model that corresponds to the maximum likelihood of generating the set of RNA sequences of a training set. Efficient algorithms exist for aligning additional RNA sequences to an initial covariance model as well as for scanning genomic sequence alignments using a covariance model (Eddy and Durbin, 1994). The RFAM database can be thought of as a database of RNA covariance models, each one describing a family of homologous RNA sequences (Burge *et al.*, 2012).

Stochastic context-free grammar

Stochastic context-free grammars (SCFGs) are probabilistic models that are well suited for representing and predicting non-pseudoknotted RNA structures. They can incorporate different sources of information (such as a biophysical energy model as well as evolutionary information). They are a probabilistic version of context free grammars (Dowell and Eddy, 2004). Context-free grammars are by definition 'context free'; applied to RNA structure it means that the score corresponding to a substructure with self-contained base pairing does not depend on the base pairing outside of that substructure.

CFGs consist of a set of nonterminal symbols (states), a finite set of terminal symbols (for RNA the symbols a, c, g, u), a set of production rules and a start state. Probabilistic CFGs have the additional property that production rules are applied using a set of probabilities. Given a start state (typically the unfolded state), and an SCFG model, one can generate RNA structures with different probabilities. An algorithm (the Cocke–Younger–Kasami or CYK algorithm) is able to find the series of production rules (called a parse tree) that corresponds to the highest probability. If the grammar has the additional property of being structurally unambiguous, this parse tree corresponds to a unique RNA secondary structure (Dowell and Eddy, 2004). This result, the highest-probability RNA secondary structure, is the desired RNA secondary structure prediction of a given sequence. It is insightful to appreciate, that the CYK algorithm for finding the highest-probability parse-tree effectively is a dynamic programming algorithm with the difference that log-probabilities instead of biophysical energy contributions are being added. Giving a detailed introduction of SCFG approaches is beyond the scope of this paper; the reader is referred to the literature (Durbin and Eddy, 1998; Dowell and Eddy, 2004).

RNA single-sequence structure prediction

Single sequence secondary structure prediction approaches will be mentioned here only briefly for several reasons: first, they have been reviewed elsewhere (Shapiro *et al.*, 2007). Secondly, the approaches used are, with some noteworthy exceptions, (Shapiro *et al.*, 2001; Ding and Lawrence, 2003; Do *et al.*, 2006; Hamada *et al.*, 2009a; Bindewald *et al.*, 2010; Schroeder *et al.*, 2011) quite homogeneous in their methodology: many single sequence secondary structure prediction programs use a biophysical energy model [often the 'Turner rules' (Xia *et al.*, 1998; Mathews *et al.*, 1999, 2004; Mathews and Turner, 2006)] in combination with a form of the dynamic programming algorithm (Zuker and Stiegler, 1981; Hofacker *et al.*, 1994; Mathews, 2004; Markham and Zuker, 2008). A recent trend is that RNA secondary structure prediction programs are increasingly able to consider pseudoknotted structures (Rivas and Eddy, 1999; Dirks and Pierce, 2004; Reeder *et al.*, 2007). Thirdly, comparative methods, have, because they have by definition multiple homologous sequences as input, more information that can be used for the structure prediction problem. This, combined with the fact that it has been shown that energetic stability alone is not highly predictive for the detection of non-coding RNAs (Rivas and Eddy, 2000) makes comparative methods more attractive for genome-scanning and 'RNA gene' detection applications (Freyhult *et al.*, 2007).

Comparative methods for secondary structure prediction

One of the key concepts of bioinformatics is that a set of aligned homologous sequences can dramatically increase the accuracy with which one can interpret genomic information. RNA structure prediction is no exception to this general observation. The object is, to computationally predict the *consensus* RNA secondary structure of a set of homologous sequences. It was a much-welcomed innovation by Gardner and Giegerich, to classify the growing variety of possible approaches into three different approaches or 'plans' (Fig. 14.3) (Gardner and Giegerich, 2004). According to this classification, 'Plan A' stands for first aligning the RNA sequences using regular sequence alignment approaches and

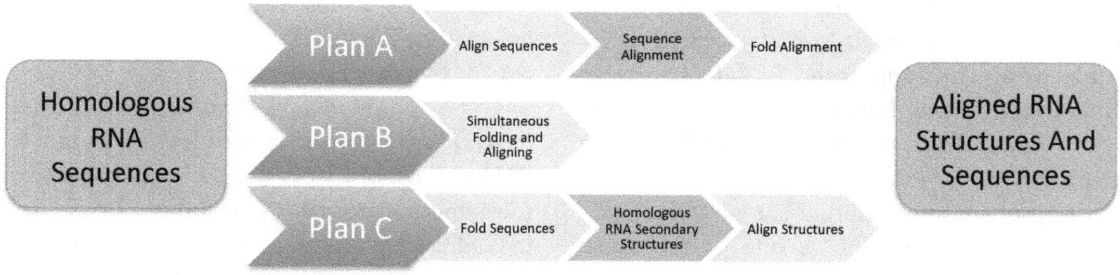

Figure 14.3 The '3 plans' for the prediction of the consensus RNA secondary structure from a set of homologous RNAs (Gardner and Giegerich, 2004).

subsequently performing structure predictions of individual sequences. 'Plan B' stands for algorithms that simultaneously align and fold RNA sequences. 'Plan C' stands for approaches in which first the predicted RNA secondary structure is computed individually for each sequence; then the set of resulting RNA secondary structures is subsequently aligned. We note, however, that the case of 'Plan B' (simultaneously aligning and folding) can indeed 'behind the scenes' be accomplished by internally following a strategy that resembles more plan A or plan C. Because it is in those cases not clear, how to assign a 'plan' to the approach, we use here a user-oriented classification: RNA consensus secondary structure prediction methods either use a set of aligned or unaligned RNA sequences as input, leading to two cases. The third case of a user input of unaligned RNA sequences that are annotated by RNA structures is not as commonly used and is not described here.

Consensus secondary structure prediction for a set of aligned RNA sequences

Feature-based machine learning approaches

The program KNetFold is based on a machine learning approach rather than an energy-based RNA folding approach (Bindewald and Shapiro, 2006). For a given set of aligned RNA sequences (length of alignment is L nucleotides), the program generates 3 $L \times L$ matrices that correspond to three different features. The first feature matrix consist of mutual information values, a measure for compensatory base changes between any two positions in the RNA sequence alignment (Bindewald *et al.*, 2006). The second feature matrix consists of values that denote the fraction of complementary base pairs (the fraction of sequences that have for a position pair *i,j* either the nucleotides G:C, A:U, or G:U). Energy-based RNA folding properties are being considered in this approach in an indirect way: The third feature matrix is a consensus of the base pairing probability matrices as produced by the RNAfold program (Hofacker *et al.*, 1994). The program is essentially a classifier, that for each position pair makes a prediction, whether or not it is more likely to be base paired. This is accomplished by using as features not only the 3 features of the current position pair, but a small 5×5 window, a 'patch' of the feature matrices. This leads to $5 \times 5 \times 3 = 75$ potential features. Important for the success of this approach was the ability to reduce the high dimensional feature-space to a set of lower-dimensional classifiers (each one being two- or three-dimensional). The hierarchical organization of the classifiers is essentially a practical implementation of a Bayesian belief network, in which a Bayesian probability is forwarded in a network until a final prediction is achieved. The information flow starting from initial feature matrices to a final matrix with classifier results (the desired structure prediction) is shown in Fig. 14.4.

A similar strategy is used by the program Bayesfold (Knight *et al.*, 2004) with key differences that the program considers nucleotide pair scores without considering local context but offers, on the other hand, the possibility of using per-position scores (for example chemical probing data) as another source of information.

Probabilistic approaches

Pfold is based on the SCGF approach (Knudsen and Hein, 1999, 2003). The clean incorporation of evolutionary information as well as compensatory base change information into one SCFG that has been trained using a set of aligned RNA sequences has been seminal. This approach is called phylogenetic SCFG or phylo-SCFG. The XRate software is a generalized software that facilitates prototyping of user-defined phylo-SCFGs (Klosterman *et al.*, 2006). The PETfold approach is an extension of the Pfold stochastic context free grammar approach (Knudsen and Hein, 1999, 2003; Seemann *et al.*, 2008). The Pfold model is extended such that the predicted consensus structure contains base pairs that have a high probability of being conserved and also are energetically favourable.

Energy-based consensus folding approaches

The RNAalifold approach is based on the idea of combining energy-based RNA folding with

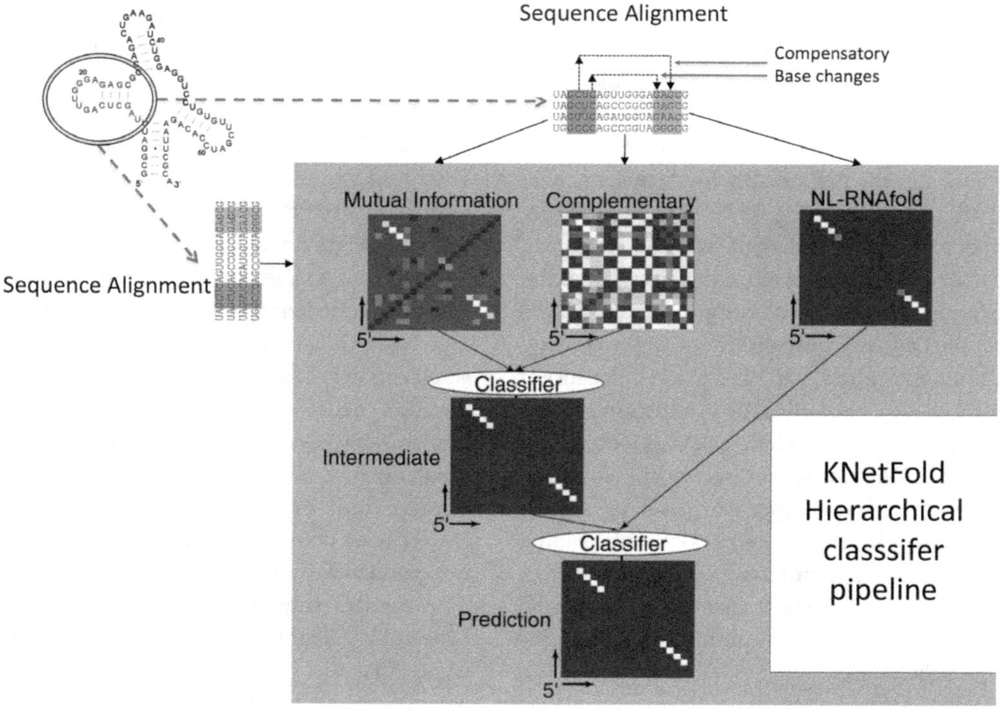

Figure 14.4 Depiction of the computational processing pipeline in KNetFold for generating a secondary structure prediction. From a set of aligned sequences, 3 feature matrices are generated (top row). A network of k-nearest neighbour classifiers is used to reduce the 'noise' inherent in these feature matrices and generate a matrix with classifier results that correspond to the predicted secondary structure (Bindewald and Shapiro, 2006).

compensatory information (Hofacker *et al.*, 2002). As a measure for covariation, not the mutual information but a novel measure (also used by the program RSPredict (Spirollari *et al.*, 2009)) that incorporates preferences of RNA base pairing is used (Hofacker *et al.*, 2002). The score assigned to each alignment column pair is a linear combination of an energy component (the sequence-average base pairing energy) and a thresholded covariation term. A dynamic programming algorithm applied to this score matrix is used to determine the predicted consensus RNA secondary structure.

Consensus secondary structure prediction for a set of *un*aligned RNA sequences

Predicting consensus secondary structure for a set of unaligned RNA sequences is the computationally harder problem, because in addition to the consensus secondary structure an alignment has to be generated. It is, ultimately, the most interesting case, in which the user does not introduce any prior knowledge, except the assumption that the input sequences are homologous. This is attractive; the flipside of this paradigm is that the involved methods are computationally significantly more costly compared to the case in which an alignment is given as input.

The approaches for simultaneous folding and aligning of RNA structures can be classified into three different classes: energy-based methods, probabilistic methods and graph-based methods. Even though a broad-scale systematic comparison between these approaches still has to be performed, it is generally viewed that energy-based methods have the advantage of higher accuracy, while probabilistic methods have the advantage of incorporating covariation information into the prediction in a less ad-hoc manner (Kiryu *et al.*, 2007).

Simultaneous folding and alignment using an 'annotated sequence approach'

Sankoff described a dynamic programming approach for a structural alignment of RNAs that corresponds to a simultaneous folding and sequence alignment (Sankoff, 1985). If L is the length of the RNA sequences and n is the number of input RNA sequences, the time-complexity of the original Sankoff algorithm is $O(L^{3n})$ and the complexity for space is $O(L^{2n})$ (Hamada *et al.*, 2009b). This complexity clearly makes it prohibitive to apply the Sankoff algorithm in its initial form to practical situations for a large number of sequences. Several programs mentioned below, take inspiration from the Sankoff model, but utilize a heuristic or a computational shortcut in order to reduce the time and space complexity of the Sankoff algorithm. One obvious and commonly used strategy, employed by programs such as Multilign (Xu and Mathews, 2011), is to compute pair wise alignments followed by progressively assembling a multiple sequence alignment. This avoids the case of applying the computationally costly Sankoff approach to more than two sequences simultaneously.

The pmcomp approach is an approach for producing, starting from unaligned RNA sequences, a set of aligned and folded RNA structures, while avoiding the computational complexity of the original Sankoff algorithm (Hofacker *et al.*, 2004). The method applied to a pair of unaligned RNA sequences starts by computing the base pairing probability matrices of the RNA sequences using a McCaskill algorithm (McCaskill, 1990). In a subsequent computational stage, an alignment of the base pairing probability matrices is performed using a dynamic programming approach. The computational complexity of this approach is $O(n^5)$. By limiting the difference of subsequence lengths that are compared with each other, a computational complexity of $O(n^3)$ is achieved. This allows for a pairwise quasi-simultaneous folding and alignment. A multiple RNA sequence-structure alignment can be computed by computing all pairwise alignments, and from that a guide tree can be computed that is used to progressively compute a multiple RNA sequence alignment (pmmulti). Extended and more efficient implementations of the pmcomp program are the programs FoldAlignM (Torarinsson *et al.*, 2007) and LocARNA (Will *et al.*, 2007). With LocARNA it is possible to align several thousand sequences.

Typical strategies for improving the computational complexity involve the use of sliding windows and limits on the maximum sequence separation of base pairing nucleotides. Examples of such approaches are the programs Dynalign (Mathews and Turner, 2002) and FoldAlign (Havgaard *et al.*, 2005). In the FoldAlign approach the computational complexity is reduced by setting an upper limit on the length differences of subsequences that are considered for potential alignments. The program Stemloc achieves reduced computational complexity by simultaneously employing constraints on the considered alignments and folds (Holmes, 2005).

The program Murlet is also based on the Sankoff approach for aligning predicted single-sequence base pair probability matrices. Extensive preprocessing (such as excluding from consideration structural alignments with base pairs that have a low probability in the single-sequence base pair probability matrices) reduces the computational cost so that sequence lengths of more than 300 nucleotides are becoming practical (Kiryu *et al.*, 2007).

The SCARNA method avoids the use of the computationally expensive Sankoff algorithm altogether (Tabei *et al.*, 2006). To align a pair of RNA sequences, a list of 5′ and 3′ regions corresponding to fixed-length stem-candidates is generated for both sequences. These fixed-length sequence regions corresponding to stem-halves are aligned using a hidden-Markov-model, analogous to a regular sequence alignment. Unlike a stochastic context free grammar, the hidden Markov model cannot represent nestedness. This can lead to inconsistent sequence regions that are being matched; those inconsistent matches are removed from the solution space using a filtering heuristic. This approach is extended to a multiple-sequence alignment version with the program MXSCARNA (Tabei *et al.*, 2008). A related method that performs a consensus secondary structure prediction by identifying stems that are supported by consistent matches between sequence pairs is Tfold (Engelen and Tahi, 2010). The program RNA Sampler uses an approach for predicting conserved stems by probabilistically sampling aligned stems based on a stem conservation score that consists of an intra-sequence component (base pairing probabilities) and an inter-sequence component (base alignment probabilities) (Xu *et al.*, 2007).

The program TurboFold uses an approach that combines sequence-intrinsic information (predicted base pairing probabilities for a given sequence) with sequence-extrinsic information (pairwise sequence alignment based probabilities of co-incidence of matching and base pairing). The extrinsic information updates the used energy model in the next iteration of computing the intrinsic base pairing prediction. This process converges after about three iterations (Harmanci *et al.*, 2011). Recently, TurboKnot, an extension that considers pseudoknots, has been reported (Seetin and Mathews, 2012).

Simultaneous folding and alignment with a probabilistic approach

CMfinder generates covariance models to describe motifs, a finite mixture model describes how motifs are distributed in sequences, and an expectation maximization algorithm is used to search motif space (Yao *et al.*, 2006). The method consists of two computational stages: (i) construction of a heuristic initial alignment (stable secondary structure elements are chosen as candidates using RNAfold; next a consensus alignment is generated based on a tree-edit model); and (ii) refinement of alignments is performed by optimizing the log-likelihood of the combination of covariance models that represents motifs.

The CARNAC approach consists of three computational stages: First, a list of all possible stems of all involved sequences is generated. In the second stage, pair wise alignments between stems corresponding to different sequences are performed. The optimal common secondary structure between sequence pairs is performed based on the list of aligned stems. In the third stage, a consensus structure is generated by subsequently placing stems according to a consensus-confidence measure the authors describe in their paper (Perriquet *et al.*, 2003; Touzet and Perriquet, 2004). Another systematic application of the stochastic context-free grammar approach to a pair of RNA sequences (pairSCFG) has been reported for the Consan method (Dowell and Eddy, 2006).

The program SimulFold is a method for predicting an RNA sequence alignment, a consensus structure (including pseudoknots) and an evolutionary tree, given as input a set of homologous unaligned RNA sequences. It uses a Bayesian approach, in which the predicted consensus RNA secondary structure is one that is best supported by a posterior probability. The involved Bayesian identities are not solved analytically, but by using a non-deterministic Markov Chain Monte Carlo sampling approach (Meyer and Miklos, 2007).

Simultaneous folding and alignment using scores and non-deterministic optimization

The MASTR program is the implementation of a simulated annealing optimization approach for simultaneous folding and aligning a set of homologous RNA sequences (Lindgreen *et al.*, 2007). The program optimizes an objective function that consists of three terms representing sequence conservation, base pairing probability matrices and covariation matrices respectively. It is an interesting case in which a non-deterministic algorithm performs competitively against deterministic algorithms that are guaranteed to find the global optimum (Lindgreen *et al.*, 2007).

Graph-based approaches

Bauer *et al.* (2007) present the program LARA. The problem of aligning two RNA sequences that are annotated by a predicted structure is converted to a graph-representation and modelled as an integer linear program. They report a time complexity of $O(L^2)$ for aligning a pair of RNA sequences of length L. For the multiple sequence case, the program computes all pairwise alignments and employs a progressive alignment software.

Genome-wide search for structured RNAs

The described arsenal of computational tools that allows one to predict a consensus RNA secondary structure from a set of aligned or unaligned RNA sequences is a good starting point for searching for RNAs with conserved secondary structure on a genome-wide scale. A common approach is to assume that functional non-coding RNAs require certain secondary and possibly tertiary structures for their function. There is, however, no inherent reason that would prohibit the possibility of non-coding RNAs that have a biological function from not having intra-molecular base pairing. In other words, the approaches described below are implementations for the problem of detecting structured RNA elements or structured RNAs; the more general problem of detecting all noncoding RNAs (with or without conserved secondary structure) will require additional or different approaches containing elements of gene detection that have been used to identify protein-coding genes. Also, the possibility that transcripts can act both as protein-coding messenger RNAs as well as non-coding RNAs should be appreciated (Dinger *et al.*, 2011). For an overview of the growing field of genomic screens for structured RNAs the reader is also referred to several recent review articles (Backofen *et al.*, 2007; Gorodkin *et al.*, 2010; Gorodkin and Hofacker, 2011).

Because it has been reported, that a stable predicted secondary structure alone is not sufficient for the prediction of non-coding RNAs (Rivas and Eddy, 2000), the employed methods typically involve the analysis of genomic alignments, often involving an energy-based and a covariation component. A commonly used (but not required) idea is to analyse genomic alignments in a sequence-window like fashion, with typical window sizes ranging from 100 nt to 200 nt (Le and Shapiro, 2010). The sheer amount of genomic material make the direct application of some of the methods (in particular the methods for unaligned sequences) listed above prohibitive.

The software RNAmotif is capable of detecting user-specific RNA motifs in nucleotide sequences (Macke *et al.*, 2001). Specialized software is available for the detection of certain classes of structured RNA genes: BRUCE for tmRNAs (Laslett *et al.*, 2002), tRNAscan-SE for tRNAs, miRseeker for microRNAs (Lai *et al.*, 2003), snoscan for box C/D snoRNAs, snoGPS for box H/ACA snoRNAs (Schattner *et al.*, 2005). If the task is to identify such specific classes of non-coding RNAs, those specialized programs are preferable compared to general-purpose RNA consensus structure detection programs.

The first covariance model based program that has been applied to genomic searches for conserved RNA structure was the SCFG based program QRNA (Rivas and Eddy, 2001; Rivas *et al.*, 2001). This lead to important insights regarding the abundance of conserved RNA structural elements as well as the realization, that single-sequence approaches that search for stable RNA secondary structure are limited in their utility for RNA gene detection.

The RNAz program has been pioneering in several aspects: first, it utilizes the strength of both the energy-based RNA structure prediction approach as well as the strength of a machine learning approach, called support-vector machines. Secondly, it was the first program, for which genome scale-results of RNA gene scanning were reported. Through this work it became clear that genomes of higher organisms contain a plethora of genomic regions that contain the 'fingerprint' of conserved nucleotide base pairing, possibly as a result of conserved RNA structure (Gruber *et al.*, 2007; Washietl *et al.*, 2005; Washietl, 2007).

The program Alifoldz combines several scores into one prediction that indicates whether a part of the genome corresponds to conserved RNA structure (Washietl and Hofacker, 2004). This program is a good example of how competence in RNA structure prediction can be built upon for the goal of detecting conserved RNA structural elements in genomes. The program takes as input aligned RNA or genomic sequences. If the program is presented with a set of aligned genomic sequences, it uses the 'sliding-window' approach: alignment segments of, for example, 150 nucleotides lengths are extracted from the genomic sequence alignment. The minimum free energy of each of the sequences of an extracted alignment is computed; from these MFE values an average energy is computed. To this average MFE value a second term is added that is a measure of compensatory base changes. This term favours, for a predicted consensus secondary structure, compensatory base changes while penalizing mismatches.

For genome scanning, where the most important question is not 'what is the structure' but 'where is the structure', it is vital to have a component that determines the statistical significance of the detected conserved structure. This can be accomplished by utilizing shuffled alignments. Great care has been taken by the developers of AliFoldz in using a shuffling procedure that does not introduce artefacts (and thus spurious 'significance') by ensuring that gap patterns and local conservation patterns are preserved and not lost 'in the shuffle'. Comparing the obtained structure/compensatory-base-change score with the scores corresponding to 100 shuffled alignments leads to one obtained *z*-score or *P*-value. Approaches that are inherently probabilistic, such as SCFG, have the advantage that they can immediately compute a P-value without having to compute the scores corresponding to hundreds of shuffled alignments.

Evofold is a genome-scanning application of the Stochastic Context Free Grammar approach pioneered by the Pfold approach. Again, a sliding window of a genomic alignment is chosen and analysed using the SCFG approach. One key advantage of this approach is that it computes a *P*-value without the necessity of sequence alignment shuffling. This makes the program well suited for genome-scanning purposes.

Discussion

The field of RNA biology is thriving: new experimental data becomes available, new classes of functional RNAs are being discovered, non-coding RNAs are being discovered and characterized, and the appreciation for the importance of RNA biology keeps increasing. This situation translates to the field of computational RNA structure prediction and analysis. Much progress has been made in the last three decades from single-sequence structure prediction, to multiple sequence consensus structure prediction.

A general important theme is that genomic screens for structured RNAs often involve the use of computational pipelines. Designing computational methods for solving biological problems almost always contain trade-offs between accuracy on one hand and computational efficiency on the other hand; RNA structure prediction approaches do not constitute an exception to this generality. Obtaining accurate results in an efficient manner is nonetheless possible; it typically is implemented in the form of computational pipelines. A program optimized for the fast screening of genomes (for example programs that take aligned sequences as input) can lead to a set of initial results; these results can be further analysed using the slower and more accurate programs that correspond to simultaneous aligning and folding. Examples of such computational pipelines can be found in recent review papers (Backofen *et al.*, 2007; Gorodkin *et al.*, 2010; Gorodkin and Hofacker, 2011).

The parameter sets for energy models keep evolving (Xia *et al.*, 1998; Mathews *et al.*, 1999, 2004; Mathews and Turner, 2006). Energy contributions of pseudoknots as well as multi-branched loops are, however, currently not as well understood. Also, the notion that 3D motifs, folding pathways, co-transcriptional folding and tertiary interactions can influence helix formation may inspire the design of prediction algorithms with increased sophistication (Behrouzi *et al.*, 2012). The arsenal of available computational tools for the prediction of consensus RNA structures has reached a certain level of maturity; further breakthroughs will likely involve various extensions of the basic structure prediction problem, such as the incorporation of experimental constraints, the characterization of RNA–RNA interactions, the incorporation of three-dimensional models and RNA design.

It is now becoming increasingly important to provide guidance to the user regarding the accuracy one can expect for a given prediction task and a certain software tool. Gardner and Giegerich performed such an evaluation using a set of 4 RNA families with alignment information (Gardner and Giegerich, 2004). They find that comparative methods (such as RNAalifold, Pfold, CARNAC, Dynalign, Foldalign) lead to a higher prediction accuracy compared to single-sequence methods. Given the small test set size, it is, however, not straightforward to draw general conclusions; the reported prediction accuracy (a variant of the Matthews correlation coefficient, for which 1.0 indicates a perfect prediction) is for most predictions well below 0.9. Given that the lists of available programs and available RNA structures have grown considerably since the date of publication, it will be interesting to perform a new round of comprehensive comparisons. This will be particularly interesting with respect to the 'chicken and egg' problem of comparative methods: compensatory base changes act as anchor points for the prediction algorithms; they are, however rare in the extreme cases of high sequence similarity (high conservation) on one hand and low sequence similarity (low alignment quality) on the other hand (Meyer and Miklos, 2007).

The growing set of available computational tools should be able continue to do its part to keep the 'exploding' field of RNA biology exciting.

Future trends

A current trend that will certainly continue into the near future is to incorporate experimental constraints into the RNA structure prediction algorithms. Constraints can originate from different types of experiments such as selective hydroxyl acetylation by primer extension (SHAPE) (Merino *et al.*, 2005; Wilkinson *et al.*, 2006), hydroxyl radical footprinting (Tullius and Greenbaum, 2005) or in-line probing (Regulski and Breaker, 2008; Wakeman and Winkler, 2009). These experiments provide information about which nucleotides of an RNA structure participate in base pairing (Bindewald *et al.*, 2011b). In other words, the mentioned approaches do not provide *direct* information about which nucleotides base pair with which, but can indirectly guide prediction programs towards solutions that are compatible with the results. A new version of the program Pfold, for example, can incorporate such additional experimental information in a probabilistic fashion (Sukosd *et al.*, 2012). It is clear that the additional experimental information is increasing the accuracy of prediction programs. Further research is needed to determine what gains in accuracy one can expect by incorporating SHAPE constraints into structure prediction approaches applied to different types of RNAs (Low and Weeks, 2010; Kladwang *et al.*, 2011). It would be desirable to have high-throughput experimental approaches that go beyond one-dimension to generate residue-residue distance constraints.

A parallel trend is to go beyond secondary structure prediction and attempt the tertiary structure prediction problem. A variety of programs, such as FARNA (Das and Baker, 2007), FARFAR (Das *et al.*, 2010), iFoldRNA (Sharma *et al.*, 2008), NAST (Jonikas *et al.*, 2009), RNA-composer (Popenda *et al.*, 2012), RNA2D3D (Martinez *et al.*, 2008), RSIM (Bida and Maher, 2012) and Mc-Sym (Parisien and Major, 2008), compute a predicted 3D structure starting from an RNA sequence that is typically augmented by additional constraints such as secondary structure. For a review of RNA 3D modelling and prediction approaches see (Shapiro *et al.*, 2007; Bailor *et al.*, 2011; Laing and Schlick, 2011). This development mirrors somewhat the approaches in protein structure prediction (Rother *et al.*, 2011a), with the exploration of coarse-grained approaches, fragment approaches, all-atom folding simulations and pipeline-style combinations thereof. Successes have also been reported for using tertiary structure prediction as a way to inform single-sequence secondary structure prediction (Parisien and Major, 2008; Bindewald *et al.*, 2010). An interesting case of that strategy is McFold that uses a dynamic programming approach (similar in spirit to the secondary structure prediction case) for the assembly of small 3D motifs that then result in a secondary structure prediction (Parisien and Major, 2008). The mentioned programs are single-sequence 3D structure predictors. Similar to the case of secondary structure prediction, one can expect future programs to exhibit higher prediction accuracy if the information from homologous sequences (such as covariation) can be used. An extreme case is one in which a 3D structure is known for an RNA sequence that is homologous to the query sequence. This leads to the RNA homology-modelling problem. The basic idea is that 'similar sequence implies similar 3D structure'. This is fruitful concept, although it is known that homologous RNA sequences can have a considerable amount of divergence in 3D structure (Rother *et al.*, 2011a). Thankfully, information obtained from covariance analysis can readily be used as base pair interaction constraints in 3D prediction programs. Software packages for RNA 3D structure homology are just emerging, with existing implementations being RNAbuilder (Flores *et al.*, 2009) and ModeRNA (Rother *et al.*, 2011a,b).

It is also increasingly appreciated, that there are cases in which RNA folding pathways can influence the assumed native structure (Behrouzi *et al.*, 2012) as well as give rise to functionally important intermediate structures as has been shown using the program MPGAfold (Shapiro *et al.*, 2001, 2006). Another example of a single-sequence based secondary structure folding pathway prediction program is KineFold (Behrouzi *et al.*, 2012). Cotranscriptional folding is a special case of considering folding pathways; it is based on the fact that RNAs are being transcribed in a linear fashion, potentially causing the nascent RNA strand to fold into a structure that differs from a structure that would result from a folding process that starts from an unfolded state of a complete RNA strand (Shapiro *et al.*, 2001; Pan and Sosnick, 2006; Linnstaedt *et al.*, 2009; Zhao *et al.*, 2012). It would be interesting to apply comparative approaches to the prediction and analysis of RNA folding pathways.

An important trend is the increasing emphasis on RNA–RNA interaction predictions. An example of a non-comparative method is RNAcofold, a variant of the RNAfold approach, in which the minimum free energy structure of two RNAs is computed (Bernhart *et al.*, 2006). The algorithm considers pseudoknot-free secondary structures of the predicted dimer complex. Simultaneous interactions between possibly more than two RNA strands can be considered by NUPACK (Zadeh *et al.*, 2010) or NanoFolder (Bindewald *et al.*, 2011a). Very few comparative methods exist for the general RNA–RNA interaction case. The PETcofold approach is a comparative method for RNA–RNA interaction prediction based on PETfold (Seemann *et al.*, 2010, 2011). In this approach the intra-molecular base pairing is in a first computational stage predicted with the PETfold method (described earlier) for each interaction partner individually. A second-stage co-folding step is used to compute predicted base pairs among the remaining accessible nucleotides. The Cova RNAsoftware can perform a search for long-range covariation in genomic alignments, thus facilitating the prediction of trans-RNA–RNA interactions (Bindewald and Shapiro, 2013). As in the case of RNA gene prediction: if special cases of RNA–RNA interactions are being considered (such as microRNA-target prediction), then programs that are especially designed for such predictions should be considered. Programs for microRNA target prediction are RegRNA (Huang *et al.*, 2006), PicTar (Krek *et al.*, 2005), miRanda (John *et al.*, 2004) and TargetScan (Lewis *et al.*, 2003). The list is growing and the reader is encouraged to consult relevant review papers (Das, 2012).

Another trend is to go beyond single-strand RNA structure prediction towards the prediction of a complex of multiple RNA structures. Several approaches have been described for prediction RNA–RNA interaction structures for the case of two sequences (Bernhart *et al.*, 2006). Programs such as NUPACK and NanoFolder can generate predicted complexes consisting of more than two RNA strands (Dirks and Pierce, 2003, 2004; Zadeh *et al.*, 2010; Bindewald *et al.*, 2011a). These approaches are, however, not comparative. The prediction of complexes consisting of 3 or more sequences is particularly important for the design of RNA nanostructures.

The RNA design problem turns the RNA secondary structure prediction problem upside down: given a target RNA base pairing, what should the RNA sequence be? This is the RNA design problem, or the inverse RNA structure prediction problem. This is of particular interest for the design of RNA nanostructures that consist of multiple RNA strands (Zadeh *et al.*, 2010; Bindewald *et al.*, 2011a).

Taken together: computational tools for RNA structure prediction have come a long way from single-sequence energy-based structure predictors to comparative methods for RNA–RNA interaction prediction. The trends in algorithm development of increasing incorporation of multiple sources of information (such as experimental information) and prediction of RNA interaction partners will lead to an increase of prediction accuracy and scope. These are exciting times for developers and users of RNA structure prediction programs alike.

Web resources

A list of current RNA secondary structure prediction programs can be found under http://en.wikipedia.org/wiki/List_of_RNA_structure_prediction_software. The CompaRNA web server is a recently published continuous benchmarking system for RNA secondary structure prediction approaches (Puton *et al.*, 2013). It is available at http://genesilico.pl/comparna/

Acknowledgments

This project has been funded in whole or in part with Federal funds from the Frederick National Laboratory for Cancer Research, National Institutes of Health, under contract HHSN261200800001E. The content of this publication does not necessarily reflect the views or policies of the Department of Health and Human Services, nor does mention of trade names, commercial products or organizations imply endorsement by the US Government. This research was supported in part by the Intramural Research Program of the National Institutes of Health, Centre for Cancer Research. Frederick National Laboratory for Cancer Research, Centre for Cancer Research.

References

Akmaev, V.R., Kelley, S.T., and Stormo, G.D. (2000). Phylogenetically enhanced statistical tools for RNA structure prediction. Bioinformatics *16*, 501–512.

Akutsu, T. (2000). Dynamic programming algorithms for RNA secondary structure prediction with pseudoknots. Discrete Appl. Math. *104*, 45–62.

Backofen, R., Bernhart, S.H., Flamm, C., Fried, C., Fritzsch, G., Hackermuller, J., Hertel, J., Hofacker, I.L., Missal, K., Mosig, A., *et al.* (2007). RNAs everywhere: genome-wide annotation of structured RNAs. J. Exp. Zool. B Mol. Dev. Evol. *308*, 1–25.

Bailor, M.H., Mustoe, A.M., Brooks, C.L. 3rd, and Al-Hashimi, H.M. (2011). Topological constraints: using RNA secondary structure to model 3D conformation, folding pathways, and dynamic adaptation. Curr. Opin. Struct. Biol. *21*, 296–305.

Bauer, M., Klau, G.W., and Reinert, K. (2007). Accurate multiple sequence-structure alignment of RNA sequences using combinatorial optimization. BMC Bioinformatics *8*, 271.

Behrouzi, R., Roh, J.H., Kilburn, D., Briber, R.M., and Woodson, S.A. (2012). Cooperative tertiary interaction network guides RNA folding. Cell *149*, 348–357.

Bernhart, S.H., Tafer, H., Muckstein, U., Flamm, C., Stadler, P.F., and Hofacker, I.L. (2006). Partition function and base pairing probabilities of RNA heterodimers. Algorithms Mol. Biol. *1*, 3.

Bida, J.P., and Maher, L.J. 3rd (2012). Improved prediction of RNA tertiary structure with insights into native state dynamics. RNA *18*, 385–393.

Bindewald, E., and Shapiro, B.A. (2006). RNA secondary structure prediction from sequence alignments using a network of k-nearest neighbor classifiers. RNA *12*, 342–352.

Bindewald, E., and Shapiro, B. A. (2013). Computational detection of abundant long-range nucleotide covariation in Drosophila genomes. RNA *19*, 1171–1182.

Bindewald, E., Schneider, T.D., and Shapiro, B.A. (2006). CorreLogo: an online server for 3D sequence logos of RNA and DNA alignments. Nucleic Acids Res. *34*, W405–411.

Bindewald, E., Kluth, T., and Shapiro, B.A. (2010). CyloFold: secondary structure prediction including pseudoknots. Nucleic Acids Res. *38*, W368–372.

Bindewald, E., Afonin, K., Jaeger, L., and Shapiro, B.A. (2011a). Multistrand RNA secondary structure prediction and nanostructure design including pseudoknots. ACS Nano. *5*, 9542–9551.

Bindewald, E., Wendeler, M., Legiewicz, M., Bona, M.K., Wang, Y., Pritt, M.J., Le Grice, S.F., and Shapiro, B.A. (2011b). Correlating SHAPE signatures with three-dimensional RNA structures. RNA *17*, 1688–1696.

Burge, S.W., Daub, J., Eberhardt, R., Tate, J., Barquist, L., Nawrocki, E.P., Eddy, S.R., Gardner, P.P., and Bateman, A. (2012). Rfam 11.0: 10 years of RNA families. Nucleic Acids Res. *41*, D226–D232.

Das, N. (2012). MicroRNA Targets – How to predict? Bioinformation *8*, 841–845.

Das, R., and Baker, D. (2007). Automated *de novo* prediction of native-like RNA tertiary structures. Proc. Natl. Acad. Sci. U.S.A. *104*, 14664–14669.

Das, R., Karanicolas, J., and Baker, D. (2010). Atomic accuracy in predicting and designing noncanonical RNA structure. Nat. Methods *7*, 291–294.

Derrien, T., Johnson, R., Bussotti, G., Tanzer, A., Djebali, S., Tilgner, H., Guernec, G., Martin, D., Merkel, A., Knowles, D.G., *et al.* (2012). The GENCODE v7 catalog of human long noncoding RNAs: analysis of their gene structure, evolution, and expression. Genome Res. *22*, 1775–1789.

Ding, Y., and Lawrence, C.E. (2003). A statistical sampling algorithm for RNA secondary structure prediction. Nucleic Acids Res. *31*, 7280–7301.

Dinger, M.E., Gascoigne, D.K., and Mattick, J.S. (2011). The evolution of RNAs with multiple functions. Biochimie *93*, 2013–2018.

Dirks, R.M., and Pierce, N.A. (2003). A partition function algorithm for nucleic acid secondary structure including pseudoknots. J. Comput. Chem. *24*, 1664–1677.

Dirks, R.M., and Pierce, N.A. (2004). An algorithm for computing nucleic acid base pairing probabilities including pseudoknots. J. Comput. Chem. *25*, 1295–1304.

Do, C.B., Woods, D.A., and Batzoglou, S. (2006). CONTRAfold: RNA secondary structure prediction without physics-based models. Bioinformatics *22*, e90–98.

Dowell, R.D., and Eddy, S.R. (2004). Evaluation of several lightweight stochastic context-free grammars for RNA secondary structure prediction. BMC Bioinformatics *5*, 71.

Dowell, R.D., and Eddy, S.R. (2006). Efficient pairwise RNA structure prediction and alignment using sequence alignment constraints. BMC Bioinformatics *7*, 400.

Dunham, I., Kundaje, A., Aldred, S.F., Collins, P.J., Davis, C.A., Doyle, F., Epstein, C.B., Frietze, S., Harrow, J., Kaul, R., *et al.* (2012). An integrated encyclopedia of DNA elements in the human genome. Nature *489*, 57–74.

Durbin, R., and Eddy, S.R. (1998). Biological Sequence Analysis: Probabilistic Models of Proteins and Nucleic Acids (Cambridge, UK, Cambridge University Press).

Eddy, S.R., and Durbin, R. (1994). RNA sequence analysis using covariance models. Nucleic Acids Res. *22*, 2079–2088.

Engelen, S., and Tahi, F. (2010). Tfold: efficient *in silico* prediction of non-coding RNA secondary structures. Nucleic Acids Res. *38*, 2453–2466.

Flores, S.C., Wan, Y., Russell, R., and Altman, R.B. (2009). Predicting RNA structure by multiple template homology modeling. Pac. Symp. Biocomput. 216–227.

Freyhult, E.K., Bollback, J.P., and Gardner, P.P. (2007). Exploring genomic dark matter: a critical assessment of the performance of homology search methods on noncoding RNA. Genome Res. *17*, 117–125.

Gao, F., Kasprzak, W., Stupina, V.A., Shapiro, B.A., and Simon, A.E. (2012). A ribosome-binding, 3′ translational enhancer has a T-shaped structure and engages in a long-distance RNA–RNA interaction. J. Virol. *86*, 9828–9842.

Gardner, P.P., and Giegerich, R. (2004). A comprehensive comparison of comparative RNA structure prediction approaches. BMC Bioinformatics *5*, 140.

Gorodkin, J., and Hofacker, I.L. (2011). From structure prediction to genomic screens for novel non-coding RNAs. PLoS Comput. Biol. *7*, e1002100.

Gorodkin, J., Hofacker, I.L., Torarinsson, E., Yao, Z., Havgaard, J.H., and Ruzzo, W.L. (2010). De novo prediction of structured RNAs from genomic sequences. Trends Biotechnol. *28*, 9–19.

Gruber, A.R., Neubock, R., Hofacker, I.L., and Washietl, S. (2007). The RNAz web server: prediction of thermodynamically stable and evolutionarily conserved RNA structures. Nucleic Acids Res. *35*, W335–338.

Gutell, R.R., Power, A., Hertz, G.Z., Putz, E.J., and Stormo, G.D. (1992). Identifying constraints on the higher-order structure of RNA: continued development and application of comparative sequence analysis methods. Nucleic Acids Res. *20*, 5785–5795.

Gutell, R.R., Lee, J.C., and Cannone, J.J. (2002). The accuracy of ribosomal RNA comparative structure models. Curr. Opin. Struct. Biol. *12*, 301–310.

Hamada, M., Kiryu, H., Sato, K., Mituyama, T., and Asai, K. (2009a). Prediction of RNA secondary structure using generalized centroid estimators. Bioinformatics *25*, 465–473.

Hamada, M., Sato, K., Kiryu, H., Mituyama, T., and Asai, K. (2009b). CentroidAlign: fast and accurate aligner for structured RNAs by maximizing expected sum-of-pairs score. Bioinformatics *25*, 3236–3243.

Harmanci, A.O., Sharma, G., and Mathews, D.H. (2011). TurboFold: iterative probabilistic estimation of secondary structures for multiple RNA sequences. BMC Bioinformatics *12*, 108.

Havgaard, J.H., Lyngso, R.B., and Gorodkin, J. (2005). The FOLDALIGN web server for pairwise structural RNA alignment and mutual motif search. Nucleic Acids Res. *33*, W650–653.

Hofacker, I.L., Fontana, W., Stadler, P.F., Bonhoeffer, M., Tacker, M., and Schuster, P. (1994). Fast folding and comparison of RNA secondary structure. Monatshefte für Chemie *125*, 167–188.

Hofacker, I.L., Fekete, M., and Stadler, P.F. (2002). Secondary structure prediction for aligned RNA sequences. J. Mol. Biol. *319*, 1059–1066.

Hofacker, I.L., Bernhart, S.H., and Stadler, P.F. (2004). Alignment of RNA base pairing probability matrices. Bioinformatics *20*, 2222–2227.

Holmes, I. (2005). Accelerated probabilistic inference of RNA structure evolution. BMC Bioinformatics *6*, 73.

Huang, H.Y., Chien, C.H., Jen, K.H., and Huang, H.D. (2006). RegRNA: an integrated web server for identifying regulatory RNA motifs and elements. Nucleic Acids Res. *34*, W429–434.

John B., Enright A.J., Aravin A., Tuschl T., Sander C., and Marks D.S. (2004). Human MicroRNA targets. PLoS Biol. *2*, e363.

Jonikas, M.A., Radmer, R.J., Laederach, A., Das, R., Pearlman, S., Herschlag, D., and Altman, R.B. (2009). Coarse-grained modeling of large RNA molecules with knowledge-based potentials and structural filters. RNA *15*, 189–199.

Kiryu, H., Tabei, Y., Kin, T., and Asai, K. (2007). Murlet: a practical multiple alignment tool for structural RNA sequences. Bioinformatics *23*, 1588–1598.

Kladwang, W., VanLang, C.C., Cordero, P., and Das, R. (2011). Understanding the errors of SHAPE-directed RNA structure modeling. Biochemistry *50*, 8049–8056.

Klosterman, P.S., Uzilov, A.V., Bendana, Y.R., Bradley, R.K., Chao, S., Kosiol, C., Goldman, N., and Holmes, I. (2006). XRate: a fast prototyping, training and annotation tool for phylo-grammars. BMC Bioinformatics *7*, 428.

Knight, R., Birmingham, A., and Yarus, M. (2004). BayesFold: rational 2 degrees folds that combine thermodynamic, covariation, and chemical data for aligned RNA sequences. RNA *10*, 1323–1336.

Knudsen, B., and Hein, J. (1999). RNA secondary structure prediction using stochastic context-free grammars and evolutionary history. Bioinformatics *15*, 446–454.

Knudsen, B., and Hein, J. (2003). Pfold: RNA secondary structure prediction using stochastic context-free grammars. Nucleic Acids Res. *31*, 3423–3428.

Krek, A., Grun, D., Poy, M.N., Wolf, R., Rosenberg, L., Epstein, E.J., MacMenamin, P., da Piedade, I., Gunsalus, K.C., Stoffel, M., *et al.* (2005). Combinatorial microRNA target predictions. Nat. Genet. *37*, 495–500.

Lai, E.C., Tomancak, P., Williams, R.W., and Rubin, G.M. (2003). Computational identification of Drosophila microRNA genes. Genome Biol. *4*, R42.

Laing, C., and Schlick, T. (2011). Computational approaches to 3D modeling of RNA. J. Phys. Condens. Matter *22*, 283101.

Laslett, D., Canback, B., and Andersson, S. (2002). BRUCE: a program for the detection of transfer-messenger RNA genes in nucleotide sequences. Nucleic Acids Res. *30*, 3449–3453.

Le, S.Y., and Shapiro, B.A. (2010). Data mining of functional RNA structures in genomic sequences. WIREs Data Mining and Knowledge Discovery *1*, 88–95.

Leontis, N.B., and Westhof, E. (2001). Geometric nomenclature and classification of RNA base pairs. RNA *7*, 499–512.

Lewis, B.P., Shih, I., Jones-Rhoades, M.W., Bartel, D.P., and Burge, C.B. (2003). Prediction of mammalian microRNA targets. Cell *115*, 787–798.

Lindgreen, S., Gardner, P.P., and Krogh, A. (2007). MASTR: multiple alignment and structure prediction of non-coding RNAs using simulated annealing. Bioinformatics *23*, 3304–3311.

Linnstaedt, S.D., Kasprzak, W.K., Shapiro, B.A., and Casey, J.L. (2009). The fraction of RNA that folds into the correct branched secondary structure determines hepatitis delta virus type 3 editing levels. RNA *15*, 1177–1187.

Low, J.T., and Weeks, K.M. (2010). SHAPE-directed RNA secondary structure prediction. Methods *52*, 150–158.

Lozada-Chavez, I., Stadler, P.F., and Prohaska, S.J. (2012). 'Hypothesis for the modern RNA world': a pervasive non-coding RNA-based genetic regulation is a prerequisite for the emergence of multicellular complexity. Orig. Life Evol. Biosph. *41*, 587–607.

Lyngso, R.B., and Pedersen, C.N. (2000). RNA pseudoknot prediction in energy-based models. J. Comput. Biol. *7*, 409–427.

Macke, T.J., Ecker, D.J., Gutell, R.R., Gautheret, D., Case, D.A., and Sampath, R. (2001). RNAMotif, an RNA secondary structure definition and search algorithm. Nucleic Acids Res. *29*, 4724–4735.

Markham, N.R., and Zuker, M. (2008). UNAFold: software for nucleic acid folding and hybridization. Methods Mol. Biol. *453*, 3–31.

Martinez, H.M., Maizel, J.V., Jr., and Shapiro, B.A. (2008). RNA2D3D: a program for generating, viewing, and comparing 3-dimensional models of RNA. J. Biomol. Struct. Dyn. *25*, 669–683.

Mathews, D.H. (2004). Using an RNA secondary structure partition function to determine confidence in base pairs predicted by free energy minimization. RNA *10*, 1178–1190.

Mathews, D.H., and Turner, D.H. (2002). Dynalign: an algorithm for finding the secondary structure common to two RNA sequences. J. Mol. Biol. *317*, 191–203.

Mathews, D.H., and Turner, D.H. (2006). Prediction of RNA secondary structure by free energy minimization. Curr. Opin. Struct. Biol. *16*, 270–278.

Mathews, D.H., Sabina, J., Zuker, M., and Turner, D.H. (1999). Expanded sequence dependence of thermodynamic parameters improves prediction of RNA secondary structure. J. Mol. Biol. *288*, 911–940.

Mathews, D.H., Disney, M.D., Childs, J.L., Schroeder, S.J., Zuker, M., and Turner, D.H. (2004). Incorporating chemical modification constraints into a dynamic programming algorithm for prediction of RNA secondary structure. Proc. Natl. Acad. Sci. U.S.A. *101*, 7287–7292.

McCaskill, J.S. (1990). The equilibrium partition function and base pair binding probabilities for RNA secondary structure. Biopolymers *29*, 1105–1119.

Merino, E.J., Wilkinson, K.A., Coughlan, J.L., and Weeks, K.M. (2005). RNA structure analysis at single nucleotide resolution by selective 2'-hydroxyl acylation and primer extension (SHAPE). J. Am. Chem. Soc. *127*, 4223–4231.

Meyer, I.M., and Miklos, I. (2007). SimulFold: simultaneously inferring RNA structures including pseudoknots, alignments, and trees using a Bayesian MCMC framework. PLoS Comput. Biol. *3*, e149.

Michel, F., and Westhof, E. (1990). Modelling of the three-dimensional architecture of group I catalytic introns based on comparative sequence analysis. J. Mol. Biol. *216*, 585–610.

Montange, R.K., and Batey, R.T. (2008). Riboswitches: emerging themes in RNA structure and function. Annu. Rev. Biophys. *37*, 117–133.

Nussinov, R., Pieczenik, G., Griggs, J.R., and Kleitman, D.J. (1978). Algorithms for loop matchings. SIAM J. Appl. Math. *35*, 68–82.

Pan, T., and Sosnick, T. (2006). RNA folding during transcription. Annu. Rev. Biophys. Biomol. Struct. *35*, 161–175.

Parisien, M., and Major, F. (2008). The MC-Fold and MC-Sym pipeline infers RNA structure from sequence data. Nature *452*, 51–55.

Perriquet, O., Touzet, H., and Dauchet, M. (2003). Finding the common structure shared by two homologous RNAs. Bioinformatics *19*, 108–116.

Popenda, M., Szachniuk, M., Antczak, M., Purzycka, K.J., Lukasiak, P., Bartol, N., Blazewicz, J., and Adamiak, R.W. (2012). Automated 3D structure composition for large RNAs. Nucleic Acids Res. *40*, e112.

Puton, T., Kozlowski, L.P., Rother, K.M., and Bujnicki, J.M. (2013) CompaRNA: a server for continuous benchmarking of automated methods for RNA secondary structure prediction. Nucleic Acids Res. *41*, 4307–4323.

Reeder, J., Steffen, P., and Giegerich, R. (2007). pknotsRG: RNA pseudoknot folding including near-optimal structures and sliding windows. Nucleic Acids Res. *35*, W320–324.

Regulski, E.E., and Breaker, R.R. (2008). In-line probing analysis of riboswitches. Methods Mol. Biol. *419*, 53–67.

Rivas, E., and Eddy, S.R. (1999). A dynamic programming algorithm for RNA structure prediction including pseudoknots. J. Mol. Biol. *285*, 2053–2068.

Rivas, E., and Eddy, S.R. (2000). Secondary structure alone is generally not statistically significant for the detection of noncoding RNAs. Bioinformatics *16*, 583–605.

Rivas, E., and Eddy, S.R. (2001). Noncoding RNA gene detection using comparative sequence analysis. BMC Bioinformatics *2*, 8.

Rivas, E., Klein, R.J., Jones, T.A., and Eddy, S.R. (2001). Computational identification of noncoding RNAs in E. coli by comparative genomics. Curr. Biol. *11*, 1369–1373.

Rother, K., Rother, M., Boniecki, M., Puton, T., and Bujnicki, J.M. (2011a). RNA and protein 3D structure modeling: similarities and differences. J Mol Model *17*, 2325–2336.

Rother, M., Rother, K., Puton, T., and Bujnicki, J.M. (2011b). ModeRNA: a tool for comparative modeling of RNA 3D structure. Nucleic Acids Res. *39*, 4007–4022.

Sankoff, D. (1985). Simultaneous solution of the RNA folding alignment and protosequence problems. SIAM J. Appl. Math. 810–825.

Schattner, P., Brooks, A.N., and Lowe, T.M. (2005). The tRNAscan-SE, snoscan and snoGPS web servers for the detection of tRNAs and snoRNAs. Nucleic Acids Res. *33*, W686–689.

Schroeder, S.J., Stone, J.W., Bleckley, S., Gibbons, T., and Mathews, D.M. (2011). Ensemble of secondary structures for encapsidated satellite tobacco mosaic virus RNA consistent with chemical probing and crystallography constraints. Biophys. J. *101*, 167–175.

Seemann, S.E., Gorodkin, J., and Backofen, R. (2008). Unifying evolutionary and thermodynamic information for RNA folding of multiple alignments. Nucleic Acids Res. *36*, 6355–6362.

Seemann, S.E., Richter, A.S., Gorodkin, J., and Backofen, R. (2010). Hierarchical folding of multiple sequence alignments for the prediction of structures and RNA–RNA interactions. Algorithms Mol. Biol. *5*, 22.

Seemann, S.E., Menzel, P., Backofen, R., and Gorodkin, J. (2011). The PETfold and PETcofold web servers for intra- and intermolecular structures of multiple RNA sequences. Nucleic Acids Res. *39*, W107–111.

Seetin, M.G., and Mathews, D.H. (2012). TurboKnot: rapid prediction of conserved RNA secondary structures including pseudoknots. Bioinformatics *28*, 792–798.

Shapiro, B.A., Bengali, D., Kasprzak, W., and Wu, J.C. (2001). RNA folding pathway functional intermediates: their prediction and analysis. J. Mol. Biol. *312*, 27–44.

Shapiro, B.A., Kasprzak, W., Grunewald, C., and Aman, J. (2006). Graphical exploratory data analysis of RNA secondary structure dynamics predicted by the massively parallel genetic algorithm. J. Mol. Graph. Model *25*, 514–531.

Shapiro, B.A., Yingling, Y.G., Kasprzak, W., and Bindewald, E. (2007). Bridging the gap in RNA structure prediction. Curr. Opin. Struct. Biol. *17*, 157–165.

Sharma, S., Ding, F., and Dokholyan, N.V. (2008). iFoldRNA: three-dimensional RNA structure prediction and folding. Bioinformatics *24*, 1951–1952.

Spirollari, J., Wang, J.T., Zhang, K., Bellofatto, V., Park, Y., and Shapiro, B.A. (2009). Predicting consensus structures for RNA alignments via pseudo-energy minimization. Bioinform. Biol. Insights *3*, 51–69.

Sukosd, Z., Knudsen, B., Kjems, J., and Pedersen, C.N. (2012). PPfold 3.0: fast RNA secondary structure prediction using phylogeny and auxiliary data. Bioinformatics *28*, 2691–2692.

Tabei, Y., Tsuda, K., Kin, T., and Asai, K. (2006). SCARNA: fast and accurate structural alignment of RNA sequences by matching fixed-length stem fragments. Bioinformatics *22*, 1723–1729.

Tabei, Y., Kiryu, H., Kin, T., and Asai, K. (2008). A fast structural multiple alignment method for long RNA sequences. BMC Bioinformatics *9*, 33.

Tinoco, I. Jr., and Bustamante, C. (1999). How RNA folds. J. Mol. Biol. *293*, 271–281.

Tinoco, I. Jr., Uhlenbeck, O.C., and Levine, M.D. (1971). Estimation of secondary structure in ribonucleic acids. Nature *230*, 362–367.

Tinoco, I. Jr., Borer, P.N., Dengler, B., Levin, M.D., Uhlenbeck, O.C., Crothers, D.M., and Bralla, J. (1973). Improved estimation of secondary structure in ribonucleic acids. Nat. New Biol. *246*, 40–41.

Torarinsson, E., Havgaard, J.H., and Gorodkin, J. (2007). Multiple structural alignment and clustering of RNA sequences. Bioinformatics *23*, 926–932.

Touzet, H., and Perriquet, O. (2004). CARNAC: folding families of related RNAs. Nucleic Acids Res. *32*, W142–145.

Tullius, T.D., and Greenbaum, J.A. (2005). Mapping nucleic acid structure by hydroxyl radical cleavage. Curr. Opin. Chem. Biol. *9*, 127–134.

Wakeman, C.A., and Winkler, W.C. (2009). Analysis of the RNA backbone: structural analysis of riboswitches by in-line probing and selective 2′-hydroxyl acylation and primer extension. Methods Mol. Biol. *540*, 173–191.

Washietl, S. (2007). Prediction of structural noncoding RNAs with RNAz. Methods Mol. Biol. *395*, 503–526.

Washietl, S., and Hofacker, I.L. (2004). Consensus folding of aligned sequences as a new measure for the detection of functional RNAs by comparative genomics. J. Mol. Biol. *342*, 19–30.

Washietl, S., Hofacker, I.L., and Stadler, P.F. (2005). Fast and reliable prediction of noncoding RNAs. Proc. Natl. Acad. Sci. U.S.A. *102*, 2454–2459.

Waterman, M.S., and Smith, T.F. (1978). RNA secondary structure: a complete mathematical analysis. Math. Biosci. *42*, 257–266.

Wilkinson, K.A., Merino, E.J., and Weeks, K.M. (2006). Selective 2′-hydroxyl acylation analyzed by primer extension (SHAPE): quantitative RNA structure analysis at single nucleotide resolution. Nat. Protoc. *1*, 1610–1616.

Will, S., Reiche, K., Hofacker, I.L., Stadler, P.F., and Backofen, R. (2007). Inferring noncoding RNA families and classes by means of genome-scale structure-based clustering. PLoS Comput. Biol. 3, e65.

Xia, T., SantaLucia, J. Jr., Burkard, M.E., Kierzek, R., Schroeder, S.J., Jiao, X., Cox, C., and Turner, D.H. (1998). Thermodynamic parameters for an expanded nearest-neighbor model for formation of RNA duplexes with Watson–Crick base pairs. Biochemistry 37, 14719–14735.

Xia, X., and Holcik, M. (2009). Strong eukaryotic IRESs have weak secondary structure. PLoS One 4, e4136.

Xu, X., Ji, Y., and Stormo, G.D. (2007). RNA Sampler: a new sampling based algorithm for common RNA secondary structure prediction and structural alignment. Bioinformatics 23, 1883–1891.

Xu, Z., and Mathews, D.H. (2011). Multilign: an algorithm to predict secondary structures conserved in multiple RNA sequences. Bioinformatics 27, 626–632.

Yao, Z., Weinberg, Z., and Ruzzo, W.L. (2006). Cmfinder – a covariance model based RNA motif finding algorithm. Bioinformatics 22, 445–452.

Zadeh, J.N., Steenberg, C.D., Bois, J.S., Wolfe, B.R., Pierce, M.B., Khan, A.R., Dirks, R.M., and Pierce, N.A. (2010). NUPACK: Analysis and design of nucleic acid systems. J. Comput. Chem. 32, 170–173.

Zhao, P., Zhang, W., and Chen, S.J. (2012). Cotranscriptional folding kinetics of ribonucleic acid secondary structures. J. Chem. Phys. 135, 245101.

Zuker, M., and Stiegler, P. (1981). Optimal computer folding of large RNA sequences using thermodynamics and auxiliary information. Nucleic Acids Res. 9, 133–148.

Context-free Grammars and RNA Secondary Structure Prediction

15

Markus E. Nebel and Anika Schulz

Abstract

For a long time, computational methods for RNA secondary structure prediction were typically based on more or less complex models of the free energy – defined by experimentally derived thermodynamic parameters and incomplete free energy rules. However, due to the problems even of comprehensive state-of-the-art thermodynamic models to capture some important, non-energetic influences on sequence folding, an attractive alternative is to use stochastic approaches with parameters estimated from growing databases of structural RNAs.

This motivated the development of a competing methodology towards computational RNA structure prediction analysis that builds on principles of probabilistic modelling of the class of possible foldings rather than on incomplete free energy models. Such probabilistic prediction approaches are generally based on more or less powerful extensions of the concept of traditional context-free grammars that are indeed able to capture the specific structural information collected in an arbitrary database of reliable RNAs.

This chapter deals with such probabilistic RNA folding approaches based on context-free modelling. Note that we here call an approach probabilistic if and only if it abstracts from general thermodynamic models and instead tries to learn about the structural behaviour of the molecules by training (a manageable number of) probabilistic parameters from trusted RNA structure databases. In that sense, partition function approaches, even if providing pairing probabilities, are not assumed to be probabilistic.

Introduction

Computational RNA secondary structure prediction methods traditionally use the dynamic programming (DP) principle and in general either rely on a comprehensive thermodynamic model like the standard Turner model (Xia *et al.*, 1998; Mathews *et al.*, 1999) or on a specific (more or less complex) probabilistic model, making use of structural information obtained from reliable RNA databases. Purely physics-based approaches implementing the minimum free energy (MFE) paradigm (see, for example Nussinov and Jacobson, 1980; Zuker and Stiegler, 1981; Steger *et al.*, 1984; Zuker, 1989) or building on partition function (PF) calculations (see, for instance, McCaskill, 1990; Ding and Lawrence, 2003) depend on experimentally measured thermodynamic parameters and on a particular collection of energy functions, which lack the ability to capture some important, non-energetic influences (e.g. that of enzymes) on sequence folding. However, since the impact of such influences can most likely be gathered from growing databases of solved RNA structures, corresponding probabilistic approaches to RNA analysis have emerged as an attractive alternative. They are characterized by an underlying probabilistic model, which is, rather than being a physics-based model, derived on the basis of structural information collected in an arbitrary database of reliable RNAs.

Probabilistic models

Basically, a probabilistic model produces different outcomes with different probabilities and can thus simulate a whole class of objects, where each object is assigned a probability. In connection with RNA folding, the objects are traditionally RNA sequences with underlying secondary structures, and a corresponding model typically describes a family of related sequences and their respective feasible foldings. Since the number of described sequences is infinite and since there usually exists an exponential (in the sequence length) number of possible structures for a single sequence, it is actually impossible to enumerate all these possibilities in order to obtain either a corresponding probability distribution or a probability model. In order to build a correct probabilistic model, we therefore first need a compact and, most importantly, finite representation of all possible foldings and their probabilities. An alternative to such a representation is given in the so-called stochastic context-free grammar (SCFG). Briefly, SCFGs are an extension of the concept of traditional context-free grammars (CFGs), which do not only model the class of objects (language, RNA sequences in our case) to be generated and their semantics (encoded in the so-called *derivation trees*, here the compatible foldings), but also define a probability distribution (usually non-uniform). Formally, the SCFG approach can be seen as a generalization of hidden Markov models (HMMs), which are widely and successfully used in the wide field of bioinformatics. A basic introduction to HMMs can be found, for instance, in Rabiner and Juang (1986) and Krogh (1998). For more information on HMMs use in bioinformatics we refer to Durbin *et al.* (1998).

As of here, it is worth mentioning that by using SCFG theory, it becomes possible to tackle some questions of RNA secondary structure analysis that can actually not be handled with analogue HMMs or other primary sequence based approaches, including RNA secondary structure prediction, structure based alignment of RNAs, and structure based database search for homologous RNAs (see, for instance, (Durbin *et al.*, 1998)). In addition, more powerful probabilistic modelling variants based on *length-dependent stochastic context-free grammars* (*LSCFGs*) (Weinberg and Nebel, 2011) or *conditional log-linear model* (*CLLMs*) (Do *et al.*, 2006) have been applied over the past years as alternative to plain SCFGs.

Parameter estimation

Any probabilistic model essentially builds on a particular (more or less comprehensive) set of *free parameters* denoted by Φ that induce the corresponding probability distribution on the considered objects (as a function of Φ). In bioinformatics, the parameters for a probabilistic model are typically estimated from sufficiently large sets of trusted sample data, the so-called *training set*, where there exist several mathematically well-defined and widely accepted training approaches. Once parameters Φ are chosen, the probabilistic model is fixed and is assumed to reflect, at least approximately, the biological reality.

Maximum likelihood training

A popular approach for estimating the parameters of a probabilistic model is to use the *maximum likelihood* (*ML*) method, which was invented by R.A. Fisher between 1912 and 1922 (Aldrich, 1997). Intuitively, this technique is applied as follows: on a fixed sample from a larger population, the free parameters of the underlying probabilistic model are tuned in such a way that the sample has maximum *likelihood*, which is defined as the total probability of all the objects in the sample given the considered model. This means that other values for the free parameters make the observation of the sample less likely. Formally, given a model with parameters Φ and a set of data D, the ML estimate for Φ is that value which maximizes the probability $\Pr(D \mid \Phi)$. Details will follow in 'Parameter Estimation' below.

However, we want to make clear already that in non-technical parlance the terms *likelihood* and *probability* may be used synonymously, but in statistical usage, a clear technical distinction is made: given specific values for Φ, $\Pr(D \mid \Phi)$ is the probability that we would observe the data represented by D given our model. This probability is a property of the data D conditioned on the parameters Φ. If we assume, on the contrary, that D is given but Φ is unknown we arrive at the goal to find a reasonable choice for Φ, i.e. a plausible choice of the free parameters given the observed data D. Then we aim at maximizing the likelihood $l(\Phi \mid D) := \Pr(D \mid \Phi)$, which, by definition, is conditioned on the observed data D and is actually a property of the unknown parameters

Φ. Anyway, note it can be proven that in certain scenarios the ML estimates for the free parameters can easily be derived. If we assume that our model generates objects step-by-step with various choices for each increment[1] (the one actually used is determined randomly), then it suffices to consider the paths along which each sample is generated, counting the relative frequencies of choices used. If this is done for the full set D, assigning the observed relative frequencies to probabilities Φ yields a maximum likelihood estimate. Under the condition that the dataset is large enough and that the training data are not systematically biased towards a peculiar structural composition, the resulting model may be expected to be well-behaved.

Finally, it needs to be mentioned that when estimating parameters for a model from a limited amount of data, there is a danger of *overfitting*. In statistics, the term overfitting means that a model describes random error or noise instead of the underlying relationship, which is caused by the fact that there are too many parameters relative to the number of observations. Hence, in our context overfitting potentially occurs when the number of free parameters to be estimated is too large for the training data, i.e. if the considered training data are not rich enough to reliably estimate the distinct parameters.

Bayesian parameter estimation

Another popular approach for deriving the parameters of a probabilistic model is known as *Bayesian parameter estimation*, as it relies on *Bayes' theorem*, which is fundamental to Bayesian statistics and is widely applied in numerous fields. Notably, Bayesian parameter estimation is often used in cases of insufficient data to reliably estimate at least some of the model parameters, as it enables the convenient use of prior knowledge to constrain the parameter estimates in order to prevent overfitting.

Nevertheless, due to the subjectiveness of issues like the choice of prior, Bayesian methods are not always the preferred framework for parameter estimation. In connection with training large parameter sets from small amounts of data, however, Bayesian methods provide a consistent formalism for bringing in additional information from previous experience with the same type of data (see Durbin *et al.*, 1998).

For more information on Bayesian approaches as part of the field of statistics, but also on Bayesian methods applied to a wide range of problems in bioinformatics and genomics, we refer to Box and Tiao (1992), Gelman *et al.*, (1997), Durbin *et al.* (1998) and Beaumont and Rannala (2004). Other textbooks dealing with statistical inference approaches in general include Jaynes (2003) and MacKay (2003).

Overview

The aim of this section is to introduce all the needed formalisms concerning SCFGs as mathematical vehicle for capturing biological sequence and folding information, as well as to provide some additional information on existing SCFG based probabilistic approaches in computational structural biology.

We start with a somewhat informal introduction to context-free grammars and languages as non-stochastic models of RNA structure, describing only the basic concepts and properties. For more formal and detailed information on CFGs, we refer to the excellent textbooks available today on this topic (for instance Harrison, 1978; Hopcroft and Ullman, 1979; Sipser, 1996). Afterwards, we give a self-contained overview of probabilistic modelling of classes of objects whose combinatorial structure can be described by corresponding CFGs, with a special emphasis on RNA secondary structures. In fact, we first present the basic formal definitions and some aspects of SCFGs that are relevant in the context of RNA folding. Then, we introduce and discuss some details on SCFG based prediction. For a fundamental introduction on stochastic context-free languages in general we refer to Huang and Fu (1971). A review of the use of SCFGs for RNA folding can be found, for example, in Durbin *et al.* (1998).

By the end of this section we give a brief introduction to the previously mentioned more powerful length-dependent variant of SCFGs. Finally, we discuss some related probabilistic approaches and especially as they apply to RNA folding.

---

[1] As we will see later, this assumption is fulfilled by SCFGs.

Context-free grammars

The formalism of context-free grammars (CFGs) was developed in the mid-1950s by Noam Chomsky (Chomsky, 1956), who also described their classification as a special type of formal grammar (which he called *phrase-structure grammars*). In principle, a CFG is given by a finite set of variables, called *nonterminal symbols, intermediate symbols* or simply *intermediates*. Each variable represents a class of strings, where any string is composed of a finite number of distinct primitive symbols, called *terminal symbols* or *terminals*. Assuming that V is a finite set of symbols (intermediates, terminals or mixed), we will write V^* to represent the set of all strings consisting of 0 or more symbols from V. The variable representing the class with maximum cardinality among all classes of the grammar is called *start variable* or *axiom*. Moreover, string classes represented by the variables of a CFG are called *context-free languages* (*CFLs*). They are described recursively in terms of each other, where rules relating the variables are called *production rules, productions* or simply *rules*.

Formal definitions

In order to introduce the basic formal framework concerning CFGs, we may rely on any textbook providing a discussion of formal language theory. Here, we basically adhere to Hopcroft and Ullman (1979), but we decided to use different names for the specific components:

Definition 15.1

A context-free grammar (CFG) is a 4-*tuple* $G=(I, T, R, S)$, where

- I is a finite set of intermediate or nonterminal symbols;
- T is a finite set of terminal symbols called alphabet;
- $R \subset I \times (I \cup T)^*$ is a finite set of production rules; and
- $S \in I$ is a distinguished intermediate symbol called *axiom*.

Additionally, I and T are required to be disjoint. In the sequel, we will write $f=A \to \alpha$ instead of $f=(A, \alpha) \in R$.

The *language generated* by a CFG G is defined as the class of strings (over the terminal symbols) represented by the axiom of G, and is denoted by L(G); the empty string, i.e. the string consisting of no symbol, will be denoted by ε in the sequel. Note that in literature one sometimes calls the left-hand side $A \in I$ of a production $A \to \alpha$ the *premise* of that rule, whereas the corresponding right-hand side $\alpha \in (I \cup T)^*$ is then called the *conclusion*. Traditionally, a word $\alpha \in (I \cup T)^*$ is called *sentential form*.

RNA Secondary structure grammars and language specification

Besides the popular planar graph representation (Waterman, 1978), many other ways of formalizing RNA structures have been described in literature. One well-established example is the so-called *dot-bracket representation* (Viennot and Chaumont, 1985), where a secondary structure is modelled as a string over the alphabet $\Sigma_s := \{(,),\circ\}$, with a dot $\circ$ representing an unpaired nucleotide and a pair of corresponding brackets ( ) representing two paired bases in the molecule. This representation directly relates to the modelling of RNA secondary structures by a corresponding CFG.

In fact, one can construct CFGs that will generate the set of all dot-bracket representations as their language while respecting classical constraints to RNA secondary structures (as, for instance, a minimal length for hairpin loops). Contrary to hidden Markov models, CFGs can express long-range interactions, i.e. hierarchically nested, pairwise correlations, which is essential to model the base pairings in secondary structures.

Example 15.1

A fairly simple CFG, call it G_1, with only one non-terminal symbol S representing the class of all dot-bracket strings related to secondary structures (under the assumptions of a minimum allowed hairpin loop length $min_{HL}=0$ and a minimum allowed helix length $min_{hel}=1$)[2] is described by the following production rules: $S \to \circ S$, $S \to S\circ$, $S \to (S)$, $S \to SS$, $S \to \varepsilon$. Basically, the first two rules generate unpaired bases, the third produces base pairs, the fourth bifurcations and the last one is needed to guarantee that the string construction eventually terminates. The dot-bracket representation ( )( )∘ can be derived by the stepwise application of production rules (used to replace within a sentential form the intermediate on the rule's left-hand side by its right-hand side) starting at the axiom in the following way[3]: $S \Rightarrow SS \Rightarrow (S)S \Rightarrow (\,)S \Rightarrow (\,)S\circ \Rightarrow (\,)(S)\circ \Rightarrow (\,)(\,)\circ$.

Note that rather different sets of variables and production rules, that is grammar designs, can be used to model the same or quite similar classes of strings:

Example 15.2

An alternative CFG, say G_2, for modelling secondary structures in dot-bracket representation is defined by the following productions (see Knudsen and Hein, 1999): $S \to LS$, $S \to L$, $L \to \circ$, $L \to (F)$, $F \to (F)$, $F \to LS$.

Briefly, the intermediate symbol S (axiom) initiates the generation of single-stranded bases and/or helices of the exterior loop. Symbol L actually 'decides' which substructure (single free base or complete helix starting with a corresponding free base pair) is created. Finally, intermediate symbol F recursively generates substructures in between a given base pair, since in any case it either extends a particular stem (by generating an additional base pair) or finishes its construction (by initializing a new loop). Note that grammar G_2 implicitly assumes minimum allowed lengths of $min_{HL}=2$ and $min_{hel}=1$ for hairpin loops and helices, respectively.

In a quite similar fashion, recursive specifications of CFLs can be derived as formal characterizations of classes of complex combinatorial objects. For example, the combinatorial class of all RNA secondary structures without pseudoknots, typically considered by physics-based folding algorithms (precisely, those structures that are not completely unpaired and meet the stereochemical constraint on hairpin loops consisting of at least $min_{HL} = 3$ unpaired nucleotides), can be modelled by the following CFL:

Definition 15.2 (Nebel and Scheid, 2011a)

The language L containing exactly all RNA secondary structures is given by $L := L_u L_{lu}^{+}$, where $L_{lu} := (L_l)L_u$, $L_u := \{\circ\}^*$ is the language of all dot-bracket representations of single-stranded regions and L_l is the language of all dot-bracket representations of other possible substructures, that is the smallest language satisfying the following conditions:

- $\{\circ\}^+ \setminus \{\circ, \circ\circ\} \subset L_l$ (dot-bracket representations of hairpin loops).
- If $w \in L_l$, then $(w) \in L_l$ (dot-bracket representation of a stacked pair).
- If $w \in L_l$, then $\{\circ\}^+(w) \subset L_l$ and $(w)\{\circ\}^+ \subset L_l$ (dot-bracket representations of bulge loops).
- If $w \in L_l$, then $\{\circ\}^+(w)\{\circ\}^+ \subset L_l$ (dot-bracket representations of interior loops).
- If $w_1, \ldots, w_n \in L_l$ and $n \geq 2$, then $L_u(w_1)\, L_u(w_2) \cdots L_u(w_n)L_u \subset L_l$ (dot-bracket representations of multi-branched loops).

Derivation and ambiguity

By definition, all strings of terminal symbols that can be generated from the start symbol (axiom) of a given grammar G constitute the language L(G) specified by that grammar. Accordingly, any grammar G can be used as a language specification mechanism by generating each string of the corresponding language L(G) in the following 'top-down' manner (see Example 15.1):

1. Start with the first variable (axiom) representing the complete language.
2. Find a variable in current sentential form and a rule describing a relation for that variable (i.e. left-hand side of rule equals variable) and replace it with the relation (i.e. by the right-hand side).

---

[2] Note that by definition, secondary structures must admit a minimum hairpin loop length of $min_{HL} \geq 1$, such that this CFG G_1 does actually only model a secondary structure related class. However, it has in any case been used in literature for describing RNA secondary structure, such as for example in Dowell and Eddy (2004), which is why we will consider it here, too.

[3] We will shortly provide a formal definition of notation $\Rightarrow$ and the underlying generative process as well as further examples of how to use it.

3 Repeat step 2 until no variables remain in the generated string (until it is either equal to the empty string ε or consists exclusively of terminals).

The sequence of substitutions utilized to obtain a string $w \in L(G)$ using the CFG G is called a *derivation* (or sometimes also *parse*). If in any step of the derivation process, the leftmost (rightmost) variable is substituted, then the corresponding derivation is called *leftmost derivation* (*rightmost derivation*). Here, we will only consider leftmost derivations. Note that by convention, one uses symbol $\Rightarrow$ to represent a single application of step 2 (immediate derivation) or sometimes $\overset{rule}{\Longrightarrow}$ to explicitly mark the production rule, considered for the corresponding substitution. Notation $\Rightarrow^+$ is used for a series of at least one consecutive derivation steps. Accordingly, $\Rightarrow^*$ is used for an arbitrary number (zero or more) of consecutive derivation steps. Anyway, derivations can also be represented as ordered trees, where each node represents an intermediate symbol, a terminal symbol or the empty string ε. More precisely, internal nodes always correspond to non-terminals and the labels of their children – in left-to-right order – have to form the right-hand side (conclusion) of a rule for the parent (the corresponding premise). The corresponding trees are conventionally named *derivation trees* or *parse trees*. There actually exists a bijection between leftmost derivations and derivation trees.

Example 15.3
The derivation of the dot-bracket string ((∘∘∘))∘ using grammar G_2 as specified in Example 15.2 is given by

$$S \overset{S\to LS}{\Rightarrow} LS \overset{L\to(F)}{\Rightarrow} (F)S \overset{F\to(F)}{\Rightarrow} ((F))S \overset{F\to LS}{\Rightarrow} ((LS))S$$
$$\overset{L\to\circ}{\Rightarrow} ((\circ S))S \overset{S\to LS}{\Rightarrow} ((\circ LS))S \overset{L\to\circ}{\Rightarrow} ((\circ\circ S))S$$
$$\overset{S\to L}{\Rightarrow} ((\circ\circ L))S \overset{L\to\circ}{\Rightarrow} ((\circ\circ\circ))S \overset{S\to L}{\Rightarrow} ((\circ\circ\circ))L$$
$$\overset{L\to\circ}{\Rightarrow}_1 ((\circ\circ\circ))\circ.$$

The corresponding parse tree is shown in Fig. 15.1.

In many applications, for instance in connection with enumeration problems, it is crucial that each structure is generated only once by a corresponding CFG. This means that there must be only *one* unique parse tree for any string of the CFL, or equivalently only one leftmost derivation. A CFG G where some string in L(G) has more than one parse trees (or leftmost derivations) is said to be *ambiguous*.

Example 15.4
Using the CFG G_1 introduced in Example 15.1, the string ((∘∘∘))∘ has more than one leftmost derivations, for instance:

$$d_1 = S \overset{S\to SS}{\Rightarrow} SS \overset{S\to(S)}{\Rightarrow} (S)S \overset{S\to(S)}{\Rightarrow} ((S))S \overset{S\to\circ S}{\Rightarrow} ((\circ S))S$$
$$\overset{S\to\circ S}{\Rightarrow} ((\circ\circ S))S \overset{S\to\circ S}{\Rightarrow} ((\circ\circ\circ S))S \overset{S\to\varepsilon}{\Rightarrow} ((\circ\circ\circ))S$$
$$\overset{S\to\circ S}{\Rightarrow} ((\circ\circ\circ))\circ S \overset{S\to\varepsilon}{\Rightarrow} ((\circ\circ\circ))\circ,$$

or

$$d_2 = S \overset{S\to S\circ}{\Rightarrow} S\circ \overset{S\to(S)}{\Rightarrow} (S)\circ \overset{S\to(S)}{\Rightarrow} ((S))\circ \overset{S\to S\circ}{\Rightarrow} ((S\circ))$$
$$\circ \overset{S\to S\circ}{\Rightarrow} ((S\circ\circ))\circ \overset{S\to S\circ}{\Rightarrow} ((S\circ\circ\circ))\circ \overset{S\to\varepsilon}{\Rightarrow} ((\circ\circ\circ))\circ,$$

or else

$$d_3 = S \overset{S\to SS}{\Rightarrow} SS \overset{S\to(S)}{\Rightarrow} (S)S \overset{S\to(S)}{\Rightarrow} ((S))S \overset{S\to SS}{\Rightarrow} ((SS))S$$
$$\overset{S\to SS}{\Rightarrow} ((SSS))S \overset{S\to\circ S}{\Rightarrow} ((\circ SSS))S \overset{S\to\varepsilon}{\Rightarrow} ((\circ SS))S$$
$$\overset{S\to\circ S}{\Rightarrow} ((\circ\circ SS))S \overset{S\to\varepsilon}{\Rightarrow} ((\circ\circ S))S \overset{S\to SS}{\Rightarrow} ((\circ\circ SS))S$$
$$\overset{S\to\circ S}{\Rightarrow} ((\circ\circ\circ SS))S \overset{S\to\varepsilon}{\Rightarrow} ((\circ\circ\circ S))S \overset{S\to SS}{\Rightarrow} ((\circ\circ\circ SS))S$$
$$\overset{S\to\varepsilon}{\Rightarrow} ((\circ\circ\circ S))S \overset{S\to\varepsilon}{\Rightarrow} ((\circ\circ\circ))S \overset{S\to SS}{\Rightarrow} ((\circ\circ\circ))SS$$
$$\overset{S\to\circ S}{\Rightarrow} ((\circ\circ\circ))\circ SS \overset{S\to\varepsilon}{\Rightarrow} ((\circ\circ\circ))\circ S \overset{S\to SS}{\Rightarrow} ((\circ\circ\circ))\circ SS$$
$$\overset{S\to\varepsilon}{\Rightarrow} ((\circ\circ\circ))\circ S \overset{S\to\varepsilon}{\Rightarrow} ((\circ\circ\circ))\circ$$

The corresponding parse trees are presented in Fig. 15.2. Another one is shown in Fig. 15.3, illustrating the differences in the degree of redundancy of some derivations characterized by the number of useless applications of production rules.

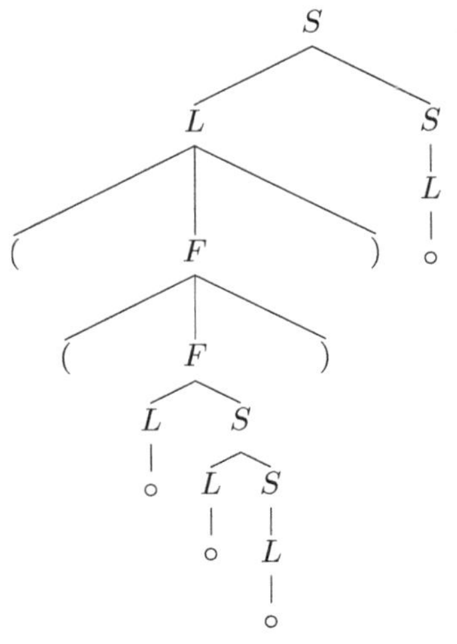

Figure 15.1 A simple derivation tree showing the unique derivation tree of the dot-bracket word ((∘∘∘))∘ using grammar G_2.

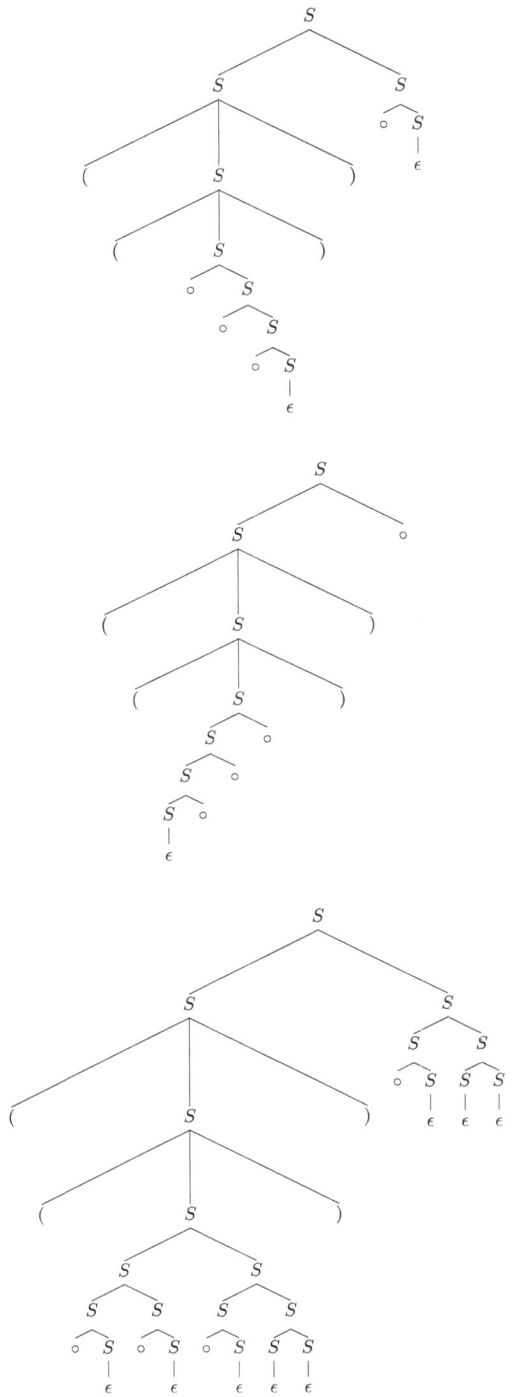

Figure 15.2 Derivation trees of different complexities. Three alternative derivation trees of the simple string ((∘∘∘))∘ using grammar G_1.

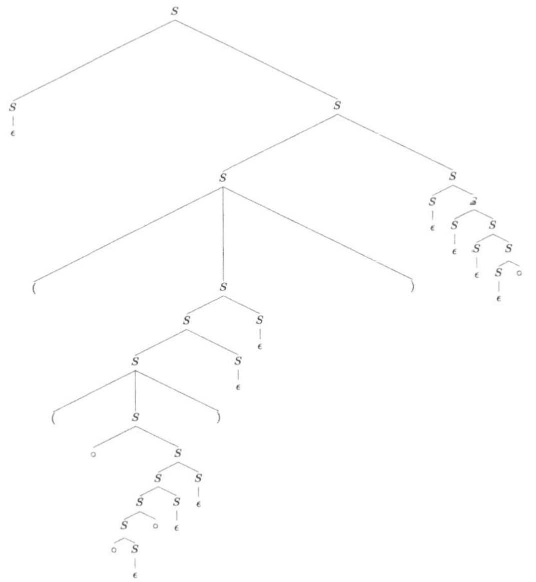

Figure 15.3 A more complex derivation tree. Another derivation tree of the simple string ((∘∘∘))∘ using grammar G_1.

Redundancy is actually caused by sequences of substitutions made with $S \rightarrow SS$ in connection with $S \rightarrow \varepsilon$ (see Figs. 15.2c and 15.3).

Moreover, considering the most reasonable (since comparably small) parse trees displayed in Figs. 15.2a and 15.2b, it seems that either $S \rightarrow \circ S$ or $S \rightarrow S\circ$ represents a needless production.

Accordingly, if for every word $w \in L(G)$, exactly *one* leftmost derivation exists, then the CFG G is called *unambiguous*.

Stochastic context-free grammars

As indicated above, SCFGs are essential for probabilistic approaches towards RNA secondary structure prediction. We already know that secondary structures can be encoded as words of a CFL and the class of all feasible structures can thus effectively be modelled via a corresponding CFG. Now the key idea for a probabilistic modelling of RNA foldings is to extend such a CFG to a corresponding SCFG by assigning probabilities to its production rules. This way, besides the mere representation of all valid foldings, we can introduce non-uniform probabilities for the various motifs of RNA secondary structures (in accordance with native folding).

Formal definitions and basic concepts

Traditional CFGs are only capable of modelling the class of all generated strings and thus inevitably induce a uniform distribution on the objects, while SCFGs additionally imply a (non-uniform) probability distribution on the considered class of objects. Basically, a particular SCFG is derived by equipping the productions of a corresponding CFG with probabilities in such a way that the induced distribution on the generated language fits, as closely as possible, the distribution of the sample data. The needed formalities are given as follows:

Definition 15.3 (Fu and Huang, 1972)
A weighted context-free grammar (WCFG) is a 5-tuple $G=(I,T,R,S,W)$, where (I,T,R,S) is a CFG and W: $R \rightarrow \mathbb{R}^+$ is a mapping such that each rule $f \in R$ is equipped with a weight $w_f := W(f)$.

If G is a WCFG, then G is a stochastic context-free grammar (SCFG) if the following additional restrictions hold:

1. For all $f \in R$, we have $W(f) \in (0, 1]$, which means the weights are probabilities.
2. The probabilities are chosen in such a way that for all $A \in I$, we have
$$\sum_{f \in R, Q(f)=A} w_f = 1,$$
where $Q(f)$ denotes the premise of the production f. In the sequel, we will write $w_f: A \rightarrow \alpha$ instead of $f=(A,\alpha) \in R$, $w_f=W(f)$.

Consequently, for any SCFG $G=(I,T,R,S,W)$, the mapping $W: R \rightarrow (0,1]$ provides a probability distribution on the production rules[4]. In many cases, the probability distribution on the productions of a SCFG G also implies a probability distribution on the words of the language L(G), such that G might adequately be used for probabilistic modelling of the generated class of objects (represented by the language); details will follow. Note that if G is a SCFG, in most cases we will write $G=(I,T,R,S,\Pr)$ and hence consider the mapping Pr: $R \rightarrow (0,1]$. The concepts of derivation and ambiguity for SCFGs are the same as for usual CFGs. This means that each word $w \in L(G)$ is generated in exactly the same way as for the corresponding CFG (I,T,R,S). However, considering $G=(I, T, R, S, \Pr)$, the mapping Pr: $R \rightarrow (0, 1]$ assigns a probability $\Pr(d)$ to each derivation d of a word $w \in L(G)$. The probability $\Pr(d)$ of a given derivation d is equal to the product of the probabilities of the production rules used in it, formally:

$$\Pr(d) = \prod_{\alpha \overset{f}{\Rightarrow} \beta \in d} P(f).$$

[4] To be precise, a distribution is implied for each subset of rules having the same left-hand side. However, it is of no importance to make this distinction in the sequel.

Furthermore, we can use the mapping Pr to compute the probability Pr(w) for each word $w \in$ L(G). To do this, we have to take into account that the SCFG G can be ambiguous, since then a word $w \in$ L(G) may have more than one derivation. In fact, if a word $w \in$ L(G) has exactly k different leftmost derivations $d_1, ..., d_k$, then we have

$$\Pr(w) = \sum_{i=1}^{k} \Pr(d_i).$$

More generally, this means that we must sum up the probabilities of all possible leftmost derivations of a particular word $w \in$ L(G), formally

$$\Pr(w) = \sum_{d=S\Rightarrow^* w} \Pr(d).$$

Thus, if the SCFG G is unambiguous, then the probability Pr(w) of a word $w \in$ L(G) can be efficiently computed, as it is actually equal to the product of the probabilities Pr(f) of the production rules $f \in R$ that have to be used to generate this word, formally

$$\Pr(w) = \prod_{\alpha \overset{f}{\Rightarrow} \beta \in d = S \Rightarrow^* w} \Pr(f).$$

Furthermore, many efficient algorithms are known to determine the unique derivation tree (left-most derivation) of a given string with respect to a given unambiguous CFG (see 'SCFG Based Algorithms' below for corresponding algorithms which are efficient even in connection with ambiguous grammars).

Example 15.5

The corresponding SCFG of grammar G_2 as introduced in Example 15.2 can be written as follows:

$$p_1 : S \to LS, p_2 : S \to L, p_3 : L \to \circ,$$

$$p_4 : L \to (F), p_5 : F \to (F), p_6 : F \to LS.$$

Hence, considering the unique leftmost derivation d of the simple string s=((∘∘∘))∘ as presented in Example 15.3, we find that

$$\Pr(s) = \Pr(d)$$
$$= \Pr(S \to LS) \cdot \Pr(L \to (F)) \cdot \Pr(F \to (F)) \cdot$$
$$\Pr(F \to LS) \cdot \Pr(L \to \circ) \cdot \Pr(S \to LS) \cdot$$
$$\Pr(L \to \circ) \cdot \Pr(S \to L) \cdot \Pr(L \to \circ) \cdot \Pr(S \to L) \cdot$$
$$\Pr(L \to \circ)$$
$$= p_1 \cdot p_4 \cdot p_5 \cdot p_6 \cdot p_3 \cdot p_1 \cdot p_3 \cdot p_2 \cdot p_3 \cdot p_2 \cdot p_3$$

Example 15.6

Consider the following SCFG based on grammar G_1 of Example 15.1:

1/15: S→∘S, 2/15: S→S∘, 3/15: S→(S), 4/15: S→SS, 5/15: S→ε.

Then, the probabilities of the different leftmost derivations given in Example 15.4 are:

$$Pr(d_1) = (1/15)^4 \cdot (2/15)^0 \cdot (3/15)^2 \cdot (4/15)^1 \cdot (5/15)^2$$
$$\approx 2.34111 \times 10^{-8},$$

$$\Pr(d_2) = (1/15)^0 \cdot (2/15)^4 \cdot (3/15)^2 \cdot (4/15)^0 \cdot (5/15)^1$$
$$\approx 4.21399 \times 10^{-6},$$

$$\Pr(d_3) = (1/15)^4 \cdot (2/15)^0 \cdot (3/15)^2 \cdot (4/15)^7 \cdot (5/15)^8$$
$$\approx 1.15480 \times 10^{-14}.$$

Note that the two more compact and thus more reasonable derivations d_1 and d_2 are significantly more probable than the comparably rather complex derivation d_3, since d_3 actually implies more rule applications (and hence more probability terms) due to the high level of redundancy. Anyway, since there are many other leftmost derivations possible, the probability of the simple dot-bracket word s=((∘∘∘))∘ is given by

$$\Pr(s) = \Pr(d_1) + \Pr(d_2) + \Pr(d_3) +$$
$$\sum_{d=S\Rightarrow^* s, d \notin \{d_1, d_2, d_3\}} \Pr(d) > \Pr(d_1) + \Pr(d_2) + \Pr(d_3)$$
$$\approx 4.2374 \times 10^{-6}.$$

Obviously, the computation of Pr(s) is quite inefficient due to the ambiguity of grammar G_1.

Note that in derivations of words according to a WCFG G = (I,T,R,S,W), both $\overset{f}{\Rightarrow}$ *and* $\overset{W(f)}{\Rightarrow}$ may be used to explicitly mark the production $f \in R$

considered for a particular immediate derivation (substitution). Finally, it needs to be mentioned that if a SCFG $G = (I, T, R, S, \Pr)$ indeed provides a probability distribution for the generated language L(G), that is if $\sum_{w \in L(G)} \Pr(w) = 1$ holds, then G is called *consistent*.

Parameter estimation

In principle, SCFGs try to learn about the typical behaviour of a particular class of objects from statistical grounds, by employing appropriate training procedures for estimating probabilities for the distinct production rules (that is, for calculating estimates for the respective grammar parameters). In fact, the probabilities of a SCFG G, which generates language L(G), can be trained from a database of words $w \in L(G)$. As indicated in 'Maximum Likelihood Training' above, SCFGs are trained according to the maximum likelihood principle on (hopefully) typical words of different sizes. Thereby, a SCFG G captures the probability distribution present in the sample set of words $w \in L(G)$ provided for the training.

The conditions for consistency of such a trained grammar have been investigated by a number of scientists. As a result, several methods for the empirical parameter estimation, which provide consistent SCFGs, have been proposed in literature. For example, assigning relative frequencies, found by counting the production rules used in the leftmost derivations of a finite sample of words $w \in L(G)$, results in a consistent SCFG G (Chi and Geman, 1998). In fact, it was shown that the maximum likelihood, the expectation maximization and a new cross-entropy minimization approach each provide a consistent SCFG without restrictions on the grammar (Nederhof and Satta, 2003; Corazza and Satta, 2006; Nederhof and Satta, 2006).

As already outlined in 'Maximum Likelihood Training' above, determining the probabilities of a SCFG by simply counting the rule's relative frequencies within all leftmost derivations actually yields a maximum likelihood estimate (Chi and Geman, 1998) (and a consistent grammar). This is especially useful in connection with unambiguous SCFGs, since then the relative frequencies can be counted efficiently, as, for every word, there is only one unique leftmost derivation to consider.

SCFGs for structure prediction

It has been known for a long time that SCFGs can be used to model RNA secondary structure (see, for instance, (Sakakibara *et al.*, 1994)). Moreover, SCFGs can be employed for deriving results on the expected structural behaviour of RNAs, which might then be used for judging the quality of predictions made by any RNA folding algorithm (Nebel, 2002b, 2004b). Such results draw a quite realistic picture compared to other attempts to describe the structure of RNA quantitatively, which, for example, assume an unrealistic combinatorial model (Waterman, 1978; Nebel, 2002a) or Bernoulli-model (Hofacker *et al.*, 1998; Nebel, 2004a) for RNA secondary structures. Furthermore, note that an SCFG mirror to the famous Turner energy model has been used in (Nebel and Scheid, 2011a) to perform the first analytical analysis of the free energy of RNA secondary structures. However, SCFGs have also been used successfully for the prediction of RNA secondary structure (Knudsen and Hein, 1999, 2003; Dowell and Eddy, 2004). In this context, the grammar's language traditionally models the set of all RNA sequences; the foldings are encoded within the derivation trees. Since the set of all possible base paired structures for a particular RNA sequence needs to be considered for calculating a corresponding prediction, any SCFG that will be useful in this context must be ambiguous, i.e. there must be more than one possible derivation tree for a given sequence representing all its feasible foldings. However, since the derivation trees are equipped with probabilities, we can ask for that of the highest probability for our prediction. To this end it is of high practical concern that each derivation tree for a sequence uniquely corresponds to one of its secondary structures. In fact, the popular CYK algorithm (details will follow in 'Cocke–Younger–Kasami algorithm', below) can be used to find the optimal (most probable) derivation tree, which is equal to the optimal folding if and only if there is a one-to-one correspondence between parse trees and secondary structures. If the same secondary structure is described by multiple valid derivation trees, the corresponding SCFG is called *structurally* ambiguous, which has been indicated to be of great disadvantage in connection with accurate DP methods (see, for instance, Giegerich, 2000; Dowell and Eddy, 2004).

Grammar design

As already indicated in 'RNA Secondary Structure Grammars and Language Specification' above (in connection with traditional CFGs), different SCFG designs can be used to model the same class of structures. Here flexibility in model design comes from the fact that basically all distinct structural motifs of RNA (like bulges or interior loops) can either be generated by distinguished rules for each motif or by a shared set of productions. With increasing number of distinguished features, the resulting SCFG gains in both explicitness and complexity which may result in a more realistic probability distribution on the modelled structure class. Principally, any grammar describing RNA secondary structures at least has to distinguish between paired and unpaired positions by using different productions to generate the corresponding symbols of the RNA sequence. For example, the SCFG G_1 only captures the simplest folding features: unpaired bases, base pairings and bifurcations.

However, attempting to construct an elaborate SCFG that not only generates secondary structures but also models structural features of the sample data as closely as possible, it is important to appropriately specify the set of production rules in order to guarantee that all substructures, which have to be distinguished, are derived from different rules. This is due to the fact that by using only one production rule f to generate different substructures (for instance, any unpaired nucleotides independent of the type of loop they belong to), there is only one weight (the probability $\Pr(f)$ of this production f) with which any of these substructures is generated. In order to distinguish between these substructures, the use of different rules $f_1, \ldots, f_k$ implies that they may be generated with different probabilities $\Pr(f_1), \ldots, \Pr(f_k)$, where $\Pr(f_1) + \ldots + \Pr(f_k) = \Pr(f)$.

This way, we guarantee that more common substructures are generated with higher probabilities than less common ones.

Example 15.7

A (rather simple) unambiguous SCFG generating the language characterized in Definition 15.2 is given by:

$$p_1 : S \rightarrow CA,$$
$$p_2 : A \rightarrow (B)C, p_3 : A \rightarrow (B)CA,$$
$$p_4 : B \rightarrow \circ\circ\circ C, p_5 : B \rightarrow CA,$$
$$p_6 : C \rightarrow \varepsilon, p_7 : C \rightarrow \circ C.$$

This grammar unambiguously generates L for the following reasons:

- Every sentential form $C(B)C(B)\cdots(B)C$ is obviously generated in a unique way; this resembles $L := L_u\, L_{lu}^{+}$ and $L_{lu} := (L_l)L_u$ of L's definition. The number of outermost pairs of brackets in the entire string uniquely determines the corresponding sentential form to be used.
- Now, B either generates a hairpin-loop from $\circ\circ\circ\, L_u$, which is possible (in an unambiguous way) by rules $B \rightarrow \circ\circ\circ C$, $C \rightarrow \circ C$ and $C \rightarrow \varepsilon$.
- Or else, B itself has to generate at least one additional pair of brackets. In this case, $B \rightarrow CA$ must be applied (only A can generate brackets) and then $A \rightarrow (B)C$ resp. $A \rightarrow (B)CA$ are used; the number of outermost brackets to be generated (from B under consideration) again uniquely determines that part of the derivation

Anyway, when changing the production p_5: $B \rightarrow CA$ used to generate any possible k-loop for $\kappa \geq 2$ (any loop that is not a hairpin loop) with probability p_5 into the two rules:

$$p_{5.1} : B \rightarrow C(B)C, p_{5.2} : B \rightarrow C(B)CA,$$

where $p_{5.1} + p_{5.2} = p_5$, it becomes possible to generate any possible 2-loop (that is, a stacked pair, a bulge (on the left or on the right), or an interior loop) and all kinds of multiloops (that is, any k-loop with $k \geq 3$) with different probabilities, which could increase the accuracy of the SCFG model. We denote the corresponding grammar by G_3'.

By additionally replacing the first of these two new rules $p_{5.1}: B \rightarrow C(B)C$, by the four productions

$$p_{5.1.1}: B \rightarrow (B), p_{5.1.2}: B \rightarrow \circ C(B),$$

$$p_{5.1.3}: B \rightarrow (B)C\circ, p_{5.1.4}: B \rightarrow \circ C(B)C\circ,$$

we obtain an even more specific grammar design, which we denote by G_3''. Notably, we then have $(p_{5.1.1} + \dots + p_{5.1.4}) + p_{5.2} = p_{5.1} + p_{5.2} = p_5$, meaning it becomes possible to distinguish between the different types of 2-loops more accurately, yielding a more realistic secondary structure model.

In fact, in the case of significant differences of the new probabilities ($p_{5.1.1}, \dots, p_{5.1.4}$ and $p_{5.2}$), we can expect a huge improvement in the model's accuracy. Note that it is not hard to see that changes to a grammar like the ones just discussed do not change the language generated. However, this is not at all obvious with respect to ambiguity of the grammar. Hence, the corresponding changes need to be performed very carefully in order to ensure that the modified grammar remains unambiguous, which indeed has been done in the presented cases.

Notably, the (structural) unambiguity of rather complex SCFG designs can often readily be proven by describing the construction of their rule sets as done in Example 15.7. In brief, one starts with a rather simple and small (so-called *lightweight*) grammar that distinguishes only the basic structural features and extend its set of productions (by replacing single productions that model one particular type of substructure by a bunch of corresponding new productions for generating the respective special types of substructures) until all substructures that need to be distinguished are represented by separate rules (and parameters). In order to avoid structural ambiguities, we only have to take care that at any point (where a more general old rule is replaced by a set of more specialized new ones) none of the considered alternative structure motifs can be constructed from more than one production. In this context, it is important to use different intermediate symbols for distinguished substructure types, thereby ensuring that any intermediate symbol of the grammar uniquely corresponds to a particular class of substructures.

It is worth mentioning that different SCFG designs generally imply differences in the induced probability distributions, as illustrated by the following example.

Example 15.8
The unique leftmost derivations of the secondary structure $s = ((\circ\circ\circ))\circ$ using the three grammars G_3, G_3' and G_3'' of Example 15.7, respectively, are given by

$$d = S \Rightarrow^* (B)C \overset{p_5}{\Rightarrow} (CA)C \overset{p_6}{\Rightarrow} (A)C \overset{p_2}{\Rightarrow} ((B)C)C$$

$$\overset{p_4}{\Rightarrow} ((\circ\circ\circ C)C)C \overset{p_6}{\Rightarrow} ((\circ\circ\circ)C)C \overset{p_6}{\Rightarrow} ((\circ\circ\circ))C$$

$$\Rightarrow^* ((\circ\circ\circ))\circ,$$

$$d' = S \Rightarrow^* (B)C \overset{p_{5.1}}{\Rightarrow} (C(B)C)C$$

$$\overset{p_6}{\Rightarrow} ((B)C)C \overset{p_4}{\Rightarrow} ((\circ\circ\circ C)C)C \overset{p_6}{\Rightarrow} ((\circ\circ\circ)C)C$$

$$\overset{p_6}{\Rightarrow} ((\circ\circ\circ))C \Rightarrow^* ((\circ\circ\circ))\circ,$$

and

$$d'' = S \Rightarrow^* (B)C \overset{p_{5.1.1}}{\Rightarrow} ((B))C \overset{p_4}{\Rightarrow}$$

$$((\circ\circ\circ C))C \overset{p_6}{\Rightarrow} ((\circ\circ\circ))C \Rightarrow^* ((\circ\circ\circ))\circ.$$

The corresponding parse trees are pictured in Fig. 15.4. Hence, we have

$$\Pr(d) = \Pr(S \Rightarrow^* (B)C) \cdot p_2^1 \cdot p_4^1 \cdot p_5^1 \cdot p_6^3 \cdot \Pr(((\circ\circ\circ))C \Rightarrow^* s),$$

$$\Pr(d') = \Pr(S \Rightarrow^* (B)C) \cdot p_4^1 \cdot p_{5.1}^1 \cdot p_6^3 \cdot \Pr(((\circ\circ\circ))C \Rightarrow^* s),$$

$$\Pr(d'') = \Pr(S \Rightarrow^* (B)C) \cdot p_4^1 \cdot p_{5.1.1}^1 \cdot p_6^1 \cdot \Pr(((\circ\circ\circ))C \Rightarrow^* s).$$

Nevertheless, standard loop-dependent thermodynamic models factor secondary structures in a more complex way, namely into terms for base pair stacking interactions (as opposed to individual base pairs) and diverse terms for different kinds of loops, which in many cases strongly depend on the lengths of the respective loops. Therefore, in order to closely mirror a particular state-of-the-art energy model, a corresponding SCFG must account for base pair stacking and

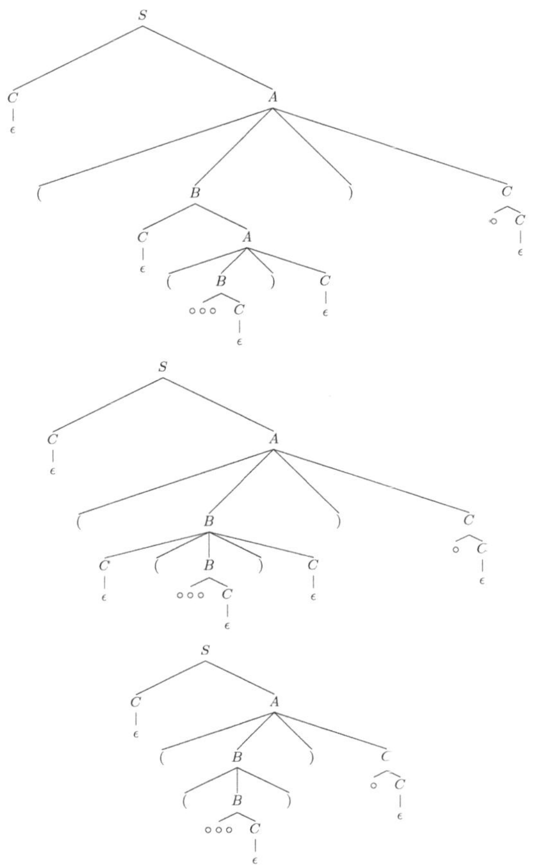

Figure 15.4 Parse trees according to different grammar designs. Figures display the unique parse trees for the dot-bracket word ((∘∘∘))∘ using the three different (more and more specialized) variants of the same lightweight grammar as discussed in Example 15.7. (a) Parse tree according to G_3. (b) Parse tree according to G_3'. (c) Parse tree according to G_3''.

explicit loop lengths. These might be considered the most important features that should and actually can be handled by a corresponding SCFG. More information on how to deal with those and similar features can be found, for instance in Dowell and Eddy (2004). Notably, as indicated by several authors (for example (Nebel and Scheid, 2011a; Nebel *et al.*, 2011)), considering SCFGs that do not or only partially model base stacking and also rarely use explicit loop lengths might actually be sufficient for obtaining a reliable probabilistic model.

Anyway, a straightforward approach for deriving a suitable grammar G_r for generating RNA sequences (where derivation trees are used to encode the underlying secondary structure) is given as follows: Initially, we construct an *unambiguous* grammar $G_s=(I, T, R, S)$ that generates the language of all possible dot-bracket representations. Afterwards, we replace any production rule of the form $X\to\alpha({}^z\beta)^z\gamma$ or $X\to\alpha\circ^z\beta$ (with $\alpha,\beta,\gamma\in(I\cup T)^*$, $X\in I$, and $({}^z)^z$ or $\circ^z$ representing $z\geq 1$ consecutive base pairs or unpaired bases) by corresponding new productions generating all considered individual base pairs and unpaired bases, respectively.

Example 15.9

For applications to structure prediction, one could, for example, use the following ambiguous yet still structurally unambiguous SCFG for RNA sequences which has been constructed in the described way on the basis of the corresponding unambiguous SCFG for RNA secondary structures from Example 15.5:

$$p_1: S\to LS, p_2: S\to L,$$
$$p_{3.1}: L\to a, p_{3.2}: L\to c, p_{3.3}: L\to g, p_{3.4}: L\to u,$$
$$p_{4.1}: L\to aFu, p_{4.2}: L\to cFg, p_{4.3}: L\to gFc, p_{4.4}: L\to uFa,$$
$$p_{5.1}: F\to aFu, p_{5.2}: F\to cFg, p_{5.3}: F\to gFc, p_{5.4}: F\to uFa,$$
$$p_6: F\to LS.$$

Note that according to this grammar, only Watson–Crick pairs are allowed in all possible base paired secondary structures for a particular RNA sequence; other parings are prohibited since there are no corresponding production rules for generating them[5]. Obviously, the transformation of the secondary structure grammar G_2 into the presented SCFG for RNA sequences implies a higher complexity by means of cardinality of the underlying rule set and hence results in a larger number of probabilistic parameters that need to be estimated by corresponding training procedures.

[5] This obviously corresponds to the case that such rules actually exist but are assigned weights or probabilities 0.

Conditional structure probabilities

The essence of SCFG based approaches towards structure prediction is that the parameters of the underlying grammar actually provide a compact representation of a *joint* probability distribution over RNA sequences and their secondary structures, which is induced by the *joint* probabilities $\Pr(r, d)$ of generating a particular leftmost derivation d and some RNA sequence r using the considered SCFG.

Example 15.10

Using the structurally unambiguous RNA grammar presented in Example 15.9, the joint probability of the secondary structure s=((∘∘∘))∘ to be generated along with the sequence *r=aucgaaug* is given by:

$$\Pr(r,s)=\Pr(r,d=S\overset{S\to LS}{\Rightarrow}LS\overset{L\to aFu}{\Rightarrow}aFuS\overset{F\to uFa}{\Rightarrow}$$

$$auFauS\overset{F\to LS}{\Rightarrow}auLSauS\overset{L\to c}{\Rightarrow}aucSauS\overset{S\to LS}{\Rightarrow}$$

$$aucLSauS\overset{L\to g}{\Rightarrow}aucgSauS\overset{S\to L}{\Rightarrow}aucgLauS\overset{L\to a}{\Rightarrow}$$

$$aucgaauS\overset{S\to L}{\Rightarrow}aucgaauL\overset{L\to g}{\Rightarrow}aucgaaug)$$

$$=\Pr(S\to LS)\cdot\Pr(L\to aFu)\cdot\Pr(F\to uFa)$$

$$\cdot\Pr(F\to LS)\cdot\Pr(L\to c)\cdot\Pr(S\to LS)\cdot$$

$$\Pr(L\to g)\cdot\Pr(S\to L)\cdot\Pr(L\to a)\cdot$$

$$\Pr(S\to L)\cdot\Pr(L\to g).$$

However, the goal of single sequence RNA secondary structure prediction is to find the best folding s for a given input sequence r. In connection with probabilistic parsing techniques, this requires a way to calculate the *conditional* probability $\Pr(s \mid r)$ of the secondary structure s given the RNA sequence r. If generative probabilistic models (like SCFGs or HMMs) are used in this context, these conditional probabilities can readily be derived from the corresponding joint probabilities. Formally, for S denoting a set of valid derivation trees sharing the same secondary structure, it follows that

$$\Pr(S\mid r)=\sum_{d\in S}\Pr(d\mid r)=\frac{\sum_{d\in S}\Pr(r,d)}{\sum_{d'\in F(r)}\Pr(r,d')},$$

where $F(r)=\{d'=S\Rightarrow^{*}r\}$ is the set of all possible derivation trees for sequence r (here S denotes the axiom of the used SCFG). Consequently, for structurally unambiguous SCFGs, the probability for generating a particular secondary structure s (with corresponding unique derivation tree d) given some RNA sequence r is equal to

$$\Pr(s\mid r)=\Pr(d\mid r)=\frac{\Pr(r,d)}{\sum_{d'\in F(r)}\Pr(r,d')},$$

where $F(r)$ then obviously defines the set of all feasible secondary structures for r (due to structural unambiguity), that is the so-called *folding space* for the considered sequence.

Parameter estimation

When using SCFGs as a model for secondary structures on RNA sequences, the grammar's parameters have to be estimated from a given sample set of real-life sequences with annotated trusted secondary structures. Like in the sequence-independent case (where only secondary structures without annotated sequences are considered), the training task involves finding exactly those probabilities for each of the production rules of particular SCFG G (that is, the set of parameters $\Phi=\{p_1,\ldots,p_n\}$ if G defines n rules) that maximize some specified objective function. However, the popular maximum likelihood technique as one of the most well-understood algorithms for parameter estimation can still be applied for this purpose.

Formally, let $D=\{(x^{(1)}, y^{(1)}),\ldots,(x^{(m)}, y^{(m)})\}$ denote the considered set of training data, composed of m pairs of RNA sequences $x^{(i)}$ with trusted (that is, in general experimentally validated or sometimes alternatively computationally derived) secondary structures $y^{(i)}$, $1\leq i\leq m$. Then, Φ is chosen to maximize the *joint likelihood* of the training sequences and their structures. Under the constraints typically imposed on the parameters of generative probabilistic models (namely that all parameters must be non-negative and certain groups of parameters must sum up to one), this likelihood[6] is given as follows:

$$l_{ML}(\Phi|D) = \prod_{i=1}^{m} \Pr(x^{(i)}, y^{(i)}; \Phi).$$

Note that the solution Φ_{ML} to this constrained optimization problem actually exists in closed form for structurally unambiguous SCFGs, being the reason why this technique is most commonly used in practice for estimating a particular set of grammar parameters. For more details, see for instance (Durbin *et al.*, 1998).

Anyway, if structural unambiguity is assured, then generative (SCFG based) training can easily and indeed efficiently be realized again by counting the observed frequencies of applications of the distinct production rules needed for generating the structures in the considered sample set, as this not only yields a consistent SCFG, but also a maximum likelihood estimator of the SCFG parameters (see 'Parameter Estimation' above). More specifically, for training a structurally unambiguous RNA grammar, we make use of the fact that for any $i \in \{1, \ldots, m\}$, we know the trusted secondary structure $y^{(i)}$ for each sequence $x^{(i)}$ of the training set D, such that the unique derivation tree that corresponds to $y^{(i)}$ can be used along with $x^{(i)}$, for $1 \leq i \leq m$, in order to determine the relative frequency of each production among all productions with the same premise. The relative frequencies that are obtained in this manner are indeed a maximum joint likelihood estimator for the probabilities that lead to the generation of the training data (see (Prescher, 2003) for details).

Choice of training data

If parameters for RNA secondary structure models are to be estimated, we basically have two different choices: First, we may consider a training set where only structures of a single biological class (for example tRNA) are contained. Then, we may expect that all structural properties (including aspects which are caused by interaction with proteins or by other *non-energetic* factors of RNA folding) that are typical to this class are trained into the respective parameter values. For a general model of RNA folding, it has to assume some kind of 'over-specialization', since we cannot expect the model to generalize well to new data from a different class. Second, we may use a rich training set of mixed biological classes. In that case, the danger of a potential lack of generalization is much smaller, but we lose the chance to capture some class-specific properties of the structures within our model.

In both cases, the main problem that comes inherently with the SCFG approach for modelling RNA structures and limits the performance of the corresponding computational prediction methods is that it is obviously highly dependent on the availability of a rich, reliable training set in order to minimize the danger of overfitting. Intuitively, this might especially be the case when using an excessively complex SCFG design that distinguishes between all different features in RNA structure aiming at a highly realistic model (for which a large number of parameters need to be determined). The reason lies in the fact that to obtain a reliable estimation result for SCFGs with large numbers of parameters, we need comprehensive training sets to ensure that enough observations are made for any structural motif modelled by one of the production rules, thereby avoiding a particular structure (or shape) being trained into the model.

In this context, it should be mentioned that recently it has been shown that for complex models (SCFGs or otherwise) performance is highly sensitive to the structural diversity of the training sample and not to the sequence diversity or sample's total size (Rivas *et al.*, 2012). This presumably implies that for constructing robustly trained models, a larger number of structurally diverse RNA families must be considered, each containing a large number of sequences with well-annotated structures.

However, despite the multitude of publicly available RNA databases resulting from the fact that the number of solved secondary structures has dramatically increased over the past years,

6 Not probability, see 'Maximum likelihood training'.

an ideal set of structural RNA data for training statistical models – according to our experience – was still to be attained. In fact, even the latest currently existing datasets like RNA STRAND (Andronescu *et al.*, 2008) do not support the satisfactory training of complex probabilistic models. Nevertheless, it should be mentioned that in an attempt to resolve that long-standing problem, a collection of four new sets of structural RNA data has been described in (Rivas *et al.*, 2012), which include rather large numbers of RNA sequences with known secondary structures (compared to previously considered sets) and where the variety and correctness of data have been crucial for inclusion.

Separation of parameters

In algorithms and applications based on SCFGs, the grammar parameters are often split into a set of *transition probabilities* and corresponding sets of *emission probabilities*. This separation into transition and emission probabilities actually corresponds to the standard treatment of (generative) model parameters as applied for example in the case of HMMs. For a particular RNA grammar G_r with underlying SCFG $G_s=(I, \Sigma_s, R, S, \Pr)$ for modelling secondary structures (i.e. generating the language of dot-bracket representations), the probabilities of the rules of G_r are thus split into transition probabilities PR_{tr} (*rule*) for *rule* $\in R$ and corresponding emission probabilities $PR_{em}(r_x)$ for $r_x \in \Sigma_r=\{a, c, g, u\}$ and $\Pr_{em}(r_{x1}r_{x2})$ for $r_{x1}r_{x2} \in \Sigma_r^2$, that is for the individual unpaired bases and possible base pairings, respectively.

Note that in cases of SCFGs for modelling RNA secondary structure, it has become custom that all emission probabilities (for the 4 individual unpaired bases and the resulting 16 distinct possible base pairings, respectively) come from the same distribution. That is, for any considered loop type, one uses the same emission probabilities for unpaired bases located within the loop and for base pairs closing a corresponding loop. Hence, this separation into rule and emission probabilities allows us to work with a smaller set of productions since e.g. all productions $L \to r_xFr_y$, r_x, r_y two paired nucleotides, are combined into a single rule $L \to (F)$ (analogous to the grammar for dot-bracket words) weighting the use of $L \to (F)$ by the emission probability $\Pr_{em}(r_xr_y)$ for the actual base pair r_x, r_y. Thus, we essentially work with the unambiguous SCFG for dot-bracket words, although we actually had to deal with the larger set of productions of the corresponding ambiguous SCFG generating any possible RNA sequence (where the derivation trees uniquely correspond to the different secondary structures for that sequence).

Example 15.11

By linking together the emissions of base pairs generated with different rules, the joint probability of generating the sequence *r=aucgaaug* and the secondary structure s=((∘∘∘))∘ (corresponding to the unique leftmost derivation *d* given in Example 15.10) is computed as follows:

$$\Pr(r,s)=\Pr_{tr}(S\to LS)\cdot\Pr_{tr}(L\to(F))\cdot$$

$$\cdot\Pr_{em}(au)\cdot\Pr_{tr}(F\to(F))\Pr_{em}(ua)$$

$$\cdot\Pr_{tr}(F\to LS)\cdot\Pr_{tr}(L\to\circ)\Pr_{em}(c)$$

$$\cdot\Pr_{tr}(S\to LS)\cdot\Pr_{tr}(L\to\circ)\Pr_{em}(g)$$

$$\cdot\Pr_{tr}(S\to L)\cdot\Pr_{tr}(L\to\circ)\Pr_{em}(a)$$

$$\cdot\Pr_{tr}(S\to L)\cdot\Pr_{tr}(L\to\circ)\Pr_{em}(g).$$

It is easily seen that for a particular SCFG, this separation might actually reduce the number of free parameters that need to be estimated by the employed training procedures in a very significant way, such that the ever-present danger of overfitting becomes less threatening.

With respect to training, separating RNA grammar parameters into transition and emission terms is quite unproblematic in practice: let $D=\{(x^{(1)}, y^{(1)}), \ldots, (x^{(m)}, y^{(m)})\}$ be the considered set of RNA data. Then, instead of using the derivation tree that corresponds to the correct secondary structure $y^{(i)}$ for sequence $x^{(i)}$, $1 \leq i \leq m$, to determine the relative frequency of each production among all productions with the same premise, we simply have to count the relative frequencies of applications of the production rules of the underlying secondary structure grammar, along with the corresponding relative frequencies of emissions of unpaired bases and base pairs observed in the training set. Using a particular parsing technique (as, for instance, *probabilistic Earley parsing*, see 'Earley Parsing' below) in order to count these frequencies might often be even more efficient in practice, due to the smaller number of production rules of the underlying secondary structure grammar. However, the relative frequencies, obtained in this manner, are still a maximum likelihood estimator for the probabilities of the more complex (by means of number of production rules) RNA grammar.

SCFG-based algorithms

When considering structure prediction as a mathematically well-defined optimization problem, given a stochastic RNA grammar G, one traditionally looks for a valid derivation tree with maximal probability among all possible derivation trees for the given input sequence[7]. This derivation tree corresponds to the most likely secondary structure using the SCFG G and is usually called *Viterbi parse*. To compute Viterbi parses, one can utilize adapted versions of well-established parsing algorithms, as we will see in 'Adapted Parsing Techniques' below. A related alternative parsing variant (following a different optimization goal) that has also been successfully applied to the RNA folding problem will be discussed in 'MEA Parsing' below. This method actually calculates a so-called *MEA parse* rather than a corresponding Viterbi parse.

Adapted parsing techniques

CFGs are simple enough to allow the construction of efficient (polynomial-time) *recognition* and *parsing* algorithms, which, for a given input string, determine whether and how (in terms of a corresponding parse tree) a tree can be generated by the considered CFG. These algorithms are usually described by recursive DP routines. Two popular examples, which we sometimes directly build on in the sequel, will be briefly discussed in this section. For a more detailed introduction to parsing techniques for CFGs and a quite exhaustive collection of existing parsing strategies in general, we refer to Grune and Jacobs (2008).

Cocke–Younger–Kasami algorithm

The most well-known and fairly simple variant due to Cocke (Cocke and Schwartz, 1970), Younger (1967) and Kasami (1965), usually called the Cocke–Younger–Kasami (CYK) algorithm, can only handle some restrictive subsets of CFGs. In fact, the CYK algorithm in its original form is only described for non-stochastic CFGs in Chomsky normal form (CNF), that is for grammars $G=(I, T, R, S)$, where any production is either of the form $A \to BC$ or $A \to a$, with $A, B, C \in I$ representing intermediate symbols and $a \in T$ denoting a single terminal symbol. However, any CFG G without ε-rules can be transformed into G' in CNF, such that $L(G)=L(G')$. For details, refer to Hopcroft and Ullman (1979). This is also possible for SCFGs without affecting the probability distribution implied. Anyway, the CYK parsing algorithm can readily be modified to compute the most probable parse tree of a given input sequence according to the probability distribution on all possible parse trees for that sequence, as induced by a given SCFG. This is typically realized by incorporating log probabilities[8] of the production rules of any considered SCFG into the DP recursions.

Since the transformation of an arbitrary CFG G into CNF may lead to an undesirable bloat in the number of productions, it has proven convenient to avoid this conversion in applications and express the CYK recursions directly in terms of the production rules of G. For instance, when considering the CFG G_1 of Example 15.1, the probabilistic variant of the CYK algorithm relies

[7] Note that when using a non-stochastic RNA grammar, all derivation trees are actually equiprobable, due to the implicit assumption of a uniform distribution.

[8] Using the logarithm of all probabilities has two main advantages: speed and accuracy. In fact, since the log of a product is equal to the sum of the logs of all factors, all products are turned into sums and addition is less expensive than multiplication. Furthermore, the use of log probabilities improves numerical stability, as the underflow problem can be essentially solved. Note that the base of the logarithm is not important as long as it is larger than 1 (for instance 2, Euler's number e, or 10).

on the following recursions for a given input sequence r:

$$M_{i,i-1} = \log(\Pr(S \to \varepsilon)), \text{ for } 1 \le i \le n, \text{ and}$$

$$M_{i,j} = \max \begin{cases} \log(\Pr(S \to r_i S)) + M_{i+1,j}, \\ M_{i,j-1} + \log(\Pr(S \to S r_j)), \\ \log(\Pr(S \to r_i S r_j)) + M_{i+1,j-1}, \\ \max_{i<k<j-1} M_{i,k} + M_{k+1,j} + \log(\Pr(S \to SS)), \end{cases}$$

for $1 \le i \le j-1$ and $1 \le j \le n$.

Obviously, there is a near-exact correspondence between probabilistic CYK parsing and standard DP algorithms for MFE calculations. In fact, SCFG based DP methods for computing the optimal (maximum probability) parse tree essentially use the same recursion scheme as physics-based DP algorithms for deriving the optimal (MFE) conformation (see, for instance Nussinov *et al.*, 1978; Nussinov and Jacobson, 1980). However, while thermodynamic methods are based on factoring the structure into a sum of energy terms or arbitrary base pair scores (according to a particular energy model), the scoring system of the CYK variant is in fact probabilistic, based on factoring the score for a structure down into a sum of log probability terms (according to the underlying SCFG model). Hence, the main difference is that in case of a sophisticated CYK recursion scheme, which distinguishes the same canonical substructures and structural motifs as a particular MFE algorithm, the corresponding scores are all derived from observations in known RNA structures rather than from experimental studies and extrapolations.

Furthermore, there is a notable difference between thermodynamic and probabilistic approaches to optimal structure prediction with respect to structural ambiguity:

- Physics-based approaches traditionally use free energy values as scores, such as the non-normalized thermodynamic scoring scheme. Hence, even if the free energy of a particular structure is scored more than once, the scores (free energies) of all considered structures effectively remain the same. This actually means that regardless of how many different ways of scoring the energy of a structure exist, the MFE structures still win. Consequently, structural ambiguity is not an issue in connection with methods for computing optimal (by means of MFE) foldings.
- Probabilistic approaches, however, use normalized probabilities for distinguished structural motifs, such as scores, resulting in a normalized scoring scheme. Thus, if a structure is considered more than once due to structural ambiguity of the underlying grammar, then the scores (structure probabilities) are inherently biased, since the probability of any secondary structure is then generally dependent on (somehow 'weighted' by) the number of its corresponding valid derivations. As a consequence, if the considered SCFG is structurally ambiguous, then the most probable parse tree represents only its own probability, but not the probabilities of the other alternatives. Hence, in connection with methods for computing the optimal (most probable) folding for a given sequence, structural ambiguity is an important practical concern (as shown for instance in Dowell and Eddy, 2004).

Earley parsing

Another widely known parsing algorithm that can actually handle *arbitrary* CFGs is known as *Earley's algorithm*, or *Earley parser* (Earley, 1970). Besides the obvious benefit that no normal form is required for applying this algorithm, it is in fact one of the most efficient known parsing techniques. Moreover, *probabilistic* Earley parsing can actually be used for determining the most probable derivation of a given input string with respect to a considered SCFG, hence providing an attractive alternative to the CYK algorithm with respect to computing Viterbi parses.

Principally, a (non-probabilistic) Earley parser operates on lists of *items*, which define position indices along with so-called *dotted rules* and are used to represent partial derivations. Specifically, a dotted rule is a production of the grammar with a dot symbol • inserted somewhere in the right-hand side of this production, where dots at the very left and right positions are explicitly allowed. For example, the production $A \to \alpha\beta$ of a grammar

$G=(I,T,R,S)$, with $A \in I$ and $\alpha, \beta \in I \cup T$, implies the three dotted rules $A \to \bullet\alpha\beta$, $A \to \alpha \bullet \beta$ and $A \to \alpha\beta\bullet$ to be considered by Earley's algorithm.

For any dotted rule, $rule = A \to \alpha \bullet \beta$, $A \in I$ and $\alpha, \beta \in (I \cup T)^*$, we can write items in the form $[i, rule, j]$, for $1 \le i,j \le n+1$, where n is the length of the input string w to be parsed. An item of this form is derived if there exists a sentential form $\gamma \in (I \cup T)^*$ such that

$$S' \Rightarrow^* w_{1,i-1} A\gamma \overset{A\to\alpha\beta}{\Rightarrow} w_{1,i-1}\alpha\beta\gamma \Rightarrow^* w_{1,i-1} w_{i,j-1}\beta\gamma.$$

Briefly, the Earley parser is initialized with the start item $[1, S' \to \bullet S, 1]$, for $S' \notin \{I \cup T\}$ and S is the axiom of the used CFG G[9]. Then, the transitive closure with respect to the following operations is computed:

- Predictor: If $\exists([i, C \to \gamma \bullet A\delta, j]$ and $A \to \alpha \in R)$, then add $[j, A \to \bullet\alpha, j]$,
- Scanner: If $\exists[i, A \to \alpha \bullet w_{j-1}\beta, j-1]$, then add $[i, A \to \alpha w_{j-1} \bullet \beta, j]$,
- Completer: If $\exists([i, A \to \alpha \bullet B\beta, k]$ and $[k, B \to \gamma\bullet, j])$, then add $[i, A \to \alpha B \bullet \beta, j]$.

Intuitively, the predictor adds all the productions that might yield a valid extension of the (nonterminal) symbol following the dot, whereas the scanner advances the dot past terminal symbols if they match the corresponding symbol of the parsed string. The completer advances the dot symbol past (nonterminal) symbols if they could actually be validly extended. The goal item $[1, S' \to S\bullet, n+1]$ is thus derived if and only if the considered grammar G generates the input string w, this is if and only if $w \in L(G)$.

Note that a corresponding probabilistic variant of Earley's algorithm simply keeps track of the probabilities of partial derivations by adding them to the corresponding items as an additional parameter. For further information on probabilistic Earley parsing, we refer to Stolcke (1995).

Inside–outside algorithm

An additional popular and highly relevant algorithm for SCFGs in CNF, which alongside others has been successfully applied to the RNA folding problem, is the so-called *inside–outside algorithm* (Lari and Young, 1990; Lari and Young, 1991). This algorithm actually represents the natural counterpart of the *forward–backward algorithm* for HMMs (for details, we refer to Durbin *et al.* (1998) and Bishop (2007)), where its computational complexity is substantially greater.

Briefly, the inside–outside algorithm has two basic steps, each realized by a corresponding DP method. In particular, for X, an intermediate symbol of the considered grammar G, the first step called *inside algorithm* (which corresponds to the forward step in case of HMMs) computes all *inside probabilities* $\alpha_X(i, j)$, $1 \le i, j \le n$, for a given word $w \in L(G)$ of size n. Afterwards, the *outside algorithm* (corresponding to the backward step for HMMs) calculates all *outside probabilities* $\beta_X(i, j)$, $1 \le i, j \le n$, for the same word w, making use of the inside values derived in step one.

Definition of inside and outside values

Specifically, for an underlying CFG $G=(I, T, R, S)$ and an input $w \in L(G) \subseteq T^n$, the inside and outside values for any two indices $i,j \in \{1, \ldots, n\}$ are defined as follows:

- $\alpha_X(i, j)$ is the probability of all leftmost derivations (derivation trees) that generates the subword $w_i \ldots w_j$ from the intermediate symbol $X \in I$, formally $\alpha_X(i, j) := \Pr(X \Rightarrow^* w_i \ldots w_j)$. Hence, this value is the probability of covering a range of the input w starting from a specified nonterminal symbol X. It might thus be computed simply as the sum of the probabilities of all derivations of the subword $w_i \ldots w_j$ starting with X.
- $\beta_X(i, j)$ is the probability of all leftmost derivations, which starting with the intermediate symbol S (the axiom of grammar G), generate the sentential form $w_1 \ldots w_{i-1} \cup w_{j+1} \ldots w_n$, formally $\beta_X(i, j) := \Pr(S \Rightarrow^* w_1 \ldots w_{i-1} X w_{j+1} \ldots w_n)$. Intuitively, this value is the probability of everything surrounding a certain nonterminal symbol X (assumed to generate $w_i \ldots w_j$) and thus in some way is dual to the corresponding inside probability.

[9] Thus, S′ is a new nonterminal symbol and S′→S a new production.

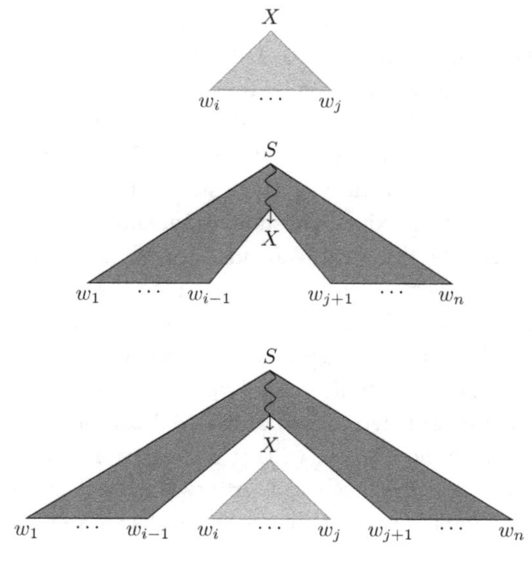

Figure 15.5 Inside and outside values. Figures illustrate the inside and outside probabilities α_x (i, j) and β_x (i, j), $1 \leq i, j \leq n$, for a particular intermediate symbol x of a SCFG G with start symbol S, as parts of a corresponding parse tree for input word $w=w_1...w_n$ using G. (a) Inside value α_x (*i, j*). (b) Outside value β_x (i, j). (c) Product α_x (i, j)·β_x (i, j).

The product of a particular inside probability and its corresponding outside probability actually covers the whole input by using a specified nonterminal symbol for a certain range, since

$$\alpha_X(i,j)\cdot\beta_X(i,j)=\Pr(X\Rightarrow^* w_i...w_j)$$

$$\cdot\Pr(S\Rightarrow^* w_1...w_{i-1}\, X w_{j+1}...w_n)$$

$$=\Pr(S\Rightarrow^* w_1...w_{i-1}\, X w_{j+1}...w_n \Rightarrow^* w_1...w_n).$$

Hence, the outside values describe a convenient relationship to the inside values. In fact, for $X \in I$ any product α_X (i, j)·β_X (i, j), $1 \leq i, j \leq n$, equals to the sum of the probabilities of all leftmost derivations using the given intermediate symbol *X* at a particular point. An illustration of individual inside and outside probabilities and their relationship is presented in Fig. 15.5.

Computation of inside values

Conceptually, the inside algorithm is strongly related to the CYK parsing algorithm. In fact, for any input sequence *r* of length *n*, the goal value of the inside algorithm is given by $\alpha_S(1,n)$, *S* the axiom of the used SCFG. This value is equivalent to the total probability of the input sequence given the underlying stochastic model and might thus be obtained by summing over all parse trees for the input. To do this, the inside algorithm replaces the max operations in the corresponding CYK recursions with sums, and additions of terms with multiplications. It does not consider log probabilities.

Just like for the CYK algorithm itself, the corresponding recursions for its inside variant may conventionally be expressed directly in terms of the productions of a particular SCFG G, albeit in its original form (Lari and Young, 1990) it can only handle grammars in CNF. For instance, given an arbitrary input sequence *r* of length *n*, then the CYK inside algorithm for the CFG G_1 of Example 15.1 can be described by the following recursion scheme:

$$\alpha_S(i,i-1)=\Pr(S\rightarrow\varepsilon), \text{ for } 1\leq i\leq n, \text{ and}$$

$$\alpha_S(i,j)=\sum\begin{cases}\Pr(S\rightarrow r_iS)\cdot\alpha_S(i+1,j),\\ \alpha_S(i,j-1)\cdot\Pr(S\rightarrow Sr_j),\\ \Pr(S\rightarrow r_iSr_j)\cdot\alpha_S(i+1,j-1),\\ \sum_{i<k<j-1}\alpha_S(i,k)\cdot\alpha_S(k+1,j)\cdot\Pr(S\rightarrow SS),\end{cases}$$

for $1\leq i\leq j-1$ and $1\leq j\leq n$.

Note that the inside algorithm is somehow analogous to the McCaskill algorithm (McCaskill, 1990) for calculating PFs on the basis of (experimentally obtained) thermodynamic parameters for RNA secondary structure. Anyway, there is also a noticeable difference between both algorithms: When calculating the equilibrium PF for a given sequence, one must be careful not to count any structure more than once in order to obtain an unbiased Boltzmann distribution of possible structures. In contrast, structural ambiguity is not an issue in connection with summed inside calculation (the analogue of the summed PF computation), since it actually sums over all possible parse trees rather than over all possible secondary structures, such that the inside algorithm gives the correct result even for structurally ambiguous SCFGs.

Computation of outside values

While inside probabilities of components of a given input string and for a particular SCFG can easily be computed with extended variants of CYK or Earley parsers, the construction of corresponding parsing algorithms for calculating the outside probabilities is in general significantly more complex. Actually, a generally valid method (of similar simplicity) corresponding to the method of (Lari and Young, 1990) that can be used in connection with arbitrary grammar specifications (not in CNF) has not yet been devised and is thus an unsolved task.

However, an attractive alternative is to use the concept of *semiring parsing* (Goodman, 1998, 1999) in order to derive a grammar specific parser that computes the outside values of all string fragments using an arbitrary SCFG. Briefly, semiring parsing represents a formalism to describe parsers for which we abstract from specific operations typically applied for applications (like summation or taking the maximum as it was done for the inside algorithm or for computing the most probable derivation tree, see above). In the general unspecified framework, a semiring is assumed to account for those operations, and the parsing algorithm is expressed by means of abstract semiring operations only. For a certain application – including recognition, derivation forests, and in case of SCFGs also Viterbi paths or inside probabilities – one has to provide the appropriate semiring and replace the abstract operations within the algorithm by those at hand. In fact, the idea of semiring parsing provides a beautifully unified view on the problem of generating parsing algorithms for calculating any of these *forward values* and also the corresponding *reverse values* which are typically much more difficult to compute[10]. As pointed out in (Goodman, 1998), this makes it especially useful for finding parsers for outside probabilities, and for parsers that can handle arbitrary grammar designs, like Earley-style parsers.

MEA parsing

In connection with RNA structure analysis, both the inside values and the outside values are of great relevance for several practical applications, including parameter re-estimation by expectation maximization[11] and a special variant of SCFG based structure prediction, called *maximum expected accuracy* (*MEA*) parsing, as opposed to Viterbi parsing.

Traditional variant

Briefly, just like Viterbi parsing, MEA parsing is also an optimization problem, but a different objective function is considered. In fact, the optimal structure is defined as the one with the highest expected number of correctly predicted positions. Thus, a *MEA parse* for a given sequence is actually a valid derivation tree which maximizes the number of correctly unpaired and paired positions with respect to the true folding of that sequence.

Among all possible derivation trees for the given input sequence, this parse can be identified by a corresponding DP method similar to the CYK algorithm. Basically, the sole difference is that the recursion scheme makes use of sequence-specific probabilities for unpaired bases and base pairs, which are traditionally calculated from the corresponding inside and outside values of subsequences of the input. Originally (see (Knudsen and Hein, 2003), supplemental material), for an input sequence of length n, the following recursive system is used for calculating the maximum expected accuracy $M_{1,n}$:

$$M_{i,i} = q_i, \text{ for } 1 \leq i \leq n, \text{ and}$$

$$M_{i,j} = \max \begin{cases} q_i + M_{i+1,j}, \\ M_{i,j-1} + q_j, \\ p_{i,j} + M_{i+1,j-1}, \\ \max\limits_{i<k<j-1} M_{i,k} + M_{k+1,j}, \end{cases}$$

$$\text{for } 1 \leq i \leq j-1 \text{ and } 1 \leq j \leq n$$

[10] Goodman (1998) introduces forward value and reverse value as generalizations of inside value and outside value, respectively, for arbitrary semirings.

[11] The inside and outside values can be used to re-estimate the probabilities (parameters) of a given SCFG by expectation maximization in a similar way as the forward and backward variables can be used for HMM training by expectation maximization, see for instance Durbin *et al*. (1998).

where q_i denotes the probability that i remains unpaired and $p_{i,j}$ denotes the probability that i pairs with j. Under the assumption that the underlying CFG $G = (I, T, R, S)$ for modelling the considered class of all feasible secondary structures exclusively uses productions of the form $A \to (B)$ for generating base pairs, where $A, B \in I$, the probability (under that model) that positions i and j, $1 \le i, j \le n$, form a base pair is effectively computed by summing over all rules forming pairs:

$$p_{i,j} = \sum_{A,B} \beta_A(i,j) \cdot \sum_{xy} \Pr(A \to xBy) \cdot \alpha_B(i+1, j-1)$$
$$= \sum_{A,B} \beta_A(i,j) \cdot \Pr(A \to (B)) \cdot \alpha_B(i+1, j-1).$$

It immediately follows that the probability for position i, $1 \le i \le n$, remaining unpaired is given by:

$$q_i = 1 - \sum_{j \ne i} p_{i,j} .$$

It should be mentioned that an efficient implementation of this MEA parsing algorithm is given by the popular Pfold tool (Knudsen and Hein, 1999, 2003). Notably, Pfold makes use of inside–outside probabilities derived on the basis of a particular stochastic variant of the (lightweight) CFG G_2 of the Example 15.2. However, Pfold has not been created for single sequence structure prediction in the first place. In fact, the underlying algorithm has originally been designed to take an alignment of related RNA sequences as input and then predict a common (consensus) structure for all sequences. Recently, a new parallelized version of the algorithm has been created, named PPfold (Sukosd *et al.*, 2011), which manages to solve the structure of much longer alignments than Pfold without breaking down.

Adjusted variant

It is worth mentioning that there exist elegant ways to incorporate a trade-off parameter γ in order to provide the user with a mechanism for controlling the *sensitivity* (Sens.) and the *positive predictive value* (PPV)[12] of the predicted foldings. These two measures were introduced in order to quantify the accuracy of RNA secondary structure prediction methods and are usually defined as follows (see, for instance (Baldi *et al.*, 2000)):

- Sens. is the relative frequency of correctly predicted pairs among all base pairs that are actually present within the native folding, whereas
- PPV is defined as the relative frequency of correctly predicted pairs among all position pairs that were predicted.

Formally, they are given by

$$\text{Sens.} = \frac{TP}{TP + FN} \text{ and}$$

$$\text{PPV} = \frac{TP}{TP + FP}, \text{ where}$$

- TP is the number of correctly predicted base pairs (*true positives*),
- FN is the number of base pairs in the native structure that were not predicted (*false negatives*) and
- FP is the number of incorrectly predicted base pairs (*false positives*).

Note that in Do *et al.* (2006), the idea of a parameter γ to control the sensitivity/PPV trade-off has been used in connection with a DP optimization scheme. According to its value, the algorithm either tends to predict only those base pairs with rather strong signals for them to belong to the native folding or it is encouraged to predict more pairings even if they might be no part of the native structure. The MEA structures for a given sequence are then defined as those of all candidate structures that maximize the number of correctly unpaired positions plus γ times the number of correctly paired positions with respect to the true folding of that sequence.

Formally, γ may take on any positive real value and the choice of $\gamma = 1$ serves as the neutral element with respect to the prediction, that is the prediction is neither biased towards a better sensitivity nor towards a better PPV. More precisely, γ may take on values in $[0, \infty)$, where for the considered sequence fragment $r_i \dots r_j$, $1 \le i, j \le n$,

[12]Note that the positive predictive value is often called specificity, as for example in Do *et al.* (2006).

- $\gamma < 1$ restricts the procedure to produce pair *i.j* only if it is extremely confident,
- $\gamma = 1$ has no impact on the decision whether *i.j* should be paired or not,
- $\gamma > 1$ encourages the algorithm to produce pair *i.j*, even if it is not confident,

that this pair belongs to the native folding.

In Do *et al.* (2006), the corresponding DP matrix M is computed according to the following recurrence:

$$M_{i,i} = q_i, \text{ for } 1 \leq i \leq n, \text{ and}$$

$$M_{i,j} = \max \begin{cases} q_i + M_{i+1,j}, \\ M_{i,j-1} + q_j, \\ \gamma \cdot 2 \cdot p_{i,j} + M_{i+1,j-1}, \\ \max\limits_{i<k<j-1} M_{i,k} + M_{k+1,j}, \end{cases}$$

for $1 \leq i \leq j-1$ and $1 \leq j \leq n$

where $p_{i,j}$ denotes the probability that i pairs with j and q_i denotes the probability that i remains unpaired. The traceback step of the corresponding DP algorithm can thus be employed to identify the MEA structures of the input sequence according to the given setting of γ. If only one MEA structure is recovered in the traceback step, the complete algorithm obviously requires $O(n^3)$ time and $O(n^2)$ space.

Note that the authors of (Do *et al.*, 2006) claim that for the default setting $\gamma = 1$, the algorithm only maximizes the expected number of correct (unpaired and paired) positions and is actually identical to the DP technique used in Pfold. This, however, is not quite right, since in the unparameterized recurrence of Pfold as introduced in Knudsen and Hein (2003), supplemental material and presented above, the pairing probability $p_{i,j}$ is actually not multiplied by factor 2. Consequently, if this factor is indeed considered, the default setting would be $\gamma = \frac{1}{2}$.

Related methods

It remains to note that principally the MEA calculations only require the posterior probabilities of any two positions being base paired, which can obviously be derived by employing the inside–outside algorithm (Lari and Young, 1990, 1991) on the basis of a particular probabilistic model. However, these pairing probabilities might equivalently be calculated with the McCaskill algorithm (McCaskill, 1990) when considering a thermodynamic model.

Furthermore, different MEA methods have been implemented for RNA folding. Originally, the posterior probabilities of base pairs were maximized (Knudsen and Hein, 1999; Do *et al.*, 2006), as described above, today it is often preferred to calculate centroid estimators (Chan *et al.*, 2005; Hamada *et al.*, 2009). Since any of these methods for finding the MEA structure have consistently improved performance over corresponding traditional algorithms that compute the most probable structure (both for thermodynamic and probabilistic models), most RNA folding packages currently prefer a MEA estimator in order to compute their predictions (Lu *et al.*, 2009).

Length-dependent stochastic context-free grammars

An essential basic fact in connection with SCFG approaches is that at any point the probability for generating a particular structure motif (as modelled by the grammar) is given by the estimated parameter value (probability) of the corresponding production rule(s). Whether a certain production rule is applicable or not does only depend on the occurrence of its premise and not on the context of this intermediate within the sentential form (thus the name context-free grammar). The same holds for the probability, which is considered for its application. However, considering the cloverleaf structure of tRNA as an example it seems obvious that the rule used to initiate the multiloop should only have a high probability if the remaining part of the input sequence is long enough to contribute 3 hairpin loops. For smaller subsequences its applications is pointless.

In order to model this behaviour of RNA molecules, it seems reasonable to additionally include *length dependencies* into traditional SCFG models. To the best of our knowledge, this idea has first been applied in Nawrocki and Eddy (2007) in connection with database similarity searching based on *covariance models* (*CMs*). Briefly, CMs are profile SCFGs (a particular SCFG architecture) for cleanly describing both the secondary structure and the primary sequence consensus of an RNA (see, for example Eddy and Durbin, 1994). They are widely used in general approaches to several RNA analysis problems, such as consensus structure prediction, multiple sequence alignment and, in fact, database similarity searching. Principally in Nawrocki and Eddy (2007), it is described how to accelerate CM searches by using a *banded* DP strategy (a standard approach in many areas of sequence analysis), which for each node calculates the probability of generating a subsequence of a particular length. Nevertheless, with respect to traditional probabilistic RNA secondary structure prediction methods, one might easily consider an appropriate *length-dependent* stochastic context-free grammar (LSCFG), as formally introduced in Weinberg and Nebel, (2011). Basically, LSCFGs exactly address the problem sketched above, that is they are defined as an extension to the concept of conventional SCFGs so that the probabilities of the productions depend on the length of the generated subword. One important benefit of this new formalism is that in general, LSCFG based algorithms can be implemented to have the same worst-case time and space requirements as their length-independent counterparts. However, due to the larger number of grammar parameters implied, algorithms implementing LSCFG models are obviously not only more explicit (due to the higher level of specialization), but unfortunately also more prone to overfitting than the corresponding traditional SCFG variants.

Formal definitions

When attempting to improve the ability of a particular stochastic model to capture typical features of a particular language within its parameters, length-dependencies can be incorporated into traditional SCFGs according to the following definition:

Definition 15.4 (Weinberg and Nebel, 2011)
A length-dependent stochastic context-free grammar (LSCFG) *is defined as a SCFG* G=(*I*, *T*, *R*, *S*, Pr) *with the following exceptions:*

- Pr: $R \times IN \to [0, 1]$ now takes a second argument (length of subword generated).
- The constraint on the probabilities changes to: $\forall A \in I\ \forall n \in N$:

$$\sum_{A \to \alpha \in R} \Pr(A \to \alpha, n) \in \{0,1\}.$$

- Additionally, we introduce a probability distribution Pr(n) on the lengths of the words in L(G), that is

$$\sum_{n \in IN : T^n \cap L(G) \neq \varnothing} \Pr(n) = 1 .$$

- Let len($A \to \alpha$) denote the length of a specific rule application $A \to \alpha$ in a parse tree, which is defined as the length of the (terminal) subword finally generated from $A \to \alpha$. Furthermore, for $\alpha \in (I \cup T)^*$ and $n \in IN$, we denote by $c_{\alpha,n}$ the number of different assignments of lengths to the symbols of α that satisfy:
  - Terminal symbols are always assigned a length of 1.
  - A non-terminal symbol B can be assigned any length l for which there is $w \in T^l$ such that $\Pr(B \Rightarrow^* w) > 0$.
  - The assigned lengths add up to n.

The probability of a parse tree for a word of length n is then Pr(n) times the product of the probabilities of all rule applications $A \to \alpha$ in the tree multiplied by $(c_{\alpha,\text{len}(A \to \alpha)})^{-1}$.

The factors $(c_{\alpha,\text{len}(A \to \alpha)})^{-1}$ and $\Pr(n)$ are necessary to ensure a probability distribution on the language that is generated by the LSCFG (see (Weinberg and Nebel, 2011) for details). In fact, considering a conventional SCFG, the probability of a parse tree δ is given by

$$\prod_{A \to \alpha \text{ applied in } \delta} \Pr(A \to \alpha),$$

whereas for the corresponding LSCFG, the probability of a parse tree δ for a terminal word $w \in T^n$ is defined by

$$\Pr(n) \cdot \prod_{A \to \alpha \text{ applied in } \delta} \Pr(A \to \alpha, \text{len}(A \to \alpha)) \cdot (c_{\alpha,\text{len}(A \to \alpha)})^{-1}.$$

Thus, when using LSCFG based models, we need to add the length (of rule applications) as an additional parameter for probability lookup in order to obtain $\Pr(A \to \alpha, \text{len}(A \to \alpha))$, and multiply each resulting probability by a corresponding factor (the reciprocal of $c_{\alpha,\text{len}(A\to\alpha)}$).

Note that in connection with RNA structure analysis, LSCFGs are interesting since they indeed make it possible to assign probabilities for a specific loop structure that depends on the size of this loop. Conventional SCFGs can model such dependencies only to a certain extent. For instance, let X denote a nonterminal symbol of some SCFG G that generates single-stranded regions. Hence, G typically defines exactly two rules with premise X, for instance p: X→∘X and $(1-p)$: X→∘. Hence, any sequence of unpaired bases is generated from X according to a geometric distribution, since $\Pr(\circ^n) = p^{n-1} \cdot (1-p)$ for $n \geq 2$. In order to induce a different distribution (characterized by a corresponding function in n), a clever modification of the underlying grammar G is needed, where most distributions will not be compactly representable. Hence, plain (length-independent) SCFGs are somewhat limited in the kind of functions in the length they realize, whereas the concept of LSCFGs embraces many more distributions.

Parameter representation

Owing to the previously mentioned facts, we can easily use the following transition probabilities for the production rules of the secondary structure grammar G_s underlying a particular RNA grammar G_r:

$$\Pr_{tr}(X \to \gamma, len = \text{len}(X \to \gamma)) :=$$

$$\begin{cases} \Pr_{tr}(X \to \gamma, len) \cdot \dfrac{1}{c_{\gamma,len}} & \text{if } G_r \text{ length-dependent,} \\ \Pr_{tr}(X \to \gamma) & \text{else.} \end{cases}$$

Accordingly, for any two positions x_1, x_2 of a given sequence r, with $1 \leq x_1 < x_2 \leq |r|$, representing a base pair $r_{x_1} r_{x_2} \in \Sigma_r^2$, we can use the corresponding emission probability

$$\Pr_{em}(r_{x_1} r_{x_2}, len = x_2 - x_1 + 1) :=$$

$$\begin{cases} \Pr_{em}(r_{x_1} r_{x_2}, len) & \text{if } G_r \text{ length-dependent,} \\ \Pr_{em}(r_{x_1} r_{x_2}) & \text{else.} \end{cases}$$

Note that any subword $r_x \in \Sigma_r$ representing a single unpaired base, $1 \leq x \leq |r|$, has length $len = x - x + 1 = 1$, such that the corresponding emission probabilities will always be the same for both the length-dependent and the traditional case. Thus, we can set $\Pr_{em}(r_x, 1) := \Pr_{em}(r_x)$.

Grouping of lengths

As proposed in Weinberg and Nebel (2011), we can confine ourselves to grouping the lengths together in several intervals which allows us to store the needed transition probabilities as a vector and thus makes it possible to retrieve them in algorithms and applications without further computational efforts. Obviously, the needed emission probabilities for base pairs (and for unpaired bases) can be stored and retrieved in the same way. As indicated in Weinberg and Nebel (2011), when choosing appropriate intervals, this restriction is – under certain circumstances – still powerful enough to yield a significant improvement over traditional SCFGs with respect to the prediction accuracy[13].

However, if lengths are to be grouped into intervals, we have to deal with the fact that not all such groupings yield a consistent grammar. Nevertheless, the following definition of consistency (for such a grouping) offers a handy sufficient condition (see (Weinberg and Nebel, 2011) for details):

[13]Note that length-dependencies can also be modelled with plain grammars, no matter if the lengths are grouped into (a finite number of) intervals or not. In fact, as stated in Weinberg and Nebel (2011), any LSCFG is equivalent to a corresponding indexed grammar (Aho, 1968; Gazdar, 1988) with significantly increased numbers of intermediate symbols and production rules. Hence, the concept of LSCFGs in principle only allows for a more compact representation.

Definition 15.5 (Weinberg and Nebel, 2011)
Let G=(*I*, *T*, *R*, *S*) a CFG and *Q* a partitioning of *IN*. We call *Q* consistent with G if it satisfies the following condition:

$\forall q \in Q, i, j \in q: \exists A \rightarrow \alpha \in R, w_i \in T^i: \alpha \Rightarrow^* w_i$ *implies* $\exists w_j \in T^j: \alpha \Rightarrow^* w_j$.

Intuitively, to satisfy this condition, we may not group lengths *i* and *j* together if there exists a production rule that can lead to a word of length *i* but not to one of length *j* (or vice versa). The partitioning into sets of one element each (which corresponds to not grouping lengths into intervals at all) is trivially always consistent. Moreover, from the point of view of parameter estimation, the interval sizes should principally grow with increasing subword lengths, since any training set will typically contain fewer observations for rule applications in connection with larger substrings. Only if the considered databases are rich enough to provide a sufficiently large number of observations per interval, we can hope for accurately estimated probabilities. Additionally, the lengths implied by the last interval should be chosen in accordance with the longest words in the training set for ensuring that the estimates for that interval do not influence the induced probability distribution.

Finally, note that as already mentioned, the danger of overfitting of the induced model is indeed much more present if length-dependent probabilities rather than their traditional length-independent counterparts are considered, since the observations made for a particular structural motif (for the application of certain productions) have to be split into distinct subsets of observations according to the corresponding lengths (or length intervals). In fact, LSCFGs typically imply significantly greater numbers of free parameters than the corresponding conventional SCFGs, where the actual numbers indeed increase with growing complexity (by means of number of distinguished length intervals) of the considered partitioning of *IN*. This is a disadvantage, since increasing the number of parameters to be estimated requires more training data. However, there is already a lack of data in connection with training (heavyweight) plain SCFGs, see 'Choice of training data' above.

Concluding remarks

Due to the limitations of standard thermodynamic models, including the inability to capture certain relevant physiological and chemical aspects in their parameters, probabilistic models derived on the basis of structural information obtained from reliable RNA databases have emerged as an alternative methodology towards single sequence RNA secondary structure prediction. In fact, several probabilistic approaches have been used so far that build on different types of models and/or ways for deriving a particular prediction. Some of them will be discussed in brief throughout this section, along with some relevant corresponding aspects.

Traditional grammar models

Originally, the class of all feasible secondary structures (that obey to certain structural constraints like for example the non-existence of isolated base pairs) was described by a corresponding SCFG, which induces a (non-uniform) probability distribution on the considered class. As regards computational structure prediction, early probabilistic approaches such as (Knudsen and Hein, 1999) seem to have chosen the structure of their SCFG rather arbitrarily; at least, there is almost no discussion about the motivation for the choice of the productions. This problem has first been addressed by Dowell and Eddy (2004), who evaluated nine different lightweight SCFGs in connection with RNA secondary structure prediction.

Notably, a structurally ambiguous SCFG mirror of modern energy-based algorithms for single sequence structure prediction has already been described by Rivas and Eddy (2000). However, not until a few years ago, several rather heavyweight (unambiguous) SCFGs mirroring state-of-the-art physics-based algorithms were applied for RNA structure analysis (Nebel *et al.*, 2011; Nebel and Scheid, 2011a,b). Those grammars actually define specific productions for all structural motifs for which there are different thermodynamic parameters or energy rules, such that the induced probabilistic models are of comparable complexity as the corresponding physics-based ones.

Similar probabilistic models

Over the years several extensions and generalizations of SCFGs have also proven useful in connection with RNA analysis, like for instance LSCFGs which naturally imply a more specific probability distribution on the modelled structure, and also the previously mentioned CLLMs. Briefly, CLLMs are more flexible than vanilla SCFGs, as they generalize upon the representation of conditional probabilities. In principle, these mathematical objects constitute a specialized class of structured classification models and a generalization of *conditional random fields* (*CRFs*). Note that CRFs are in fact a specialized class of CLLMs whose probability distributions are defined in terms of graphical models. For details, we refer to Lafferty *et al.* (2001) and Sutton and McCallum (2007).

While SCFGs (like HMMs) are generative probabilistic models, which are intuitive and allow convenient (generative) parameter training via maximum *joint* likelihood techniques, CLLMs are *discriminative* probabilistic models, where the parameters are learned by (discriminative) training methods maximizing the *conditional* likelihood.

As a consequence, CLLMs have the power to represent more complex scoring schemes than the corresponding SCFG can represent, for instance a Mfold-like energy based one as considered in Do *et al.* (2006). However, generative probabilistic models are generally easier to train and to use than discriminative variants: principally, generative (SCFG based) training can easily be realized by counting the observed frequencies of applications of the distinct production rules of an unambiguous SCFG (yielding a maximum likelihood estimator), by expectation maximization or similar methods from machine-learning. That way, the resulting estimates of the grammar parameters are adapted to a particularly considered dataset. For details, we refer to Do *et al.* (2006). Popular software packages implementing a particular SCFG and CLLM model, respectively, are Pfold (Knudsen and Hein, 1999, 2003) and CONTRAfold (Do *et al.*, 2006).

Overfitting and parameter estimation

In any case, there are two important issues that inherently arise when estimating the corresponding set of probabilistic parameters: the resulting model may suffer from overfitting or lack of generalization. In fact, the features of CLLMs used for structure prediction usually model the (energy contributions of) various structural motifs that occur in RNA secondary structure, such that a rich and reliable training set providing enough structural information is required in order to avoid overfitting, just like in case of SCFGs of comparable complexity.

In many supervised learning methods, overfitting is actually controlled through the use of regularization penalties for limiting model complexity. For instance, methods for statistical estimation of thermodynamic parameters, for example Bayesian inference approaches (see 'Bayesian Parameter Estimation' above), use the experimentally obtained Turner values (Mathews *et al.*, 1999) as an a priori specification. However, in connection with common probabilistic approaches that completely abstract from free energies, there is no *lab-based* prior to the grammar parameters like the standard Turner model for physics-based approaches. Therefore, the corresponding distribution (on the modelled structure class) has to be derived exclusively from a collection of sample data, more specifically real-life RNA data (given by pairs of RNA sequences with annotated secondary structures).

Nevertheless, it should be clear by now that the dependence of the resulting accuracy on the used model parameters is not only a major problem in connection with probabilistic approaches that rely exclusively on structural information from reliable databases, but it indeed also limits the performance of physics-based methods. In fact, the (usually many hundreds of) thermodynamic parameters used in standard state-of-the-art energy models for RNA folding are mostly estimated from experimental results (on the basis of diverse structural RNAs) and are far from being perfect, since folding processes of RNA molecules are usually to a large part controlled by a number of additional non-energetic effects. Those might only be accounted for by estimating the thermodynamic parameters from reliable sets of real-life data, but then the same problems concerning parameter estimation arise as in case of probabilistic prediction methods based on SCFGs or similar models.

Standard predictions and statistical sampling extension

Independent on the type of probabilistic model used, the corresponding DP algorithms traditionally compute either the Viterbi parse (corresponding to the most likely structure) or the MEA parse (corresponding to the structure with highest expected number of correctly predicted positions). However, independent on the objective function used, a major problem of all these DP algorithms is that due to the strict deterministic optimization applied for determining the (sub)optimal structure(s), it may happen that only such foldings might be predicted that are actually impossible to show up in nature.

In order to address this problem, several (L) SCFG based statistical sampling methods have been invented over the past years (Nebel and Scheid, 2011b, 2012; Scheid and Nebel, 2012) which in contrast to the sampling extension of the physics-based PF approach (Ding and Lawrence, 2003) randomly sample secondary structures from the ensemble induced by the probabilistic model rather than from the Boltzmann low-energy ensemble. These statistical sampling approaches do not only enable the generation of more diverse sets of candidate structures, but they can also be used for characterizations of the complete ensemble of feasible foldings. Accordingly, there are several reasonable ways for deriving predictions from a generated statistically representative and reproducible sample set. These include the calculation of specific MEA or centroid structures (Ding *et al.*, 2005) as described in Nebel and Scheid (2011b), which reflect the structures in the overall sample rather then entire structure ensemble.

Notably, statistical sampling usually requires the same time and space complexities as the standard DP algorithms, specifically $O(n^3)$ time and $O(n^2)$ storage for an input sequence of length n. However, the worst-case time complexity can be reduced by a linear factor to $O(n^2)$ by heuristic approximation of the preprocessing step, making it the most efficient (L)SCFG based prediction approach to date – but generally not without sacrificing some accuracy (see Nebel and Scheid, 2012, 2013).

Pseudoknot prediction

Finally, it should be noted that all the approaches discussed within this chapter only consider secondary structures without pseudoknots (see, for example (Pleij and Bosch, 1989; Abrahams *et al.*, 1990; Gutell and Woese, 1990; Dam *et al.*, 1992; Pleij, 1994)). This means they only deal with the mutual planar arrangements of base paired helices and are restricted to exclusively generating non-crossing base pairs (crossing pairs are prohibited). In fact, non-planar base pair interactions cannot be modelled by traditional context-free grammar formalisms.

Besides, the problem of RNA secondary structure prediction including arbitrary pseudoknots is NP complete – even for rather simple energy models underlying the corresponding algorithms (see (Akutsu, 2000; Lyngso and Pedersen, 2000)). Therefore, much effort has been put into the development of polynomial-time algorithms for predicting RNA secondary structures that allow certain restricted classes of pseudoknots. However, owing to their quite high time (polynomials of degree 5 or 6) and space complexities, most of these algorithms are principally not applicable for long input sequences. Notably, some of these algorithms are grammar based variants (for instance (Matsui *et al.*, 2005; Kato *et al.*, 2006)) that make use of various more powerful grammar models capable to model RNA secondary structures with pseudoknots (and also RNA–RNA interactions). One example are so-called *multiple context-free grammars* (*MCFGs*), which are defined by a weakly context-sensitive grammar formalism that deals with tuples of strings (see, for instance (Seki *et al.*, 1991; Kato *et al.*, 2006, 2009; Kato, 2007; Wild, 2010)). Notably, MCFGs for all common classes of RNA pseudoknots have been presented in Nebel and Weinberg (2012).

References

Abrahams, J., van den Berg, M., van Batenburg, E., and Pleij, C. (1990). Prediction of RNA secondary structure, including pseudoknotting, by computer simulation. Nucleic Acids Res. *18*, 3035–3044.

Aho, A. (1968). Indexed grammars – an extension of context-free grammars. J. ACM *15*, 647–671.

Akutsu, T. (2000). Dynamic programming algorithms for RNA secondary structure prediction with pseudoknots. Discrete Appl. Math. *104*, 45–62.

Aldrich, J. (1997). R.A. Fisher and the making of maximum likelihood 1912–1922. Stat. Sci. *12*, 162–176.

Andronescu, M., Bereg, V., Hoos, H., and Condon, A. (2008). RNA STRAND: the RNA secondary structure and statistical analysis database. BMC Bioinformatics *9*, 340.

Baldi, P., Brunak, S., Chauvin, Y., Andersen, C., and Nielsen, H. (2000). Assessing the accuracy of prediction algorithms for classification: an overview. Bioinformatics *16*, 412–424.

Beaumont, M., and Rannala, B. (2004). The Bayesian revolution in genetics. Nat. Rev. Genet. *5*, 251–261.

Bishop, C. (2007). Pattern Recognit. and Machine Learning (Springer, New York).

Box, G., and Tiao, G. (1992). Bayesian Inference in Statistical Analysis (John Wiley and Sons, Inc., New York).

Chan, C., Lawrence, C., and Ding, Y. (2005). Structure clustering features on the Sfold Web server. Bioinformatics *21*, 3926–3928.

Chi, T., and Geman, S. (1998). Estimation of probabilistic context-free grammars. Comput. Ling. *24*, 299–305.

Chomsky, N. (1956). Three models for the description of language. IEEE Trans. Inf. Theory 2, 113–124.

Cocke, J., and Schwartz, J. (1970). Programming Languages and Their Compilers: Preliminary notes (Technical report Courant Institute of Mathematical Sciences, New York).

Corazza, A., and Satta, G. (2006). Cross-Entropy and estimation of probabilistic context-free grammars. In Proceedings of HLT-NAACL'06. Association for Computational Linguistics, Stroudsburg, PA, pp. 335–342.

Dam, E., Pleij, K., and Draper, D. (1992) Structural and functional aspects of RNA pseudoknots. Biochemistry *31*, 11665–11676.

Ding, Y., and Lawrence, C. (2003). A statistical sampling algorithm for RNA secondary structure prediction. Nucleic Acids Res. *31*, 7280–7301.

Ding, Y., Chan, C., and Lawrence, C. (2005). RNA secondary structure prediction by centroids in a Boltzmann weighted ensemble. RNA *11*, 1157–1166.

Do, C., Woods, D., and Batzoglou, S. (2006). CONTRAfold: RNA secondary structure prediction without physics-based models. Bioinformatics *22*, e90–e98.

Dowell, R., and Eddy, S. (2004). Evaluation of several lightweight stochastic context-free grammars for RNA secondary structure prediction. BMC Bioinformatics *5*, 71.

Durbin, R., Eddy, S., Krogh, A., and Mitchison, G. (1998). Biological Sequence Analysis: Probabilistic Models of Proteins and Nucleic Acids. (Cambridge University Press, Cambridge, UK).

Earley, J. (1970). An efficient context-free parsing algorithm. Comm. ACM *13*, 94–102.

Eddy, S., and Durbin, R. (1994). RNA sequence analysis using covariance models. Nucleic Acids Res. *2*, 2079–2088.

Fu, K., and Huang, T. (1972). Stochastic grammars and languages. Int. J. Comput. Inform. Sci. *1*, 135–170.

Gazdar, G. (1988). Applicability of indexed grammars to natural languages. In Natural Language Parsing and Linguistic Theories, Reyle, U., and Rohrer, C., eds. (Reidel Dordrecht, Dordrecht, Holland), pp. 69–94.

Gelman, A., Carlin, J., Stern, H., and Rubin, D. (1997). Bayesian Data Analysis (Chapman & Hall/CRC, New York).

Giegerich, R. (2000). Explaining and Controlling Ambiguity in Dynamic Programming. In Proceedings of the 11th Annual Symposium on Combinatorial Pattern Matching (SpringerVerlag, London, UK), pp. 46–59.

Goodman, J. (1998). Parsing Inside-Out. PhD thesis, Harvard University, Cambridge, MA.

Goodman, J. (1999). Semiring parsing. Comput. Ling. *25*, 573–605.

Grune, D., and Jacobs, C. (2008). Parsing Techniques – A Practical Guide. 2nd edition (Springer, New York).

Gutell, R., and Woese, C. (1990). Higher order structural elements in ribosomal RNAs: Pseudo-knots and the use of noncanonical pairs. Proc. Natl. Acad. Sci. U.S.A. *87*, 663–667.

Hamada, M., Kiryu, H., Sato, K., Mituyama, T., and Asai, K. (2009). Prediction of RNA secondary structure using generalized centroid estimators. Bioinformatics *25*, 465–473.

Harrison, M. (1978). Introduction to Formal Language Theory. (Addison-Wesley, Reading, MA).

Hofacker, I., Schuster, P., and Stadler, P. (1998). Combinatorics of RNA secondary structures. Discrete Appl. Math. *88*, 207–237.

Hopcroft, J., and Ullman, J. (1979). Introduction to Automata Theory, Languages, and Computation. (Addison-Wesley, Reading, MA).

Huang, T., and Fu, K. (1971). On stochastic context-free languages. Inf. Sci. *3*, 201–224.

Jaynes, E. (2003). Probability Theory: The Logic of Science. (Cambridge University Press, Cambridge, UK).

Kasami, T. (1965). An efficient recognition and syntax analysis algorithm for context-free languages. Technical Report AFCRL-65-758 Air Force Cambridge Research Laboratory.

Kato, Y. (2007). Formal grammars for describing RNA pseudoknotted structure and their application to structure analysis. PhD thesis, Department of Information Processing, Graduate School of Information Science, Nara Institute of Science and Technology.

Kato, Y., Seki, H., and Kasami, T. (2006). RNA pseudoknotted structure prediction using stochastic multiple context-free grammar. IPSJ Trans. Bioinform. *47*, 12–21.

Kato, Y., Akutsu, T., and Seki, H. (2009). A grammatical approach to RNA–RNA interaction prediction. Pattern Recognit. *42*, 531–538.

Knudsen, B., and Hein, J. (1999). RNA secondary structure prediction using stochastic context-free grammars and evolutionary history. Bioinformatics *15*, 446–454.

Knudsen, B., and Hein, J. (2003). Pfold: RNA secondary structure prediction using stochastic context-free grammars. Nucleic Acids Res. *31*, 3423–3428.

Krogh, A. (1998). An introduction to hidden Markov models for biological sequences. In Computational Biology: Pattern Analysis and Machine Learning Methods, Salzberg, S.L., Searls, D.B., and Kasif, S., eds. (Elsevier, Amsterdam), pp. 45–63.

Lafferty, J., McCallum, A., and Pereira, F. (2001). Conditional random fields: probabilistic models for segmenting and labelling sequence data. In 18th Proceedings of the International Conference in Machine Learning, pp. 282–289.

Lari, K., and Young, S. (1990). The estimation of stochastic context-free grammars using the Inside–outside algorithm. Comput. Speech Lang. *4*, 35–56.

Lari, K., and Young, S. (1991). Applications of stochastic context-free grammars using the Inside–outside algorithm. Comput. Speech Lang. *5*, 237–257.

Lu, Z., Gloor, J., and Mathews, D. (2009). Improved RNA secondary structure prediction by maximizing expected pair accuracy. RNA *15*, 1805–1813.

Lyngso, R., and Pedersen, C. (2000). RNA pseudoknot prediction in energybased models. J. Comput. Biol. *7*, 409–427.

MacKay, D. (2003). Information Theory, Inference, and Learning Algorithms. (Cambridge University Press, Cambridge, UK).

Mathews, D., Sabina, J., Zuker, M., and Turner, D. (1999). Expanded sequence dependence of thermodynamic parameters improves prediction of RNA secondary structure. J. Mol. Biol. *288*, 911–940.

Matsui, H., Sato, K., and Sakakibara, Y. (2005). Pair stochastic tree adjoining grammars for aligning and predicting pseudoknot RNA structures. Bioinformatics *21*, 2611–2617.

McCaskill, J. (1990). The equilibrium partition function and base pair binding probabilities for RNA secondary structure. Biopolymers *29*, 1105–1119.

Nawrocki, E., and Eddy, S. (2007). Query-dependent banding (QDB) for faster RNA similarity searches. PLoS Comput. Biol. *3*, e56.

Nebel, M. (2002a). Combinatorial properties of RNA secondary structures. J. Comput. Biol. *9*, 541–574.

Nebel, M. (2002b). On a Statistical Filter for RNA Secondary Structures. Technical report Frankfurter Informatik-Berichte.

Nebel, M. (2004a). Investigation of the Bernoulli-Model of RNA secondary structures. Bull. Math. Biol. *66*, 925–964.

Nebel, M. (2004b). Identifying good predictions of RNA secondary structure. In Proceedings of the Pacific Symposium on Biocomputing, pp. 423–434.

Nebel, M., and Scheid, A. (2011a). Analysis of the free energy in a stochastic RNA secondary structure model. IEEE/ACM Trans. Comput. Biol. Bioinformat. *8*, 1468–1482.

Nebel, M., and Scheid, A. (2011b). Evaluation of a sophisticated SCFG design for RNA secondary structure prediction. Theory Biosci. *130*, 313–336.

Nebel, M., and Scheid, A. (2012). A n^2 RNA secondary structure prediction algorithm. In Proceedings of Bioinformatics 2012 – International Conference on Bioinformatics Models, Methods and Algorithms, Vilamoura, Portugal, pp. 66–75.

Nebel, M., and Scheid, A. (2013). Fast RNA secondary structure prediction using fuzzy stochastic models. Comm. Comput. Inf. Sci. *357*, pp. 175–194.

Nebel, M., and Weinberg, F. (2012). Algebraic and combinatorial properties of common RNA pseudoknot classes with applications. J. Comput. Biol. *19*, 1134–1150.

Nebel, M., Scheid, A., and Weinberg, F. (2011). Random generation of RNA secondary structures according to native distributions. Algorithms Mol. Biol. *6*.

Nederhof, M.-J., and Satta, G. (2003). Probabilistic parsing as intersection. In Proceedings of the 8th IWPT, Nancy, France, pp. 137–148.

Nederhof, M.-J., and Satta, G. (2006). Estimation of consistent probabilistic context-free grammars. In Proceedings of HLT-NAACL'06, New York, pp. 343–350.

Nussinov, R., and Jacobson, A. (1980). Fast algorithms for predicting the secondary structure of single-stranded RNA. Proc. Natl. Acad. Sci. U.S.A. *77*, 6309–6313.

Nussinov, R., Pieczenik, G., Griggs, J., and Kleitman, D. (1978). Algorithms for loop matchings. SIAM J. Appl. Math. *35*, 68–82.

Pleij, C. (1994). RNA pseudoknots. Curr. Opin. Struct. Biol. *4*, 337–344.

Pleij, C., and Bosch, L. (1989). RNA pseudoknots: structure, detection, and prediction. Methods Enzymol. *180*, 289–303.

Prescher, D. (2003). A tutorial on the expectation maximization algorithm including maximum-likelihood estimation and EM training of probabilistic context-free grammars. 15th European Summer School in Logic, Languages and Information (ESSLL 1–03), Vienna, Austria.

Rabiner, L., and Juang, B. (1986). An introduction to hidden Markov models. IEEE ASSP Magazine 3, 4–16.

Rivas, E., and Eddy, S. (2000). Secondary structure alone is generally not statistically significant for the detection of noncoding RNAs. Bioinformatics *16*, 583–605.

Rivas, E., Lang, R., and Eddy, S. (2012). A range of complex probabilistic models for RNA secondary structure prediction that include the nearest neighbour model and more. RNA *18*, 193–212.

Sakakibara, Y., Brown, M., Hughey, R., Mian, I., Sjölander, K., Underwood, R., and Haussler, D. (1994). Stochastic context-free grammars for tRNA modelling. Nucleic Acids Res. *22*, 5112–5120.

Scheid, A., and Nebel, M. (2012). Statistical RNA Secondary Structure Sampling Based on a Length-Dependent SCFG Model. Technical report, University of Kaiserslautern.

Seki, H., Matsumura, T., Fujii, M., and Kasami, T. (1991). On multiple context-free grammars. Theor. Comput. Sci. *88*, 191–229.

Sipser, M. (1996). Introduction to the Theory of Computation (PWS Publishing, Boston, MA).

Steger, G., Hofmann, H., Förtsch, J., Gross, H., Randles, J., Sänger, H., and Riesner, D. (1984). Conformational transitions in viroids and virusoids: comparison of results from energy minimization algorithm and from experimental data. J. Biomol. Struct. Dyn. *2*, 543–571.

Stolcke, A. (1995). An efficient probabilistic context-free parsing algorithm that computes prefix probabilities. Comput. Ling. *21*, 165–201.

Sukosd, Z., Knudsen, B., Vaerum, M., Kjems, J., and Andersen, E. (2011). Multithreaded comparative RNA secondary structure prediction using stochastic context-free grammars. BMC Bioinformatics *12*.

Sutton, C., and McCallum, A. (2007). An introduction to conditional random fields for relational learning. In Introduction to Statistical Relational Learning, Getoor, L., and Taskar, B., eds. (MIT Press, Cambridge, MA), pp. 93–128.

Viennot, G., and Chaumont, M.V.D. (1985). Enumeration of RNA Secondary Structures by Complexity. Paper presented at Mathematics in Medicine and Biology, Lecture Notes in Biomathematics 57, 360–365.

Waterman, M. (1978). Secondary structure of single-stranded nucleic acids. Advances in Mathematics Supplementary Studies *1*, 167–212.

Weinberg, F., and Nebel, M. (2011). Applying length-dependent stochastic context-free grammars to RNA secondary structure prediction. Algorithms *4*, 223–238.

Wild, S. (2010). An Earley-style parser for solving the RNA–RNA interaction problem. Bachelor Thesis, University of Kaiserslautern, Kaiserslautern, Germany.

Xia, T., SantaLucia, Jr., J., Burkard, M., Kierzek, R., Schroeder, S., Jiao, X., Cox, C., and Turner, D. (1998). Thermodynamic parameters for an expanded nearest-neighbor model for formation of RNA duplexes with Watson–Crick base pairs. Biochemistry *37*, 14719–14735.

Younger, D. (1967). Recognition and parsing of context-free languages in time n^3. Inf. Control *10*, 372–375.

Zuker, M. (1989). On finding all suboptimal foldings of an RNA molecule. Science *244*, 48–52.

Zuker, M., and Stiegler, P. (1981). Optimal computer folding of large RNA sequences using thermodynamics and auxiliary information. Nucleic Acids Res. *9*, 133–148.

Stochastic Context-free Grammars and RNA Secondary Structure Prediction

16

James W.J. Anderson

Abstract

Prediction of RNA secondary structure from a single sequence, or an alignment of sequences, is a core problem in bioinformatics. Many approaches to RNA secondary structure prediction have been attempted, and probabilistic methods using stochastic context-free grammars (SCFGs) have been one of the more successful tries. In particular, SCFGs can be combined with a molecular evolution model to produce consensus structure predictions which more accurately predict RNA secondary structure than when considering single-sequence prediction. The use of SCFGs in RNA secondary structure prediction, and the potential for further developments make for a truly interesting topic.

In this chapter we discuss the application of SCFGs to RNA secondary structure prediction, from a single sequence, or a single fixed alignment. An introduction to RNA secondary structure prediction is given, and some technical issues for SCFGs, such as normal forms and grammar design, are discussed, as is estimating SCFG parameters. Methods are shown for predicting RNA secondary structures, and some measures are given for analysing SCFG variability. Finally, a brief discussion concerning their predictive quality is had, with some suggestions for further work and web resources given.

Introduction

RNA secondary structure

Ribonucleic acid (RNA) secondary structure prediction is an important problem in molecular biology; almost as soon as RNA started to be sequenced, methods have been established to determine the structure from the sequence of nucleotides. Many models have been developed for RNA secondary structure prediction, and in this chapter one in particular will be examined: SCFGs.

Firstly, we might like to know what defines RNA secondary structure, and why it is important that it is able to be predicted. To look at secondary structure, we start by looking at the structure of deoxyribonucleic acid (DNA), as it is from this RNA forms. DNA is made up of a double helix of nucleotides, adenine (A), thymine (T), cytosine (C) and guanine (G); the helices held together by hydrogen bonds. The nucleotides bond in pairs due to their chemical structure, A bonds with T, C with G. When DNA is copied, only a single strand is copied, using A, C, G, and uracil (U) in place of T, which form the nucleotides of RNA. Single strands of RNA then fold back on themselves to increase thermodynamic stability. It has been found experimentally that AU, CG are base pairs (echoing the AT, CG pairs found in DNA), but U and G bond also, although with less thermodynamic stability than the AU and CG base pairs. At this point we introduce 'dot-parenthesis' notation, where a dot represents an unpaired nucleotide, and corresponding parentheses indicated paired nucleotide. Fig. 16.1 shows an example of a sequence, secondary structure, and graphical representation of the secondary structure, created using the VARNA applet (Darty *et al.*, 2009).

Unsurprisingly, there are many biological applications of RNA secondary structure prediction. The most obvious one is in looking at gene

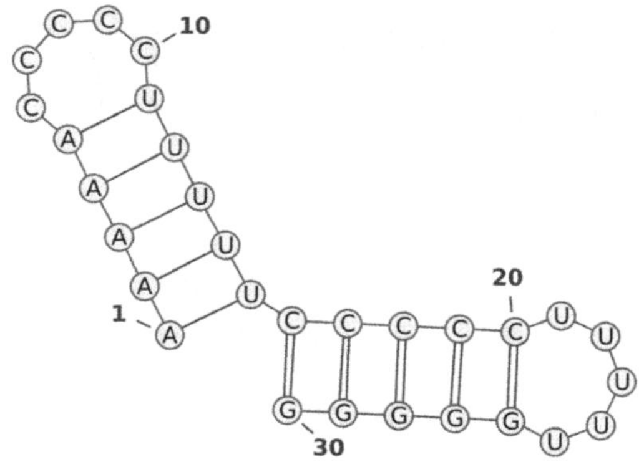

Figure 16.1 Secondary structure (((((.....)))))(((((.....))))) of sequence AAAAACCCCCUUUUUCCCCCUUUUUGGGGG.

expression. RNA can play key roles in gene expression, encouraging or discouraging expression of a given gene, or making post-transcriptional or -translational modifications. Structural flexibility can be important in roles like chaperoning (de Almeida Ribeiro *et al.*, 2011), splice site recognition (Gahura *et al.*, 2011), translation (Loh *et al.*, 2011), and as riboswitches (Vitreschak *et al.*, 2004). Naturally then, by predicting RNA secondary structure, we will have a better idea of the function of the RNA. Similarly, with the advent of next-generation sequencing and transcriptomics, regulatory effects can be looked for by genome-wide prediction of RNA secondary structure (Washietl *et al.*, 2005).

Similarly, we would like to understand RNA folding for RNA sequence design. If there is a specific RNA secondary structure we require in some cellular mechanism, for example in building biochemical structures like the DNA box (Andersen *et al.*, 2009), we will need to know what sequences could be synthesized in a laboratory to fold into the required RNA structure. Computational applications like Frankenstein (Lyngsø *et al.*, 2012) use RNA folding programs to generate sequences likely to fold into these desired structures. RNA folding methods have also been used in RNA gene finding (Meyer 2010), as structural components can be detected at the genomic level.

The problem changes slightly when homologous RNA sequences are considered. RNA secondary structure is conserved in evolution, so whilst the sequence content might change with mutation and crossover events, in general secondary structure does not, as this would likely change the function of the RNA. So, when considering the secondary structure of our homologous RNAs, it is important that we think about the alignment of the RNA sequences. Equally, when aligning homologous RNA sequences, we must consider the potential secondary structure of the sequences. We would like, therefore, a method, which, given a set of homologous RNA sequences, will give an alignment and a common structure.

Many approaches have been tried in an attempt to produce such an alignment and structure. Early attempts include summing over all possible secondary structures and evaluating them with respect to free-energy functions (Pipas and McMahon, 1975). Since RNA folding helps stability, and less free energy encourages stability, minimizing the free energy of the structure was appealing. Another approach formulated the folding of RNA as a loop matching problem, and developed dynamic programming algorithms to find an optimal secondary structure with respect to a simple score function (1 if the two nucleotides can form a base pair, 0 otherwise), hence looking to maximize the number of base pairs in the structure (Nussinov *et al.*, 1978). Biological and thermodynamical principles have since been used to advanced these energy functions to get more accurate predictions, which have been used to great effect in algorithms such as Mfold (Zuker and Stiegler, 1981; Zuker and Sankoff, 1984; Zuker, 2003), UNAfold (Markham and Zuker, 2008), and RNAfold (Hofacker *et al.*, 1994).

The use of SCFGs in RNA secondary structure prediction was based on the success of Hidden Markov Models (HMMs) in protein and gene modelling (Krogh *et al.*, 1993). It was desired to apply HMMs to annotate RNA secondary structure, but in an HMM, one cannot model global dependencies, which are certainly required for RNA secondary structure. Thus SCFGs, as generalisations of HMMs, were used, initially by Sakakibara *et al.* (1994). Notably the Pfold algorithm (Knudsen and Hein, 1999, 2003), which used comparative modelling techniques, which was shown to be effective in comparison with other SCFGs (Dowell and Eddy, 2004; Anderson *et al.*, 2012). For reviews on RNA secondary structure prediction, see Gardner and Giegerich (2004), Shapiro *et al.* (2007), or Xu *et al.* (2012).

However, one key feature of the basic SCFG method is the inability to predict pseudoknotted structures. A pseudoknot is an RNA secondary structure containing at least two stem-loop structures in which half of one stem is intercalated between the two halves of another stem, for example the structure '((((... [[[...)))) ...]]]'. RNA structures with pseudoknots in them are highly common (Andronescu *et al.*, 2008). They occur in virtually all classes of RNA, and have important biological function (Chen and Greider, 2005). Pseudoknotted structures are reasonably common in RNA secondary structure, and methods have been taken to adapt the SCFG method to pseudoknotted structures, but we cannot predict them with the straightforward SCFG approach. Consequently, throughout this chapter, I will be referring specifically to non-pseudoknotted structures, unless otherwise stated.

Data sources

There are many data sources for RNA sequences with known secondary structures. Two of the main databases are discussed here, but there are a large number of sources available.

RNASTRAND (Andronescu *et al.*, 2008) is an online database of RNA sequences with known secondary structures, which itself takes data from many other sources (Berman *et al.*, 1992; Brown, 1999; Cannone *et al.*, 2002; Westbrook *et al.*, 2003; Griffiths-Jones *et al.*, 2005; Sprinzl and Vassilenko, 2005; Andersen *et al.*, 2006). RNASTRAND is mainly a database of single-sequence entries. At the time of writing, it contains over 4500 structures, of many different types of RNA, such as tmRNA, 16S ribosomal RNA, tRNA, and ribonuclease P RNA. RNASTRAND describes itself as containing 'simple yet powerful ways of analysing, searching and updating' the database. It is the power of the search which proves particularly useful: being able to search for structural motifs like stems or pseudoknots, and easily filter searched RNA makes RNASTRAND easy to use to obtain a required RNA dataset.

On the other hand, Rfam (Griffiths-Jones *et al.*, 2005) is an online database more focused on alignments of RNA families. Noteworthy are the 'seed' alignments, manually curated alignments of RNA sequences designed to be particularly accurate with respect to structure conservation across sequences. So, if you need some trusted alignments for a dataset, this is likely where you will find these. At writing, Rfam contains over 2000 families of RNA sequences, each family with a wiki-type documentation to allow for updating of information concerning a given family. Rfam offers searching by family type, and many families actually contain three-dimensional structures too. Both Rfam and RNASTRAND allow users to upload new sequences and new structures to keep data banks dynamic.

Combinatorics of RNA secondary structures

So why is RNA secondary structure prediction a difficult problem? Well, firstly, we need a model for how the RNA folds, which is not trivial in itself. However, even when we have our model, the combinatorics of RNA secondary structures start to become overwhelming, and consequently prediction methods have typically resorted to a function minimization or maximization approach by recursing over structures. Here, we will outline the combinatorics of (non-pseudoknotted) RNA secondary structures so that when considering SCFGs, we have better intuition as to why the method has been so successful.

The typical strategy for counting RNA secondary structures is to create a recursion based on the length of the structure concerned. Here we consider all structures which can be represented in dot-parenthesis format, with the paired nucleotides having at least two unpaired nucleotides between them. Note particularly that these do not allow for pseudoknots. However, if we wanted to place additional constraints upon what we considered a valid structure, the counting procedure would still be the same.

For a sequence of length one, the only possible structure is a single, unpaired nucleotide, represented by a dot. We only allow base pairs exist with at least two nucleotides in between, so for a sequences of lengths two and three, again there is only one allowed structure, all unpaired bases. For a sequence of length 4, we could have either .... or

(..) as our structures. This then yields to a general recursion for the number of secondary structures S_n for a sequence of length n, by considering whether the first base is unpaired or paired to some other base k:

$$S_{n+1} = S_n + \sum_{k=0}^{n-2} S_k S_{n-k-1}$$

$$S_1 = S_2 = S_3 = 1$$

For larger secondary structures, we can use a large sample approximation to get a asymptotic result for the number of secondary structures

$$S_n \approx 1.104 \times n^{3/2} \times 2.618^n \text{ for large } n$$

It can clearly been seen from this that the number of RNA secondary structures grows exponentially as n becomes large, and in fact it becomes computationally unfeasible to evaluate models on all individual possible structures for longer nucleotide sequences. However, in a similar way to the method used to enumerate RNA secondary structures, dynamic programming, particularly with SCFGs, can be used to recurse over all valid RNA secondary structures in prediction algorithms.

In this chapter we look to explore RNA secondary structure prediction, mainly in the framework of SCFGs. Firstly, single-sequence prediction is considered, then prediction from a single fixed alignment. Upon discussion of the measures of variability of SCFGs, a brief introduction to the thermodynamic approach of RNA secondary structure prediction is given. Note the alignment part of the comparative prediction method is not discussed, as it is beyond the scope of what is done here.

Context-free grammars and SCFGs

So why is it that SCFGs are used to model RNA secondary structure? As mentioned in the introduction, using SCFGs was inspired by the use of Hidden Markov Models (HMMs) in protein and gene modelling. The strength of using HMMs was that they could model the coding and non-coding regions of DNA, and something similar was desired for RNA, for example with base pairing and unpaired regions. However, secondary structures in RNA contain global dependencies, and so it was necessary to use a more complex model than an HMM. SCFGs, simply a generalization of HMMs, can describe a number of long-range interactions, including several in RNA secondary structure.

To explore the use of SCFGs in RNA secondary structure prediction, we begin by defining context-free grammars and stochastic context-free grammars (Chomsky, 1959). A grammar G is a 4-tuple (N, V, P, S) consisting of the following components:

- a finite set N of non-terminal symbols, throughout denoted 'non-terminals';
- a finite set V of terminal symbols that is disjoint from N, throughout denoted 'terminals';
- a finite set P of production rules;
- a distinguished symbol $S \in N$ that is the start symbol.

Furthermore, a grammar is said to be context free if

- production rules of the form $A \to aA$ and $A \to a$ for $A \in N$ and $a \in V$ are allowed (requiring solely this condition would give us a regular grammar);
- the left-hand side of each production rule consists of at most a single non-terminal.

Lastly, we can define our context-free grammars to be stochastic: a grammar is said to be stochastic if there is a probability distribution on the implementation of production rules for each $n \in N$.

Given this definition, it is not immediately obvious why SCFGs might be used for RNA secondary structure prediction. To see this, we will work with an example. Consider the following SCFG:

- Non-terminals $N = \{S\}$
- Terminals $V = \{(,)\}$
- Production rules with probabilities $P = \{(S \to (S), 0.5), (S \to SS, 0.3), (S \to (), 0.2)\}$
- Start symbol S

This grammar might typically be represented in the following shorthand:

$$S \rightarrow \begin{matrix} (S) & | & SS & | & (\,) \\ 0.5 & | & 0.3 & | & 0.2 \end{matrix}$$

To produce strings of symbols, we start with the start variable S, then apply a production rule, chosen accordingly to its probability, to change S to, for instance, (S). We then continue applying production rules until we are left with only terminals. For example, to generate the sequence ()() we will need to proceed through a sequence containing at least two variables at some point, e.g. as in the derivation $S \rightarrow SS \rightarrow (S)S \rightarrow ()S \rightarrow ()(S) \rightarrow ()()$. The descendant sequences to each of the two occurrences of S in SS, each has to be a balanced sequence of parentheses. However, due to the context-free limitation, once the initial S has been split into *SS* there is no possibility of coordinating the descendant sequences of each S. Where HMMs allow the modelling of sequential dependencies – the next state only depends on its predecessor – CFGs allow the modelling of hierarchical dependencies, i.e. dependencies that have a tree-like structure.

The derivation tree of the string ()() by the grammar can be seen in Fig. 16.2. A derivation tree is a tree with internal nodes labelled by variables and leaves labelled by terminal symbols. The children of an internal node are the symbols (variables and terminal symbols) of the right hand side of the production rule used to replace the variable in the derivation. The final sequence of the derivation is read off the leaves from left to right. Any derivation in a CFG can be represented by a derivation tree, as there is an equivalence between trees and sequences with balanced parentheses. The parenthesis example captures exactly the capabilities and limitations of CFGs. In this way, SCFGs generate strings which correspond to RNA secondary structures.

At this point, it is worth discussing a technical issue- the difference between sequence-generating grammars and structure-generating grammars. We define a sequence-generating grammar to be one where the terminals are the nucleotides A, C, G and U (and possibly the gap character '-' if considering alignments). The advantage of sequence-generating grammars is that we can have different probability distributions over nucleotides in certain regions of RNA structures (Knudsen and Hein, 1999), for example making stem base pairs more C-G rich. The disadvantage is that many more parameters are required, and these cannot always be accurately estimated from data. Often sequence generating grammars are abbreviated, giving rules denoted as $S \rightarrow s$, where s denotes a single unpaired nucleotide, so this is in fact four production rules expressed as one.

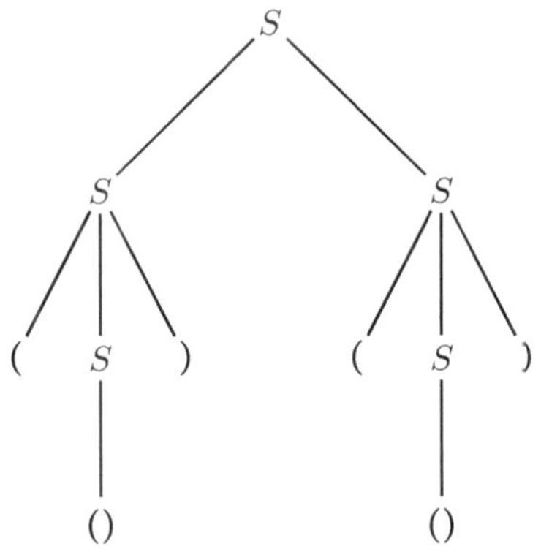

Figure 16.2 Derivation of the string ()(). The grammar has two uses of rule S→(S), one use of rule S→SS, and two uses rule of S→( ).

Structure generating grammars, on the other hand, generate strings of dots and parentheses, and we have already seen an example of this type of grammar. They are simply sequence generating grammars with the nucleotide probabilities constant. This characterization, though, offers additional mathematical flexibility, as will be seen later. To calculate the probability of, for example, non-terminal S generating a single unpaired A, we take the probability S generates a dot, and multiply it by the nucleotide probability for an A. In this way, the probability of generating a string of nucleotides becomes the product of the grammar probabilities and the nucleotide probabilities. Throughout this chapter, I will mainly use examples from structure-generating grammars, for simplicity, but results will apply equally to both types.

Now that we have concepts of how SCFGs generate strings of nucleotides, and how these correspond to RNA secondary structures, it is worth introducing some measures of success. There have been many metrics on RNA secondary structures designed (Shapiro and Zhang, 1990; Moulton *et al.*, 2000; Freyhult *et al.*, 2005; Hamada *et al.*, 2010), but the ones commonly used are sensitivity, positive predictive value (PPV), and *F*-score – all of which are defined over predicting base pairs correctly.

Before considering sensitivity, PPV, and *F*-score, we must define some intermediaries. Firstly, we define true positives (TP) to be the number of base pairs which are in the correct structure and also the predicted structure. We define false positives (FP) to be the number of base pairs which are in the predicted structure, but not in the correct structure. Lastly, we define false negatives (FN) to be the number of base pairs which are in the correct structure but not in the predicted structure. This then gives us values for sensitivity, PPV, and *F*-score:

$$\text{Sensitivity}=\frac{\text{TP}}{\text{TP}+\text{FN}}$$

$$\text{PPV}=\frac{\text{TP}}{\text{TP}+\text{FP}}$$

$$\text{F-score}=\frac{2*\text{TP}}{2*\text{TP}+\text{FN}+\text{FP}}$$

Note that sensitivity, PPV, and *F*-score are all between 0 and 1. For sensitivity, a 0 shows none of the base pairs predicted are there, and a 1 indicates all base pairs in the true structure have been predicted. For PPV, a 0 shows that all base pairs have been predicted incorrectly, whilst a 1 indicates all predicted base pairs are correct (by convention, if no base pairs are predicted we use a PPV of 1).

These measures are hardly ideal measures for RNA secondary structure prediction accuracy. Firstly, note that these measures were originally developed with binary classifiers in mind, where predictions made were independent. However, with base pairs, if one prediction is wrong, it can affect the correctness of other predicted pairs, by blocking them, for instance. Once you start to go wrong with a prediction, the sensitivity and PPV suffer quickly. Secondly, though, note that these measures have little to do with biological function, and do not reflect the functional landscape. An RNA which has mis-folded by, say, missing one particular base pair, may keep the same function, but if a different base pair is missing, the function may completely change. Of course, it is not possible to create a measure with this taken account for, but additional realism with accuracy measures would greatly benefit RNA secondary structure prediction method design.

Normal forms

This section forms a more technical section discussing CFG normal forms, null variables, null productions, and pseudoknotted structures. Readers more interested in the practical side of RNA secondary structure prediction using SCFGs may therefore want to skip this section, at least on first reading.

Chomsky normal form

The definition of context-free grammars allow arbitrary right hand sides of production rules – for example, we can have the rule $S \rightarrow x$, where x is any collection of non-terminals and terminals. Just as there were several possible ways we could restrict regular grammars without changing expressibility (left or right, extended or not), there are several ways the right hand side of the productions of a grammar can be restricted without losing expressibility. These are known as normal forms, and requiring grammars to be on normal form simplifies algorithms as only the restricted set of production types need to be considered.

We will use the Chomsky normal form (CNF) (Chomsky, 1959). Many of the algorithms used are originally designed to work on grammars in CNF, so it is important to understand how to change the grammar being used to CNF. A grammar is said to be in CNF if all of its production rules are of the form:

- $A \rightarrow BC$ for $A \in N$ $B,C \in N \setminus \{S\}$
- $A \rightarrow \alpha$ for $A \in N, \alpha \in V$
- $S \rightarrow \varepsilon$, where S is the start symbol and ε the empty string

Any context-free grammar can be converted to a CNF grammar. For right hand sides of production rules with a sequence of more than two symbols we can break the sequence into the constituent symbols by introducing auxiliary variables. If a symbol in a right-hand side of length two is a terminal symbol we can replace it with an auxiliary variable that can only be replaced with this terminal symbol, and if it is the special start variable S we can replace it with an auxiliary variable with exactly the same production rules as S. However, when the right hand side is shorter than two symbols but the production rule is not in concordance with the CNF restrictions, we need to proceed with a bit more caution.

An empty sequence right hand side, or null production, is only allowed for the special start variable S. To eliminate all other null productions we first identify all nullable variables, i.e. variables that can in one or more steps be replaced with the empty sequence. A variable U is nullable if

- $U \rightarrow \varepsilon \in \mathrm{P}$
- $U \rightarrow V_1 ... V_k$ and $V_1, ..., V_k$ are all nullable

Based on this recursive rule we can formulate Algorithm 16.1 to efficiently determine the set of nullable variables. Once these have been identified, for each nullable variable, with the exception of S, we remove the null production for this variable, and for every production where it occurs on the right hand side we make two copies of this production: one where the variable still occurs and one where it has been removed from the right hand side sequence. When the right hand side consists of just a single variable – remember that right-hand sides with just a single symbol are required to be a terminal symbol for grammars in CNF – the transformation is even less complicated. All we need to do is copy all the productions of the right hand side variable to the left hand side variable.

Algorithm 16.1 Nullable variables

```
N=0
repeat
  for U → x → P do
    if x does not contain terminal symbols or
    variables not in N then
      Add U to N
until no further variables are added to N in this
iteration
```

This all appear simple enough, so why the warning to proceed with caution? As long as only the language of a grammar, i.e. the set of finite sequences it can generate, is of concern, the last two types of transformations do not require special attention. However, in bioinformatics mostly parameterized, in particular stochastic, context-free grammars are used.

Manipulation of probabilities can easily be included in the first set of transformations mentioned: breaking a long right hand side sequence into pairs, replacing a terminal symbol in a pair with a variable, and adding a copy of S for right hand side use. The only part not strictly trivial would be tying of the probabilities between the productions of S and the copy of S when inferring probabilities from data.

The manipulations eliminating null productions and replacement productions, on the other hand, require more complicated manipulation of probabilities. For an already parameterized SCFG the new probabilities, after elimination of replacement productions, can be described by a system of linear equations. This, in turn, can be solved to find the probabilities of the transformed grammar. Inferring probabilities from data, however, can become a difficult if not impossible task, if parameters are required to reflect the structure of the original grammar (as will usually be the case). Under maximum likelihood estimation we would need to maximize an equation system similar to the one for parameterized grammars, but with the parameters of the original grammar occurring polynomially, rather than linearly, as variables.

Elimination of null productions is potentially even more complicated when the grammar is stochastic. Even when the grammar is already parameterized, determining probabilities of the transformed grammar may not be straightforward. If the grammar has a production $U \rightarrow V\ W$ where both V and W are nullable, the probability that U is eventually replaced by the empty sequence depends on the product of the probabilities that V and W are eventually replaced by the empty sequence. If there are no cycles in these dependencies, the nullability probability can easily be determined for every variable. However, with cyclic dependencies the nullability probabilities may have to be specified by a quadratic equation system, which in general can be hard to solve (Håstad *et al.*, 1993). Because of these complications, when designing a SCFG the best advise is, if at all possible, to avoid introducing replacement productions with cyclic structure and in general null productions (Algorithm 16.2).

Algorithm 16.2 Eliminating replacement productions

```
repeat
  for U → V ∈ P do
    for V → x ∈ P do
      Add U → x to P
until no updates where required in this iteration
Eliminate all U → V productions from P
```

Double emission normal form

However, while CNF is a convenient form for SCFGs in terms of algorithmics, it does not help particularly for RNA secondary structure prediction. To see why, consider the following sequence-generating grammar in CNF:

$$S \rightarrow SS \mid AC \mid BC$$
$$A \rightarrow BS$$
$$B \rightarrow s$$
$$C \rightarrow d$$

This grammar can then produce strings of *s* and *d*. However, there is no way of knowing which *d* are intended to be paired together. Equally, the grammar can produce strings which do not correspond to RNA secondary structures, like *sd*- the *d* must be paired to something to create a valid structure.

Consequently, when considering RNA secondary structure prediction, double emission normal form (Anderson *et al.*, 2012) is much more convenient to use. We restrict rules to be of the following form

$$U \rightarrow VW$$
$$U \rightarrow (V)$$
$$U \rightarrow .$$

for non-terminals *U*, *V*, *W*. Double emission normal form maintains all of the flexibility of CNF, as, similarly, any RNA grammar (that is, one that does not produce parentheses that do not match) in CNF can be expressed in double emission normal form, and vice versa, with the exception of being able to generate the empty string. Given the similarity of the two forms, null and replacement productions can be dealt with in a similar fashion.

Furthermore, note that the rules $U \rightarrow \varepsilon$ and $U \rightarrow V$ are not allowed. This was originally to prevent cyclical productions from existing, that is having both of the rules $U \rightarrow V$ and $V \rightarrow U$, as many of the algorithms for SCFGs will become stuck in infinite loops if both of these rules exist. Similarly, cyclical productions could come about if we have the rules $U \rightarrow VW$, $V \rightarrow UW$ and $W \rightarrow \varepsilon$. However, when we are designing a grammar by hand, we are unlikely to put in cyclical productions, so these rules could be allowed, if we wanted the extra flexibility. However, double emission normal form in practice allows enough flexibility to predict all types of RNA secondary structures seen.

Pseudoknots

As discussed previously, all grammars and normal forms constructed have not been able to predict pseudoknotted structures. The basic structure of a pseudoknot can be represented as $\{w \mid w$ has the form $[i\ (j]i)j\}$. These can, in fact, never be constructed with SCFGs since the form of a pseudoknot is not context-free by the Pumping Lemma (Brown and Wilson, 1996). Hence one cannot use SCFGs in their pure form to predict pseudoknots. Various approaches have been tried to address this problem.

Intersections of SCFGs can be used to produce pseudoknotted structures (Brown and Wilson, 1996; Lefebvre, 1996). Brown and Wilson (1996) use an intersection of two SCFGs, where the structure was classified as a pseudoknot if it can be identified as having one helix which has high probability under one statistical model and simultaneously having high probability under the other model. Lefebvre (1996) uses 'm-tape context-free grammars', a similar extension in which multiple rows of grammars can evolve and combinations of these grammars taken to represent secondary structure. Although the method uses SCFGs to predict pseudoknots, the intersecting approach is a heuristic approach, and guaranteed optimality of the resulting parse is unfortunately sacrificed, leading to less accurate predictions.

Alternatively, one can leave the framework of SCFGs, whilst still using grammars (Rivas and Eddy, 2000). A formal grammar structure was developed which had a one-to-one correspondence with a pseudoknot folding algorithm (Rivas and Eddy, 1999), which used minimum free-energy structures to find optimal secondary structure. Although this method included context-free grammars, it extended the grammar by using special non-terminal symbols, which dictated specific rearrangements of substrings, taking the method out of a SCFG. Similarly, another method of pseudoknot prediction from SCFGs is to use the posterior probabilities found by SCFGs, but decode them in a different manner, such as iterative helix production. There is scope for further research to be done here.

Ambiguity, completeness and grammar design

Ambiguity and completeness

Grammar ambiguity is concerned with SCFGs producing more than one derivation for a given structure. We say a CFG is ambiguous if the grammar has two or more different derivations for a given RNA secondary structure. With an unambiguous grammar, there is a one-to-one correspondence between derivations and RNA secondary structures. Consequently, statements we make about the probability distribution on derivations correspond to statements about the probability distribution on RNA secondary structures. However, with ambiguous grammars, we cannot make that interpretation of the probability distribution on derivations.

For example, recall the example grammar used previously. Three different derivation trees for the sequence ()()()() are shown as an example in Fig. 16.3. These correspond to the leftmost derivations $S \rightarrow SS \rightarrow ()S \rightarrow ()SS \rightarrow ()()S \rightarrow ()()SS \rightarrow ()()()S \rightarrow ()()()()$, $S \rightarrow SS \rightarrow SSS \rightarrow ()SS \rightarrow ()()S \rightarrow ()()SS \rightarrow ()()()S \rightarrow ()()()()$ and $S \rightarrow SS \rightarrow SSS \rightarrow SSSS \rightarrow ()SSS \rightarrow ()()SS \rightarrow ()()()S \rightarrow ()()()()$. Observe that though the trees are similar and utilizes each of the three production rules the same number of times, they are still distinctly different.

Given that it is hard to parse an ambiguous grammar's probability distribution, efforts have been made to avoid ambiguity in RNA secondary structure prediction grammars (Reeder *et al.*, 2005; Branrand *et al.*, 2007), despite determining ambiguity being an undecidable problem (Hopcroft and Ullman, 1979). However, it seems that, practically, CFG ambiguity is less of a problem than it might look from a theoretical standpoint. Small ambiguous grammars have been found to perform poorly (Dowell and Eddy, 2004), but this was perhaps due to fundamental limitations in the grammar design than the ambiguity. However, many larger ambiguous grammars have been found that perform as well, or better, than the best unambiguous grammars (Anderson *et al.*, 2012).

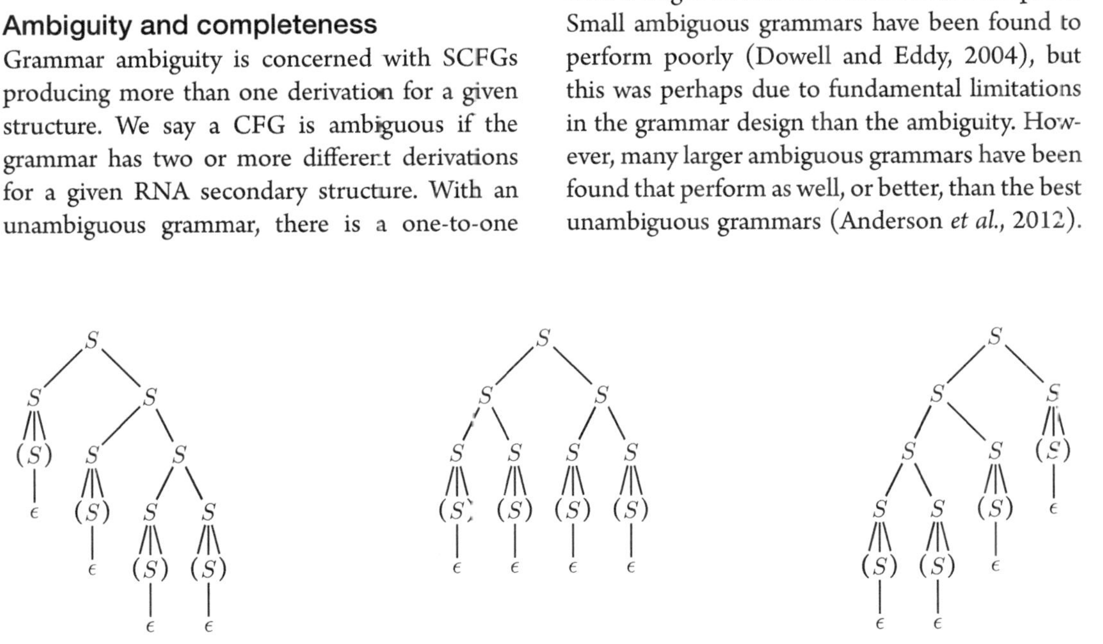

Figure 16.3 Three different derivations for the string ()()()()

Even though potentially suboptimal predictions were chosen, the predictions were still good. However, ambiguity is still something that should be considered in SCFG design.

Completeness can be seen as the complement to ambiguity, considering instead whether or not a grammar can form enough structures, as opposed to too many. For RNA secondary structure, we say a grammar is complete if it has a derivation for all possible structures, which have no hairpins shorter than length two, that is, it can generate all structures that we consider valid. Again, we might consider that completeness is a desirable property for a grammar to have, but again there are examples of incomplete grammars still performing well (Anderson *et al.*, 2012). However, the structures which these grammars could not generate were rarely seen in biology, suggesting that the set of structures they could generate was perhaps a good approximation to the set of structures we consider valid.

Note a complete, unambiguous grammar cannot be simply modified without compromising one of the properties. Adding production rules, if they are ever used, will create ambiguity by providing additional derivations. Equally, removing production rules will create incompleteness, unless the rule is never used in a derivation, as the original grammar is assumed unambiguous. Moreover, grammars that are unambiguous and complete are vastly outnumbered by grammars that are not, so designing grammars which are both unambiguous and complete can be practically difficult.

Lightweight grammar design

The last thing to discuss before we move onto considering algorithms for SCFGs is the design of the grammar. Up to this point, we have been dealing with example grammars only, for simplicity. However, when using a SCFGs to predict RNA secondary structure optimally, we will need more than this. Several attempts have been made to find out which set of non-terminals and production rules will create the best CFG for RNA secondary structure prediction. Firstly, we focus on the design of lightweight SCFGs, that is grammars with relatively few non-terminals, production rules, and therefore parameters.

Initially the main CFG used was the Knudsen and Hein grammar (Knudsen and Hein, 1999, 2003), hereafter denoted KH99. KH99 was a remarkable simple grammar:

$$S \rightarrow \begin{matrix} L & LS \\ .131 & .869 \end{matrix}$$

$$L \rightarrow \begin{matrix} s & dF\hat{d} \\ .895 & .105 \end{matrix}$$

$$F \rightarrow \begin{matrix} dF\hat{d} & LS \\ .788 & .212 \end{matrix}$$

with d, $\hat{d}$ indicating paired nucleotides, and s indicating an unpaired nucleotide. The grammar generated stem and loop regions, and, although originally designed to be paired with an evolutionary model, performs well on single sequences too (Anderson *et al.*, 2012). Particularly, it has only three production rule parameters to be estimated from data, which means it is less likely to suffer from overfitting.

While KH99 was effective, it seems to have been chosen relatively arbitrarily, in that there is little discussion about what motivated the choice of production rules. This problem was addressed by Dowell and Eddy, where nine different 'lightweight' (i.e. a small number of parameters) SCFGs were evaluated for single sequence structure on a benchmark set of RNA secondary structures (Dowell and Eddy, 2004). Lightweight grammars were considered here for their computational advantages and simplicity, since if one would like to extend SCFGs to model multiple sequences, a combinatorial explosion of parameters can occur if parameterization is not kept small. The grammars tested were all hand designed, that is that people sat down and tried to figure out which grammars would best produce secondary structure.

Dowell and Eddy (2004) considered two types of grammar: 'Stacking' and 'Simple'. Stacking grammars have production rules such as $P(\mathrm{b},\hat{\mathrm{b}}) \rightarrow \mathrm{a}P(\mathrm{a},\hat{\mathrm{a}})\hat{\mathrm{a}}$ for nucleotides $\mathrm{a},\hat{\mathrm{a}},\mathrm{b},\hat{\mathrm{b}}$ which gives a larger set of production rules in the grammar. Simple grammars like KH99 do not have these production rules. One might expect the stacking grammars to have better predictive accuracy due to the added strength of the model, but it is quite possible that as there are many more parameters, which must be estimated from training data, that this is not the case. They found that several of the grammar performed comparably on their test dataset, and the stacking grammars generally outperformed the non-stacking grammars, but KH99 performed the best.

This search for good grammars was further extended by use of an evolutionary algorithm to enable a computational search (Anderson *et al.*, 2012). A population of small, very simple grammars was used for an initial population, then grammars were mutated, for example by adding a production rule or non-terminal, and bred, for example by combining production rules. Grammars were then selected for, with a fitness based on the ability to correctly predict RNA secondary structures. This process of mutation, breeding and selection was then iterated for a fixed number of generations, and the grammars tested on an independent dataset, to avoid over-fitting. Furthermore, a brute force search of all small grammars that are with at most two non-terminals was done, but most performed poorly, perhaps as expected.

Many grammars, however, were found from this evolutionary search, which performed as well, or slightly better, than KH99. In particular, a visual inspection of the grammars found yielded no suggestions as to why they were so good at secondary structure prediction – they seemed relatively random in choices of non-terminals and production rules. For example, the largest, most complicated, and best grammar is shown below.

$$A \rightarrow DE \mid AB \mid BA \mid AH \mid . \mid (F) \mid (H)$$
$$B \rightarrow .$$
$$C \rightarrow (H)$$
$$D \rightarrow BB \mid AC$$
$$E \rightarrow . \mid (H)$$
$$F \rightarrow FB \mid CF$$
$$G \rightarrow GH \mid (H) \mid (C)$$
$$H \rightarrow FA \mid AF \mid HH \mid (B) \mid (H)$$

Overall, this gives an interpretation of grammars as similar to machine learning approaches, in style – simply structures that learn a probability distribution from data and regurgitate it. There seem to be many grammars possible that perform similarly for the RNA secondary structure prediction problem, and it is the method of SCFG prediction, via either inside–outside or CYK, as opposed to grammar design, which seems to be the methodological bottleneck.

Heavyweight grammar design

We can, of course, look to design more complicated grammars to more closely mimic RNA secondary structure. In particular, one might try and mimic the thermodynamic model in SCFGs, for example to use the advantages of comparative modelling.

Most approaches using heavyweight SCFGs have been to try to reproduce the thermodynamic model (Nebel and Scheid, 2011; Rivas *et al.*, 2012). We can create SCFGs that have production rules which mirrors the energy rules, and can calculate the full partition function, to aid suboptimal folding, or simply to incorporate the evolutionary model. However, given results (Nebel and Scheid, 2011) produced by these grammars, they seem no better than the smaller, lightweight SCFGs used in Pfold, or the thermodynamic methods themselves. This gives further credence to the idea that it is not the SCFG design which is important, but the method of decoding, given the probability distribution.

Algorithms for SCFGs, parameter estimation, structure prediction

Algorithms for SCFGs

We now move on to some of the main algorithms used for SCFG prediction. Throughout, we assume that SCFGs are in double emission normal form. There are two main sets of algorithms which are used for RNA secondary structure prediction with SCFGs. I will first introduce the algorithms, and then discuss how each one can be applied.

The Cocke–Younger–Kasimi (CYK) algorithm (Younger, 1967) was originally used to determine whether or not a context-free grammar could derive a given string. However, it is easily adapted to be useful for SCFGs, and determines the maximum probability of a derivation of the SCFG for a given string. It is a dynamic programming algorithm, so backtracking through the dynamic programming table will give the SCFG derivation which produced this maximum probability. This derivation can then be used to find the structure with the maximum probability derivation.

For each non-terminal U, string *s*, and string indices *i* and *j*, we define

$$C(U,i,j)=\begin{cases} P[U\rightarrow .]P[s[i]=.] & \text{if } i=j \\ \max\left\{\max\limits_{U\rightarrow VW}\max\limits_{i\le k<j} P[U\rightarrow VW]\,C(V,i,k)C(W,k+1,j),\right. \\ \left.\max\limits_{U\rightarrow (V)} P[U\rightarrow (V)]C(V,i+1,j-1)P[s[i],s[j]=()]\right\} & \text{if } i<j \\ 0 & \text{otherwise} \end{cases}$$

$C(U, i, j)$ is then the maximum probability of U generating the subsequence $s[i] \ldots s[j]$, and in particular, $C(S,1,|s|)$ gives the maximum probability of the SCFG generating the sequence. To compute this table, start with subsequences of length one, then length two, and so on, until you have computed the CYK table for the entire string. Pseudocode for the algorithm can be found in Algorithm 16.3. As can be seen in the pseudocode, we can find $C(S, 1, |s|)$ in time $O(|P||s|^3)$ and space $O(|V||s|^2)$, as we compute the $C(U, i, j)$ values in order of subsequences of s of increasing length.

The inside and outside algorithms (Lari and Young, 1990) are similar to the CYK algorithm, but instead of calculating the maximum probability of a derivation, they calculate the total probability of generating subsequences. First we define the inside table, for each non-terminal U, string *s*, and string indices *i* and *j*:

$$I(U,i,j)=\begin{cases} P[U\rightarrow .]P[s[i]=.] & \text{if } i=j \\ \sum\limits_{U\rightarrow VW}\sum\limits_{i\le k<j} P[U\rightarrow VW]I(V,i,k)I(W,k+1,j) \\ +\sum\limits_{U\rightarrow (V)} P[U\rightarrow (V)]I(V,i+1,j-1)P[s[i],s[j]=()]\} & \text{if } i<j \\ 0 & \text{otherwise} \end{cases}$$

and similarly, the outside table:

$$O(U,i,j)=\begin{cases} 1 & \text{if } U=S,\ i=1,\ j=|s| \\ \sum\limits_{V\rightarrow UW}\sum\limits_{k>j} P[V\rightarrow UW]O(V,i,k)I(W,j+1,k) \\ +\sum\limits_{V\rightarrow WU}\sum\limits_{k<i} P[V\rightarrow WU]O(V,k,j)I(W,k,i-1) \\ +\sum\limits_{V\rightarrow (U)} P[V\rightarrow (U)]O(V,i-1,j+1)P[s[i-1],s[j+1]=()] & \text{if } i<j \\ 0 & \text{otherwise} \end{cases}$$

$I(U, i, j)$ then gives the total probability of U generating the subsequence $s[i] \ldots s[j]$, and $O(U, i, j)$ the total probability of generating everything except the subsequence $s[i] \ldots s[j]$. Hopefully, considering the derivation trees required to produce the subsequence $s[i] \ldots s[j]$, it is intuitive why the tables are named 'inside' and 'outside'. The pseudocode for the inside algorithm is analogous to the CYK algorithm pseudocode. The outside is computed in a similar manner, but instead starting from subsequences of length $|s|$, and working down to subsequences of length one. Again, the inside and outside algorithms are computed in time $O(|P||s|^3)$ and space $O(|V||s|^2)$.

Parameter estimation

For parameter estimation, we require a set of RNA sequences with known secondary structures, that is, in the language of machine learning, we

Algorithm 16.3 CYK algorithm

for i = 1 **to** $|s|$ **do**
  for $U \in V$ **do**
    $C(U, i, i) = P(U \rightarrow s[i])$
for l = 1 **to** $|s|-1$ **do**
  for i = 1 **to** $|s|-l$ **do**
    for $U \in V$ **do**

$$C(U,i,i+l) = \max\left\{\max\limits_{U\rightarrow VW}\max\limits_{i\le k<l} P[U\rightarrow VW]C(V,i,k)C(W,k+1,l),\ \max\limits_{U\rightarrow (V)} P[U\rightarrow (V)]C(V,i+1,l-1)P[s[i],s[l]=(\ )]\right\}$$

use supervised training. Interestingly, if you try to train the SCFG in an unsupervised manner, that is just using RNA sequences, the SCFG will choose parameters which enable it to produce long strings at lost 'cost'. To illustrate this, consider the start variable for KH99. The probabilities for the production rules $S \rightarrow LS$ and $S \rightarrow L$ given are 0.869 and 0.131 respectively (Knudsen and Hein, 1999). However, consider also a SCFG with rules $S \rightarrow LS$, $S \rightarrow L$ and $L \rightarrow .$, with rule probabilities 0.99, 0.01 and 1. This latter SCFG will generate strings solely of unpaired bases with very high probability – much higher than KH99 does – and this is, in fact, the sort of SCFG you get when using unsupervised training, as you are looking to maximize the probability of the data. For this reason, we need known secondary structures for parameter estimation.

We need also the nucleotide probabilities for SCFG parameter estimation. When using SCFGs over alignments, an evolutionary model is used, discussed later in this chapter. However, then considering only single sequences, we simply take the frequency at which nucleotides appear in paired and unpaired positions and use these frequencies as probabilities. For example, if we see 100 AU base pairs out of 500 total base pairs, then the probability of seeing an AU base pair would be 1/5.

Now that we have our nucleotide probabilities and our SCFG algorithms, we are interested in determining the production rule probabilities which will be useful in producing the most accurate structure predictions. We can do this parameter estimation using either the CYK algorithm, or by using the inside and outside algorithms.

For parameter estimation with the CYK algorithm, we first consider unambiguous grammars. Recall that if an unambiguous grammar can derive a string, then it will only have one way of doing so. So here, we use the CYK algorithm instead to generate the known structure, initially using uniform rule probabilities, and an indicator function in the algorithm to determine whether or not the algorithm is generating the known structure. Thus, by keeping track of the rule frequencies used in the derivation to generate this known structure, we will be able to count the total number of rules used over the whole training dataset. Again, we just take the frequencies of rules observed. Taking KH99 as an example, if we saw 200 $S \rightarrow LS$ rules used, and 100 $S \rightarrow L$ rules used, our rule probabilities would be 2/3 and 1/3 respectively.

For ambiguous grammars, we no longer have a single derivation for each structure. So, if we want to estimate parameters on the dataset, we must choose at random a single derivation. This is done in the backtracking step of the CYK algorithm: if there is not a single production rule which generates a given subsequence, one is chosen at random. Of course, implementing this does not require knowing whether or not a SCFG is ambiguous *a priori*, since picking randomly will still be correct for unambiguous grammars. Again, by keeping track of production rules, you can determine production rule probabilities.

For the inside and outside algorithms, we used inside–outside training (Lari and Young, 1990). This is an implementation of expectation maximization training (Demster *et al.* 1977). First we randomly initialise the rule probabilities, then use these to calculate the expected number of rule uses. From the expected number of rule uses, we re-estimate the rule probabilities. This process is then iterated until convergence of rule probabilities. The inside and outside algorithms can be used to calculate the expected rule frequencies. If we let P_{old} be the old SCFG probabilities, then the new probabilities, P_{new} are estimated as follows:

$$P_{new}[U \rightarrow VW] = \frac{E[\#U \rightarrow VW \text{ is used}]}{E[\#U \text{ is used}]}$$

$$= \frac{\sum_{i=1}^{m-1}\sum_{j=i+1}^{m}\sum_{k=1}^{j-1} P_{old}[U \rightarrow VW] I(V,i,k) I(W,k+1,j) O(U,i,j)}{\sum_{i=1}^{m}\sum_{j=1}^{m} I(U,i,j) O(U,i,j)}$$

There are good reasons why one might want to use one variety of training over the other. The CYK training method is significantly quicker: not only does the algorithm take less time (practically, not in complexity) than inside-outside, but there is only a single iteration required. Thus, for very large SCFGs, or if many are being trained, this may be of benefit. On the other hand, inside–outside training deals with ambiguous SCFGs much better, considering all derivations they can make instead of a random one. Similarly, inside–outside training is more focused on getting the overall probability distribution correct, as opposed to the simply a maximum-likelihood style estimator the CYK algorithm uses.

Similarly, we might wonder how stable the parameter estimation procedure is, that is, if we choose slightly different parameters, might the SCFG predict entirely different structures? This was in particular an important question when considering designing grammars with an automated approach (Anderson *et al.*, 2012). When the dataset for parameter estimation was made to be a random subset of the whole dataset, the stability of parameterization could be tested. In fact, with only a handful of sequences, the parameters of small SCFGs varied very little, but with slightly larger SCFGs, the variation was more – not surprising, given there are more production rules to estimate. With a dataset of around 100 sequences, though, the variation was very small. This idea has been further confirmed by looking at disturbances in the sampling probabilities of a heavyweight SCFG, and it's corresponding prediction accuracy, and finding that the structures produced do not vary particularly (Scheid and Nebel, 2012). Thus, we can be reasonably confident when estimating SCFG probabilities that we will have reached a reasonable parameterization.

Secondary structure prediction

Now we have our SCFG with a set of parameters found from known structures, we want to begin predicting structures for other sequences. Again, the CYK, inside, and outside algorithms are used in a similar way to the parameter estimation.

Prediction with the CYK algorithm is straightforward. Firstly, we fill in the CYK table using the algorithm above. The value $C(S,1,|s|)$ gives the maximum probability of the SCFG generating the sequence, and it is the derivation corresponding to this probability which we use for the predicted structure. This derivation, as with the parameter estimation, is found by backtracking through the CYK table. We could also use a modified version CYK algorithm to determine the n-most probable structures (Chappelier and Rajman, 1998).

Prediction with the inside and outside algorithms is a little more difficult. From the inside and outside tables, one can derive the base pair probability matrix $B(i, j)$, the probability of i pairing with j:

$$B(i,j)=\frac{\sum_{U\to(V)} O(U,i,j)P[U\to(V)]P[s[i],s[j]=()]I(V,i+1,j-1)}{I(S,1,|s|)}$$

that is, by summing over all rules of type $U\to(V)$.

Once we have the base pair probability matrix, we have a probability distribution over secondary structures, and we can do posterior decoding to obtain our predicted structure. At this point we must decide what we are looking to maximize. Typically, expected positions correctly predicted or expected base pairs correctly predicted are chosen as the criterion to maximize over. For example, if we are looking to maximize the expected number of positions correctly predicted, we consider

$$E(i,j)=\max\begin{cases}\frac{1}{2}\left(1-\sum_{1\le k\le|s|}B(i,k)\right)+E(i+1,j) & (1)\\ B(i,j)+E(i+1,j-1) & (2)\\ \max_{i+1<k<j}[B(i,k)+E(i+1,k-1)+E(k+1,j)] & (3)\end{cases}$$

$E(i, j)$ is the maximum expected number of positions correctly predicted for the subsequence $i \dots j$. Case (1) considers this maximum having been achieved from i being unpaired, case (2) considers this maximum having been achieved from i being paired with j, and case (3) considers this maximum having been achieved from i being paired with some k in between i and j. $E(1,|s|)$ is then the maximum expected number of positions correctly predicted for the entire sequence, and the structure which produces this maximum, found by backtracking through the $E(i, j)$ table, is the structure predicted.

However, there is no fundamental reason why the posterior base pair probability matrix should be decoded in this way – we are simply maximizing over a function. We could define alternative functions to maximize over to simulate, for instance suboptimal folding, or to include more biophysical constraints. One simple idea would be to implement iterative helix production, as has been done for the thermodynamic model (Harmanci *et al.*, 2011), to simulate the quicker forming on helices observed biochemically (Craig *et al.*, 1971). This is an avenue for further research for SCFG methods.

There are some key differences between the CYK and inside–outside methods of prediction (Anderson *et al.*, 2012). The CYK algorithm, predicting structures with the highest probability, predicts structures with a larger proportion of base pairs. This is in some sense unsurprising when considering the double emission normal form, as forming a base pair uses only a single rule for creating two positions. Hence the CYK algorithm can predict the structure 'quicker', which typically comes with a higher probability, and so the sensitivity of CYK predictions tends to be relatively higher. Inside-outside, on the other hand, will consider probability contributions from all derivations. Therefore, structures predicted by the inside–outside algorithm tend to have relatively higher PPV, only predicting base pairs where the majority of derivations (weighted by probability mass) 'agree' on a base pair being present. In particular, when considering predictions in the context of comparative modelling, this difference can be more pronounced.

Example

At this point many algorithms have been presented, as well as methods of secondary structure prediction for single sequences. Here, I provide an example of implementation of the algorithms and calculations throughout.

Here we will use sequences from Rfam (Griffiths-Jones *et al.*, 2005). In particular, we will consider RF00390, an upstream pseudoknot domain found in viruses like the tobacco mosaic virus (Felden *et al.*, 1996). The first sequence is as follows:

UAAGUUCUCGAUCUUUAAAAUCG

In particular, this sequence has a secondary structure, which is pseudoknotted,

. [[[.... ((((]]] ....))))

which means the SCFG method will not be able to predict it correctly. However, this will give an opportunity to show some of the strengths and weaknesses of the method, and demonstrate the accuracy metrics.

Firstly, we will need to convert KH99 to double emission normal form so that we can use the algorithms developed above:

$$S \rightarrow \underset{.117}{\cdot} \quad \underset{.014}{(F)} \quad \underset{.869}{LS}$$

$$L \rightarrow \underset{.896}{\cdot} \quad \underset{.105}{(F)}$$

$$F \rightarrow \underset{.788}{(F)} \quad \underset{.212}{LS}$$

with parameters for the SCFG are taken from the Pfold model. Next, we will need unpaired and paired probabilities. For these, we simply take the frequencies of nucleotides observed in Knudsen and Hein (1999). We have the probability of (A,C,G,U) being unpaired as (0.268, 0.214, 0.267, 0.251), and the probability of (AU,UA,CG,GC,GU,UG) being paired as (0.180, 0.180, 0.272, 0.272, 0.048, 0.048), allowing only canonical base pairs for simplicity.

Firstly, we consider the CYK algorithm. Now we have the SCFG and nucleotide probabilities, we can calculate the full CYK table for each nonterminal. The value of $C(S,1,|s|)$ here is 9.35e–18, and, backtracking through the CYK table, we then get the structure:

CYK predicted structure: ... . ((((... ....))))

True structure: . [[[.... ((((]]] ....))))

Inside–outside is very similar. Fig. 16.4 shows the base pair probability matrix.

Note that there is practically no probability mass associated with the pseudoknotted base pairs AAG, CUU. Since the SCFG cannot produce both sets of base pairs, it must 'choose' between the two sets, and we can see the choice is for the 'regular' pairs. Looking at the base pair probability matrix produced by inside-outside, we might expect the

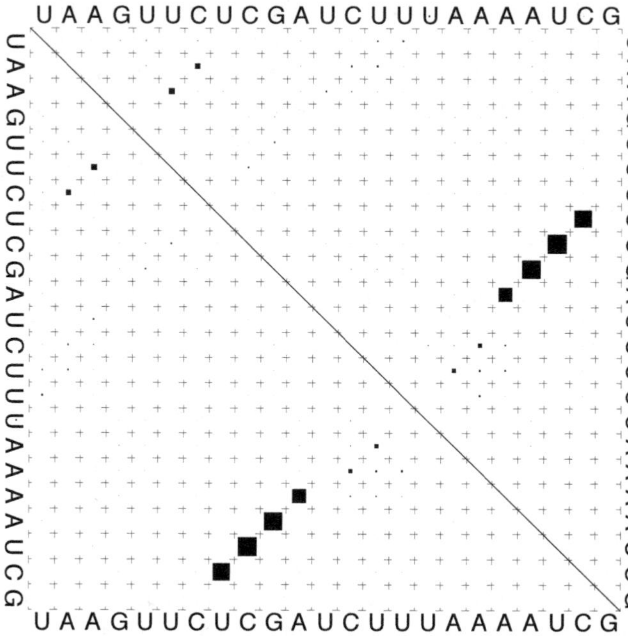

Figure 16.4 An example of a base pair probability matrix.

same structure produced by CYK, and this is in fact what the MEA produces:

MEA predicted structure:	... . ((((........))))
True structure:	.[[[....((((]]]....))))

It is not uncommon that CYK and MEA produce the same secondary structure prediction. Furthermore, when they do produce different structures, they often share the same 'core' structure, that is that CYK will produce the same structure as inside-outside, but with a few more base pairs predicted, and vice versa. This can be seen by considering the normal form and production rules: producing a base pair will produce two non-terminals. By the SCFG predicting many base pairs, it will use fewer production rules, and so, in general, a higher probability structure will result. This is the type of structure that CYK will choose. Inside–outside cares more about probabilities over all derivations, and so it is more selective about the base pairs it chooses.

Finally, we can calculate the accuracy measures of sensitivity, PPV, and *F*-score. We have $TP=4$, $FP=0$, $FN=3$. So the sensitivity is $4/7=0.571$, the PPV is $4/4=1$, and the *F*-score, the harmonic mean of sensitivity and PPV, is 0.727. For smaller sequences, this is approximately what we might expect, in terms of accuracy.

Comparative modelling

The main strength of SCFG prediction methods, at least, over other prediction methods, is their ease to combine with an evolutionary model. As RNA secondary structure has function, we might expect it to be conserved. Therefore, by considering an alignment of homologous sequences, the extra information provided by the homology should be expected to contribute to the model's ability of structure prediction.

We consider evolution in the setting of having one, single alignment. There are methods of dealing with multiple alignments and averaging, but they are beyond the scope of this chapter. Here, we begin by discussing the typical evolutionary model (Knudsen and Hein, 1999, 2003) used in RNA secondary structure prediction, then the application of this evolutionary model to SCFGs, and finally by discussing some more technical issues, such as considering gaps in alignments and tree estimation.

Evolutionary model and column probabilities

Recall that for single sequences we had that the probability of non-terminal U generating an unpaired A was $P[U \rightarrow .]$ * P[unpaired A]. However, when we are considering an alignment of RNA sequences, we now have columns of nucleotides instead of single nucleotides. If we have a column, say, A and U, the probability of U generating the unpaired column of *A* and *U* is now $P[U \rightarrow .]$ * P[unpaired column *A*, *U*]. So the SCFG rule probabilities stay the same, and the same algorithms can be used for prediction, but we just need an expression for the column probabilities.

As with many evolutionary models, we will need simplifying assumptions. We assume that RNA sequences evolved from a common ancestor, and that lineages are independent. Furthermore, we require evolutionary processes in different unpaired or paired positions of the sequence to be independent, so the probability for the whole alignment is the product of the probabilities of individual patterns. We also need the evolutionary processes to be the same within all paired and the same (although possibly distinct) within all unpaired positions. Lastly, we want the model to be time-reversible, so that evolution looks similar forward and backward in time.

A very simple model for column probabilities might be to say that, for example, the probability of an unpaired column of A, A, and U is

$$\text{P}[\text{unpaired column A,A,U}] =$$

$$\text{P}[\text{unpaired A}]^2 \times \text{P}[\text{unpaired U}]$$

Why might we want this model? Well, if we consider that secondary structure is likely to be conserved. If there is an A in position 1, and a G in position 6, then it is very unlikely that position 1 and 6 pair, even if all the other sequences look like 1 and 6 should pair. The reason for this is that is 1 and 6 are an A and G in one sequence, that sequence will not fold to have a base pair in positions 1 and 6. Consequently, we might think that this first sequence has a different function, due to its different secondary structure – but this contradicts our assumption that structure is conserved. Thus, requiring base pairs in all positions where we pose a base pair is necessary for our model.

However, what this model does assume is that all of the homologous RNA sequences are independent. Our evolutionary assumptions, though, assume that they evolved from a common ancestor, so we would like to include this co-variation in the structure prediction model.

To do this, first we must define a model for nucleotide evolution. We follow the Pfold evolutionary model (Knudsen and Hein, 1999, 2003), which assumes that nucleotides evolution follows a general reversible model (Tavaré, 1986), that is according to a continuous-time Markov chain. However, the evolutionary model used for RNA secondary structures is slightly different to the classic evolutionary models like the Jukes-Cantor model (Jukes and Cantor, 1969), in that we want paired columns to evolve together, to maintain base pair complementarity.

We consider first unpaired bases. To model the evolution of an unpaired nucleotide, we need evolutionary rates, that is our Q-matrix in the continuous time matrix. Thus, for unpaired bases, the probability that, say, A evolves to G in time t is

$$\text{P}[\text{A evolves to G in time } t] = (\exp(tQ))_{1,3}$$

where Q is the 4×4 Q-matrix of the continuous time Markov chain, and we are taking the 1,3 entry to correspond to the first nucleotide A, and the third nucleotide, G. The calculation for paired columns is similar, but instead we have a 16×16 Q-matrix to correspond to all possible base pairs. The probability that an AU base pair evolves to a CG base pair is

$$\text{P}[\text{AU evolves to CG in time } t] = (\exp(tQ))_{4,7}$$

where similarly, we are taking the 4,7 entry to correspond to the fourth base pair AU, and the seventh base pair CG.

Implementation with SCFGs

Now we have our nucleotide substitution model, we want to calculate the column probabilities. To do so, we follow the Felsenstein pruning algorithm

(Felsenstein, 1981). For unpaired columns, we define a vector $p_i = (p_{i1}, \dots p_{i4})$ for each node in the phylogenetic tree, where p_{i1} is the probability that the sequence at node i is an A, p_{i2} is the probability that the sequence at node i is a C, and so on. We begin by setting the leaf probabilities, that is, for the nucleotides we observe. So, if for sequence 1 we observe an A, the probability vector at that node is (1,0,0,0). For a node n_3 with two children n_1 and n_2 with branch lengths t_1 and t_2 respectively, we have that the probability that n_3 is an A is

$$P[n_3 = A] = \sum_{x_1 \in \{1,2,3,4\}} \sum_{x_2 \in \{1,2,3,4\}} \exp(-t_1 Q)_{1,x_1} \cdot$$

$$P[n_1 = x_1] \exp(-t_2 Q) 1,_{x_2} P[n_2 = x_2]$$

where x_1 and x_2 represent the nucleotides of n_1 and n_2 respectively. To get the probability of a column, we also need a prior on nucleotides (pi_A, pi_C, pi_G, pi_U). This is usually set to the frequencies at which nucleotides are observed in the dataset. Now we can finally calculate the column probability:

$$P[\text{column unpaired}] = \sum_{x_1 \in \{A,C,G,U\}} \pi_{x1} P[n_a = x_1]$$

where n_a is the ancestor node. Column probabilities for those columns which form base pairs are calculated in a similar manner. We define a vector $p_i = (p_{i1}, \dots p_{i16})$ for each node in the phylogenetic tree, where p_{i1} is the probability that the column pair at node i is an AA, p_{i2} is the probability that the sequence at node i is a AC, and so on. We begin by setting the leaf probabilities, that is, for the nucleotides we observe. So, if for sequence 1 we observe an AU, the probability vector at that node is (0, 0, 0, 1, 0, 0, …, 0). For a node n_3 with two children n_1 and n_2 with branch lengths t_1 and t_2 respectively, we have that the probability that n_3 is an AU is

$$P[n_3 = AU] = \sum_{x_1 \in \{1,\dots,16\}} \sum_{x_2 \in \{1,\dots,16\}} \exp(-t_1 Q)_{4,x_1}$$

$$P[n_1 = x_1] \exp(-t_2 Q)_{4,x_2}$$

$$P[n_2 = x_2]$$

where x_1 and x_2 again represent the nucleotides of n_1 and n_2 respectively. Then finally, with our prior on base pairs $(pi_{AA}, pi_{AC}, pi_{AG}, pi_{AU}, \dots, pi_{UU})$, we have the paired-column probability:

$$P[\text{column paired}] = \sum_{x_1 \in \{AA,AC,AG,AU,\dots,UU\}} \pi x_1 P[n_a = x_1]$$

If we permit only canonical base pairs, that is AU, CG, and GU pairs, it is easy to see there are many fewer terms in this sum.

Once one has these probabilities for paired and unpaired columns, one can calculate the CYK and inside–outside algorithms in the normal manner (see Extended Example for details), and structure prediction done in the normal manner. This is in some sense one of the big advantages of the SCFG method – one can decouple the evolutionary model from the SCFG model – the total probability is simply the product of the evolutionary probabilities and the SCFG probabilities.

The main implementation of this evolutionary approach is in Pfold (Knudsen and Hein, 1999, 2003), which has later been parallelised into PPfold (Sükösd *et al.*, 2012). The evolutionary model produces considerable better predictions than for single sequences. This is not only due to additional structural constraints seen through many sequences, but if the predictions from the simple evolutionary model described at the beginning of this section are not as good as those with the nucleotide substitution model.

Estimating rates

To estimate mutation rates, we follow the process in Knudsen and Hein, (1999). Columns with over 85% sequence identity were taken, so as to suggest only single mutations. Counts were recorded of differences in base pairs and single nucleotides over the sequences. Gaps were not counted. The rates could then be calculated as a function of the frequencies of nucleotide and base pair mutations observed, and the evolution time between sequences. In practice, though, since this estimation procedure requires a reasonably big dataset, and the mutation rates are made available in the manuscript, it is easier to use the parameters found there. The full rate matrices are available to download from the Pfold website (see Web Resources section).

Insertion–deletion events

At this point, little has been mentioned with respect to how to model insertion–deletion events. One approach is to model the gap as an unknown nucleotide, but requiring it to be unpaired, as we cannot have a base pair involving a nucleotide that is not present. To implement this, we simply put $(1, 1, 1, 1)$ in the leaf-node probability vector. This also has the advantage that the phylogenetic probabilities are the same with or without completely gapped columns, so columns with many gaps can be pruned from the alignment without affecting the prediction quality. This is the approach used by Pfold and PPfold, which eliminate columns with greater than 75% gaps.

Alternatively, one can treat the gap as a fifth nucleotide. One might expect insertion–deletion events to require a different evolutionary model than substitution events, something which is not considered by treating gaps as unknown nucleotides. Similarly, frequency of gaps will depend on evolutionary distances and number of sequences in the alignment. This approach has been successfully used in RNA gene finding (Rivas and Eddy, 2001).

Finally, one can change the evolutionary model, for example by combining the above model with an HMM to better take account of insertion–deletion events (Fang *et al.*, 2008). The HMM considers evolution in structured and unstructured regions, and results suggest that this yields better accuracy than the SCFG approach with a simple evolutionary model. Again, this is an area where more research could be done, particularly with correlated evolution models.

Example

Now that we have developed the full comparative model, an example of its use is provided. Consider an alignment of the following RNAs, assumed to have a consensus secondary structure (((....))):

```
>RNA_1
GGGCGUGCCC
>RNA_2
AGGCGUGCCU
>RNA_3
GGGCGAACCC
```

Furthermore, we can use a predicted evolutionary tree. Fig. 16.5 shows an examples evolutionary tree, with each node having two leaf children, and the corresponding branch lengths.

Here we will only focus on calculating the phylogenetic probabilities for columns, since the other algorithms for SCFG prediction have been demonstrated prior to this. First, we look at the phylogenetic probability that column 1 (G,A,G) is unpaired. We will assume a prior on unpaired nucleotides of (0.268, 0.214, 0.267, 0.251), the overall frequencies seen in Knudsen and Hein (1999), and use the rate matrix for stems in Knudsen and Hein (1999). For each node, we can now calculate the associated probability vector:

Node	Likelihood of (A, C, G, U)
1	(0,0,1,0)
2	(1,0,0,0)
3	(0,0,1,0)
4	(0.162, 0.011, 0.105, 0.016)

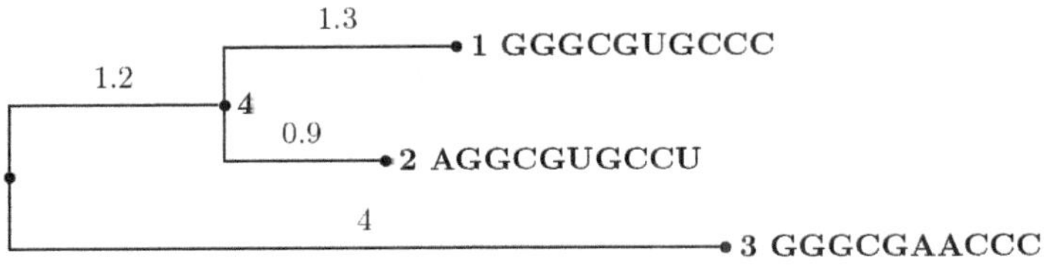

Figure 16.5 Evolutionary tree, with nodes labelled and branch lengths given, for our RNA alignment.

Ancestor $(0.021, 0.010, 0.019, 0.011)$

and then the probability of the column being unpaired is simply the dot product of the prior on unpaired nucleotides, and ancestor nucleotide likelihood:

$$0.00563 + 0.00214 + 0.00507 + 0.00271 = 0.01555$$

Similarly, we consider the phylogenetic probability of the columns 1 (G,A,G) and 10 (C,U,C) being paired. For simplicity, we will assume that only canonical base-pairs (AU, UA, CG, GC, GU, UG) are allowed, and that the prior on base pairs is (0.180, 0.180, 0.272, 0.272, 0.048, 0.048), the normalized stem frequencies from Knudsen and Hein (1999). For each node, we can now calculate the associated probability vector:

Node	Likelihood of (AU, UA, GC, CG, UG, GU)
1	(0,0,1,0,0,0)
2	(1,0,0,0,0,0)
3	(0,0,1,0,0,0)
4	(0.121, 0.011, 0.079, 0.007, 0.077, 0.095)
Ancestor	(0.019, 0.003, 0.040, 0.003, 0.003, 0.029)

and then the probability of the two columns being paired

$$0.00342 + 0.00054 + 0.01088 + 0.00082 + 0.00014 + 0.00139 = 0.01719$$

When we have, for each column, the probability of the column being unpaired, and, for each pair of columns, the probability of the pair of columns being paired, we can proceed to structure prediction via CYK or inside-outside.

As with the single-sequence structure prediction problem, once we have the probability distribution over RNA secondary structures, we can decode this distribution in a number of ways. Pfold uses an MEA approach, but other approaches have been implemented as well. A Bayesian sampling approach, which samples helices according to the distribution from SCFGs has been implemented (Doose and Metzler, 2012), which also allows the prediction of pseudoknots, since helices can be nested. Alternatively, an approach which simultaneously aligns and folds (Bradley *et al.*, 2008) using an evolutionary model and SCFG allows us to also take into account the uncertainty in the alignment. However, there is significant room for developments here in novel ways to decode the posterior distribution to gain better predictive accuracy.

Measures of SCFG variability

As we have seen, SCFGs place a probability distribution on the space of RNA secondary structures. So, we might like to use some results from probability theory to get additional measures to help us understand the space of SCFGs. This is particularly relevant to consider aspects of RNA folding like the evolution of folding pathways – not just looking at a single secondary structure, but the space as a whole.

One of the main problems, though, is computational. As mentioned in the Introduction section, the number of RNA secondary structures of length n grows exponentially as n becomes large. So we cannot simply evaluate the probability of each structure, and then get measures of structural variability, because this would be computationally unfeasible. Similarly, there aren't easy ways to average over the whole space to get global measures on the space. So, to get anything computationally feasible, we must make use of the dynamical programming approaches used in the CYK and inside–outside algorithms, to sum over the whole space in cubic time.

Information entropy

The main method which can be calculated in this way is information entropy. For a probability distribution P with events X, we define the information entropy $H(P)$:

$$H(P) = -\sum_{x \in X} P(x)\log_2(P(x))$$

Informally, the information entropy (or Shannon entropy) measures the spread of probability mass over the probability space. We can see from the above equation that if there is only a single event, with probability one, then the information entropy is 0. Similarly, if we have a uniform distribution, that is n events with probability $1/n$ each, we have an entropy of $\log_{2(n)}$. So 0 and $\log_{2(n)}$ provide the minimum entropy and maximum entropy respectively. When the base of the logarithm is 2, the entropy is measured in bits.

Entropy has been calculated previously for RNA secondary structure space. In particular, there are many approaches for calculating thermodynamic entropy for the thermodynamic model (Mathews *et al.*, 1999; Markham and Zuker, 2008), considering positional thermodynamic entropy (Hofacker *et al.*, 1994) and thermodynamic entropy in response to base pair mutation (Kiryu and Asai, 2012).

Firstly, we can improve maximum entropy bound of $\log_{2(\text{number of structures})}$, since we know the number of structures of length n. Recall the asymptotic approximation for the number of structures of length n:

$$Sn \approx 1.104 \times n^{3/2} \times n^{3/2} \times 2.618^{n}$$

So our maximum entropy bound is approximately:

$$H_{\max}(n) \approx 0.142 - \frac{3}{2}\log_2(n) + 1.388n$$

Now, to move on to calculating the information entropy for single sequences (Wang *et al.*, 2011). We begin by noting the information entropy can be written neatly as a function of expected rule frequencies, summing over the set of production rules P (Nedefhof and Satta, 2004).

$$H(G) = -\sum_{r \in P} \log_2(\mathrm{P}[r])\mathrm{E}[\text{uses of } r]$$

This is advantageous as expected rule frequencies can be computed using the inside–outside algorithm in cubic time. Once we have computed inside and outside values, $I(U, i, j)O(U, i, j)$ gives the expectation of $s[i..j]$ being derived from an occurrence of U and

$$O(U,i,j)\sum_{k=i}^{j-1} \mathrm{P}[U \to VW]I(V,i,k)I(W,k+1,j)$$

the expectation of this happening by an initial application of the $U \to VW$ production. Summing over all subsequences of s we can find the expected number of uses of $U \to VW$. Similarly we can find the expected number of uses of terminal production rules $U \to .$ from $O(U, i, i)P[U \to .]$ for i with $s[i]$ unpaired., and in the same manner the paired rule types.

However, this information entropy is only useful to calculate for unambiguous SCFGs. We want to be able to make statements about RNA secondary structure space, and it is only for unambiguous SCFGs that there is a one-to-one correspondence between SCFG derivations and RNA secondary structures. With ambiguous SCFGs, we can still calculate the information entropy, but instead we calculate the spread of mass over SCFG derivation space, not over RNA secondary structure space.

Information entropy can also be calculated with phylo-SCFGs, that is SCFGs which generate alignment columns, instead of single nucleotides (Sukosd *et al.*, 2012). We have

$$\begin{aligned} H(G) &= -\sum_{d \in \Phi} \mathrm{P}[d]\log_2(\mathrm{P}[d]) \\ &= \log_2(T) - \frac{1}{T}\sum_{d \in \Phi} \mathrm{P}[d]\log_2(\mathrm{P}_G[d]) \\ &\quad - \frac{1}{T}\sum_{d \in \Phi} \mathrm{P}[d]\log_2(\mathrm{P}_T[d]) \end{aligned}$$

where Φ is the set of all derivations, T represents the total probability of the alignment, and P_G and P_T represent the SCFG probabilities and evolutionary probabilities respectively. We can furthermore calculate the individual terms

$$\sum_{d \in \Phi} \mathrm{P}[d]\log_2(\mathrm{P}_G[d]) =$$

$$\sum_{d \in \Phi} \mathrm{P}[d]\log_2(\mathrm{P}_T[d]) = \sum_{i \neq j} \log_2(P_T(i,j \mid \widehat{i,j})) \sum_{d \in \Phi} 1^{d}($$

$$\sum_{i \neq j} \log_2(P_T(i,j \mid \widehat{i,j}) \sum_{d \in \Phi} 1^{d}(i,j)\mathrm{P}[d] +$$

$$\sum_{i} \log_2(P_T(i \mid \dot{i})) \sum_{d \in \Phi} 1^{s}(i)\mathrm{P}[d]$$

with 1^{d} the indicator function for a given derivation d. Note that once we have calculated the inside–outside algorithms, we can calculate the expected rule frequencies in quadratic time, and so the entropy is calculated in quadratic time. This has been implemented in PPfold 3.0.

Once we have calculated information entropy, we are interested in knowing what it can tell us about the space of RNA secondary structures. In some sense, information entropy can be viewed as signal. When the model has lower entropy, the model has a less spread out probability distribution, and so we might expect the SCFG model is more 'confident' in the prediction. Similarly, when the probability mass is spread out, many secondary structures might look likely, and so we might expect the SCFG model is less 'confident' in making the prediction.

Similarly, we would not expect information entropy would correlate with predictive accuracy. A model might not be very confident about the prediction it makes, but it might still be correct, and similarly, the model might allocate a lot of probability to one structure, but it could be totally wrong. The limitation of information entropy to explain RNA secondary structure space is fundamentally limited by the SCFG model it is part of.

However, there are some useful results that can be found (Sükösd *et al.*, 2012). Examples can be found where information entropy can be used to distinguish between alignments with larger information content than others, and this happens to explain the difference in predictive accuracy. Similarly, the amount of information added by the evolutionary model can be quantified against the amount of information in a single sequence for a given alignment. Finally, by considering a large number of random sequences, one can make statements about the relative information content of an alignment. However, with all statistical models, we must view this as only part of the picture, as a tool which can help shed light on how the model views the space.

Reliability scores

Pfold and PPfold also give reliability scores obtained from posterior probabilities. Once the base pair probability matrix is obtained via inside-outside, we will predict a structure using a simple MEA algorithm. But how sure are we in our choice of structure? Imagine that we have pairing probabilities of one for each base pair, then we would be very confident. But if we have pairing probabilities of 0.7 for each base pair, we may get the same structure from MEA, but be less confident about it. This reliability score gives a measure of how close the pairing probabilities are, by calculating a function of the distance the pairing probabilities are from one, if positions are predicted paired, and zero if positions are unpaired.

However, as with entropy, this reliability measure is only part of the picture. It can indicate if there are other structures the model might prefer, which are close in probability, which could be useful if you do not, for some biological reason, have confidence in the prediction. On the other hand, it is useful in sequence design programs like Frankenstein (Lyngsø *et al.*, 2012), as it can emulate the idea of folding stability. Overall though, it is limited by the SCFG model as a framework, and does not correlate particularly well with predictive accuracy.

Thermodynamic methods

At this point, it is worth briefly discussing thermodynamic methods of RNA secondary structure prediction, for comparison. The method is based on energy parameters, which must be gained experimentally (Mathews *et al.*, 1999). In particular, we must consider energy contributions from different combinations of base pairs. For example, if an AU base pair is next to a GU base pair, there will be a different energy contribution than if the AU base pair is next to another AU base pair. These so-called stacking energies mean that a large number of parameters must be found, over a hundred, which is why many of the thermodynamic models are based on the same energy models.

Once we have all of these energy parameters, again we run into the problem that we cannot evaluate the free energy of every possible secondary structure. So, what is found, is the structure with the minimum free energy, through dynamic programming, and this is the structure predicted.

Since, for each secondary structure, we have an associated energy, this naturally leads to a Boltzmann probability distribution over secondary structures. To calculate the probability of a secondary structure S, we have

$$P(S) = \frac{1}{Q} e^{-F(S)/kT}$$

where $F(S)$ is the free energy of S, T is the temperature, k a constant, and Q a normalizing factor. To calculate Q, one needs to calculate the partition function (McCaskill, 1990), and to find the free-energy of a given secondary structure, one simply sums over all the loops and stems, finding the free energy of the smaller components, using the free energy parameters determined experimentally. However, if one is only interested in calculating the minimum free energy structure, this can be done without calculating the full partition function (Hofacker *et al.*, 1999). In particular, it can be done in cubic time, the same complexity as the SCFG method.

The main implementations of the thermodynamic are UNAfold (Markham *et al.*, 2008) and RNAfold, as part of the Vienna package (Hofacker *et al.*, 1994). They are essentially the same method, but with slightly different energy models. They predict the minimum free energy structure, in the same way that SCFGs can predict the maximum probability structure. Also like SCFG methods, they place a probability distribution on RNA secondary structures, so various different posterior decoding methods have been implemented to obtain slightly different prediction methods. Sfold (Ding *et al.*, 2000) samples from the Boltzmann distribution generated by the free energy model, and generates a consensus structure from the samples. Kinefold (Xayaphoummine *et al.*, 2005) uses kinetic folding concepts in combination with the free energy model to try and more closely echo biological folding mechanisms.

The main thermodynamic method has a couple of weaknesses. Firstly, it does not consider non-canonical base pairs, those are non-AU. -CG, or -GU base pairs. Part of the reason for this is practical – the number of parameters that would need to be determined from experiments would become significantly larger, and perhaps obtaining these would not be feasible. However, many RNA secondary structures do contain non-canonical base pairs. Secondly, the method cannot be adapted to an evolutionary model. The thermodynamic methods that do consider alignments (Bernhart *et al.*, 2008) only gain the information concerning additional structure constraints from having multiple sequences with a consensus structure, but this is often considerable. Pseudoknots, though, can be considered by adapting the energy model, for example in pknots (Rivas and Eddy, 1999). For further reading on the thermodynamic approach, see Janssen *et al.*, (2011).

Discussion

As can be seen by the number of approaches attempted at the RNA secondary structure prediction problem, even within the framework of SCFGs, there has not been overwhelming success at the problem. Methods continue to operate at the same kind of sensitivity on datasets, at an average of around 0.55 for single sequences, and 0.75 for alignments. However, this accuracy is highly variable, it may be 1 on one sequence, and 0 on another, without much warning.

Thermodynamic methods and SCFG method continue to compete, with thermodynamic methods having a slight edge on single-sequence structures. The methods do not predict identical structures, or even similar structures (Anderson *et al.*, 2012) when the methods go wrong, and there does not seem to be a way of knowing *a priori* which method you might want to predict your structure.

Furthermore, although many methods have been developed to measures entropy, or folding pathways (Shapiro *et al.*, 2001), these are all fundamentally limited by the models which produce them. They give us useful statistics concerning what the models think is going on in the RNA folding, and how the RNA might folding differently, but, due to the model accuracy, cannot be completely trusted.

However, it is important that there is continued work on the methods. With developments in next-generation sequencing techniques, particular the fact that methods like RNA-seq are now easily available, and the data being generated in transcriptomics is considerable. RNA folding methods will want to be used in genome-wide studies, considering RNA structure-based effects on gene expression and regulation. Cumulative error from poor prediction will create statistical problems in these studies, and possibly invalidate conclusions. Since RNA is so fundamental in molecular dynamics, we must have a better understanding on the mechanisms for folding, and be able to translate that understanding into modelling principles.

Future trends

RNA is as interesting as ever. Particularly as biological technologies advance, the role of RNA will be come more and more crucial to understand.

Three-dimensional structure prediction techniques are still in their infancy but will no doubt develop as technology and data advance. Particularly, the possibility of extending SCFG prediction to three-dimensional structures will be an interesting task. Similarly, as RNA roles in gene expression become better understood, structural features of RNA may well be used in combination with network inference techniques to better understand gene regulatory networks.

Of course, there are many additional applications of RNA-secondary structure prediction which will benefit from improved methodology: RNA gene finding, annotation, functional prediction, RNA–protein interactions, the list is long. Improved RNA molecular evolution models or alignments may enable prediction quality to improve. However, there has been little investigation into which components of the RNA secondary structure prediction model are important to predictive success.

Given that continued work is important in the field of RNA secondary structure prediction, the question is how to go about it. It seems unlikely that many improvements will come from an improved alignment or evolutionary model, partly because these models are close to being as good as they will get, and partly because the single-sequence structure prediction problem for RNA is still unsolved. The probabilistic and thermodynamic approaches seem to cover all the bases. One might imagine a machine-learning approach could work, but in my opinion, SCFGs are really just that. It seems unlikely to me that a fundamentally different approach will arise, especially given how much work has been done to develop the probabilistic and thermodynamic approaches. Models need to become more biologically realistic, particularly SCFGs. As mentioned previously, SCFGs are little more than a machine-learning approach, and there are many more biological processes going on in RNA folding than a simple sequence to structure map. RNA-specific, co-transcriptional effects, as well as kinetic effects, could be considered, or perhaps something so simple as being able to incorporate some notion of temperature-dependence might prove useful. The probability distribution generated by SCFGs is can still be used, but perhaps novel methods of decoding could be used to more closely echo biological processes. There is clearly much still to do.

Web resources

There are many web resources available for RNA secondary structure prediction. In particular, most of the methods cited throughout will have made open source programs available, with links in the manuscripts. Similarly, most studies which use datasets will have made them publicly available, with links in the manuscripts. However, a few of the main resources are as follows.

SCFG based approaches

Pfold: http://www.daimi.au.dk/~compbio/pfold/
PPfold: http://www.daimi.au.dk/~compbio/pfold/downloads.html
PhyloQFold (pseudoknot sampling SCFG model): http://evol.bio.lmu.de/_statgen/software/phyloqfold/
Evolgram (evolving SCFGs): http://www.stats.ox.ac.uk/~anderson/Code/evolvingscfgs.html

Thermodynamic approaches

UNAfold: http://mfold.rna.albany.edu/
Vienna RNA Package (including RNAfold): http://www.tbi.univie.ac.at/~ivo/RNA/
Sfold (Boltzmann sampling): http://sfold.wadsworth.org/cgi-bin/index.pl
Kinefold (kinetic model): http://kinefold.curie.fr/
Pknots (pseudoknot model): http://selab.janelia.org/software.html

Data sources

Rfam: http://rfam.sanger.ac.uk/

RNA STRAND http://www.rnasoft.ca/strand/

References

de Almeida Ribeiro, E., Beich-Frandsen, M., Konarev, P.V., Shang, W., Veerek, B., Kontaxis, G., Hmmerle, H., Peterlik, H., Svergun, D.I., Blsi, U., *et al.* (2012). Structural flexibility of rna as molecular basis for hfq chaperone function. Nucleic Acids Res. *40*, 8072–8084.

Andersen, E., Dong, M., Nielsen, M., Jahn, K., Subramani, R., Mamdouh, W., Golas, M., Sander, B., Stark, H., Oliveira, C., *et al.* (2009). Self-assembly of a nanoscale DNA box with a controllable lid. Nature *459*, 73–76.

Andersen E.S., Rosenblad, M.A., Larsen, N., Westergaard, J.C., Burks, J., Wower, I.K., Wower, J., Gorodkin, J., Samuelsson, T., and Zwieb, C. (2006). The tmRDB and SRPDB resources. Nucleic Acids Res., *34*, 163–168.

Anderson, J.W.J., Tataru, P., Staines, J., Hein, J., and Lyngso, R. (2012). Evolving stochastic context-free grammars for RNA secondary structure prediction. BMC Bioinformatics *13*, 78.

Andronescu, M., Bereg, V., Hoos, H., and Condon, A. (2008). RNA STRAND: The RNA secondary structure and statistical analysis database', BMC Bioinformatics 9, 340.

Berman, H.M., Olsonm, W.K., Beveridgem, D.L., Westbrook, J., Gelbin, A., Demeny, T., Hsieh, S.H., Srinivasan, A.R., and Schneider, B. (1992). The nucleic acid database. A comprehensive relational database of three-dimensional structures of nucleic acids. Biophys. J. *63*, 751–759.

Bernhart, S., Hofacker, I., Will, S., Gruber, A., and Stadler, P. (2008). Rnaalifold: improved consensus structure prediction for rna alignments. BMC Bioinformatics 9, 474.

Brabrand, C., Giegerich, R., and Møller, A. (2007). Analyzing ambiguity of context-free grammars. In Proc. 12th International Conference on Implementation and Application of Automata, CIAA '07, volume 4783 of LNCS. (Berlin Heidelberg, Springer-Verlag).

Bradley, R.K., Pachter, L., and Holmes, I. (2008). Specific alignment of structured RNA: stochastic grammars and sequence annealing. Bioinformatics *24*, 2677–2683.

Brown, J. (1999). The ribonuclease P database. Nucleic Acids Res. *27*, 314–314.

Brown, M., and Wilson, C. (1996). RNA pseudoknot modeling using intersections of stochastic context free grammars with applications to database search. Pacific Symposium on Biocomputing 109–125.

Cannone, J., Subramanian, S., Schnare, M., Collett, J., D'Souza, L., Du, Y., Feng, B., Lin, N., Madabusi, L., Muller, K., *et al.* (2002). The comparative RNA web (CRW) site: an online database of comparative sequence and structure information for ribosomal, intron, and other RNAs. BMC Bioinformatics 3, 15.

Chappelier, J., and Rajman, M. (1998). A generalized CYK algorithm for parsing stochastic CFG', Proc. of 5eme conference sur le Traitement Automatique du Langage Naturel, pp. 153–157.

Chen, J., and Greider, C. (2005). Functional analysis of the pseudoknot structure in human telomerase RNA. Proc. Natl. Acad. Sci. U.S.A., *102*, 8080–8085.

Chomsky, N. (1956). Three models for the description of language. IRE Trans. Inf. Theory 2, 113–124.

Craig, M.E., Crothers, D.M., and Doty, P. (1971). Relaxation kinetics of dimer formation by self complementary oligonucleotides. J. Mol. Biol. *62*, 383–401.

Darty, K., Denise, A., and Yann, P. (2009). VARNA: Interactive drawing and editing of the RNA secondary structure. Bioinformatics *25*, 1974–1975.

Dempster, A.P., Laird, N.M., and Rubin, D.B. (1977). Maximum likelihood from incomplete data via the EM algorithm. J. R. Stat. Soc. Series B Stat. Methodol. *39*, 1–38.

Ding, Y., Chan, C.Y., and Lawrence, C.E. (2005). RNA secondary structure prediction by centroids in a boltzmann weighted ensemble. RNA *11*, 1157–1166.

Doose, G., and Metzler, D. (2012). Bayesian sampling of evolutionarily conserved RNA secondary structures with pseudoknots. Bioinformatics *28*, 2242–2248.

Dowell, R., and Eddy, S. (2004). Evaluation of several lightweight stochastic context-free grammars for rna secondary structure prediction. BMC Bioinformatics *5*, 71.

Fang, X.Y., Luo, Z.G., and Wang, Z.H. (2008). Predicting RNA secondary structure using profile stochastic context-free grammars and phylogenic analysis. J. Comput. Sci. Technol. *23*, 582–589.

Felden, B., Florentz, C., Giegé, R., and Westhof, E. (1996). A central pseudoknotted three-way junction imposes tRNA-like mimicry and the orientation of three 5′ upstream pseudoknots in the 3′ terminus of tobacco mosaic virus RNA. RNA 2, 201–12.

Felsenstein, J. (1981). Evolutionary trees from dna sequences: A maximum likelihood approach. J. Mol. Evol. *17*, 368–376.

Freyhult, E., Gardner, P., and Moulton, V. (2005). A comparison of RNA folding measures. BMC Bioinformatics 6, 241.

Gahura, O., Hammann, C., Valentov, A., Fta, F., and Folk, P. (2011). Secondary structure is required for 3 splice site recognition in yeast. Nucleic Acids Res. *39*, 9759–9767.

Gardner, P., and Giegerich, R. (2004). A comprehensive comparison of comparative RNA structure prediction approaches. BMC Bioinformatics 5, 140.

Griffiths-Jones, S., Moxon, S., Marshall, M., Khanna, A., Eddy, S.R., and Bateman, A. (2005). Rfam: annotating non-coding rnas in complete genomes. Nucleic Acids Res. 33 (Suppl 1), D121–D124.

Hamada, M., Sato, K., and Asai, K. (2010). Prediction of RNA secondary structure by maximizing pseudo-expected accuracy. BMC Bioinformatics *11*, 586.

Harmanci, A., Sharma, G., and Mathews, D. (2011). Turbofold: Iterative probabilistic estimation of secondary structures for multiple RNA sequences. BMC Bioinformatics *12*, 108.

Håstad, J., Phillips, S., and Safra, S. (1993). A well characterized approximation problem. Inf. Process. Lett. *47*, 301–305.

Hofacker, I., Fontana, W., Stadler, P., Bonhoeffer, L., Tacker, M., and Schuster, P. (1994). Fast folding and comparison of RNA secondary structures. Chem. Monthly *125*, 167–188.

Hopcroft, J., and Ullman, J. (1979). Introduction to Automata Theory, Languages, and Computation. (Reading, MA, Addison-Wesley).

Janssen, S., Schudoma, C., Steger, G., and Giegerich, R. (2011). Lost in folding space? comparing four variants of the thermodynamic model for RNA secondary structure prediction. BMC Bioinformatics *12*, 429.

Jukes, T.H., and Cantor, C.R. (1969). Evolution of Protein Molecules. (New York, Academic Press), pp. 21–132.

Kiryu, H., Asai, K. (2012). Rchange: Algorithms for computing energy changes of RNA secondary structures in response to base mutations. Bioinformatics *1*, 1–2.

Krogh, A., Brown, M., Mian, I., Sjölaner, K., and Haussler, D. (1993). Hidden Markov models in computational biology: applications to protein modeling. J. Mol. Biol. *235*, 1501–1531.

Knudsen, B., and Hein, J. (1999). RNA secondary structure prediction using stochastic context-free grammars and evolutionary history. Bioinformatics *15*, 446–454.

Knudsen, B., and Hein, J. (2003). Pfold: RNA secondary structure prediction using stochastic context-free grammars. Nucleic Acids Res. *31*, 3423–3428.

Krogh, A., Brown, M., Mian, I., Sjolander, K., and Haussler, D. (1993). Hidden Markov Models in computational biology: Applications to protein modeling. J. Mol. Biol. *235*, 1501–1531.

Lari, K., and Young, S.J. (1990). The estimation of stochastic context-free grammars using the inside–outside algorithm. Comput. Speech Lang. *4*, 35–56.

Lefebvre, F. (1996). A grammar-based unification of several alignment and folding algorithms. ISMB-96 Proceedings 2, 143–154.

Loh, E., Memarpour, F., Vaitkevicius, K., Kallipolitis, B.H., Johansson, J., and Sondn, B. (2011). An unstructured 5-coding region of the prfa mrna is required for efficient translation. Nucleic Acids Res. *40*, 1818–27.

Lyngsø, R.B., Anderson, J.W.J., Badugu, A., Hyland, T., Sizikova, E., and Hein, J. (2012). Frnakenstein: multiple target inverse rna folding. BMC Bioinformatics *13*, 260.

Markham, N.R., and Zuker, M. (2008). UNAFold: software for nucleic acid folding and hybridization. Methods Mol. Biol. *453*, 3–31.

Mathews, D., Sabina, J., Zuker, M., and Turne, D. (1999). Expanded sequence dependence of thermodynamic parameters improves prediction of RNA secondary structure. J. Mol. Biol. *288*, 911–940.

McCaskill, J.S. (1990). The equilibrium partition function and base pair binding probabilities for RNA secondary structure. Biopolymers *29*, 1105–1119.

Meyer, I. (2007). A practical guide to the art of RNA gene prediction. Brief. Bioinform. *8*, 396–414.

Moulton, V., Zuker, M., Steel, M., Pointon, R., and Penny, D. (2000). Metrics on RNA secondary structures. J. Comput. Biol. *7*, 277–292.

Nebel, M., and Scheid, A. (2011). Evaluation of a sophisticated SCFG design for RNA secondary structure prediction'. Theory Biosci. *130*, 313–336.

Nederhof, M., and Satta, G. (2004). Kullback-Leibler distance between probabilistic context-free grammars and probabilistic finite automata. In Comput. Speech Lang. Assoc. Comput. Ling. *2004*, 71.

Nussinov, R., Pieczenik, G., Griggs, J., and Kleitman, D. (1978). Algorithms for loop matchings. SIAM J. Appl. Math. *35*, 68–82.

Pipas, J., and McMahon, J. (1975). Method for predicting RNA secondary structure. Proc. Natl. Acad. Sc. U.S.A. *72*, 2017–2021.

Reeder, J., Steffen, P., and Giegerich, R. (2005). Effective ambiguity checking in biosequence analysis. BMC Bioinformatics *6*, 153.

Rivas, E., and Eddy, S.R. (1999). A dynamic programming algorithm for RNA structure prediction including pseudoknots. J. Mol. Biol. *285*, 2053–2068.

Rivas, E., and Eddy, S.R. (2000). The language of RNA: a formal grammar that includes pseudoknots. Bioinformatics *16*, 334–340.

Rivas, E., and Eddy, S.R. (2001). Noncoding RNA gene detection using comparative sequence analysis. BMC Bioinformatics *2*, 8.

Rivas, E., Lang, R., and Eddy, S.R. (2012). A range of complex probabilistic models for rna secondary structure prediction that include the nearest neighbor model and more. RNA *18*, 193–212.

Sakakibara, Y., Brown, M., Hughey, R., Mian, I., Sjolander, K., Underwood, R., and Haussler, D. (1994). Stochastic context-free grammars for tRNA modeling. Nucleic Acids Res. *22*, 5112–5120.

Scheid, A., and Nebel, M. (2012). Evaluating the effect of disturbed ensemble distributions on SCFG based statistical sampling of RNA secondary structures. BMC Bioinformatics *13*, 159.

Shapiro, B.A., and Zhang, K. (1990). Comparing multiple RNA secondary structures using tree comparisons. Comput. Appl. Biosci. *6*, 309–318.

Shapiro, B.A., Bengali, D., Kasprzak, W., and Wu, J.C. (2001). RNA folding pathway functional intermediates: their prediction and analysis. J. Mol. Biol. *312*, 27–44.

Shapiro, B.A., Yingling, Y.G., Kasprzak, W., and Bindewald, E. (2007). Bridging the gap in RNA structure prediction. Curr. Opin. Struct. Biol. *17*, 157–165.

Sprinzl, M., and Vassilenko, K. (2005). Compilation of tRNA sequences and sequences of tRNA genes. Nucleic Acids Res. *33*, 139–140.

Stein, P.R., and Waterman, M.S. (1978). On some new sequences generalizing the Catalan and Motzkin numbers. Discrete Math. *26*, 261–272.

Sükösd, Z., Knudsen, B., Kjems, J., and Pedersen, C.N.S. (2012). PPfold 3.0: fast RNA secondary structure prediction using phylogeny and auxiliary data. Bioinformatics *28*, 2691–2692.

Sükösd, Z., Knudsen, B., Anderson, J.W.J., Novák, Kjems, J., and Pedersen, C.N.S. (2012). Characterising RNA secondary structure space using information entropy. BMC Bioinformatics *14*(Suppl 2), S22.

Tavaré, S. (1986). Some probabilistic and statistical problems in the analysis of DNA sequences. Lectures on Mathematics in the Life Sciences *17*, 57–86.

Vitreschak, A.G., Rodionov, D.A., Mironov, A.A., and Gelfand, M.S. (2004). Riboswitches: the oldest mechanism for the regulation of gene expression? Trends Genet. *20*, 44–50.

Wang, Y., Manzour, A., Shareghi, P., Shaw, T.I., Li, Y.W., Malmberg, R.L., and Cai, L. (2011). Stable stem enabled Shannon entropies distinguish non-coding RNAs from random backgrounds. Paper presented at: IEEE 1st International Conference on Computational Advances in Bio and Medical Sciences (ICCABS), 184.

Washietl, S., Hofacker, I.L., Lukasser, M., Huttenhofer, A., and Stadler, P.F. (2005). Mapping of conserved RNA secondary structures predicts thousands of functional noncoding RNAs in the human genome. Nat. Biotechnol. *23*, 1383–1390.

Westbrook, J., Feng, Z., Chen, L., Yang, H., and Berman, H. (2003). The protein data bank and structural genomics. Nucleic Acids Res. *31*, 489–491.

Xayaphoummine, A., Bucher, T., Thalmann, F., and Isambert, H. (2003). Prediction and statistics of pseudoknots in rna structures using exactly clustered stochastic simulations. Proc. Natl. Acad. Sci. *100*, 15310–15315.

Xayaphoummine, A., Bucher, T., and Isambert, H. (2005). Kinefold web server for rna/dna folding path and structure prediction including pseudoknots and knots. Nucleic Acids Res. *33*(Suppl 2), W605–W610.

Xu, Z., Almudevar, A., and Mathews, D.H. (2012). Statistical evaluation of improvement in RNA secondary structure prediction. Nucleic Acids Res. *40*, e26.

Younger, D. (1967). Recognition and parsing of context-free languages in time n3. Inf. Control *10*, 189–208.

Zuker, M. (2003). Mfold web server for nucleic acid folding and hybridization prediction. Nucleic Acids Res. *31*, 3406–3415.

Zuker, M., and Sankoff, D. (1984). RNA secondary structures and their prediction. Bull. Math. Biol. *46*, 591–621.

Zuker, M., and Stiegler, P. (1981). Optimal computer folding of large RNA sequences using thermodynamics and auxiliary information. Nucleic Acids Res. *9*, 133–148.

Index

U

V

W

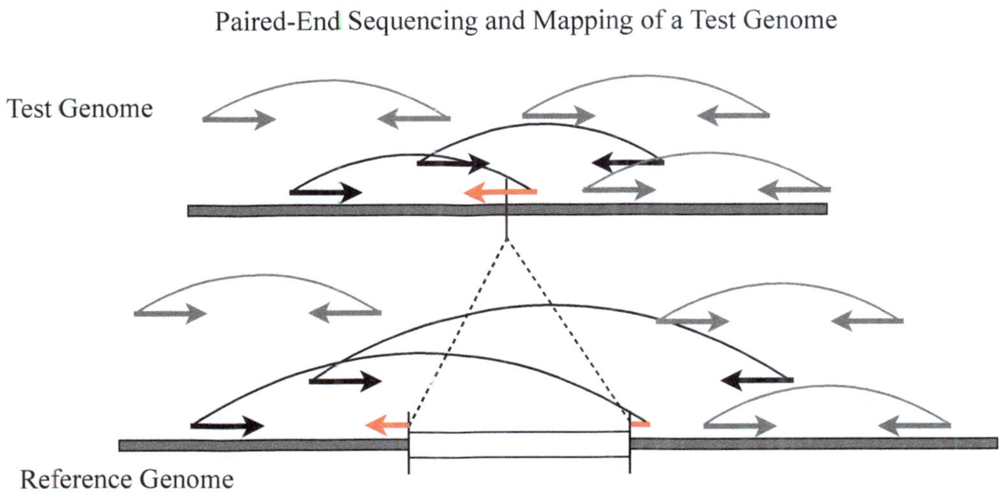

Figure 1.2 To identify structural variants (SVs) from a test genome, fragments from a test genome are sequenced from both ends and the paired-reads mapped to a reference. Evidence of a structural variant (in this case a deletion) comes from three signals in the mappings: (1) changes in read depth (RD), the depth of mapped reads (black and grey); (2) discordant paired-read mappings (PR) (black reads); and (3) split-read mappings (SR) (red read).

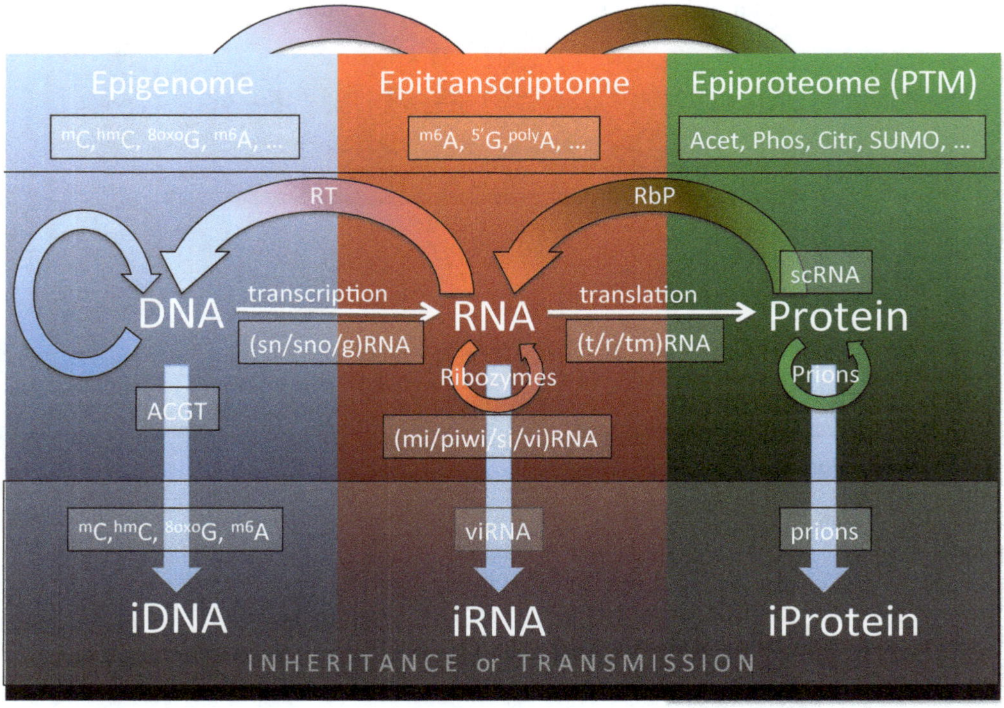

Adapted from Saletore *et al*., Genome Biology, 2012

Figure 2.1 The increasing complexity of the central dogma of molecular biology. (Top) The originally proposed central dogma, with unidirectional information flow. (Bottom) The current view of the central dogma, wherein information content can flow 'backwards' or 'sideways' with reverse transcriptases (RT), RNA-binding proteins (RbPs) and RNA editing. Also, information can be copied within each of the three realms: genetic (blue) copying such as with transposable elements, transcriptional (red) copying with ribozymes and rich levels of RNA regulation using small RNAs (micro, piwi, si, vi RNAs), and proteomic (green) copying using prions.

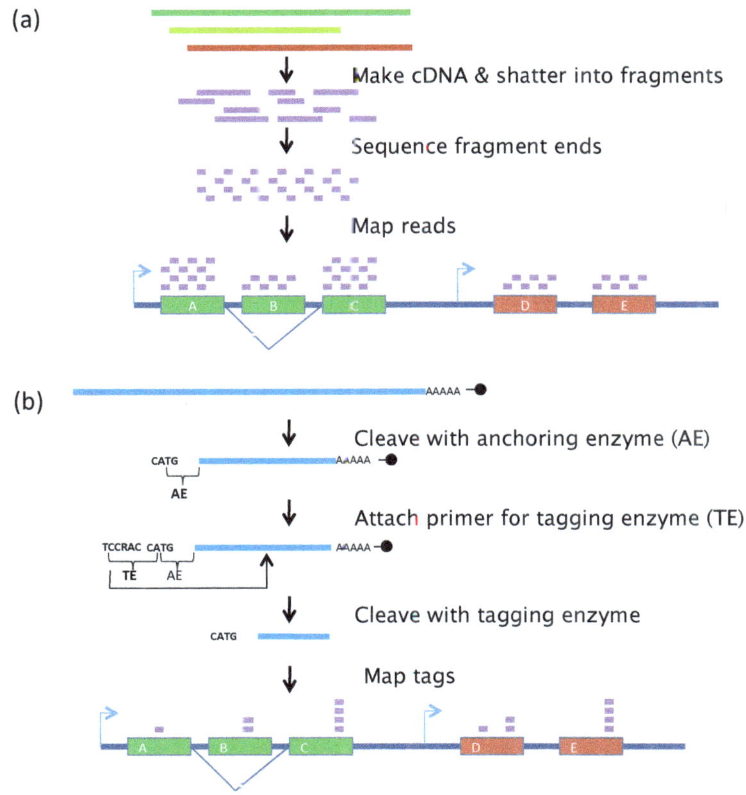

Figure 3.1 A schematic representation of the RNA sequencing protocols. (a) RNA-Seq (b) DGE.

Key: Pink 15-19 nt;Red 20-21 nt; Green 22 nt; Blue 23-24 nt;Grey 25-27 nt;Black 28-35 nt; Orange other;

Figure 4.4 Post-analysis visualization of a locus in Gbrowse showing differential expression between two sequencing libraries. Visualization in this way allows comparison with nearby annotation features; in this case, a known coding sequence, identification of strandedness of individual siRNAs, and, if properly configured, the length of sequenced siRNAs in the region.

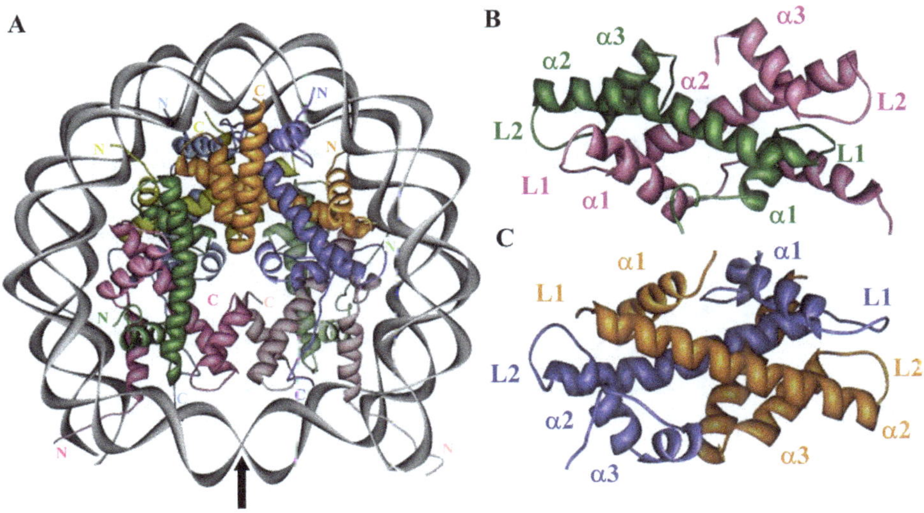

Figure 7.1 Architecture of the histone octamer core in the nucleosome. **(A)** Mutual positioning of the histone octamer and **DNA** in the nucleosome. The dyad axis is shown by an arrow at the middle of the nucleosome. H3 (pink), H4 (green), H2A.Z (blue) and H2B (orange), and DNA (grey) in the nucleosome. Dark and light colours distinguish each copy of the monomers. The N- and C-termini are indicated. The arrow indicates the direction of the dyad axis. The structure has been constructed using PDB file 1F66 for the nucleosome with the histone variant H2A.Z, deposited by Luger and coworkers (Suto *et al.*, 2000). (**B**) Handshake motifs formed by H3-H4 and H2A-H2B. The monomers H3, H4, H2Z, and H2B are collared pink, green, blue, and orange, respectively, consistent with (**A**). Adopted from (Ramaswamy *et al.*, 2005).

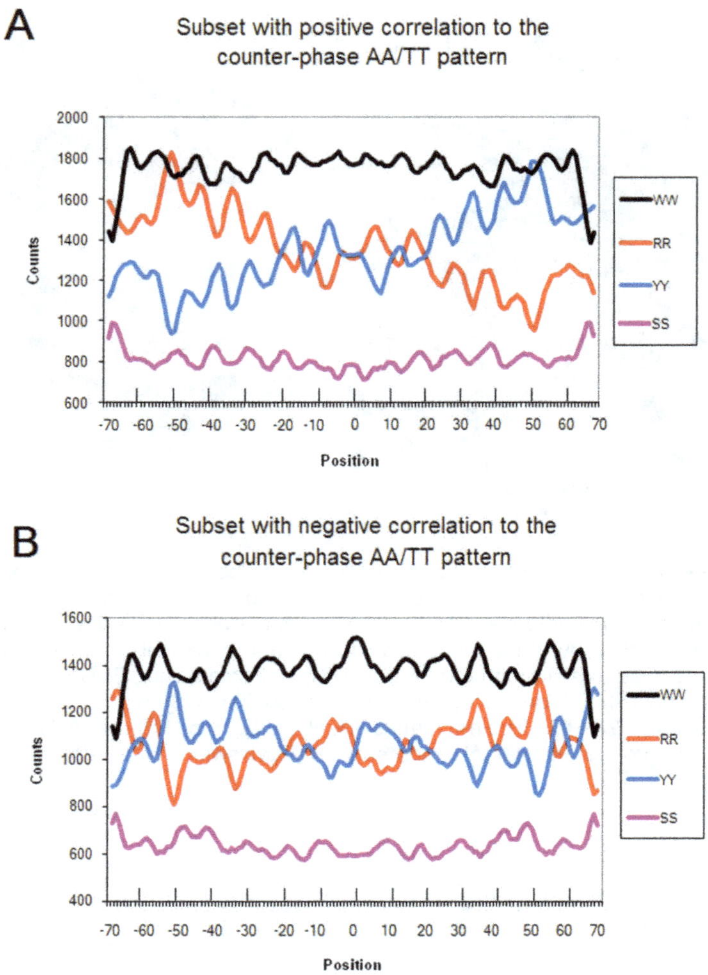

Figure 7.6 Combined dinucleotide distributions (smoothed by 3) for nucleosome subsets with positive (A) and negative (B) AA/TT correlation to the counter-phase AA/TT pattern from (Ioshikhes *et al.*, 1996). Notice opposite phases for RR patterns at A and B, for YY patterns at A and B, and steep gradients for RR and YY patterns at A (no obvious gradients at B). The opposite phases for the patterns in the A and B panels are related to inverse positioning of respective sequence elements in conventional (major) patterns and respective anti-patterns (presented in Figures 7.2 and 7.3, left and right sides, respectively). Adopted from Ioshikhes *et al.* (2011).

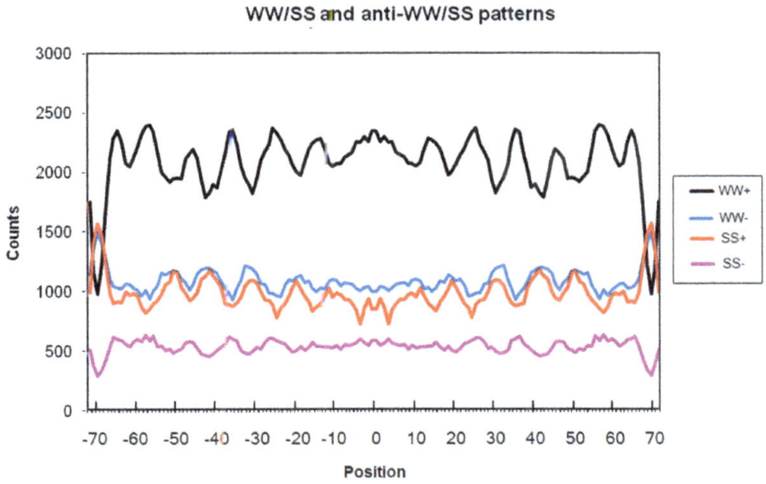

Figure 7.7 WW and SS patterns (smoothed by 3) for subsets with positive (+) and negative (–) WW/SS correlation to the major WW/SS patterns from Albert *et al.* (2007). Notice identical phases for the WW– and SS+ and for the WW+ and SS– patterns. Adapted from Ioshikhes *et al.* (2011).

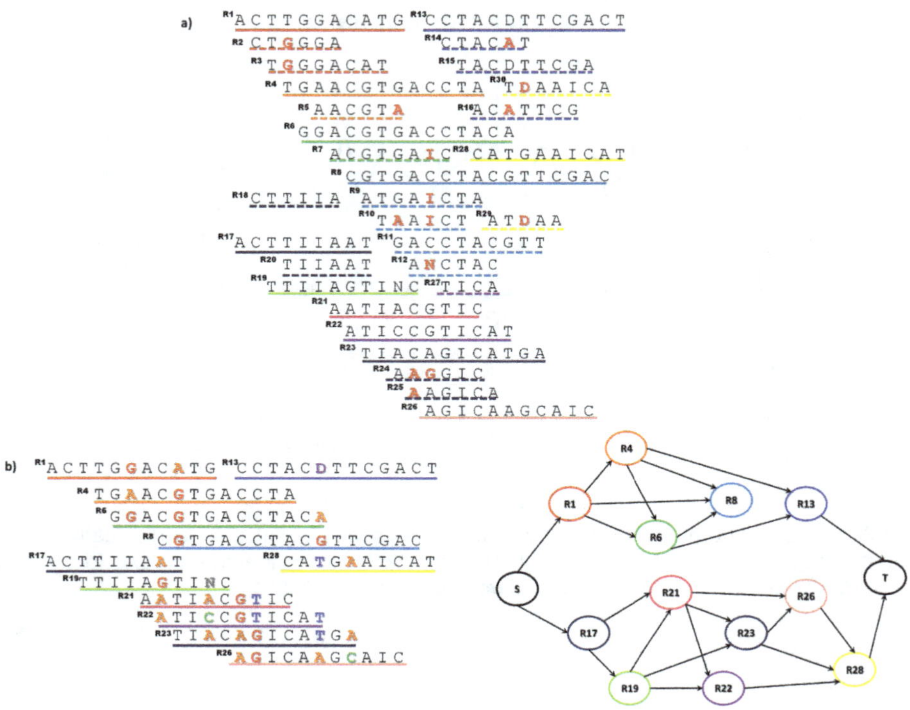

Figure 12.4 (adapted from (Astrovskaya *et al.*, 2011)) (a) Finding superreads from the aligned reads ($n=2$). The reads are divided into the superreads (underlined by a bold line) and the subreads (underlined by a dashed line of the corresponding colour). A mismatch between subread and its superread is in the red colour. For example, *R2* and *R3* are the subreads of *R1*, both subreads have one difference with the superread *R1*. (b) A read graph example. At the left, the aligned superreads ($n=2$) are given. The positions with several allelic values in overlap are highlighted (red (*G*), orange (*A*), green (*C*), blue (*T*), grey (*N*) and purple (*D*)). For simplicity of the presentation, differences are marked separately across reads *R1-R13* and *R17-R28*. At the right, a read graph ($n=2$, $m=2$) is drawn (the overlaps that are less than 3 bases are not considered).

www.ingramcontent.com/pod-product-compliance
Lightning Source LLC
LaVergne TN
LVHW062216200926
848484LV00006B/1409
9781908230294